Medicinal Chemistry

Medicinal Chemistry

A Molecular and Biochemical Approach

THIRD EDITION

Thomas Nogrady

Donald F. Weaver

OXFORD

UNIVERSITY PRESS

2005

OXFORD
UNIVERSITY PRESS

Oxford University Press, Inc., publishes works that further
Oxford University's objective of excellence
in research, scholarship, and education.

Oxford New York
Auckland   Cape Town   Dar es Salaam   Hong Kong   Karachi
Kuala Lumpur   Madrid   Melbourne   Mexico City   Nairobi
New Delhi   Shanghai   Taipei   Toronto

With offices in
Argentina   Austria   Brazil   Chile   Czech Republic   France   Greece
Guatemala   Hungary   Italy   Japan   Poland   Portugal   Singapore
South Korea   Switzerland   Thailand   Turkey   Ukraine   Vietnam

Published by Oxford University Press, Inc.
198 Madison Avenue, New York, New York 10016

www.oup.com

Oxford is a registered trademark of Oxford University Press

Library of Congress Cataloging-in-Publication Data
Nogrady, Th. Medicinal chemistry: a molecular and biochemical
approach/Thomas Nogrady, Donald F. Weaver—3rd ed.
        p. cm
Includes bibliographical references and index
ISBN 13 978-0-19-510455-4; 978-0-19-510456-1 (pbk.)
ISBN 0-19-510455-2; 0-19-510456-0 (pbk.)
    1. Pharmaceutical chemistry. I Weaver, Donald F., 1957-II. Title.
[DNLM: 1. Chemistry, Pharmaceutical. 2. Drug Design. 3. Receptors,
Drug. QV 744 N777m 2005]

RS403.N57 2005
615'.7—dc22   2004058105

10045 05 294

9 8 7 6 5 4 3 2 1
Printed in the United States of America
on acid-free paper

To Heather and Cheryl
Inspiration, Critic, Sustenance

Preface

There have been many changes in medicinal chemistry and molecular pharmacology since the second edition of this book was published in 1988. Accordingly, it has required extensive updating. This process was initiated in the Chemistry Department of Queen's University, Kingston, Canada where Dr. Nico van Gelder, an Adjunct Emeritus Professor, introduced the now retired Thomas Nogrady to Donald Weaver, a medicinal chemist and clinical neurologist. Together, Weaver and Nogrady undertook the challenge of updating this book. In this way the third edition of *Medicinal Chemistry* was started and the two authors have worked together to ensure a continuity in the style and content that has made this book popular among students and researchers alike.

The areas of change in this new edition are many and varied. Since molecular modeling has assumed an increasingly prominent role in drug discovery, we have expanded the discussion of modeling techniques. Description of other new techniques such as high throughput screening and applications of genomics in drug design have also been added. In terms of medicinal chemistry applications, neuropharmacology has enjoyed many advances in the past decade; much new information from this field has been included. In concert with these advances, new therapies have been introduced for Alzheimer's disease, Parkinson's disease, multiple sclerosis and epilepsy – these new therapies are explicitly discussed in the third edition. Emerging disorders like those of protein folding (e.g. Creutzfeldt-Jakob disease and other prion disorders) are also considered. Chapter 5 on hormonal therapies has been thoroughly updated and re-organized. An entire new chapter on the immune system has been added (chapter 6), reflecting the increased interest in therapeutic molecular manipulation of immunity. Emerging enzyme targets in drug design (e.g. kinases, caspases) are discusses in this edition. Recent information on voltage-gated and ligand-gated ion channels has also been incorporated. The sections on antihypertensive, antiviral, antibacterial, anti-inflammatory, antiarrhythmic, and anti-cancer agents, as well as treatments for hyperlipidemia and for peptic ulcer, have been substantially expanded.

Despite these many changes, the overall structure and philosophy of the book remain unchaged. Therapeutic agents are organized according to their targets – the conceptual centerpiece of the first two editions. The nine chapters of the third edition are grouped in two parts: the basic principles of medicinal chemistry (chapters 1–3), and applications of medicinal chemistry from a target-centered viewpoint (chapters 4–9). Given this organizational structure, the book is not a catalogue of drugs. It does not present medicinal chemistry in a "telephone directory of drugs" way. Rather, it emphasizes the understanding of mechanisms of drug action, which includes drug and receptor structure. The book's target-centered philosophy facilities a clear, mechanistic understanding of how and why drugs work. This should give students a conceptual framework that will enable them to continue learning about drugs and drug action long after they have left school.

As with the first and second editions, this text is aimed primarily at students of pharmacy, pharmacology and chemistry who are interested in drug design and development. It provides the core of biochemical- and molecular-level thinking about drugs needed for a basic medicinal chemistry course. Another new feature of this edition is designed to enhance the book's appeal to all readers: the multiple sections on the "Clinical–Molecular Interface." These sections strengthen the book's clinical relevance by making it easier to understand the treatment of human disease at a molecular level.

Many co-workers, colleagues and reviewers have given their time, expertise and insights to aid the development of this third edition. Dr. Chris Barden (Department of Chemistry, Dalhousie University) provided detailed remarks on the entire book. Joshua Tracey checked molecular structures for accuracy, providing extensive assistance with molecular formulae; Vanessa Stephenson checked references and the suggested reading citations; and Dawnelda Wight provided clerical assistance with tables. Cheryl Weaver, Felix Meier, Vanessa Stephenson, Valerie Compagna-Slater, Michael Carter, Buhendwa Musole, Kathryn Tiedje, and Colin Weaver provided additional assistance with figures and diagrams. To all of them we offer our thanks.

In addition, one of us (DFW) wishes to express his gratitude to Dr. R. A. Purdy, Head, Division of Clinical Neurology, Dalhousie University, for his generous "protection of time" to provide the many hours necessary for the revision of this book. We also thank the editorial staff of Oxford University Press, Jeffrey House in particular, and Edith Barry, for working on the second and third edition, and for their never-ending patience. As with previous editions, we look forward to a continuing dialogue with our readers so that future editions can be further improved.

T. N.
Kingston, Ontario

D. F. W.
Halifax, Nova Scotia

April, 2005

Contents

3. Basic Principles of Drug Design III—Designing Drug Molecules to Fit Receptors   106

II BIOCHEMICAL CONSIDERATIONS IN DRUG DESIGN: FROM DRUGGABLE TARGETS TO DISEASES

Introduction to Part II   185

4. Messenger Targets for Drug Action I: Neurotransmitters and Their Receptors   193

Medicinal Chemistry

I
GENERAL MOLECULAR PRINCIPLES
OF DRUG DESIGN

Introduction to Part I

Designing drug molecules to alleviate human disease and suffering is a daunting yet exhilarating task. How does one do it? How does a researcher sit down, paper in hand (or, better yet, a blank computer screen), and start the process of creating a molecule as a potential drug with which to treat human disease? What are the thought processes? What are the steps? How does one select a target around which to design a drug molecule? When a researcher does design a molecule, how does she or he know if it has what it takes to be a drug?

These are important questions. The previous century ended with an explosion of activity in gene-related studies and stem cell research; the new one is emerging as the "Century of Biomedical Research." We have now witnessed the global spectre of SARS (Severe Acute Respiratory Syndrome) and avian flu, which has emphasized the looming importance of infectious disease to global health. Concerns about the capacity of "Mad Cow" disease to infect humans have focused attention on the safety of our food supply. AIDS and obesity-related disorders have not gone away, but rather are increasing in incidence and prevalence. Long-recognized diseases, such as stroke and Alzheimer's dementia, are becoming more common as a greater proportion of the human population reaches old age. Not surprisingly, the need for drug discovery to address these important diseases is increasingly being recognized as a societal priority.

Not only is drug discovery important to the medical health of humankind, it is also an important component of our economic health. New chemical entities (NCEs) as therapeutics for human disease may become the "oil and gas" of the 21st century. As the world's population increases and health problems expand accordingly, the need to discover new therapeutics will become even more pressing. In this effect, the design of drug molecules arguably offers some of the greatest hopes for success.

DRUG DESIGN: A CONCEPTUAL APPROACH

Successful drug design is multi-step, multidisciplinary and multi-year. Drug discovery is not an inevitable consequence of fundamental basic science; drug design is not merely a technology that generates drugs for humans on the basis of biological advances—if it were that simple, more and better drugs would already be available. Medicinal chemistry is a science unto itself, a central science positioned to provide a molecular bridge between the basic science of biology and the clinical science of medicine (analogous to chemistry being the central science between the traditional disciplines of biology and physics). From a very broad perspective, drug design may be divided into two phases:

1. Basic concepts about drugs, receptors, and drug–receptor interactions (chapters 1–3).
2. Basic concepts about drug–receptor interactions applied to human disease (chapters 4–9).

The first phase comprises the essential building blocks of drug design and may be divided into three logical steps:

1. Know what properties turn a molecule into a drug (chapter 1).
2. Know what properties turn a macromolecule into a drug receptor (chapter 2).
3. Know how to design and synthesize a drug to fit into a receptor (chapter 3).

Knowledge of these three steps provides the necessary background required for a researcher to sit down, paper in hand, and start the process of creating a molecule as a potential drug for treating human disease.

Step 1 involves knowing what properties turn a molecule into a drug. All drugs may be molecules, but all molecules are certainly not drugs. Drug molecules are "small" organic molecules (molecular weight usually below 800 g/mol, often below 500). Penicillin, acetylsalicyclic acid, and morphine are all small organic molecules. Certain properties (geometric, conformational, stereochemical, electronic) must be controlled if a molecule is going to have what it takes even to emerge as a drug-like molecule (DLM). When designing a molecule to be a drug-like molecule and, hopefully, a drug, the designer must have the ability to use diverse design tools. Now, computer-aided molecular design (CAMD) is one of the most important design tools available. CAMD incorporates various rigorous mathematical techniques, including molecular mechanics and quantum mechanics. When using CAMD to design a drug, one must remember that a drug molecule is complex and has sub-unit parts. Some of these parts enable the drug to interact with its receptor, while other parts permit the body to absorb, distribute, metabolize, and excrete the drug molecule. Once a drug-like molecule successfully becomes a candidate for the treatment of a disease, it has graduated to the status of drug molecule.

Step 2 involves knowing what properties turn a macromolecule into a receptor. All receptors may be macromolecules, but all macromolecules are certainly not receptors. Receptor macromolecules are frequently proteins or glycoproteins. Certain properties must be present if a macromolecule is going to have what it takes to be a druggable target. The receptor macromolecule must be intimately connected with the disease in question, but not integral to the normal biochemistry of a wide range of processes.

Step 3 involves designing a specific drug-like molecule to fit into a particular drug-gable target. During this task many molecules will be considered, but only one (or two) will emerge as promising starting points around which to further elaborate the design process. This prototype compound is referred to as the *lead compound*. There is a variety of ways of identifying a potential lead compound, including rational drug design, random high throughput screening, and focused library screening. Once a lead compound has been successfully identified, it must be optimized. Optimization may be achieved using quantitative structure–activity relationship (QSAR) studies. Synthetic organic chemistry is a crucial component of this step in drug development. The process of drug design must be validated by actually making and testing the drug molecule. An ideal synthesis should be simple, be efficient, and produce the drug in high yield and high purity.

Once the basics of drug design are in place, the drug designer next focuses upon the task of connecting a drug–receptor interaction to a human disease—this is the goal of the second phase. For example, how does one design a drug for the treatment of cancer or Alzheimer's disease? This phase of drug design requires an understanding of biochemistry and of the molecular pathology of the disease being treated.

The human body normally moves through time with its various molecular processes functioning in a balanced, harmonious state, called *homeostasis*. When disease occurs, this balance is perturbed by a pathological process. For a drug molecule, the goal is to rectify this perturbation (via the action of molecular therapeutics) and to return the body to a state of healthy homeostasis. Logically, there are many approaches to attaining this therapeutic goal. First, one may ask what are the body's normal inner (*endogenous*) control systems for maintaining homeostasis through day-to-day or minute-to-minute adjustments? These control systems (for example, neurotransmitters, hormones, immunomodulators) are the first line of defense against perturbations of homeostasis. Is it possible for the drug designer to exploit these existing control systems to deal with some pathological process? If there are no endogenous control systems, how about identifying other targets on endogenous cellular structures or macromolecules that will permit control where endogenous control has not previously existed? Alternatively, instead of pursuing these endogenous approaches, it is sometimes easier simply to attack the cause of the pathology. If there is a harmful microorganism or toxin in the environment (*exogenous*), then it may be possible to directly attack this exogenous threat to health and inactivate it. Accordingly, this phase of drug development, which connects the drug–receptor interaction to human disease, may be divided into three logical approaches:

1. Know how to manipulate the body's endogenous control systems (chapters 4–6).
2. Know how to manipulate the body's endogenous macromolecules (chapters 7 and 8).
3. Know how to inactivate a harmful exogenous substance (chapter 9).

A full understanding of the three steps of phase 1 and the three approaches of phase 2 will enable the researcher to design drugs.

DRUG DESIGN: A PRACTICAL APPROACH

This book aims to put forth a strategy to facilitate the insightful design of new chemical entities as therapies for human disease—a strategy that will foster the ability to sit down in front of a blank computer screen and draw molecules that may help cure the various maladies that afflict humankind. This strategy uses a molecular-level understanding of human biochemistry and pathology to drive the design of drug-like molecules engineered to fit precisely into targets of drug action (druggable targets).

A Drug as a Composite of Molecular Fragments

For the practical implementation of this idealistic strategy, drug molecules are conceptualized as being assembled from biologically active building blocks (*biophores*) that are covalently "snapped together" to form an overall molecule. Thus, a drug molecule is a *multiphore*, composed of a fragment that enables it to bind to a receptor (*pharmacophore*), a fragment that influences its metabolism in the body (*metabophore*), and one or more fragments that may contribute to toxicity (*toxicophores*). The drug designer should have the ability to optimize the pharmacophore while minimizing the number of toxicophores. To achieve this design strategy, these fragments or building blocks may be replaced or interchanged to modify the drug structure. Certain building blocks (called *bioisosteres*), which are biologically equivalent but not necessarily chemically equivalent, may be used to promote the optimization of the drug's biological properties.

DRUG DESIGN: THE HUMANITARIAN APPROACH

In traditional medicine there are two major therapeutic approaches to the treatment of human disease: surgical and medical. Surgical procedures are labour intensive and time demanding; they help a limited number of individuals, one at a time, mostly in rich or developed nations. Medical therapy, on the other hand, is based on drug molecules and thus has the capacity to positively influence the lives of more people, often over a shorter time frame. Medical therapeutics offer hope in both developed and developing parts of the world—hopefully to rich and poor alike.

After public health measures (e.g., safe drinking water, hygienic disposal of waste water), the discovery of drugs has had one of the largest beneficial effects on human health. Penicillin has saved countless lives through the effective treatment of devastating infectious diseases. Before penicillin, a diagnosis of meningococcal meningitis was invariably a death sentence. Penicillin reduced bacterial meningitis to a treatable disorder. Similarly, drugs for the treatment of high blood pressure have substantially reduced the impact of this "silent killer" that leads to myocardial infarction (heart attack) or cerebral infarction (stroke).

It can be awe-inspiring to witness the effects of a seemingly trivial amount of drug. The panic-stricken child who cannot breathe because of an asthma attack gets prompt relief from the inhalation of a mere 100 micrograms of salbutamol sulphate. Uncontrolled and potentially life-threatening seizures (*status epilepticus*) in a young adult are quickly brought under control with the intravenous administration of 2 mg of lorazepam. The terrified older adult with crushing chest pain from a myocardial infarction gains rapid relief from 8 to 10 mg of morphine. Drugs are truly amazing molecules.

A medicinal chemist can help thousands or even millions of people with a carefully designed new drug molecule. The practice of science is a very human activity; medicinal chemistry is a humanitarian science.

1

Basic Principles of Drug Design I

Drug molecules: structure and properties

Most drugs are molecules, but most molecules are not drugs. Every year, millions of new molecules are prepared, but only a very small fraction of these are ever considered as possible drug candidates. A chemical compound must possess certain characteristics if it is to cross the hurdle from being an organic molecule to becoming a drug molecule. *Medicinal chemistry* is the applied science that is focused on the design (or discovery) of new chemical entities (NCEs) and their optimization and development as useful drug molecules for the treatment of disease processes. In achieving this mandate, the medicinal chemist must design and synthesize new molecules, ascertain how they interact with biological macromolecules (such as proteins or nucleic acids), elucidate the relationship between their structure and biological activities, determine their absorption and distribution throughout the body, and evaluate their metabolic transformations. Not surprisingly, medicinal chemistry is multidisciplinary, drawing on theoretical chemistry, organic chemistry, analytical chemistry, molecular biology, pharmacology, and biochemistry. Despite these complexities, medicinal chemistry has a clear "bottom line"—the design and discovery of drug molecules.

1.1 DEFINITION AND PROPERTIES OF A DRUG MOLECULE

1.1.1 What Is a Drug Molecule? What Is a Drug-Like Molecule?

A molecule is the smallest particle of a substance that retains the chemical identity of that substance; it is composed of two or more atoms held together by chemical bonds (i.e., shared electron pairs). Although molecules are highly variable in terms of structure, they may be organized into families on the basis of certain groupings of atoms called functional groups. A functional group is an assembly or cluster of atoms that generally reacts in the same way, regardless of the molecule in which it is located; for example, the carboxylic acid functional group (-COOH) generally imparts the property of acidity to any molecule in which it is inserted. It is the presence of functional groups that determines the chemical and physical properties of a given family of molecules. A functional group is a centre of reactivity in a molecule.

A *drug molecule* possesses one or more functional groups positioned in three-dimensional space on a structural framework that holds the functional groups in a defined geometrical array that enables the molecule to bind specifically to a targeted biological macromolecule, the *receptor*. The structure of the drug molecule thus permits a desired biological response, which should be beneficial (by inhibiting pathological processes) and which ideally precludes binding to other untargeted receptors, thereby minimizing the probability of toxicity. The framework upon which the functional groups are displayed is typically a hydrocarbon structure (e.g., aromatic ring, alkyl chain) and is usually chemically inert so that it does not participate in the binding process. The structural framework should also be relatively rigid ("conformationally constrained") to ensure that the array of functional groups is not flexible in its geometry, thus preventing the drug from interacting with untargeted receptors by altering its molecular shape. To be successful in countering a disease process, however, a drug molecule must have additional properties beyond the capacity to bind to a defined receptor site. It must be able to withstand the journey from its point of administration (i.e., the mouth for an orally administered drug) until it finally reaches the receptor site deep within the organism (i.e., the brain for a neurologically active drug).

A *drug-like molecule* (DLM) possesses the chemical and physical properties that will enable it to become a drug molecule should an appropriate receptor be identified (see figure 1.1). What are the properties that enable a molecule to become a drug–like molecule? In general, a molecule should be small enough to be transported throughout the body, hydrophilic enough to dissolve in the blood stream, and lipophilic enough to cross fat barriers within the body. It should also contain enough polar groups to enable it to bind to a receptor, but not so many that it would be eliminated too quickly from the body via the urine to exert a therapeutic effect. Lipinski's Rule of Five does a good job of quantifying these properties. According to this rule, a drug-like molecule should have a molecular weight less than 500, a logP (logarithm of its octanol–water partition coefficient) value less than 5, fewer than five hydrogen bonding donors, and less than 10 hydrogen bonding acceptors.

1.1.2 Structural Integrity of a Drug Molecule: Pharmaceutical, Pharmacokinetic and Pharmacodynamic Phases

Although a drug molecule may be administered in many different formulations, oral administration as a tablet is the most common form. Following oral administration, the drug molecule journeys from the gastrointestinal tract throughout the body until it reaches the drug receptor. During this journey "from gums to receptor," the drug molecule traverses many phases (pharmaceutical, pharmacokinetic, and pharmacodynamic) and is subjected to multiple assaults on its structural and chemical integrity (see figure 1.2).

1.1.2.1 Pharmaceutical Phase

The pharmaceutical phase is the time from the point of administration of the drug molecule until it is absorbed into the circulation of the body. For an orally administered drug, the pharmaceutical phase starts in the mouth and ends when the drug is absorbed across the intestinal wall. A drug may be administered either "systemically," which involves

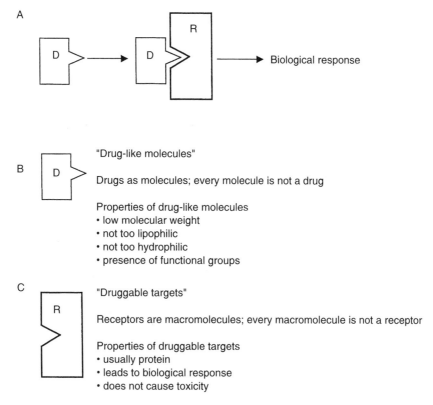

Figure 1.1 Drug-like molecules and druggable targets. Certain properties permit a molecule to become a drug-like molecule and certain properties permit a macromolecule to become a druggable target. When a drug-like molecule interacts with a druggable target to give a biological response, it becomes a drug molecule and the druggable target becomes a *receptor*. When a drug molecule is successfully and beneficially distributed to people with a disease, it becomes a useful drug molecule.

the drug entering the bloodstream and being distributed throughout the entirety of the body, or "locally," which involves site-specific administration directly onto the region of pathology. Systemic administration may be achieved by the following routes: (1) via the gastrointestinal tract (usually orally, sometimes rectally); (2) parenterally, using intravenous, sub-cutaneous, intramuscular, or (rarely) intra-arterial injection; (3) topically, in which the drug is applied to the skin and is absorbed transdermally into the body to be widely distributed via the bloodstream; or (4) by direct inhalation into the lungs.

The most frequent route of administration is oral. From the perspective of a drug designer who is endeavoring to engineer drug molecules, many factors must be taken into consideration when designing a drug for oral administration. On its journey from the mouth (the point of first administration) to the drug's receptor deep within the organ systems of the body, the drug molecule undergoes a variety of potential assaults to the integrity of its chemical structure. This attack begins in the mouth where saliva contains digestive enzymes such as ptyalin or salivary α-amylase. The drug molecule next enters

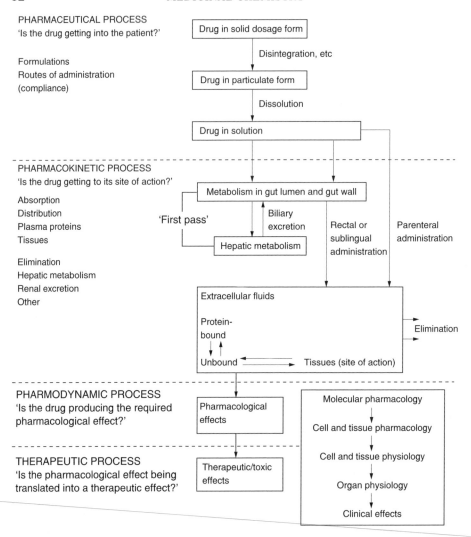

Figure 1.2 The three phases of drug processing. The journey from the point of administration to the microenvironment of the receptor is a complex and arduous journey for the drug molecule. (Adapted from D. G. Grahame-Smith, J. K. Aronson (2002). *Clinical Pharmacology and Drug Therapy*, 3rd Edn. New York: Oxford University Press. With permission.)

the stomach at which point it is subjected to a pH of 1.8–2.2, as well as to a variety of pepsin enzymes. Under such acidic conditions, certain functional groups, such as esters, are vulnerable to hydrolysis—an important point of consideration during drug design. From the stomach, the drug molecule sequentially enters the three portions of the small intestine: duodenum, jejunum, and ileum. Within the small intestine the pH is alkalinized to 7.8–8.4, and the drug molecule is subjected to a complex array of intestinal and pancreatic enzymes including peptidases, elastase, lipases, amylase, lactase, sucrase,

Table 1.1 pH Values for Tissue Fluids

Fluid	pH
Aqueous humor (eye)	7.2
Blood, arterial	7.4
Blood, venous	7.4
Blood, maternal umbilical	7.3
Cerebrospinal fluid	7.4
Duodenum	4.5–7.8
Intestine	6.0–8.3
Lacrimal fluid (tears)	7.4
Milk, breast	7.0
Nasal secretions	6.0
Prostatic fluid	6.5
Saliva	6.4
Semen	7.2
Stomach	1.8
Sweat	5.4
Urine	5.6–7.0
Vaginal secretions, premenopause	4.5
Vaginal secretions, postmenopause	7.0

phospholipase, ribonuclease, and deoxyribonuclease. The drug designer must consider these environments of varying pH combined with digestive enzymes when selecting functional groups to be incorporated into a drug molecule. Table 1.1 presents pH values for a variety of tissue fluids.

The pharmaceutical phase also includes the process of drug absorption from the gastrointestinal tract into the body fluids. In general, little absorption of a drug molecule occurs in the stomach since the surface area is relatively small. Absorption takes place mainly from the intestine where the surface area is greatly expanded by the presence of many villi, the small folds in the intestinal surface. Drug absorption across the gastrointestinal lining (which may be regarded functionally as a lipid barrier) occurs mainly via passive diffusion. Accordingly, the drug molecule should be largely un-ionized at the intestinal pH to achieve optimal diffusion/absorption properties. The most significant absorption occurs with weakly basic drugs, since they are neutral at the intestinal pH. Weakly acidic drugs, on the other hand, are more poorly absorbed since they tend to be un-ionized in the stomach rather than in the intestine. Consequently, weakly basic drugs have the greatest likelihood of being absorbed via passive diffusion from the gastrointestinal tract. Table 1.2 provides ionization constants for a variety of weakly basic and weakly acidic drugs.

A final point of consideration (at the pharmaceutical phase) when designing drugs for oral administration concerns product formulation. A pill is not simply a compressed mass of drug molecules. Rather, it is a complicated mixture of fillers, binders, lubricants, disintegrants, colouring agents, and flavoring agents. If a drug molecule is biologically

Table 1.2 Ionization Constants of Common Drugs

Weak acids	pK_a	Weak bases	pK_a
Acetaminophen	9.5	Alprenolol	9.6
Acetazolamide	7.2	Amiloride	8.7
Ampicillin	2.5	Amiodarone	6.5
Aspirin	3.5	Amphetamine	9.8
Chlorpropamide	5.0	Atropine	9.7
Cromolyn	2.0	Bupivacaine	8.1
Ethacrynic acid	2.5	Chlordiazepoxide	4.6
Furosemide	3.9	Chlorpheniramine	9.2
Levodopa	2.3	Chlorpromzaine	9.3
Methotrexate	4.8	Clonidine	8.3
Penicillamine	1.8	Codeine	8.2
Pentobarbital	8.1	Desipramine	10.2
Phenobarbital	7.4	Diazepam	3.0
Phenytoin	8.3	Diphenhydramine	8.8
Propylthiouracil	8.3	Diphenoxylate	7.1
Salicylic acid	3.0	Ephedrine	9.6
Sulfadiazine	6.5	Epinephrine	8.7
Sulfapyridine	8.4	Ergotamine	6.3
Theophylline	8.8	Hydralazine	7.1
Tolbutamide	5.3	Imipramine	9.5
Warfarin	5.0	Isoproterenol	8.6
		Kanamycin	7.2
		Lidocaine	7.9
		Methadone	8.4
		Methamphetamine	10.0
		Methyldopa	10.6
		Metoprolol	9.8
		Morphine	7.9
		Norepinephrine	8.6
		Pentazocine	7,9
		Phenylephrine	9.8
		Pindolol	8.6
		Procainamide	9.2
		Procaine	9.0
		Promazine	9.4
		Promethazine	9.1
		Propranolol	9.4
		Pseudoephedrine	9.8
		Pyrimethamine	7.0
		Scopolamine	8.1
		Terbutaline	10.1
		Thioridazine	9.5

active at an oral dose of 0.1 mg, then fillers are necessary to ensure that the pill is large enough to be seen and handled. Additional *excipient* additives are required to permit the pill to be compressed into a tablet (binders), to pass through the gastrointestinal tract without sticking (lubricants), and to burst open so that it can be absorbed in the small intestine (disintegrants). Fillers include dextrose, lactose, calcium triphosphate, sodium chloride, and microcrystalline cellulose; binders include acacia, ethyl cellulose, gelatin, starch mucilage, glucose syrup, sodium alginate, and polyvinyl pyrrolidone; lubricants include magnesium stearate, stearic acid, talc, colloidal silica, and polyethylene glycol; disintegrants include starch, alginic acid, and sodium lauryl sulphate. The importance of this design consideration follows a 1968 Australasian outbreak of phenytoin drug toxicity caused by the replacement of an excipient in a marketed formulation of an anti-seizure drug called phenytoin; the new excipient chemically interacted with the phenytoin drug molecule, ultimately producing toxicity.

1.1.2.2 Pharmacokinetic Phase

Once the drug molecule has been released from its formulation, it enters the pharmacokinetic phase. This phase covers the time duration from the point of the drug's absorption into the body until it reaches the microenvironment of the receptor site. During the pharmacokinetic phase, the drug is transported to its target organ and to every other organ in the body. In fact, once absorbed into the bloodstream, the drug is rapidly transported throughout the body and will have reached every organ in the body within four minutes. Since the drug is widely distributed throughout the body, only a very small fraction of the administered compound ultimately reaches the desired target organ—a significant problem for the drug designer. The magnitude of this problem can be appreciated by the following simple calculation. A typical drug has a molecular weight of approximately 200 and is administered in a dose of approximately 1 mg; thus, 10^{18} molecules are administered. The human body contains almost 10^{14} cells, with each cell containing at least 10^{10} molecules. Therefore, each single administered exogenous drug molecule confronts some 10^6 endogenous molecules as potential available receptor sites—the proverbial "one chance in a million."

In addition to this statistical imbalance, the drug molecule also endures a variety of additional assaults during the pharmacokinetic phase. While being transported in the blood, the drug molecule may be bound to blood proteins. The degree of protein binding is highly variable. Highly lipophilic drugs do not dissolve well in the aqueous serum and thus will be highly protein bound for purposes of transport. If a person is taking more than one drug, various drugs may compete with each other for sites on the serum proteins. Human serum albumin (HSA) is one of the proteins commonly involved in drug transportation. Table 1.3 gives the percentage protein binding for a diversity of common drugs.

During this transport process, the drug is exposed to metabolic transformations that may chemically alter the integrity of its chemical structure. This metabolic attack is most likely to occur during passage through the liver. In fact, some drug molecules are completely transformed to biologically inactive metabolites during their first pass through the liver; this is the so-called *first pass effect*. A complete first pass effect renders a drug molecule useless since it is metabolically transformed to an inactive form prior to reaching

Table 1.3 Percentage Protein Binding for Common Drugs

99%	95–99%	90–95%	50–90%	<50%
Levothyroxine	Amitriptyline	Diazoxide	Aspirin	Alcohol
Phenylbutazone	Chlorpromazine	Disopyramide	Carbamazepine	Aminoglycosides
Triiodothyronine	Clofibrate	Phenytoin	Chloramphenicol	Digoxin
Warfarin	Diazepam	Propranolol	Chloroquine	Paracetamol
	Furosemide	Tolbutamide	Lidocaine	Procainamide
	Gold salts	Valproate	Quinidine	
	Heparin		Simvastatin	
	Imipramine		Sulfonamides	

(Adapted from D. G. Grahame-Smith, J. K. Aronson (2002). *Clinical Pharmacology and Drug Therapy*, 3rd Edn. New York: Oxford University Press. With permission.)

any possible receptor site. Due to the anatomical arrangement of blood vessels in the abdomen, all orally administered drugs must immediately pass through the liver following absorption from the small intestine. Accordingly, a drug molecule that is susceptible to a first pass effect should in theory be designed and formulated in a manner that minimizes small intestine absorption. One method of reducing a first pass effect is to administer the drug sublingually so that it is absorbed under the tongue and has an opportunity of avoiding the initial pass through the liver. See figure 1.3 for anatomical details of the three phases that a drug must endure in traveling to its site of action.

Like the liver, the kidney is another organ system that may influence the effectiveness of a drug molecule during the pharmacokinetic phase. Small, hydrophilic, and highly polar molecules (e.g., sulphonates, phosphonates) run a significant chance of being rapidly excreted via the renal system. Such molecules have short half-lives (the period of time during which one-half of the drug molecules is excreted). A short half-life reduces the effectiveness of a drug molecule because it shortens the time duration available to the drug for distribution and binding to its receptor. In addition, as a general rule, a drug is administered at least once every half-life; a drug with a half-life of 24 hours may be administered once per day whereas a drug with a 12 h half-life must be given at least twice per day. If a drug has a half-life of 20 minutes it would be impractical to administer it three times per hour. Table 1.4 presents the half-lives for a variety of drug molecules.

The final impediment to drug molecule effectiveness during the pharmacokinetic phase is the existence of barriers. In order to reach its target organ, the drug molecule must traverse a variety of membranes and barriers. This is particularly true if the drug is destined to enter the brain, which is guarded by the blood–brain barrier. This is a lipid barrier composed of endothelial tight junctions and astrocytic processes. The blood–brain barrier can be exploited for purposes of drug design. Molecules can be designed not to cross this barrier. This design feature is highly desirable if one wishes to develop drug molecules for non-neurologic indications that will have no neurologic side effects. On the other hand, the existence of the blood–brain barrier must be explicitly considered when designing drugs for neurological indications. Another highly relevant barrier is the maternal–placental barrier. This must be considered when designing drugs for women of childbearing age. The maternal–placental barrier is a lipid barrier much like

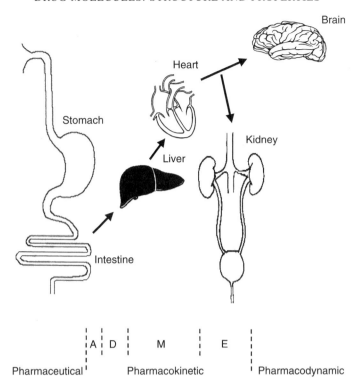

Figure 1.3 The three phases of drug processing. Different organ systems inflict varying degrees of assault on the integrity of the drug molecule during its journey to the receptor. Stomach acid initiates the assault. Liver enzymes may destroy the drug in a first pass effect. If the drug is too polar, the kidney will rapidly excrete it.

the blood–brain barrier and most drugs designed to enter the brain will likewise traverse the maternal–placental barrier.

1.1.2.3 Pharmacodynamic Phase

After a drug molecule has surmounted the barriers of the pharmacokinetic phase and has been distributed throughout the body, it ultimately reaches the microenvironment of the receptor where its biological effect will be exerted. Once the drug molecule has entered the region of its receptor, it is in the pharmacodynamic phase. During this phase, the molecule binds to its receptor through the complementarities of their molecular geometries. The functional groups of the drug molecule interact with corresponding functional groups of the receptor macromolecule via a variety of interactions, including ion–ion, ion–dipole, dipole–dipole, aromatic–aromatic, and hydrogen bonding interactions. The binding of the drug molecule to its receptor enables the desired biological response to occur. The nature of the drug–receptor interaction is described in detail in chapter 2.

Table 1.4 Half-Lives of Selected Drugs in Patients (with Normal Renal Function)

<1 h	1–4 h	4–12 h	12–24 h	1–2 days	>2 days
Adenosine	Aminoglycosides	Acetazolamide	Bromocriptine	Amlodipine	Amiodarone
Cocaine	Atropine	Acyclovir	Chlorpromazine	Carbamazepine	Bisphosphonates
Dobutamine	Azathioprine	Amiloride	Clonidine	Chlorpropamide	Chloroquine
Dopamine	Captopril	Caffeine	Doxycycline	Clonazepam	Phenobarbital
Iloprost	Cephalosporins	Chloramphenicol	Fluvoxamine	Diazoxide	Thyroxine
Naloxone	Cimetidine	Clomethiazole	Haloperidol	Digoxin	
Nitroprusside	Ciprofloxacin	Clozapine	Minocycline	Triiodothyronine	
Penicillins	Colchicine	Diltiazem	Ouabain	Warfarin	
Succinylcholine	Diclofenac	Gabapentin			
	Erythromycin	Hydralazine			
	Ethambutol	Ketoconazole			
	Furosemide	Metronidazole			
	Ibuprofen	Quinidine			
	Isoniazid	Theophylline			
	Isosorbide	Tolbutamide			
	mononitrate	Trimethoprim			
	Levodopa	Valproate			
	Lidocaine	Vigabatrin			
	Morphine				
	Ondansetron				
	Pravastatin				
	Procainamide				
	Ranitidine				
	Sumatriptan				

(Adapted from D. G. Grahame-Smith, J. K. Aronson (2002). *Clinical Pharmacology and Drug Therapy*, 3rd Edn. New York: Oxford University Press. With permission.)

1.1.3 Structural Fragments of a Drug Molecule: Pharmacophore, Toxicophore, Metabophore

As previously defined, a drug molecule consists of functional groups displayed in a defined geometric array that permits a binding interaction with a receptor during the pharmacodynamic phase of drug action. The three-dimensional arrangement of atoms within a drug molecule that permits a specific binding interaction with a desired receptor is called the *pharmacophore*. The atoms that constitute the pharmacophore are a subset of all the atoms within the drug molecule. The pharmacophore is the bioactive face of the molecule and is that portion of the molecule that establishes intermolecular interactions with the receptor site. (In principle, the term pharmacophore is an abstract concept. A pharmacophore is the assembly of geometric and electronic features required by a drug molecule to ensure both an optimal supramolecular interaction with its target receptor and the elicitation of a biological response. The term pharmacophore does not represent a single real molecule but a portion of a molecule. It is incorrect to name a structural skeleton, such as a phenothiazine or a prostaglandin, as a pharmacophore. It is correct, however, to regard a pharmacophore as the common structural denominator shared by a set of bioactive molecules; the pharmacophore accounts for the shared molecular interaction capabilities of a group of structurally diverse drug molecules toward a common target receptor.)

Depending on which face it puts forward, a single drug molecule may interact with more than one receptor and thus may have more than one *pharmacophoric pattern*. For example, one bioactive face of acetylcholine permits interaction with a muscarinic receptor, while another bioactive face of acetylcholine permits interaction with a nicotinic receptor (section 4.2). Similarly, the excitatory neurotransmitter glutamate may bind to a range of different receptors, such as the NMDA and AMPA receptors (section 4.7), depending upon the pharmacophoric pattern displayed by the glutamate molecule toward the receptor with which it is interacting.

The other portions of the drug molecule that are not part of the pharmacophore constitute *molecular baggage.* The role of this molecular baggage is to hold the functional group atoms of the pharmacophore in a fixed geometric arrangement (with minimal conformational flexibility) to permit a specific receptor interaction while minimizing *both* interactions with toxicity-mediating receptors *and* the metabolic (via liver) and rapid excretion (via kidney) problems associated with the pharmacokinetic phase.

Two other less frequently discussed fragments of a drug molecule are the *toxicophore* and the *metabophore*. Conceptually, these two types of fragment are analogous to the pharmacophore. (Collectively, pharmacophores, toxicophores, and metabophores may be referred to as *biophores*.) The toxicophore is the three-dimensional arrangement of atoms in a drug molecule that is responsible for a toxicity-eliciting interaction. If a drug molecule has multiple toxicities arising from several undesirable interactions, then it may possess more than one toxicophore. From the perspective of drug design, if a toxicophore does not overlap with the pharmacophore in a given drug molecule, then it may be possible to redesign the molecule to eliminate the toxicity. However, if the pharmacophore and toxicophore are congruent molecular fragments, then the toxicity is inseparable from the desired pharmacological properties. The metabophore is the three-dimensional

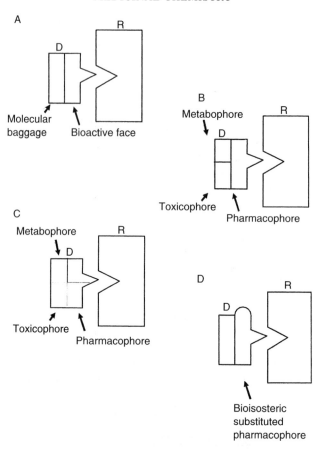

Figure 1.4 A drug molecule contains many parts. The bioactive face is the portion of the drug molecule that interacts with the receptor; the remainder of the molecule, called molecular baggage, holds the bioactive face in a desired geometry. The pharmacophore is the arrangement of molecules that permits the bioactive face to interact with the receptor. The toxicophore is the fragment that is responsible for toxicity; the metabophore is the fragment that is responsible for metabolism. If these various fragments are separate (as in B), then toxicity can be "designed out of the drug molecule"; if they overlap (as in C), then it may be impossible to separate the toxicophore from the pharmacophore. It is sometimes possible to replace all or part of the pharmacophore with a biologically equivalent fragment called a bioisostere.

arrangement of atoms in a drug molecule that are responsible for the metabolic properties. Since functional groups are responsible not only for drug–receptor interactions but also for metabolic properties, the metabophore and the pharmacophore tend to be inextricably overlapped. Nevertheless, from the viewpoint of drug design, it is sometimes possible to manipulate the structure of either the pharmacophore or the molecular baggage portions of the drug molecule to achieve a metabophore that overcomes problems with liver-mediated first pass effects or that either hastens or delays renal excretion (see figure 1.4).

1.1.4 Structural Fragments of Drug Molecules: Interchangeable Bioisosteres

A drug molecule may be conceptualized as a collection of molecular fragments or building blocks. The most important fragment is the pharmacophore, with the functional groups of the pharmacophore being displayed on a molecular framework composed of metabolically inert and conformationally constrained structural units. These structural units may be an alkyl chain, an aromatic ring, or a section of peptide chain backbone. When designing or constructing a drug molecule, one can thus pursue a fragment-by-fragment building block approach. In conceptualizing this approach, one sees that certain molecular fragments, although structurally distinct from each other, may behave identically within the biological milieu of the receptor microenvironment. These structurally distinct yet biofunctionally equivalent molecular fragments are referred to as *bioisosteres*. (A bioisosteric drug is a drug molecule that arises from the replacement of either an atom or a group of atoms with a biologically equivalent atom or group of atoms to create a new molecule with pharmacological properties similar to those of the parent molecule.)

There are many examples of bioisosteric substitutions. For example, a drug that contains a sulphonate functional group (SO_3^-) within its pharmacophore may interact with a receptor via an electrostatic interaction, whereby the negatively charged sulphonate group interacts with a positively charged ammonium within the receptor. In designing analogs of this drug, it would be possible to replace the sulphonate with a bioisosterically equivalent carboxylate group. The carboxylate group would be able to interact electrostatically with the ammonium functional group in a fashion analogous to the sulphonate moiety. This bioisosteric substitution would bring additional advantages such as a prolonged half-life for the drug molecule since the carboxylate is less polar than the sulphonate and is thus less susceptible to rapid renal excretion. There are many other examples of bioisosteric substitutions. For example, H- may be replaced by F-; a carbonyl group (C=O) may be replaced by a thiocarbonyl group (C=S); a sulphonate may be replaced by a phosphonate.

Bioisosteric substitutions may be categorized as *classical* or *non-classical*. Classical bioisosteres are functional groups that possess similar valence electron configurations. For example, oxygen and sulphur are both in column VI of the periodic table; thus, a thio–ether (-C-S-C-) is a classical bioisosteric substitution for an ether (-C-O-C-) functional group. Non-classical bioisosteres are functional groups with dissimilar valence electron configurations; for instance, a tetrazole moiety may be used to replace a carboxylate since many biological systems are unable to differentiate between these two very structurally distinctive functional groups (see figure 1.5).

A consideration of bioisosterism is important in drug design. A systematic exploration of bioisosteres when constructing drug molecules as collections of molecular fragments enables a rigorous structural consideration of varying pharmacophores and their properties during the pharmaceutical, pharmacokinetic, and pharmacodynamic phases of drug action.

1.1.5 Structural Properties of Drug Molecules

A drug molecule is a collection of molecular fragments held in a three-dimensional arrangement that determines and defines all of the properties of the drug molecules.

Classical bioisosteres

1. Monovalent atoms or groups:

OH NH$_2$ CH$_3$ Cl F H

SH PH$_2$

Br i-Pr

I

2. Bivalent atoms or groups:

—CH$_2$— —NH— —O— —S—

—COCH$_2$R —CONHR —CO$_2$R —COSR

3. Trivalent atoms:

—CH═ —N═

—P═ —As═

4. Tetravalent atoms:

—C— —Si— —N$^+$— —P$^+$—

Nonclassical bioisosteres

1. Functional group replacements:

a. Halogens:

X CF$_3$ CN N(CN)$_2$ C(CN)$_3$

b. Hydroxyl group:

OH CH$_2$OH CH(CN)$_2$

NHCOR NHSO$_2$R NHCN NHCONHR

c. Carbonyl group:

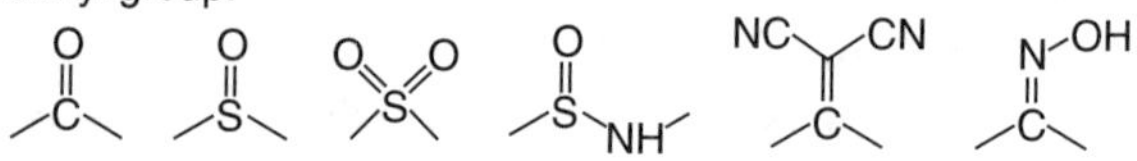

d. Carboxylic acid group:

COOH CONH$_2$ CONHSO$_2$R CONHCN

SO$_3$H SO$_2$NH$_2$ SO$_2$NHR

PO$_3$H$_2$ PO$_2$HNH$_2$ PO$_2$HOEt

e. Amide:

—NHCO— —CONH— —NHCH$_2$— —COCH$_2$—

—NHCS— —NHCO$_2$— ═NHCH$_2$═ —NHCOS—

f. Thioether:

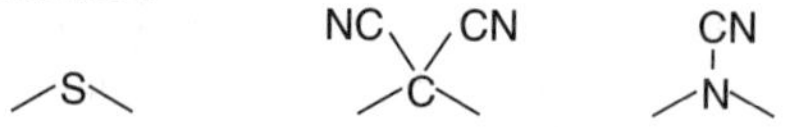

g. Thiourea:

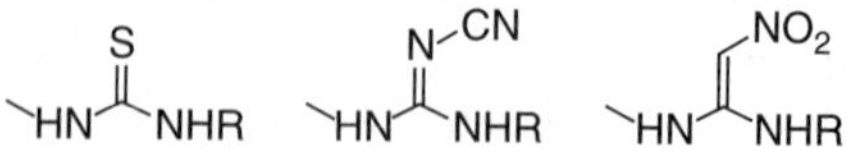

(Continued)

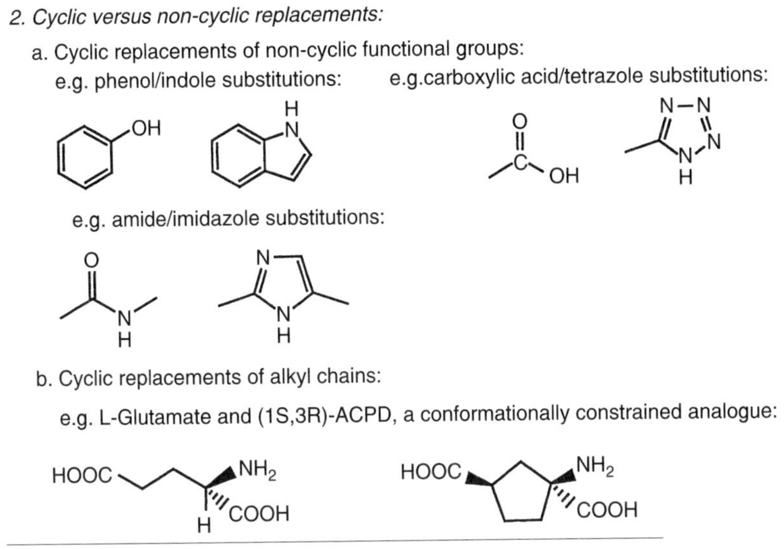

2. Cyclic versus non-cyclic replacements:

a. Cyclic replacements of non-cyclic functional groups:

e.g. phenol/indole substitutions: e.g.carboxylic acid/tetrazole substitutions:

e.g. amide/imidazole substitutions:

b. Cyclic replacements of alkyl chains:

e.g. L-Glutamate and (1S,3R)-ACPD, a conformationally constrained analogue:

Figure 1.5 Bioisosteres. These are biologically equivalent molecular fragments that can be used to replace portions of a drug molecule.

These properties dictate the therapeutic, toxic, and metabolic characteristics of the overall drug molecule. These properties also completely control the ability of the drug to withstand the arduous journey from the point of administration to the receptor site buried deep within the body. These physical properties of drug molecules may be categorized into the following major groupings:

1. Physicochemical properties
2. Shape (geometric, steric, conformational, topological) properties
3. Stereochemical properties
4. Electronic properties

Physicochemical properties are crucial to the pharmaceutical and pharmacokinetic phases of drug action; the other three properties are fundamental to the pharmacodynamic interaction of the drug with its receptor. Physicochemical properties (section 1.2) reflect the solubility and absorption characteristics of the drug and its ability to cross barriers, such as the blood–brain barrier, on its way toward the receptor. Geometric, steric, and topological properties (section 1.3) and stereochemical properties (section 1.4) describe the structural arrangement of the atoms within the drug molecule and influence the geometry of approach as the drug molecule enters the realm of the receptor. Electronic properties (section 1.5) reflect electron distribution within the drug molecule and determine the nature of the precise binding interaction between the drug and its receptor (by hydrogen bonding and various other forms of electrostatic interaction). From the perspective of the drug designer, the electronic properties are among the most

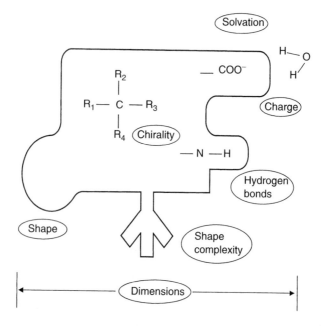

Figure 1.6 Properties of a drug molecule. A drug has many properties (size, shape, topology, polarity, chirality) that influence its ability to interact with a receptor. Each of these properties is required for the unique pharmacological activity of a drug molecule.

difficult to predict and to engineer with insight into a molecule. Accordingly, extensive use is now made of quantum mechanics and classical mechanics force field calculations (section 1.6) to determine electronic and structural properties of drug molecules (before the molecules are ever synthesized). Figure 1.6 summarizes the structural characteristics of a drug molecule.

1.2 PHYSICOCHEMICAL PROPERTIES OF DRUG MOLECULES

All drug molecules interact with biological structures (e.g., biomembranes, the cell nucleus), biomolecules (e.g., lipoproteins, enzymes, nucleic acids) and other small molecules on their way "from the gums to the receptor." Only by first unraveling the relatively simple primary interactions between a drug molecule and the various molecular structures that it encounters during its journey to the receptor can we understand drug activity at the cellular and molecular level. Since all biological reactions take place in an aqueous medium or at the interface of water and a lipid, the properties of water and this boundary layer must be studied as part of a comprehensive understanding of the interaction of a drug molecule with its receptor. Physicochemical properties reflect the solubility characteristics of a drug (in both aqueous and lipid environments) and help to determine the ability of a drug to penetrate barriers and gain access to receptors throughout the body.

1.2.1 Role and Structure of Water: Influence on Drug Structure

Life is based on water, the major constituent of living organisms and their cells. Drugs are transported within the aqueous bloodstream and most receptor sites are bathed in water molecules. The water molecule is thus central to the structure and function of most drugs and their associated receptors. Besides being a universal solvent, water participates in many reactions, and its role is therefore much more than that of an inert medium. Water is a very reactive and unusual chemical compound. Solubility, surface activity, hydrogen bonding, hydrophobic bonding, ionization, acidity, and solvation effects on macromolecular conformation all involve water.

Water structure is the consequence of the unique and unusual physical properties of the H_2O molecule. Water has a higher melting point, boiling point, and heat of vaporization than hydrides of related elements, such as H_2S, H_2Se, and H_2Te, or related isoelectronic compounds such as HF, CH_4, or NH_4. These properties are all a measure of the strong intermolecular forces that act between individual water molecules. These strong forces do not permit the ice crystal to collapse or water molecules to leave the surface of the liquid phase easily when heated. The forces result from the high polarity of water caused by the orientation of the H-O-H bond angle, which is 104.5°. The more electronegative oxygen attracts the electron of the O-H bond to a considerable extent, leaving the H atom with a partial positive charge (δ^+), while the O atom acquires a partial negative charge (δ^-). Since the molecule is not linear, H_2O has a *dipole moment*. The partial positive and negative charges of one water molecule will electrostatically attract their opposites in other water molecules, resulting in the formation of *hydrogen bonds*. Such noncovalent bonds can also be formed between water and hydroxyl, carbonyl, or NH groups.

In ice, each oxygen atom is bonded to four hydrogen atoms by two covalent and two hydrogen bonds. When ice melts, about 20% of these hydrogen bonds are broken, but there is a strong attraction between water molecules even in water vapor. Liquid water is therefore highly organized on a localized basis: the hydrogen bonds break and re-form spontaneously, creating and destroying transient structural domains, the so-called "flickering clusters." The half-life of any hydrogen bond between two water molecules is only about 0.1 nanosecond.

Water can interact with ionic or polar substances and may destroy their crystal lattices. Since the resulting hydrated ions are more stable than the crystal lattice, solvation results. Water has a very high dielectric constant (80 Debye units [D] versus 21 D for acetone), which counteracts the electrostatic attraction of ions, thus favoring further hydration. The dielectric constant of a medium can be defined as a dimensionless ratio of forces: the force acting between two charges in a vacuum and the force between the same two charges in the medium or solvent. According to Coulomb's law,

$$F = q_1q_2/Dr^2 \qquad (1.1)$$

where F is the force, q_1 and q_2 are the charges, and r is the distance separating them. D, the dielectric constant, is a characteristic property of the medium. Since D appears in the denominator, the higher the dielectric constant, the weaker the interaction between the two charges.

Polar functional groups such as aldehydes, ketones, and amines, possessing free electron pairs, form hydrogen bonds readily with water. Compounds containing such functional groups dissolve to a greater or lesser extent, depending on the proportion of polar to apolar moieties in the molecule. Solutes cause a change in water properties because the hydrate "envelopes" (which form around solute ions) are more organized and therefore more stable than the flickering clusters of free water. As a result, ions are water-structure *breakers*. The properties of solutions, which depend on solute concentration, are different from those of pure water; the differences can be seen in such phenomena as freezing-point depression, boiling-point elevation, and the increased osmotic pressure of solutions.

Water molecules cannot use all four possible hydrogen bonds when in contact with *hydrophobic* (literally, "water-hating") molecules. This restriction results in a loss of entropy, a gain in density, and increased organization. So-called "icebergs"—water domains more stable than the flickering clusters in liquid water—are formed. Such icebergs can form around single apolar molecules, producing inclusion compounds called *clathrates*. Apolar molecules are thus water-structure *formers*. The interaction between a solute and a solid phase—for example, a drug with its lipoprotein receptor—is also influenced by water. Hydrate envelopes or icebergs associated with one or the other phase will be destroyed or created in this interaction and may often contribute to conformational changes in macromolecular drug receptors and, ultimately, to physiological events. *Hydrophilic* ("water-loving") molecules are also relevant to the process of drug design.

1.2.1.1 Water: The Forgotten Molecule of Drug Design

Although under-appreciated and understudied, water plays a fundamental and crucial role in determining the properties of drug molecules. Water is one of the most important players in determining the pharmacokinetic properties of a drug molecule. It directly influences the conformation of the receptor macromolecule. Water bathes every drug molecule, hydrogen bonding to important functional groups of the molecule (see figure 1.7). Water strongly influences the drug–receptor interaction.

Despite these obvious facts, water has frequently been forgotten during the drug design process. In the past (and still to this day), many computer-aided studies on drugs calculated the drug properties *in vacuo*, completely neglecting the influence of water. In future, comprehensive studies of drug structure must include an equally comprehensive evaluation of the role of water.

1.2.2 Solubility Properties of Drug Molecules

Since a large percentage of all living structures consists of water, all biochemical reactions are based on small molecules dissolved in an aqueous phase (like the cellular cytoplasm) or on macromolecules dispersed in this phase—usually both. However, the equally important nonaqueous structures of cells, such as plasma membranes or the membranes of organelles, are of a lipid nature, and prefer to dissolve nonpolar hydrophobic (*lipophilic*) molecules. Accordingly, a highly significant physical property of all physiologically and pharmacologically important drug molecules is their solubility (in both aqueous *and* non-aqueous environments), because only in solution can they interact with the cellular and subcellular structures that carry drug receptors, thus triggering

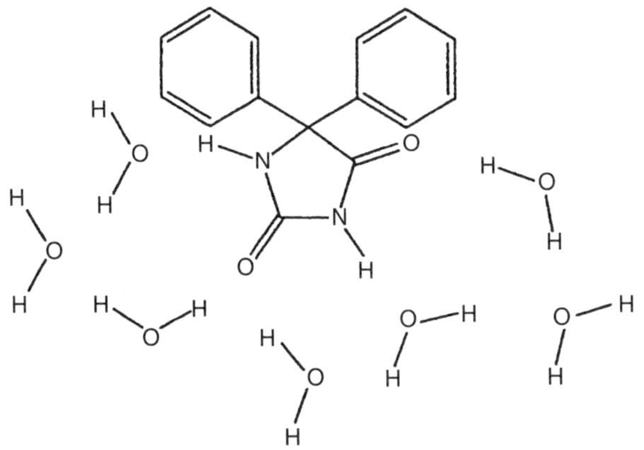

Figure 1.7 Effects of hydration on a drug molecule: A drug molecule does not exist in a vacuum; it is hydrated. In order to interact with its receptor, it must be dehydrated. Water molecules hydrogen-bond to the functional groups of the drug molecule. Additional water molecules then hydrogen-bond to these inner water molecules. The overall result consists of many layers of hydration.

pharmacological reactions. Theoretically, there are no absolutely insoluble compounds; every molecule is soluble in both the aqueous and nonaqueous lipid "compartments" of a cell. The degree of solubility, however, differs between each compartment. The proportion of these concentrations at equilibrium—or the ratio of solubilities—is called the *partition coefficient;* partition coefficients are extremely important when understanding the properties of drug molecules. Most successful drugs exhibit solubility to some extent in both water and lipid environments.

Solubility is a function of many molecular parameters. Ionization, molecular structure and size, stereochemistry, and electronic structure all influence the basic interactions between a solvent and solute. As discussed in the previous section, water forms hydrogen bonds with ions or with polar nonionic compounds through -OH, -NH, -SH, and -C=O groups, or with the nonbonding electron pairs of oxygen or nitrogen atoms. The ion or molecule will thus acquire a hydrate envelope and separate from the bulk solid; that is, it dissolves. The interaction of nonpolar compounds with lipids is based on a different phenomenon, the hydrophobic interaction, but the end result is the same: formation of a molecular dispersion of the solute in the solvent.

Although successful drugs tend to exhibit solubility in both aqueous and lipid environments, there are a few examples in which solubility in only one of these phases correlates with pharmacological activity. One such example is the local anesthetic activity of *p*-aminobenzoic acid esters, which is partly proportional to their lipid solubility. Another thoroughly investigated example is the bactericidal activity of aliphatic alcohols. In the homologous series beginning with *n*-butanol and ending with *n*-octanol, the bactericidal activity changes with increasing molecular weight. Whereas *n*-butanol and *n*-pentanol are active against *Staphylococcus aureus*, higher members of the series fail to kill the bacteria because the necessary concentration cannot be reached, arising from solubility considerations.

partition coefficients, and to simplify the determination of P values for small molecules. The fragmental constants are determined statistically by regression analysis; they are additive, and their sum provides a reasonable value for logP. Detailed tables of the f values for various functional groups have been published by Rekker (1977) and are sometimes used in computer program algorithms that calculate logP values. Somewhat analogous to the fragmental constants are the atomic constants put forth by Ghose and Crippen (1986); these assign logP values for every atom in a molecule and then determine the logP for the overall molecule by summing these values. Currently, there are a number of computer programs available (e.g., cLogP) for calculating logP values. In addition to these theoretical methods for calculating logP, there are a number of experimental protocols, including the classical "shake flask" method, and various chromatographic techniques, including HPLC methods.

1.2.4 Surface Activity Effects of Drug Molecules

Although a capacity to cross biological membranes and barriers is important for most drugs, there are also pharmaceutical agents that display mechanisms of action that are more dependent upon activities at surfaces. Pharmacologic reactions may occur on biological surfaces and interfaces. The energy situation at a surface differs markedly from that in a solution because special intermolecular forces are at work; therefore, surface reactions require specific consideration. In living organisms, membranes comprise the largest surface, covering all cells (the plasma membrane) and many cell organelles (the nucleus, mitochondria, and so forth). Dissolved macromolecules such as proteins also account for an enormous surface area (e.g., 1 ml of human blood serum has a protein surface area of $100 \, m^2$). Biological membranes also (i) serve as a scaffold that holds a large variety of enzymes in proper orientation, (ii) provide and maintain a sequential order of enzymes that permits great efficiency in multistep reactions, and (iii) serve as the boundaries of cells and many tissue compartments. In addition, many drug receptors are bound to membranes.

It is therefore apparent why the physical chemistry of surfaces and the structure and activity of surface-active agents are also of interest to the medicinal chemist. Antimicrobial detergents and many disinfectants exert their activity by interacting with biological surfaces and are important examples of surface-active drug effects.

1.2.4.1 Surface Interaction and Detergents

All molecules in a liquid phase interact with each other and exert a force on neighboring molecules. We have already discussed the hydrogen-bonding interaction of water molecules that creates clusters. The water molecules at a gas–liquid interface, however, are exposed to unequal forces, and are attracted to the bulk water of the liquid phase because no attraction is exerted on them from the direction of the gas phase. This accounts for the surface tension of liquids.

Because the dissolution of a solid is the result of molecular interaction between a solvent and the solid (which, once dissolved, becomes a solute), polar compounds capable of forming hydrogen bonds are water soluble, whereas nonpolar compounds dissolve only in organic solvents as the result of van der Waals and hydrophobic bonds. Compounds

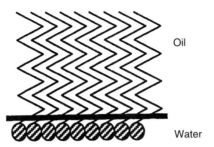

Oil

Water

Figure 1.8 Micellar structure of a soap molecule on an oil–water interface. The nonpolar alkyl chains are in the nonpolar phase; the polar carboxylate head groups are in the aqueous phase.

that are *amphiphilic* (i.e., containing hydrophobic as well as hydrophilic groups) will concentrate at surfaces and thereby influence the surface properties of these interfaces. Only in this way can amphiphilic detergents, through their hydrogen bonding with water and nonpolar interaction with a nonpolar (organic) phase or with air, maintain an orientation that ensures the lowest potential energy at an interface. A classic example of such behavior is given by soap, a mixture of alkali-metal salts of long-chain fatty acids. Figure 1.8 shows the interaction of soap molecules at an oil–water boundary; the circle symbolizes the anionic carboxylate or the polar "head group," and the zigzag line represents the hydrophobic alkyl chain.

A detergent-like soap forms a colloidal solution. At a very low concentration, soap molecules will be dissolved individually. At a higher concentration, the molecules find it more energy efficient to "remove" their hydrophobic tails from the aqueous phase and let them interact with each other, thus forming a miniature "oil drop" or nonpolar phase, with the polar heads of the soap molecules in the bulk water. At a concentration that is characteristic for a given individual detergent, molecular aggregates, known as *micelles,* are formed. They are often spherical colloidal particles, but can also be cylindrical. The concentration at which such micelles are formed is called the *critical micellar concentration*, and can be determined by measuring the light diffraction of the solution as a function of detergent concentration. The diffraction will show a sudden increase when micelles begin to form.

When soap is dispersed in a nonpolar phase, inverted micelles are formed in which the nonpolar tails of the soap molecules interact with the bulk solvent while the hydrophilic heads interact with each other. This behavior of amphiphilic molecules explains how they can disperse nonpolar particles in water: the hydrocarbon tail of the amphiphile interacts with the particle, such as an oil droplet, dirt, or a lipoprotein membrane fragment, covers the particle, and then presents its hydrophilic head groups to the aqueous phase.

1.2.5 The Clinical–Molecular Interface: Bioavailability and Drug Hydration

One of the authors recently encountered a 21-year-old male presenting to the emergency room in *status epilepticus* (prolonged, uncontrolled seizures). This patient had a seven–year history of epilepsy, well controlled with the drug phenytoin at a dose of 300 mg/day. Indeed, he had not experienced a seizure in more than a year. In the emergency room his serum level of phenytoin was "undetectable"; six months earlier, a routine measurement had revealed a serum phenytoin level of 68 μmol/L. He was given diazepam 5 mg intravenously (IV) and phenytoin 1,000 mg IV. His seizures soon abated. When asked why he had stopped taking his phenytoin he stated that he had not, but had been taking the same dose for years. His mother confirmed this story. She stated that he took his daily dose of phenytoin every morning at breakfast and that she had witnessed his doing so, every day for the past six years.

Upon questioning, it was discovered that this individual purchased his phenytoin in bottles of 1,000 capsules, in order to save money. He routinely stored this phenytoin in the basement of his home—a relatively damp and cool location—for months at a time; one week earlier he had started to use the capsules from one of these old stored bottles. When thirty "old" capsules from this recently opened bottle of 1000 capsules were weighed, all had masses in excess of 295 mg. When thirty capsules from a recently purchased supply of "new" phenytoin were weighed, all had masses less than 280 mg. An examination of an old capsule revealed that the contents were hardened and slightly discoloured. Subsequent analyses revealed that the phenytoin within the old capsules had become excessively hydrated from the ambient humidity of their storage conditions. The resulting hardened mass of drug material was less soluble within the gastrointestinal tract and was thus less *bioavailable* for absorption.

Many factors can influence the bioavailability of a drug molecule following oral absorption. Damp storage conditions of the drug can cause increased molecular hydration with concomitant altered solubility. When a drug molecule is crystallized using different solvents or different conditions, the resulting change in crystal morphology can influence bioavailability and thus alter biological results.

1.3 SHAPE (GEOMETRIC, CONFORMATIONAL, TOPOLOGICAL, AND STERIC) PROPERTIES OF DRUG MOLECULES

Physicochemical properties are important in determining the ability of a drug molecule to survive the pharmacokinetic phase and to reach the region of the receptor. The interaction of the drug molecule's pharmacophore with its complementary receptor during the pharmacodynamic phase of drug action is dependent upon a geometrically precise and accurate intermeshing of two molecular fragments. A rigorous control of molecular geometry and shape is crucial to the drug design process. Knowledge of molecular geometry also plays an important role in understanding quantitative structure–activity relationships during drug optimization (see section 3.3.2).

1.3.1 Conformational Isomerism and Drug Action

The concept of *conformational isomerism* is central to any consideration of molecular shape. Molecules that are flexible may exist in many different shapes or conformers. Conformational isomerism is the process whereby a single molecule undergoes transitions from one shape to another; the physical properties of the molecule have not changed, merely the shape. Conformational isomerism is demonstrated by compounds in which the free rotation of atoms around chemical bonds is not significantly hindered. The energy barrier to the transition between different conformations is usually very low

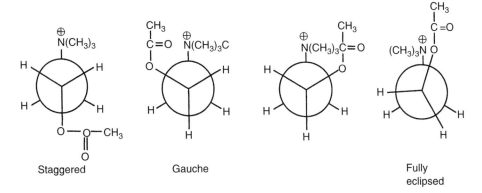

Figure 1.9 Newman projections of acetylcholine. Rotation around torsional angles permits many different conformers (shapes) of a molecule.

(on the order of 4–8 kJ/mol), and is easily overcome by thermal motion unless the molecule is made rigid or because nonbonding interactions between functional groups of the molecule favor one conformer over an infinite number of others. The concept and biophysical reality of "preferred" drug conformations and their potential role in receptor binding are currently important issues among drug designers.

For aliphatic compounds, the well-known Newman projection is used to show the relative position of the substituents on two atoms connected to each other (as in ethane derivatives). For example, figure 1.9 shows several possible conformers of acetyl-choline. When the trimethylammonium-ion and acetoxy functional groups are as far removed as possible, we speak of a *fully staggered* conformation (erroneously and confusingly also called a *trans* conformation). When the two groups overlap, they are *eclipsed.* Between these two extremes are an infinite number of conformers called *gauche* (or skew) *conformers* or *rotamers* (rotational isomers). The potential interaction energy of the trimethylammonium-ion and acetoxy groups is lowest in the staggered conformation and highest when the two groups are eclipsed. The stability of these rotamers is normally opposite. An exception to this exists when two functional groups show a favorable nonbonding interaction (e.g., hydrogen-bond formation).

Since the transition between rotamers occurs very rapidly, the existence of any one conformer can be discussed in statistical terms only. For example, in acyclic hydrocar-bon molecules, it has been assumed that long hydrocarbon chains exist in the staggered, fully extended, zigzag conformation. There is, however, a considerable probability of their also existing in skew conformations, effectively reducing the statistical length of the carbon chain. Such considerations become important if one wishes to calculate effective intergroup distances in drugs, which play a role in the geometric fit and bind-ing to receptors. For instance, in the anticholinergic agents hexamethonium (**1.1**) and decamethonium (**1.2**), the two quaternary trimethylammonium groups are connected by six and ten methylene (-CH2-) groups, respectively.

Observations emphasize the need for extreme caution in proposing geometry-based hypotheses when dealing with drug conformations and their correlations with receptor structure, especially when the drug contains flexible acyclic hydrocarbon segments and

Hexamethonium (1.1)   Decamethonium (1.2)

is not conformationally constrained. Many publications have proposed receptor mapping techniques based on the distances between assumed key atoms (usually heteroatoms) or functional groups in drugs, determined by prolonged quantum-chemical calculations of "preferred" conformers. Similarly, the design of a number of drugs has been based on questionable assumptions about drug–receptor binding, all founded on conformational analysis. These oversimplifications are subject to criticism. Such caveats, however, do not detract from the utility of conformational analysis of drugs, from the importance of calculating intergroup geometric distances, or from the potential value of these methods in drug design and molecular pharmacology.

Flexible molecules that lack conformational constraints may assume a variety of different conformations, thus increasing the likelihood of drug toxicity by enabling interactions with undesirable receptor sites. The drug designer may address these problems by using various methods to decrease the degrees of conformational freedom. For instance, within the hydrocarbon skeleton of a drug, an alkane moiety could be replaced by either an alkene or an alkyne; the increased barrier to rotation around double or triple bonds (as compared to single bonds) adds a considerable measure of conformational constraint. However, one of the most popular techniques for decreasing conformational freedom is to replace acyclic hydrocarbon fragments with cyclic fragments, such as cyclohexane rings or aromatic rings.

The conformational analysis of cyclohexane and its derivatives has been well explored. The cyclohexane ring itself can assume several conformations. The chair conformation is more stable than either the boat or twist form because it permits the maximum number of substituents to exist in a staggered conformation relative to their neighbors. The substituents can assume two conformations relative to the plane of the ring (defined by carbon atoms 2, 3, 5, and 6): *axial* (a), in which they point up or down; and *equatorial* (e), in which they point in the direction of the ring's circumference. As the cyclohexane ring keeps flipping back and forth between many chair forms, the substituents on the ring alternate between axial and equatorial conformations unless stabilized (see figure 1.10).

Although cyclohexane is more conformationally rigid than an acyclic hydrocarbon, there are several ways to additionally stabilize or "freeze" the conformation of a cyclohexyl ring.

1. By electrostatic repulsion of two adjoining substituents (e.g., in 1,2-dichlorocyclohexane, a diaxial conformation is forced).
2. By steric repulsion.
3. By using a bulky substituent like the *tert*-butyl group, which always maintains an equatorial position.
4. By using multiple cyclohexyl rings adjoined to each other.

The use of multiple adjoined rings is an effective means of locking conformation. Polycyclic structures, such as decaline or the steroids, are rigid and maintain stable conformations. In such rigid systems, the axial and equatorial substituents can display

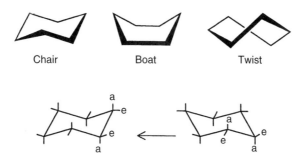

Figure 1.10 The six-membered cyclohexane ring can adopt a series of different conformations. The chair conformation is more stable than the boat or twist because it permits a maximal number of substituents to exist in a staggered conformation relative to their neighbors.

cis/trans isomerism without the presence of a double bond; the restriction on their rotation is ensured by the ring system itself. Diastereomerism can also occur in these molecules. In substituted cyclohexanes, or their heterocyclic analogs, 1,2–diaxial or the equivalent diequatorial substituent pairs are considered to be *trans*, while the axial–equatorial pair is regarded as *cis*. 1,3–diequatorial substituents are, however, *cis*.

The axial or equatorial nature of a substituent has a bearing on its reactivity, or ability to interact with its environment. Equatorial substituents are more stable and less reactive than their axial counterparts. For example, equatorial carboxyl groups are stronger acids than axial ones because of the higher stability of the carboxylate ion, whereas equatorial esters are hydrolyzed more slowly than axial ones because they are less accessible to protons or hydroxyl ions during acid- or base-catalyzed hydrolysis.

Even better than an acyclic saturated hydrocarbon such as cyclohexane is the use of aromatic rings, especially polyaromatic systems. The unsaturated structure of the aromatic ring imparts planarity and rigidity. Drugs in which the functional groups are appended to an aromatic ring have marked conformational rigidity. In the realm of neurologic drug design, the use of tricyclic structures containing aromatic rings is extremely common in major antipsychotics (e.g., chlorpromazine (**1.3**)), antidepressants (e.g., amitriptyline (**1.4**)), and anticonvulsants (e.g., carbamazepine (**1.5**)).

Although they are superb for achieving planarity and rigidity, polyaromatic systems may come accompanied with the risk of a side-effect—*carcinogenicity*. Such rigid,

Chlorpromazine
(1.3)

Amitriptyline (1.4)

Carbamazepine
(1.5)

planar compounds sometimes have the capacity to insert themselves within nucleic acids, potentially inducing cancer-causing changes.

When contemplating the effect of drug conformation on drug–receptor interactions, one must not forget that the receptor macromolecule also undergoes changes in its molecular geometry, as postulated by the Koshland induced-fit hypothesis (see chapter 2). Owing to the enormously more complex nature of macromolecular structure, less is known about such changes. Many examples of conformational changes of enzymes during their reactions with substrates have been well studied and described in the literature, including those of carboxypeptidase, dihydrofolate reductase, and acetylcholinesterase (see section 7.1.2).

1.3.2 Steric Effects on Drug–Receptor Interactions

During the geometrical interaction between a drug and its receptor, steric factors frequently emerge as extremely important considerations. At times, a large, bulky substituent appended to a fragment within a drug molecule may physically impede the geometry of interaction between a drug and its receptor. Historically, the first attempt to include steric effects in relationship studies between the structure and pharmacological activity of a molecule was the *Taft steric parameter* (E_s). This parameter was defined as the difference between the logarithm of the relative rate of the acid-catalyzed hydrolysis of a carboxymethyl-substituted compound, and the logarithm of the rate of hydrolysis of methyl acetate as a standard:

$$E_{SX} = \log K_{XCOOCH_3} - \log K_{CH_3COOCH_3} \qquad (1.3)$$

where X is the molecule or molecular fragment in question to which a carboxy-methyl group has been attached. With some corrections suggested by other authors, E_s has proven to be useful in quite a few structure–activity correlations.

Another classical measure of the molecular geometry of substituents is the *Verloop steric parameter*. This is calculated from bond angles and atomic dimensions—primarily the lengths of substituent groups and several measures of their width. Trivial as this may sound, the consideration of molecular "bulk" is an important and often neglected factor in making multiple quantitative correlations of structure and pharmacological activity. Balaban et al. (1980) devised several related methods that are still in use today.

1.4 STEREOCHEMICAL PROPERTIES OF DRUG MOLECULES

Since drugs interact with optically active, asymmetric biological macromolecules such as proteins, polynucleotides, or glycolipids acting as receptors, many of them exhibit stereochemical specificity. This means that there is a difference in action between stereoisomers of the same compound, with one isomer showing pharmacological activity while the other is more or less inactive. In 1860, Louis Pasteur was the first to demonstrate that molds and yeasts can differentiate between (+)- and (−)-tartarates, utilizing only one of the two isomers.

Therefore, complementarity between an asymmetric drug and its asymmetric receptor is often a criterion of drug activity. The effects of highly active or highly specific drugs depend more upon such complementarity than do those of weakly active drugs. Occasionally, the stereoselectivity of a drug is based on a specific and preferential metabolism of one isomer over the other, or on a biotransformation that selectively removes one isomer. Such

Diazepam (1.6)

(S)-*N*-Methyloxazepam
(1.7)

stereoselective biotransformations may have far-reaching consequences. For instance, microsomal hydroxylation of the tranquilizer diazepam (**1.6**) occurs stereoselectively, yielding (S)-*N*-methyloxazepam (**1.7**). Since this hydroxylated metabolite is pharmacologically active, the stereochemical circumstances of the activation process are crucial, not only for the extent of the activation but also for the rate of elimination of the metabolite.

1.4.1 Optical Isomers of Drugs

Optical isomerism is the result of a dissymmetry in molecular substitution. The basic aspects of optical isomerism are discussed in various textbooks of organic chemistry. Optical isomers (*enantiomers*) *may* have different physiological activities from each other *provided* that their interaction with a receptor or some other effector structure involves the asymmetric carbon atom of the enantiomeric molecule *and* that the three different substituents on this carbon atom interact with the receptor. The *Easson–Stedman hypothesis* assumes that a three-point interaction ensures stereospecificity, since only one of the enantiomers will fit; the other one is capable of a two-point attachment only, as shown in figure 1.13 for the reaction with a hypothetical planar receptor. However, it is reasonable to assume that receptor stereospecificity can also undergo a change when the receptor conformation is altered by a receptor–drug interaction.

The difference in pharmacological action between enantiomers can be considerable. (−)-Levorphanol (**1.8**), a synthetic analgesic, has a binding equilibrium constant (K_D) of 10^{-9} M. (K_D) is a dissociation constant, indicating that this drug will occupy half of all accessible morphine receptors at a nanomolar concentration. (+)-Dextrorphan (**1.9**), the optical antipode of (−)-levorphanol, has a K_D of 10^{-2} M, reflecting a high and nonphysiological concentration. Qualitatively, dextrorphan is not an analgesic at all, but a very effective antitussive (cough suppressant), an action entirely different from analgesia. (+)-Muscarine (**1.10**) is about three orders of magnitude more effective as a cholinergic neurotransmitter than (−)-muscarine (**1.11**). A very large body of data is available on the selectivity of enantiomeric drugs.

It should be emphasized that the mere sign (+ or −) of the optical rotation produced by an enantiomer is not biochemically decisive to the action of such a molecule. The *absolute* configuration of the compound in question must be considered; in modern organic chemistry the Cahn–Ingold–Prelog sequence rules are followed, and have increasingly replaced the ambiguous and obsolescent D and L designations for *relative* configuration. Again, the reader is referred to standard organic chemistry texts for details. Figure 1.11 shows two types of stereoisomerism relevant to drug action.

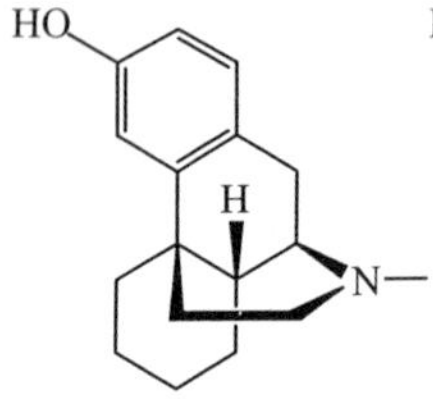

(−)-Levorphanol
(1.8)

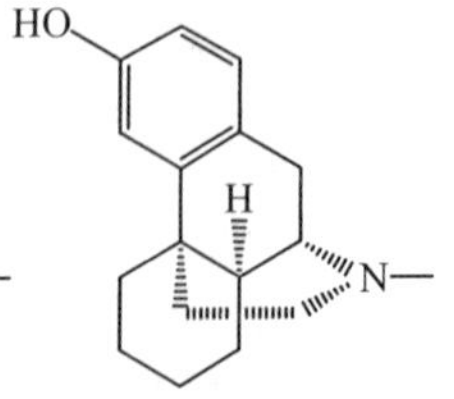

(+)-Dextrorphan
(1.9)

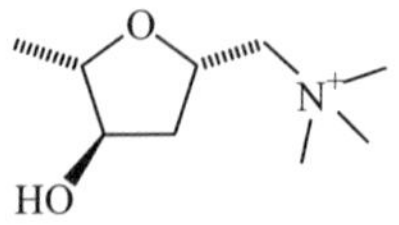

(+)-Muscarine
(1.10)

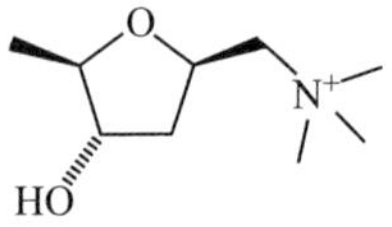

(−)-Muscarine
(1.11)

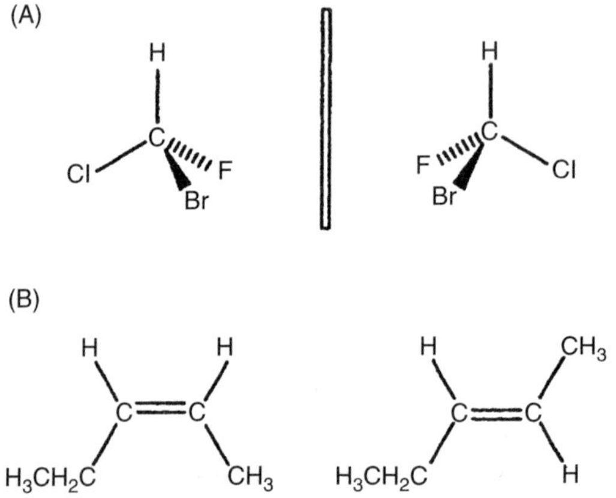

(A)

(B)

Figure 1.11 Stereoisomers: a carbon atom bonded to four different substituents in a chiral carbon or a stereogenic center. Such molecules cannot be superimposed upon their mirror image. A receptor will recognize one stereoisomer but not another. Such stereoisomers are designated as either R or S. Stereoisomerism may also occur around double bonds, producing *cis* or *trans* orientations of the substituents on either face of the double bond.

Even though enantiomeric drug pairs quite often show different potencies, they are seldom antagonists of each other, since the differences in their action are due to differences in their binding properties; antagonists (see section 2.4) usually bind more strongly than agonists, and the less active enantiomer of a pair is typically incapable of displacing the more active one from the receptor.

Diastereomeric drugs—those having two or more asymmetric centers—are usually active in only one configuration. Unlike enantiomers, which have identical physicochemical properties, the absorption, distribution, receptor binding, metabolism, and every other aspect that influences the pharmacological activity of a drug are different for each diastereomer.

1.4.2 Enantiomers and Pharmacological Activity

Lehman et al. (1986) stated the definitions of stereoselectivity in the following manner: the better fitting enantiomer (the one with higher affinity for the receptor) is called the *eutomer*, whereas the one with the lower affinity is called the *distomer*. The ratio of activity of the eutomer and distomer is called the *eudismic ratio*; the expression of the *eudismic index* is

$$EI = \log \text{affinity}_{\text{Eu}} - \log \text{affinity}_{\text{Dist}} \tag{1.4}$$

In a series of agonists and antagonists (for definitions, see section 2.4), the eudismic affinity quotient can also be defined as a measure of stereoselectivity. Because of widespread misconceptions, the distomer of a racemate is often considered "inactive" and of no consequence to pharmacological activity, an idea reinforced by the fact that resolution (i.e., separation) of racemates is economically disadvantageous. In the 1980s, Ariëns and his associates (Ariëns et al., 1983; Ariëns, 1984, 1986) published a series of influential books and papers that showed the fallacy of this concept and pointed out the necessity of using pure enantiomers in therapy and research; thankfully, this message has now been learned.

The distomer should therefore be viewed as an impurity constituting 50% of the total amount of a drug—an impurity that in the majority of cases is by no means "inert." Possible unwanted effects of a distomer are as follows:

1. It contributes to side effects.
2. It counteracts the pharmacological action of the eutomer.
3. It is metabolized to a compound with unfavorable activity.
4. It is metabolized to a toxic product.

However, there are instances in which the use of a racemate has advantages; sometimes it is more potent than either of the enantiomers used separately (e.g., the antihistamine isothipendyl), or the distomer is converted into the eutomer *in vivo* (the anti-inflammatory drug ibuprofen).

Admittedly, the separation of enantiomers is often difficult and expensive. However, now that we are in the 21st century, the need for optically active drugs capable of stereospecific interactions with drug receptors is a recognized prerequisite in drug design.

1.4.3 Geometric Isomers of Drugs

Cis/trans isomers are the result of restricted rotation along a chemical bond owing to double bonds or rigid ring systems in the isomeric molecule. These isomers are not mirror images and have very different physicochemical properties, as reflected in their pharmacological activity. Because the functional groups in these molecules are separated by different distances in the different isomers, they cannot as a rule bind to the same receptor. Therefore, geometric isomerism as such may be of interest to the medicinal chemist.

In biological systems, there are a number of examples of the importance of *cis/trans* isomerization. The human eye contains one of the most important examples. Rod cells and cone cells are the two types of light-sensitive receptor cells in the human retina. The three million rod cells enable vision in dim light; the 100 million cone cells permit colour perception and vision in bright light. Within the rod cells, 11-*cis*-retinal (**1.12**) is converted into rhodopsin a light-sensitive molecule. When rod cells are exposed to light, isomerization of the C11–C12 double bond occurs, leading to the production of a *trans*-rhodopsin, called metarhodopsin, that contains all-*trans*-retinal (**1.13**). This *cis/trans* isomerization of rhodopsin leads to a change in molecule geometry, which in

contain many ionic species (phospholipids, proteins) that can repel or bind ionic drugs; and ion channels, usually lined with polar functional groups, can act in an analogous manner. Ionic drugs are also more hydrated; they may therefore be "bulkier" than nonionic drugs.

As a rule of thumb, drugs pass through membranes in an undissociated form, but act as ions (if ionization is a possibility). A pK_a in the range of 6–8 would therefore seem to be most advantageous, because the nonionized species that passes through lipid membranes has a good probability of becoming ionized and active within this pK_a range. This consideration does not relate to compounds that are actively transported through such membranes.

A high degree of ionization can prevent drugs from being absorbed from the gastrointestinal tract and thus decrease their systemic toxicity. This is an advantage in the case of externally applied disinfectants or antibacterial sulfanilamides, which are meant to remain in the intestinal tract to fight infection. Also, some antibacterial aminoacridine derivatives are active only when fully ionized. These now obsolete bacteriostatic agents intercalate (position or interweave themselves) between the base pairs of DNA. The cations of these drugs, obtained by protonation of the amino groups, then form salts with the DNA phosphate ions, anchoring the drugs firmly in position. Ionization can also play a role in the electrostatic interaction between ionic drugs and the ionized protein side chains of drug receptors. Therefore, when conducting experiments on drug–receptor binding, it is advisable to regulate protein dissociation by using a buffer. The degree of ionization of any compound can be easily calculated from the Henderson–Hasselbach equation:

$$\% \text{ ionized} = 100/(1 + \text{antilog } [\text{pH} - pK_a]) \tag{1.6}$$

1.5.3 Electron Distribution in Drug Molecules

More recently, a variety of other methods has been developed to describe the electronic distribution properties of drug molecules. The electron distribution in a molecule can be estimated or determined by experimental methods such as dipole-moment measurements, NMR methods, or X-ray diffraction. The latter method provides very accurate electron-density maps, but only of molecules in the solid state; it cannot be used to provide maps of the nonequilibrium conformers of a molecule in a physiological solution.

To provide easily obtained yet rigorous assessments of electron distribution properties, quantum mechanics calculations are now employed (see section 1.6). Molecular quantum mechanics calculations provide several methods for calculating the orbital energies of atoms, combining the individual atomic orbitals into molecular orbitals, and deriving from the latter the probability of finding an electron at any atom in the molecule— which is tantamount to determining the electron density at any atom. There are several methods for doing this, with varying degrees of sophistication, accuracy, and reliability. These calculations permit quantification of the charge density on any atom in a drug molecule. Such atomic electron density values may be used when correlating molecular structure with biological activity during the drug molecular optimization process.

In addition to providing values for charge densities on individual atoms, quantum mechanics calculations may also be used to determine the energies of delocalized orbitals; such energy values may also be used when correlating molecular structure with pharmacologic activity. The energies of delocalized orbitals have attracted considerable interest since the early 1960s, when Szent-Györgyi (1960), in his brilliant pioneering book on submolecular biology, directed attention to charge-transfer complexes (see section 2.3.5). The energies of the highest occupied molecular orbital (HOMO) and the lowest unoccupied molecular orbital (LUMO) are a measure of electron-donor and electron-acceptor capacity, respectively, and consequently determine donors and acceptors in charge-transfer reactions. HOMO and LUMO are also reliable estimates of the reducing or oxidizing properties of a molecule. They are expressed in β units (a quantum-chemical energy parameter whose value varies from 150 to 300 U/mol). The smaller the numerical value of HOMO (a positive number), the better the molecule is as an electron donor, since the small number indicates that less energy is required to remove an electron from it. Likewise, the smaller the magnitude of the LUMO (a negative number), the more stable the orbital for the incoming electron, which favors electron-acceptor characteristics. Thus, by examining the numerical values of the HOMO and LUMO of a pair of drug molecules, one can often decide whether a charge-transfer complex can be formed, and which compound will be the donor and which the acceptor.

In addition to providing insights concerning correlation of molecular structure with pharmacologic bioactivity, quantum mechanics calculations of electron distribution may also be employed to understand the molecular basis of drug toxicity. For instance, overall p-electron density of polycyclic hydrocarbons has traditionally been assumed to correlate with the carcinogenicity of these compounds. According to this hypothesis, defined reactive regions on the molecule undergo metabolism to form reactive intermediates such as epoxides, which react with cell constituents such as the basic nitrogen atoms in nucleic acids. Although this model has been widely cited in the literature, it is appropriate to warn the reader that, however attractive, it is seriously questioned. However, p-electron density is very important in the chemical reactivity of aromatic rings.

1.6 PREDICTING THE PROPERTIES OF DRUG MOLECULES: QUANTUM MECHANICS AND MOLECULAR MECHANICS

When confronted with the task of designing drugs, it would be wonderful to have a method for predicting the properties of drug molecules before having to actually synthesize and purify them. The synthetic preparation of new molecules is challenging, time consuming, and expensive. Theoretical chemistry, combined with modern computational methods, offers a powerful solution to this prediction dilemma.

The docking of a drug with its receptor site is a precise interaction between two molecules. The success of this interaction is dependent upon the geometry, conformation and electronic properties of the two molecules. Designing drugs requires techniques for determining and predicting the geometry, conformation, and electronic properties of both small molecules (i.e., drugs with molecular weights less than 800) and macromolecules (i.e., receptor proteins.) Quantum pharmacology and molecular modeling calculations are such techniques. Molecular modeling is the evaluation of molecular properties and structures using computational chemistry and molecular graphics to provide three-dimensional visualization and representation of molecules. Quantum pharmacology is the application of the methods of modern computational chemistry to understanding drug action at the molecular and atomic level of structural refinement. CADD (computer-aided drug design) and CAMD (computer-aided molecular design) are the employment of computer-aided techniques to design, discover, and optimize bioactive molecules as putative drugs.

Molecular modeling and quantum pharmacology calculations have emerged as extremely important techniques in modern medicinal chemistry. A review of drug design papers in the *Journal of Medicinal Chemistry* and of pharmaceutically relevant papers in the *Journal of the American Chemical Society*, covering the year 2000, reveals that 43% of these papers included computational chemistry techniques in their design and analyses of drug molecule action. Clearly the dawn of the 21st century has emphasized the exponentially growing importance of molecular modeling and quantum pharmacology in drug design. Accordingly, a basic understanding of medicinal chemistry in the modern era requires an appreciation of the fundamentals of quantum mechanics, molecular mechanics, and the other techniques of computational chemistry as applied to drug design. The medicinal chemist who uses commercially available computer programs to design drugs should not treat them as merely "black boxes," and should have some insight into their conceptual basis.

1.6.1 Methods of Quantum Pharmacology for Molecular Geometry Optimization: Quantum Mechanics, Molecular Mechanics, QM/MM Calculations

The first and foremost goal of quantum pharmacology is to predict and determine the optimal geometry of drug molecules and drug receptors. This is best achieved by using a "mechanics" method that permits the geometry of a molecule to be expressed as a function of energy. By minimizing this energy function, one can ascertain the optimal geometry of the molecule. Quantum mechanics and molecular mechanics are the dominant "mechanics" methods in quantum pharmacology (see figure 1.12).

1.6.1.1 Quantum Mechanics

The *Schrödinger equation* is the centrepiece of quantum mechanics and lies at the heart of much of modern science. In its simplest form, the Schrödinger equation may be represented as

$$H\psi = E\psi \tag{1.7}$$

where ψ is the wave function, E is the energy of the system, and H (the "Hamiltonian operator") is the shorthand notation for a mathematical operator function that operates on other mathematical functions. Once the wavefunction is known for a particular system, then any physical property may in principle be determined for that system. However, ψ is just a normal mathematical function; it has no special mathematical properties.

If the system being studied is a simple hydrogen atom with a single electron outside of a positively charged nucleus, the Schrödinger equation may be solved exactly. The wavefunctions which satisfy the Schrödinger equation for this simple hydrogen atom are called orbitals; a hydrogenic atomic orbital is therefore the three-dimensional mathematical function from which one may calculate the energy and other properties of a single electron. For single atoms that contain multiple electrons (polyelectronic mono-atomic systems), the wavefunction for the atom (ψ) is a product of one-electron wavefunctions (χ_i), one for each electron. For molecules that contain multiple atoms (polyelectronic, polyatomic molecular systems) the wave function for the molecule (Ψ) is a product of one-electron wavefunctions (ϕ_i), where ϕ is a three-dimensional mathematical function

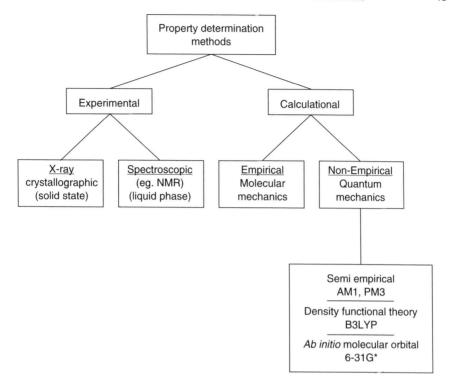

Figure 1.12 Determining the properties of drug molecules. Drug molecules may have their properties ascertained by either experimental or theoretical methods. Although experimental methods, especially X-ray crystallography, are the "gold standard" methods, calculational approaches tend to be faster and do provide high quality information. Nonempirical techniques, such as ab initio quantum mechanics calculations, provide accurate geometries and electron distribution properties for drug molecules.

representing the energy and properties of an individual electron within the molecule, and where these unknown molecular orbitals ϕ may be represented as a linear combination of known atomic orbital functions (χ_i).

In quantum pharmacology, the goal is to determine the wavefunction Ψ for the drug molecule so that the energy and properties of the drug may be calculated. However, Schrödinger's equation may be exactly solved only for the hydrogen atom. It is not possible to provide an exact mathematical solution for the wavefunction of an entire molecule. Accordingly, quantum mechanics calculations that provide approximate, but not exact, solutions for the drug molecule wavefunction are employed; these approximate methods are called *molecular orbital calculations*.

In molecular orbital calculations, the molecular orbitals ϕ are represented as a linear combination of atomic orbital functions (χ_i). A variety of different mathematical functions may be used to represent these atomic orbital functions. If a very sophisticated mathematical function is used, then the resulting answer is higher in quality, providing very accurate energies and geometries for the drug molecule being studied; however, such calculations may be extremely expensive in terms of computer time required. If a simpler mathematical function is used, it may be calculated more rapidly, but will provide a cruder approximation of the numerical properties of the atomic orbital function that it endeavours to represent. Classically, the most obvious types of mathematical functions used to represent atomic orbitals are called *Slater-type orbitals*. If one uses Slater-type functions for each atomic orbital which is filled (e.g., for carbon using 1s, 2s, and 2p atomic orbitals), then the resulting set of functions is termed the *basis set*. The term basis set applies to a set of mathematical functions used to describe the shape of the orbitals in an atom.

Molecular orbital calculations may be broadly divided into two types: *ab initio* and semi-empirical. The term *ab initio* is an unfortunate choice of words since it gives a mistaken idea of quality; nevertheless, it is used universally for molecular orbital wavefunction calculations that explicitly consider all electrons within the drug molecule. Ab initio calculations may be done at a variety of basis set levels. The higher the basis set level, the more likely will the calculation reproduce experimental observations, such as bond lengths determined from X-ray crystallographic methods. Not surprisingly, the current medicinal chemistry literature contains numerous examples in which quantum pharmacology calculations using ab initio methods have been employed to understand the properties of drug molecules.

Despite the length of time required for their completion, ab initio calculations are themselves not always successful in reproducing experimental observations and do require prolonged calculational times. To address these potential problems, considerable effort has been expended to devise the so-called *semi-empirical* molecular orbital calculations. Semi-empirical methods employ a variety of approximations and assumptions to reduce the complexity of the mathematics and thus the time required for a calculation. Typically, semi-empirical calculations consider only valence shell electrons. Core electrons, such as 1s electrons, are ignored under the assumption that they play little if any role in biological and biochemical processes. To compensate for ignoring core electrons, empirically derived parameters are incorporated into the calculations; these "fudge factors" help the semi-empirical calculation to reproduce experimental results while neglecting to calculate a number of difficult integral equations that would be present in the ab initio mathematical formulation. It is not uncommon for semi-empirical calculations to run 2 to 3 times faster than ab initio calculations. There are a number of types of semi-empirical calculations. Historically, CNDO (complete neglect of differential overlap) and INDO (intermediate neglect of differential overlap) parameterizations were used. These first methods were somewhat crude and proved to be of little value in medicinal chemistry and quantum pharmacology. More recently, parameterizations such as AM1 and PM3 yield impressive results when compared with a range of experimental observables. AM1 and PM3 semi-empirical calculations have been used successfully over the past 5 to 10 years to model a variety of drugs and drug–receptor interactions, and their utility in quantum pharmacology calculations continues to be explored and expanded.

1.6.1.2 Molecular Mechanics

Molecular mechanics is based on the principles of classical mechanics, rather than those of quantum mechanics. Quantum mechanics is based on an explicit consideration of electrons and electron properties. Molecular mechanics, on the other hand, does not consider electrons explicitly. In molecular mechanics, atoms are regarded as distensible balls, bearing charge, and connected to other distensible balls via springs. The mathematics of molecular mechanics is thus rapid and trivial, which makes the technique ideal for the treatment of pharmaceutically relevant macromolecules.

The term *molecular mechanics* refers to a heavily parameterized calculational method that leads to accurate geometries and accurate relative energies for different conformations of molecules. The molecular mechanics procedure employs the fundamental equations of vibrational spectroscopy, and represents a natural evolution of the notions that atoms are held together by bonds and that additional interactions exist between nonbonded atoms. The essential idea of molecular mechanics is that a molecule is a collection of particles held together by elastic or harmonic forces, which can be defined individually in terms of potential energy functions. The sum of these various potential energy equations comprises a multidimensional energy function termed the *force field*, which describes the restoring forces acting on a molecule when the minimal potential energy is perturbed. The force field approach supposes that bonds have natural lengths and angles, and that molecules relax their geometries to assume these values. The incorporation of van der Waals potential functions and electrostatic terms allows the inclusion of steric interactions and electrostatic effects. In strained systems, molecules will deform in predictable ways, with strain energies that can be readily calculated. Thus molecular mechanics uses an empirically derived set of simple classical mechanical equations, and is in principle well suited to provide accurate a priori structures and energies for drugs, peptides, or other molecules of pharmacological interest.

Molecular mechanics lies conceptually between quantum mechanics and classical mechanics, in that data obtained from quantum mechanical calculations are incorporated into a theoretical framework established by the classical equations of motion. The *Born–Oppenheimer approximation*, used in quantum mechanics, states that Schrödinger's equation can be separated into a part that describes the motion of electrons and a part that describes the motion of nuclei, and that these can be treated independently. Quantum mechanics is concerned with the properties of electrons; molecular mechanics is concerned with the nuclei, while electrons are treated in a classical electrostatic manner.

The heart of quantum mechanics is the Schrödinger equation; the heart of molecular mechanics is the force field equation. A typical molecular mechanics force field is shown below:

General form of a force field equation:

$$V = V_r + V_\theta + V_\omega + V_{inv} + V_{nb} + V_{hb} + V_{\text{cross}} \qquad (1.8)$$

Specific force field equation (AMBER):

$$V_r = \sum k_r (r - r_0)^2 \qquad (1.9)$$

$$V_\theta = \sum k_\theta (\theta - \theta_0)^2 \qquad (1.10)$$

$$V_\omega = \sum \frac{V_n}{2}(1 + \cos(n\phi - \gamma)) \tag{1.11}$$

$$V_{nb} = \sum_{i<j} \left(\frac{A_{ij}}{r_{ij}^{12}} - \frac{B_{ij}}{r_{ij}^6} + \frac{q_i q_j}{\varepsilon r_{ij}} \right) \tag{1.12}$$

$$V_{hb} = \sum \left(\frac{C_{ij}}{r_{ij}^{12}} - \frac{D_{ij}}{r_{ij}^{10}} \right) \tag{1.13}$$

Or

$$V = \sum k_r(r - r_0)^2 + \sum k_\theta(\theta - \theta_0)^2 + \sum \frac{V_n}{2}(1 + \cos(n\phi - \gamma))$$
$$+ \sum_{i<j} \left(\frac{A_{ij}}{r_{ij}^{12}} - \frac{B_{ij}}{r_{ij}^6} + \frac{q_i q_j}{\varepsilon r_{ij}} \right) + \sum \left(\frac{C_{ij}}{r_{ij}^{12}} - \frac{D_{ij}}{r_{ij}^{10}} \right) + V_{\text{cross}} \tag{1.14}$$

where V_r represents bond length energies, V_θ represents bond angle energies, V_ω represents dihedral angle energies, and V_{nb} represents non-bonded interaction energies (van der Waals and electrostatic), and V_{hb} represents hydrogen bonding interactions. Typically, the bond stretching and bending functions are derived from Hooke's law harmonic potentials; a truncated Fourier series approach to the torsional energy permits accurate reproduction of conformational preferences.

The molecular mechanics method is extremely parameter dependent. A force field equation that has been empirically parameterized for calculating peptides must be used for peptides; it cannot be applied to nucleic acids without being re-parameterized for that particular class of molecules. Thankfully, most small organic molecules, with molecular weights less than 800, share similar properties. Therefore, a force field that has been parameterized for one class of drug molecules can usually be transferred to another class of drug molecules. In medicinal chemistry and quantum pharmacology, a number of force fields currently enjoy widespread use. The MM2/MM3/MMX force fields are currently widely used for small molecules, while AMBER and CHARMM are used for macromolecules such as peptides and nucleic acids.

1.6.1.3 QM/MM Calculations

Both quantum mechanics and molecular mechanics permit optimization of the geometry of a molecule. However, each method has its strengths and weaknesses. Molecular mechanics calculations are extremely fast and efficient in providing information about the geometry of a molecule (especially a macromolecule); unfortunately, molecular mechanics provides no useful information about the electronic properties of a drug molecule. Quantum mechanics, on the other hand, provides detailed electronic information, but is extremely slow and inefficient in dealing with larger molecules. For detailed calculations on small molecules, high level ab initio molecular orbital quantum mechanics calculations are preferred. For calculations on larger molecules, including peptidic drugs or drug receptors, molecular mechanics is preferred. For a small molecule that is extremely flexible, one may wish to calculate many different conformations of the same molecule. For such a problem, a preliminary series of molecular mechanics calculations to identify a smaller number of low energy conformers, prior to performing a quantum mechanics calculation, may be indicated.

At other times, quantum mechanics and molecular mechanics may be used together in harmony. These are the so-called QM/MM calculations that have become popular over the past several years. If one wishes to use quantum pharmacology calculations to simulate a drug interacting with a site on a receptor protein, such calculations have both small molecule and large molecule components. To approach this problem, one uses QM calculations "nested" within MM calculations. The overall protein is studied using molecular mechanics calculations; however, the small region around the receptor site (and the drug interacting with that receptor via electrostatic interactions) is studied using ab initio quantum mechanics calculations. Regions intermediate between these two zones and at the interface between the molecular mechanics optimized region and the quantum mechanics optimized region may be studied using intermediate semi-empirical molecular orbital calculations.

1.6.1.4 Energy Minimization Algorithms

Whether one is using Schrödinger's equation of quantum mechanics or the force field equation of molecular mechanics, both approaches must be used in conjunction with an energy minimization algorithm. These two mechanics approaches provide a single energy for a single given geometry of the molecule; that is, they express geometry as a function of energy—this function defines an energy surface such that all possible geometries of the molecule are defined by a point on the energy surface. To obtain the optimal geometry, one must minimize the energy function (as defined by either the Schrodinger equation or a force field); that is, one must find the lowest point or deepest well on the energy surface. This is a multi-dimensional problem complicated by the presence of many local energy troughs on the energy surface which are minima in a mathematical sense, but which are higher in energy than the one single global energy minimum. Many of the minimization algorithms in current use are based on either a *steepest descent method* or a *Newton–Raphson method*, which require first and second derivative information about the energy surface, respectively. The steepest descent method is superior if the starting geometry of the drug molecule under consideration is far from the global minimum on the energy surface. The Newton–Raphson method, on the other hand, is superior when fine-tuning the geometry of the drug molecule within the depths of the energy surface well. The two methods are frequently used in sequence, with the steepest descent method being used prior to final optimization by a Newton–Raphson method.

1.6.2 Methods of Quantum Pharmacology for Conformational Analysis: Monte Carlo Methods, Molecular Dynamics, Genetic Algorithms

The ability of a drug molecule to interact with its receptor is dependent not only on the geometry of the drug molecule (as defined by bond lengths, bond angles, and interatomic distances), but also on the conformation of the molecule (as defined by rotations around torsional angles). If the drug molecule under study is "large and floppy" (i.e., it is conformationally labile and exists in a family of low energy conformers), it is difficult to identify the lowest energy conformer using quantum pharmacology calculations. For example, if the putative drug being studied is a hexapeptide, it will exist in a multiplicity of low energy shapes; the hexapeptide's potential energy surface will have many, many low energy wells, and trying to identify the global energy minimum (the lowest energy well) is a challenging task. Such energy surfaces may have billions of low energy wells, and trying to identify the single lowest energy well is a computationally demanding problem. This problem is sometimes referred to as the *multiple minima problem*. The multiple minima problem also explains our inability to predict protein folding when our only starting information is the primary amino acid sequence of a protein.

There exist a number of techniques for addressing the multiple minima problem when trying to identify the lowest energy conformer for a flexible drug or for a receptor protein. These techniques are computational chemistry methods that enable one to "search the conformational space" of the floppy drug molecule or protein under study. The *Monte Carlo method* was one of the first methods used to search conformational space, having been adapted from classical statistical mechanics. Using this method, random moves are made to the rotatable bonds of an isolated molecule. Then, using a *Metropolis sampling procedure*, it is possible to generate a large number of suitable conformations. The spectrum of acceptable conformations is then energy minimized (using a quantum mechanics or molecular mechanics approach, as discussed above), and ranked by energy. Although it is necessary to generate a large number of conformations, in principle it is possible, within a user-defined timeframe, to achieve a representative sample from low-energy conformational space.

A second, widely used method for searching conformational space is through *molecular dynamics* calculations. A simple definition of molecular dynamics is that it simulates the motions of a system of atoms with respect to the forces that are present and acting on the molecule. This collection of forces causes the system to change, but by collective motion of atoms over time, in a way that is described by integrating Newton's second law of motion ($F = ma$, where F is the force acting on an atom, m is its mass, and a is its acceleration). If one can calculate the next configuration of the collection of atoms, it is possible to follow the evolution of the atomic movements within the molecule over time. This is different from the Monte Carlo method, which requires outside intervention to produce change by a random move; in molecular dynamics, all changes result without external intervention and arise from within the system itself. In a molecular dynamics calculation, the molecule is "heated" by assigning velocities randomly to the atoms for a given temperature. Once the first velocities have been assigned, the molecular dynamics simulation is self perpetuating. As the simulation of the atomic movements progresses, the new atomic positions are calculated. By "heating" the molecule and permitting it to cool, it is possible to explore the conformational space of the molecule, thereby identifying low energy shapes.

The *genetic algorithm* method is a technique that has very recently gained attention for searching conformational space. Genetic algorithms may be applied to the multiple minima problem of molecular conformational analysis via a variety of methods. In one such method, the torsional angles within a given molecule are designated as "genes."

Then, two randomly selected starting conformations for this molecule are generated; one conformation is termed "mother," the other is "father." One-half of the genes (torsional angles) is selected from each of the parents and combined to produce an offspring conformation. If this offspring has a lower energy than its parents (as determined using either molecular mechanics or quantum mechanics calculations), the conformation is said to have "fitness" and is permitted to survive. The "most fit" conformations are permitted to propagate by exchanging their genes with their sibling conformers. A mathematical procedure, termed a "mutation operator," is used to incorporate greater diversity amongst the genes as successive generations are created. Genetic algorithm calculations permit families of low energy conformers to be identified.

Monte Carlo methods, molecular dynamics calculations, and genetic algorithm methods are all techniques for searching conformational space; each has strengths and weaknesses. The techniques are complementary rather than competitive, and may be used together in a concerted attempt to identify low energy conformers of drug molecules. Since these methods are simply techniques for skipping across the conformational space of a molecule, they must be used in conjunction with a mechanics method (e.g., quantum mechanics or molecular mechanics) to provide values of energy for the varying conformations that they generate.

One final issue, which confounds the use of these methods for identifying the elusive global energy minimum, concerns the biological relevance of this lowest energy conformation once it has been identified. Just because a detailed quantum mechanics calculation has identified a given conformation as the lowest energy shape for a drug molecule, this does not mean that this is the *bioactive conformation*. The interaction of a drug with its receptor is a dynamic process in which each molecule flexes to fit the other. It is entirely possible that the drug molecule may assume a higher energy conformation (by several kcal/mol) in order to achieve this fit, thereby rendering the search for a global energy minimum somewhat irrelevant.

For someone who has never taken a course in quantum mechanics, this discussion of quantum pharmacology may have been somewhat confusing. However, understanding these basic principles is important because of the important and ever-increasing role of molecular modeling in drug design and discovery. The diverse concepts of section 1.6 presented thus far may be summarized as follows. The "mechanics" methods (quantum mechanics [section 1.6.1.1], molecular mechanics [section 1.6.1.2]) provide a single value of energy for a single shape or conformation of a drug molecule. Since a molecule may have an almost infinite variety of shapes, the infinite number of single energy values corresponding to these shapes define a surface (termed the potential energy hypersurface). The lowest point on this surface (global minimum) is assumed to represent the most probable shape of the molecule. However, finding the lowest point on the surface is difficult. Methods such as Monte Carlo, molecular dynamics, and genetic algorithms (section 1.6.2) permit one to "hop and skip" across the potential energy hypersurface to sample it in a point-by-point fashion which may identify a point (i.e., a single conformation of the molecule) which lies in a low energy region of the surface. Once a low energy region of the surface has been so identified, then energy minimization algorithms (e.g., Newton-Raphson, section 1.6.1.4) may be used to "fine tune" the geometry and conformation of the molecule to ensure that the lowest energy structure has been identified. These concepts are diagrammatically illustrated in figure 1.13.

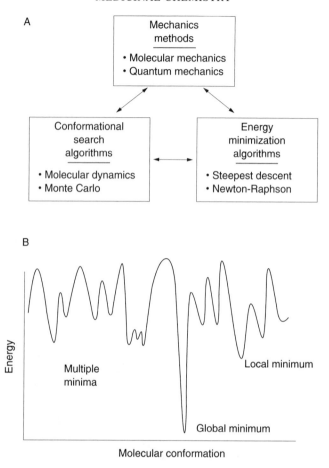

Figure 1.13 Mechanics methods assign an energy value to all possible shapes and geometries of a drug molecule or receptor macromolecule. The set of all possible shapes of a molecule defines the potential energy hypersurface for that molecule, shown in two dimensions in Part B. When attempting to identify the most probable shape of a molecule, it is necessary to search the hypersurface for the lowest energy shape (global minimum). Because of the many millions of valleys on such a surface (multiple minima), it is difficult to find the true global minimum and not just one particular local minimum. Conformational search algorithms permit an approach to "hopping" across the hypersurface in an attempt to sample it and, hopefully, find the region of the global minimum. Once the region of a minimum has been identified, energy minimization algorithms permit the bottom of the minimum energy well to be attained.

1.6.3 Applications of Quantum Pharmacology Calculations to "Small Molecule" Drug Studies

Most drug molecules are classed as "small molecules", that is, molecules with a molecular weight less than 800. Such molecules are ideally suited to having their structures probed and understood using quantum pharmacology calculations (see figure 1.14). More

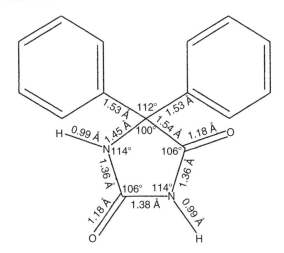

Figure 1.14 Ab initio quantum mechanics calculations can be employed to rigorously provide geometries (bond lengths, bond angles, torsional angles) for a drug molecule. This figure shows such values for the anticonvulsant drug phenytoin.

than merely providing structural information about geometries and conformations, quantum pharmacology calculations may be employed to provide useful data central to the drug design process.

An important application of quantum pharmacology calculations to small molecules is in the area of *quantitative structure-activity relationships* (QSAR). QSAR will be discussed at length in chapter 3. Over the past 30 years, QSAR has progressed from the regression equations of Hansch, through 2D QSAR, to modern 3D QSAR methods. The applications of quantum pharmacology calculations are well exemplified in 2D QSAR studies. Typically, these studies start with 10–20 analogs of a bioactive molecule. These analogs range from biologically active to inactive. Each analog, regardless of its bioactivity, undergoes extensive calculations and is described by a series of descriptors. *Geometric descriptors* reflect properties such as bond lengths, bond angles, and interatomic distances within the analogue series. *Electronic descriptors* represent properties such as atomic charge densities, molecular dipoles, and energy of the highest occupied molecular orbital. *Topological descriptors* encode aspects of molecular shape and branching and are frequently represented by graph theory indices, such as the Randic indices. *Physicochemical descriptors* reflect properties related to the ability of the molecules to traverse biological barriers such as the blood–brain barrier, and include values such as the octanol–water partition coefficient. These descriptors, especially the geometric and electronic descriptors, may be ascertained using quantum mechanics calculations. Once the descriptors have been determined, a data array is constructed with descriptors along one axis of the array and biological activity along the other axis. Statistical methods are then used to search the array and to identify the minimal descriptor set capable of differentiating between biological activity and inactivity. As a corollary to this, it is possible to deduce the bioactive face of the molecule, thereby identifying the pharmacophore.

A second application of quantum pharmacology is in the process of *pseudoreceptor mapping*. If, the pharmacophore for a family of drug molecules has been determined by means of QSAR calculations, then it is possible to deduce what the corresponding receptor should look like. For instance, if the pharmacophore has a positively charged ammonium located 6 Å from a hydrogen bonding acceptor, then the corresponding receptor site may have a negatively charged carboxylate and a hydrogen bonding donor located in an appropriate geometric orientation. Establishing the geometry of the model receptor (or pseudoreceptor map) can be achieved using either quantum mechanics or molecular mechanics calculations. Therefore, even though the structure of the actual receptor is unknown, the nature of the molecular properties that it should have can be ascertained. In principle, this pseudoreceptor map can be used to design or identify other molecules capable of docking with it.

A third important application of quantum pharmacology/molecular modeling is its use in *de novo* drug design of novel molecular shapes that will fit into a known receptor site. If the molecular structure of a receptor protein has been solved by experimental methods such as X-ray crystallography, and if the location of a potential receptor site within this protein has been deduced, then it may be possible to design small molecules to fit into this receptor site. By identifying hydrogen bonding donors or acceptors and other points for intermolecular interactions on the receptor site, it is possible to design complementary molecules to fit into this site. In short, this is the process of designing a pharmacophore and then designing the molecular baggage around the pharmacophore to ensure that the functional groups are held in an appropriate three-dimensional arrangement. Molecular mechanics and quantum mechanics are well suited to this task of designing new molecules as putative drugs.

1.6.4 Applications of Quantum Pharmacology to Large Molecule Studies

Although quantum pharmacology calculations are more rigorous and robust when applied to small molecules, such calculations may also be applied to macromolecules. There are few drug molecules that are macromolecules; peptides, such as insulin, are the exception. Usually, it is the receptor that is the macromolecule. Although receptors are discussed in detail in chapter 2, the role of quantum pharmacology in optimizing the structure of macromolecules will be presented here.

The most important potential application of quantum pharmacology to macromolecular modelling is in the area of protein structure prediction. Protein structure may be considered at multiple levels of refinement: primary structure refers to the amino acid sequence; secondary structure is defined by the local conformations induced by hydrogen bonding along the peptide backbone (e.g., α-helix, β-sheet, β-turn); tertiary structure concerns the three-dimensional structure of the protein arising from hydrogen bonding, electrostatic interactions, and other intramolecular interactions involving either side-chain or backbone functional groups; quaternary structure refers to the three-dimensional structure of proteins composed of more than one peptide chain. From this hierarchical system of structure arises the fundamental question (called the *protein folding problem*) in applying computational chemistry to protein structure: does the primary amino acid sequence determine the three-dimensional structure of a protein, and, if so, what are the rules that will permit us to predict tertiary structure with only a knowledge of primary structure? To date, the protein folding problem remains unsolved—we cannot predict the overall three-dimensional structure of a protein.

There have been many attempts to solve the protein folding problem. The first attempts initiated the process by focusing on secondary structure prediction, starting from the amino acid sequence. The first commonly used method was the *Chou and Fasman* method. From an analysis of known structures, Chou and Fasman devised propensity tables, which gave the probability of a given secondary structure for each individual amino acid. The *Garnier, Osguthorpe and Robson* (GOR) method is an extension of this statistical approach. However, extending beyond secondary structure prediction has proven difficult. One method of attempting to predict three-dimensional structure is via *sequence alignment* and *homology modeling*. In this process, the calculations begin with the crystal structure of a known protein. Then, a protein with an unknown three-dimensional structure is "aligned" with the structure of the known protein. Similar amino acids are aligned with each other; for example, a glutamate in one protein may be aligned with an aspartate in the other protein. Regions of the two proteins with similar amino acids are aligned against each other and are said to have sequence homology. The three-dimensional structure of the unknown protein is then set to be analogous to the three-dimensional structure of the known protein. Although useful, this procedure still does not solve the protein folding problem and it does require a similar protein with an experimentally solved structure.

Another important application of large-scale quantum pharmacology calculations to drug molecule design is the process of *docking simulation*. Either molecular mechanics calculations in isolation or QM/MM calculations may be used to simulate a drug molecule interacting with a proposed receptor site in a macromolecule such as a protein. Such simulations may be of value in understanding a drug's mechanism of action at a molecular or atomic level of refinement and may also be of utility in designing improved analogs of the drug molecule. These simulations (sometimes referred to as *in silico* [preferred] or *in computo* experiments to distinguish them from *in vitro* and *in vivo* experiments) may be made more physiologic by including solvation effects. Sometimes, it is possible to add hundreds if not thousands of explicit water molecules around the docking simulation about the drug and its receptor. The presence of solvating waters influences the conformation and reactive properties of the drug and its receptor. The task of adding many water molecules dramatically increases the computational intensity of this work.

Through the consideration of large molecules, quantum pharmacology may someday make the jump to *quantum medicine*. More than simply permitting an elucidation of optimal geometries for purposes of drug design, quantum medicine will enable a detailed molecular and submolecular understanding of human disease at a rigorous and quantitative level of conceptual refinement.

1.6.5 The Clinical–Molecular Interface: "Butterfly Angles" in Tricyclic Drugs

Tricyclic molecules are frequently used in drug design and as drugs to treat a diversity of disorders. Tricyclic drugs contain three rings fused together. Molecules belonging to this structural class are routinely used for the treatment of psychosis, schizophrenia, depression, epilepsy, headache, insomnia, and chronic pain. In treating these many disorders, tricyclic drugs demonstrate an ability to bind to a plethora of different (and structurally quite distinct) receptors, including many types of dopamine receptors, serotonin receptors, acetylcholine receptors, and even the voltage-gated sodium channel protein. Beyond this, tricyclic molecules have been suggested as a treatment for prion-based dementias such as Creutzfeldt–Jakob disease or the human variant "mad cow disease." A review of the patent literature also discloses suggestions that tricyclics may be useful in the treatment of Alzheimer's disease. Consequently, the tricyclic moiety is regarded as a *preferred platform*—a chemical structure that can be successfully exploited to produce a wide range of drugs for very diverse clinical indications.

Case 1.1. A 19-year old female attempted "suicide by tricyclics" by ingesting 13.2 g of carbamazepine (an anticonvulsant drug). She then had multiple seizures and was taken to an emergency room in coma (secondary to the interaction of the drug with voltage-gated Na^+ channels). By 18 h post-overdose she was awake, but was experiencing writhing movements in her arms (possibly secondary to the drug interacting with dopamine receptors). Between 20 and 40 h post-overdose she had no "bowel sounds" (i.e., no gastrointestinal peristalsis, secondary to the drug interacting with acetylcholine receptors in the bowel). With supportive care, she ultimately made an uneventful recovery (see Weaver et al., 1988). This not uncommon case exemplifies the range of receptors available for binding to a tricyclic drug molecule.

The clinically relevant relationship between tricyclic structure and bioactivity can be assessed using quantum pharmacology calculations. It is possible to quantify the spatial relationships (butterfly angles) between the planes defined by the "aromatic wings" of the tricyclic molecules. A series of angular descriptors (figure 1.15) can be used as measures of these spatial relationships; these descriptors can be accurately calculated, using molecular orbital calculations. Anticonvulsant effects are mediated primarily through the voltage-gated Na^+ channel; antipsychotic effects are mediated primarily through dopamine receptors.

1.6.6 Experimental Alternatives to Quantum Pharmacology Calculations for Small Molecules: X-Ray Crystallography and NMR Spectroscopy

Over the past decade, quantum pharmacology calculations have proven to be an immensely powerful tool in drug design and medicinal chemistry. However, several experimental techniques offer the same information.

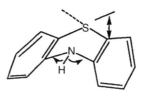

Figure 1.15 The "Butterfly Angle". Tricyclic drugs consist of anticonvulsants (carbamazepine), antidepressants (amitriptyline), and antipsychotics (chlorpromazine). Although all three families consist of three interconnected ring systems, the orientation between the rings varies, imparting a different spectrum of bioactivity.

1.6.6.1 X-Ray Crystallography

When it comes to determining the geometry of a drug molecule, X-ray crystallography remains "the gold standard." It is used extensively to study the structure of molecules and is the most powerful method available for the determination of molecular structure. To apply X-ray crystallography to drug molecules, the compound must first be crystallized into a solid form; within this crystalline solid, many drug molecules lie stacked together. X-rays have wavelengths of approximately 1 nm, a scale of atomic dimensions. When X-rays strike a crystalline solid, the X-rays interact with electrons in the atoms and are scattered in different directions, with varying intensities due to interference effects. When this interference is constructive, in-phase waves combine to produce a wave of greater amplitude that can be indirectly detected by exposing a spot on a photographic film. When the interference is destructive, the waves cancel each other such that a decreased X-ray intensity is recorded. These interference effects arise because the different atoms within the molecule of the crystalline solid scatter the X-rays in different directions. This scattered radiation produces maxima and minima in various directions, generating a diffraction pattern. The quantitative aspects of the diffraction pattern are dependent on the distances between planes of atoms within the crystal and on the X-ray wavelength; these relationships may be mathematically analyzed by means of the Bragg equation

$$n\lambda = 2d \sin \theta \qquad (1.15)$$

where n is an integer, λ is the X-ray wavelength, d is the spacing between atomic layers, and θ is the angle of scattering. By analyzing the angles of reflection and the intensities of diffracted X-ray beams, it is possible to determine the location of atoms within the molecule. Thus, determining the molecular structure of a crystalline solid is equivalent to determining the structure of one molecule. This in turn provides detailed information about the structure of the drug molecule (i.e., bond lengths, bond angles, interatomic distances, molecular dimensions).

X-ray crystallography has a long history of contributions to medicinal chemistry. Perhaps first and foremost is the work of Dorothy Hodgkin who transformed X-ray crystallography into an indispensable scientific method. Her first major achievement was the crystallographic determination of the structure of penicillin in 1945; in 1964 she received the Nobel Prize in Chemistry for determining the structure of Vitamin B12. Myoglobin and hemoglobin were the first proteins (in 1957 and 1959) to be subjected to a successful X-ray analysis. This was achieved by J. C. Kendrew and Max Perutz at Cambridge University; they received the 1962 Nobel Prize. In what is perhaps the most famous application of X-ray crystallography, James Watson and Francis Crick in 1953 used X-ray data from Rosalind Franklin and Maurice Wilkins to deduce the double-helix structure of DNA. Watson, Crick, and Wilkins received the 1962 Nobel Prize in Medicine for this work; Franklin was already deceased.

Clearly, in its infancy X-ray crystallographic determination of molecular structure was a challenging task. Nowadays, this is no longer the case. Automated X-ray diffractometers, direct methods for structure determination, and increasingly sophisticated computers and more efficient software have permitted X-ray crystallography of small drug molecules to become almost routine. Rather than solving single molecule structures, it is now possible to study families of compounds. Such X-ray studies provide valuable experimental information about the precise dimensions of drug molecules. In addition to providing structural insights into small drug molecules, X-ray crystallography can also provide data concerning drug–macromolecule interactions when the drug and its receptor are co-crystallized.

1.6.6.2 Nuclear Magnetic Resonance Spectroscopy

Although, historically, X-ray crystallography was the only practical experimental technique for structural elucidation of molecules, nuclear magnetic resonance (NMR) spectroscopy has been making significant inroads for many years. NMR is a spectroscopic technique that enables "visualization" of nuclei within a drug molecule. However, not all atomic nuclei can give rise to an NMR signal; only nuclei with values of I (the spin quantum number) other than zero are "NMR active". The spin number of a nucleus is controlled by the number of protons and neutrons within the nucleus; the nuclear spin varies from element to element and also varies among isotopes of a given element. A nucleus with a spin quantum number I may take on $2I+1$ energy levels when it is placed in an applied magnetic field of strength H. The amount of energy separating these levels increases with increasing H; however, the amount of energy separating adjacent levels is constant for a given value of H. The specific amount of energy separating adjacent levels, ΔE, is given by

$$\Delta E = (H \gamma h)/(2\pi) \tag{1.16}$$

where γ is the magnetogyric ratio for a given isotope, H is the strength of the applied magnetic field, and h is Planck's constant. The creation of an NMR spectrum for a drug molecule is related to this difference in energy (ΔE) between adjacent energy levels. In the NMR experiment, a nucleus is energetically excited from one energy level into a higher level. Since the exact value for ΔE is related to the molecular environment of the nucleus being excited, there now exists a way of relating the value of ΔE to the molecular structure; this enables the molecular structure to be determine.

Nuclear magnetic resonance (NMR) is based on the fact that a number of important nuclei (e.g., ^1H, ^2H, ^{13}C, ^{19}F, ^{23}Na, ^{31}P, ^{35}Cl) show the atomic property called magnetic momentum; their nuclear spin quantum number I is larger than zero (for ^1H, ^{13}C, ^{19}F, and ^{31}P, $I = 1/2$). When such a nucleus (or an unpaired electron) is put into a strong magnetic field, the axis of the rotating atom will describe a precessional movement, like that of a spinning top. The precessional frequency ω_0 is proportional to the applied magnetic field H_0: $\omega_0 = \gamma H_0$, where γ is the magnetogyric ratio, which is different for each nucleus or isotope. Since the spin quantum number of the nucleus can be either $+1/2$ or $-1/2$, there are two populations of nuclei in any given sample, one with a higher energy than the other. These populations are not equal: the lower-energy population is slightly more abundant. The sample is then irradiated with the appropriate radiofrequency. At a certain frequency, the atom population with the lower energy will absorb the energy of the radiofrequency and be promoted to the higher energy level, and will be in resonance with the irradiating frequency. The energy absorption can be measured with a radio receiver (just as in the case of any other electromagnetic radiation such as ultraviolet or infrared, using the appropriate detectors) and can be displayed in the form of a spectrum of absorption versus the irradiating frequency. The great information content of this spectrum derives from the fact that each nucleus of a molecule (e.g., each proton) will have a slightly different resonance frequency, depending on its "environment" (the atoms and electrons that surround it). In other words, its magnetic momentum will be "shielded" differently in different functional groups. This makes it easy to distinguish, for example, the protons on a C-CH group from an O-CH$_3$ group or an N-CH$_3$ group, aliphatic or aromatic protons, carboxylic acid or aldehyde protons, and so on, because they absorb at different frequencies. In the same fashion, every carbon atom in a molecule can be distinguished by ^{13}C magnetic resonance spectroscopy.

The only drawback to NMR is its low sensitivity. Concentrations in the millimolar range are sometimes required, although with computer enhancement techniques (such as Fourier transform) signals at 10^{-6}–10^{-5} M concentrations can be detected. This is especially important for nuclei that have a low natural abundance, such as ^{13}C (1.1%) or deuterium, ^{2}H (0.015%).

Fourier-transform (pulsed) proton NMR techniques allow an even more sophisticated assignment of resonances to specific protons. If the single high-frequency pulse is replaced by two pulses of variable pulse separation, the introduction of a second time parameter yields a *two-dimensional* NMR spectrum, with two frequency axes. Resonances on the diagonal are the normal, one-dimensional spectrum, but off-diagonal resonances show the mutual interaction of protons through several bonds. This allows the assignment of all protons even in very large molecules; recently, the three-dimensional spectrum of a small protein has been deduced by use of a three-pulse method.

Nuclear magnetic resonance permits counting of the protons in a molecule. The area under each NMR resonance peak is proportional to the protons contained in that functional group. One of the easily identifiable groups in the spectrum is used as a relative standard; electronic integration of the peak areas will give the number of protons in each group of signals, clarifying the assignment of resonances to specific structural features.

The detection of relaxation rates is a further application of NMR spectroscopy. When a particular nucleus, such as a methyl proton, is irradiated by a strong radiofrequency and absorbs it, the populations of protons in the high- and low-spin states are equalized and the signal disappears after a while. It will be recalled that the NMR signal is based on energy absorption; if all of the nuclei of a given type are in the high-spin state, absorption is not possible and "saturation" occurs. Upon removal of the strong irradiating frequency, the high- and low-spin populations will once again become unequal by transferring energy either to the solvent (spin–lattice relaxation, T_1) or to another spin in the molecule (spin–spin relaxation, T_2), and the appropriate spectrum line will assume its original amplitude. The time necessary for this recovery is called the *relaxation time,* whereas its reciprocal is the *relaxation rate.* We shall see in some later examples how relaxation rates can be used in elucidating molecular interactions.

Another tool in NMR spectral analysis is the observation of slight shifts of the various peaks. Hydrogen bonding and charge-transfer complex formation will shift resonances downfield (to lower frequencies) and upfield, respectively. On the other hand, the *coupling constant,* or separation distance between the sublines of doublets or triplets, is a result of line splitting by neighboring protons. Thus, line multiplicity (in addition to line position) is used in determining the nature of a proton and its neighbor. An ethyl group, for instance, gives a triplet (—CH$_3$ split by the adjacent —CH$_2$– group into three peaks) and a quartet (—CH$_2$— split into four peaks by the —CH$_3$). The magnitude of the coupling constant for two protons is also influenced by the dihedral angle of the X—Y bond in an H—X—Y—H structure, and can be used in conformational analysis.

In peptides, the coupling constants of the —CH— and —NH— protons show a correlation with the dihedral angle. This, however, can be ambiguous, since some coupling constants can be assigned to four different dihedral angles. Additional structural information can be obtained from the coupling constant of the H—^{13}C—N—H structure or H—C—C—^{15}N arrangement, giving correlations that do not overlap with the H—X—Y—H curve.

Much NMR work has been done on the interaction of small molecules with macromolecules, which is obviously of great interest in drug–receptor binding studies as well as in enzymology. In principle, the small-molecule resonances are easy to follow, provided they are not overlapped by the very complex and broad spectra of the macromolecules in the same solution. This technique was used to gain information on drug binding to serum albumin, and in some cases the binding moieties of the small molecule could be recognized by increased relaxation rates of some of the protons. It is much more difficult to obtain data on the dynamics of the binding of a macromolecule, such as an enzyme.

The advent of high magnetic field spectrometers and two-dimensional spectroscopy techniques has facilitated the utility of NMR spectroscopy in determining molecular structure. In a typical experiment, the key to using NMR spectroscopy in determining molecular structure lies in the information obtained from interactions among protons in the drug molecule. As part of this process, it is mandatory to assign all of the individual resonances in the NMR spectrum to specific protons in the drug molecule. The assignment of resonances is greatly simplified by two-dimensional NMR experiments: COSY (*correlated spectroscopy*), which provides information about through-bond interactions between protons; and NOESY (*nuclear Overhauser enhancement spectroscopy*), which provides information about through-space interactions. These techniques are permitting NMR spectroscopy to provide valuable structural information about drug molecules and even about the macromolecular peptidic receptors with which the drug molecules interact.

1.6.6.3 Comparison of Experimental Techniques

In comparing the use of experimental techniques such as NMR and X-ray crystallography against theoretical techniques such as quantum mechanics and molecular mechanics, a number of strengths and weaknesses must be considered. In general, experimental techniques have the strength of being applied to "real molecules" and being "less abstract." X-ray crystallography is the "gold standard" and the "method of choice" for determining the structure of a drug molecule. However, before X-ray crystallography can be used, the compound must be synthesized, purified, and crystallized. Also, X-ray crystallography provides structural information about a solid-state form of the drug molecule. This solid-state geometry and conformation may bear no resemblance whatsoever to the solution phase structure of the drug. (Determining how a drug interacts with its receptor by using solid state X-ray data is analogous to determining how geese fly by studying a single frozen goose stacked amongst other frozen geese in a grocery store freezer.) NMR spectroscopy at least permits solution-phase structural studies, but still does not supply the detailed data provided by either X-ray or quantum pharmacology studies. Quantum pharmacology calculations are inexpensive, fast, and do not require that the compound has already been synthesized. Thus, quantum pharmacology calculations may be used in a predictive manner to determine which molecules should next be synthesized. Furthermore, many theoretical chemists would argue that current quantum mechanics calculations provide structural data on small drug molecules that is equivalent to an X-ray structure; others might dispute this assertion.

1.6.7 Bioinformatics and Cheminformatics

Bioinformatics and cheminformatics are significant, new, rapidly evolving techniques focused upon the management of information. They are exerting an important influence on the future of medicinal chemistry, drug design, and quantum pharmacology calculations. *Bioinformatics* refers to the tools and techniques (usually computational) for storing, handling, and communicating the massive and seemingly exponentially increasing amounts of biological data emerging mainly from genomics research but also from other areas of biological research. Bioinformatics has the goal of enabling and accelerating biological and pharmacological research. It encompasses a diverse range of activities including data capture, automated data recording, data storage, data access, data analysis, data visualization, and the use of search engines and query tools for probing multiple databases. Bioinformatics also endeavors to draw correlations between biological data from multiple sources in an attempt to identify novel information that may have utility in drug design; the use of bioinformatics in drug design is now ubiquitous and all pervasive.

Bioinformatics attempts to combine data from the following three principal types of study:

1. Gene discovery studies
   - High-throughput genetic sequencing
   - Genetic linkage studies
   - Genetic maps
   - Polymorphism studies
2. Gene function studies
   - Gene chips and microarrays
   - Gene expression profiles
   - Functional genomics
   - Proteomics
   - Metabolic pathways
3. Clinical trial studies
   - Clinical trial efficacy data
   - Pharmacokinetic studies
   - Pharmacogenetics
   - Pharmacogenomics
   - Toxicology studies
   - Patient data

Most recently, bioinformatics is being employed to understand the organization and function of the human genome. However, the use of bioinformatics to characterize smaller, less complex genomes, notably bacteria and yeast, has preceded studies of the human genome. For example, the *Saccharomyces cerevisiae* genome project, which delivered the first complete eukaryotic genome with 16 chromosomes and 6200 genes, provided a model for ways in which DNA sequence information could be used in the systematic study of biochemical and functional processes.

Bioinformatics is proving invaluable in harnessing the power to study bacterial genomes in the search for new antibiotics. Over the past four decades, the search for new antibiotics has been essentially restricted to a relatively small number of well-known classes of compounds. Although this approach yielded numerous effective compounds, clinical resistance (i.e., antibiotic-resistant "superbugs") ultimately arose because of insufficient chemical variability. Bioinformatics-aided exploration of bacterial genomes is providing opportunities to expand the range of potential drug targets and to facilitate a shift from direct antimicrobial screening programs to rational target-based strategies. By comparing the genes of a given type of bacteria with the human genome it is possible to identify genes unique to the bacteria which may be targeted in such a way as to reduce potential toxicity in humans. Moreover, by determining the function of these bacteria-specific genes, it is possible to ascertain their usefulness as targets in designing drugs that will be lethal to those bacteria. Thus, bioinformatics is an extremely powerful tool for the future of theoretical drug design.

Cheminformatics is the chemistry equivalent to bioinformatics and involves the tools and techniques (usually computational) for storing, handling, and communicating the massive and ever-increasing amounts of data concerning molecular structures. Like bioinformatics, cheminformatics attempts to combine data from varying sources:

1. Molecular modelling studies
2. High-throughput screening results for molecules
3. Structure-based drug design studies
4. Small molecule compound libraries
5. Virtual chemical libraries

There are many examples of applying cheminformatics to drug design. For instance, if the pharmacophore for a particular class of compounds has been identified through QSAR studies, then it is possible to search other families of molecules to ascertain whether this pharmacophore is present in other classes of molecules. Various mathematical algorithms are in place to permit overlapping of structurally different molecules to see whether a common pharmacophore exists. In short, this is using cheminformatics to discover other molecules with the same pharmacophore but with different "molecular baggage" portions. A technique that is somewhat analogous to this pharmacophore search application of cheminformatics is to use a docking algorithm to systematically insert all molecules within a compound library into a known receptor site. By this strategy, the three-dimensional structure of a receptor has been determined by X-ray crystallography. Next, each molecule within an extensive library of molecules is docked with this receptor via computer simulation. Molecules that fit into the receptor can be identified and subsequently explored in an experimental setting.

Cheminformatics is also used extensively in "combichem" approaches to drug discovery (see chapter 3). If a high throughput assay is available for a particular disease, then it is possible to screen a large library of small-molecule compounds through this screen to identify a potential lead candidate. A problem central to this approach is to verify that the library of small molecules possesses true molecular diversity and that the molecules contained within the library contain all possible functional groups displayed systematically in three-dimensional space. Cheminformatics calculations based on molecular modeling and quantum pharmacology methods may be used to verify that the library of compounds truly has comprehensive molecular diversity.

When used in harmony, bioinformatics and cheminformatics are a powerful combination of computer-intensive techniques which will grow in power over the coming decade as information-handling technologies improve in sophistication. Currently, these two informatics techniques represent the most rapidly growing technology in the future of drug design.

Selected References

General Organic Chemistry Textbooks

W. Brown, C. Foote (2002). *Organic Chemistry,* 3rd ed. New York: Harcourt.
J. Clayden, N. Greeves, S. Warren, P. Wothers (2001). *Organic Chemistry.* New York: Oxford.
M. A. Fox, J. K. Whitesell (2004). *Organic Chemistry,* 3rd ed. Toronto: Jones and Bartlett.
G. Solomons, C. Fryhle (2002). *Organic Chemistry,* 7th ed. New York: John Wiley.

General Medicinal Chemistry Textbooks

J. N. Delgado, W. A. Remers (1998). *Wilson and Gisvold's Textbook of Organic, Medicinal and Pharmaceutical Chemistry*, 10th ed. New York: Lippincott, Williams and Wilkins.
D. G. Grahame-Smith, J. K. Aronson (2002). *Textbook of Clinical Pharmacology and Drug Therapy.* 3rd ed. New York: Oxford University Press.
R. B. Silverman (2004). *The Organic Chemistry of Drug Design and Drug Action*, 2nd ed. San Diego: Academic Press.
D. A. Wilkins, T. L. Lemke (2002). *Foye's Principles of Medicinal Chemistry*, 5th ed. New York: Lippincott, Williams and Wilkins.

PROPERTIES OF DRUG MOLECULES

A. Albert (1985). *Selective Toxicity,* 7th ed. London: Chapman and Hall.
A. Albert, E. Serjeant (1984). *The Determination of Ionization Constants,* 3rd ed. London: Chapman and Hall.
E. J. Ariëns, W. Soudijn, P. Timmermans (Eds.) (1983). *Stereochemistry and Biological Activity of Drugs.* Oxford: Blackwell.
E. J. Ariëns (1984). Stereochemistry, a basis for sophisticated nonsense in pharmacokinetics and clinical pharmacology. *Eur. J. Clin. Pharmacol. 26*: 663–668.
E. J. Ariëns (1986). Chirality in bioactive agents and its pitfalls. *Trends Pharmacol. Sci. 7*: 200–205.
A. Balaban, A. Chiriac, J. Motoc, Z. Simon (1980). *Steric Fit in Quantitative Structure–Activity Relations.* Berlin: Springer.
E. Blackwood, C. L. Gladys, K. L. Leering, A. E. Petrarca, J. K. Rush (1968). Unambiguous specification of stereoisomerism about a double bond. *J. Amer. Chem. Soc. 90*: 509–510.

M. Charton, J. Motoc (Eds.) (1983). *Steric Effects in Drug Design.* Berlin: Springer.

K. C. Chu (1980). The quantitative analysis of structure–activity relationships. In: M. Wolff, ed. *The Basis of Medicinal Chemistry*, 4th ed. *Part 1.* New York: Wiley-Interscience. Pp.393–418.

G. M. Crippen, T. F. Havel (1988). *Distance Geometry and Molecular Conformation.* New York: Wiley.

J. M. Daniels, E. R. Nestmann, A. Kerr (1997). Development of stereoisomeric drugs: a brief review of scientific and regulatory considerations. *Drug Inf. J. 31*: 639.

P. M. Dean (1990). *Concepts and Applications of Molecular Similarity.* New York: Wiley.

P. M. Dean (1995). *Molecular Similarity in Drug Design.* New York: Chapman and Hall.

J. N. Delgado, W. A. Remers (Eds.) (1998). *Textbook of Organic Medicinal and Pharmaceutical Chemistry.* New York: Lippincott Williams and Wilkins.

F. Franks (Ed.). (1972). *Water. A Comprehensive Treatise, vol. 1. The Physics and Physical Chemistry of Water.* New York: Plenum.

N. P. Franks, W. R. Lieb (1984). Do general anesthetics act by competitive binding to specific receptors? *Nature 310*: 599–601.

A. K. Ghose, G. M. Crippen (1986). Atomic Physicochemical Paramotors for 3-D Quantitative Structure–Activity Relationships. J. Comput. Chem. 7: 565–577.

M. Gumbleton, W. Sneader (1994). Pharmacokinetic considerations in rational drug design. *Clinical Pharmacokinetics 26*: 161.

L. P. Hammet (1970). *Physical Organic Chemistry,* 2nd ed. New York: McGraw-Hill.

C. Hansch, A. Leo (1995). *Exploring QSAR: vol. 1, Fundamentals and Applications in Biology and Chemistry.* Washington: American Chemical Society.

E. Hines (2000). What's new in excipients? *Pharm. Formul. and Qual. 4*: 24.

H. Kubinyi (1993). Hansch analysis and related approaches. In: R. Mannhold (Ed.). *Methods and Principles of Medicinal Chemistry, vol. 1.* Weinheim: VCH.

I. D. Kuntz, E. Meng, B. Shoichet (1994). Structure based molecular design. *Acc. Chem. Res. 27*: 117.

P. A. Lehman (1986). Stereoisomerism and drug action. *Trends Pharmacol. Sci. 7*: 281–285.

A. Leo, C. Hansch, D. Hoekman (1995). *Exploring QSAR: Vol. 2, Hydrophobic, Electronic and Steric Constants.* Washington: American Chemical Society.

C. A. Lipinski (1986). Bioisosterism in drug design. *Ann. Rep. Med. Chem. 21*: 283.

B. Longoni, R. W. Olsen (1992). Studies of the mechanism of interaction of anesthetics with GABA-A receptors. In: G. Biggio, E. Costa (Eds.). *GABAergic Synaptic Transmission.* New York: Raven.

L. R. Low, N. Castagnoli (1978). Enantioselectivity in drug metabolism. *Ann. Rep. Med. Chem. 13*: 304–315.

R. F. Rekker (1977). *The Hydrophobic Fragmental Constant.* New York: Elsevier.

R. F. Rekker, R. Mannhold (1992). *Calculation of Drug Lipophilicity.* Weinheim: VCH.

R. B. Silverman (1992). *The Organic Chemistry of Drug Design and Drug Action.* New York: Academic Press.

S. C. Stinson (2001). Chiral pharmaceuticals. *Chem. Eng. News.* Oct. *1*: 79.

C. R. Symous (1981). Water structure and reactivity. *Acc. Chem. Res. 14*: 179–187.

A. Szent-Györgyi (1960). *Introduction to Submolecular Biology.* New York: Academic Press.

J. Uppenbrink, J. Mervis (2000). Drug discovery. *Science 287*: 1951.

A. Verloop (1987). *The STERIMOL Approach to Drug Design.* New York: Marcel Dekker.

C. G. Wermuth, C. R. Ganellin, P. Lindberg, L. A. Mitscher (1998). Terms used in medicinal chemistry: IUPAC recommendations: 1997. *Ann. Rep. Med. Chem. 33*: 385.

M. E. Wolf (Ed.) (1995). *Burger's Medicinal Chemistry and Drug Discovery.* New York: Wiley Interscience.

Selected References in Quantum Pharmacology

Note: Over the past five years, the role of molecular modeling, bioinformatics, and cheminformatics in drug design has grown exponentially. Traditionally, these areas have not been discussed in medicinal chemistry, and many students of medicinal chemistry have not taken background courses in these areas. Accordingly, we have included a more comprehensive list of references.

GENERAL QUANTUM MECHANICS TEXTBOOKS

P. W. Atkins, R. S. Friedman (1997). *Molecular Quantum Mechanics*. Oxford: Oxford.
J. E. House (1998). *Fundamentals of Quantum Mechanics*. San Diego: Academic Press.
J. Simons, J. Nichols (1997). *Quantum Mechanics in Chemistry*. Oxford: Oxford.
A. Szabo, N. Ostlund (1996). *Modern Quantum Chemistry*. New York: Dover.

GENERAL COMPUTATIONAL CHEMISTRY TEXTBOOKS

T. Clark (1985). *A Handbook of Computational Chemistry*. New York: John Wiley.
D. B. Cook (1998). *Handbook of Computational Quantum Chemistry*. Oxford: Oxford.
F. Jensen (1999). *Introduction to Computational Chemistry*. New York: John Wiley.
A. R. Leach (1996). *Molecular Modelling Principles and Applications*. Essex: Longman.
D. C. Young (2001). *Computational Chemistry*. New York: John Wiley.

REVIEWS ON AB INITIO QUANTUM MECHANICS

J. Cioslowski (1993). *Rev. Comput. Chem. 4*: 1.
E. R. Davidson (1998). *Encycl. Comput. Chem. 3*: 1811.
S. Shaik (1998). *Encycl. Comput. Chem. 5*: 3143.

REVIEWS ON SEMI-EMPIRICAL QUANTUM MECHANICS

A. J. Holder (1998). *Encycl. Comput. Chem. 1*: 8 (Review on AM1).
K. Jug, F. Neumann (1998). *Encycl. Comput. Chem. 1*: 507.
J. J. P. Stewart (1998). *Encycl. Comput. Chem. 3*: 2080 (Review on PM3).
J. J. P. Stewart (1998). *Encycl. Comput. Chem. 3*: 2000.
W. Thiel (1996). *Adv. Chem. Phys. 93*: 703.
M. C. Zerner (1991). *Rev. Comput. Chem. 2*: 313.

REVIEWS ON MOLECULAR MECHANICS

N. L. Allinger (1998). *Encycl. Comput. Chem. 2*: 1013.
U. Burkert, N. Allinger (1982). *Molecular Mechanics*. Washington: American Chemical Society.
A. G. Csaszar (1998). *Encycl. Comput. Chem. 1*: 13.
P. Kollman (1987). *Ann. Rev. Phys. Chem. 38*: 303.
B. Reindl (1998). *Encycl. Comput. Chem. 1*: 196.

MOLECULAR DYNAMICS AND MONTE CARLO METHODS

P. Balbuena, J. Seminario (1999). *Molecular Dynamics*. Amsterdam: Elsevier.
D. L. Beveridge (1998). *Encycl. Comput. Chem. 3*: 1620.
R. A. Marcus (1997). *Adv. Chem. Phys. 101*: 391.
J. J. de Pablo, F. A. Escobedo (1998). *Encycl. Comput. Chem. 3*: 1763.
D. C. Rapaport (1997). *The Art of Molecular Simulation*. Cambridge: Cambridge.

X-RAY CRYSTALLOGRAPHY AND NMR

M. R. Jefson (1998). Applications of NMR spectroscopy to structure determination. *Ann. Rev. Med. Chem. 23*: 275.

J. J. Stezowski, K. Chandrasekhar (1986). X-ray crystallography as an aid to drug design. *Ann. Rev. Med. Chem. 21*: 293.

COMPUTATIONAL CHEMISTRY OF BIOMOLECULES

P. S. Charifson (Ed.) (1997). *Practical Application of Computer-Aided Drug Design.* New York: Marcel Dekker.

H. van de Waterbeemd (1995). *Advanced Computer Assisted Techniques in Drug Discovery.* New York: John Wiley.

BIOINFORMATICS AND CHEMINFORMATICS

F. A. Bisby (2000). The quiet revolution: biodiversity informatics. *Science 289*: 2309.

A. Edwards, C. Arrowsmith, B. des Pallieres (2000). Proteomics: new tools for a new era. *Mod. Drug Discov. 3*: 34.

B. R. Jasny, L. Roberts (2001). Unlocking the genome. *Science 294*: 81.

E. S. Lander, R. Weinberg (2000). Genomics: journey to the center of biology. *Science 287*: 1777.

K. Pal (2000). The keys to chemical genomics. *Mod. Drug Discov. 3*: 46.

S. Pollock, H. Safer (2001). Bioinformatics in the drug discovery process. *Ann. Rep. Med. Chem. 36*: 201.

M. Sanchez (2001). Bioinformatics: the wave of the future. *Biotechnol. Focus.* March: 16.

A. Sugden, E. Pennisi (2000). Bioinformatics and biodiversity. *Science 289*: 2305.

A. M. Thayer (2000). Bioinformatics. *Chem. Eng. News.* Feb. 7: 19.

K. J. Watkins (2001). Bioinformatics. *Chem. and Eng. News.* Feb. 19: 29.

CLINICAL-MOLECULAR INTERFACE

D. F. Weaver, P. Camfield, A. Fraser (1988). Massive Carbamazepine overdose: clinical and pharmacological observations in five episodes. *Neurology 38*: 755–759.

2

Basic Principles of Drug Design II

Receptors: structure and properties

2.1 THE RECEPTOR CONCEPT AND ITS HISTORY

Along with improved methods for understanding and designing drug molecules, the other central theme of medicinal chemistry over the past 40 years has been the elucidation of the structure and function of drug receptors, an endeavour that continues unabated in the 21st century. This is not surprising, given the importance of the receptor to the pharmacodynamic phase of drug action. Specific drugs (i.e., those that act at very low concentrations) exert their effects by interacting with a specific macromolecular structure (the receptor) in the living cell. This results in the brief formation of a reversible drug–receptor complex. This, in turn, triggers a secondary mechanism such as the opening of an ion channel, or catalyzes the formation of a second messenger, often cyclic AMP (cAMP). Other molecular participants within this chain reaction, such as kinases, are then activated. This cascade of events finally results in the physiological (and hopefully therapeutic) change attributed to the drug. The same mechanisms also operate with endogenous agents such as hormones and neurotransmitters.

It is generally accepted that endogenous or exogenous agents interact specifically with a *receptor site* on a specialized *receptor molecule*. Drug interaction with this site of binding, which has chemical recognition properties, may or may not trigger the sequence of biochemical events discussed above; therefore, one must distinguish carefully between sites of action (true receptors) and sites of binding (silent receptors or, occasionally, separate allosteric antagonist-binding sites).

The receptor concept dates back to 1878. The notion was initially formulated by John Langley, a British physiologist who worked on the biological properties of atropine (**2.1**) and pilocarpine (**2.2**) (see section 4.2). However, the actual term *receptor* was first introduced in 1907 by Paul Ehrlich, the famous pioneer of chemotherapy and immunochemistry. His concepts of receptor binding (corpora non agunt nisi fixata—"compounds do not act unless bound"), bioactivation, the therapeutic index, and drug resistance are still valid in principle, though they have undergone considerable expansion and refinement. The early history of the receptor concept is recounted by Parascandola (1980).

Table 2.1 Chemical Bonds and Average Bond Energies

Bond type	Example	Total interaction energy, $-\Delta E$ (kJ/mol)	Electrostatic energy, $-\Delta E_{es}$ (kJ/mol)	Charge-transfer energy, ΔE_{et} (kJ/mol)
Dispersion (van der Waals)	Xe...Xe	1.9	0	0
Hydrophobic	$C_6H_6...C_6H_6$	4.2	$\neq 0$	$\neq 0$
Hydrogen	$H_2O...H_2O$	37	38	9
Charge transfer	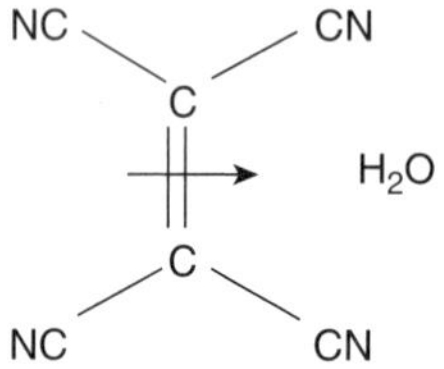	17	16	4
Dipole–dipole	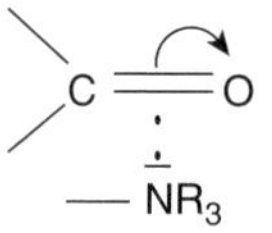	~5		
Ion–dipole	$F^{\ominus}...H_2O$	171	154	75
Ionic	$NH_4^{\oplus} F^{\ominus}$	685	757	149
	$H^{\oplus} Cl^{\ominus}$	450		
Covalent	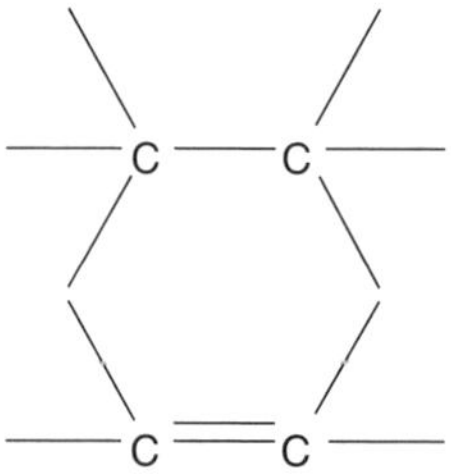	346		
		614		

(Modified from Stenlake (1979) and Kollman (1980).)

2.2.1 Covalent Bond Interactions

Although very important in traditional organic chemistry, covalent bonds are less important in drug–receptor binding than noncovalent interactions. It is generally not desirable to have a drug covalently linked to its receptor, since such an interaction would persist for a long period of time. Such prolonged interactions tend to lead to difficulties with lengthy drug half-lives and potentially to toxicity problems. Accordingly, the only receptors to which covalent binding is desirable are those that belong to exogenous (or "non-self") targets, including viruses, bacteria, parasites, or tumours (see sections 9.1 and 9.2). In short, it is okay for a drug to covalently bind to a disease-causing bacterium, but it is not okay for a drug to covalently bind to a diseased liver.

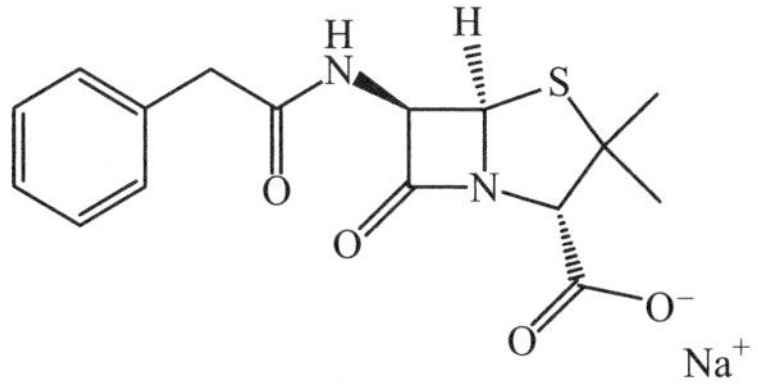

Penicillin (2.3)

While most drugs do not covalently attach themselves to their receptors, a few do. Penicillin (**2.3**), one of the most important antibacterial agents of the past century, functions via the formation of covalent bonds. It acts by acylating a bacterial transpeptidase enzyme that is vital to cell-wall synthesis within the bacterium; by structurally disrupting the cell wall, penicillin leads to death of the bacterial cells. Bonds to receptor sites are also formed by antiparasitic agents that inactivate the thiol enzymes of a parasite through bonding of a heavy metal (e.g., As, Bi, Sb) to the sulphur atoms in the enzyme's thiol groups. Finally, antitumor nitrogen mustards alkylate the amino groups of guanine bases in DNA and crosslink the two strands of the DNA double helix, preventing gene replication and transcription.

2.3.2 Ionic Bond Interactions

Ionic bonds are formed between ions of opposite charge. Their electrostatic interaction is very strong:

$$E = e_1 e_2 / Dr \tag{2.1}$$

with a bonding energy (E) that can approach or even exceed the energy of a covalent bond. Ionic bonds are ubiquitous and, since they act across long distances, play an important role in the actions of ionizable drugs. The interaction between a negatively charged carboxylate and a positively charged ammonium is a prototypic example of an ionic interaction. The use of charged groups within a drug molecule can be used to influence the pharmacokinetic properties of the molecule. For example, incorporating highly polar charged groups, such as sulphonates, will decrease a drug's half-life by increasing the rate of renal excretion. Also, charged groups can be used to preclude a drug molecule from traversing the blood–brain barrier.

2.3.3 Dipole–Dipole Interactions

Molecules in which there is a partial charge separation between adjacent atoms or functional groups can interact either with each other (via a *dipole–dipole* interaction) or with ions. Dipole moments are bond moments resulting from charge differences and the distance between charges within a molecule; they are vectorial quantities and are expressed in Debye units (about 10^{-20} esum, or *e*lectro*s*tatic *u*nits per *m*eter). Linear group moments (as in *p*-dichlorobenzene) can cancel one another out; nonlinear ones (e.g., *m*-dichlorobenzene) are added vectorially. Since so many functional groups have dipole moments, dipole–dipole interactions are frequent. A carbonyl (C=O) functional group, for example, constitutes a dipole since the carbon is electropositive and the oxygen is electronegative. The energy of dipole–dipole interactions can be calculated from the following expression:

$$E = \frac{2\mu_1\mu_2 \cos\theta_1 \cos\theta_2}{Dr^3} \qquad (2.2)$$

where μ is the dipole moment, θ is the angle between the two poles of the dipole, D the dielectric constant of the medium, and r the distance between the charges involved in the dipole. Thus, this interaction occurs over a fairly long range, declining only with the third power of the distance between the dipole charges.

Ion–dipole interactions are even more powerful, with energies that can reach 100–150 kJ/mol. The energy of such an interaction can be calculated from

$$E = e\mu \cos \hbar/D(r^2 - d^2) \qquad (2.3)$$

where e is the fixed charge and d the length of the dipole. Because the bond energy in this interaction declines only with the square of the distance between the charged entities, it is consequently very important in establishing the initial interaction between two ligands. A classic example of a dipole–ion interaction is that of hydrated ions which, in the process of hydration, become different from the same ions in a crystal lattice.

2.3.4 Hydrogen Bonding Interactions

Hydrogen bonding has considerable importance in stabilizing structures by *intra*molecular bond formation. Classical examples of such bonding occur in the protein α-helix and in the base pairs of DNA. Surprisingly, hydrogen bonds are probably less important in *inter*molecular bonding between two structures (i.e., the drug and its receptor) in aqueous solution because the polar groups of such structures form hydrogen bonds with the solvating water molecules. There is no advantage in exchanging hydrogen bonding with water molecules for hydrogen bonding with another molecule unless additional, stronger bonding brings the two molecules into sufficient proximity.

Hydrogen bonding is based on an electrostatic interaction between the nonbonding electron pair of a heteroatom (N, O, and even S) as the donor, and the electron-deficient hydrogen atom of —OH, —SH, and —NH groups. Hydrogen bonds are strongly directional, and linear hydrogen bonds are energetically preferred to angular bonds. Hydrogen bonds are also somewhat weak, having energies ranging from 7 to 40 kJ/mol.

2.3.5 Charge Transfer Interactions

The term *charge transfer* refers to a succession of interactions between two molecules, ranging from very weak donor–acceptor dipolar interactions to interactions that result in the formation of an ion pair, depending on the extent of electron delocalization. Charge transfer (CT) complexes are formed between electron-rich donor molecules and electron-deficient acceptors. Typically, donor molecules are p-electron-rich heterocycles (e.g., furan, pyrrole, thiophene), aromatics with electron-donating substituents, or compounds with free, nonbonding electron pairs. Acceptor molecules are p-electron-deficient systems such as purines and pyrimidines or aromatics with electron-withdrawing substituents (e.g., picric acid).

A classic example of CT complex formation occurs in the solution of iodine (an acceptor) in cyclohexene (a donor), when the solution assumes a brown color due to a shift in its absorption spectrum. The brown is not a color in the physical sense, but rather the result of a very broad absorption band encompassing about 200 nm in the visible spectrum and evolving as a result of electronic changes in the CT complex. In contrast, a solution of iodine in CCl_4—an inert solvent—is purple.

Drug–receptor interactions often involve CT complex formation. Examples include the reactions of antimalarials with their receptors and of some antibiotics that intercalate with DNA. The CT energy is proportional to the ionization potential of the donor and the electron affinity of the receptor, but is usually no higher than about 30 kJ/mol.

2.3.6 Dispersion and Van der Waals Interactions

Van der Waals bonds exist between all atoms, even those of noble gases, and are based on polarizability—the induction of asymmetry in the electron cloud of an atom by a nucleus of a neighboring atom (i.e., a positive charge). This is tantamount to the induced formation of a dipole. However, although the interaction: between induced dipoles sets up a temporary local attraction between the two atoms, this noncovalent interaction decreases very rapidly, in proportion to $1/R^6$, where R is the distance separating the two molecules. Such van der Waals forces operate within an effective distance of about 0.4–0.6 nm and exert an attractive force of less than 2 kJ/mol; therefore, they are often overshadowed by stronger interactions. While individual van der Waals bonds make a very low energy contribution to a system, a large number of van der Waals forces can add up to a sizable amount of energy.

2.3.7 Hydrophobic Interactions

Hydrophobic binding plays an important role in stabilizing the conformation of proteins, in the transport of lipids by plasma proteins, and in the binding of steroids to their receptors, among other examples. The concept of these indirect forces, first introduced by Kauzman in the field of protein chemistry, also explains the low solubility of hydrocarbons in water. Because the nonpolar molecules of a hydrocarbon are not solvated in water, owing to their inability to form hydrogen bonds with water molecules, the latter become more ordered around the hydrocarbon molecule, forming a molecular level interface that is comparable to a gas–liquid boundary. The resulting increase in solvent structure leads to a higher degree of order in the system than exists in bulk water, and therefore a loss of entropy. When the hydrocarbon structures—whether two protein side chains or hexane molecules dispersed in water—come together, they will "squeeze out" the ordered water molecules that lie between them (figure 2.1). Since the displaced water is no longer part of a boundary domain, it reverts to a less ordered structure, which results in an entropy gain. This change is sufficient to improve the free energy of the system by about 3.4 kJ/mol for every methylene group, and is tantamount to a bonding energy because it favours the association of hydrophobic structures. Naturally,

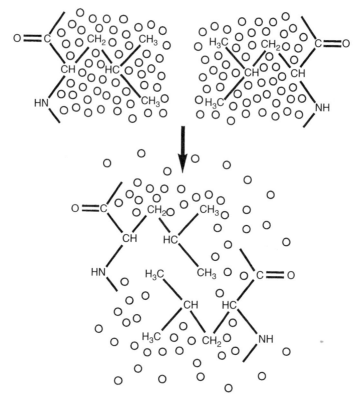

Figure 2.1 Schematic diagram of the hydrophobic interaction between two leucine side chains of a protein. By displacing part of the hydrate envelope, the two alkyl side chains occupy the same water "cavity" while many of the water molecules (represented by circles) become randomized. Thus the entropy of the system increases, resulting in a favorable stabilization.

once the hydrocarbon chains are in sufficient proximity, van der Waals forces become operative between them.

2.3.8 Selection of Drug–Receptor Binding Forces in Drug Design

When designing a drug for a particular therapeutic application, the drug designer has the opportunity to engineer functional groups capable of specific drug–receptor binding interactions into the drug molecule. As discussed in chapter 1, a drug molecule is a collection of geometrically arranged functional groups displayed on a molecular framework. These functional groups establish interactions with the drug receptor by one or more of the various binding forces discussed above.

When designing a drug, the designer wishes to have an energetically favorable and geometrically optimal interaction with the receptor site. This may be achieved in two strategies: (i) by having multiple points of contact between the drug molecule and the receptor (i.e., the pharmacophoric pattern in the drug matches to several complementary sites on the corresponding receptor site); and (ii) by having each individual point of contact between the drug and the receptor as energetically strong as possible.

The first strategy concerns the optimal number of contact points between the drug and the receptor. If the drug molecule has only two functional groups capable of binding to a receptor, then the interaction lacks specificity; such a drug could interact with too many putative receptors and would probably demonstrate unwanted toxicities. On the other hand, if the drug molecule has too many functional groups capable of interaction with a receptor, the molecule tends to be too polar and is thus too poorly absorbed and too rapidly excreted. Also, such a highly polar molecule will not likely cross the blood–brain barrier. Therefore, when designing a drug, an average of 3–5 points of contact between the drug and the receptor tends to be optimal; this corresponds to the drug molecule containing 3–5 functional groups capable of establishing binding interactions with the receptor macromolecule. If the drug is to cross the blood–brain barrier, then fewer contact points may be required; if the drug is to stay confined to the gastrointestinal tract and not absorbed, then more contact points may be tolerated.

The second strategy concerns the selection of functional groups capable of enabling the most energetically desirable interaction with the receptor site. As stated, polar groups tend to give the most energetically favorable binding interactions. Ionic interactions, for example, are among the strongest. However, desirable though they may be, too many polar groups make the drug molecule too hydrophilic, causing poor absorption, rapid excretion, and poor distribution. Usually, a mixture of varying functional groups with varying properties is desirable. If the drug is to cross the blood–brain barrier, incorporating lipophilic groups (such as aromatic rings capable of both lipophilic interactions and charge transfer interactions) into the drug molecule satisfies the twofold role of adding a point of contact between the drug and the receptor and of increasing the lipophilicity of the drug so that it can diffuse into the brain.

The drug designer must select functional groups from the following interaction types to be incorporated into the drug molecule: ionic interactions (e.g., carboxylate, sulphonate, phosphate, ammonium); dipole–dipole interactions (e.g., carbonyl, thiocarbonyl, hydroxyl, thiol, amine); hydrogen-bonding donors and acceptors (e.g., carbonyl, thiocarbonyl, hydroxyl, thiol, amine); charge transfer interaction (e.g., heteroaromatics, aromatics), or hydrophobic interactions (e.g., *tert*-butyl, *sec*-butyl). Initially, these groups are selected to enable an optimal pharmacodynamic interaction with the drug receptor macromolecule. However, these functional groups may also be selected to influence the pharmacokinetic and pharmaceutical properties of the drug molecule. Highly polar functional groups will facilitate renal excretion; lipophilic functional groups will promote passive diffusion across the blood–brain barrier.

2.4 DEFINITIONS OF CLASSICAL BINDING TERMS FOR DRUG–RECEPTOR INTERACTIONS

The findings of classical pharmacology serve as a basis for a discussion of drug–receptor interactions at a biological level. To aid in this discussion, some classical pharmacological binding terms are briefly defined. The traditional *dose–response curve* is central to these discussions, and a representative example is given in figure 2.2.

An *agonist* is a substance that interacts with a specific cellular constituent, the receptor, and elicits an observable biological response. An agonist may be an endogenous physiological substance such as a neurotransmitter or hormone, or it can be an exogenous substance such as a synthetic drug.

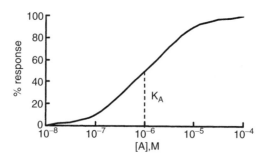

Figure 2.2 A dose-response curve on a semilogarithmic scale.

There are also *partial agonists* that act on the same receptor as other agonists in a group of ligands (binding molecules) or drugs. However, regardless of their dose, they cannot produce the same maximal biological response as a full agonist. This behavior necessitates introducing the concept of *intrinsic activity* of an agonist. This is defined as a proportionality constant of the ability of the agonist to activate the receptor as compared to the maximally active compound in the series being studied. The intrinsic activity is a maximum of unity for full agonists and a minimum of zero for antagonists. The intrinsic activity is comparable to the K_m of enzymes.

An *antagonist* inhibits the effect of an agonist but has no biological activity of its own in that particular system. It may compete for the same receptor site that the agonist occupies, or it may act on an allosteric site, which is different from the drug–receptor site. In *allosteric inhibition*, antagonist binding distorts the receptor, preventing the agonist from binding to it; that is, the antagonist changes the affinity of the receptor for the agonist. In a different system, it may have an independent pharmacological activity.

An *inverse agonist* is a drug which acts at the same receptor as an agonist yet produces an opposite effect. Inverse agonists are also sometimes called *negative antagonists*. However, an inverse agonist must be differentiated from an antagonist. Inverse agonists produce biological effects opposite to those of agonists; antagonists produce no biological effect.

An *autoreceptor* (a macromolecule typically found in a nerve ending) is a receptor that regulates, via either positive or negative feedback processes, the synthesis and/or release of its own physiological ligand. Thus, a neurotransmitter receptor may, upon binding with the neurotransmitter, either increase or decrease the biosynthesis of that neurotransmitter. This is distinct from a *heteroreceptor*, which is a receptor that regulates the synthesis and/or release of chemical mediators other than its own ligand.

Receptor *down-regulation* is a phenomenon whereby an agonist, after binding to a receptor, actually induces a decrease in the number of those receptors available for binding. Receptor *up-regulation* is the opposite and involves an agonist-induced increase in the number of receptors.

Affinity is the ability of a drug to combine with a receptor; it is proportional to the binding equilibrium constant K_D. A ligand of low affinity requires a higher concentration to produce the same effect as a ligand of high affinity. Both agonists and antagonists have affinity for the receptor.

Efficacy describes the relative intensity with which agonists vary in the response they produce when occupying the same number of receptors and with the same affinity. Efficacy and intrinsic activity are different concepts.

Potency refers to the dose of a drug required to produce a specific effect of given magnitude as compared to a standard reference. Potency is dependent upon both affinity and efficacy.

The *median effective dose* (ED_{50}) is the amount of a drug required for half-maximal effect, or to produce an effect in 50% of a group of experimental animals. It is usually expressed as mg/kg body weight. The *in vitro* ED_{50} should be expressed as a molar concentration (EC_{50}) rather than as an absolute amount. The *median inhibitory concentration* (IC_{50}) is the concentration at which an antagonist exerts its half-maximal effect. The *median toxic dose* (TD_{50}) is the dose required to produce a particular toxic effect in 50% of animals or subjects. If that toxic effect is death, then a *median lethal dose* (LD_{50}) may be defined. The *therapeutic index* is the ratio of TD_{50} to the ED_{50}.

The term pD_2 refers to the negative logarithm of the molar concentration of an agonist necessary for half-maximal effect. It is thus a measure of affinity under ideal conditions (i.e., a linear dose–response relationship). The pA_2 is the negative logarithm of the molar concentration of an antagonist that necessitates the doubling of the agonist dose to counteract the effect of that antagonist and restore the original response.

Since the drug–receptor interaction ultimately leads to a biological or clinical response, several other terms should also be defined at this point. With some drugs, the intensity of the response to a given dose may decrease over a period of time; this is the phenomenon of *tolerance*. An individual patient may be either *hyporeactive* or *hyperreactive* to a drug in that the patient's unique receptor-mediated response to a given dose of that drug may be either decreased or increased relative to the general population. Sometimes, individuals experience an *idiosyncratic* response to a drug, that is, a response that is infrequently observed in most patients.

Figure 2.3 shows how the concepts of affinity, efficacy, and agonist can be interpreted within the context of a classical dose–response curve.

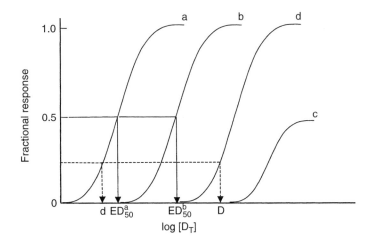

Figure 2.3 Schematic dose–response curves. Curves a and b (with different ED_{50} values) show the actions of drugs in the same series acting on the same receptor site with different intrinsic activities. Curve c represents a partial agonist of the same series. Thus, a and b are agonists; c is a partial agonist. Curve d is the action of a in the presence of a competitive antagonist. Compounds represented by curves a and b have the same efficacy.

2.5 CLASSICAL THEORIES OF DRUG–RECEPTOR BINDING INTERACTIONS

The classical theories of drug action were developed by Gaddum and Clark in the 1920s and extended to antagonists by Schild. These ideas were expanded by Stephenson (1956) and by Ariëns and his school from 1960 to 1980. It is not possible to appreciate and critically appraise current and rapidly changing ideas on the molecular nature of drug–receptor interactions without reviewing the classical pharmacological theories. Since about 1970, progress in methodology has made direct measurement of drug binding to receptors a routine procedure. The classical theories were of necessity based on measurement of the final effect of drug action—an effect that is many steps removed from the drug–receptor binding process. Therefore, the modern approach more closely follows molecular lines whereas the older classical pharmacological methodology operates at the cellular and organismic level. Naturally, both avenues have advantages and disadvantages. We shall deal first with the dose–response relationship before reviewing current receptor models.

The classical *occupation theory* of Clark rests on the assumption that drugs interact with independent binding sites and activate them, resulting in a biological response that is proportional to the amount of drug–receptor complex formed. The response ceases when this complex dissociates. Assuming a bimolecular reaction, one can write

$$D + R \rightleftharpoons DR \tag{2.4}$$

where D = drug and R = receptor. The dissociation constant at equilibrium is

$$K_D = \frac{[D][R]}{[DR]} \tag{2.5}$$

The effect (E) is directly proportional to the concentration of the drug–receptor complex:

$$E = \alpha[DR] \tag{2.6}$$

The maximum effect (E_{max}) is attained when all of the receptors are occupied:

$$E_{max} = \alpha[R_T] \tag{2.7}$$

where the total receptor concentration $[R_T]$ is

$$[R_T] = [R] + [DR] \tag{2.8}$$

and α is a proportionality factor. Therefore, from (2.5) and (2.7):

$$\frac{[DR]}{[R_T]} = \frac{[D]}{K_D + [D]} \tag{2.9}$$

Dividing (2.6) by (2.7):

$$\frac{[DR]}{[R_T]} = \frac{E}{E_{max}} \tag{2.10}$$

Combining equations (2.9) and (2.10) yields

$$E = \frac{E_{\max}[D]}{K_D + [D]} \tag{2.11}$$

Equation (2.11) indicates a hyperbolic relationship between the effect and the concentration of free drug. The ED_{50} is therefore equal to K_D. Incidentally, equation (2.11) is identical to the Michaelis–Menten relationship in enzyme kinetics, with $E_{\max}$ representing $V_{\max}$. Dose–response curves usually show effect versus the logarithm of the total drug concentration $[D_T]$, assuming that the concentration of bound drug is so small as to be negligible and that $[D_T] \cong [D]$. However, if the receptor concentration $[R_T]$ becomes large relative to K_D, then

$$ED_{50} = K_D + 0.5[R_T] \tag{2.12}$$

meaning that, at a high bound-drug concentration, the total concentration of drug may exceed K_D by an amount equal to one-half the total receptor concentration. It seems that the case $ED_{50} = K_D$ is rather exceptional. If occupation of some of the receptors is sufficient for a maximal response, as often happens, *spare receptors* will be present and

$$\frac{ED_{50}}{K_D} < 1 \tag{2.13}$$

and the true value of K_D (and thus the affinity of the drug for the receptor) will be underestimated. This case may be an indication that an "induced fit" takes place, since it seems that a small number of agonist molecules can trigger a conformational change in many receptors, leading to the activation of a larger number of receptors than seems to be warranted. The "spare" receptor concept can be tied to the idea of efficacy or intrinsic activity, meaning that some drugs may have to activate fewer receptors than others to elicit a full pharmacological effect, and are thus said to be more efficacious.

Agonists that yield parallel dose–response curves with the same maximum are assumed to act on the same site but with different affinities. Nonreceptor binding to a "site of loss" (sometimes called a "silent" receptor) can thus be distinguished from relevant binding.

Schild extended these ideas to the description of effects when a *competitive antagonist* (A) is present. If Y is the proportion of receptors occupied, that is, if

$$Y = \frac{[DR]}{[R_T]} \tag{2.14}$$

and the AR complex is inactive, then

$$\frac{K_D}{[D]} = \frac{(1 + [A]K_A)Y}{1 - Y} \tag{2.15}$$

where K_A, is the association constant of the antagonist. If the same biological response is achieved at a lower drug concentration $[d]$ in the absence of the antagonist, then

$$K_D[d] = \frac{Y}{1 - Y} \tag{2.16}$$

Dividing (2.15) by (2.16):

$$\frac{[D]}{[d]} = 1 + [A]K_A \qquad (2.17)$$

The Schild equation is thus obtained, where $[D]/[d]$ is the "dose ratio." From equation (2.17), $[A]_2$, the antagonist concentration necessitating a doubling of the agonist concentration to achieve the pure agonist effect, is

$$[A]_2 = \frac{1}{K_A} \qquad (2.18)$$

and

$$pA_2 = -\log K_A \qquad (2.19)$$

which provides a convenient experimental method for measuring the "activity" of an antagonist. This is, of course, analogous to the $pD_2 = -\log K_D$ concept.

The effect of a competitive inhibitor can also be expressed as an *inhibitor affinity constant* (K_I) by plotting the inhibitor concentration versus the reciprocal of the reaction velocity, or versus the reciprocal concentration of a labeled ligand (the isotopically radiolabeled agonist that is displaced by the antagonist). The intersect of the lines so generated is $-K_I$.

$$K_I = \frac{IC_{50}}{\frac{1+[L^*]}{K^*}} \qquad (2.20)$$

where IC_{50} is the inhibitor concentration that displaces 50% of the labeled ligand, $[L^*]$ is the concentration of the labelled ligand, and K^* is its dissociation constant. This is a method that is particularly suited for *in vitro* binding experiments; however, it is not suitable for organ preparations or whole-animal studies. Rapidly growing experimental evidence that takes into account the latest *in vitro* binding experiments favors a modified form of the occupation theory of drug activity. There are, however, phenomena that are unexplained by the occupation theory·

1. The inability of partial agonists to elicit a full response while blocking the effect of more active agents.
2. The existence of drugs that first stimulate and then block an effect.
3. *Desensitization* or *tachyphylaxis*—diminution of the effect of an agonist with repeated exposure to or higher concentrations of that agonist.
4. The concept of spare receptors.

To accommodate some or all of these phenomena, several alternatives to the occupation theory have been proposed. None of them is entirely satisfactory, and some have no physicochemical basis.

The *rate theory* of Paton, as modified by Paton and Rang, rejects the assumption that the response is proportional to the number of occupied receptors, and instead proposes a relationship of response to the *rate* of drug–receptor complex formation. According to this view, the duration of receptor occupation determines whether a molecule is an agonist, partial agonist, or antagonist. Accordingly, the concept of intrinsic activity becomes unnecessary. The rate theory offers an adequate explanation for the ability of some antagonists to trigger a response before blocking a receptor, and also accounts for desensitization. However, it lacks a plausible physicochemical basis and conflicts with some experimentally established facts (e.g., the slow dissociation rate of agonists).

The *induced-fit theory*, developed by Koshland primarily for enzymes, states that the morphology of a binding site is not necessarily complementary to the conformation— even the preferred conformation—of the ligand. According to this theory, binding produces a mutual plastic molding of both the ligand and the receptor as a dynamic process. The conformational change triggered by the mutually induced fit in the receptor macromolecule is then translated into the biological effect. Although this model does not lend itself to the mathematical derivation of binding data, it has altered our ideas on ligand–receptor binding in a revolutionary way, eliminating the rigid and obsolete "lock and key" concept of earlier times.

There are other theories of drug–receptor binding relationships. Belleau's *macromolecular perturbation theory* suggests that when a drug–receptor interaction occurs, one of two general types of macromolecular perturbation is possible: a specific conformational perturbation leads to a biological response (agonist), whereas a nonspecific conformational perturbation leads to no biological response (antagonist). Changeux's *activation–aggregation theory* is an extension of the macromolecular perturbation theory and suggests that a drug receptor (in the absence of a drug) still exists in an equilibrium between an activated state (bioactive) and an inactivated state (bio-inactive); agonists bind to the activated state while antagonists bind to the inactivated state.

Interesting though these theories may be, most are of limited practical use to the medicinal chemist who is about to design a drug. To the drug designer, a receptor is a flexible macromolecule (usually a protein) capable of a dynamic "hand-in-glove" (rather than "lock-in-key"), geometrically precise, stereospecific interaction with a flexible drug molecule, mediated via two or more specific intermolecular binding forces; this interaction, in turn, leads to an alteration in some biological or biochemical process. Because of this latter requirement, the receptor site should be somehow linked to the functional domain of the protein, so that drug binding may influence protein biological function.

2.6 EXPERIMENTAL QUANTIFICATION OF DRUG–RECEPTOR BINDING INTERACTIONS

The drug–receptor interaction may be quantified through the use of *binding constants*, which are derived from *in vivo* pharmacological experiments or from the *in vitro* use of labeled ligands.

As shown above, in the reaction of a drug with a single population of noninteracting sites,

$$D + R \rightarrow DR \tag{2.21}$$

where

$$K_D = \frac{[D][R]}{[DR]} \tag{2.22}$$

and where the fraction of occupied sites v is

$$v = \frac{[DR]}{[DR] + [R]} \tag{2.23}$$

2.6.1 Direct Plot

In the direct plot, equation (2.22) is solved for [DR] and its value substituted into equation (2.23):

$$v = \frac{[D]}{K_D + [D]} \tag{2.24}$$

K_D can then be obtained from a plot of v against [D] if the receptor concentration is constant. This is, of course, the same as the direct plot of enzyme activity shown in every biochemistry textbook. As with all hyperbolic relationships, there are several drawbacks to this technique: many data points are needed at the beginning of the curve, at low [D] values, where accuracy is limited. Also, determination of the maximum effect is almost impossible, since we are dealing with an asymptotic curve.

2.6.2 Titration Plot

In the titration plot, equation (2.24) is solved for K_D:

$$K_D = [D]\frac{(1 - v)}{v} \tag{2.25}$$

and, obtaining $\log_{10}$ of both sides,

$$\log K_D = \log [D] + \log \frac{1 - v}{v} \tag{2.26}$$

where one can use the negative logarithms

$$- \log [D] = pD \tag{2.27}$$

and

$$- \log K_D = pK_D \tag{2.28}$$

and arrive at

$$pD = pK_D + \log \left(\frac{1 - v}{v} \right) \tag{2.29}$$

In acid–base titrations, pD is pH and pK_D is pK_a because $[D] \approx [H^+]$. This is the reason for the name of the curve, well known from analytical chemistry. The drawback of this plot is that many points are needed in the vicinity of the inflection point.

2.6.3 Double Reciprocal Plot

The hyperbolic direct plot can easily be straightened out, as analogies from classical enzymology teach us. The most popular data treatment for yielding straight lines is the double reciprocal plot, also known as the Lineweaver–Burke or Benesi–Hildebrand plot. Here, we take the reciprocal of equation (2.24):

$$\frac{1}{v} = 1 + \frac{K_D}{[D]} \tag{2.30}$$

If one plots $1/v$ against $1/[D]$, K_D and v can be obtained directly with good precision.

2.6.4 Dixon Plot

Another useful method of data reduction is the Dixon plot, where $1/v$ is plotted against [I], the inhibitor concentration, at a fixed [D]. This allows for the determination of K_D without the need to determine the absolute concentration of [D]—a great advantage in cases in which the substrate is a polynucleotide or a protein, as is often the case in chemotherapy.

2.6.5 Scatchard Plot

Perhaps the most widely used method for extracting binding data is the *Scatchard plot*. This is obtained from

$$\frac{v}{[D]} = \frac{1}{K_D} - \frac{v}{K_D} \tag{2.31}$$

Thus, plotting $v/[D]$ against v or, alternatively, $[D]_{bound}/[D]_{free}$ against $[D]_{bound}$ gives a straight line. The slope is $1/K_D$ (the binding constant); the abscissa intercept is the number of binding sites if v is shown as mol v/mol R.

The great advantage of the Scatchard plot is its linearity (i.e., all data are weighted equally). Errors in measurement register on both axes and are therefore eliminated. The Scatchard plot is most useful when there are multiple binding sites; in that case, however, the plot is not linear. This most commonly occurs in the case involving a small population of receptors with a high affinity, accompanied by a large population with a low affinity. The Scatchard plot is an excellent way of extracting binding-site constants and numbers from experimental findings. However, it must be remembered that these properties of the Scatchard plot are valid only if the binding sites are independent (that is, if there is no cooperative interaction between them). Indeed, downward curvature of the Scatchard plot may result from the underestimation of nonspecific binding, simultaneous binding on two sites, and other factors beside negative cooperativity and differences in affinity.

2.6.6 Hill Plot

The cooperativity of receptor sites can be recognized from binding data. Cooperativity means that binding to one receptor site facilitates binding to subsequent receptor sites in the same population. The classical example is oxygen binding by hemoglobin, which is treated in every biochemistry text. In a direct plot, the curve of cooperative binding is sigmoidal instead of hyperbolic.

If equation (2.24) is modified to incorporate n theoretical sites on the receptor, then

$$v = \frac{n[D]^n}{K_D + [D]^n} \tag{2.32}$$

from which

$$\frac{v}{n - v} = \frac{[D]^n}{K_D} \tag{2.33}$$

If F is the fraction of occupied active sites, then the number of occupied sites becomes

$$v = nF \tag{2.34}$$

and therefore

$$\frac{v}{n - v} = \frac{[D]^n}{K_D} \tag{2.35}$$

In logarithmic form, this becomes

$$\log \frac{F}{1 - F} = n \log[D] - \log K_D \tag{2.36}$$

and a straight line, the Hill plot, results if $\log F/(1 - F)$ is plotted against $\log[D]$. The slope is n. Thus the Hill plot gives, approximately, the number of interacting sites. If the slope of the Hill plot is less than unity, negative cooperativity is suspected (i.e., the binding of the first ligand inhibits subsequent binding). The insulin receptor shows such behavior. Positive or negative cooperativity would indicate a conformational change that increases or decreases the affinity of the receptor site for the drug.

2.7 GENERAL MOLECULAR CONCEPTS
OF DRUG RECEPTOR ACTION

The preceding sections have explored classical pharmacological concepts based on the dose–response relationships in tissue or organ preparations. The enormous complexity of living systems and the remoteness of cause from effect (i.e., drug administration from pharmacological action) introduce many complications and artefacts into the study of such relationships.

Molecular pharmacologists and physical scientists have therefore sought to simplify the experimental system as much as possible. This objective has been increasingly realized as the methodology of quantitative binding experiments on membrane preparations (and, later, on isolated receptors) has become more sophisticated, precise, and simple. Isotopically labelled compounds of very high activity have made it possible to work with physiological ligand concentrations down to the picomole level (10^{-12} M). This has allowed direct experimental access to receptor binding sites and has led to the development of several complementary receptor models. Physical chemistry techniques (X-ray crystallography, NMR conformational analysis, molecular modeling) are now enabling precise molecular pictures of receptor structure. The reshaping of our ideas on drug receptors is an ongoing process.

2.7.1 Functional Molecular Properties of Drug Receptors

Early receptor models, based on pharmacological data rather than direct ligand-binding measurements, postulated that agonists and their competitive antagonists became bound to the same receptor site and competed for it. This view was partly based on findings in enzymology, in which this concept is generally valid for metabolite–antimetabolite competition as well as for activity studies of vitamins and hormones. An antimetabolite is a molecular analog of an intermediate in a physiologically relevant metabolic pathway that replaces a natural substrate. In doing so, it prevents the biosynthesis of physiologically important substances within the organism. The close structural resemblance of agonists and antagonists in these categories constitutes direct proof that they have identical binding sites. The lack of structural correlations between many neurotransmitters and their blocking agents, however, initiated a review of the competitive binding hypothesis.

It is generally accepted that there is complementarity between a ligand (either endogenous [e.g., hormone or neurotransmitter] or exogenous [e.g., drug molecule]) and its receptor site in the sense of the induced-fit concept; this suggests a mutual molding of the drug and macromolecule to take full advantage of stereoelectronic interactions. Under optimal conditions, the energies liberated in binding can reach 40–50 kJ/mol, a figure equivalent to binding equilibrium constants of about 10^{-8}–10^{-9}, which is considered to represent a high affinity.

Complementarity in the context of induced fit implies a plasticity of the receptor macromolecule in terms of an ability to undergo conformational changes and associate with ligands. In its activated state (i.e., a different conformation), the receptor can interact with effector molecules, which then transmit a nerve impulse or other signals to other structures. The complementarity also determines the selectivity of the receptor. For stereospecific binding, it is generally assumed that a ligand must have three unequal substituents; this is considered sufficient for great selectivity. The discrete forms of a receptor site are, of course, the result of receptor plasticity.

Recognizing this capacity of the receptor to assume different molecular geometries without a significant change in function is probably essential to achieving some understanding of the pluralistic nature of many receptors. It is physiologically and structurally unreasonable to assume that a given type of receptor—probably a complex, multisubunit structure that is part of an even more complex membrane framework—is absolutely identical throughout an organism. Mautner pointed out in 1967, long before the structure of any drug receptor was known in any detail, that the medicinal chemist would have to deal with an isoreceptor concept in the same matter-of-fact way that an enzymologist accepts isozymes. Despite recent advances in molecular biology, our present knowledge of receptor structure is still evolving.

This confusion is complicated even further by receptor multiplicity. Consider, for example, the presence of opiate receptors in both the central nervous system and the ileum. Not only do they have different roles as participants in neuromodulation and peristaltic

Norepinephrine
(2.4)

regulation, respectively, but they are also probably different in a morphological sense: the neuromodulatory receptors are assumed to be *presynaptic*, or situated on the presynaptic terminal membrane ahead of the synaptic gap, whereas other receptors are the classical *postsynaptic* receptors, embedded in the postsynaptic membrane of the effector cell or of the next neuron. In the first case, the receptor modulates neurotransmitter release; in the second, it may activate an enzyme such as adenylate cyclase, or trigger an action potential. As we shall see later, almost all neurotransmitters show receptor multiplicity, and medicinal chemists deal with multiple adrenergic receptor subtypes and many different opiate receptors, just to name two examples.

Receptor plasticity could be invoked as the underlying common trait of multiple receptors. For example, although the multiple adrenergic isoreceptors are similar, they react to the common neurotransmitter norepinephrine (**2.4**) in a quantitatively different manner. They also show a drug specificity that varies from organ to organ and differs in various species of animals. In subsequent chapters of this book, receptor multiplicity as the rule rather than the exception will become amply evident. It is to be hoped that, in time, the comparison of isoreceptor molecular structures will provide precise criteria for their differentiation.

The multiplicity of receptor or recognition sites for agonists and antagonists is well documented. One may distinguish (i) agonist binding sites, (ii) competitive antagonist binding sites (accessory sites), and (iii) noncompetitive antagonist or regulatory binding sites (allosteric sites).

The *agonist binding site* is the subject of continuous discussion throughout this book, ranging from a purely physical approach to the treatment of its biochemical characteristics, where these are known. In this discussion, it is implicit that we are dealing with discrete loci on the receptor macromolecule: specific amino acids, lipids, or nucleotides held in just the right geometric configuration by the scaffold of the rest of the molecule, as well as by its supramolecular environment such as a membrane.

Competitive antagonists were originally assumed to bind to the agonist binding site and, in some way, displace and exclude the agonist as a result of their very high affinity but lack of intrinsic activity. This behavior would result in displaced but parallel dose–response curves. Our present views are at variance with such simplistic older ideas. The mere fact of great chemical dissimilarity between agonists and competitive antagonists in the vast field of neurotransmitters precludes identity of the two receptor sites. It is evident at a glance that a careful analysis is needed to discern correlations between agonist–antagonist pairs or even between antagonists of the same class. As always, there are notable exceptions to this. For example, opiate analgesics and their

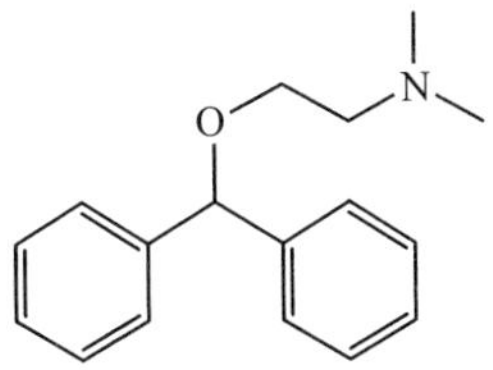

Diphenhydramine
(2.5)

antagonists are very similar in structure, but if one considers the relationship between the endogenous peptide opiates known as enkephalins and the opiate antagonists, the conspicuous dissimilarity of the two groups is once again apparent.

The most remarkable property of antagonists is their great receptor affinity, which is sometimes two to four orders of magnitude greater than that of the agonists. Many antagonists have large nonpolar moieties, usually aromatic rings. Therefore, *accessory binding sites* must exist on the receptor to accommodate these large hydrophobic groups. What is even more remarkable is that there are some compounds that are antagonistic in more than one system. Diphenhydramine (**2.5**), for example, has an antihistaminic as well as an anticholinergic action.

Competitive antagonists can be viewed in two ways. In one of these, the antagonist binding site is considered to be topically close to the agonist site and may even partially overlap it. The antagonist will therefore interfere with agonist access to the receptor, even though it need not necessarily occupy both the agonist and the accessory sites. On the other hand, the antagonist may functionally deny agonist accessibility by altering the receptor affinity. This would be closely analogous to allosteric inhibition.

Allosteric sites are at a distance from the agonist site and may even be on a different receptor protomer in the receptor–effector complex. Their occupation by allosteric inhibitors results in a conformational change that is propagated to the agonist site and changes its affinity. There is thus a mutual exclusion between the agonist and an allosteric antagonist. Moreover, classical pharmacological models cannot distinguish between competitive and allosteric inhibition. Allosteric effectors are not necessarily inhibitors. Just as in enzymology, some may activate whereas others deactivate one or another state of a receptor.

2.7.2 Molecular-Level Conceptual Models of Receptors

The transition from classical pharmacological theories of drug–receptor interactions to real, physical models of receptors is crucial for drug design. As part of this transition, a number of molecular-level conceptual models of receptors have been put forth over the years. The *two-state receptor model* and the *mobile receptor model* are two examples of such models. Although these models have limited direct utility for the medicinal chemist involved in drug design, they are extremely instructive for a number of reasons. These models emphasize the fact that many receptors are not just simple macromolecules, which interact with a drug in a "hand-in-glove" fashion. On the contrary, some receptors are extremely dynamic, existing as a family of low-energy conformers existing in equilibrium with each other. Other receptors have complex multi-unit structures, being composed of more than one protein; facilitatory and inhibitory interactions exist between these subunits and may alter the drug–receptor interaction. Finally, some receptors are not only dynamic in terms of their shape, but also mobile, drifting in the membrane like an iceberg in the ocean.

2.7.2.1 The Two-State Receptor Model

The two-state receptor model (also known as the Monod–Wyman–Changeux model) was developed on the basis of the kinetics of competitive and allosteric inhibition as well as through interpretation of the results of direct binding experiments. This model postulates that, regardless of the presence or absence of a ligand, the receptor exists in two distinct states, the R (relaxed, active, or "on") and T (tense, inactive, or "off") states, which are in equilibrium with each other. An agonist (drug, D) has a high affinity for the R state and will shift the equilibrium to the right; an antagonist (inhibitor, I) will prefer the T state and will stabilize the TI complex. Partial agonists have about equal affinity for both forms of the receptor.

Some members of a receptor population are in the R state, even in the absence of any agonist. Thus, the receptor can be thought of having a "tone" like a resting muscle. The ratio of states is defined by the equilibrium constants K_L, K_T, and K_R (for drug D or inhibitor I), and gives true physicochemical meaning to the concept of *intrinsic activity*.

In contrast to the assumption made in the classical occupation theory, the agonist in the two-state model does not activate the receptor but shifts the equilibrium toward the R form. This explains why the number of occupied receptors does not equal the number of activated receptors.

2.7.2.2 The Receptor Cooperativity Model

Receptor cooperativity, which has largely been studied on hormone receptors, is explained by further extension of the two-state model. It is assumed that the cooperation of several receptor protomers is necessary for an effect like the opening of an ion channel, with all of these protomers having to attain an R or a T state to open or close a pore. This means that the binding sites or the receptor protomers on which these sites are situated must interact, and, as they do so, their affinity changes as a function of the proportion of R-state receptors in the assembly. This also means that a drug–receptor complex can trigger the transition of an unoccupied neighboring receptor from the T to the R state. If a ligand facilitates binding or the effect of the receptor, the cooperativity is positive; if it hinders these, the cooperativity is negative (e.g., in the insulin receptor). Negative cooperativity could also account for the spare receptors (receptor reserve) seen in many systems. As receptors cluster during their own metabolic cycle, low ligand occupancy in such clusters may still lead to a large change in the cluster configuration, resulting in a full effect without a 1:1 ratio of ligand–receptor binding. Scatchard plots of ligand binding will be concave for positive and convex for negative cooperativity. Hill plots can also indicate the type of cooperativity involved.

As is already evident from foregoing discussions, effector or amplifier systems are the parts of the receptor oligomer which convey the fact that a drug has become bound to the receptor (or, to be more precise, that there has been a conformational change to the

R or T state), signaling subsequent links in the receptor–effector chain that ultimately trigger the biological effect. Besides initiating the effector step of the drug action, effectors are often amplifiers, magnifying an inconspicuous initial event like the binding of a few thousand ligand molecules at 10^{-9}–10^{-10} M concentration. Amplification can take the form of a *cascade,* as in the well-known case of epinephrine or glucagon: these hormones initiate glycogenolysis through a series of enzyme activation steps, causing the initial effect to be magnified approximately 100 million fold. Indeed, one is struck by the omnipresence of the enzyme adenylate cyclase (AC), which serves as first amplifier or effector in a large number of drug-initiated cascades as well as in numerous biochemical chains of events in enzymology. The cAMP that is subsequently produced activates kinases, which phosphorylate different proteins acting as final effectors. Since the majority of receptors are localized in cell membranes, this sequence of events constitutes *inter*cellular communication.

Another type of effector is the ion channel of an excitable membrane, which in its R (open) conformation allows the passage of about 10,000–20,000 ions in a single impulse, resulting in either membrane depolarization or polarization and a multitude of possible physiological phenomena.

2.7.2.3 *The Mobile Receptor Model*

The mobile receptor model was proposed by Cuatrecasas and by De Haën in an attempt to explain why so many different drugs, hormones, and neurotransmitters can activate adenylate cyclase. According to classical concepts, a recognition site is permanently associated with an effector site, and will regulate its operation on a one-to-one or some other stoichiometric basis. The recognition site is, of course, specific.

If this hypothesis is applied to the case of adenylate cyclase, one of two conditions would have to be assumed: that there are either as many adenylate cyclase isozymes as there are receptors acting through them, or that adenylate cyclase would need an enormous variety of specific recognition sites that can answer to many ligands. The latter possibility would imply a lack of selectivity. However, there is no evidence for either assumption.

The mobile receptor concept is an attempt to offer a solution to this problem, in recognizing that the lipid membrane is a two-dimensional liquid in which the embedded proteins can undergo rapid lateral movement or translation at a rate of 5–10 μm/min—an enormous distance on a molecular scale. The recognition protomer of a receptor complex therefore need not be permanently associated with an effector molecule, and thus no stoichiometric relationship is required. Instead, the recognition protomer can undergo rapid lateral movement and, when activated to the R state, can engage in what has been dubbed a "collision coupling." The R state of the receptor has the appropriate conformation to trigger effector activity, which could be the opening of an ionophore or the activation of adenylate cyclase. Therefore, different recognition sites can activate the same adenylate cyclase molecule at different times through the same mechanism. By the same token, a single recognition site could activate several adenylate cyclase molecules or other effector systems during its active lifetime. Such multiple collision couplings can be seen as the molecular explanation of positive cooperativity and the concept of receptor reserve. There is no need to invoke multiple recognition sites on the enzyme or a multitude of isoenzymes, only the physical separation of recognition and effector sites and their multipotential association. However, it is probably true that a recognition site that operates an ion gate is more permanently associated with the ionophore than is the drug or hormone receptor that acts through adenylate cyclase or the phosphatidylinositol system.

2.8 RECEPTOR ACTION: REGULATION, METABOLISM, AND DYNAMICS

Like all proteins, receptors or receptor subunits are coded by appropriate genes, transcribed to mRNA, translated, and further post-translationally processed in the rough endoplasmic reticulum. After the receptor protein is packaged in the Golgi apparatus, various carbohydrates are removed and others are added to the branched oligosaccharide "antenna" structures; this process is referred to as "capping." The oligosaccharides of the receptor glycoprotein seem to serve as recognition units necessary for high-affinity ligand binding, and also as protection from premature proteolytic degradation. The assembled supramolecular receptor is then inserted into the cell membrane as an *intrinsic protein,* that is, one that usually spans the width of the lipid bilayer. It can therefore communicate with the extracellular space as well as with the inside of the cell, thus fulfilling its role as a transmembrane signal transducer.

Membrane-bound receptors undergo dynamic processes that serve as regulatory mechanisms. This has led to the idea that such receptor regulation is just as important in the overall response of the system as is the response of the target organ (e.g., a muscle cell or secretory cell). There are several categories of regulatory mechanisms, and they differ primarily in the rate of response: some are very fast (milliseconds to seconds), whereas others are much slower and delayed. At this point, we know of the following mechanisms:

1. Regulation at the genetic level
2. Regulation by ligand
3. Covalent modification
4. Noncovalent modification
5. Receptor clustering
6. Migration of receptors and receptor internalization
7. Internalization and proteolytic degradation

2.8.1 Receptor Regulation at the Genetic Level

Regulation at the genetic level is often observed for hormones that can regulate the rate of synthesis either of their own receptor or of other functionally related receptors (e.g., regulation of oxytocin receptor synthesis in the uterus by estrogens).

2.8.2 Receptor Regulation by Ligand

Ligand regulation of receptors can be either self-regulatory (*homospecific*) or transregulatory (*heterospecific*). In the first case, ligand binding may initiate internalization of the ligand–receptor complex, thus removing receptors from the cell surface and decreasing the number of available receptors (as in the case of the insulin receptor). Heterospecific regulation is shown, for instance, by histamine, which at high concentrations can activate acetylcholine receptors; and by benzodiazepine anxiolytics (tranquilizers), which regulate GABA receptors.

2.8.3 Covalent Modification of Receptors

Covalent modification of receptors occurs by phosphorylation, sulfhydryl–disulfide redox reactions, and proteolytic cleavage, in the same manner as occurs for many enzymes. Upon ligand binding, the receptor may phosphorylate itself on a tyrosine or serine residue, or the ligand-induced conformational change may make the receptor a substrate for a phosphorylase kinase. Sulfhydryl redox reactions, seen in the nicotinic cholinoceptor, result in alteration of relative ligand sensitivities; in the insulin receptor, they lead to affinity changes. Thus covalent modifications of receptors, whether homospecific or heterospecific, lend biochemical significance to the pharmacological terms affinity and intrinsic activity.

2.8.4 Noncovalent Modification of Receptors

Noncovalent modifications of receptors can involve interactions with small ligands (ions, nucleotides) or macromolecules (as in the mobile receptor model), leading to allosteric changes. They can also influence the receptor environment, causing a change in membrane potential or receptor distribution (clustering, patching). A notable example is the effect of Na^+ ions on the relative affinity of opiate receptors toward agonists and antagonists. The lateral mobility of the ligand–receptor complex, a phenomenon still not well understood, is further regulated by an alteration in membrane fluidity triggered by the ligand–receptor complex itself.

2.8.5 Receptor Clustering

Receptor clustering, although a noncovalent interaction, is really an entirely separate regulatory mechanism. Peptide hormone receptors in particular are known to form clusters that are observable microscopically by use of fluorescent receptor probes. Clustering is a necessary but insufficient prerequisite for the pharmacological effect. Ligand binding to clustered receptors is still necessary for cell activation in such instances as insulin receptor-mediated lipolysis in adipocytes (fat cells). As implied earlier, clustering could explain receptor cooperativity in a positive sense, as well as in a negative sense.

2.8.6 Receptor Internalization

Receptor clustering also accounts for receptor internalization. The basis of this phenomenon is endocytosis via coated pits. These pits are apparent in electron micrographs as membrane invaginations coated on the inner (cytoplasmic) side with a web of the protein clathrine. It has been suggested that certain receptor proteins have structural domains that allow them to react with coated pits. Receptor clusters in coated pits are rapidly endocytosed, resulting in the formation of vesicles (endosomes) that are then transported to the interior of the cell. Perhaps the most important role of internalization is the removal of receptors from the plasma membrane, the down-regulation of a receptor population.

2.9 RECEPTOR TYPES AS DETERMINED BY MOLECULAR MODE OF ACTION

The goal of a drug molecule is to influence cellular function via a receptor-mediated mechanism. Cellular function may be conceptualized into three very broad categories of activity:

1. Transmitting information from one cell to an adjacent cell (via voltage-gated and ligand-gated ion channels).
2. Transmitting information from the exterior of the cell to the interior of the cell (via G-protein coupled receptors).
3. Biosynthetic activity within the interior of the cell (in the nucleus and cytosol via enzyme-catalyzed activities).

Accordingly, receptor types may be categorized into five types, which address these broad categories of cellular function:

1. Voltage-gated ion channels
2. Ligand-gated ion channels
3. G-protein coupled receptors
4. Enzymes and ligand-operated enzymes
5. Protein synthesis-regulating receptors

Voltage-gated ion channels (Na^+, Ca^{2+}, and K^+ ion channels) are large transmembrane proteins whose conformation is dependent upon the transmembrane voltage gradient. By controlling transmembrane transport of ions they control electrical activity (e.g., the neuronal action potential) and thus the transmission of information from cell to cell. Ligand-gated ion channels are large transmembrane proteins whose conformation is dependent upon the presence or absence of particular ligands bound to the protein; ligand binding causes the channel to open and ions to cross the cellular membrane, thereby influencing cellular function. The acetylcholine, GABA-A, NMDA, and glycine receptors are ligand-gated ion channels. G proteins (discussed in detail in the next section) are proteins that permit a ligand binding to the exterior of the cell to influence metabolic processes, such as enzymatic activity, within the cell. Since enzymes are catalysts that promote biosynthesis within the cell, they are logical receptors for drug action. Finally, protein synthesis-regulating receptors are found in both the cytosol and the nucleus and are capable of binding steroid and thyroid hormones. The hormone binds to a domain on the receptor protein, which in turn binds to a particular nucleotide sequence on a gene, thereby regulating its transcription.

This list of potential receptors encompasses the majority of receptors that permit drug regulation of endogenous biochemical processes. However, this list of receptors is not comprehensive for all drugs available for the treatment of human disease. Infections produced by bacteria such as *Staphylococcus aureus* respond to antibiotics such as penicillin. Penicillin binds to "exogenous receptors" located on the bacterium.

2.10 RECEPTOR ACTION: MECHANISMS IN RECEPTOR SIGNAL TRANSDUCTION

From the genome sequence of humans and other organisms deduced in the past several years, it is now appreciated that there are literally thousands of different receptors. However, it is likewise appreciated that Nature is efficient and has organized these thousands of receptors into a mere handful of "superfamilies." One of these superfamilies consists of the voltage-gated ion channels. Recent work by MacKinnon and co-workers has provided groundbreaking structural data on this superfamily. Another, and probably more important, superfamily consists of the highly conserved seven-transmembrane domain G-protein coupled receptors. These receptors are seemingly omnipresent in a diversity of disease states, and are so adaptable that they can detect ligands as big and complex as peptide hormones or as small and subtle as photons of light.

Binding of an agonist or antagonist by a receptor is the first step in a long cascade of events leading to the ultimate, macroscopic physiological effect of the drug or endogenous substance. In the case of receptors that operate on ion channels (see section 8.1), the recognition site and the ion channel are part of the same supramolecular receptor oligomer, and the ion channel will operate in direct response to ligand binding on different parts of the recognition subunits. A more complex chain of events takes place in the vast majority of receptors—those utilizing chemical signaling, such as the G-protein coupled receptors, for transmembrane chemical signaling.

The general scheme of transmembrane chemical signaling begins with the arrival of an extracellular first messenger—a neurotransmitter, hormone, or another endogenous substance, or an exogenous ligand such as a drug or bacterial toxin. The receptor–ligand interaction takes place outside the cell, and in most instances the ligand does not enter the cytoplasm. There are, however, some exceptions, as discussed in the previous section on receptor internalization. Generally, the signal delivered by the ligand is conveyed to the cell interior by the receptor–ligand complex, which interacts with a *transducer*. The receptor–ligand–transducer ternary complex then interacts with an *amplifier*, usually an enzyme, which produces a substance that activates an *internal effector* (usually a phosphorylase kinase); the effector kinase then phosphorylates—and thereby activates or deactivates—a site-specific enzyme that regulates the final cellular response. Three systems, using different second messenger transducers, are known:

1. The adenylate cyclase system
2. The guanylate cyclase system
3. The inositol triphosphate–diacylglycerol system

2.10.1 The Adenylate Cyclase System

This system has been elucidated by a number of investigators over a relatively long period. Sutherland and Rall discovered cAMP in 1958, Rodbell and co-workers showed the need for GTP in the process in 1971, and the complete sequence of events was

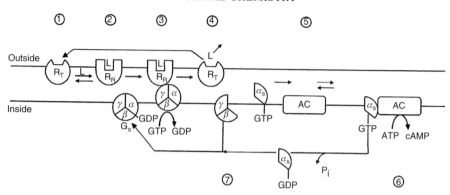

Figure 2.4 Model of adenylate cyclase activation. (1) The receptor, in the tense (off) conformation (R_T), binds ligand (L) to form (2) the activated ligand–receptor complex (R_R–L), which can now undergo collision coupling with the stimulatory guanyl-nucleotide binding protein trimer (G_S). (3) The ternary complex (L–R_R–GS) is activated by an exchange of the GDP bound on the G protein for a GTP. (4) The ternary complex dissociates into inactive receptor (R_T), the ligand (L), the β_γ subunits of the G protein, and (5) the activated α_S subunit of the G protein. (6) The active α_S subunit binds to adenylate cyclase (AC) and activates it, initiating cAMP synthesis from ATP. (7) The α_S subunit is inactivated by hydrolysis of GTP to GDP and inorganic phosphate (Pi); the α_S subunit–GDP complex recycles by reassociating with the β_γ subunits.

mapped by Gilman and his group. This chain of reactions is shown in figure 2.4. Membrane receptors that operate through adenylate cyclase can do so either by activating the amplifier (see below) or by inhibiting it. When the receptor is occupied by its ligand, it forms a transient complex with a guanyl nucleotide binding protein, which is occupied by GDP. These protein transducers—either stimulatory (G_s) or inhibitory (G_i)—become activated in the binding process. In the ternary ligand–receptor–G_{GDP} complex, the GDP is exchanged for a GTP, which triggers the release of the α_s subunit of the $\alpha\beta\gamma$ trimer G_s protein. The β and γ subunits are also released. The active α_s subunit then combines with the adenylate cyclase (AC) enzyme (the amplifier), which produces cAMP, the second messenger. The active G_s state is terminated by a ligand-activated GTPase, which hydrolyzes the bound GTP to GDP. Presumably, the G protein is then reconstituted from the three subunits in the inactive form, ready for the next binding cycle with an occupied receptor. It must be kept in mind that the receptors, the G proteins, and the cyclase interact in a mobile system by collision coupling, and thus a large diversity of receptors can activate the same population of G proteins and cyclase.

The final step in signal transduction is the action of cAMP on the regulatory subunit of the enzyme, protein kinase A. This ubiquitous enzyme then phosphorylates and activates enzymes with functions specific to different cells and organs. In fat cells, protein kinase A activates lipase, which mobilizes fatty acids; in muscle and liver cells, it regulates glycogenolysis and glycogen synthesis.

The molecular properties of G proteins and their subunits, as well as the structural basis of the interactions among the α, β, and γ subunits of G proteins and between these subunits and the associated receptor, has been immensely facilitated by X-ray crystallographic and molecular modeling studies of these proteins. Early work suggested that there were three types of heterotrimeric G protein: G_s, G_i, and G_t; where G_s and G_i effected stimulation and inhibition of adenyl cyclase, respectively, and G_t coupled rhodopsin to regulation of photoreceptor cell function. Currently, approximately 40 heterotrimeric G protein subunits have been identified, and for most of these G proteins multiple subtypes show unique distributions in the brain and peripheral tissues. Analogous studies have also shown that each α unit has two domains: the GTPase activity domain and the GTP-binding site domain.

G proteins play a role in various diseases and in the long-term response to various drugs. If the GTP is replaced by a nonhydrolyzable synthetic analog, guanyl-5'-yl-imidodiphosphate (Gpp(NH)p; the anhydride oxygen is replaced by an NH group), the reaction cannot be terminated, and cAMP will be produced continuously. The GTPase, which normally terminates the active state, can also be inactivated by cholera toxin. The potentially fatal diarrhea and electrolyte loss that occur in cholera reflect the fact that cAMP is an activator of fluid secretion in the intestine. G proteins are also involved in other diseases. Neurofibromatosis type 1, a familial disorder characterized by multiple benign tumours of glial cells, is due to a genetic mutation that alters GTPase activity, which in turn leads to abnormal cellular growth. In addition to such involvement in specific disease states, G proteins are also involved in the body's response to chronic exposure to psychoactive drugs. Drug-induced alterations in G protein subunit concentrations influences signal transduction pathways in the brain, contributing to both the addictive and therapeutic properties of these drugs.

An exciting development in G-protein-mediated signal transduction research has been the realization that proteins produced by *oncogenes* (cancer-causing genes) are also GTP-binding proteins.

2.10.2 The Guanylate Cyclase System

The second type of signal transduction utilizes cyclic GMP (cGMP) instead of cAMP as second messenger. It plays a role not only in insect behavior but also in the human retina and in the functioning of atrial natriuretic factor, a hormone produced by the heart which regulates blood pressure. It is quite likely that cGMP can also act through Ca^{2+} as a third messenger in activating Ca-dependent protein kinases.

2.10.3 The Inositol Triphosphate–Diacylglycerol System

The third widely utilized signaling pathway is based on phosphatidylinositol, a normal constituent of the cell membrane. The extracellular signal is received by a membrane-bound receptor that interacts with a G_s protein, activating phospholipase C (phosphatidylinositol diphosphate [PIP_2] phosphodiesterase), an enzyme that cleaves phosphate diesters. The two products of this cleavage reaction are inositol triphosphate (IP_2) and diacylglycerol (DG), both of which act as second messengers, but in different cellular compartments. The structure of this G-protein receptor is similar to that of the other G proteins (see figure 2.5); a comparison of this DG/IP_3 G protein to the previously described adenylate cyclase system is shown in figure 2.6.

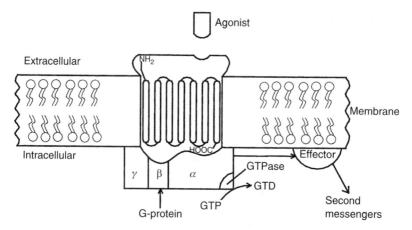

Figure 2.5 Generalized model of the G-protein coupled receptor.

Inositol triphosphate is water soluble and therefore diffuses into the cytoplasm, where it mobilizes calcium from its stores in microsomes or the endoplasmic reticulum. The Ca^{2+} ions then activate Ca-dependent kinases (like troponin C in muscle) directly or bind to the ubiquitous Ca-binding protein calmodulin, which activates calmodulin-dependent kinases. These kinases, in turn, phosphorylate cell-specific enzymes.

Diacylglycerol, on the other hand, is lipid soluble and remains in the lipid bilayer of the membrane. There it can activate protein kinase C (PKC), a very important and widely distributed enzyme which serves many systems through phosphorylation, including neurotransmitters (acetylcholine, α_1- and β-adrenoceptors, serotonin), peptide hormones (insulin, epidermal growth hormone, somatomedin), and various cellular functions (glycogen metabolism, muscle activity, structural proteins, etc.), and also interacts with guanylate cyclase. In addition to diacylglycerol, another normal membrane lipid, phosphatidylserine, is needed for activation of PKC. The DG–IP$_3$ limbs of the pathway usually proceed simultaneously.

The phosphatidylinositol pathway is completed by regeneration of the phospholipid from IP$_3$ and DG. It is remarkable that IP$_3$ is successively dephosphorylated to mositol. The last step of this sequence is inhibited by Li$^+$ ions, which block phosphatidylinositol synthesis. Li salts are used to control the symptoms of manic-depressive illness, an affective mental disorder (see section 3.5.4), and it is thus tempting to implicate the last reaction of the PI pathway in the etiology of this disorder.

2.11 SELECTING A RECEPTOR APPROPRIATE FOR DRUG DESIGN

In developing new chemical entities as therapeutics, the medicinal chemist or drug designer is confronted with the task of identifying and selecting a receptor against which to target the drug to be designed; this is not a trivial task. Philosophically, there are many basic approaches to receptor site selection that may be pursued when tackling this task.

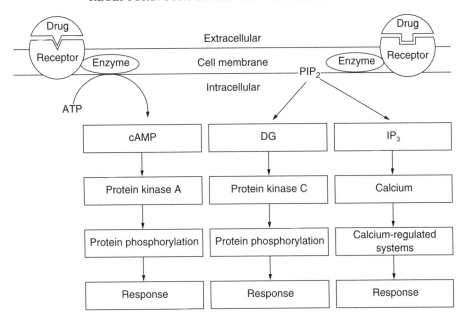

Figure 2.6 Second messenger systems mediate effects of drugs acting on G-protein coupled receptors. The drug stimulates an enzyme upon binding to the membrane-bound receptor. If the enzyme is adenylate cyclase, this stimulates the production of cyclic AMP, which in turn stimulates protein kinase A, causing protein phosphorylation and ultimately a biological response. If the enzyme is phospholipase C, this stimulates the production of the phosphoinositide cycle, which in turn stimulates two mechanisms: (i) increased protein phosphorylation via stimulation of protein kinase C by diacylglycerol (DG); and (ii) the activation of calcium-regulated cellular systems. (Adapted from D. G. Grahame-Smith, J. K. Aronson (2002). *Clinical Pharmacology and Drug Therapy*, 3rd Edn. New York: Oxford University Press. With permission.)

2.11.1 Disease-Centered Receptor Selection

In this approach, several possible receptors for a single disease entity (e.g., Alzheimer's disease, stroke, rheumatoid arthritis) are explored. If the objective is Alzheimer's disease, for example, then drugs may be designed to target one or more of the following potential receptor sites: acetylcholine esterase, β-amyloid peptide, or tau protein—each being a protein, functionally and structurally distinct from one another, that may (or may not) play a central role in the aetiology, pathogenesis, or symptomatology of Alzheimer's disease. The strength of a disease-centered approach is that it permits the designer to pursue whatever target is necessary to fight the disease, without being confined to a particular class of receptor.

2.11.2 Systems-Centered Receptor Selection

Physiologically, the human body may be considered as a collection of various functional systems: nervous, endocrine, immune, cardiac, respiratory, gastrointestinal, genitourinary, musculoskeletal, and dermatological. These physiological systems may be categorized into three larger groupings: control systems (nervous, endocrine, immune), support systems (cardiac, respiratory, gastrointestinal, genitourinary) and structural systems (musculosketelal, dermatological). The control systems can exert immense control over the support systems: the nervous system, through its autonomic division, can influence heart rate, the diameter of blood vessels, respiratory rate, the diameter of bronchioles within the lung, gastrointestinal motility and secretions, and bladder contractility. Drug design expertise obtained within one of these systems frequently extends from one disease to another within that same, single physiological system. For example, in drug design for the nervous system, designing drugs to cross the blood–brain barrier is useful for many diseases of the brain. Also, a system-centered approach enables a single receptor to be evaluated in many disease states. If drugs are being developed as antiglutamatergic NMDA antagonists, their potential utility in a range of medical problems including epilepsy, stroke, or psychiatric disorders, can be evaluated.

2.11.3 Pathological Process-Centered Receptor Selection

Finally, the drug designer could pursue a pathological process-centered (e.g., vascular atherosclerosis, neoplasia, inflammation, infection, apoptosis) approach. Drugs designed to target one of these pathological processes may be used against different diseases in different physiological systems. For example, drugs designed to treat infections may be used for infections extending from a sinusitis in the facial region to an abscess of the foot; likewise, a drug designed to treat neoplasia may be used for cancers in the lungs, bowel, or liver. Drugs developed for vascular disorders can be used to treat medical problems as diverse as myocardial infarction (heart attack), cerebral infarction (stroke), intermittent claudication (leg pain while walking, due to decreased blood supply), or erectile dysfunction. Finally, drugs that target apoptosis (i.e., suppressing genetically encoded, preprogrammed cell death) could in theory be employed against a wide variety of degenerative and neurodegenerative disease states.

2.11.4 Molecular Process-Centered Receptor Selection

The above-mentioned three time-honored approaches to receptor site selection are based upon a conceptualization of human disease at either a gross anatomical or histopathological degree of structural refinement. The pharmacist or drug designer, however, must always exploit atomic- and molecular-level thinking organized within a biochemical framework. Drugs are therapeutic molecules that alter the biochemistry of the human state; accordingly, they must be designed at a molecular level. To enable this biochemical conceptualization of drug action, this book identifies six drug design targets that facilitate therapeutic molecule design and drug receptor selection at an atomic and molecular level of structural refinement (reflected in Part II):

1. Messenger targets: drugs that target neurotransmitters and their receptors (chapter 4)
2. Messenger targets: drugs that target hormones and their receptors (chapter 5)
3. Messenger targets: drugs that target immunomodulators and their receptors (chapter 6)
4. Nonmessenger targets: drugs that target endogenous cellular structures (chapter 7)
5. Nonmessenger targets: drugs that target endogenous macromolecules (chapter 8)
6. Nonmessenger targets: drugs that target exogenous structures (chapter 9)

2.11.4.1 Targets 1–3

The first three categories target endogenous messengers. An endogenous messenger is a molecule synthesized in one or more cells or organs within the body and transported to other cells or organs within the body, enabling the transmission of information and effecting an alteration in biochemical function in the receiving cell or organ. In a systems-centered approach to drug design, the human body is regarded as a collection of physiological systems; the role of the three control systems (nervous, endocrine, immune) is to maintain homeostasis (i.e., a balanced and regulated internal electrical/chemical/cellular milieu). As a simplified generalization, the nervous system controls short-term homeostasis via electrical biochemical processes (using neurotransmitters), the endocrine system controls intermediate-term homeostasis via chemical biochemical processes (using hormones), and the immune system controls long-term homeostasis via cellular biochemical processes (using immunomodulators). (Keep in mind that there is actually significant overlap between the nervous and endocrine systems, the endocrine and immune systems, and even the nervous and immune systems.) Nevertheless, messenger molecules are ideal candidates around which to design therapeutic molecules, since they permit the drug designer to have molecular-level access to the body's own endogenous control systems.

Many human disease states arise directly from abnormalities of messenger molecules. The symptoms of Parkinson's disease arise from an underactivity of the dopamine neurotransmitter, whilst psychosis arises from overactivity of the dopamine neurotransmitter. The symptoms of Alzheimer's disease involve underactivity of the acetylcholine neurotransmitter, and the symptoms of Huntington's disease involve defective metabolism of the GABA neurotransmitter. At the hormonal level, diabetes results from either an absolute or a functional deficiency of the insulin hormone, whereas hypothyroidism is produced by a deficiency of thyroid hormone. Although the mechanistic relationship is somewhat less direct, many other human pathological conditions involve abnormalities in homeostasis (e.g., systemic arterial hypertension [high blood pressure], cardiac arrhythmias [chest palpitations], bronchospasm [asthma], abnormal gastric secretions [peptic ulcer disease], altered gastric motility [irritable bowel syndrome], abnormal bladder contractions [spastic bladder]) and thus can be treated via appropriate "tweaking" of one or more of the three control systems. Finally, some diseases produce end-organ pathology that in turn affects homeostatic processes. A stroke, for example, may enhance the activity of glutamate, which then produces excitotoxicity by binding to ligand-gated ion channels (thus augmenting and enlarging the neuropathology of the stroke). Once again, such pathological states may be treated, in theory and in practice, by altering the control systems.

Messenger targets are ideal for drug design. Most neurotransmitters, many hormones, and a number of immunomodulators are small molecules of low molecular weight. By designing and synthesizing analogs of these molecules, it is possible to produce agonists and antagonists that enable therapeutic modulation over endogenous control systems.

2.11.4.2 Targets 4–6

The next three categories are nonmessenger targets. Not all pathological processes can be treated by adjustments of endogenous control systems; therefore, nonmessenger targets have to be considered. Like the messenger target, these nonmessenger targets can be divided into three logical groups.

The first category of nonmessenger targets consists of cellular structures that are not directly influenced by neurotransmitter, hormonal, or immunomodulatory control. A cell consists of a genetic apparatus (the nucleus), surrounded by biosynthetic machinery (cytoplasmic structures such as rough endoplasmic reticulum), encased in a cellular membrane. This structural arrangement affords a plethora of receptors as targets for drug design. The outer delineating membrane contains numerous proteins which enable biological information to be transmitted from one cell to the next cell (via voltage-gated ion channels) or from the outside of a cell to the inside of that same cell (via G-protein mechanisms); these membrane-bound proteins are superb candidate receptors for drug design and have been successfully exploited in developing drugs for epilepsy, cardiac rhythm problems, and local anesthetics. Within the cell, the nucleus and its associated nucleic acids offer a rich assortment of drug targets (DNA replication, transcription, translation, mitosis) that may be targeted for the treatment of cancer (sarcomas, carcinomas, leukaemia).

The second group consists of the endogenous macromolecules. The most important of these macromolecules are the enzymes. Enzymes are biological catalysts that enhance a wide range of biochemical processes. Enzyme inhibitors offer an approach to therapeutics for a variety of disease processes. More recently, lipids and carbohydrates are also being recognized as viable target receptors in drug design.

The final category of nonmessenger targets includes exogenous pathogens such as prions, viruses, bacteria, fungi, and parasites. These pathogens produce numerous, clinically common localized infections (e.g., abscesses, meningitis, encephalitis, sinusitis, pneumonia, gastroenteritis, cystitis), less common infections (e.g., myocarditis, osteomyelitis), and well-recognized systemic infections (e.g., syphilis, AIDS). Apart from obvious infections caused by such agents, infections are also implicated in the indirect causation of other pathologies. For example, bacteria have been implicated as a cause of peptic ulcer disease and may even play a role in arterial wall damage related to atherosclerosis. Infectious agents have also been speculated to exert an effect in the etiology of diseases such as multiple sclerosis (for which attempts to identify a causative virus have been in progress for decades) and even type 1 diabetes. The most recently appreciated pathogens, prions, have been implicated in the devastating neurological disorder of Creutzfeldt–Jakob disease and bovine spongiform encephalopathy (mad cow disease and its human variant). These prion-based neurodegenerative diseases produce a rapidly progressive dementia associated with the onset of rapid, lightning-like seizures (myoclonic seizures) early in the course of the disease.

Drug design that focuses on targets 4–6 is different from drug design around targets 1–3. For the nonmessenger targets, the presence of a small molecule ligand is less frequent. Accordingly, it is necessary to find a molecule that influences the nonmessenger receptor target either via rational drug design (requiring three-dimensional structural knowledge of the receptor) or via high throughput screening (requiring combinatorial chemistry).

An identification of the pathological process being addressed, combined with an appreciation of which one of the six biochemical approaches (chapters 4–9) is to be pursued, enables the task of molecular-level drug design to be undertaken. In designing the drug to fit the receptor (discussed in detail in chapter 3), the molecular properties that make a molecule a drug molecule and not just an organic molecule (chapter 1) and the molecular properties that make a receptor molecule viable as a target (chapter 2) must always be kept in mind. The drug molecule must be able to withstand the pharmaceutical and pharmacokinetic phases of drug action and must have the necessary geometric, conformational, stereochemical, electronic, and physicochemical properties necessary to specifically bind with the receptor at the pharmacodynamic phase of action. The receptor molecule should be unique to the pathological process under study, accessible to the drug, and capable of stereospecific, saturable binding with a binding equilibrium constant in the nanomolar range. The ultimate realization of the successful drug candidate will require geometrically precise drug design (using quantum pharmacology calculations or experimental methods such as X-ray crystallography) and efficient drug synthesis (using synthetic organic chemistry).

2.12 THE CLINICAL–MOLECULAR INTERFACE: THE CONCEPT OF RATIONAL POLYPHARMACY

People who are afflicted with chronic diseases frequently find themselves taking many drugs. Hopefully, these drugs complement each other in terms of their mechanism of action and are not competing against each other. The capacity of a drug to either augment or diminish the bioactivity of a co-administered drug is frequently determined at the level of the receptor. A drug designer who is developing drugs for a disease for which therapeutics are already available may wish to consider developing an agent with the capacity for *rational polypharmacy* (also called *rational polytherapy*). Rational polypharmacy is usually achieved by designing drugs that work at different receptors, but which ultimately are of benefit to treatment of the same disease. The treatment of Alzheimer's disease may ultimately provide good examples of this approach: the co-administration of a cholinesterase enzyme inhibitor with an anti-amyloid agent would be an example of rational polypharmacy, whereas the co-administration of two competitive cholinesterase inhibitors simultaneously would be an example of irrational polypharmacy.

As a general rule, one drug in higher doses is better than two drugs in lower doses. The notion that two drugs can be given together, in lower doses, to improve efficacy while decreasing toxicity is usually a fallacy.

Case 2.1. A 76-year-old female is brought to an outpatient clinic. The family is convinced that their mother has Alzheimer's disease. On examination, she is definitely confused and disoriented. She does not know the date, does not know the name of the city in which she lives, cannot perform simple arithmetic, cannot draw simple diagrams, cannot identify a watch, and cannot spell the word "WORLD" backwards. However, it is also revealed that she is on lorazepam (for agitation), carbamazepine (for trigeminal neuralgia), oxazepam (for insomnia), amitriptyline (for depression), and propranolol for high blood pressure. When the administration of these medications was stopped, her mental status returned to normal. She had a drug-induced reversible delirium, rather than an irreversible dementia. She is an example of the "do not diagnose dementia while the patient is on a dozen drugs" rule.

2.12.1 Drug–Drug Interactions in Drug Design

The problem of drug–drug interactions is closely related to the concept of rational polypharmacy. Drug–drug interactions frequently occur secondary to molecular

Table 2.2 Cimetidine Drug Interactions

Amitriptyline	Phenytoin
Carbamazepine	Propafenone
Chloroquine	Propranolol
Diazepam	Quinidine
Doxepin	Quinine
Labetalol	Sulfonylurea
Lidocaine	Theophylline
Metoprolol	Triamterene
Metronidazole	Valproic acid
Moricizine	Verapamil
Oxazepam	Warfarin

interactions between two co-administered drugs, and are a common clinical problem. Table 2.2 shows a partial listing of drugs with which cimetidine interacts; a number of these interactions are clinically relevant.

When designing drugs for a chronic disease, the possibility of drug–drug interactions should be taken into consideration: some may be beneficial, but most are not. Drug–drug interactions may be classified as follows:

1. Pharmacodynamic drug–drug interactions
   a. Competitive homotopic molecular targets
      Same site on the same receptor (e.g., diazepam and lorazepam are both benzodiazepines working at the same site on the GABA-A receptor).
   b. Non-competitive homotopic molecular targets
      Different sites on the same receptor (e.g., diazepam and phenobarbital both bind to the GABA-A receptor, but at different sites: benzodiazepine site versus barbiturate site).
   c. Convergent heterotopic molecular targets
      Different receptors targeting the same biochemical process (e.g., diazepam plus vigabatrin: diazepam is an agonist for the GABA-A receptor, upregulating GABA function; vigabatrin is a GABA transaminase enzyme inhibitor that upregulates GABA function by increasing concentrations of GABA in the brain).
   d. Divergent heterotopic molecular targets
      Different receptors targeting different biochemical processes, but affecting the same disease process (e.g., diazepam plus phenytoin: diazepam is an agonist for the GABA-A receptor, while phenytoin is an antagonist of the voltage-gated Na^+ channel receptor; both drugs work to prevent seizures, but by entirely different mechanisms).
2. Pharmacokinetic drug–drug interactions
   a. A—Absorption competition (similar structures compete for absorption in gut).
   b. D—Distribution competition (competitive binding on albumin within bloodstream).

    c. M—Metabolism competition (competition for same enzymes in liver).

    d. E—Elimination competition (similar structures are competitive for kidney excretion).

3. Pharmaceutical drug–drug interactions

    Chemical reaction in gut (e.g., co-administration of valproic acid with an antacid).

4. Adjunctive polypharmacy

    Two different drugs targeting completely different aspects of a common disease (e.g., giving an antiplatelet agent and an antihypertensive agent to a person with a stroke; one agent treats platelet clots that may cause stroke, the other agent treats the high blood pressure that also may cause strokes).

Selected References

General References on Receptors

D. R. Burt (1985). Criteria for receptor identification. In: H. T. Yamamura, S. J. Enna, M. J. Kuhar (Eds.). *Neurotransmitter Receptor Binding,* 2nd ed. New York: Raven Press, pp. 41–60.

P. Ehrlich (1897). *Klin Jahr. 6*: 299.

R. Flower (2002). Drug receptors: a long engagement. *Nature 415*: 587.

B. G. Katzung (Ed.) (2001). *Basic and Clinical Pharmacology*, 8th ed. New York: Lange.

A. Korolkovas (1970). *Essentials of Molecular Pharmacology.* New York: John Wiley.

J. Langley (1878). *J. Physiol.* (London) *1*: 367.

H. Lullmann, K. Mohr, A Ziegler, D. Bieger (2000). *Color Atlas of Pharmacology.* New York: Thieme.

J. Parascandola (1980). Origins of the receptor theory. *Trends Pharmacol. Sci. 1*: 189–192.

D. F. Smith. (Ed.) (1989). *CRC Handbook of Stereoisomers: Therapeutic Drugs.* Boca Raton: CRC Press.

Drug–Receptor Interaction Forces

P. Andrews (1986). Functional groups, drug–receptor interactions and drug design. *Trends Pharmacol. Sci. 7*: 148–151.

A. Ben-Nairn (1980). *Hydrophobic Interactions.* New York: Plenum.

P. H. Doukas (1975). The role of charge-transfer processes in the action of bioactive materials. In: E. J. Ariëns (Ed.). *Drug Design*, vol. 5. New York: Academic Press, pp. 133–167.

P. A. Kollman (1980). The nature of the drug–receptor bond. In: M. F. Wolff (Ed.). *The Basis of Medicinal Chemistry,* 4th ed. Part 1. New York: Wiley-Interscience, pp. 313–329.

J. B. Stenlake (1979). *Foundations of Molecular Pharmacology,* vol. 2. London: Athlone Press, chapter 2.

Quantifying Receptor Binding

A. De Lean, D. Rodbard (1979). Kinetics of cooperative binding. In: R. D. O'Brien (Ed.). *The Receptors.* New York: Plenum Press, pp. 140–192.

D. G. Haylett (2003). Direct measurement of drug binding to receptors. In: *Textbook of Receptor Pharmacology*, 2nd ed. Boca Raton: CRC Press, pp. 153–180.

J. M. Klotz, D. L. Hunston (1971). Properties of graphical representation of multiple classes of binding sites. *Biochemistry 10*: 3065–3069.

P. M. Laduron (1982). Towards a unitary concept of opiate receptor. *Trends Pharmacol. Sci. 3*: 351–352.

K. E. Light (1984). Analyzing non-linear Scatchard plots. *Science 223*: 76–77.

Receptor Binding Terms

E. J. Ariëns (1983). Intrinsic activity: partial agonists and partial antagonists. *J. Cardiovasc. Pharmacol. 5*: S8–SI5.

F. J. Barrantes (1979). Endogenous chemical receptors: some physical aspects. *Annu. Rev. Biophys. Bioeng. 8*: 287–321.

J. M. Boeynaems, J. E. Dumont (Eds.) (1980). *Outlines of Receptor Theory.* New York: Elsevier/North Holland.

J. P. Changeux (1995). The acetylcholine receptor: A model for allosteric membrane proteins. *Biochem. Soc. Trans. 23*: 195.

D. Colquhoun (1973). The relation between classical and cooperative models for drug action. In: H. P. Rang (Ed.). *Drug Receptors.* Baltimore: University Park Press, pp. 149–182.

A. De Lean, P. J. Munson, D. Rodbard (1979). Multivalent ligand binding to multisubsite receptors: application to hormone–receptor interactions. *Mol. Pharmacol. 15*: 60–70.

A. De Lean, D. Rodbard (1979). Kinetics of cooperative binding. In: R. D. O'Brien (Ed.). *The Receptors.* New York: Plenum Press, pp. 143–192.

J. DiMaio, F. R. Ahmed, P. Shiller, B. Belleau (1979). Stereo-electronic control and decontrol of the opiate receptor. In: F. Gualtieri, M. Gianella, and C. Melchiorre (Eds.). *Recent Advances in Receptor Chemistry.* New York: Elsevier/North Holland, pp. 221–234.

M. D. Hollenberg, P. Cuatrecasas (1979). Distinction of receptor from non-receptor interaction in binding studies. In: R. D. O'Brien (Ed.). *The Receptors.* New York: Plenum Press, pp. 193–214.

P. M. Laduron (1984). Criteria for receptor sites in binding studies. *Biochem. Pharmacol. 33*: 833–839.

D. E. Macfarlane (1984). On the enzymatic nature of receptors. *Trends Pharmacol. Sci. 5*: 11–15.

H. G. Mautner (1980). Receptor theories and dose–response relationships. In: M. E. Wolff (Ed.). *The Basis of Medicinal Chemistry,* 4th ed., vol. 1. New York: Wiley-Interscience, pp. 271–284.

M. Poo (1985). Mobility and localization of proteins in excitable membranes. *Annu. Rev. Neurosci. 8*: 369–406.

D. Rodbard (1980). Agonist versus antagonist. *Trends Pharmacol. Sci. 1*: 222–225.

K. Starke (1981). Presynaptic receptors. *Annu. Rev. Pharmacol. Toxicol. 21*: 7–30.

R. P. Stephenson (1956). A modification of receptor theory. *Br. J. Pharmacol. 11*: 379–393.

D. J. Triggle, C. R. Triggle (1976). *Chemical Pharmacology of the Synapse.* New York: Academic Press, chapter 2.

Receptor Metabolism

J. L. Carpentier, P. Gorden, A. Roberts, L. Orci (1986). Internalization of polypeptide hormones and receptor cycling. *Experientia 42*: 734–744.

R. B. Dickson (1985). Endocytosis of polypeptides and their receptors. *Trends Pharmacol. Sci. 6*: 164–167.

I. A. Hanover, R. B. Dickson (1985). The possible link between receptor phosphorylation and internalization. *Trends Pharmacol. Sci. 6*: 457–459.

M. D. Hollenberg (1985). Receptor regulation, Parts 1 and 2. *Trends Pharmacol. Sci. 6*: 242–245; 299–302.

J. A. Koenig, J. Edwardson (1997). Endocytosis and recycling of G protein receptors. *Trends Pharmacol Sci. 18*: 276.

Molecular Structure of Receptors

A. A. Abdel-Latif (1986). Calcium-mobilizing receptors, polyphosphoinositides, and the generation of second messengers. *Pharmacol. Rev. 38*: 227–272.

M. J. Berridge (1985). The molecular basis of communication within the cell. *Sci. Am. 253*: 142–152.

H. R. Bourne (1997). How receptors talk to trimeric G proteins. *Curr. Opin. Cell Biol. 9*: 134.

R. Flower (2002). Drug receptors: a long engagement. *Nature 415*: 587.

J. C. Garrison (1985). Possible roles of protein kinase C in cell function. *Annu. Rep. Med. Client. 20*: 227–236.

A. G. Gilman (1995). G proteins and regulation of adenylyl cyclase. *Biosci. Rep. 15*: 65–97.

D. G. Grahame-Smith, J. K. Aronson (2002). *Textbook of Clinical Pharmacology and Drug Therapy*, 3rd ed. New York: Oxford University Press.

H. Hamm, A. Gilchrist (1996). Heterotrimeric G proteins. *Curr. Opin. Cell Biol. 8*: 189–196.

L. E. Hokin (1985). Receptors and phosphoinositide-generated second messengers. *Annu. Rev. Biochem. 54*: 205–235.

P. W. Majerus, T. M. Conolly, H. Deckmyn, T. S. Ross, T. E. Bross, H. Ishii, V. S. Bansal, D. B. Wilson (1986). The metabolism of phosphoinositide-derived messenger molecules. *Science 234*: 1519–1526.

S. R. Nahorski, D. A. Kendall, I. Batty (1986). Receptors and phosphoinositide metabolism in the central nervous system. *Biochem. Pharmacol. 35*: 2447–2453.

E. J. Neer (1995). Heterotrimeric G proteins: organizers of transmembrane signals. *Cell 80*: 249–257.

Y. Nishizuka (1986). Studies and perspectives of protein kinase C. *Science 233*: 305–312.

P. J. Parker, L. Coussens, N. Totty, L. Rhea, S. Young, E. Chen, S. Stabel, M. D. Waterfield, A. Ullrich (1986). The complete primary structure of protein kinase C, the major phorbol ester receptor. *Science 233*: 853–859.

E. Pfeuffer, R. M. Dreher, H. Metzger, T. Pfeuffer (1985). Catalytic unit of adenylate cyclase: purification and identification by affinity cross binding. *Proc. Natl. Acad. Sci. USA 82*: 3086–3090.

T. Schneidere, P. Igelmund, J. Hescheler (1997). G protein interaction with K^+ and Ca^{2+} channels. *Trends Pharmacol. Sci. 18*: 8–11.

H. Schulman (1984). Calcium-dependent protein kinases and neuronal function. *Trends Pharmacol. Sci. 5*: 188–192.

R. K. Sunahara, C. W. Dessauer, A. Gilman (1996). Complexity and diversity of mammalian adenylyl cyclases. *Ann. Rev. Pharmacol. Toxicol. 36*: 461–480.

K. Wickman, D. Clapham (1995). Ion channel regulation by G proteins. *Physiol. Rev. 75*: 865–885.

3

Basic Principles of Drug Design III

Designing drug molecules to fit receptors

3.1 OVERALL STRATEGY: THE MULTIPHORE
METHOD OF DRUG DESIGN

The aim of this book is to provide a conceptual framework for medicinal chemistry. Chapter 1 dealt with the properties necessary to transform a molecule into a drug-like molecule. Chapter 2 described the properties that determine whether a macromolecule could be a receptor. It is now necessary to develop a method of designing drug molecules to fit into receptor molecules. The *multiphore method* of drug design is such a method.

The multiphore method conceptualizes a drug as being constructed in a modular fashion from bioactive subunits, or *biophores*. Since a drug is invariably composed of many biophores, it is a *multiphore*. The most important biophore within the drug structure is the *pharmacophore*, the subset of atoms within the drug that permits energetically favorable binding to the receptor site with the elucidation of a subsequent beneficial biological response. Other portions of the molecule determine the metabolic and toxicological properties of the drug; these are the *metabophores* and *toxicophores*, respectively.

In the design of drugs using the multiphore method it is important to remember that there is nothing special about any particular drug molecule. A successful drug molecule is merely a collection of "hetero-atom rich" functional groups appropriately positioned on the three-dimensional space of a hydrocarbon framework in a fixed geometrical relationship that enables a desirable interaction with a receptor macromolecule. When the medicinal chemist knows the *bioactive zone* of the receptor macromolecule, he or she identifies multiple functional groups, usually within 15 Å of each other, within that bioactive zone. The selection of these receptor-based functional groups is a crucial step. For example, if the receptor zone were within the brain, it would be inadvisable to select many charged (anionic or cationic) functional groups, since a drug capable of binding to these receptor-based functional groups via electrostatic interactions would be too polar to diffuse across the blood–brain barrier and enter the brain.

Next, complementary functional groups capable of energetically favorable intermolecular interactions with the receptor-based functional groups are selected. These complementary functional groups will ultimately form part of the drug that is being designed and will constitute the pharmacophore. These drug-based functional groups are then "clicked together" in three-dimensional space by being covalently attached to a relatively rigid hydrocarbon frame. The number of functional groups determines the number of contact points between the drug molecule and the receptor macromolecule. A *three-point pharmacophore* will have three different intermolecular interactions between the drug and the receptor. The number of points of contact is also an important consideration. A large number of points of contact is favorable from a pharmacodynamic perspective since it enables a more specific and unique drug–receptor interaction, concomitantly decreasing the likelihood of toxicity. However, a large number of points of contact is unfavorable from a pharmacokinetic perspective, since the resulting increased polarity of the drug molecule tends to decrease the pharmacological half-life and also to decrease the ability of the drug to diffuse across membranes during its distribution throughout the body. In general, most neuroactive drugs have 2–4 points of contact, while most non-neuroactive drugs have 3–6 points of contact.

Once the pharmacophore has been designed, the remainder of the molecular fragments (individually composed of metabophores or toxicophores or inert bioinactive spacers, but collectively referred to as *molecular baggage*) are assembled. One of the primary goals of the molecular baggage component is to hold the pharmacophore in a desired conformation such that it can interact with its receptor. However, these additional molecular fragments also serve other functions. For instance, a metabophore can be inserted into this portion of the molecule. If one is designing an intravenous drug with a short half-life, one may want to include an ester moiety. This would constitute a metabophore since the ester would be hydrolyzed, resulting in rapid inactivation of the whole molecule.

Once the prototype drug molecule has been prepared and biologically evaluated, various toxicities may become apparent during preliminary testing in animals. Fragments of the molecule that are responsible for these toxicities (i.e., toxicophores) can then be deduced. If the toxicophore is separate and distinct from the pharmacophore, a new toxicity-free molecule can be engineered. If there is too much overlap between the toxicophore and the pharmacophore, it may not be possible to "design out" the toxicity. It is important to emphasize that a drug molecule may have many different toxicophores, reflecting different toxicities. A toxicophore is merely a pharmacophore that permits an undesirable interaction with an "untargeted" receptor.

The multiphore method is versatile and is not restricted to *de novo* drug design, as the above discussion might imply. For example, if the drug molecule is discovered by accident or in a random screening process, the multiphore conceptualization is still applicable. Through structure–activity studies (discussed below) it is still possible to discern fragments that constitute the pharmacophore and potential toxicophores, and thus it is still possible to re-engineer the molecule for improved performance. The strength of the multiphore method is its treatment of drug molecules as collections of bioactive fragments. If one fragment is giving problems, it is possible to simply insert another biologically similar fragment (*bioisostere*) that will hopefully overcome the identified problem.

Clearly, the multiphore method of drug design is an iterative process. It takes repeated rounds of re-evaluation and redesign before a final candidate drug molecule is developed. This iterative process has the following five steps:

1. Design or identification of a lead (prototype) compound (section 3.2)
2. Synthesis and initial biological evaluation of the lead compound (section 3.3)
3. Optimization of the lead compound for the pharmacodynamic phase (section 3.4)
4. Optimization of the lead compound for the pharmacokinetic and pharmaceutical phases (section 3.5)
5. Pre-clinical and clinical evaluation of the optimized lead compound analog (section 3.6)

The remainder of this chapter will discuss these five steps in drug optimization within the context of the multiphore method of drug design.

3.2 IDENTIFICATION OF A LEAD COMPOUND

Lead compounds are still a sine qua non of drug design. A lead compound (pronounced "lēd" and not to be mistaken for a salt of element 82) is invariably an organic molecule that acts as a prototype drug around which future optimization is centered and focused. Identifying a lead compound is the key to starting the drug discovery engine. There are several well-tested methods for uncovering or identifying lead compounds as prototype agents around which to design and optimize a drug molecule:

1. Serendipity
2. Endogenous sources (drug candidates from molecules within us; e.g., insulin for diabetes)
3. Exogenous sources: ethnobotany (drug candidates from natural product sources)
4. Rational drug design
5. High throughput screening programs
6. Genomics/Proteomics

Elegant though some of these may sound, serendipity has historically been the most successful discoverer of drugs.

3.2.1 Lead Compound Identification—A Short History of Drug Discovery

"How have drugs historically been discovered? Can these traditional techniques be further exploited to help discover new and better drugs?" The answer to the latter question is probably "no"!

Since the dawn of humankind, efforts have been made to discover remedies for the ailments of life. Although there are numerous examples of the trials and tribulations associated with these efforts, the story of epilepsy affords many instructive anecdotes.

The failure of premodern physicians to develop adequate therapies reflected their inability to gain a viable mechanistic understanding of epilepsy. In primitive times, surgical "therapies" for epilepsy included *trephining* holes through the patient's skull in order to release "evil humours and devil spirits." The ancients also employed a varied and bizarre assortment of ad hoc "medical" therapies, ranging from rubbing the body of an epileptic with the genitals of a seal, to inducing episodes of sneezing at sunset. In early Roman times human blood was widely regarded as curative, and people with epilepsy frequently sucked the blood of fallen gladiators in a desperate attempt to find a cure. Charlatans would also offer (for a generous fee, of course) to massage the heads of people with epilepsy, thereby realigning the bone plates of the skull, taking pressure off the brain, and alleviating the curse of epilepsy.

By the Middle Ages, alchemy and astronomy formed the scientific foundations of epilepsy therapy. The use of "magical prescriptions" flourished. These remedies ranged from grotesque therapies, such as the ingestion of dog bile or human urine, to the use of somewhat more innocuous precious stone amulets. During the Renaissance, these magical treatments were rejected by the medical profession in favor of "rational and scientific" Galenic therapies. These treatments relied extensively upon forced vomiting and bowel purging with concomitant oral administration of peony extracts. Also, during this time, the notion of epilepsy being secondary to hypersexuality emerged, and castration, circumcision, or clitoridectomy were widely advocated. Occasionally, various pharmacologically active organic molecules (e.g., strychnine, curare, atropine, valerian, picrotoxin, and quinine) were also used, but all ultimately failed—sometimes killing the unfortunate patient.

Inorganic salts were also considered as putative therapies during the late Renaissance. Copper-based therapy flourished during the eighteenth and nineteenth centuries. Aretaios had introduced the use of antiepileptic copper salts in the first century AD, and had described significant therapeutic successes. These reported successes with copper therapy were embraced during the 1700s, leading to other therapeutic attempts with lead, bismuth, tin, silver, iron, and mercury, thus giving rise to *metallotherapy*. The subsequent widespread failure of metallotherapy, due to lack of efficacy and excessive toxicity, led to its abandonment during the late pre-modern era.

Thus, in the millennia extending from antiquity to the mid-nineteenth century, epilepsy remained a medical condition surrounded by mystique—permitting charlatanism, superstition, and quackery to prosper. In general, the therapies of this time were without merit, as demonstrated by the detailed but disturbing description of King Charles II's death, which provides a comprehensive summary of the complexity and futility of seizure therapy during the pre-modern era.

> In 1685, the king fell backward and had a violent convulsion. Treatment was begun immediately by a dozen physicians. He was bled to the extent of 1 pint from his right arm. Next, his shoulder was incised and cupped, depriving him of another 8 oz. of blood. After an emetic and 2 purgatives, he was given an enema containing antimony, bitters, rock salt, mallow leaves, violets, beet root, chamomile flowers, fennel seed, linseed, cinnamon, cardamom seed, saffron and aloes. The enema was repeated in 2 hours and another purgative was given. The king's head was shaved and a burn blister was raised on his scalp. A sneezing powder of hellebore root and one of cowslip flowers were administered to strengthen the king's brain. Soothing drinks of barley water, licorice and sweet almond were given, as well as extracts of mint, thistle leaves, rue, and angelica. For external treatment, a plaster of Burgundy pitch and pigeon dung was liberally applied to the king's feet. After continued bleeding and purging, to which were added melon seed, manna, slippery elm, black cherry water, and dissolved pearls, the king's condition did not improve and, as an emergency measure, 40 drops of human skull extract were given to allay convulsions. Finally, bezoar stone was forced down the king's throat and into his stomach. As the king's condition grew increasingly worse, the grand finale of Raleigh's antidote, pearl julep, and ammonia water were pushed into the dying king's mouth. (Swinyard, 1980)

During an era in which Isaac Newton and Charles Darwin were making colossal advances in physics and biology, medical therapy and drug development was still mired in primitive, unprogressive and cruel rituals. Fortunately, by the mid 1800s times began to change, and over the subsequent 150 years substantial progress was made. Table 3.1 lists a number of important drug discoveries between 1842 and 2000. Although advances were obviously being made, much of this drug discovery relied on serendipity rather than rational drug design.

Table 3.1 Drug Discoveries, 1842–2000

1842	Long introduces ether as an anesthetic
1857	Locock accidentally discovers bromides as anticonvulsants
1867	Lister pioneers use of phenol as a surgical antiseptic
1869	Liebreich discovers hypnotic effects of chloral hydrate
1876	Stricker uncovers analgesic properties of salicylic acid
1882	Guthzeit and Conrad synthesize a series of barbiturates
1891	Erlich pioneers concepts of "receptor" and "chemotherapy"
1899	Meyer and Overton discover effect of lipid solubility on anesthetic action
1903	Fischer and von Mering identify hypnotic properties of barbiturates (see 1882)
1906	Hunt and Taveau synthesize and study acetylcholine
1912	Hauptmann accidentally discovers barbiturates as anticonvulsants (see 1903, 1882)
1921	Loewi demonstrates that acetylcholine is a neurotransmitter
1922	Banting and Best purify insulin as treatment for diabetes
1927	Szent-Gyorgyi isolates ascorbic acid (Vitamin C)
1929	Fleming serendipitously discovers antibacterial properties of penicillin
1932	Mietzach, Klarer, Domagk introduce first anti-streptococcal drug
1934	Ruzicka first synthesizes progesterone
1938	Merritt and Putnam use screening to identify hydantoins as anticonvulsants
1940	Chain and Florey introduce manufactured penicillin
1942	Ehrhard and Schauman produce synthetic analgesics (meperidine, methadone)
1945	Woodward and Doering synthesize quinine
1947	Lands introduces isoproterenol as a bronchodilator
1952	Charpentier identifies tricyclic phenothiazines as antipsychotics
1953	Watson and Crick deduce structure of DNA
1959	Beecham Laboratories develops semisynthetic penicillins
1959	Searle introduces the birth control pill
1960	Hoffmann-La Roche tests benzodiazepines as anxiolytics (Librium, Valium)
1962	Hansch develops principle of quantitative structure–activity relationships
1962	Pullman introduces quantum mechanics to drug design
1967	Cotzias pioneers the use of L-DOPA for treatment of Parkinson's disease
1975	Biochemically driven rational drug design begins to flourish as method
1985	Improved computers enable computer-aided drug design to advance
1995	Advances in combinatorial chemistry advance high throughput screening
2000	Widespread use of cholinesterase inhibitors for symptomatic treatment of Alzheimer's disease

3.2.2 Lead Compound Identification by Serendipity

Many of the significant advances in drug discovery over the past one and a half centuries owe their success to serendipity; the discovery of penicillin (discussed in chapter 9) is a legendary example of its importance. However, many other drugs were also discovered by accident. Continuing the theme of epilepsy from section 3.2.1 provides additional examples of serendipity and drug discovery.

3.2.2.1 The Benzodiazepines: Diazepam and Serendipity

The benzodiazepines are one of the most widely used (and abused) classes of drugs; they are widely used in the treatment of epilepsy, insomnia, anxiety, movement disorders, and a variety of other neurological disorders. The discovery of the benzodiazepines is a good example of the importance of serendipity.

In the early 1950s, following the accidental discovery of phenothiazines as antipsychotic agents, interest in tricyclic molecules as potential therapeutic agents for neurological and psychiatric disorders became widespread. Accordingly, in 1954, Sternbach of Hoffmann–La Roche laboratories decided to reinvestigate the synthetic chemistry of a series of tricyclic benzheptoxdiazine compounds upon which he had worked more than twenty years earlier. By reacting an alkyl halide with a variety of secondary amines, he prepared forty analogs, all of which were inactive as muscle relaxants and sedatives. When additional chemical studies revealed that these compounds were really quinazoline-3-oxides rather than benzheptoxdiazines, the project was summarily abandoned in 1955.

Nearly two years later, a colleague at Hoffmann–La Roche discovered an untested crystalline sample while cleaning and tidying cluttered laboratory benches. Although many other compounds had been simply discarded, this compound, later called chlordiazepoxide, was submitted for biological evaluation. Chlordiazepoxide (7-chloro-2-(methylamino)-5-phenyl-3H-1,4-benzodiazepine-4-oxide; Librium) demonstrated profound anti-anxiety properties. Additional chemical studies revealed that this single compound was a 1,4-benzodiazepine, unlike its forty quinazoline-3-oxide predecessors—the chance use of methylamine, a primary rather than a secondary amine, had resulted in a different synthetic pathway. Numerous analogs, including diazepam (Valium), were soon prepared and the anticonvulsant properties of this class of compounds were quickly discerned. The benzodiazepines soon emerged as one of the most important, and lucrative, classes of drug molecules.

The discovery of benzodiazepines is a story of serendipity and certainly one that is difficult to predictably reproduce as part of a drug discovery program. Regrettably (or fortuitously), this story of the benzodiazepines is not an isolated example. Valproic acid, an agent used to treat epilepsy, migraine, chronic pain, and bipolar affective disorder, was also discovered by accident.

3.2.2.2 Aliphatic Carboxylates: Valproic Acid and Serendipity

In postwar France, the Berthier Pharmaceutical Company in Grenoble began to pursue a sideline project of producing soothing liquid bismuth preparations for acute tonsillitis. Being dissatisfied with the commonly used oils, they elected to use the supposedly physiologically inert valproic acid as a solvent for their bismuth compounds. In doing so, they gained immense practical experience with the handling and manipulation of this solvent.

In 1962, Pierre Eymard, a graduate student at the University of Lyon, synthesized a series of khellin derivatives. Khellin is a biologically active substance that occurs in the fruit of the wild Arabian *Khell* plant and which has been used for centuries by herbalists for the treatment of kidney stones. Eymard arranged to have his new compounds biologically evaluated at the École de Médecine et de Pharmacie in nearby Grenoble. When attempts to produce a solution of these khellin compounds failed, advice was sought from H. Meunier of the nearby Laboratoire Berthier. In view of Berthier's recent peripheral interest in valproic acid as a solvent for bismuth compounds, Meunier recommended valproic acid as a nontoxic inert solvent.

Eymard's khellin derivatives were dissolved in valproic acid and, following the practice of submitting all such compounds for evaluation in an antiepileptic screening model, they were studied for anticonvulsant activity. These preliminary studies revealed profound anticonvulsant activity. Shortly after this, Meunier serendipitously decided to use valproic acid as a solvent for an unrelated coumarin compound and, although chemically dissimilar to Eymard's khellins, this coumarin exhibited identical anticonvulsant properties. The fact that both compounds had been dissolved in the same solvent was realized immediately. The antiepileptic action of valproic acid was thus discovered completely by accident, with the first successful clinical trial occurring in 1963.

Although serendipity has been quite successful in drug design, it is a method that is difficult to reproduce. Accordingly, over the past fifty years, a variety of other drug discovery methods have been pioneered.

3.2.3 Lead Compound Identification from Endogenous Sources

In the attempt to identify logical or rational methods for designing and discovering lead compounds, the notion of exploiting endogenous molecules quickly comes to mind. Human disease arises from perturbations of normal biochemical processes. A logical therapeutic approach involves the administration of one or more of these naturally occurring endogenous biochemical molecules, or analogs thereof. In addition, certain human diseases seem to arise from a deficiency of a certain endogenous molecule. It is reasonable to assume that such diseases could be cured or at least helped by the administration of the missing molecule.

Medicinal chemistry has many examples of the development of successful therapeutics based on an exploration of endogenous compounds. The treatment of *diabetes mellitus*, for example, is based upon the administration of insulin, the hormone that is functionally deficient in this disease. The current treatment of Parkinson's disease is based upon the observation that the symptoms of Parkinson's disease arise from a deficiency of dopamine, an endogenous molecule within the human brain. Since dopamine cannot be given as a drug since it fails to cross the blood–brain barrier and enter the brain, its biosynthetic precursor, L-DOPA, has been successfully developed as an anti-Parkinson's drug. Analogously, the symptoms of Alzheimer's disease arise from a relative deficiency of acetylcholine within the brain. Current therapies for Alzheimer's-type dementia are based upon the administration of cholinesterase enzyme inhibitors that prolong the effective half-life of remaining acetylcholine molecules within the brain.

As discussed in chapter 1, the human body contains many different molecules and thus offers many opportunities for the discovery of lead compounds based on endogenous molecules. Nowhere is this opportunity more apparent than in the area of peptide neu-rotransmitters and peptide hormones (see chapters 4 and 5). Neurotransmitters and hor-mones are endogenous messengers, controlling diverse biochemical processes within the body. Not surprisingly, they have the capacity to be ideal starting points in the drug discovery process. However, there are a number of major problems that must be con-fronted when exploiting peptides or proteins as lead compounds for drug discovery. Peptidomimetic chemistry is an attempt to address these problems.

3.2.3.1 Peptidomimetic Chemistry as a Source of Lead Compounds

Although they are potent endogenous bioactive molecules, peptides rarely make good drugs. There are several reasons for the failings of peptides as drugs:

1. Peptides are too big (molecular weight frequently over 1000 dalton).
2. Peptides are often too flexible (thus binding with too many receptors, leading to toxicity).
3. Peptides contain amide bonds that can be metabolized by hydrolysis.
4. Peptides cannot be given orally as drugs (they tend to be digested).
5. Peptides do not readily cross the blood–brain barrier to enter the brain.

Despite these obvious deficiencies, peptides have a number of properties that make them attractive as starting points in drug design:

1. Peptides contain numerous stereogenic (chiral) centers (an excellent starting point when designing stereoselective drugs).
2. Peptides contain many functional groups (e.g., carboxylate, ammonium, hydroxyl, thiol) that can readily constitute functional groups within a pharmacophore; since receptors are usually proteins, peptides are good starting points for designing a molecule to interact with a receptor, owing to the energetically favorable nature of peptide–peptide interactions.
3. Peptides are easily synthesized and many analogs can be readily produced.
4. Peptides can have their conformation and geometries easily optimized by energy minimization calculations using current computational methods (e.g., molecular mechanics); this makes subsequent modeling studies easy.
5. Peptides function as neurotransmitters and hormones and thus are good starting materials when designing bioactive molecules.

Since peptides are ideal starting molecules that cannot be turned into successful peptidic drugs, the specialty area of *peptidomimetic chemistry* has emerged. The goal of pep-tidomimetic chemistry is to design small, conformationally constrained, non-peptidic organic molecules that possess the biological properties of a peptide. Hopefully, this will retain the strength of the peptide as a putative drug while eliminating the problems. There are two approaches whereby peptidomimetic chemistry can achieve this design goal.

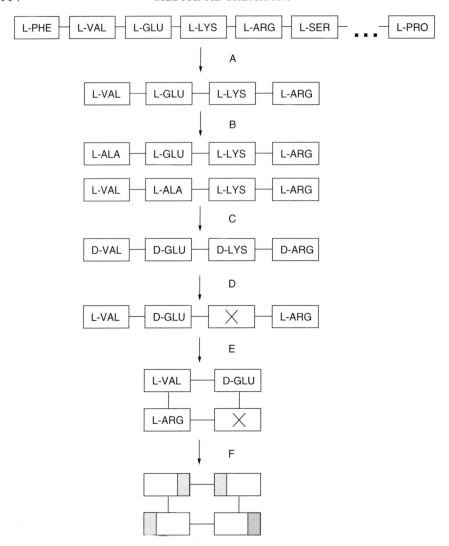

Figure 3.1 Peptidomimetic chemistry attempts to produce a non-peptidic drug to mimic a bioactive peptide. In Step A, the smallest bioactive fragment of the larger peptide is identified; in Step B, a process such as an alanine scan is used to identify which of the amino acids are important for bioactivity; in Step C, individual amino acids have their configuration changed from the naturally occurring L-configuration to the unnatural D-configuration (in an attempt to make the peptide less "naturally peptidic"); in Step D, individual amino acids are replaced with atypical unnatural amino acids and amino acid mimics; in Step E the peptide is cyclized to constrain it conformationally; finally, in Step F, fragments of the cyclic peptide are replaced with bioisosteres in an attempt to make a non-peptidic organic molecule.

The first approach is shown in figure 3.1. This approach uses various techniques (e.g., alanine scanning) to identify the smallest peptide segment with biological activity within the overall peptide. This "minimal bioactive segment" may be cyclized or have its stereochemistry altered in order to attain restriction of conformational freedom and to render the molecule less like a naturally occurring peptide. Next, this segment is then rebuilt isosteric fragment by isosteric fragment, gradually replacing each portion of the molecule in a stepwise fashion. For example, the amide bond may be replaced by a bioisosterically equivalent amide bioisostere. In this fashion, an equivalent but non-peptidic organic molecule drug eventually emerges.

An alternative approach is a little less plodding and perhaps a little more elegant. The three-dimensional structure of the peptide is determined using either theoretical (molecular mechanics, molecular orbital calculations) or experimental (X-ray crystallographic, NMR spectroscopic) methods. Next, an educated guess (hopefully based on some experimental data) is made to suggest which portion of the peptide is the pharmacophore. The geometries of the functional groups within the pharmacophore are then measured from the theoretical and experimental studies of the peptide's geometry and conformation. For example, these data may show that the peptide pharmacophore contains a carboxylate group located 4.6 Å from a hydroxyl group, which in turn is 5.1 Å from a phenyl group. Using these precise data, databases of known organic molecules are then computationally searched to identify an organic molecule with similar functional groups held in the same position in three-dimensional space. Hopefully, this will yield a non-peptidic but bioactive organic molecule drug.

3.2.3.2 Other Endogenous Drug Lead Platforms: Carbohydrates, Nucleic Acids

Although peptides have been studied the most extensively, there are other endogenous molecules within the human body worthy of exploitation as drug discovery platforms. These include nucleic acids, lipids, and carbohydrates, which are discussed in detail in chapter 8. These molecules share the same potential strengths and weaknesses as do peptides. Likewise, there is a need to develop small organic molecules as mimetics of these other endogenous molecules. Although not as clearly defined as peptidomimetic chemistry, ultimately, "nucleotidomimetic" or "carbohydromimetic" chemistries may eventually emerge as new design strategies for lead compound identification.

3.2.4 Lead Compound Identification from Exogenous Sources: Ethnopharmacology

In section 3.2.3, the utility of naturally occurring molecules, which are endogenous to the human body, was discussed as a source of bioactive lead compounds. An alternative is to exploit molecules that are endogenous to other life forms (animal or plant) but do not naturally occur within humans. Such molecules would be classed as exogenous from the perspective of drug design for humans.

Historically, plant-based natural products have been a source of useful drugs. The analgesic opiates come from the poppy plant. Digitalis for congestive heart failure was first isolated from the foxglove plant. Various antibiotics (penicillin) and anticancer agents (taxol) are derived from natural product sources. There are numerous other examples.

There is good reason to be optimistic about the potential future usefulness of such exogenous compounds as a continuing source of potential lead compounds. With many thousands of years of trial-and-error by evolution on her side, Mother Nature is a vastly superior experimentalist to any mere human organic chemist. Insect evolution has permitted the biosynthesis of anticoagulant molecules to ensure that a bite will provide a good supply of blood to the biting insect; such compounds could be useful in the treatment of stroke secondary to blood clots within the brain. Amphibian evolution has enabled the biosynthesis of antibacterial peptides on the skins of frogs so that they can avoid infections as they swim through stagnant swamp waters; peptides such as these could be a good starting point for the peptidomimetic design of novel antibacterial agents. Reptile evolution has culminated in the biosynthesis of neuroactive venoms for purposes of hunting and defense; these molecules have been fine-tuned by evolution as agents specific for neurotransmitter receptors. Plant evolution has culminated in a wide variety of biomolecules that affect any animal that may choose to eat them: it is biologically advantageous for some plants to be eaten so that their seeds can be dispersed in the stool of the animal that ate them; conversely, it is biologically advantageous for other plants to produce noxious chemicals to decrease the likelihood of their being eaten. Because of these diverse biological activities, any of these non-human biosynthetic molecules could, in principle, be a lead compound for human drug discovery.

Another promising feature of animal- or plant-based natural products is that they are a superb source of molecular diversity. As a synthetic chemist, Nature is much more creative and is not constrained to the same finite number of synthetic reactions typically employed by human synthetic organic chemists. When designing new innovative therapies, molecular diversity is important. Furthermore, when developing compound libraries for high throughput screening (see section 3.2.6), it is important to have libraries that capture molecular diversity and are not merely large collections of structurally similar analogs.

Although ethnopharmacology, the scientific investigation of natural products, folk medicine, and traditional remedies, has led to some bona fide drugs (e.g., reserpine (**3.1**), quinine (**3.2**), ephedrine (**3.3**)), it has not proven to be a reliable or efficient source of leads. However, natural products have always been and still are an inexhaustible source of drug leads as well as drugs. Renewed interest in natural products and the novel structures they provide is especially noticeable in marine pharmacology, a practically virgin

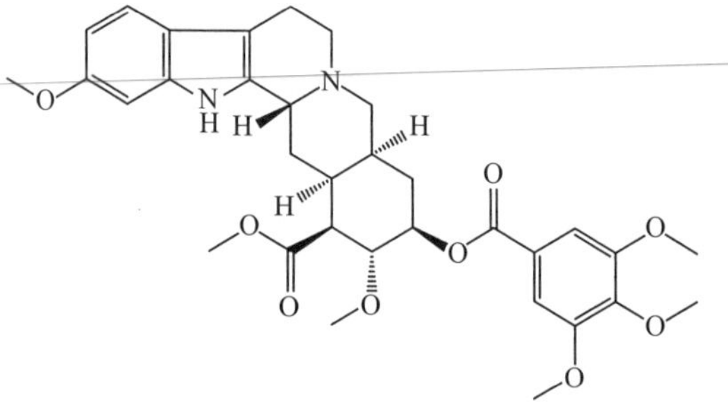

Reserpine (3.1) 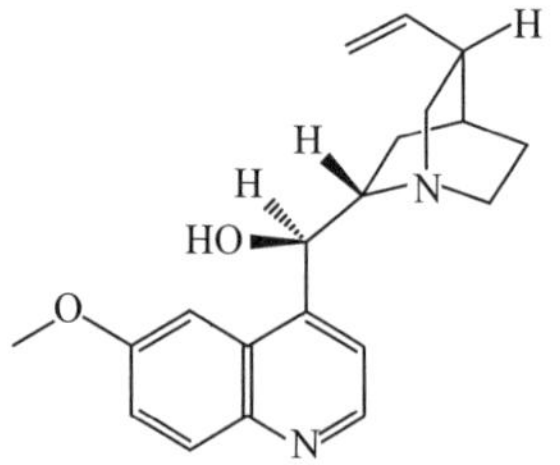   Quinine (3.2)

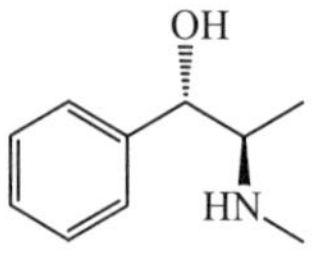

Ephedrine (3.3)

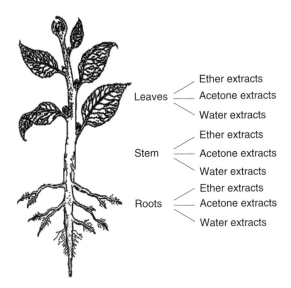

Figure 3.2 Drugs from natural sources: different molecules can be isolated from the leaves, stems, and roots. From each of these sources, extracts conducted with solvents with different polarities will yield different natural products. This complex extraction system ensures the identification of all possible candidate molecules from a plant source.

territory. Several research institutes and well-established groups (notably the Scripps Institute of Oceanography and the University of Hawaii) are producing some very promising results in this field. The isolation of prostaglandins from a coral was one of the more startling recent discoveries in marine pharmacology. Figure 3.2 shows how a plant can be "treated" for medicinal extracts.

An extension of natural products chemistry is the biochemical information derived from the study of metabolic pathways, enzyme mechanisms, and cell physiological phenomena; this research has revealed exploitable differences between host and parasite (including malignant cells), and between normal and pathological function in terms of these parameters. The large and fertile area of *antimetabolite* (metabolic inhibitors) and *parametabolite* (metabolic substitutes) chemistry is based on such stratagems, and has found use in the field of enzyme inhibition and in conjunction with nucleic acid metabolism. The design of drugs based on biochemical leads remains a highly sophisticated endeavor, light-years removed from the random screening of sulfonamide dyes in which it has its origin.

3.2.5 Lead Compound Identification by Rational Drug Design

The previous two methods of lead compound discovery exploited naturally occurring endogenous or exogenous compounds. However, of the approximately 10^{200} "small" organic molecules that could theoretically exist in our world (10^{52} of which are drug-like molecules), many would be purely synthetic substances that do not occur naturally. The lead compound discovery methods discussed in sections 3.2.5 and 3.2.6 afford an opportunity to explore these non-naturally occurring synthetic compounds as potential lead compounds.

The concept of rational drug design (in contrast to its logical counterpart, irrational drug design) implies that the disease under consideration is understood at some fundamental molecular level and that this understanding can be exploited for purposes of drug design. Such an understanding would facilitate the design of purely synthetic molecules as putative drugs. Although this ideal of rational drug design has been pursued for many years (see section 3.2.5.1 for a historical example), it really only began to emerge as a viable drug discovery strategy in the last three decades of the 20th century.

3.2.5.1 Bromides for Epilepsy: An Early Example of "Rational" Drug Design

Bromine was discovered in seawater in 1826. Recognizing its chemical similarity to iodine, French physicians immediately exploited it as an iodine alternative for the treatment of numerous conditions, including syphilis and thyroid goitre. Although no beneficial effects were reported for either bromine or its potassium salt, their widespread use persisted and eventually the depressant effect of potassium bromide on the nervous system, so-called *ivresse bromurique*, was recognized. However, it was a report in the German literature concerning bromide's ability to induce impotence and hyposexuality, rather than *ivresse bromurique*, which lead to its discovery as an anticonvulsant.

In 1857, Sir Charles Locock, the physician accoucheur to Queen Victoria, ascribing to the then prevalent view that epilepsy arose from excessive sexuality, introduced bromide as an anaphrodisiac to suppress the supposed hypersexuality of epileptics. The bromide salts (e.g., potassium bromide, sodium bromide) were administered in substantial doses, ranging from 0.3 g/day in children to a staggering 14 g/day in adults. Although side effects had been considerable (and included psychoses and serious skin rashes), bromides were successful in 13 of the 14 patients treated. On 11 May 1857, at a meeting of the Royal Medical and Chirurgical Society, Locock proudly reported his success in treating "hypersexual" epilepsies with bromides. He argued that logical and rational drug development had finally been achieved for the time: epilepsy arises from excessive sexuality; potassium bromide suppresses sexuality; therefore, potassium bromide successfully treats epilepsy. In principle it seemed like a major success of rational drug design. In reality, it was little more than yet another serendipitous discovery, since hypersexuality has absolutely nothing to do with epilepsy. Regardless of the flawed reasoning, bromides were a major step forward in the treatment of epilepsy and their use persisted until the introduction of phenobarbital in 1912.

3.2.5.2 A Modern Definition of Rational Drug Design

Thankfully, the science of rational drug design has improved substantially since Victorian times. In modern times, rational drug design is defined as a method whereby a disease is understood at a molecular level such that the biological macromolecules involved in the disease process have been identified, purified, characterized, and have had their three-dimensional structure elucidated, thereby enabling the intelligent and insightful engineering of organic molecules to dock to that macromolecule and alter its functional properties in a therapeutically beneficial manner. This is a design method that is lofty and difficult to attain.

Rational drug design is an iterative process, dependent upon feedback loops and new information. When the drug designer makes the first prototype molecule, this molecule becomes a probe with which to test the drug design hypotheses. The molecule can then be further designed and refined to better improve its ability to dock with the receptor site and elicit a biological response. This cycle of "design–test–redesign–retest" can go on for several iterations until the optimized molecule is achieved.

Rational drug design is a very intellectually satisfying activity. The successful rational design of a drug is similar to solving a major puzzle using your wits and wisdom. The macromolecules involved in the disease have been determined; the structures of these macromolecules have been ascertained using X-ray crystallography and/or computer-aided molecular design; a small organic molecule capable of binding to the macromolecule has been cleverly designed; a synthesis for this small organic molecule has been devised; and biological testing has confirmed the bioactivity of the small organic molecule.

Despite protestations from the "naysayers" who despondently claim that all drugs are discovered by serendipity, there is an increasing number of examples that exemplify the successes and practical utility of rational drug design. Perhaps one of the earliest examples is the discovery of cimetidine, an H2-antagonist drug used for the treatment of peptic ulcer disease. Even though the complete structure of the receptor was not fully appreciated, the careful manipulation of the molecule's physicochemical properties (based in part upon an understanding of the underlying histamine molecule) led to the discovery of cimetidine. More recently, the design of drugs for the treatment of AIDS has provided superb examples of rational drug design. The HIV virus encodes for a protease, called HIV-1/PR, that is required for its replication. Using X-ray crystallographic studies of recombinant and synthetic HIV-1/PR helped to identify the active site of the protein. Next, a complex between the HIV-1/PR protein and a prototype inhibitor was also structurally solved using X-ray crystallography. These structural studies greatly facilitated the process of rational drug design, ultimately leading to six rationally designed therapeutics: amprenavir, indinavir, nelfinavir, ritonavir, saquinavir, and lopinavir.

As evidenced by the aforementioned examples, structural chemistry is front and center in enabling rational drug design. Molecular modeling, also called quantum pharmacology, has been instrumental to many of the advances in rational drug design. Some cynics are quick to pontificate that there are no drugs that have been designed by computers. Strictly speaking, this is true; likewise, no drugs have been designed by a nuclear magnetic resonance spectrometer. Computers do not design drugs, people do. However, computers are immensely valuable in advancing and progressing the art and science of rational drug design.

3.2.5.3 Role of Quantum Pharmacology in Rational Drug Design

Quantum pharmacology calculations have, are, and will continue to revolutionize rational drug design. Until the dawning of the 20th century, it was believed that all reality was eminently describable through Newton's Laws. These laws, the foundation of Newtonian mechanics, appeared able to rigorously represent the physical properties of reality, including natural objects as large as planets or as small as billiard balls. However, through the work of Bohr and others, it was eventually realized that classical Newtonian mechanics failed at the atomic level of reality—atoms did not behave like billiard balls. An alternative approach was needed for the quantitative evaluation of molecular phenomena. Molecular quantum mechanics was to be such an approach.

In the first three decades of the 20th century, there occurred many significant advances in theoretical physics and physical philosophy. Planck showed that energy is emitted in the form of discrete particles or quanta; Einstein expanded upon this theory with the proposal that an atom emits radiant energy only in quanta, and that this energy is related to the mass and to the velocity of the light; Schrödinger incorporated these evolving ideas of the new quantum theory into an equation that described the wave behavior of a particle (wave mechanics); Heisenberg formulated a complete, self-consistent theory of quantum physics, known as matrix mechanics; and Dirac showed that Schrödinger's wave mechanics and Heisenberg's matrix mechanics were special cases of a larger operator theory. The capacity for a robust, mathematical description of molecular-level phenomena seemed to be at hand.

Since the Schrödinger equation (which lies at the mathematical heart of quantum mechanics) permitted quantitative agreement with experiment at the atomic level, the physicists of the 1930s predicted an end to the experimental sciences, including biology, suggesting that they would merely become a branch of applied physics and mathematics. These hopes were excessively optimistic and soon proved groundless. Although in principle the Schrödinger equation afforded a complete description of Nature, in practice it could not be solved for the large molecules of medical and pharmacological interest. Early hopes that quantum mechanics would solve the problems of drug design were dashed in despair.

Over the past thirty years, however, three advances have changed the practical usefulness of molecular quantum mechanics:

1. The advent of semi-empirical molecular orbital calculations and density functional theory, which employ mathematical assumptions to simplify the application of quantum mechanics to drug molecules of intermediate to large size.
2. The development of molecular mechanics, which incorporates quantum mechanical data into a simplified mathematical framework derived from the classical equations of motion to permit reasonable calculations on biomolecules of large size.
3. The construction of "supercomputers" capable of performing the massive calculations necessary for considering very large biomolecules. Accordingly, quantum pharmacology has become an attainable goal and calculational computer modeling permits large molecules to be studied meaningfully.

Computer-assisted molecular design (CAMD) employs these powerful computational techniques to engineer molecules for desired receptor site geometries. CAMD had gained widespread acceptance and is beginning to prove its usefulness in many realms of pharmacological endeavor. It has the potential to profoundly influence the future of rational drug design. CAMD enables rigorous modeling of drug molecules, of receptor macromolecules, and of complex drug–receptor interactions. All of these calculations are immensely important to rational drug design.

The most important aspect of computer-assisted modeling is the capability to perform three-dimensional docking experiments, a tool used by all major drug companies and biotechnology firms. Starting with the X-ray structure of the macromolecule, a space-filling molecular model is created, including hydrate envelopes around it. By separately generating the three-dimensional model of a hypothetical drug, a modeler can manipulate the two by modern fast computers and can directly examine the fit of the ligand in the active site; the investigator can change the substituents, conformation, and rotamers of the drug on the screen, and can repeat the docking.

This enormous progress in computer hardware and software, elucidation of macromolecular structure and ligand–receptor interactions, crystallography, and molecular modeling is hopefully bringing us to the threshold of a breakthrough in drug design. We are now able to design lead compounds de novo on the basis of the structure of the receptor macromolecule. However, computerized drug design is still only an instrument that reduces empiricism in an experimental science; the inherent approximations of innumerable conformers and molecular parameters of drug and receptor, and the methodological inaccuracies and difficulties of comparison, will never allow the elimination of insight and trial.

3.2.6 Lead Compound Identification by Combinatorial Chemistry with High Throughput Screening

Random screening, while seemingly wasteful, has an important place in developing lead compounds in areas in which theory lags. Screening for antitumor activity has been carried on for more than 30 years by the U.S. National Cancer Institute, with tens of thousands of compounds being tested on tumors *in vivo* and *in vitro*. More recently, a computerized prescreening method has been applied to this process, saving time and expense, and hence the screening is not as random as it used to be. A successful random search for antibacterial action was conducted by several pharmaceutical companies in the 1950s. They tested soil samples from all over the world, which resulted in the discovery of many novel structures and some spectacularly useful groups of antibiotics, notably the tetracyclines (**3.4**). In fact, microbial sources have supplied an enormous number of new drug prototypes, sometimes of staggering complexity. Recently, the large-scale automated testing of microbial mutants has been realized and combined with recombinant DNA techniques to speed up the efficient discovery and production of new antibiotics.

Some would argue that drug discovery through screening provides the "irrational" counterpart to rational drug design. This remark is unjustifiably harsh and is somewhat

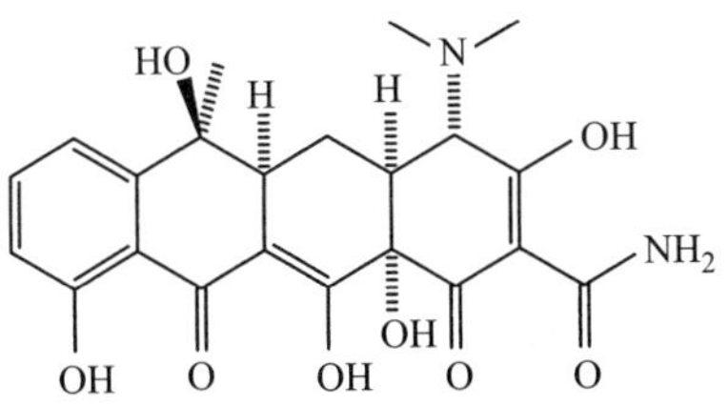

Tetracycline (3.4)

facetious. As mentioned, screening of compounds has a long and rather illustrious history and has produced many useful anticancer and antibiotic drugs. The discovery of the anticonvulsant drug phenytoin provides an early example of drug discovery through screening.

3.2.6.1 Drug Discovery by Screening: Diphenylhydantoin, An Early Example

As a chemical species, the hydantoins have been known since the 1860s. By the latter half of the nineteenth century numerous hydantoin analogs had been synthesized, but only one, 5-ethyl-5-phenylhydantoin (nirvanol), demonstrated any clinical utility. Wernecke introduced nirvanol in 1916 as a "less toxic hypnotic"; however, enthusiasm rapidly waned when its chronic toxicity became recognized. Not surprisingly, a second hydantoin, 5,5-diphenylhydantoin (phenytoin), which had long remained on the laboratory shelf, appeared doomed to obscurity; phenytoin had been first synthesized by Biltz in 1908, through a condensation of urea with benzil which exploited a pinacolone rearrangement.

In the late 1930s, T. Putnam initiated a screening programme to search for new anticonvulsants, using protection against electroshock-induced convulsions as a selection criterion. A makeshift apparatus to execute these experiments was assembled using a commutator salvaged from a World War I German aircraft. Having studied the structure of phenobarbital, Putnam randomly requested a diverse selection of heterocyclic phenyl-containing compounds from a variety of chemical manufacturers. He also communicated with a number of pharmaceutical companies. The Parke-Davis Company provided nineteen heterocyclic phenyl-substituted compounds that had been deemed "worthless hypnotics." Phenytoin was one of these nineteen compounds. Putnam screened hundreds of compounds but only phenytoin combined high activity with low toxicity. In 1936, Putnam's colleague, Houston Merritt, initiated a clinical evaluation of phenytoin, which soon led to its widespread marketing as an anticonvulsant drug.

The pioneering screening techniques that heralded the discovery of phenytoin profoundly influenced subsequent antiepileptic drug discovery. Hundreds of hydantoin analogs were synthesized and screened for biological activity; hundreds of other pentaatomic heterocyclic compounds (e.g., succinimides, oxazolidinediones) were likewise synthesized and screened for biological activity. Many of these new compounds found their way into the market place, with varying degrees of therapeutic success.

3.2.6.2 Drug Discovery by Screening: A Modern Definition

Thankfully, the science of drug discovery by screening has advanced since the time of Merritt and Putnam. Modern drug discovery by screening is more of a systematic technological tour de force than a hit-or-miss gamble. The reasons for these advances are obvious. Although rational drug design is elegant, it is also slow and thus time-inefficient. It takes a long time to identify the proteins that are involved in a disease, then crystallize them and design drugs to bind to them. Worse, some proteins, especially membrane- bound proteins, seem to defy crystallization, while some diseases do not even have identifiable proteins involved in their pathogenesis and etiology. Screening methods attempt to address all of these deficiencies in drug discovery.

If a reliable bioassay is available, it is possible to screen thousands or even millions of compounds against this bioassay. The crystal structures of key protein receptors do not have to be known; indeed, the proteins do not even have to be identified. If the bioassay is fast and efficient, and if the library of compounds being screened is diverse and comprehensive, then in principle it should be possible to identify a lead compound years before the practitioner of rational drug design. However, the key to success lies in the "goodness of the library of compounds" (i.e., a combinatorial chemistry library) and in the "goodness of the screening bioassay" (i.e., high throughput screening methods).

3.2.6.3 Combinatorial Chemistry and Drug Discovery by Screening

A key to success in drug discovery by screening is the availability of a large and structurally diverse library of compounds. If the library contains a million compounds that are all analogs of each other, then it may be large but it is probably not sufficiently diverse. The library should have the full range of functional groups (cations, anions, hydrogen bond donors, hydrogen bond acceptors, lipophilic, aromatic, etc.) displayed in all possible permutations and combinations in three-dimensional space. Creating such a library is not a trivial task.

Combinatorial chemistry is both the philosophical and the practical method with which to create structurally diverse compound libraries. Combinatorial chemistry is defined as that branch of synthetic organic chemistry that enables the concomitant synthesis of large numbers of chemical variants in such a manner as to permit their evaluation, isolation, and identification. Combinatorial chemistry affords techniques for the systematic creation of large but structurally diverse libraries. From a technical perspective, there are several avenues of approach to library creation:

1. Libraries of oligomers of naturally occurring monomers
    a. Oligopeptide libraries
    b. Oligonucleotide libraries
2. Libraries of oligomers of non-naturally occurring monomers
    a. Oligocarbamate libraries
    b. Oligourea libraries
    c. Oligosulfone libraries
    d. Oligosulfoxide libraries
3. Libraries of monomers with multiple sites for substituents
    a. Synthetic ease privileged structure
        Dioxapiperazines
    b. Pharmacological activity privileged structures
        Benzodiazepines
        Dihydropyridines
        Hydantoins
    c. Novel template structures
        Dihydrobenzopyrans

Historically, the first major libraries were oligomers of naturally occurring monomers. A good example would be a library of all possible tripeptides. Using the twenty naturally occurring amino acids, it is possible to produce 8000 different tripeptides. If atypical amino acids and amino acids in the unnatural D configuration are included, it is possible to achieve 125,000 different compounds with relative ease. Peptide libraries are easy to synthesize and, since amino acid side chains possess a wide variety of different functional groups, it is possible to achieve a good measure of structural diversity. However, in general, peptides are not drugs and a peptide lead would have to be modified into a drug-like molecule. In addition to oligopeptides, other naturally occurring oligomeric libraries are possible, including oligonucleotide libraries.

A step beyond the naturally occurring oligomeric libraries is to create libraries from non-naturally occurring monomeric building blocks. The medicinal chemistry literature contains a fair number of examples of such libraries, including oligocarbamates and oligoureas. Although these libraries overcome the limitations of naturally occurring oligomeric libraries, most drugs are not polymers.

To address this problem, new libraries emerged in which the central moiety was a small organic molecule. The diversity library was then constructed by attaching many different substituents to this central moiety. Some of these moieties were selected because they were very simple to synthesize. For example, dioxapiperazines are cyclic dipeptides and thus are relatively trivial to prepare. Other monomers were selected because they had a good track record for being drug-like molecules. Benzodiazepines are a good example of such libraries.

In preparing these various libraries, extensive use is made of *solid phase synthetic methods*. These methods are all derived from the *solid phase peptide synthesis* (SPPS) method developed by Merrifield in 1963. When performing a large number of syntheses, it is preferable to perform the synthetic steps on a solid bead rather than completing the entire synthesis in the solution phase. The solid-phase technique makes byproduct removal and final compound purification easier. The organic chemistry literature contains a wealth of different types of solid-phase supports and novel linkers for attaching the synthetic substrate to the bead.

3.2.6.4 *High Throughput Screening and Drug Discovery*

If a large, chemically diverse library is available, the next problem is to evaluate these compounds in a time-efficient manner. If a 200,000 compound library is available, the biological evaluation assay must be rapid and reliable. If the assay were capable of testing five compounds per day, it would take 110 years to evaluate the entire library. Clearly, this is not the time for elaborate *in vivo* testing. Fast, efficient *in vitro* assays are required. The ability to inhibit an enzyme is a good example of a potentially useful assay for high throughput screening.

A variety of high throughput assays have been developed and perfected over the past 10–20 years. These include the following basic types of assay:

1. Microplate activity assays (assay is in solution in a well; the result of the assay, such as enzyme inhibition, is linked to some observable, such as color change, to enable identification of bioactivity)
2. Gel diffusion assays (biological target is mixed in soft agar and spread as a thin film; the compound library is spread on the surface of the film; after allowing for compound

diffusion, an appropriate developing agent is sprayed on the agar surface and areas in which bioactivity has occurred will show up as distinct zones)
3. Affinity selection assays (compound library is applied to a protein target receptor; all compounds that do not bind are removed; compounds that do bind are then identified)

Of these, microplate assays are probably the most widely used. Screening combinatorial libraries in 96- or even 384-well microplates is time and cost efficient. Using modern robotic techniques, it is possible to perform more than 100,000 bioassays per week in a microplate system (permitting the above-described 200,000 compound library to be screened in two weeks, rather than over a century).

In addition to selecting an appropriate assay, it is also necessary to have a pooling strategy. It is more efficient to test many compounds per well on the microplate, rather than one. If one could test 100 compounds per well, then the standard 96-well plate would enable almost 10,000 compounds to be evaluated in one experiment. (Currently, multiwell plates containing more than 96 wells are routinely being used.) To facilitate effective pooling, the library of compounds is usually divided into a number of nonoverlapping subsets.

The synthetic strategy employed during the combinatorial syntheses can be used to assist in determining these pooling strategies. In *random incorporation syntheses*, a single bead could contain millions of different molecular species. In *mix and split syntheses* (also called *pool and divide syntheses* or *one bead–one compound syntheses*) only one compound is attached to any given solid-phase synthetic bead.

The evolution of methods for combinatorial syntheses and high throughput screening will be necessary to address the explosion of druggable targets soon to be identified by the genomics and proteomics revolutions. Genomics and proteomics represent the future of lead compound identification.

3.2.7 Lead Compound Identification through Genomics and Proteomics

Conservative estimates suggest that more than 4000 human diseases are influenced by distinct and potentially targetable (or druggable) genetic factors. Current drug design strategies are struggling with fewer than 500 druggable receptor proteins. Endeavoring to identify lead compounds for an additional 3500 targets will overwhelm present-day drug design technologies. Drug design has come a long way since the unfortunate demise of King Charles II, but clearly it has a very long way to go. Genomics and proteomics represent a possible pathway to enhanced future drug discovery.

3.2.7.1 Genomics and Lead Compound Discovery

Genomics is the study of genes and their functions. On June 26, 2000–the dawning of the present century–a historic milestone in genomic science was attained when researchers involved with the Human Genome Project jointly announced that they had sequenced 97–99% of the human genome–the all-encompassing collection of human genes. The human genome consists of 23 pairs of chromosomes with three billion base pair codes for approximately 24,000–30,000 functional genes (original estimates of 100,000–120,000 genes seem to have been incorrectly high). The genomic gold rush that followed the human genome project produced a flood of 50,000–75,000 sequences as potential targets for future drug design. Despite the size of this flood, its flow has not filled the drug discovery pipeline with winning candidates.

Determining gene structure and function through genomics definitely does illuminate the path for deciphering human biochemistry and for linking specific genes to specific diseases. Although genomics did deliver phenomenal masses of raw information, the genomics technologies have so far failed to deliver the more than 10,000 anticipated druggable targets predicted by the early hyperbole of the genomics era. Such success will require post-genomics technologies. Taking genomics one step further for the purpose of drug discovery will require linking specific proteins to those specific genes. Clearly, there exists a vast gap between genomics and drug discovery. Bridging this gap will ultimately be a daunting task that lies within the domain of proteomics.

3.2.7.2 Proteomics and Lead Compound Discovery

Proteomics is a protein-based science that seeks to provide new, fundamental information about proteins on a genome-wide scale; to date, proteomics is still more of a concept that a neatly defined smooth-running biotechnology. More concretely, proteomics is the molecular biology discipline that seeks to elucidate the structure and function profiles of all proteins encoded within a specific genome; this collection of proteins is termed the *proteome*. The proteomes of multicellular organisms present an immense challenge in that more than 75% of the predicted proteins have no apparent cellular function. Furthermore, although the human proteome has more than 100,000 proteins, only a fraction of these proteins are expressed in any individual cell type. If specific diseases are to be linked to specific proteins, it is imperative that ways be developed to deduce which individual protein is expressed in which individual cell.

Since protein and mRNA concentrations tend to be correlated, *DNA microarray technology* is a powerful technique with which to monitor the relative abundance of a specific mRNA in an individual cell and to correlate this with a specific protein. (Regrettably, since mRNA and protein levels do not perfectly correlate, a direct measurement of protein abundance would be preferable.) In addition to these microarray technologies, many other technologies will be required if proteomics is to deliver the drugs promised by genomics. For example, drug design requires much more than merely knowing the primary amino acid sequence of a protein; it requires a precise knowledge of the protein's three-dimensional structure, down to the level of the ångström. To date, science has no technology that enables one to use the information coded in a protein's primary amino acid sequence to deduce the overall tertiary structure of the protein. This is the *multiple minima problem* (also called the *protein folding problem*) referred to in chapter 1. The need to solve this problem has given rise to the subdiscipline of *structural proteomics*, a technology that is based upon the principle that structure underlies function and that endeavors to provide three-dimensional structural information for all proteins.

Another evolving subdiscipline is *interaction proteomics*. Protein–protein interactions are a key element of almost all cellular processes. These interactions underlie the events of cell-cycle regulation, cellular architecture, intracellular signal transduction, nucleic acid metabolism, lipid metabolism, and carbohydrate metabolism. A rigorous characterization of the complexities of cellular protein–protein interactions will afford a robust understanding of the integrated networks that drive cellular function. Furthermore, many human diseases, including cancer and neurodegenerative diseases, seem to arise from aberrant protein–protein association mechanisms. Interaction proteomics seeks to elucidate the complete set of interactions that define protein–protein associations.

Even when the technologies of structural proteomics and interaction proteomics have evolved to maturity, the pathway to the awaiting plethora of drugs is still not paved and perfect. Drugs are small organic molecules. Obtaining these drug molecules will require yet another step in the "from genomics–to proteomics–to disease" cascade. Just as proteomics is a crucial bridge uniting genomics to disease, so too will an equally crucial bridge be needed to unite proteomics with therapeutics. Hopefully, the bioinformatics/cheminformatics spectrum will be that bridge.

3.2.7.3 Bioinformatics and Cheminformatics in Lead Compound Discovery

Bioinformatics and cheminformatics constitute an *in silico* science that endeavors to predict the phenomenology of cellular physiology and pharmacology at a molecular level using computational methods. Using databases of compounds and other theoretical molecular design techniques, bioinformatics and cheminformatics will attempt to identify novel molecules to alter the function of various proteins defined by the genome-based proteome. Bioinformatics/cheminformatics will apply knowledge-discovery and pattern-recognition algorithms to the genome-wide and proteome-wide experimental data, thereby facilitating drug design. If structural proteomics has identified the functional portion of an important protein, cheminformatics will search large databases of drug-like molecules to identify one that has the right shape and properties to dock with the protein. Because of the importance of bioinformatics and cheminformatics to the future of drug design, these topics are discussed in greater detail in chapter 1.

An interesting and recent advance in cheminformatics is *chemogenomics*. In conventional cheminformatics, a single drug is designed for a single protein target; in chemogenomics, multiple drugs will be designed to target multiple-gene families. Data gleaned for one protein can be applied to structurally similar proteins coded by the same gene family. Chemogenomics represents a new conceptual approach to target identification and drug development.

3.2.8 Pharmacogenomics and the Future of Lead Compound Discovery

The spectrum of genomics/proteomics/bioinformatics/cheminformatics is defining the future of lead compound discovery and drug design; *pharmacogenomics* is defining the future beyond that! Conventional drug design attempts to discover drugs to treat particular diseases; pharmacogenomics attempts to design individualized drugs to treat particular people with particular diseases. On the basis of a variety of genetic testing, a physician would be able to predict how an individual patient would respond to a specific drug and if this patient will experience any specific side effects. On the basis of person-to-person variability in pharmacokinetics and pharmacodynamics, pharmacogenomics will study how genetic variations affect the ways in which particular people respond to specific drug molecules. Traditionally, drug design has developed drugs for "everyone" with a given disease; pharmacogenomics will enable the tailor-made design of chemotherapies for specific populations or individuals with diseases.

In attempting to achieve this lofty ideal, pharmacogenomics will rely upon genetic data such as *single nucleotide polymorphism maps*. The assembly of a single nucleotide polymorphism (SNP) map for the genome will represent a set of characterized bio-markers spread throughout the human genome; this SNP map will highlight individual variations within particular genes, some of which may be associated with particular dis-eases, thus identifying the genetic variability inherent in human populations that is cru-cial to the task of individualized drug design.

The emergence of pharmacogenomics will also enhance the interaction of medicinal chemistry as a discipline with other disciplines, including the social sciences, ethics, and economics. Society has a difficult enough time paying for currently available drug therapies. Who will pay for individualized therapies? Will this further widen the chasm between "have" and "have-not" populations, between "developed" and "developing" nations?

3.3 SYNTHESIS OF A LEAD COMPOUND

Although drug design and computer-aided molecular design are powerful techniques, ultimately the lead compound must be made and evaluated; it is imperative that we make the jump from *in silico* to *in vitro* and *in vivo*. This is the point at which medici-nal chemistry overlaps heavily with synthetic organic chemistry. Organic synthesis (from the Greek, *synthetikos*, "to put together") is the preparation of complicated organic molecules from other, simpler, organic compounds. Because of the ability of carbon atoms to form chains, multiple bonds, and rings, an almost unimaginably large number of organic compounds can be conceived and created.

In planning a synthetic route for the preparation of a desired molecule (termed the *target molecule*) the organic chemist devises a *synthetic tree*–an outline of multiple available routes to get to the target molecule from available starting materials. An organic synthesis may be either *linear* or *convergent*. A linear synthesis constructs the target molecule from a single starting material and progresses in a sequential step-by-step fashion. A convergent synthesis creates multiple subunits through several parallel linear syntheses, and then assembles the subunits in a single final step. Since overall yield is a function of the number of steps performed, convergent syntheses have higher yields and are preferred over linear syntheses.

In creating synthetic routes for the development of drug molecules, the synthetic chemist wants to create a molecular entity in which functional groups (carbonyls, amines, etc.) are correctly positioned in three-dimensional space; this will enable the creation of functional biophoric fragments such as the pharmacophore. The synthetic chemist has ten general classes of reactions available for such synthetic tasks:

1. Aliphatic nucleophilic substitution
2. Aromatic electrophilic substitution
3. Aliphatic electrophilic substitution
4. Aromatic nucleophilic substitution
5. Free-radical substitution

6. Additions to carbon–carbon multiple bonds
7. Additions to carbon–heteroatom multiple bonds
8. Elimination reactions
9. Rearrangements
10. Oxidations and reductions

These ten reactions provide the capacity to construct a molecular framework and then to position functional groups precisely on this framework. Accordingly, these ten reactions permit two fundamental construction activities:

1. Creation of C-C/C=C/C-H bonds (for the purpose of building a structural framework)
2. Creation of functional groups (to give functionality to the framework)

Appendix 3.1 at the end of this chapter lists 100 fundamental reactions used by synthetic medicinal chemists to create C-C bonds and functional groups during drug molecule preparation. Detailed discussion and mechanisms for these reactions are not provided, but are available in many textbooks of basic or advanced organic chemistry.

3.3.1 Synthon Approach to Drug Molecule Synthesis

To use these 100 reactions (and additional ones) for the purposes of drug molecule synthesis requires an organized and systematic approach. The synthesis of a complicated drug molecule from simple starting materials must be approached in a rigorous and systematic fashion. The *synthon approach* provides such a scheme. This approach is based upon the notion that it is easier to work backwards from the target molecule (i.e., the drug to be synthesized) to the starting materials. This backwards-thinking process is referred to as *retrosynthetic analysis.* In retrosynthetic analysis, a procedure known as *disconnection* is used to dissect a molecule into progressively smaller and smaller fragments until readily available starting materials are obtained. A disconnection involves breaking a bond to a carbon atom. The fragments that result from this disconnection are referred to as *synthons.* Typically, a bond is broken and the electron pair is assigned to one of the fragments, resulting in a positively charged synthon and a negatively charged synthon. The next task is to find readily available starting materials that can actually be used as sources of these synthons. These available materials that are equivalent to a synthon are referred to as *synthetic equivalents*, and are generally commercially available compounds that represent the nucleophile and the electrophile that must react. Sometimes, prior to a disconnection, one functional group is converted to another "synthetically equal" functional group through the process of a *functional group interconversion* (FGI). This FGI may produce a molecule which is easier to disconnect and thus easier to synthesize.

Figure 3.3 shows a simple example of a retrosynthetic synthesis of cyclohexanol. Cyclohexanol may be disconnected to a hydride ion and a hydroxycarbocation (the synthons). Sodium borohydride and cyclohexanone are the synthetic equivalents of these two synthons. Thus, reacting cyclohexanone with $NaBH_4$ will produce cyclohexanol.

It is apparent that this system of disconnections can be applied to molecules of significant complexity to deduce a synthetic route.

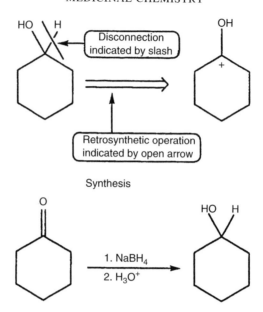

Figure 3.3 The synthon approach: example of a retrosynthetic approach to the synthesis of cyclohexanol.

3.3.2 Need for Efficient Synthetic Methods

Typically, medicinal chemists work in either academic or industrial research laboratories. Accordingly, they tend to develop "research level" syntheses. At the research level, it is acceptable for a synthetic scheme to employ unusual catalysts or large amounts of organic solvents. However, this is not acceptable at the "scale-up level." For a drug molecule to be successful, it must also be affordable. If a drug costs $85 per milligram to synthesize and will be administered at 60 mg/day for many months, then it will not be a commercially viable drug. If a drug is to be successful it will have to be produced in large quantities in an industrial setting, necessitating the implementation of syntheses that are "scalable," efficient, and environmentally friendly. A synthetic scheme with fewer steps and employing water as a preferred solvent has distinct advantages over a technically complicated, low-yield synthesis that uses large quantities of organic solvents that are difficult to dispose of.

Thus, there are some differences between a classical synthetic organic chemist and a synthetic medicinal chemist. The classical synthetic organic chemist is proud of a complex multistep synthesis which may, regrettably, have a low yield. The synthetic medicinal chemist is pleased to design a molecule that can be synthesized in as few steps as possible, hopefully with a high yield and few by-products. Figures 3.4.–3.6 present syntheses of three common drug molecules.

3.3.3 Need for a Robust Biological Model for Compound Evaluation

Once the lead compound has been synthesized, it is next necessary to biologically evaluate it. A number of very important characteristics go into making a biological assay useful. Ideally, the assay should be rapid, cost-effective, efficient, and easy to implement.

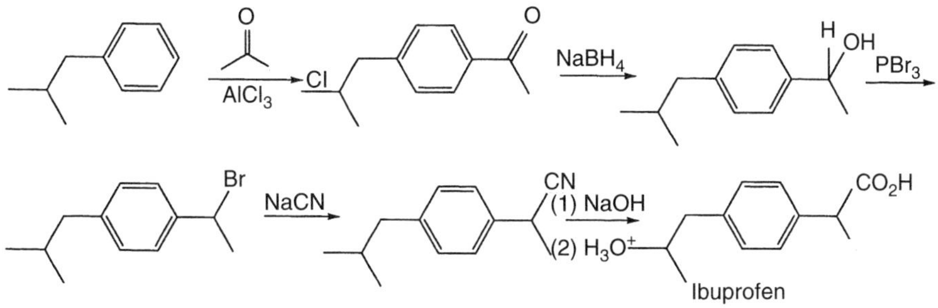

Figure 3.4 The synthesis of ibuprofen is initiated by a Friedel-Crafts acylation of an alkyl-substituted benzene ring. The resulting ketone is then reduced to an alcohol with sodium borohydride. The alcohol functionality then undergoes a functional group interchange by conversion to a bromide. In turn, this permits the introduction of an additional carbon atom in the form of a nitrile introduced via an S_N2 nucleophilic displacement. This is then hydrolyzed to give the target molecule.

This is particularly true if one is pursuing lead compound discovery by high throughput screening of millions of compounds.

More importantly, the biological model should accurately reflect the human disease for which it is being used as a drug-screening tool. Just because an animal model produces a similar disease as is found in humans, there is no guarantee that it will truly reflect the corresponding human disease. For example, occluding the middle cerebral artery in a gerbil will produce a "gerbil stroke." However, are strokes in gerbils identical to strokes in humans? A drug that successfully treats strokes in gerbils may not necessarily treat strokes in humans. Alternatively, there may be diseases that are unique to humans or primates, thereby making it difficult to develop meaningful biological assays in species such as rodents. Alzheimer's disease is a superb example of this dilemma. For years, drug design in Alzheimer's disease was delayed by the absence of a reasonable animal model. Modern advanced molecular biology techniques are addressing this issue by enabling the engineering of rats that over-express the β-amyloid protein that seems to cause Alzheimer's disease in humans.

Biological assays for compound evaluation may be broadly categorized as follows:

1. *In silico*—theoretical evaluation achieved by computing simulation
2. *In vitro*—"test-tube" evaluation done without a whole animal
    a. Binding studies (e.g., competition studies using radioactive ligands)
    b. Functional assay studies (e.g., functional assays using enzymes)
3. *In vivo*—evaluation done with a whole animal

Each of these models has its strengths and weaknesses.

The *in silico* methods are the most rapid and cost effective. They are also the ones that are most divorced from reality. Consequently, *in silico* methods are acceptable for preliminary screens, but are completely unacceptable for the advanced assessment of a candidate compound.

In vitro methods are a definite step up. Radiolabeled competitive binding studies can be used to ascertain whether a drug binds to a receptor. Such studies simply give information

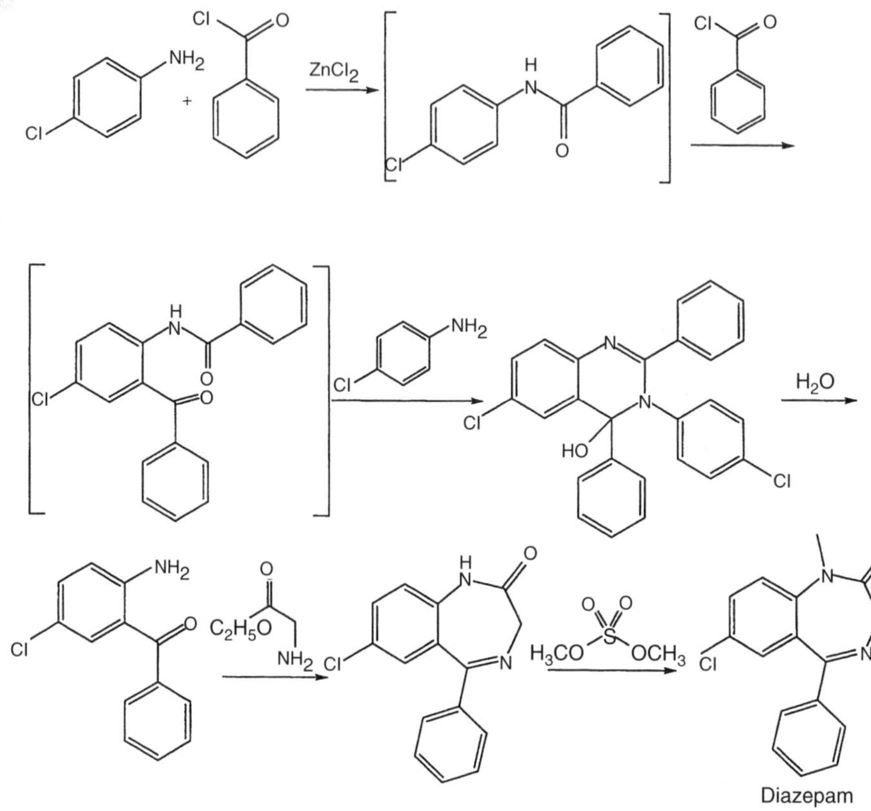

Figure 3.5 The synthesis of diazepam is initiated by the double acylation of an aromatic amine with an aromatic acid chloride. A second equivalent of the *p*-chloroaniline leads to a six-membered ring with two nitrogens. This is hydrolytically opened to expose a free amino group which reacts with an aminoester to yield a seven-member ring. The amide nitrogen is then methylated.

about the binding process; they do not tell if the drug is an agonist or antagonist. Functional *in vitro* assays with a measurable biological outcome are required to tell whether a compound is functioning as an agonist or an antagonist.

By far the most useful assay is the *in vivo* assay. Regrettably but understandably, this assay is the most labor-intensive and costly. *In vivo* assays give the highest quality information about the efficacy of a lead compound. *In vivo* models are accurate animal models of a human disease. Ideally, a candidate drug molecule should not be advanced in the development process unless it demonstrates good to excellent efficacy in an appropriate *in vivo* model.

3.3.4 Deciding to Optimize a Lead Compound

Once a biologically active lead compound has been identified (i.e., designed, synthesized, evaluated), the next decision is whether to proceed with the optimization of the lead compound. If the disease is important and if the lead compound has promising

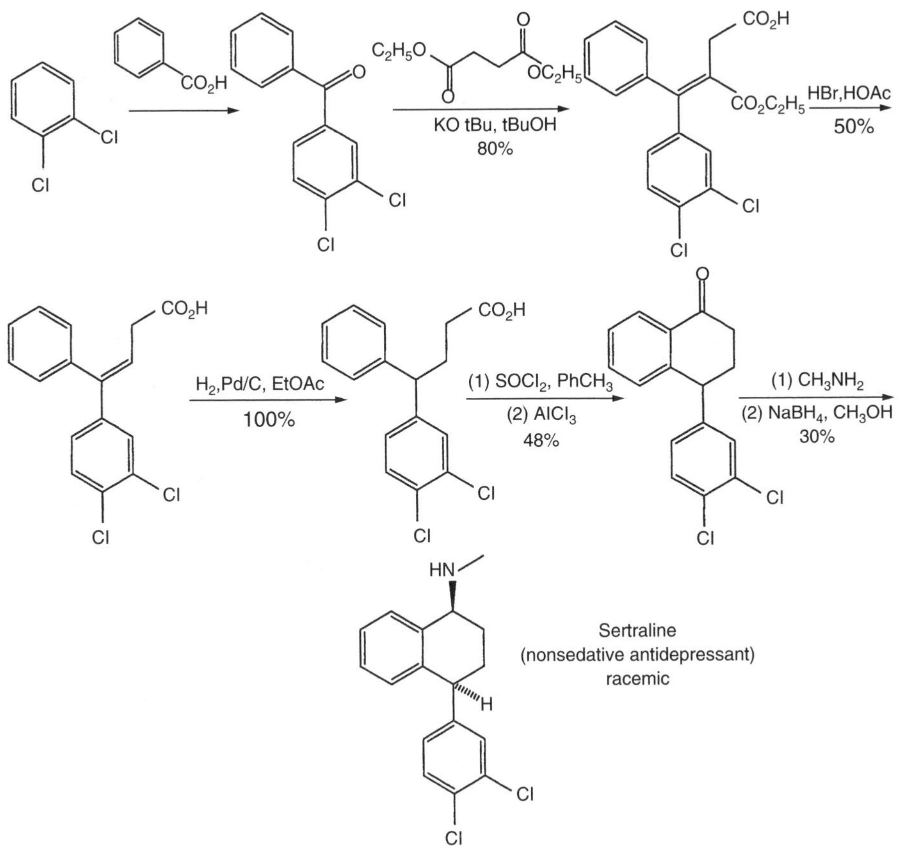

Figure 3.6 The synthesis of sertraline.

activity, then this decision may be relatively straightforward. Nevertheless, the optimization of a lead compound is a lengthy and expensive undertaking, fraught with a frighteningly high rate of failure.

Even in the 21st century, drug design is more of a hope than an achievement. It means the application of previously recognized correlations of biological activity with physico-chemical characteristics in the broadest sense, in the hope that the pharmacological success of a not yet synthesized compound can be predicted. Few drugs in use today were discovered entirely in this way. One of the principal difficulties in this approach is that the available — and very sophisticated—methods for predicting drug action cannot foretell toxicity and side effects, nor do they help in anticipating the transport characteristics or metabolic fate of the drug *in vivo*.

Although some practicing biologists and pharmacologists still regard efforts at drug design with some condescension and ill-concealed impatience, a slow but promising development gives renewed hope that progress in this area will not be less rapid than in the application of biology and physical chemistry to human and animal pathology. The explosive development of computer-aided drug design and bioinformatics (see chapter 1) promises to lead to the era of true rational drug design.

Until the early 1960s, lead compound optimization was an intuitive endeavor based on long experience, keen observation, serendipity, sheer luck, and a lot of hard work. The probabilities of finding a clinically useful drug were not good; it was estimated that anywhere from 3000 to 5000 compounds were synthesized in order to produce one optimized drug. With today's even stricter drug safety regulations, the proportions are even worse and the costs skyrocket, retarding the introduction of new drugs to an almost dangerous extent. The classical method usually applied in lead compound optimization was molecular modification — the design of analogs of a proven active "lead" compound. The guiding principle was the paradigm that minor changes in a molecular structure lead to minor, quantitative alterations in its biological effects. Although this may be true in closely related series, it depends on the definition of "minor" changes. The addition of two seemingly insignificant hydrogen atoms to the Δ^8 double bond of ergot alkaloids eliminates their uterotonic activity, but replacement of the N-CH$_3$ substituent by the large phenethyl group in morphine increases the activity less than tenfold. Extension of the side chain of diethazine by only one carbon atom led to the serendipitous discovery of chlorpromazine and the field of modern psychopharmacology.

There are two conclusions to be drawn from these random examples. First, a merely structural change in an organic molecule is meaningless as long as its physicochemical consequences remain unexplored and the molecular basis of its action remains unknown. Structure, in the organic chemical sense, is only a repository, a carrier of numerous parameters of vital importance of drug activity, as is amply illustrated in the first chapter of this book.

The second conclusion to be drawn from the above examples—and innumerable others—is that the discovery of qualitatively new pharmacological effects is often a discontinuous jump in an otherwise monotonous series of drug analogs and is hard to predict, even with fairly sophisticated methods.

Although a beginning has been made, drug design is far from being either automatic or foolproof. The decision to optimize a proper lead compound—a necessity in drug design and development—is still based on experience, serendipity, and luck, given our basic ignorance of molecular phenomena at the cellular level. Now, however, in the 21st century we can at least have some confidence (thanks in no small part to computer-aided molecular design and bioinformatics) that the optimization of lead compounds and the corresponding discovery of new drugs will be able to keep pace with the progress of biomedical research.

3.4 OPTIMIZING THE LEAD COMPOUND:
THE PHARMACODYNAMIC PHASE

Once a lead compound has been identified by one of the techniques described above and the decision has been made to optimize this compound, the next task is to determine an approach to compound optimization. Typically, this approach is achieved via a two-step strategy:

1. Optimizing the compound for pharmacodynamic interactions
2. Optimizing the compound for pharmacokinetic and pharmaceutical interactions

This two-step strategy enables the drug to be optimized first for its ability to bind to a receptor site and then for its ability to actually travel to that receptor site.

Optimizing the lead compound for the pharmacodynamic phase is likewise approached via a two-step strategy:

1. Synthesizing analogs of the lead compound
2. Correlating analog structure with bioactivity via quantitative structure–activity relationship (QSAR) studies

This two-step strategy starts with the synthesis of numerous analogs of the lead compound. This is meant to explore the structural diversity of the lead compound. Next, QSAR studies are employed to correlate these structural diversities with bioactivity. The results of the QSAR studies are then used to design new and, hopefully, improved analogs.

3.4.1 Analog Synthesis

Classical structure modifications have been the mainstay of drug optimization since the earliest days. As emphasized earlier at several points, structural modifications expressed in organic chemical terms are really only symbols for modification of the physicochemical properties of various structures. Nevertheless, the medicinal chemist usually thinks in terms of structure, since that is the language of organic synthesis. It is therefore appropriate to deal with such an approach, provided one keeps in mind that it is somewhat obsolete because it is twice removed from the arena of drug–receptor interactions.

3.4.1.1 Variation of Substituents via Homologation

The first approach is to vary substituents. The variation of substituents can follow many directions. It can be used to increase or decrease the polarity, alter the pK_a, and change the electronic properties of a molecule. Exploration of homologous series is one of the most often used strategies in this regard, because the polarity changes that are induced are very gradual. Homologation is a standard first approach to substituent variation; indeed, a "standard joke" in medicinal chemistry is to pursue the "methyl, ethyl, propyl, … futile" series of analogs. Despite the somewhat facetious nature of this statement, there are many examples in which homologation is important to drug design.

The case of the antibacterial action of aliphatic alcohols has been known in detail for many years; an increase in chain length leads to increased activity, with a sudden cutoff point at C_6–C_8, due to insufficient solubility of these homologs in an aqueous medium because of their high lipophilicity. On the other hand, local anesthetics depend on lipid solubility in the membrane, and the duration of anesthesia produced by the nupercaine derivatives varies between 10 and 600 minutes for a series of alkyl substituents ranging from a simple -H to -n-pentyl. Another well-known example is the profound qualitative change in action between promethazine (**3.5**), an H_1-antihistaminic drug in which two -CH$_2$- groups separate the ring and side-chain nitrogens, and promazine (**3.6**, an analog of promethazine), which has three methylene groups and predominantly exhibits tranquilizing properties. Higher homologs can, on occasion, become antagonists of the lower members of a series.

3.4.1.2 Introduction of Double Bonds

The introduction of double bonds changes the stereochemistry of a molecule and decreases the flexibility of carbon chains. The E and Z isomers may show very different binding properties; for example, the Δ^1-double bond of prednisone (**3.7**) increases its activity against rheumatoid arthritis over that of its parent compound cortisol (**3.8**)

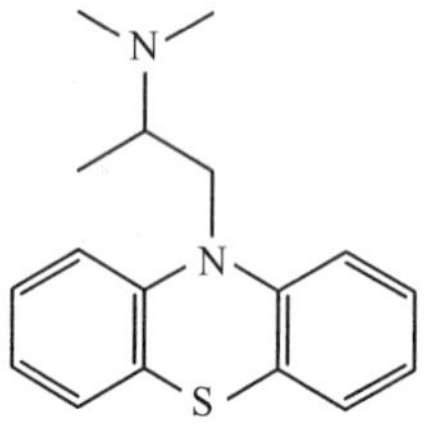

Promethazine (3.5)

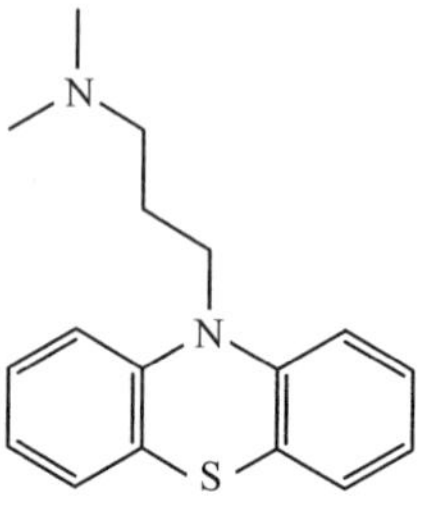

Promazine (3.6)

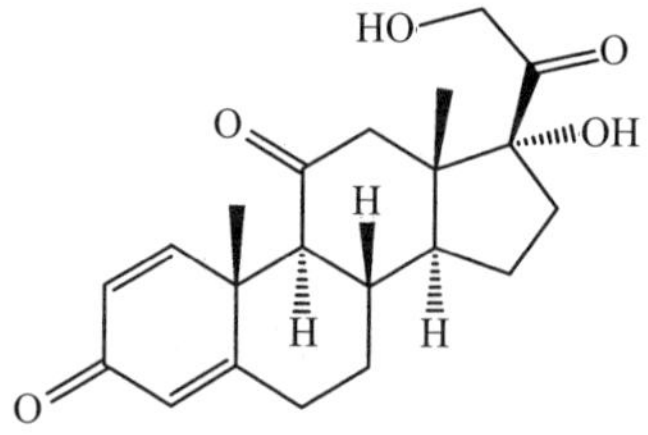

Prednisone (3.7)

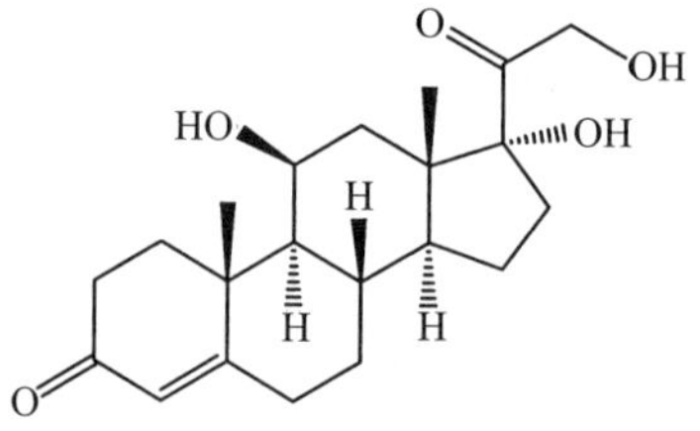

Cortisol (3.8)

by about 30-fold. Decreasing molecular flexibility is also important in drug design. If a drug molecule is too flexible, it will be able to fit into too many different receptors, leading to undesirable effects and toxicity.

3.4.1.3 Variations in Ring Structure

Variations in ring structure are endless in drug synthesis, and are often used in the service of some other change or are introduced simply for patent-right purposes. Inspection of

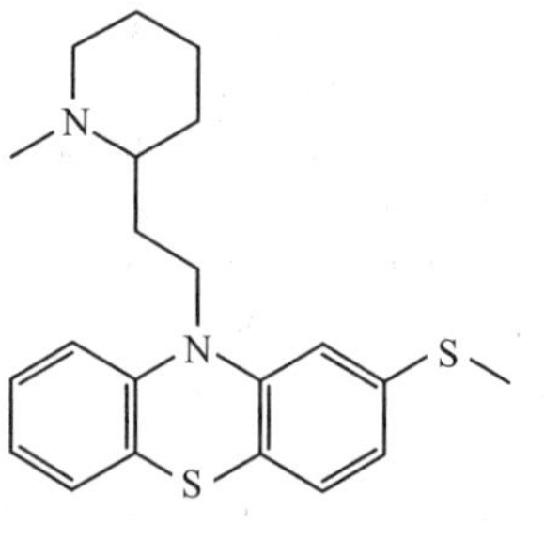

Thioridazine (3.9)

some of the bewildering variations of rings in the older H_1 antihistamines reveals them to be simply variations on the ethylenediamine structure, differing only quantitatively in their effect. Sometimes, however, relatively minor changes in ring structure lead to profound qualitative changes. The most famous example of this occurs in the transition from the neuroleptic dopamine-blocking phenothiazine drugs to the antidepressant dibenzazepines such as imipramine, in which the replacement of -S- by -CH$_2$-CH$_2$- changes the molecular geometry.

Ring opening or closure usually leads to subtle changes in activity, provided that nothing else changes. Three examples (among many possibilities) come to mind: incorporation of the N-methyl substituents of chlorpromazine (1.3) into a closed piperazine ring in prochlorperazine tremendously increases the antiemetic effect while the neuroleptic activity declines. Of course, this may be due to the introduction of a new basic center. In thioridazine (**3.9**), the neuroleptic effect increases with the introduction of a closed ring, but extrapyramidal side effects (tremor, stooped posture, slow and shuffling gait) become noticeable.

The inclusion of rings helps to conformationally constrain a molecule and make it less flexible. As discussed in the previous section, this is a desirable design strategy. Incorporating alkyl rings may change the solubility of the molecule, increasing lipophilicity. The incorporation of an aromatic ring may change the pharmacokinetics of the drug.

3.4.1.4 Structure Pruning and Addition of Bulk

As noted earlier, the pharmacophore of a drug is usually confined to a few functional groups or parts of the whole molecule, which can be a large one. In the case of such complex natural products as alkaloids, which may be difficult or impractical to synthesize (e.g., tubocurarine (**3.10**)), the first design attempt is usually directed at simplification of the molecule, pruning away those structural elements that are not part of the pharmacophore and do not serve to hold crucial binding groups in their appropriate positions. The most successful dissection of a molecule can be seen in the case of morphine. Starting with morphine (**3.11**), the oxygen bridge (i.e., the furan ring) is first removed, resulting in levorphanol (**3.12**), a morphinan. By eliminating ring C, the benzomorphan series is obtained. Its most successful member is pentazocine (**3.13**), which retains only the two methyl groups from ring C and has a lower addiction liability. The simplest (and, incidentally, oldest) modification of the morphine molecule is seen in meperidine (**3.14**),

Tubocurarine (3.10)   Morphine (3.11)

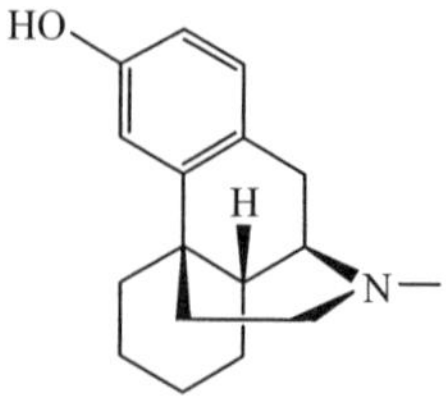

Levorphanol
(3.12)

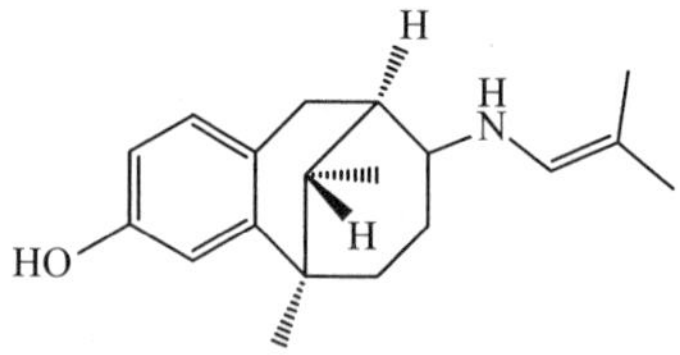

Pentazocine (3.13)

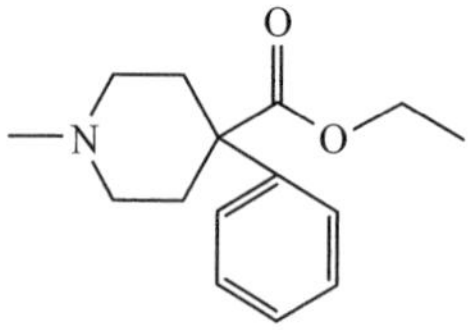

Meperidine (3.14) 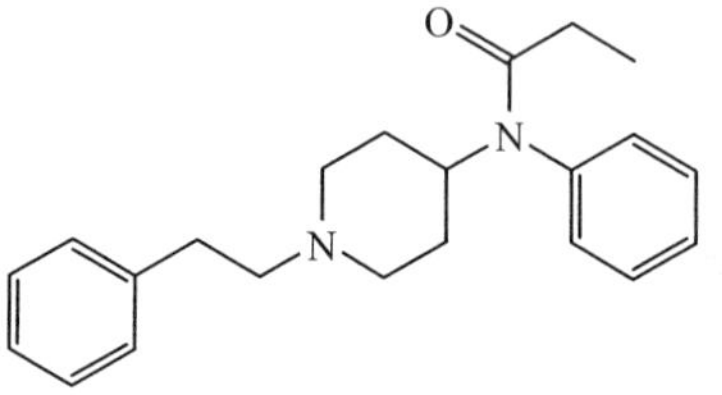   Fentany (3.15) 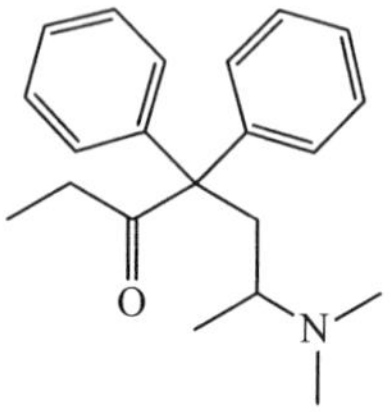   Methadone
(3.16)

a phenylpiperidine that has many congeneric analogs. Fentanyl (**3.15**) is designed along similar lines and has some tremendously active analogs, such as sufentanil. Even in the methadone (**3.16**) molecule, the remnants of the piperidine ring are discernible. On the basis of these and other analogs, the opiate pharmacophore consists of:

1. A nonbonding N electron pair
2. A phenyl ring, three carbons removed from the N
3. A quaternary carbon, next to the phenyl ring

Basically, the same criteria apply to the enkephalins.

The addition of bulky substituents to a drug molecule often results in the emergence of antagonists, since it permits the utilization of auxiliary binding sites on the receptor. This trend is especially noticeable among the neurotransmitters. For example, the anticholinergics, β-adrenergic blocking agents, and some serotonin antagonists show this correlation.

Large substituents often prevent enzymatic attack on a drug, thereby prolonging its useful life. This technique was used to impart resistance to β-lactamase to the semi-synthetic penicillins. The need for the proximity of the phenyl group to the lactam is quite interesting: phenylbenzyl penicillin (8–26) is inactive as an enzyme inhibitor because the phenyl group no longer hinders access of the enzyme to the lactam bond.

3.4.1.5 Physicochemical Alterations

Alteration of the physicochemical characteristics within a drug series is, of course, a result of structural modification; it is just our point of view that changes. It is rather difficult to change only a single parameter with any specific modification, with the potential exception of lipophilicity, which increases with the addition of "inert" hydrocarbon groups. The degree of lipophilicity — so important in drug action and quantitative SAR (QSAR) investigations — is otherwise subject to change together with the Hammet σ-constant, a descriptor of the electron–donor or electron–acceptor capability of a substituent.

3.4.1.6 Isosteric Variations

The *isosteric replacement* of atoms or groups in a molecule is widely used in the design of antimetabolites or drugs that alter metabolic processes. Isosteric groups, according to Erlenmeyer's definition, are isoelectronic in their outermost electron shell. However, since their size and polarity may vary, the term isostere is somewhat misleading. Isosteres are classified according to their valence (i.e., the number of electrons in the outer shell):

Class I: halogens; OH; SH; NH_2; CH_3
Class II: O, S, Se, Te; NH; CH_2
Class III: N, P, As, CH
Class IV: C, Si, N^+, P^+, S^+, As^+
Class V: -CH=CH-, S, O, NH (in rings)

Thus, for instance, the exchange of OH for SH in hypoxanthine gives the antitumor agent 6-mercaptopurine (**3.17**). Fluorine, the smallest halogen, replaces hydrogen well, giving, for instance, fluorouracil (**3.18**), which is also an antitumor antimetabolite. Interchanges of -CH- and nitrogen are common in rings, as seen in the antiviral agent ribavirin (**3.19**).

The oldest example of the use of "nonclassical" isosteres involves the replacement of the carboxamide in folic acid by sulfonamide, to give the sulfanilamides. Diaminopyrimidines, as antimalarial agents, are also based on folate isosterism, in addition to the exploitation of auxiliary binding sites on dihydrofolate reductase. This concept of nonclassical isosteres or *bioisosteres* — that is, moieties that do not have the same number of atoms or identical electron structure — is really the classical structure modification approach.

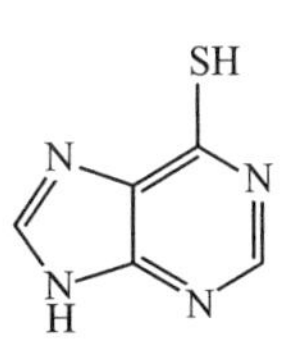

6-Mercaptopurine (3.17) 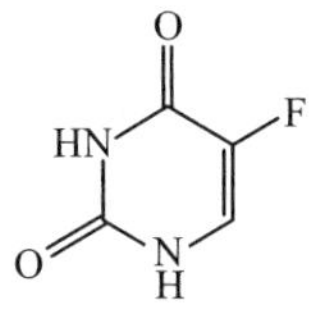   Fluorouracil   Ribavirin (3.19)
                                      (3.18)

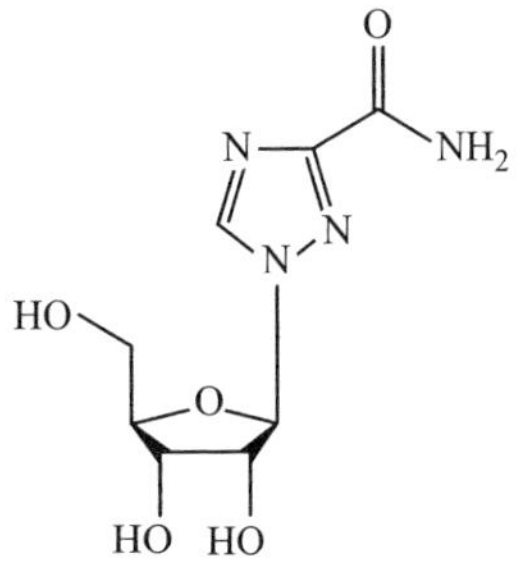

3.4.1.7 Correlating Analog Structure with Bioactivity

The approach involving the design of analogs of an active lead compound has remained unchanged for decades, and the expertise of the synthetic medicinal chemist is as much in demand as ever; however, the intuitive process of selecting structural modifications for synthesis becomes circumspect in this approach, and models based on multiple regression analysis and pattern recognition methods, using very powerful computer techniques, are increasingly being employed as aids. It is obviously much faster and cheaper to calculate the required properties of novel compounds from a large pool of data on their analogs than to synthesize and screen all such new compounds in the classical fashion. Only promising candidates are investigated experimentally. The results gained this way are incorporated into the database, expanding and strengthening the theoretical search. Eventually, sufficient material accumulates to aid in making a confident decision about whether the "best" analog has been prepared or whether the series should be abandoned. Quantitative structure–activity relationship studies represent a systematic approach to this correlation of structure with pharmacological activity.

3.4.2 Quantitative Structure–Activity Relationship (QSAR) Studies

The relationship between chemical structure and biological activity has always been at the center of drug research. In the past, drug structures were modified intuitively and empirically, depending on the imagination and experience of the synthesizing chemist, and were based on analogies. Surprisingly, the results were often gratifying, even if obtained only serendipitously or on the basis of the wrong hypothesis. However, this hit-or-miss approach, practiced even now, is enormously wasteful. Considering that only one of several thousand synthesized compounds will reach the pharmacy shelves, and that the development of a single drug can cost millions of dollars, it is imperative that rational short-cuts to drug design be found. Quantitative structure–activity relationship (QSAR) studies represent this important rational short-cut. QSAR endeavors to elevate drug design from an art to a science.

QSAR methods are in part retrospective as well as predictive, since a "training set" of compounds of known pharmacological activity must first be established. The purpose of such methods is to increase the probability of finding active compounds among those eventually synthesized, thus keeping synthetic and screening efforts within reasonable limits in relation to the success rate. There are three main classifications of QSAR methods:

1. 1D-QSAR (e.g., Hansch analysis)
2. 2D-QSAR (e.g., pattern recognition analysis)
3. 3D-QSAR (e.g., comparative molecular field analysis)

Each method has its own strengths and weaknesses.

3.4.2.1 1D-QSAR — Hansch Analysis

Historically, this is the most popular mathematical approach to QSAR. The major contribution of Hansch analysis is in recognizing the importance of log$P$, where P is the octanol–water partition coefficient. Log$P$ is perhaps the most important measure of a drug molecule's solubility. It reflects the ability of the drug to partition itself into the lipid surroundings of the receptor microenvironment.

Introduced by Corwin Hansch in the early 1960s, Hansch analysis considers both the physicochemical aspects of drug distribution from the point of application to the point of effect and the drug–receptor interaction. In a given group of drugs that have analogous structures and act by the same mechanism, three parameters seem to play a major role:

1. The *substituent hydrophobicity constant,* based on partition coefficients analogs to Hammet constants:

$$\pi_X = \log P_X - \log P_H \tag{3.1}$$

where P_X is the partition coefficient of the molecule carrying substituent X, and P_H is the partition coefficient of the unsubstituted molecule (i.e., substituted by hydrogen *only*). More positive π values indicate higher lipophilicity of the substituent. Since these values are additive, P values measured on standard molecules permit prediction of hydrophobicity of novel molecules.

2. The *Hammet substituent constant σ*

3. *Steric effects,* described by the Taft E_s values

The σ and π constants of substituents are often useful when correlated to biological activity in the statistical procedure known as multivariate regression analysis. As is well known from pharmacological testing of various drug series, such correlations can be either linear or parabolic. The linear relationship is described by the equation

$$\log 1/C = a\pi + bE_S + c\sigma + d \tag{3.2}$$

where C is the drug concentration for a chosen standard biological effect, and a, b, c, and d are regression coefficients to be determined by iterative curve fitting. The parabolic relationship fits the equation

$$\log 1/C = -a\pi^2 + b\pi + cE_S + d\sigma + e \tag{3.3}$$

The coefficients a, b, c, d, and e are fitted to the curve by the least-squares procedure, using regression methods for which computer programs are readily available. The extent of the fit is judged by the correlation coefficient r or the multiple regression coefficient r^2, which is proportional to the variance. A perfect fit gives $r^2 = 1.00$. Once the best fit has been achieved and r or r^2 has been maximized by using a reasonable number of known compounds (15–20 is an advisable number, depending on the number of variables tested, with even more compounds being even better), the curve can be used to predict the biological activity of compounds that have not been tested or, indeed, have not even been synthesized. This requires only the substitution of the optimized regression coefficient constants into the equation, and the use of π, σ, and E_s values, which are usually available for just about any substituent. Naturally, independent variables other than π or σ—including ionization constants, activity coefficients, molar volumes, or molecular orbital parameters—can also be used.

To achieve these various "best fits," statistical methods are employed. A regression analysis of the effects of various substituents on a molecule using the Hansch approach is very useful, saving much time and effort in the synthesis and testing of new drugs. Hundreds of examples of such analyses are available in the literature; many show positive predictive values for drug activity, whereas some other drug series cannot be interpreted by this method.

Regression analysis is currently the most widely used correlative method in drug design. This is because it simplifies problems within a set of compounds by using a limited number of descriptors, notably the Hansch hydrophobic constant π, Hammet constants, or other electronic characteristics of substituents, and the Taft steric constant E_s.

Nevertheless, there are several difficulties and pitfalls in using the Hansch method. First, the inherent disadvantage of regression analysis is that one can obtain good fits ($r^2 > 0.9$) simply by manipulating the constants. Therefore, curve fitting must be done for a relatively large number of compounds to ensure that all predictors are considered. Second, the mode of action may change for drugs within a seemingly continuous series, invalidating the comparison of some compounds in the series with the predictor compounds. The Hansch method cannot anticipate such a change.

Other problems with the Hansch method are that biological systems are often too crude as models for its application, or the electronic effects operative in a drug molecule are not sufficiently understood or precise. Finally, the method requires considerable time and expense, even in the hands of an expert. Difficulties notwithstanding, the Hansch approach took both chemists and pharmacologists out of the dark age of pure empiricism and allowed them to consider simultaneously the effects of a large number of variables of drug activity—a feat unattainable with classical methods.

Nevertheless, Hansch analysis revolutionized drug molecule optimization and directly led to two other strategies for molecule optimization: the Free–Wilson method and the Topliss decision tree.

The Free–Wilson Method. This method also assumes that biological activity can be described by the additive properties of the substituents on a basic molecular structure. In the Fujita–Ban modification of this method

$$\log 1/C = \sum a_i X_i + \mu_0 \tag{3.4}$$

where C is the drug concentration for a standardized effect, a_i is the group contribution of the ith substituent to the pharmacological activity of the substituted molecule, X is unity if substituent i is present and zero otherwise, and $\mu_0 = 1/C$ for the parent compound. Regression analysis is used to determine a_i and μ. In the Fujita–Ban modification of the Free–Wilson method, no assumptions are made about the relevance of the model parameters to the biological activity of the molecule. The effect of each substituent is considered to be independent of any other, and each makes a constant contribution to the overall activity of the molecule. Therefore the method is applicable to compounds with more than one variable group. The result is a data matrix that shows the contribution of each substituent in each position to the overall biological effect of the molecule. The Free–Wilson equation bears close similarities to the linear Hansch equation, and the results of the two can be comparable. The Free–Wilson method, however, cannot predict the activities of compounds that have substituents not included in the matrix. Consequently, this method has found only limited application in drug series where many close analogs are already available but physicochemical data are lacking.

Topliss Decision Tree Method. This method is quicker and easier to use than the Hansch method. The *Topliss scheme* is an empirical method in which each compound is tested before an analog is planned, and is compared in terms of its physical properties with analogs already planned. Like the Free–Wilson method, the Topliss decision tree is no longer extensively used. The 2D- and 3D-QSAR methods are gradually supplanting the 1D methods.

3.4.2.2 2D-QSAR — Pattern Recognition Analysis

2D-QSAR is a somewhat more advanced method for correlating activity and structure. The first step in performing a 2D-QSAR is to select the *training set*. This is a subset of molecules that are diverse in terms of both structure and bioactivity. Ideally, the compounds that are available cover the full spectrum of bioactivity, ranging from active (fully and partially, covering a 10^3-fold range in receptor binding affinities) to inactive. It is difficult to determine what makes a molecule bioactive (or conversely what makes a molecule bioinactive) if all of the compounds tested have similar bioactivities. The more molecules the better, but a reasonable start can be made with as few as ten compounds. It is important not to use all available molecules, since another subset is held back and retained as a *test set*. This test set will ultimately be used to validate any prediction algorithm that is developed through the study of the training set.

Next, every molecule in the training set, regardless of its pharmacological activity, is characterized by a series of *descriptors*:

1. Geometric descriptors
   Bond lengths
   Bond angles
   Torsional angles
   Interatomic distances
2. Electronic descriptors
   Charge densities on individual atoms
   Energy of the highest occupied molecular orbital
   Energy of the lowest unoccupied molecular orbital
   Molecular dipole
3. Topological descriptors
   Graph theory indices
     Randic indices
     Kier–Hall indices
   Ad hoc indices
     Number of rings in the molecule
     Number of aromatic rings in the molecule
4. Physicochemical descriptors
   Octanol–water partition coefficients
     Log*P*
     (Log*P*)^2
   Hydrogen bonding number
     Number of hydrogen bonding donor sites
     Number of hydrogen bonding acceptor sites

These various descriptors may be calculated using various molecular mechanics and quantum mechanics approaches, as discussed in chapter 1.

The geometric descriptors reflect molecular geometry and are conceptually straight-forward. Electronic descriptors reflect properties arising from variations in electron distribution throughout the drug molecule framework. Topological descriptors endeavor to describe molecular branching and complexity through the notion of *molecular connectivity*. The concept of molecular connectivity, introduced by Kier and Hall in 1976, describes compounds in topological terms. Branching, unsaturation, and molecular shape are all represented in the purely empirical connectivity index $^1\chi$, which correlates surprisingly well with a number of physicochemical properties including the partition coefficients, molar refractivity, or boiling point. These graph theory indices are useful to differentiate between an *n*-butyl substituent and a *tert*-butyl substituent. The physico-chemical indices reflect the ability of the drug to partition itself into the lipid surroundings of the receptor microenvironment.

All of these descriptors are calculated for every compound within the training set. Next, a 2D data array is constructed. Along the vertical axis, all of the training set compounds are listed in descending order of bioactivity. Along the horizontal axis, all of the descriptors are arranged for every training set compound. This data array is then probed with statistical calculations to ascertain the minimum number of descriptors that differentiate active compounds from inactive compounds. In order to probe the data array, several methods are available. *Pattern recognition* and *cluster analysis,* two recent quantitative methods, make use of sophisticated statistics and computer software.

Pattern recognition can be used to deal with a large number of compounds, each characterized by many parameters. First, however, these raw data must be processed by scaling and normalization—the conversion of diverse units and orders of magnitude from many sources — so that the chosen parameters become comparable. Feature selection methods exist for weeding out irrelevant "descriptors" and obtaining those that are potentially most useful. By using "eigenvector" or "principal component" analysis algorithms, these multidimensional data are then projected two-dimensionally onto a plot whose axes are the two principal components or two (transformed and normalized) parameters that account for most of the variance; these are the two eigen-vectors with the highest values. Previously unrecognized relational patterns between large numbers of compounds characterized by multidimensional descriptors will thus emerge in a new, comprehensible, two-dimensional plot. The projection of unknowns onto this eigenvector plot will determine their relationship to active and inactive compounds.

Cluster analysis is similar in concept to pattern recognition. It can define the similarity or dissimilarity of observations or can reveal the number of groups formed by a collection of data. The distance between clusters of data points is defined either by the distance between the two closest members of two different clusters or by the distances between the centers of clusters.

Once the data array has been probed and the minimum number of descriptors that differentiate activity from inactivity has been ascertained, a *prediction algorithm* is deduced. This algorithm attempts to quantify the bioactivity in terms of the relevant descriptors. The predictive usefulness of this algorithm is then validated by being applied to the test set compounds. If the prediction algorithm is sufficiently robust, it can be used to direct the syntheses of optimized compounds.

3.4.2.3 3D-QSAR — Comparative Molecular Field Analysis

Like other forms of QSAR, 3D-QSAR starts with a series of compounds with known structures and known biological activities. The first step is to align the molecular structures. This is done with alignment algorithms that rotate and translate the molecule in Cartesian coordinate space so that it aligns with another molecule. The work starts with the most rigid analogs and then progresses to conformationally flexible molecules that are aligned with the more rigid ones. The end result is that all the molecules are eventually aligned, each on top of another.

Once the molecules of the training set have been aligned, a *molecular field* is computed around each molecule, based upon a grid of points in space. Various molecular fields are composed of *field descriptors* that reflect properties such as steric factors or electrostatic potential. The *field points* are then fitted to predict the bioactivity. A *partial least-squares algorithm* (PLS) is used for this form of fitting. Based upon this PLS calculation, two pieces of information are deduced for every region of space within the molecular field about the molecule: the first piece of information states whether that region of space correlates with biological activity; the second piece of information determines whether the functional group on the molecule within that region of space should be bulky, aromatic, electron donating, electron withdrawing, and so on. The predictions from these molecular field calculations are then validated by being applied to a test set of compounds.

3.4.2.4 Pharmacophore Identification — A Corollary of QSAR

All drugs have pharmacological activity as a result of stereoelectronic interaction with a receptor. The receptor macromolecule "recognizes" the arrangement of certain functional groups in three-dimensional space and their electron density. It is the recognition of these groups rather than the structure of the entire drug molecule that results in an interaction, normally consisting of noncovalent binding. The collection of relevant groups responsible for the effect is the pharmacophore, and their geometric arrangement is called the pharmacophoric pattern, whereas the position of their complementary structures on the receptor is the *receptor map*. Over the years, many attempts have been made to define the pharmacophores and their pattern on many drugs. The first attempts were rather naive and simplistic, but the recent use of QSAR has contributed greatly to the evolution of sophisticated methods of practical significance.

The identification of the pharmacophore is a logical corollary of a QSAR calculation. If the minimum number of descriptors that differentiate activity from inactivity is known, it is possible to deduce the *bioactive face* of the molecule — that part of the molecule around which all of the relevant descriptors are focused. This bioactive face logically defines the pharmacophoric pattern of the bioactive molecules.

When using QSAR calculations to optimize a drug for the pharmacodynamic phase, it is important to use relevant biological activities. If *in vivo* activities are used, the bioactivities will be influenced by pharmacokinetic and pharmaceutical factors. In order for QSAR calculations to reflect the pharmacodynamic phase, the bioactivities should be based on *in vitro* data — optimally, receptor binding studies.

QSAR studies are not restricted to the optimization of biological activity at the pharmacodynamic phase. Since toxicity also arises from drug–receptor interactions, the QSAR method can be used to identify the *biotoxic* face of the molecule (i.e., the toxicophore), which could then be engineered out of the molecular structure.

Once QSAR calculations have been used to optimize the pharmacodynamic interactions of the drug molecule, the next step is to optimize the pharmacokinetic and pharmaceutical phases of drug action.

3.5 OPTIMIZING THE LEAD COMPOUND: PHARMACOKINETIC AND PHARMACEUTICAL PHASES

Once the lead drug molecule has been optimized for the pharmacodynamic phase, it must next be optimized for the pharmacokinetic and pharmaceutical phases. If a drug molecule cannot withstand the trip from the gut to the receptor microenvironment, it makes no difference whether the drug actually binds to the receptor.

Many factors must be taken into consideration when optimizing for the pharmacokinetic and pharmaceutical phases. Will the drug be annihilated within the gastrointestinal tract? Will the drug be absorbed? Can the drug molecule be distributed throughout the body? Will the drug be destroyed in the liver? Will the drug's half-life be too short, or too long? Will the drug be excreted too rapidly? If the drug is destined for a brain-based receptor, can the drug cross the blood–brain barrier?

3.5.1 ADME Considerations

When optimizing for the pharmacokinetic/pharmaceutical phases, considerations of ADME (absorption, distribution, metabolism, excretion) are among the most important. (Sometimes ADME is extended to ADMET because of the inclusion of toxicity.) The drug designer must optimize the drug so that it can remain structurally intact during its absorption and distribution. This can be a daunting task, since the body inflicts many metabolic chemical reactions upon the drug molecule during the processes of absorption and distribution. Understanding these metabolic reactions is crucial to the continuing optimization of the drug molecule.

3.5.1.1 Overview of Metabolic Reactions Affecting Drug Molecules

The body uses its usual array of chemical reactions to attack the structural integrity of a drug molecule and to promote its excretion from the body. These reactions can be categorized into Phase I and Phase II reactions. Phase I transformations (oxidation, reduction, hydrolysis) initiate the chemical modification, frequently adding functional groups and increasing polarity; Phase II transformations (conjugations) promote the aqueous solubility of the drug metabolite so that it may be excreted from the body. A more detailed delineation of these transformations is as follows:

Phase I transformations:
  Oxidations
    Oxidation of aliphatic carbon atoms
    Oxidation of carbons adjacent to sp^2 hybridized atoms
    Oxidation of C=C (alkene) systems
    Oxidation of C-O systems

Oxidations of C-N systems
Oxidation of C-S systems
Oxidations of aldehyde and alcohol
Aromatic hydroxylations
  Reductions
  Carbonyl reductions
  Nitro reductions
  Azo reductions
  Reductive dehalogenation
  Hydrolyses
  Ester hydrolysis
  Amide hydrolysis
Phase II transformations
  Glucuronic acid conjugation
    O-glucuronidation
    N-glucuronidation
    S-glucuronidation
    C-glucuronidation
  Sulfuric acid conjugation
  Amino acid conjugation
  Acetyl conjugation
  Methyl conjugation
  Glutathione conjugation
  Hydration or water conjugation

3.5.1.2 Selected Examples of Metabolic Reactions

The previous section listed the various reactions that the body can inflict upon a drug molecule. This section presents brief examples of a number of these reactions.

Oxidation: Aliphatic Carbon Atoms. Oxidation at the terminal carbon atom of an alkyl substituent is ω-oxidation; oxidation of the carbon atom located second from the end is ω-1 oxidation. Unless specifically catalyzed by an enzyme, ω-1 oxidation tends to occur more frequently. The anticonvulsant drug ethosuximide is metabolized at both the ω and ω-1 position.

Oxidation: Alkene Epoxidation. Alkenes may react to produce epoxides (alternatively, sometimes, the alkenes do not react and are metabolically stable). The epoxide is unstable and is subject to ring opening via a nucleophilic attack. The anticonvulsant drug carbamazepine is metabolized via epoxidation to yield carbamazepine-10,11-epoxide; in turn, this is rapidly opened to yield carbamazepine-10,11-diol.

Oxidation: Carbons Adjacent to sp^2 Atoms. Carbon atoms that are situated adjacent to imine, carbonyl, or aromatic groups are frequently oxidized. This reaction appears to be catalyzed by the cytochrome P-450 enzyme system. Typically, a hydroxyl group is attached to the carbon as part of the oxidation process.

Oxidation: Aromatic Hydroxylation. Since many drugs contain aromatic rings, this is a very common metabolic transformation. The process tends to be species specific, with human showing a strong tendency to hydroxylation in the *para* position. This reaction proceeds via an arene epoxide intermediate. The anticonvulsant drug phenytoin is metabolized by being *para*-hydroxylated in its aromatic rings.

Oxidation: Carbon–Nitrogen Systems. Primary amines may be hydroxylated at the nitrogen atom (*N-oxidation*) to yield the corresponding hydroxylamine. Alternatively, primary alkyl or arylalkyl amines may undergo hydroxylation at the α-carbon to give a carbinolamine that decomposes to an aldehyde and ammonia (through the process of *oxidative deamination*). Secondary aliphatic amines may lose an alkyl group first (*N-dealkylation*) prior to oxidative deamination.

Oxidation: Carbon–Oxygen Systems. Molecules containing ether linkages may undergo *oxidative O-dealkylation.* In this process, the carbon atom located α to the oxygen atom is hydroxylated, followed by cleavage of the C-O bond.

Oxidation: Alcohols and Aldehydes. The oxidation of alcohols to aldehydes and of aldehydes to carboxylic acids is routine, and is catalyzed by alcohol dehydrogenase and aldehyde dehydrogenase, respectively.

Oxidation: Carbon–Sulfur Systems. The most common metabolic process that affects a C-S system is *S-oxidation*. The S atom is oxidized to a sulfoxide. In the case of thioketones, the C=S double bond is converted to a C=O bond. For thioethers, oxidative *S-dealkylation* is a possibility.

Reduction: Carbonyl groups. The carbonyl group (-(C=O)-) is reduced through a reaction that is catalyzed by an aldo-keto reductase requiring NADH as a cofactor. A large number of aromatic and aliphatic ketones are reduced to the corresponding alcohols; these reductions are frequently stereospecific. α,β-Unsaturated ketones are typically metabolized to saturated alcohols.

Reduction: Nitro Groups. Nitro reduction is catalyzed by the cytochrome P450 system in the presence of the NADPH cofactor; it is a multistep process in which the reduction of the nitro group to a nitroso group is a rate-limiting step. The metabolic conversion of the nitro group in clonazepam to an amine is a representative example.

Hydrolyses: Esters and Amides. The plasma, liver, kidney, and intestines contain a wide variety of nonspecific amidases and esterases. These catalyze the metabolism of esters and amides, ultimately leading to the formation of amines, alcohols, and carboxylic acids. Kinetically, amide hydrolysis is much slower than ester hydrolysis. These hydrolyses may exhibit stereoselectivity.

Conjugations. Conjugation reactions are Phase II metabolic reactions that are enzymatically catalyzed and involve the attachment of small polar molecules (glucuronic acid, sulphate, amino acids) to the drug. This, in turn, makes the drug more water soluble and

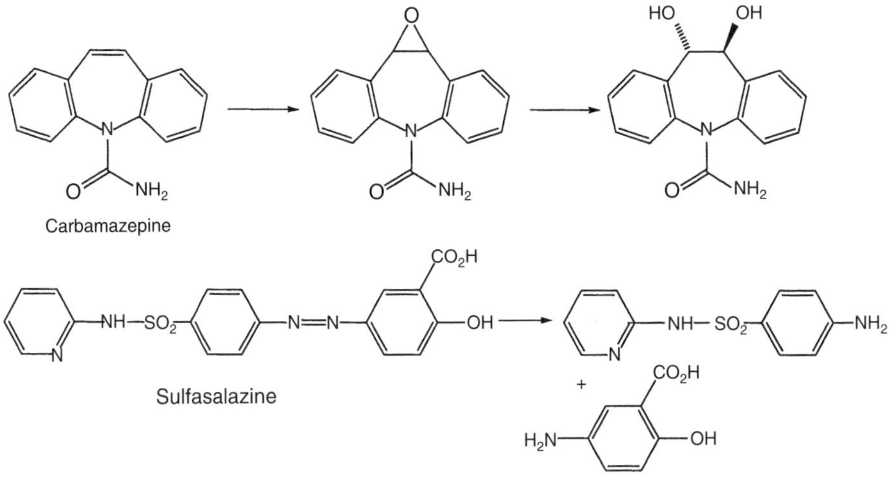

Carbamazepine

Sulfasalazine

Figure 3.7 Examples of drug metabolism by oxidative and reductive pathways.

thus more readily excreted in the urine or bile. The enzymes that catalyze conjugations are transferases such as glucuronosyltransferase, sulfotransferase, glycine N-acyltransferase, and glutathione S-transferase. The conjugation reactions normally target hydroxyl, carboxyl, amino, or thiol groups. Glucuronidation is the most common conjugation method. There are four classes of glucuronide metabolites: O-, N-, S-, and C-glucuronides. Sulphate conjugations occur less frequently.

Figure 3.7 depicts several of these reactions.

3.5.1.3 Vulnerability of Molecular Building Blocks to Metabolic Reactions

When designing and optimizing a drug molecule within the multiphore conceptualization, it is important to remember that the drug molecule is constructed from molecular building blocks that add function and structure to the molecule. It is important that the vulnerability of each of these building blocks to metabolic attack be appreciated during the drug design process. This section lists the major molecular building blocks and briefly outlines their susceptibility to metabolism.

Alkanes. Alkyl functional groups tend to be metabolically nonreactive and to be excreted unchanged. Therefore, alkanes can be used to build the framework of a molecule or as lipophilic functional groups. There are a few exceptions to the rule of alkanes being metabolically inert. Rarely, a linear alkyl group will be oxidized in a process that is catalyzed by a mixed-function oxidase enzyme. When this occurs, it does so either at the end of the hydrocarbon chain or adjacent to the final carbon (the "omega-minus-one carbon"). The metabolism of butylbarbital is a good example of this (see figure 3.8).

Cycloalkanes. Cycloalkanes are conformationally restricted alkanes. Three rings are employed in drug design: cyclopropane, cyclopentane, cyclohexane (the latter two are

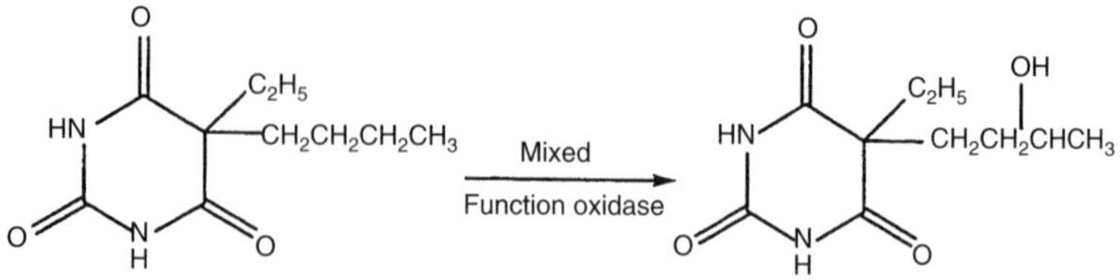

Figure 3.8 Metabolism of butylbarbital by mixed-function oxidase.

used quite commonly). While cyclopropane may be reactive, due to ring strain, cyclopentane and cyclohexane are metabolically inert. These substituents are useful for increasing lipid solubility.

Alkenes. Alkenes are, in general, metabolically stable. The majority of alkene-containing drugs do not exhibit significant rapid metabolism at the double bond. There are some isolated examples of alkene-containing compounds that undergo epoxidation, catalyzed by mixed-function oxidase, or that add water across the double bond to give an alcohol.

Halogenated Hydrocarbons. Halogenated hydrocarbons are not easily metabolized and show significant stability *in vivo*. The addition of halogens tends to increase the lipophilicity and to prolong the half-life of the drug.

Aromatic Hydrocarbons. Aromatic rings are very susceptible to oxidation, in particular to aromatic hydroxylation. The oxidation of aromatic rings frequently proceeds via an epoxide intermediate, which may actually be stable enough to be isolated. The hydroxylation of an aromatic ring increases hydrophilicity, thus promoting renal excretion and slightly decreasing the half-life of the drug. Aromatic hydrocarbons are oxidized in a number of organs, but the liver is a preferred location.

Alcohols. The alcohol functional group tends to be metabolized by a variety of enzymes. Primary and secondary alcohols are oxidized to carboxylic acids or ketones. Tertiary alcohols are substantially more stable to oxidation. Rather than oxidase-mediated oxidation, alcohols can also be conjugated. If the alcohol is conjugated with glucuronic acid, a glucuronide forms; if it is conjugated with sulfuric acid, a sulfate is formed. Regardless, both of these conjugations increase hydrophilicity and decrease the half-life of the drug molecule.

Phenols. Phenols are oxidized via hydroxylation to yield a diphenolic molecule. This hydroxylation is either *ortho* or *para* to the primary alcohol. Phenols may also be conjugated with either glucuronic acid or sulfuric acid.

Ethers. In general, ethers are very stable and are excreted unchanged. Sometimes, an ether that involves a small alkyl group (occasionally a methyl, rarely an ethyl) will be dealkylated, with the small alkyl group being excreted as an aldehyde; the remainder of the drug molecule is left as an alcohol.

Aldehydes and Ketones. Many metabolic routes are possible, including both oxidation and reduction. However, oxidations are more common. Aldehydes are very susceptible to oxidation, which is catalyzed by various enzymes including aldehyde oxidase and aldehyde dehydrogenase; this oxidation yields a carboxylic acid. Ketones, on the other hand, tend to be stable to oxidation. Conversely, aldehydes are seldom metabolized by reduction. Ketones, however, frequently undergo reduction to a secondary alcohol; this is particularly true for α,β-unsaturated ketones.

Carboxylic Acids. Metabolism of carboxylic acids is relatively straightforward. First and foremost, carboxylic acids undergo a diverse variety of conjugations. Carboxylic acids conjugate with glucuronic acid, glutamine, and glycine; the resulting conjugates are water soluble and more easily excreted. Alternatively, carboxylic acids may be oxidized, especially beta to the carboxyl group.

Esters. Not surprisingly, esters are not metabolically stable. They are readily converted to the corresponding free acid and alcohol via hydrolysis, a process that may be either base- or acid-catalyzed. Esters are rapidly hydrolyzed throughout many locations within the body. Orally administered ester-based drugs are quickly hydrolyzed within the stomach.

Amides. Although similar to esters in terms of being a functional derivative of a carboxylic acid, amides, unlike esters, are relatively metabolically stable. In general, amides are stable to acid- and base-catalyzed hydrolysis. This stability is related to the overlapping electron clouds within the amide functionality and the corresponding multiple resonance forms. Amidases are enzymes that can catalyze the hydrolysis of amides. Nevertheless, amides are much more stable than esters.

Carbonates, Carbamates. The ester portion of both carbonates and carbamates is hydrolyzed to give the monosubstituted carbonic acid, which is unstable and decomposes with loss of carbon dioxide. Thus, carbonates are hydrolyzed to give alcohol and carbon dioxide; carbamates are hydrolyzed to yield an alcohol, an amine, and carbon dioxide.

Ureides. Compounds that contain urea functionalities are stable and are not commonly metabolized or hydrolyzed.

Amines. An important metabolic route for primary and secondary amines is conjugation with either glucuronic acid or sulfuric acid to yield the corresponding water-soluble glucuronides and sulfates. Another common reaction is dealkylation of secondary and tertiary amines. The alkyl groups are sequentially removed and then "lost" as either aldehydes or ketones. The amine is thus converted from a tertiary amine to a secondary amine and then to a primary amine. This form of metabolism occurs most favorably if the alkyl group is small, such as methyl, ethyl, or propyl. Finally, amines may be acetylated, with the resulting amide undergoing typical amide metabolism.

Nitro. Typically, the nitro group is attached to an aromatic ring. Nitro groups are reduced to the corresponding amine.

Thio Ethers. Thio ethers are commonly oxidized to a sulfoxide or sulfone. The conversion of chlorpromazine to chlorpromazine sulfoxide is a good example of this.

Sulfonic Acids. Drugs that contain sulfonic acids are quite acidic. They have high water solubility and are rapidly excreted, leading to a rather short half-life. They can also be irritating to the gastrointestinal tract during the pharmaceutical phase.

Sulfonamides. In drug design, sulfonamides are frequently attached to aromatic rings. Thus many drugs are constructed around a benzenesulfonamide nucleus. This nucleus is quite stable to acid-, base-, or enzyme-catalyzed hydrolysis.

Epoxides. Epoxides or oxiranes are quite reactive metabolically. These three-membered rings will readily open in the presence of most nucleophiles. They will react with most biopolymers, including proteins and nucleic acids. Compounds with epoxides have very short half-lives and tend to be toxic.

Aziridines. These are similar to epoxides. They are extremely reactive to most nucleophiles and open readily. Aziridines tend to behave as alkylating agents and thus have carcinogenic effects. In harmony with this, drugs containing aziridines have been used to treat cancer.

Tetrahydrofuran and Furan. Some drugs containing tetrahydrofurans (oxalanes) tend to be readily air oxidized, and thus have a poor shelf life and are unstable even prior to being administered; other tetrahydrofuran-containing drugs are metabolically stable. Furans (oxoles) behave more like aromatic compounds than ethers; accordingly, they undergo aromatic hydroxylation.

Pyrrole and Pyrrolidine. Pyrroles (azoles) undergo aromatic hydroxylation; pyrrolidines (azolidines) undergo conjugation with either glucuronic acid or sulfuric acid.

Thiophenes and Tetrahydrothiophenes. Thiophenes (thioles) are subject to aromatic hydroxylation; tetrahydrothiophenes (thiolanes) undergo oxidation of the sulfur to give sulfoxides or sulfones.

Oxazole and Isoxazole. These are five-membered ring heterocycles containing oxygen and a nitrogen. Two common forms are oxazole (1,3-oxazole) and isoxazole (1,2-oxazole), both of which undergo aromatic hydroxylation.

Pyrazole and Imidazole. These are five-membered ring heterocycles containing two nitrogens, one basic and one neutral. Two common forms are pyrazole (1,2-diazole) and imidazole (1,3-diazole); both are prone to aromatic hydroxylation.

Pyridine and Piperidine. These are six-membered heterocyclic rings containing one nitrogen. Piperidine (azine) is aromatic and undergoes aromatic hydroxylation. Piperidine functions as a secondary amine and undergoes hydroxylation with either glucuronic acid or sulfuric acid.

Indole. This is a bicyclic (one five-membered ring plus one six-membered ring) aromatic heterocycle. Since most indole-containing drugs are substituted with the remainder of the drug being positioned at the 3 position, the aromatic hydroxylation tends to occur at the 4–7 position.

Coumarin. The coumarin moiety is found in a number of important drugs. This is a bicyclic heterocycle containing two six-membered rings and two oxygens, one endocyclic, one exocyclic. Since the coumarin contains an intramolecular lactone ester, it undergoes hydrolysis to yield a carboxylic acid and a phenol.

3.5.2 Site of Delivery Considerations

The next major consideration during optimization for the pharmacokinetic/pharmaceutical phase concerns the design of drugs to overcome barriers during their distribution. Of these barriers, the blood–brain barrier is by far the most important to the drug designer.

3.5.2.1 The Blood–Brain Barrier

The blood–brain barrier (BBB) is both a friend and a foe to the drug designer. When designing a drug for a non-neurologic condition, the BBB is a friend in that it can be used to preclude entry of the drug into the brain, thereby decreasing side effects. However, when designing a drug for a neurological indication, the BBB is a true barrier that must be crossed. Moreover, if a drug has been designed to cross the BBB, it typically will also cross the maternal–fetal placental barrier — a consideration when designing drugs that may be used in women of childbearing age.

The notion of a BBB first emerged in the early years of the twentieth century when it was observed that organic dyes injected into animals stained all tissues except the brain. There appeared to be some invisible barrier that prevented certain molecules from entering into the brain. Over the past twenty to thirty years, the structural basis of the BBB has been more carefully delineated. Many different structural components contribute to the BBB.

The small blood vessels, or capillaries, are the first structural level of the BBB. Drug molecules are distributed throughout the body by the bloodstream and the capillary is the point at which a drug leaves the bloodstream to bind to a receptor. Within the brain, capillaries are composed of cells, called *endothelial cells*, that are connected to each other by *tight junctions*. These junctions are a first-line impediment, slowing the journey of the drug molecule from within the capillary to a receptor site on a neuron. The next component of the BBB is the *astrocyte*. The astrocyte is a *glial cell* (helper cell) within the brain. The astrocyte wraps itself around the capillary to provide yet another line of defense between the drug in the capillary and the neuronal receptor to which it is traveling. The tight junction–astrocyte barrier is unique to the brain and forms the BBB (see figure 3.9).

There are two methods for crossing the BBB: passive diffusion or active transport. Passive diffusion is the route preferred by most neuroactive drugs. Lipid solubility is a desired chemical property for a molecule to diffuse across the BBB. For a molecule to cross the BBB by passive diffusion it should have a molecular weight less than 650 g/mol and should have a log$P$ (logarithm of the octanol–water partition coefficient) value

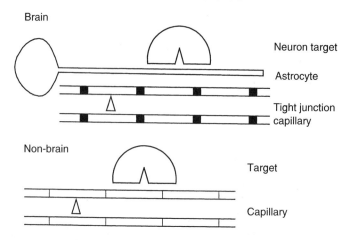

Figure 3.9 The blood–brain barrier (BBB) is a major impediment to the delivery of drugs to the brain. In the brain, in order for a drug molecule to leave a capillary and successfully journey to a neuronal receptor, it must traverse multiple barriers. The walls of capillaries in the brain are different from those in non-brain tissues. Tight junctions prevent the drugs from readily crossing the capillary. Next, in the brain, another type of cell, called an astrocyte, forms an additional barrier that must be traversed. Astrocytes are not present outside of the brain.

between 1.5 and 3.0; in accord with these requirements, the molecule should not be too polar, probably having no charged groups and fewer than three hydrogen-bonding donors or acceptors.

The second route for traversal across the BBB is by active transport. This is an energy-dependent process whereby a protein carrier physically shuttles the drug molecule across the BBB. There are a number of molecular substrates that the brain requires for its normal functioning; these substances are not biosynthesized within the brain and are not able to enter the brain by passive diffusion. Because of their importance to normal brain neurochemistry, evolution has resulted in the existence of protein carriers to transport them into the brain. D-glucose and L-phenylalanine are two such molecules, and there are a number of others. Therefore, if a drug is designed as a structural analog of D-glucose or L-phenylalanine, there is a possibility that it will be actively transported across the BBB.

3.5.3 Prodrug and Softdrug Design Considerations

Sometimes, ADME considerations can be used to the advantage of the drug designer. This is particularly true when applied to the concept of a *prodrug*. A prodrug is a drug molecule that is biologically inactive until it is activated by a metabolic process. The active compound is released as a metabolite.

The purpose of a pro-drug can be to:

1. Increase or decrease the metabolic stability of a drug
2. Interfere with transport characteristics

3. Mask side effects or toxicity
4. Improve the flavor of a drug

An ester, for example, can be used to "mask" a carboxylate. Within the body, the ester is hydrolyzed, releasing the drug in its bioactive carboxylate form.

3.5.3.1 Regulation of Drug Stability

The regulation of drug stability can take two directions: a prodrug can increase the *in vivo* stability of an active compound and prolong its action, or it can automatically limit its duration and prevent potential toxicity.

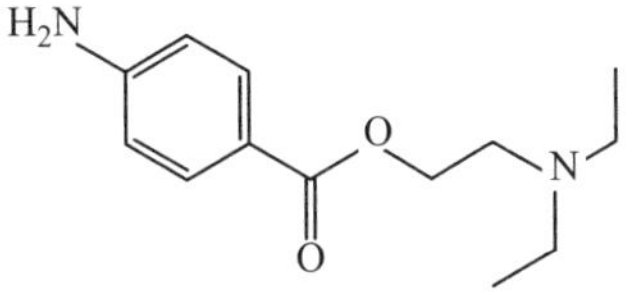

Procaine (3.20) 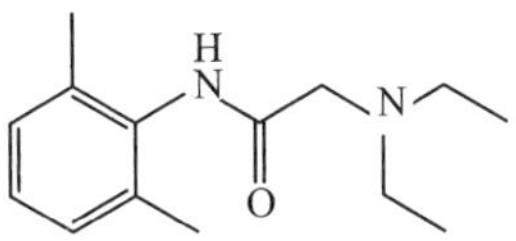   Lidocaine (3.21) 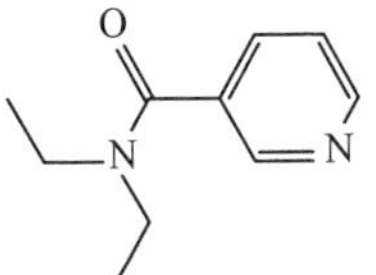   Tocainide (3.22)

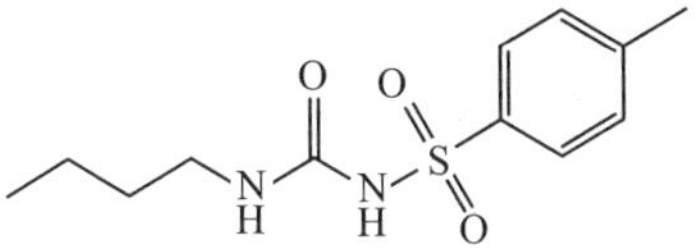

Tolbutamide (3.23) 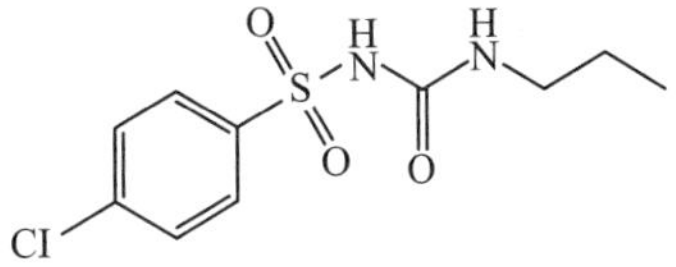   Chlorpropamide (3.24)

There are many examples of drug stabilization. Among local anesthetics, procaine (**3.20**) is an ester and is therefore very easily hydrolyzed by esterases. By conversion of the ester into an amide in lidocaine (**3.21**), the duration of action is increased several fold. Lidocaine is also used intravenously as an antiarrhythmic agent. In that application, it must pass through the liver — the principal drug-metabolizing organ — in which it loses an N–ethyl group to become a convulsant and emetic. To minimize these unwanted and toxic effects, tocainide (**3.22**) — whose α-methyl group prevents degradation, and which lacks the vulnerable N-ethyl groups — was prepared. This compound is not a prodrug in the strict sense, but rather represents a molecular modification.

Replacement of a "vulnerable moiety" such as a methyl group by a less readily oxidized chlorine was used to transform the short-acting tolbutamide (**3.23**), an oral antidiabetic, into the long-acting chlorpropamide (**3.24**), with a half-life sixfold greater than its parent.

A *decrease* in stability is often a desirable modification. For example, succinylcholine (suxamethonium; **3.25**) — a neuromuscular blocking agent used in surgery — has a self-limiting activity, since the ester is hydrolyzed in about 10 minutes, preventing the potential for overdose, which could be fatal with more stable curarizing agents.

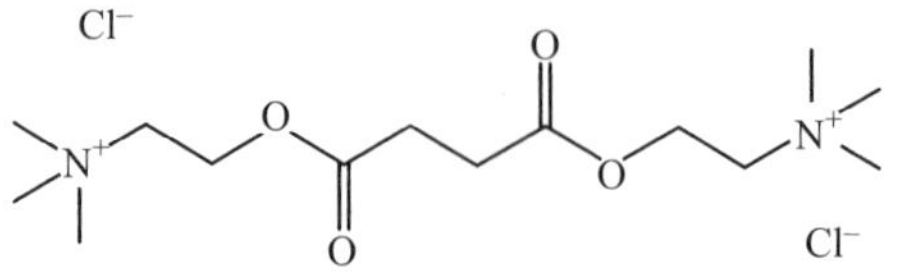

Succinylcholine (3.25)

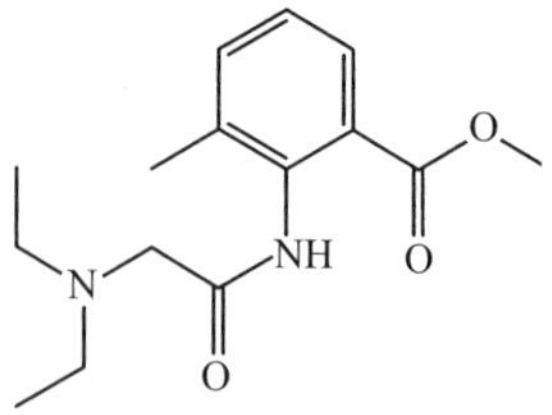

Tolycaine (3.26)

An ester group can be introduced into a local anesthetic, such as tolycaine (**3.26**), to prevent the drug from reaching the CNS if it is injected intravascularly by accident or abuse. The ester group is fairly stable in the tissues but is very rapidly hydrolyzed in the serum to the polar carboxylic acid, which cannot penetrate the blood–brain barrier.

3.5.3.2 Interference with Transport Characteristics

Interference with transport characteristics can serve many purposes. The introduction of a hydrophilic "disposable moiety" can restrict a drug to the gastrointestinal tract and prevent its absorption. Such a type of drug is represented by the intestinal disinfectant succinyl-sulfathiazole (**3.27**). On the other hand, lipophilic groups can ensure peroral activity, as in the case of the penicillin derivative pivampicillin (**3.28**), which enters the circulation and then slowly releases the antibiotic in its free acid form, producing high blood levels of the latter.

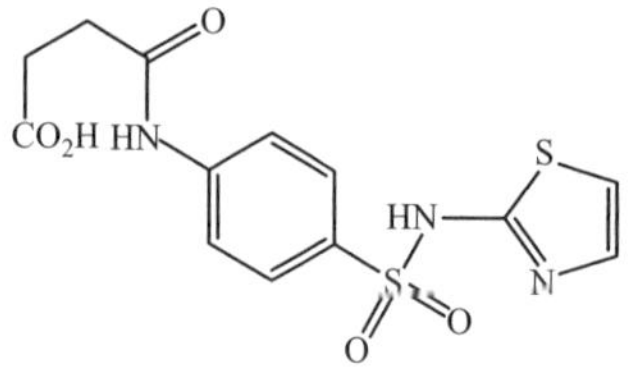

Succinyl-sulfathiazole (3.27)

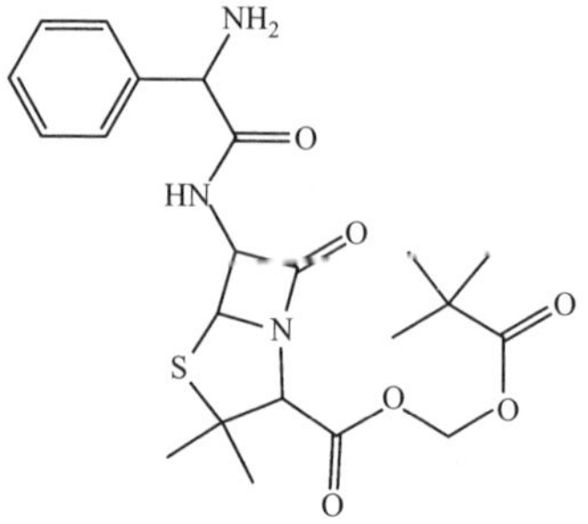

Pivampicillin (3.28)

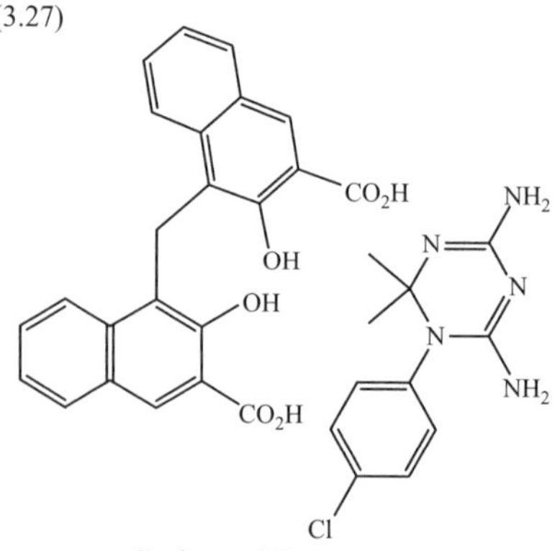

Cycloguanil Pamoate (3.29)

Lipophilic groups that are not easily hydrolyzed are used extensively for *depot preparations*, which liberate the active drug molecule slowly, for a period of days or weeks. Steroid hormone palmitates and pamoates, and antimalarial esters (e.g., cycloguanil pamoate, **3.29**), can deliver the active drugs over a prolonged time; cycloguanil, for example, is released over a period of several months. This can be a great convenience for the patient, especially in areas with remote medical facilities.

Drug designers have attempted for many years to use selective drug-transport moieties, and have met with moderate success. The idea is to attach a drug, such as an antitumor agent, to a natural product that will accumulate selectively in a specific organ and act as a "Trojan horse" for the drug. The attachment of alkylating agents to estrogens has been tried in the treatment of ovarian cancer, and amino acids have also been used as drug carriers. A recent ingenious application of the carrier concept is the utilization of *antibodies* — which can, at least in principle, be tailored to any site — as drug carriers. In this regard, antitumor agents such as adriamycin (**3.30**) and methotrexate (**3.31**) have been linked covalently to leukemia antibodies and melanoma antibodies, with some initial success. The large-scale preparation of antibodies is, of course, a major difficulty in this approach; however, the new monoclonal antibodies hold great promise.

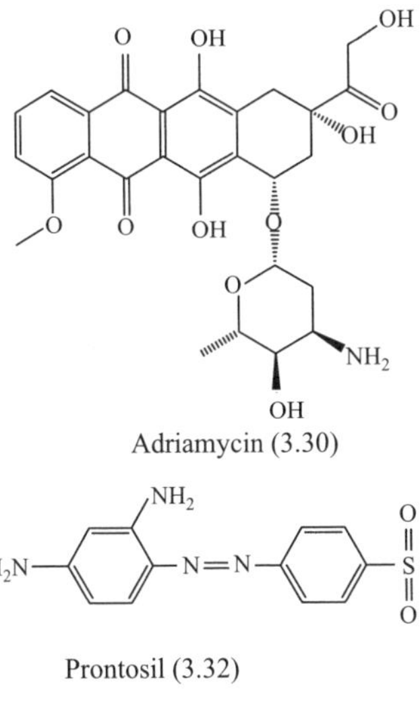

Adriamycin (3.30)

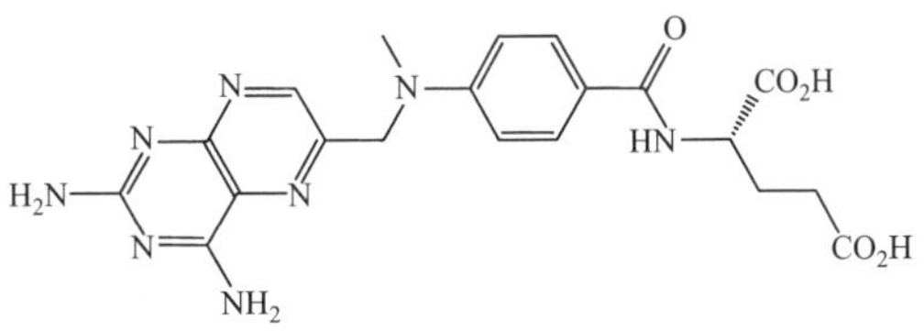

Methotrexate (3.31)

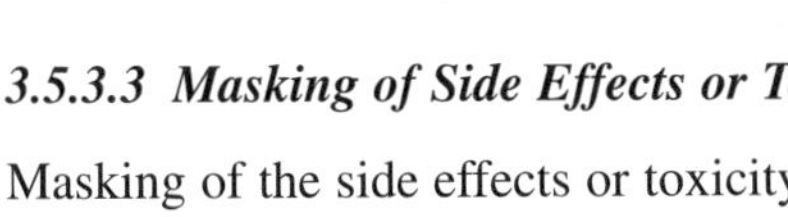

Prontosil (3.32)

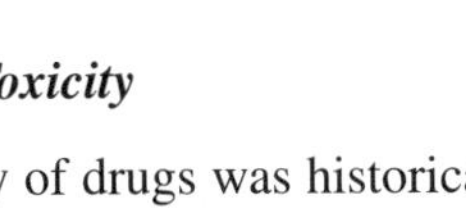

Phenacetin (3.33)

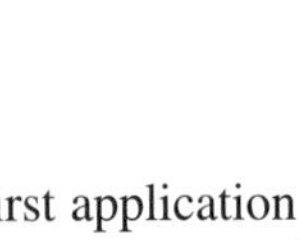

Aspirin (3.34)

3.5.3.3 *Masking of Side Effects or Toxicity*

Masking of the side effects or toxicity of drugs was historically the first application of the prodrug concept. This concept goes back to the turn of the twentieth century, and in fact many prodrugs were not at the time really recognized as such. For instance, castor oil is a laxative because it is hydrolyzed intestinally to the active ricinoleic acid. However, the classical example is prontosil (**3.32**), which undergoes a reduction to sulfanilamide. The analgesic phenacetin (**3.33**) acts in the form of its hydrolysis product, p-acetaminophenol. Another classical example of side-effect masking occurs in aspirin (**3.34**) and its many analogs — the result of a considerable effort to eliminate the gastric bleeding caused by salicylic acid.

Selective bioactivation (toxification) is illustrated in the case of the insecticide malathion (**3.35**). This acetylcholinesterase inhibitor is desulfurized selectively to the toxic malaoxon, but only by insect and not mammalian enzymes. Malathion is therefore relatively nontoxic to mammals (LD_{50} = 1500 mg/kg, rat; p.o.). Higher organisms rapidly detoxify malathion by hydrolyzing one of its ester groups to the inactive acid, a process not readily available to insects. This makes the compound doubly toxic to insects since they cannot eliminate the active metabolite.

3.5.3.4 Improvement of Taste

Taste improvement is quite an important aspect of drug modification, especially in pediatric medicine. The extremely bitter taste of some antibiotics, such as clindamycin (**3.36**) or chloramphenicol (**3.37**), can be masked successfully by preparing esters or pamoate salts of these drugs, which are very insoluble and therefore have no taste.

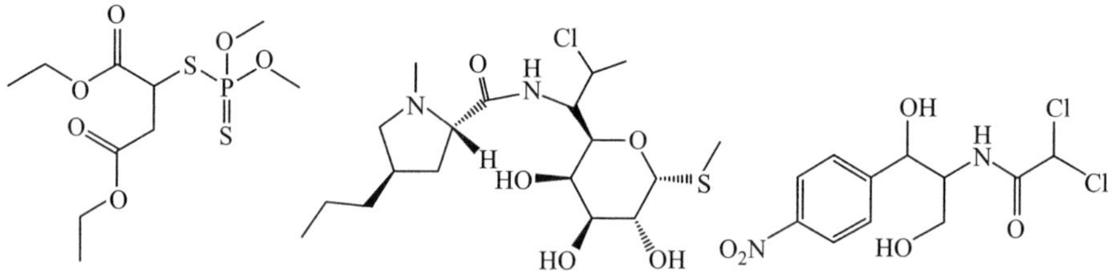

Malathion (3.35)   Clindamycin (3.36)   Chloramphenicaol (3.37)

3.5.4 Innovations in Drug Delivery

Novel drug delivery systems can also have a profound effect on pharmacokinetics, even if they do not involve the use of prodrugs in the classical sense. Novel polymers have permitted the development of membranes with controlled diffusion rates. For example,

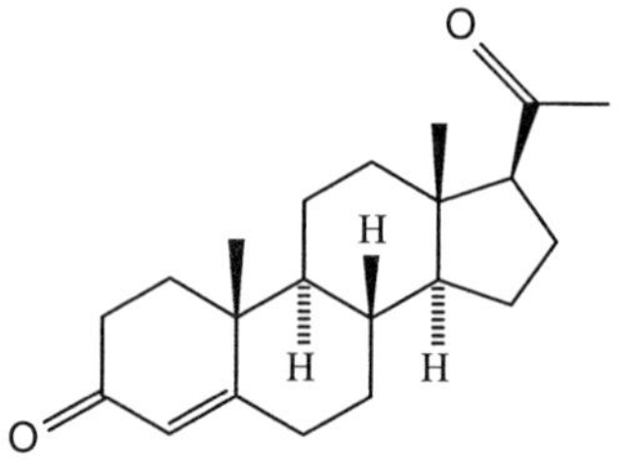

Progesterone (3.38)

pilocarpine (**2.2**), used in the treatment of glaucoma, can be applied in a steady-release ocular insert that lasts for a week. The intrauterine release of progesterone (**3.38**) as a contraceptive has also been achieved, with a single insert lasting a year. The great advantage of this is that the constant release rate of 65 µg/day means that much less drug is released than with the use of oral contraceptive tablets. The transdermal delivery of scopolamine as an antiemetic for motion sickness represents another successful application of microporous membrane technology. Here the drug is applied in a plastic strip similar to a "Band-Aid," usually behind the ear. Low-density lipoproteins and liposomes (drug-filled lipid–cholesterol vesicles measuring a fraction of a micrometer) are also being used to protect drugs from enzymatic destruction during transport in the bloodstream.

Osmotic minipumps — cylinders measuring about 25×5 mm — are widely used to deliver constant amounts of drug solutions to experimental animals. They require surgical implantation. The osmotic compartment swells in contact with tissue fluid and squeezes the drug reservoir, displacing the drug solution in a continuous flow. The rate of delivery is specified by the size of the opening in the container and the swelling rate of the osmotic "syringe."

The great advantage of these systems is the uniform drug delivery they permit, as opposed to the enormous drug level fluctuations inherent with the traditional oral or injected parenteral modes of drug administration. Patient compliance and convenience of use are also ensured. Although these interesting developments in bioengineering are not, strictly speaking, in the realm of drug design or even medicinal chemistry, they can nevertheless contribute substantially to the success of drug therapy.

3.6 FROM LEAD DISCOVERY TO CLINICAL TRIALS: THE CONCEPT OF A "USEFUL DRUG"

The ultimate long-term goal of medicinal chemistry is to design a "useful drug". A useful drug is more than a molecule that is safe and efficacious. It is more than a compound that produces papers in scholarly journals. A useful drug is a drug molecule that is not only safe and efficacious, but also one that can pass government regulations, pass through multiple levels of human clinical trials, be economically produced in large quantities, be successfully marketed, and can ultimately help people with disease. Successfully treating humans with disease is the "bottom line" in drug design.

Perhaps the greatest hurdle along the pathway of a molecule becoming a useful drug is the need to sequentially pass *clinical trials*. However, before a drug can be evaluated in human clinical trials, it must first successfully negotiate *preclinical testing*. This frequently involves five or six types of test, and is completed in non-human animals:

1. Acute toxicity — acute dose that is lethal in 50% of animals; usually two species, usually two routes of administration
2. Subacute toxicity — physiology, histology, autopsy studies; two species, sometimes with dosings over a 6 month time period
3. Chronic toxicity — detailed organ evaluation; two species, sometimes studied for 1–2 years

4. Mutagenic potential — effects on genetic stability of bacteria (Ames test) of mammalian cells in culture
5. Carcinogenic potential — required if drug is to be administered for prolonged periods of time
6. Reproductive performance effects — effects on animal progeny, production of birth defects

Once a molecule successfully passes the preclinical testing, it is ready for human clinical trials. There are four phases of human clinical trial.

Phase 1 — the effects of the drug as a function of dose are measured in a small number (25–45) of healthy volunteers who do not have the disease under study; safety is primarily evaluated.

Phase 2 — the drug is studied in a small number of people (20–150) who have the disease under study; both safety and efficacy are evaluated.

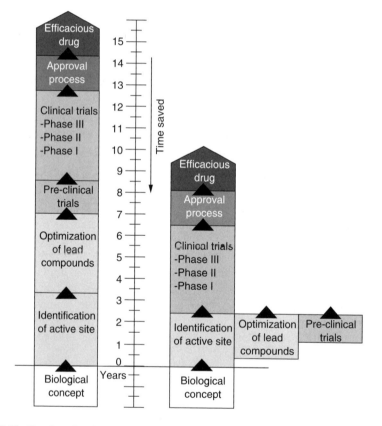

Figure 3.10 The drug development process. The timeline for drug discovery and development is long, adding to the high cost of drug development. Computational methods have helped to shorten this timeline.

Phase 3 — the drug is studied in a large number of people (hundreds to thousands) who have the disease under study, typically using a multi-center, double-blind, placebo–controlled, randomized clinical trial (RCT) protocol.

Phase 4 — once the drug has been approved for market, vigilant post-marketing surveillance is done to ascertain the possible appearance of previously undetected toxicities or problems.

The drug development process is long, as is shown in figure 3.10. During the drug development phases, toxicity is one of the most important hurdles to the success of a drug molecule. Toxicity can affect the person who is taking the medication (causing skin rashes, liver problems, bone marrow failure, etc.) or, in the case of women, can affect their developing fetus if they are pregnant. Table 3.2 lists some of the drugs

Table 3.2 Drugs Producing Adverse Effects on the Fetus

Drug	Effect
ACE inhibitors	Kidney damage
Amphetamines	Abnormal developmental patterns
Androgens	Masculinization of female
Busulfan	Congenital malformations
Carbamazepine	Neural tube defects affecting brain formation
Cocaine	Stroke in fetus
Cyclophosphamide	Congenital malformations
Cytarabine	Congenital malformations
Diethylstilbestrol	Vaginal adenocarcinoma in child
Ethanol	Risk of fetal alcohol syndrome
Etretinate	High risk of multiple congenital malformations
Iodine	Congenital goiter, hypothyroidism
Isotretinoin	High risk of face, ear, and other malformations
Methotrexate	Multiple congenital abnormalities
Methylthiouracil	Hypothyroidism in child
Metronidazole	May be mutagenic (animal studies show no evidence for mutagenic or teratogenic effects in humans)
Penicillamine	Congenital skin malformations
Phenytoin	Fetal hydantoin syndrome
Propylthiouracil	Congenital goiter
Streptomycin	Eighth nerve toxicity (deafness) in child
Tamoxifen	Increased risk of spontaneous abortion or fetal damage
Tetracycline	Discoloration and defects of teeth and altered bone growth
Thalidomide	Phocomelia (shortened bones of the limbs)
Trimethadione	Multiple congenital abnormalities
Valproic acid	Neural tube defects of the brain

Table 3.3 Approved Drugs Withdrawn Because of Toxicity

Drug	Year	Adverse reaction
Astemizole	1998	Interactions (e.g., with grapefruit juice)
Benoxaprofen	1982	Liver damage
Centoxin	1993	Increased mortality
Cerivastatin	2001	Muscle breakdown
Cisapride	2000	Cardiac arrhythmias
Clioquinol	1975	Optic neuropathy (eye problem)
Dexfenfluramine	1997	Cardiac valve abnormalities
Fenfluamine	1997	Cardiac valve abnormalities
Flosequinan	1993	Increased mortality
Indoprofen	1984	Gastrointestinal bleeding/perforation
Metipranolol 0.6% eyedrops	1990	Anterior uveitis (eye problem)
Mibefradil	1998	Many drug interactions
Nomifensine	1986	Hemolytic anemia
Noscapine	1991	Gene toxicity
Remoxipride	1994	Aplastic anemia
Sertindole	1998	Cardiac arrhythmias
Suprofen	1987	Renal impairment
Temafloxacin	1992	Various serious adverse effects
Terodiline	1991	Cardiac arrhythmias
Tolcapone	1998	Hepatobiliary disorders
Triazolam	1991	Psychiatric disorders
Troglitazone	1997	Hepatic disorders
Zimeldine	1983	Hypersensitivity
Zomepirac	1983	Anaphylaxis

that produce an adverse effect on the fetus. Toxicity problems have resulted in many drugs being withdrawn from the market, as shown in table 3.3. As shown in figure 3.11, toxicity problems are a major cause of drug rejection during the development process.

Successful completion of the four phases of human clinical trials enables a drug to be widely distributed for the treatment of human disease. However, a simple reality of drug discovery is that drugs are developed by industry. The lead compound may have been identified in an academic university-based laboratory, but the clinical trials are invariably completed by the industrial sector. Academic institutions, governments, or international organizations (e.g., the World Health Organization, WHO) do not develop drugs. Because of this, drug molecules tend to be developed only if they have a good prospect for being profitable. In order to be profitable, a drug molecule should be patented so that the vendor can enjoy exclusive rights to its marketing. Although a discussion of the criteria for patentability is beyond the scope of this book, the drug

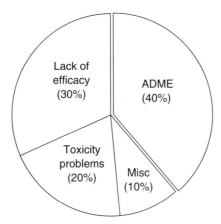

Figure 3.11 Failure of drug molecules. Many drugs that are successful in the pharmacodynamic phase ultimately fail to become useful drugs. This pie-chart presents the reasons for failure at this stage of the development process. Toxicity is an important cause of failure.

molecule should be chemically unique, without evidence of previous pharmacological development.

Selected References

Designing a Lead Compound

D. Baker, A. Sali (2001). Protein structure prediction and structural genomics. *Science 294*: 93–96.

F. A. Bisby (2000). The quiet revolution: biodiversity informatics and the internet. *Science 289*: 2309–2312.

S. Borman (2000). Combinatorial chemistry: redefining the scientific method for drug discovery. *Chem. Eng. News 78*: 53–66.

A. S. V. Burgen and G. C. K. Roberts (Eds.) (1986). *Molecular Graphics and Drug Design. Topics in Molecular Pharmacology*, vol. 3. Amsterdam: Elsevier.

N. C. Cohen (1983). Towards the rational design of new leads in drug research. *Trends Pharmacol. Sci. 4*: 503–506.

J. Drews (2000). Drug discovery: a historical perspective. *Science 287*: 1960–1964.

A. Edwards, C. Arrowsmith, B. Pallieres (2000). Proteomics: new tools for a new era. *Mod. Drug Discov. 3*: 34–45.

C. Ezzell (2002). Proteins rule. *Sci. Amer. 286*: 40–47.

J. Gasteiger (2003). *Handbook of Chemoinformatics*. New York: Wiley.

C. Henry (2001). Pharmacogenomics. *Chem. Eng. News 79*: 37–45.

A. J. Hopfinger (1985). Computer-assisted drug design. *J. Med. Chem. 28*: 1133–1139.

P. N. Kant, P. Daftari (1986). Marine pharmacology: bioactive molecules from the sea. *Annu. Rev. Pharmacol. Toxicol. 26*: 117–142.

P. Krogsgaard-Larsen, S. B. Christensen, H. Kofod (Eds.) (1984). *Natural Products and Drug Development*. Copenhagen: Munksgaard.

E. S. Lander, R. A. Weinberg (2000). Genomics: journey to the center of biology. *Science 287*: 1777–1782.

K. Pal (2000). The keys to chemical genomics. *Mod. Drug Discov. 3*: 46–58.

P. J. Rosenthal (2003). Antimalarial drug discovery: old and new approaches. *J. Exp. Biol. 206*: 3735–3744.

A. M. Rouhi (2003). Pharmaceuticals: rediscovering natural products. *Chem. Eng. News 41*: 77–90.

P. A. Singer, A. Daar (2001). Harnessing genomics and biotechnology. *Science. 294*: 87–89.

N. Sleep (2000). Sorting out combinatorial chaos. *Mod. Drug Discov. 3*: 37–46.

N. J. de Souza, B. N. Ganguli, J. Reden (1982). Strategies in the discovery of drugs from natural sources. *Annu. Rep. Med. Chem. 17*: 301–310.

J. J. Stezowski, K. Chandrasekhar (1986). X-ray crystallography of drug molecule–macromolecule interactions as an aid to drug design. *Annu. Rep. Med. Chem. 21*: 293–302.

E. A. Swinyard (1980). History of the Antiepileptic Drugs. In: G. Glaser, J. Penry, D. Woodbury (Eds.). *Antiepileptic Drugs: Mechanism of Action*. New York: Raven Press. p. 1.

A. S. Verkman (2004). Drug discovery in academia. *Am. J. Physiol. 286*: C465–C474.

Synthesis of a Lead Compound

M. A. Fox, J. K. Whitesell (2004). *Organic Chemistry* 3rd ed. Toronto: Jones and Bartlett, (excellent description of syntheses of diazepam, ibuprofen, sertraline).

N. J. Hrib (1986). Recent development in computer-assisted organic synthesis. *Annu. Rep. Med. Chem. 21*: 303–311.

J. MacCoss, T. Baillie (2004). Organic chemistry in drug discovery. *Science 303*: 1810–1813.

S. L. Schreiber (2000). Target-oriented and diversity-oriented organic synthesis in drug discovery. *Science 287*: 1964–1969.

Optimizing a Lead Compound

T. M. Allen, P. R. Cullis (2004). Drug delivery systems: entering the mainstream. *Science 303*: 1818–1821.

S. C. Basak, D. K. Harris, V. R. Magnuson (1984). Comparative study of lipophilicity versus topological molecular descriptors in biological correlations. *J. Pharm. Sci. 73*: 429–437.

N. Bodor (1982). Soft drugs: strategies for design of safer drugs. In: J. A. Keverling–Buisman (Ed.). *Strategy in Drug Design*. Amsterdam: Elsevier.

N. Bodor (1984). Novel approaches to the design of safer drugs: soft drugs and site–specific delivery systems. In: B. Testa (Ed.). *Advanced Drug Research*, vol. 13. New York: Academic Press, pp. 255–331.

D. E. Clark (2003). *In silico* prediction of blood–brain barrier permeation. *Drug Discov. Today 8*: 927–933.

P. V. Desai, E. C. Coutinho (2001). QSAR in drug discovery and development. *Asian Chem. Lett. 5*: 77–86.

O. Dror, A. Shulman-Peleg, R. Nussinov, H. J. Wolfson (2004). Predicting molecular interactions *in silico*: I. A guide to pharmacophore identification and its applications to drug design. *Curr. Med. Chem. 11*: 71–90.

N. L. Henderson (1983). Recent advances in drug delivery system technology. *Annu. Rep. Med. Chem. 18*: 275–284.

W. L. Jorgensen (2004). The many roles of computation in drug discovery. *Science 303*: 1813–1818.

G. M. Keseru (2003). Structure/function/SAR and molecular design. Mol. Divers. 7: 1.

L. B. Kier, L. H. Hall (1976). *Molecular Connectivity in Structure–Activity Studies.* Letchworth: Research Studies Press.

G. L. Kirschner, B. R. Kowalski (1978). The application of pattern recognition to drug design. In: E. J. Ariëns (Ed.). *Drug Design*, vol. 9. New York: Academic Press, pp. 73–131.

G. Klopman, R. Contreras (1985). Use of artificial intelligence in structure–activity relationships of anticonvulsant drugs. *Mol. Pharmacol. 27*: 86–93.

C. M. Krejsa, D. Horvath, S. L. Rogalski, J. E. Penzotti, B. Mao, F. Barbosa, J. C. Migeon (2003). Predicting ADME properties and side effects: the BioPrint approach. *Curr. Op. Drug Discov. Develop. 6*: 470–480.

H. Kubinyi (1979). Lipophilicity and drug activity. In: E. Jacket (Ed.). *Drug Research*, vol. 23. Basel: Birkhäuser, pp. 97–198.

C. A. Lipinski (1986). Bioisosterism in drug design. *Annu. Rep. Med. Chem. 21*: 283–291.

P. P. Mager (1980). The MASCA model of pharmacochemistry, I. Multi-variate statistics. In: E. J. Ariëns (Ed.). *Drug Design*, vol. 9. New York: Academic Press, pp. 187–236.

G. R. Marshall (1987). Computer-aided drug design. *Annu. Rev. Pharmacol. Toxicol. 27*: 193–213.

E. F. Meyer, Jr (1980). Interactive graphics in medicinal chemistry. In: E. J. Ariëns (Ed.). *Drug Design*, vol. 9. New York: Academic Press, pp. 267–289.

W. M. Pardridge (1985). Strategies for delivery of drugs through the blood–brain barrier. *Annu. Rep. Med. Chem. 20*: 305–313.

R. Perkins, H. Fang, W. Tong, W. J. Welsh (2003). Quantitative structure–activity relationship methods: perspectives on drug discovery and toxicology. *Environ. Toxicol. Chem. 22*: 1666–1679.

M. J. Poznansky, R. L. Juliano (1984). Biological approaches to the controlled delivery of drugs: a critical review. *Pharmacol. Rev. 36*: 277–336.

A. K. Rappe, C. J. Casewit (1996). *Molecular Mechanics Across Chemistry.* Sausalito: University Science Books.

J. N. Simpkins, J. McCormack, K. S. Estes, M. E. Brewster, E. Sheck, N. Bodor (1986). Sustained brain-specific delivery of estradiol causes long-term suppression of luteinizing hormone secretion. *J. Med. Chem. 29*: 1809–1812.

A. A. Sinkula (1975). Prodrug approach in drug design. *Annu. Rep. Med. Chem. 10*: 306–316.

J. G. Topliss (Ed.) (1983). *Quantitative Structure–Activity Relationships in Drugs.* New York: Academic Press.

J. G. Topliss, J. Y. Fukunaga (1978). QSAR in drug design. *Annu. Rep. Med. Chem. 13*: 292–303.

J. G. Topliss, Y. C. Martin (1975). Utilization of operational schemes for analog synthesis in drug design. In: E. J. Ariëns (Ed.). *Drug Design*, vol. 5. New York: Academic Press, pp. 1–21.

F. Torrens (2003). Structural, chemical topological, electrotopological and electronic structure hypotheses. *Comb. Chem. HTS 6*: 801–809.

J. Vaya, S. Tamir (2004). The relation between the chemical structure of flavonoids and their estrogen-like activities. *Curr. Med. Chem. 11*: 1333–1343.

D. A. Winkler (2003). The role of quantitative structure–activity relationships (QSAR) in bio-molecular discovery. *Brief. Bioinf. 3*: 73–86.

A. Xie, C. Liao, Z. Li, Z. Ning, W. Hu, X. Lu, L. Shi, J. Zhou (2004). Quantitative structure–activity relationship study of histone deacetylase inhibitors. *Curr. Med. Chem.: Anti-Cancer Agents 4*: 273–299.

APPENDIX 3.1: BASIC REACTIONS FOR DRUG
MOLECULE SYNTHESIS

A. Building Blocks for the Framework Structure:
C-C, C=C, C≡C, and C-H Bond Formation

C-C Bond Formation

1. Acetoacetic ester synthesis

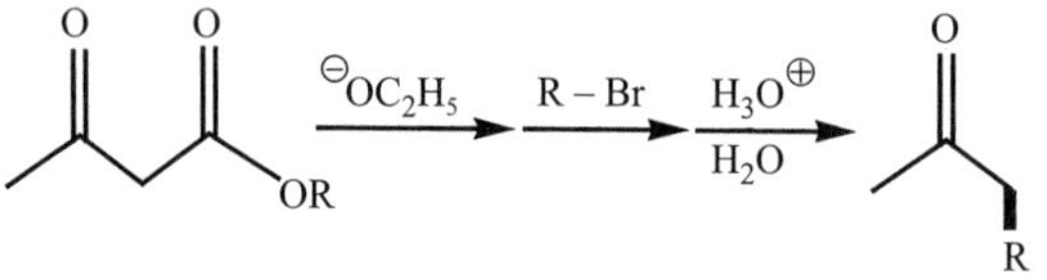

2. Aldol reaction

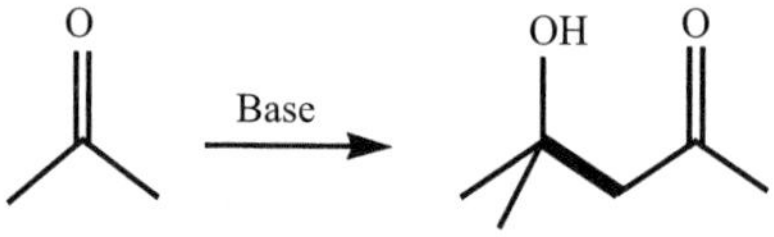

3. Alkylation of ketone enolate anion

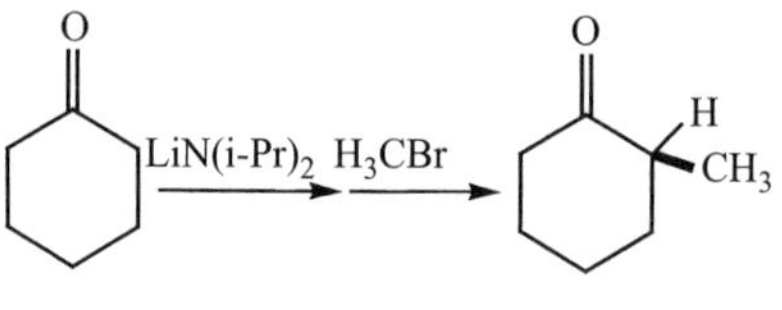

4. Claisen condensation

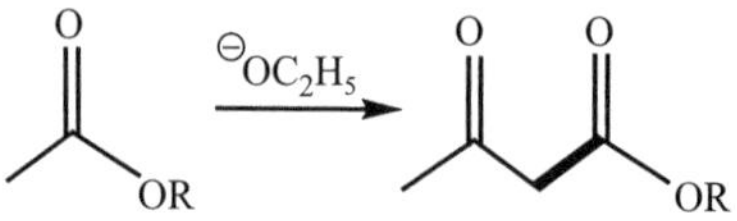

5. Claisen rearrangement

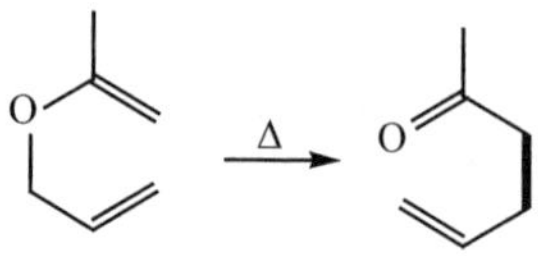

6. Conjugate addition to an α,β-unsaturated carbonyl group

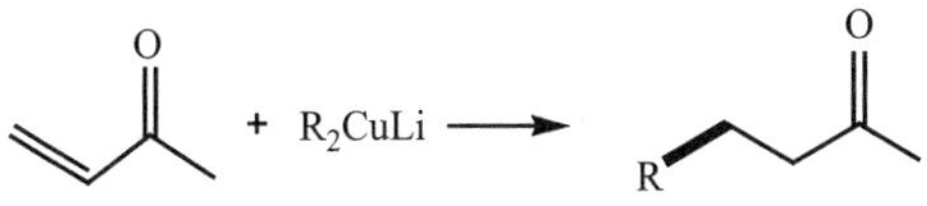

7. Cope rearrangement

8. Diels–Alder reaction

9. Friedel–Crafts alkylation

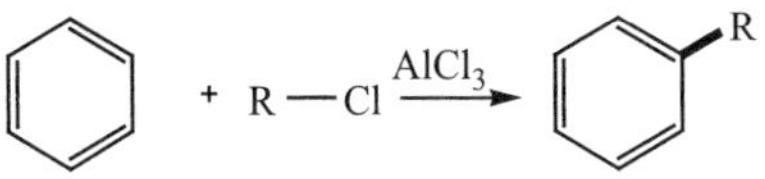

10. Grignard addition

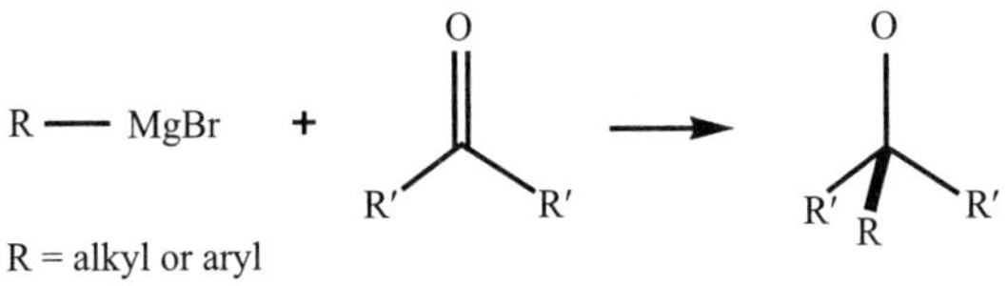

R = alkyl or aryl

11. Malonic ester synthesis

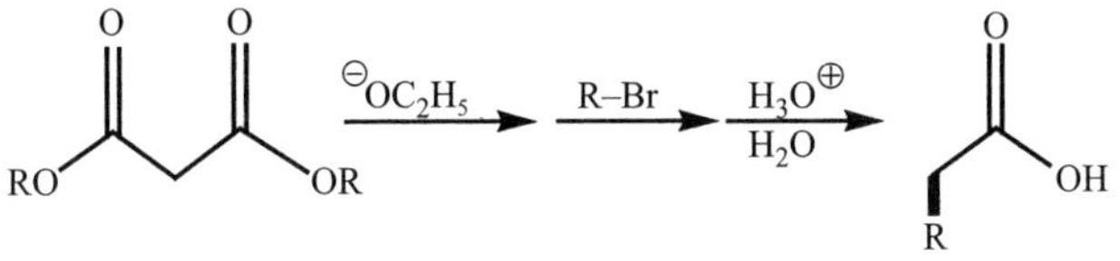

12. Michael reaction

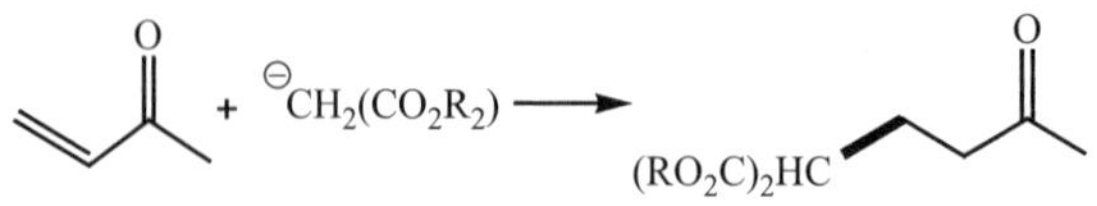

$$\text{(structure)} + {}^{\ominus}\text{CH}_2(\text{CO}_2\text{R}_2) \longrightarrow (\text{RO}_2\text{C})_2\text{HC}\text{(structure)}$$

13. S$_N$2 displacement by cyanide

$$\text{R}\text{—Br} \quad \xrightarrow{{}^{\ominus}\text{C}\equiv\text{N}} \quad \text{R}\text{—C}\equiv\text{N}$$

14. S$_N$2 displacement by acetylide anion

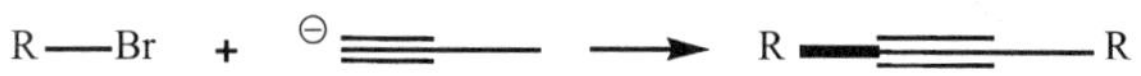

$$\text{R}\text{—Br} \quad + \quad {}^{\ominus}\equiv\text{—} \quad \longrightarrow \quad \text{R}\text{—}\equiv\text{—R}$$

C=C Bond Formation

15. Aldol condensation

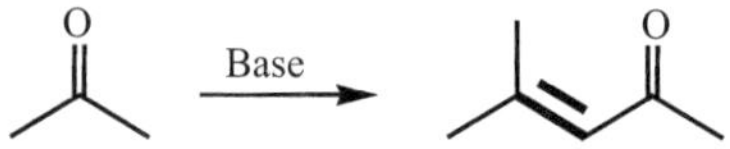

$$\text{(structure)} \xrightarrow{\text{Base}} \text{(structure)}$$

16. Catalytic hydrogenation of an alkyne

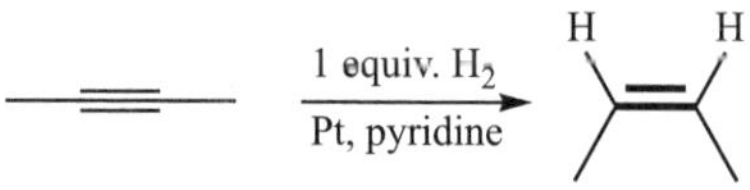

$$\text{—}\equiv\text{—} \quad \xrightarrow[\text{Pt, pyridine}]{\text{1 equiv. H}_2} \quad \text{(structure)}$$

17. Dehydration

$$\text{(structure, OH, H)} \xrightarrow{\text{H}_3\text{O}^{\oplus}} \text{(structure)}$$

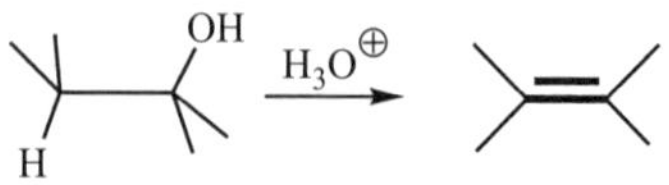

18. Dehydrohalogenation

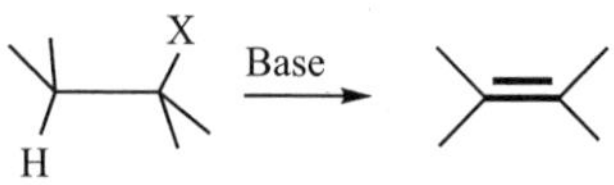

$$\text{(structure, X, H)} \xrightarrow{\text{Base}} \text{(structure)}$$

19. Dissolving metal reduction of an alkyne

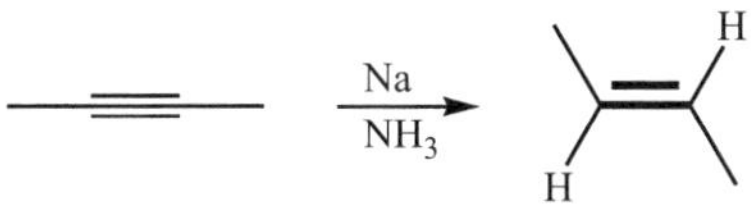

20. Hoffman elimination

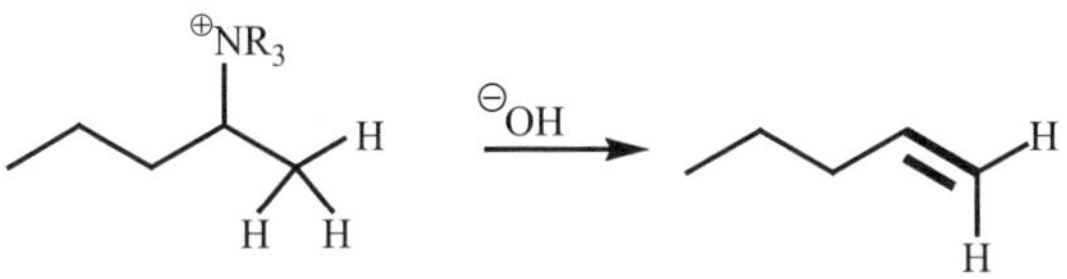

21. Reductive elimination of a vicinal dihalide

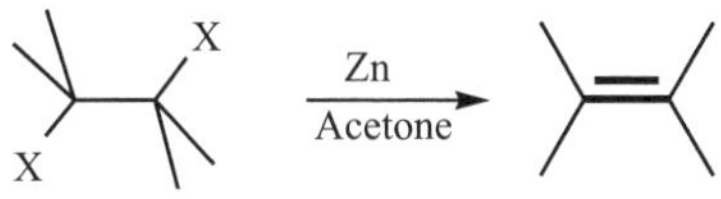

22. Wittig reaction

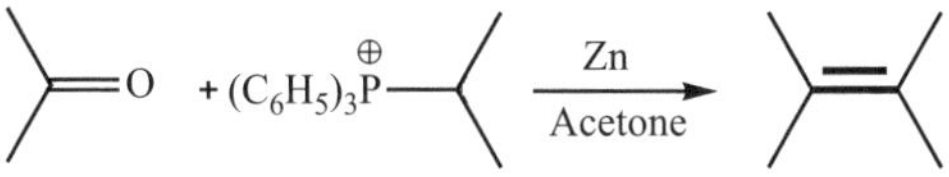

C≡C Bond Formation

23. Dehydrohalogenation

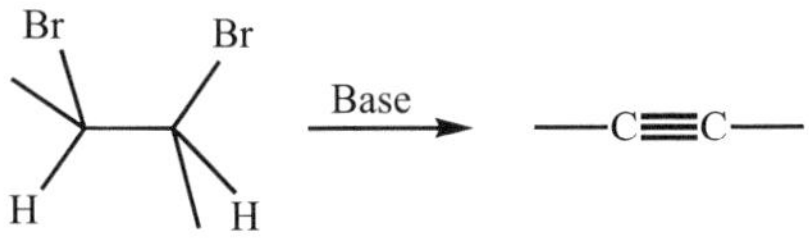

24. S_N2 displacement by an acetylide anion

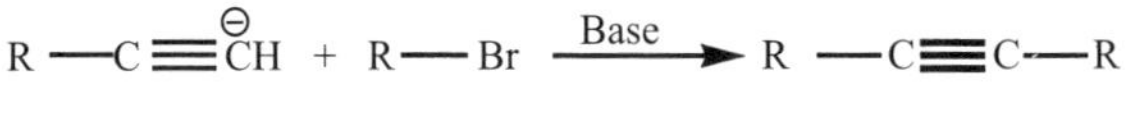

C-H Bond Formation

25. Catalytic hydrogenation of an alkene (or alkyne)

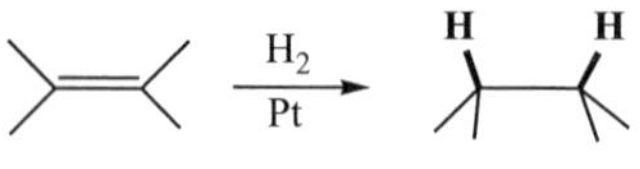

26. Clemmensen reduction

27. Decarboxylation of a β-ketoacid

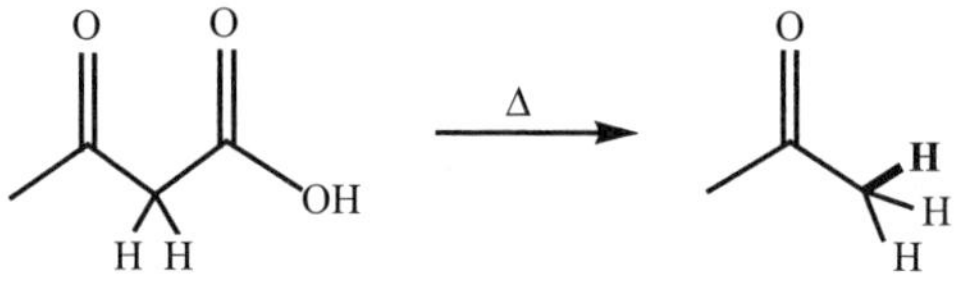

28. Dissolving metal reduction of an alkyne

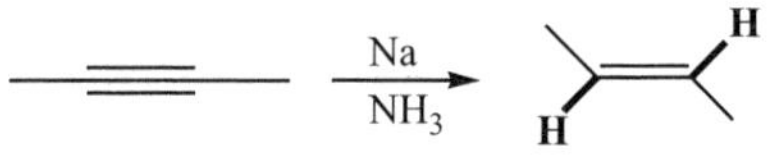

29. Hydrolysis of a Grignard reagent

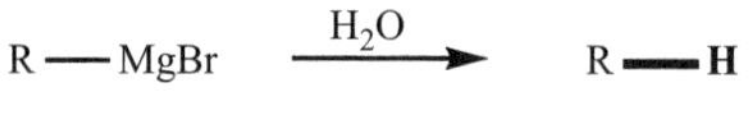

B. Functional Groups for Biophore Creation

-X (halogen) Functional Group

30. Addition of H-X

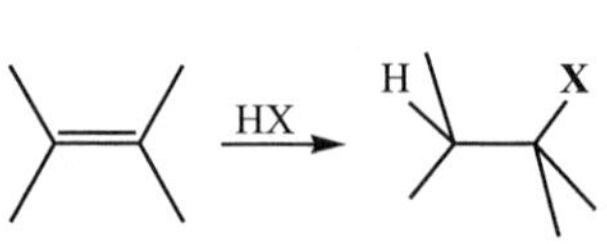

31. Addition of X_2

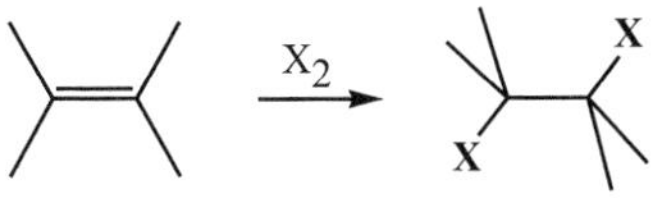

32. Bromination of an alkene

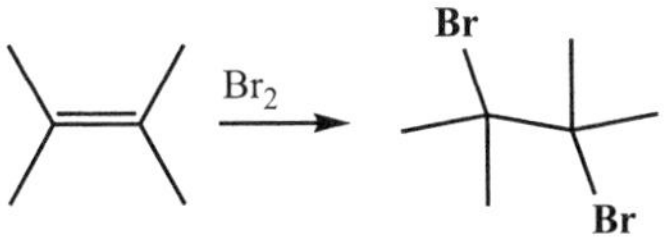

33. Chlorination of an alkene

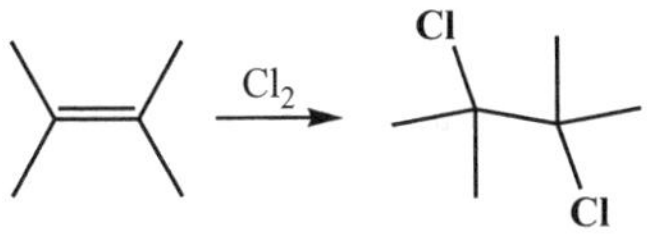

34. Conversion of alcohol to an alkyl halide

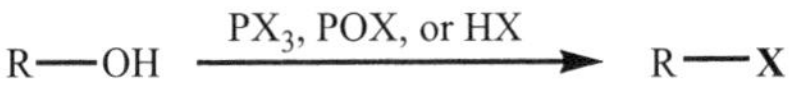

$$R\text{—}OH \xrightarrow{\text{PX}_3,\ \text{POX, or HX}} R\text{—}X$$

35. Electrophilic aromatic substitution

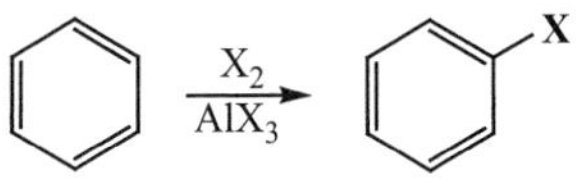

36. Free-radical halogenation

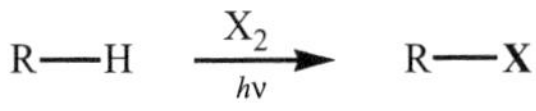

$$R\text{—}H \xrightarrow[h\nu]{\text{X}_2} R\text{—}X$$

37. α-Halogenation of a ketone

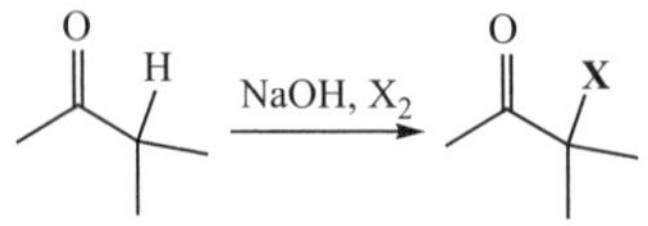

38. Hell–Volhard–Zelinski reaction

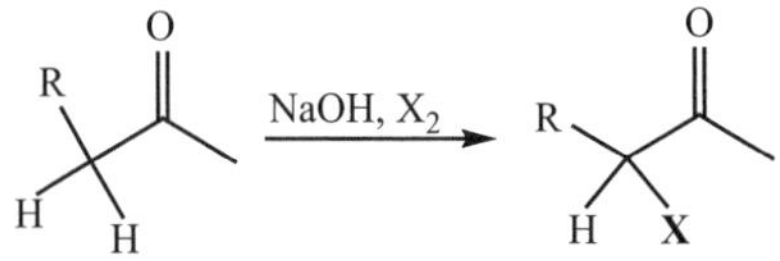

-OH Functional Group

39. Aldol reaction

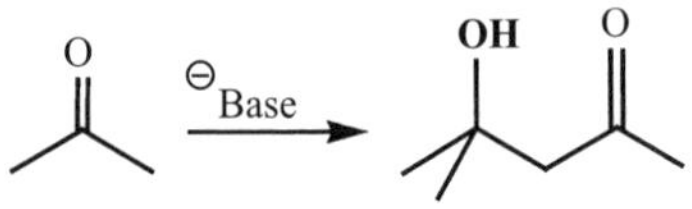

40. Cannizzaro reaction

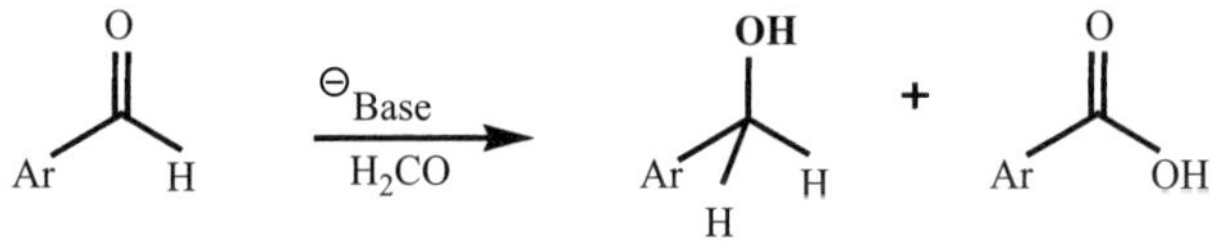

41. Cyanohydrin formation

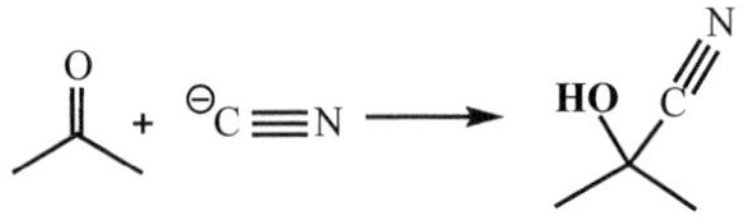

42. Grignard reaction of an aldehyde or ketone

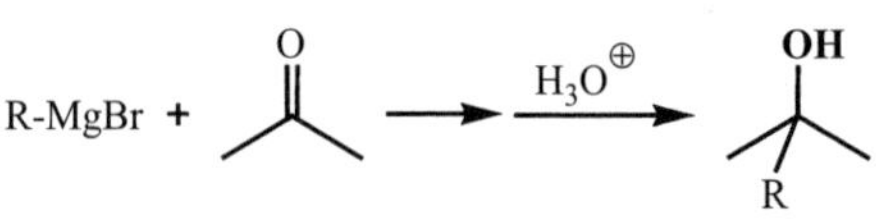

43. Grignard reaction of an ester

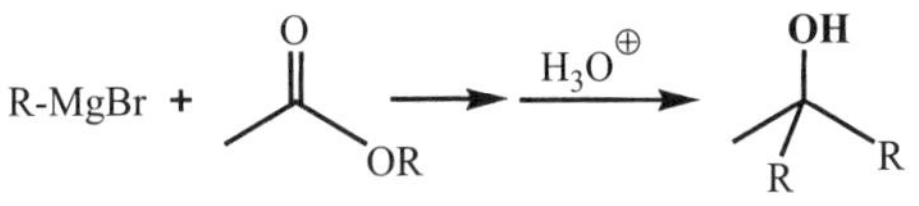

44. Hydration of an alkene (Markovnikov regiochemistry)

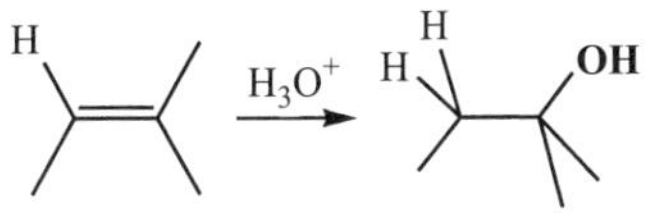

45. Hydroboration–oxidation (anti-Markovnikov regiochemistry)

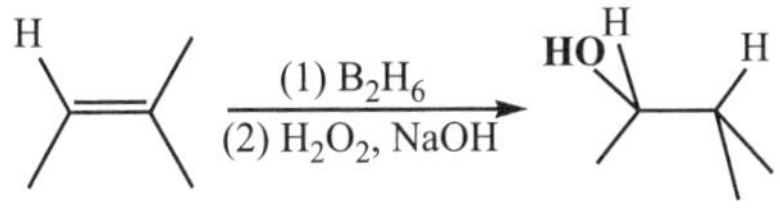

46. Hydrolysis of an alkyl halide

47. *cis*-Hydroxylation

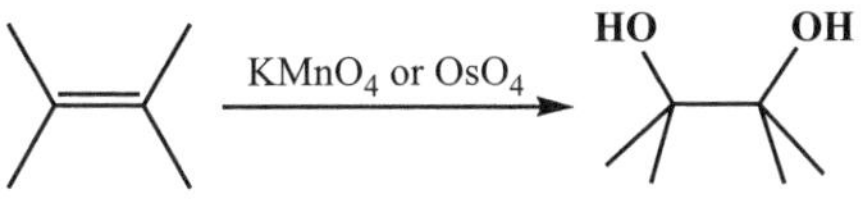

48. Metal hydride reduction of an aldehyde or ketone

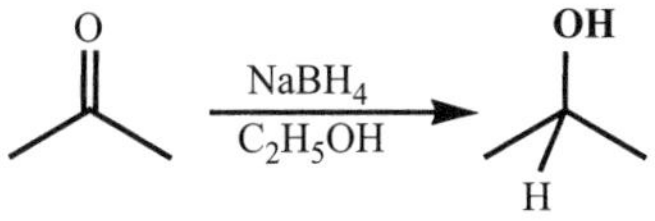

49. Metal hydride reduction of an ester

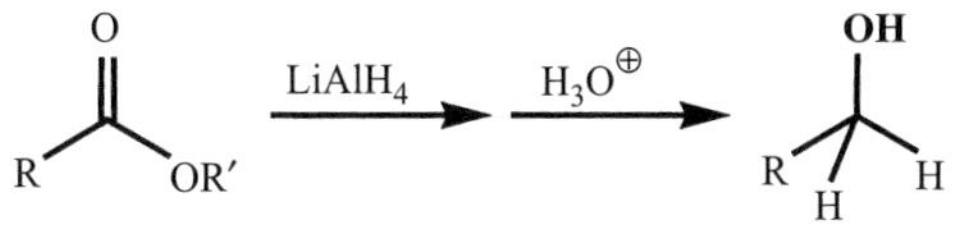

50. Nucleophilic opening of an epoxide

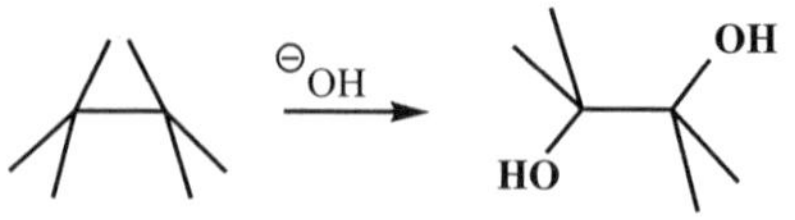

51. Oxymercuration–demercuration (Markovnikov regiochemistry)

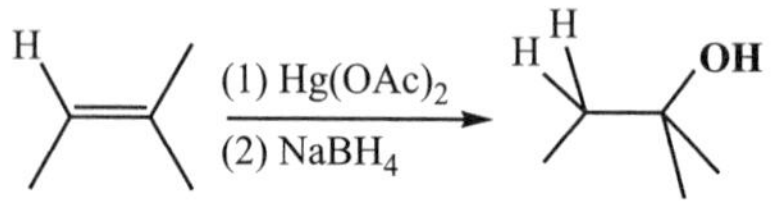

-C≡N Functional Group

52. Cyanohydrin formation

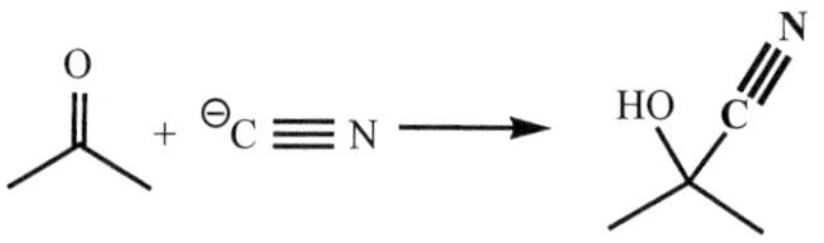

53. Dehydration of an amide

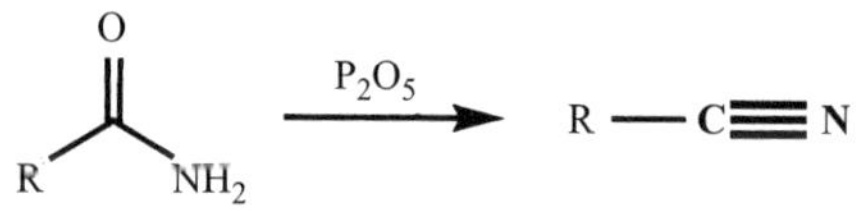

54. S_N2 displacement by cyanide

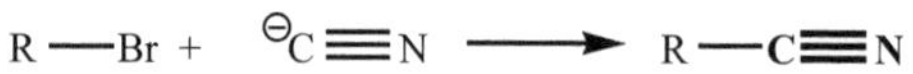

$$R-Br + \ ^{\ominus}C\equiv N \longrightarrow R-C\equiv N$$

-NH₂ Functional Group

55. Aminolysis of an alkyl halide

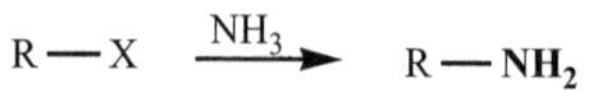

$$R-X \ \xrightarrow{NH_3} \ R-NH_2$$

56. Gabriel synthesis

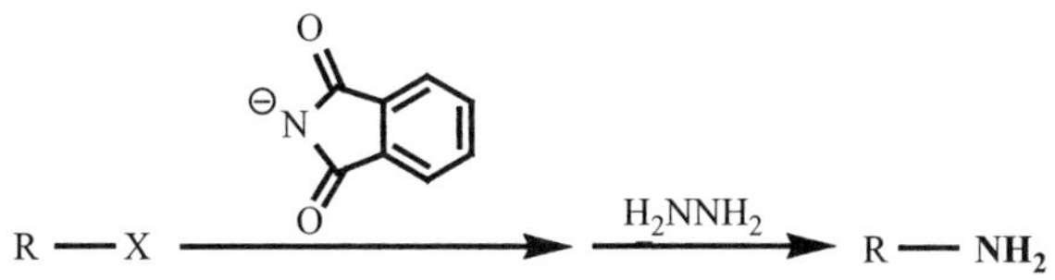

R — X $\xrightarrow{\hspace{2cm}}$ $\xrightarrow{H_2NNH_2}$ R — NH$_2$

57. Hoffmann rearrangement

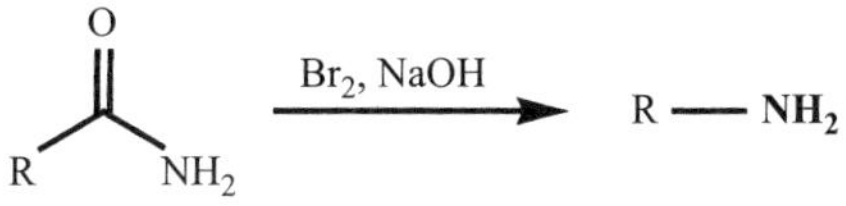

$\xrightarrow{Br_2, NaOH}$ R — NH$_2$

58. Lithium aluminum hydride reduction of an amide

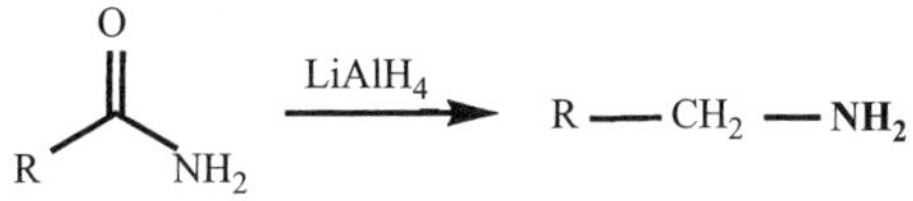

$\xrightarrow{LiAlH_4}$ R — CH$_2$ — NH$_2$

59. Reduction of an aromatic nitro compound

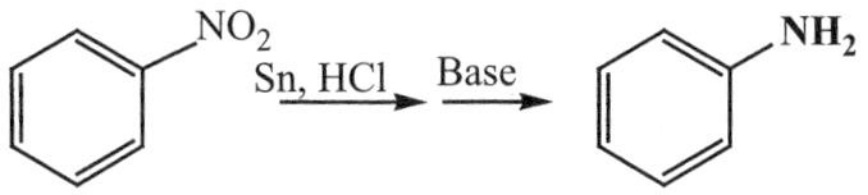

$\xrightarrow{Sn, HCl}$ $\xrightarrow{Base}$

60. Reductive amination of a ketone

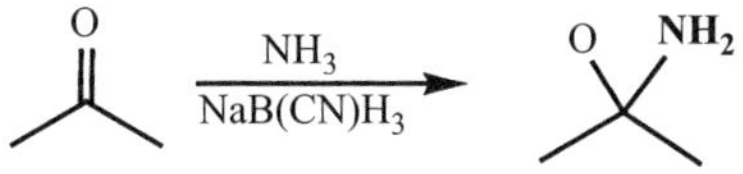

$\xrightarrow[NaB(CN)H_3]{NH_3}$

-C=O Functional Group

61. Claisen condensation

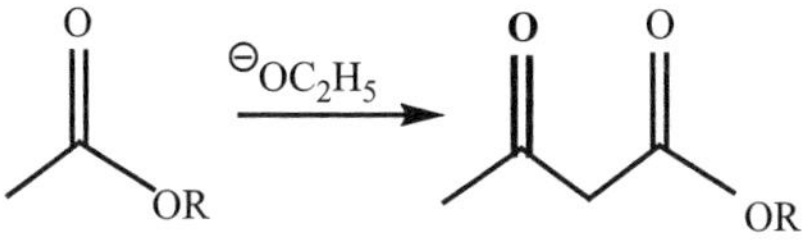

$\xrightarrow{\ominus OC_2H_5}$

62. Claisen rearrangement

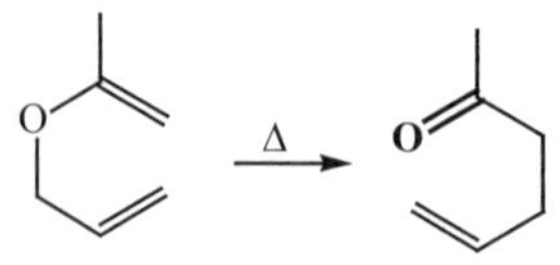

63. Chromate oxidation of a secondary alcohol

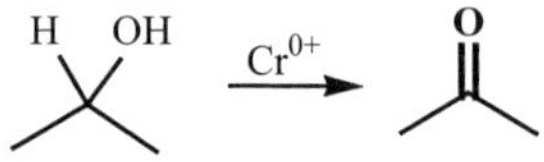

64. Freidel–Crafts acylation

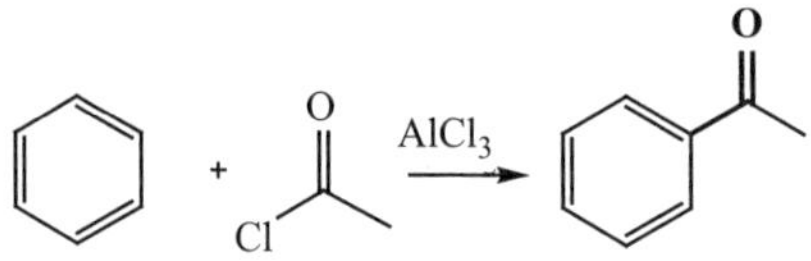

65. Hydrolysis of an acetal

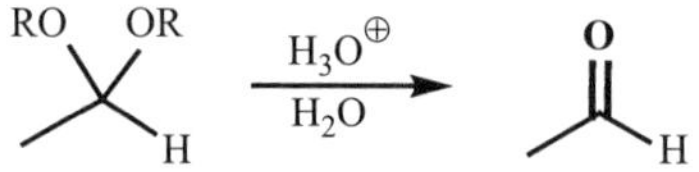

66. Hydrolysis of a ketal

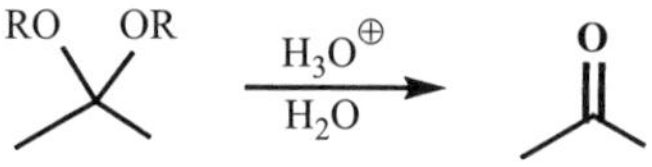

67. Hydrolysis of a terminal alkyne

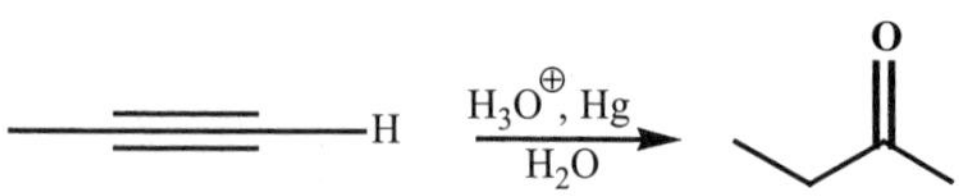

68. Oxidation of a primary alcohol

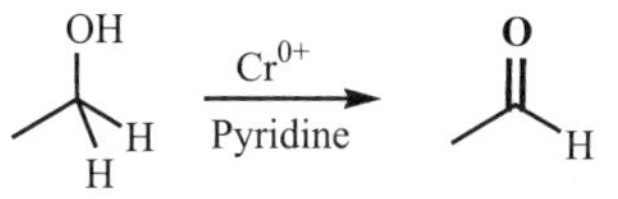

$$\underset{\text{H}}{\overset{\text{OH}}{\lambda_{\text{H}}}} \xrightarrow[\text{Pyridine}]{\text{Cr}^{0+}} \overset{\text{O}}{\lambda_{\text{H}}}$$

69. Ozonolysis of an alkene

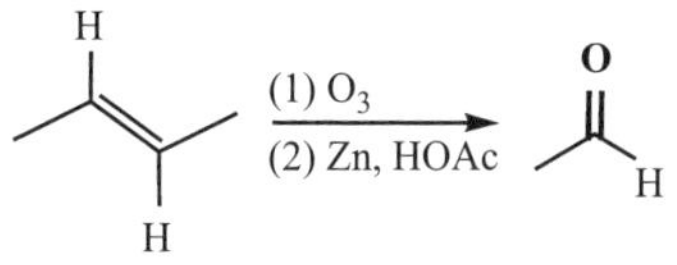

$$\xrightarrow[\text{(2) Zn, HOAc}]{\text{(1) O}_3} \overset{\text{O}}{\lambda_{\text{H}}}$$

70. Pinacol rearrangement

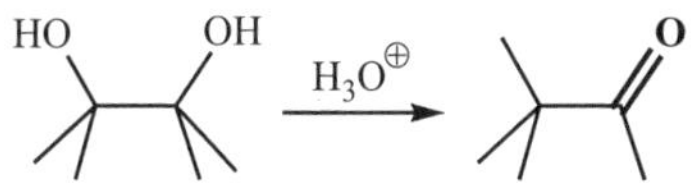

$$\underset{}{\overset{\text{HO} \quad \text{OH}}{\lambda}} \xrightarrow{\text{H}_3\text{O}^{\oplus}} \overset{\text{O}}{\lambda}$$

-COOH Functional Group

71. Benzilic acid rearrangement

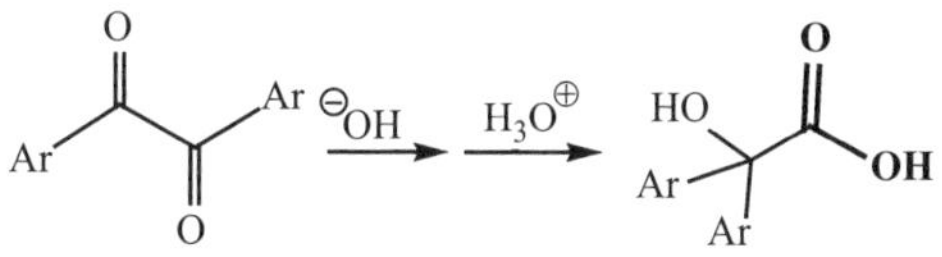

$$\underset{\text{Ar}}{\overset{\text{O}}{\lambda}}\overset{\text{Ar}}{\underset{\text{O}}{\lambda}} \xrightarrow{\ominus\text{OH}} \xrightarrow{\text{H}_3\text{O}^{\oplus}} \underset{\text{Ar}}{\overset{\text{HO}}{\underset{\text{Ar}}{\lambda}}}\overset{\text{O}}{\underset{}{\lambda_{\text{OH}}}}$$

72. Carboxylation of a Grignard reagent

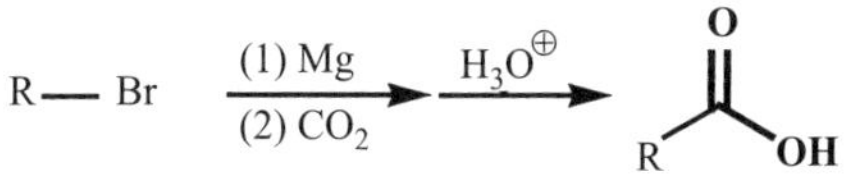

$$\text{R}\text{---}\text{Br} \xrightarrow[\text{(2) CO}_2]{\text{(1) Mg}} \xrightarrow{\text{H}_3\text{O}^{\oplus}} \underset{\text{R}}{\overset{\text{O}}{\lambda_{\text{OH}}}}$$

73. Hydrolysis of a carboxylic acid derivative

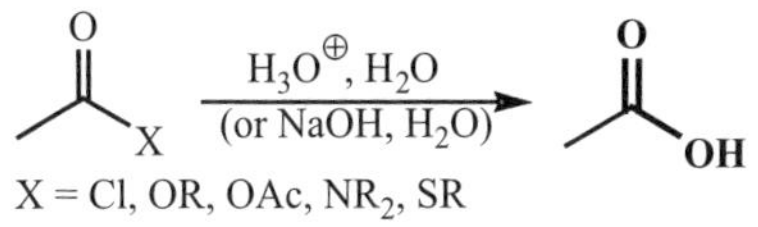

$$\underset{\text{X}}{\overset{\text{O}}{\lambda}} \xrightarrow[\text{(or NaOH, H}_2\text{O)}]{\text{H}_3\text{O}^{\oplus}, \text{H}_2\text{O}} \overset{\text{O}}{\lambda_{\text{OH}}}$$

X = Cl, OR, OAc, NR$_2$, SR

74. Hydrolysis of a nitrile

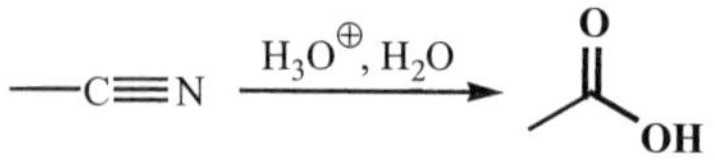

$$-C\equiv N \xrightarrow{H_3O^{\oplus}, H_2O} \overset{O}{\underset{OH}{\bigcup}}$$

75. Iodoform reaction

$$\overset{O}{\underset{CH_3}{\bigcup}} \xrightarrow{I_2,\ NaOH} \xrightarrow{H_3O^{\oplus}} \overset{O}{\underset{OH}{\bigcup}}$$

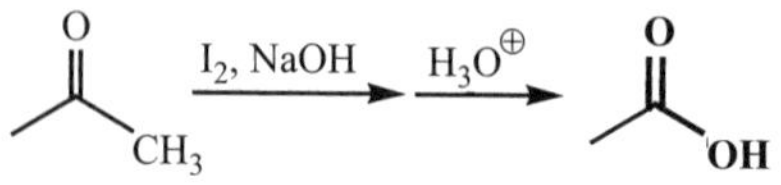

76. Oxidation of a primary alcohol

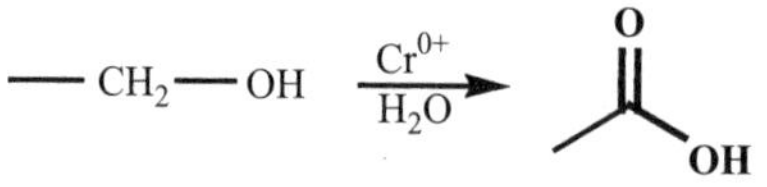

$$-CH_2-OH \xrightarrow[H_2O]{Cr^{0+}} \overset{O}{\underset{OH}{\bigcup}}$$

77. Oxidation of an aldehyde

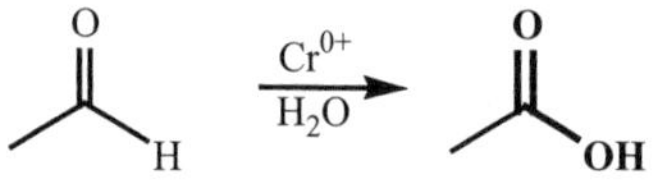

$$\overset{O}{\underset{H}{\bigcup}} \xrightarrow[H_2O]{Cr^{0+}} \overset{O}{\underset{OH}{\bigcup}}$$

78. Permanganate oxidation of an alkyl side chain of an arene

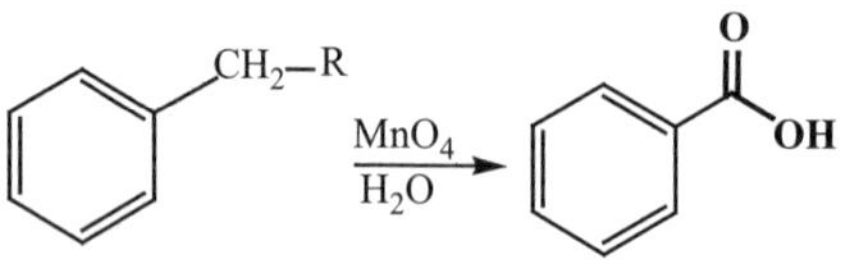

$$\underset{}{\text{C}_6\text{H}_5-CH_2-R} \xrightarrow[H_2O]{MnO_4} \underset{}{\text{C}_6\text{H}_5-\overset{O}{\underset{OH}{\bigcup}}}$$

-COOR Functional Group

79. Baeyer–Villiger oxidation

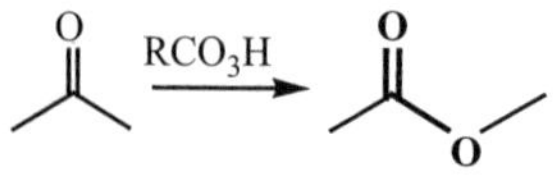

$$\overset{O}{\underset{}{\bigcup}} \xrightarrow{RCO_3H} \overset{O}{\underset{O}{\bigcup}}$$

80. Esterification of a carboxylic acid

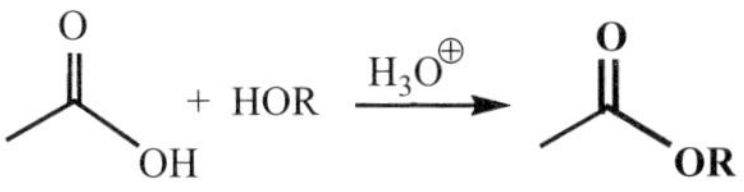

81. Transesterification

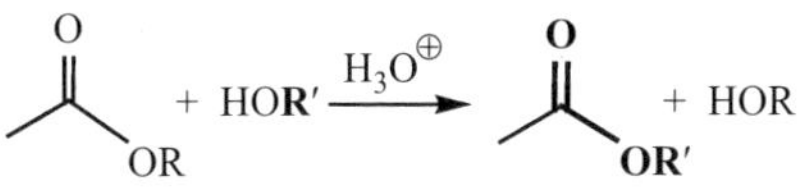

-COCl Functional Group

82. Treatment of an acid with thionyl chloride

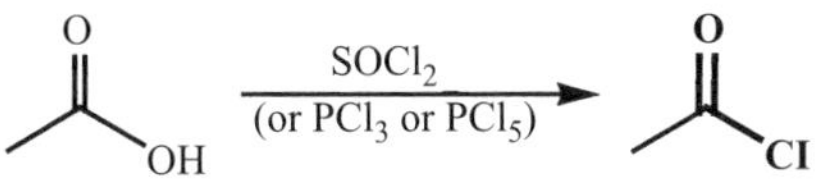

-CONH₂ Functional Group

83. Amidation of a carboxylic acid derivative

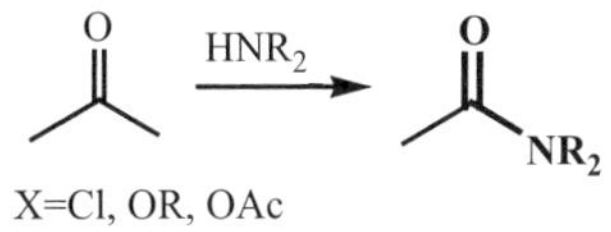

X=Cl, OR, OAc

84. Beckmann rearrangement

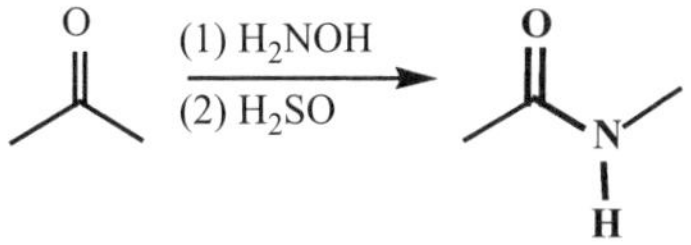

R-OR′ Functional Group

85. Peracid oxidation of an alkene

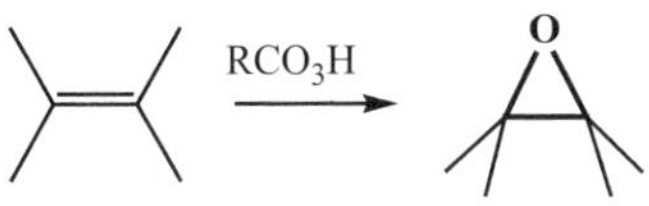

86. Williamson ether synthesis

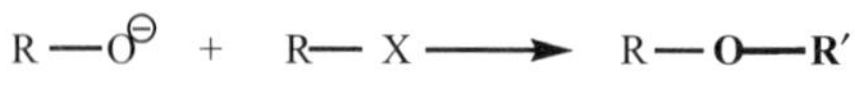

$$R - O^{\ominus} \quad + \quad R - X \longrightarrow \quad R - O - R'$$

Aromatic Functionalities

87. Clemmensen reduction or Wolff–Kishner reduction

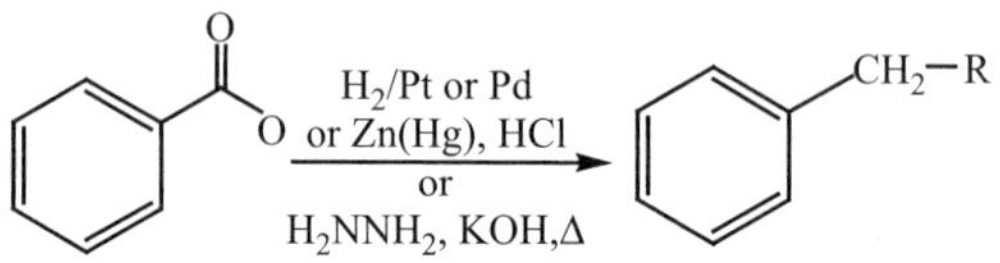

88. Cyanation via diazomium ion

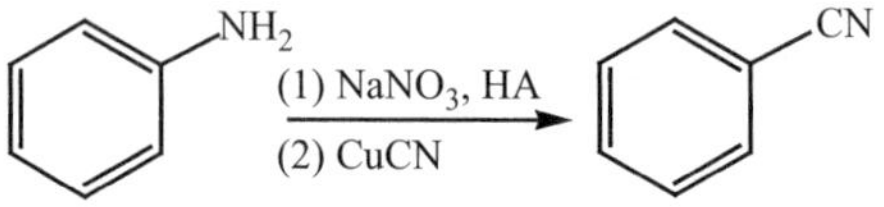

89. Amidation of a carboxylic acid derivative

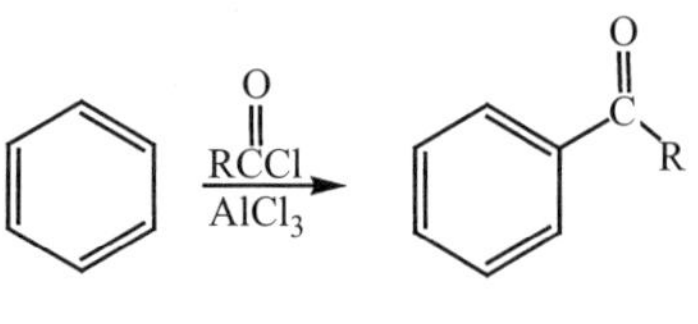

90. Halogenation

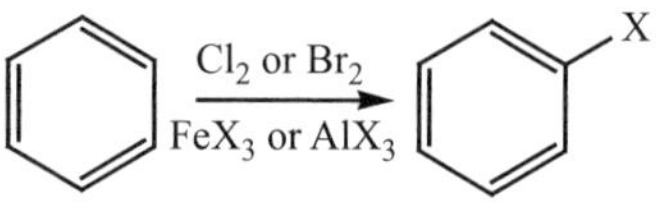

91. Reduction via diazonium ion

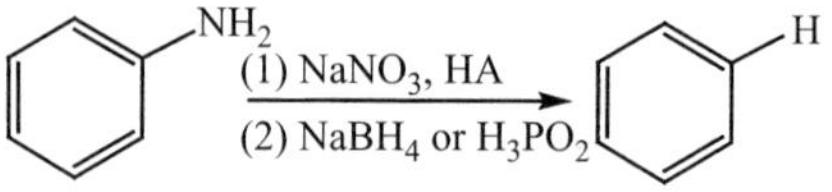

92. Reduction of nitro groups

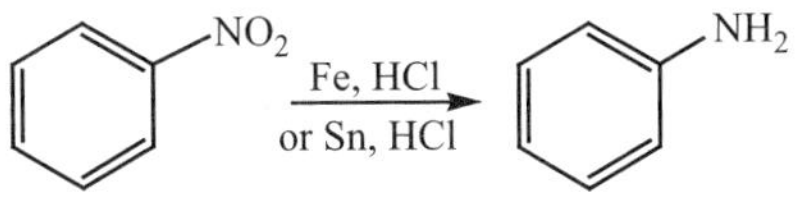

93. Nitration

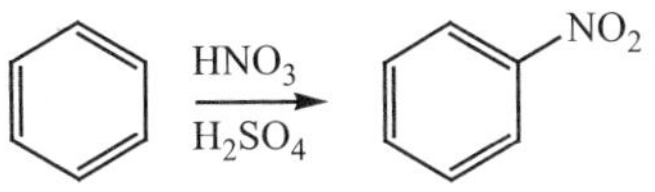

94. Oxidation of alkyl group

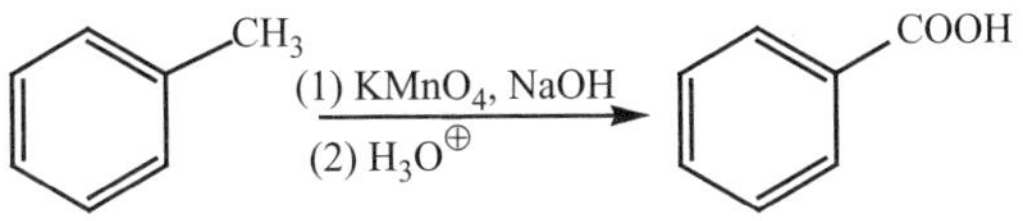

95. Sulfonation

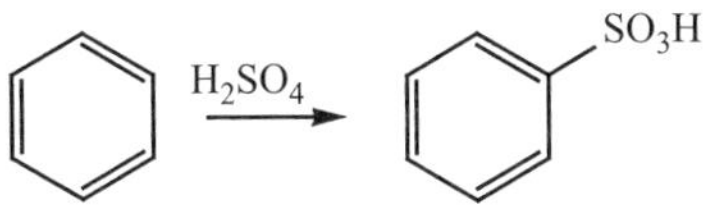

96. Nucleophilic aromatic substitution via diazonium ion

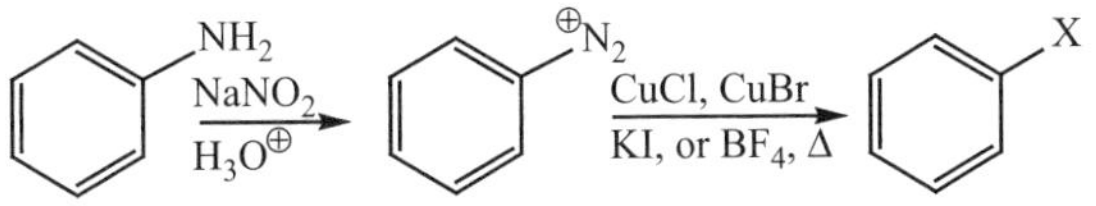

Other Functional Groups

97. Acid dehydration

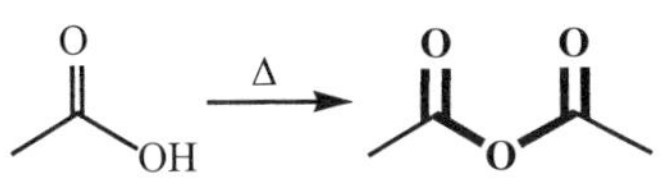

98. Enamine formation

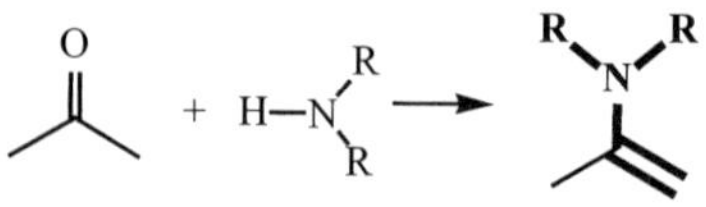

99. Imine formation

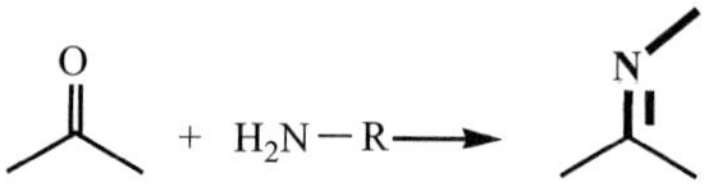

100. Ketal (acetal) formation

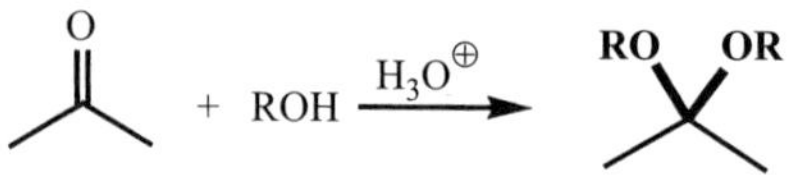

II
BIOCHEMICAL CONSIDERATIONS IN DRUG DESIGN

From Druggable Targets to Diseases

Introduction to Part II

Part I described the basics of medicinal chemistry and drug design. These principles follow three basic activities:

1. Design a molecule with drug-like properties.
2. Select a receptor that is a potential drug target.
3. Use design methods to identify and optimize a lead compound as a prototype drug.

From this sequence of activities, a number of observations can be made. Not all molecules are drugs, but certain properties enable a molecule to be a drug-like molecule and potentially a drug. Analogously, not all macromolecules are receptors, but certain properties enable a macromolecule to be a druggable target. With these two basic facts in place, it is then necessary to understand how to design drug-like molecules that can specifically interact with drug targets. This process involves lead compound identification (via rational drug design or high throughput screening, using either random or focused libraries) followed by lead compound optimization (via quantitative structure–activity relationship studies). Throughout the full spectrum of this design process, computer-aided drug design with molecular mechanics and molecular orbital calculations is an important design tool.

In order to discover therapeutic compounds, it is next necessary to connect these general design principles to the specific reality of human disease. This means that the medicinal chemist must have an organized knowledge of relevant biology and biochemistry and must be able to integrate this knowledge with the principles of drug design, thereby enabling the development of a therapeutic molecule. But how does one select a potential drug target for the disease being studied? This requires a conceptual approach for drawing relationships between druggable targets and specific human diseases. Such an approach is essential to enable the mechanistic connection between a disease and a molecule. There are many approaches by which this conceptual connection can be developed.

THE PHYSIOLOGICAL SYSTEMS APPROACH

One approach for connecting diseases to molecules is to focus on the ten fundamental physiological systems of the human body and the particular diseases associated with these systems:

1. Cardiovascular system (angina, myocardial infarction, arrhythmias, arterial hypertension, valvular heart disease)
2. Dermatological system (erythroderma, icthyosis, Stevens–Johnson syndrome, Behcet's disease, acute blistering diseases)
3. Endocrine system (Cushing's disease, Addison's disease, carcinoid syndrome, diabetes, hyperthyroidism, Grave's disease, hypothyroidism)
4. Gastrointestinal system (inflammatory bowel disease [ulcerative colitis, Crohn's disease], peptic ulcer, pancreatitis, cholecystitis, hepatitis, choledocholithiasis)
5. Genitourinary system (nephrologic—glomerulonephritis, chronic renal failure; urological—benign prostatic hypertrophy, prostatitis)
6. Hematological system (anemia, polycythemia, thrombocytopenia, leukemia, lymphoma, multiple myeloma)
7. Immune system (allergic rhinitis, polymyositis, autoimmune diseases [systemic lupus erythmatosus], graft vs. host disease)
8. Musculoskeletal system (rheumatoid arthritis, ankylosing spondylitis, Sjogren's syndrome, osteoporosis)
9. Nervous system (dementia, stroke, epilepsy, extrapyramidal diseases [Parkinson's], demyelinating diseases [multiple sclerosis], neuropathy, myasthenia gravis, psychosis, schizophrenia)
10. Respiratory system (chronic obstructive pulmonary disease [COPD; emphysema, chronic bronchitis], acute obstructive lung disease [asthma], chronic restrictive lung disease [connective tissue lung disease])

This is a "top-down" classification system. It starts with the disease and works down to the biochemical and molecular processes involved in the disease. There are advantages and disadvantages to such a classification system. It has the same organizational lines as conventional medicine. When designing drugs for the nervous system for example, the rules for designing the drug to enter the brain will apply to all molecules being designed. Nevertheless, this classification system is not ideal in connecting "disease to molecule." For example, when designing drugs for the cardiovascular system, many different receptors (adrenergic, cholinergic) and many different pathological processes (atherosclerosis, inflammation) are involved.

THE PATHOLOGICAL PROCESS APPROACH

A second conceptual approach is a histopathological classification system. This system categorizes disease processes in terms of damage at the tissue and cellular level. This approach focuses on ten fundamental pathological processes (conveniently designated with the THIND[2] mnemonic):

1. Traumatic (pathology from injury)
   External source (destructive physical injury; e.g., motor vehicle accident)
   Internal source (effects of high blood pressure on arterial walls)
2. Toxic (pathology from poisons)
   Biological toxins (snake venom)
   Chemical/physical toxins (toxic chemicals, radiation)
3. Hemodynamic/vascular (pathology from disorders of blood vessels)
   Ischemic (decreased blood supply to an organ or tissue; e.g., stroke or heart attack)
   Hemorrhagic (excessive bleeding from a ruptured blood vessel; e.g., ruptured aneurysm)
4. Hypoxic (pathology from inadequate supply or excessive demand for oxygen by a tissue)
   Generalized/organ specific (lung disease, anemia, decreased blood supply)
   Cellular hypoxia (cyanide poisoning of electron transport chain in mitochondria)
5. Inflammatory (pathology from abnormal inflammatory response in the body)
   Autoimmune and/or chronic diseases (systemic lupus erythmatosus, rheumatoid arthritis)
   Response to environmental triggers (asthma)
6. Infectious (pathology from microbes or infectious agents)
   Prions, viruses, bacteria, fungi, parasites (pneumonia, meningitis, gastroenteritis)
7. Neoplastic (pathology from tumors, cancer)
   Carcinoma (adenocarcinoma of the breast)
   Sarcoma (osteogenic sarcoma)
8. Nutritional (pathology from too much/too little food intake)
   Deficiency (vitamin deficiency secondary to reduced intake)
   Excess (obesity leading to diabetes)
9. Developmental (pathology in the chemistry of heredity)
   Inborn errors of metabolism (Fanconi's syndrome, cystinuria)
   Genetic diseases (Huntington's disease)
10. Degenerative (pathology from age-related tissue breakdown)
   Protein misfolding diseases (Alzheimer's dementia, prion diseases)
   Apoptosis (pre-programmed cell death)
   Mechanical "wear-and-tear" (osteoarthritis [or, more correctly, osteoarthropathy])

This classification system is based on a traditional pathology approach to disease with emphasis on *etiology* (causative factors) and *pathogenesis* (mechanism of disease, particularly at a cellular level). There are strengths and weaknesses in this system as well. The strengths are many. For example, drug design that targets neoplasia may lead to drugs with many applications, including lung cancer, bowel cancer, or brain cancer. Likewise, drug design that targets inflammation could have applications to many different diseases, affecting many organ systems. Nevertheless, this approach focuses more on cellular targets than on molecular targets.

THE MOLECULAR MESSENGER AND NONMESSENGER
TARGET SYSTEM

A third conceptual approach, the one favored in this book, is to focus on the biochemical and molecular processes of human disease. The authors call this system the *Messenger and Non-Messenger Target System* (MANMETS). It may be classified as follows:

1. Messenger targets I—Neurotransmitters
   a. Acetylcholine and cholinergic receptors (nicotinic, muscarinic [M1, M2, ...])
   b. Norepinephrine and adrenergic receptors (α, β)
   c. Dopamine and dopaminergic receptors (D1, D2, ...)
   d. Serotonin and serotonergic receptors (HT1, HT2, ...)
   e. Histamine and histaminergic receptors (H1, H2, ...)
   f. Gamma-aminobutyric acid and GABAergic receptors (GABA-A, GABA-B, ...)
   g. Glutamate and glutamatergic receptors (NMDA, AMPA, kainate)
   h. Glycine and glycinergic receptors
   i. Neuropeptides and peptidergic receptors
       Opioid peptides
       Neurokinins
       Neuropeptide Y
       Galanin
       Cholecystokinin
   j. "Neurogases" and associated receptors
       NO
       CO
   k. Taurine and beta-alanine
   l. Purines and adenosine
2. Messenger targets II—Hormones
   a. Steroid hormones and their receptors
       Estrogens
       Progestins
       Androgens
       Adrenal steroids
   b. Peptide hormones and their receptors
       Pituitary neurohormones
       Oxytocin and vasopressin
       Insulin and glucagon
       Renin–angiotensin hormones
3. Messenger targets III—Immunomodulators
   a. Immunosuppressants and their receptors
   b. Immunomodulators/immunostimulants and their receptors
4. Non-messenger targets I—Endogenous cellular structures
   a. Membrane targets
       Membrane lipids
       Membrane proteins

Voltage-gated ion channels
Ligand-gated ion channels
G-proteins
  b. Cytoplasmic organelle targets
    Mitochondrial targets
    Rough endoplasmic reticulum
    Smooth endoplasmic reticulum
  c. Nuclear targets
    Targeting DNA replication
    Targeting transcription
    Targeting translation
    Targeting mitosis/meiosis
5. Non-messenger targets II—Endogenous macromolecules
  a. Proteins
    Enzyme proteins
    Hydrolases
      Amidases (proteases)
      Esterases (lipases)
    Ligases
      Carboxylases
      Synthetases
    Lyases
      Decarboxylases
      Dehydrases
    Oxidoreductases
      Oxidases
      Reductases
      Dehydrogenases
    Transferases
      Kinases
    Transaminases
    Enzyme cofactors
      Vitamins
    Non-enzyme proteins
      Abnormal folding proteins (amyloid)
      Growth factors (nerve growth factor)
      Endogenous proteins from other animals (snail conotoxins)
  b. Nucleic acids
  c. Lipids
  d. Carbohydrates
  e. Heterocycles
  f. Inorganics
6. Non-messenger targets III—Exogenous pathogens
  a. Microbes
    Prions
    Viruses

         Bacteria
         Fungi
         Parasites
     b. Environmental toxins
         Biological
         Chemical
             Organic
             Inorganic
         Physical

   Within each of these categories there is a further refinement of targets. As discussed
in chapter 9, for example, possible druggable targets for antifungal drug design may be
subdivided as follows:

1. Fungal membrane disruptors via mechanical insertion
2. Ergosterol biosynthesis inhibitors via 14α-demethylase enzyme inhibition
3. Ergosterol biosynthesis inhibitors via squalene epoxidase enzyme inhibition
4. Ergosterol biosynthesis inhibitors via Δ^{14}-reductase enzyme inhibition
5. Fungal cell wall disruptors

   These subclassifications are given in detail in the corresponding chapters (4–9). (As
an extension to the MANMETS system, the authors devised a further classification
system based entirely on molecular structure. Drug molecules were divided into acyclic
and cyclic structures, which were then further subdivided. For example, the cyclic molecules were categorized into steroids, heterocycles, and so on. These were then subcategorized; for instance, heterocycles had many subcategories including benzodiazepines,
imidazolidinediones, dihydropyridines, etc. Regrettably, this classification system is too
extensive and too cumbersome to be useful. Furthermore, the connection to biological
intuition is lost in such a system. Accordingly, it will not be presented in this book.)
   The MANMETS system is a "bottom-up" classification system. It starts at the level
of the biomolecule and works up to the pathological processes (traumatic, toxic, …),
then to the physiological systems (cardiovascular, endocrine, …) and ultimately to the
diseases affecting these systems. Because of its molecular-based approach, it offers definite advantages for drug design. MANMETS classifies the molecular targets that are
biochemically relevant to human disease and to drug design.
   The goal of medicinal chemistry is to design novel chemical compounds that will
favorably influence ongoing biochemistry in the host organism in some beneficial manner.
As discussed in chapters 4–6, one of the most obvious approaches is to either mimic or
block endogenous messengers used by the organism itself to control or alter its own biochemistry. These endogenous messengers may be neurotransmitters (fast messengers),
hormones (intermediate), or immunomodulators (slow), working at the electrical, molecular, or cellular levels, respectively. However, not all pathologies that afflict the human
organism can be addressed by manipulation of these messengers. Accordingly, it becomes
necessary to directly target other cellular components and/or endogenous macromolecules
that are not normally directly controlled through binding to endogenous messengers.

The identification of such cellular targets for drug design necessitates an appreciation of the anatomy of cellular structure. These endogenous macromolecules are the catalysts and molecular machinery that enable the cell to perform its normal metabolic functions; accordingly, they afford numerous druggable targets. Finally, if that approach is not sufficient, it would next be necessary to attack the agent causing the disease process, perhaps a bacterium or virus.

MANMETS thus provides a sequential step-by-step working algorithm with which to design a drug for a particular disease state:

1. Design the drug to manipulate endogenous messengers that would normally respond to the disease process.
2. Design the drug to influence endogenous targets involved in the disease but not influenced by messenger systems.
3. Design the drug to attack exogenous causes of the disease.

For example, when confronted with the task of designing drugs for systemic arterial hypertension and atherosclerosis, there are many targets. Following Step 1, drugs could be designed to interact with messenger neurotransmitter (adrenergic) receptors (e.g., "beta-blockers") or messenger hormonal (renin–angiotensin) receptors to lower blood pressure. Following Step 2, drugs could be designed to interact with non-messenger protein targets such as enzymes involved in fluid homeostasis (e.g., diuretics targeting the carbonic anhydrase enzyme). Following Step 3, drugs could be designed to interact with exogenous targets (e.g., chlamydia infection) that some workers hypothesize are involved with augmenting the arterial wall damage initiated by high blood pressure.

The MANMETS system also facilitates logical methods for remembering the side effects of drugs, without having to resort to "brute memorization." For example, drugs used for the treatment of psychosis may cause the movement disorder known as parkinsonism as a side effect because of their interactions with dopamine receptors. Similarly, drugs used for the treatment of epilepsy may produce untoward events in individuals susceptible to heart arrhythmias because seizures and cardiac arrhythmias are both mediated by voltage-gated ion channels.

A final advantage of MANMETS is that it is not merely a catalogue of information pertinent to drug design and medicinal chemistry. It provides a structural *and* conceptual framework that enables this knowledge and information to be stored and logically manipulated in a meaningful way for purposes of practical drug design. As an information storage, processing, and utilization framework, MANMETS endeavors to provide a conceptual outline that logically interfaces the practice of medicinal chemistry with bioinformatics and cheminformatics. MANMETS is a molecular-level system that provides a comprehensive organization of druggable targets—past, present and future; future targets emerging from genomics/proteomics research can be readily integrated into the MANMETS system.

4

Messenger Targets for Drug Action I

Neurotransmitters and their receptors

4.1 OVERVIEW OF RELEVANT NEUROANATOMY AND NEUROPHYSIOLOGY

This chapter deals with endogenous messenger molecules (neurotransmitters) within the human central and peripheral nervous systems (CNS, PNS), their targets, and the drugs that affect them. Since most of these messengers act on nerve cells (neurons), it is appropriate to review the anatomy and physiology of the nervous system and to discuss briefly the neuronal networks that can be manipulated therapeutically. Since the nervous system also influences homeostasis throughout the entire body, neurotransmitters are ideal messengers to target when designing drugs within a rational biochemical conceptual framework.

Neurons offer some of the most important targets in drug design (see figure 4.1).

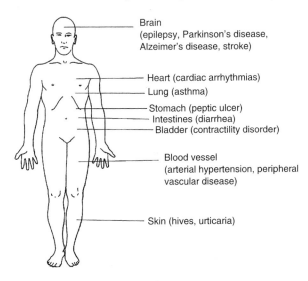

Figure 4.1 Neurotransmitters and neurotransmitter receptors as messenger targets: Neurotransmitters affect organs and disease processes throughout the entire body, affording a wealth of targets for drug design. Drugs based on neurotransmitters are not restricted to brain diseases; they can be used for the treatment of diseases affecting virtually any organ system within the body.

193

4.1.1 The Neuron

The human brain is composed of more than 100 billion neurons with widely differing molecular properties. Neurons are highly specialized cells that trigger and conduct bioelectric impulses, communicate with each other through intricate networks, and regulate all tissues and organs within the body. The membrane of the nerve cell is "excitable" because it can undergo changes in its permeability, mediated by trans-membrane ion channel proteins and triggered by small, endogenous neurotransmitter molecules.

Figure 4.2 shows the organization of a neuron. The cell body carries short, branching *dendrites*, which receive and transfer incoming signals to the cell; these signals are then transmitted to the next neuron (or to a tissue) by the long *axon*. The axon of a neuron is insulated by the lipid *myelin sheath*, which is interrupted by the *nodes of Ranvier*. These gaps allow the exchange of ions between the axon and its surroundings. The axon terminates in a nerve ending, which may be a *neuromuscular endplate* that communi-cates with the membranes of muscle cells. In other neurons, the nerve ending can be a knoblike *synaptic bouton* in contact with the dendrites, axon, or cell body of another nerve cell, with chemical signals rather than electric impulses being used for transmis-sion. The synaptic end of a neuron contains mitochondria and one or more types of *synaptic vesicles*—spheres of 0.3–0.9 μm diameter, surrounded by a membrane and filled with a neurotransmitter that is often complexed with protein and ATP. The presy-naptic membrane seems to have an inner grid composed of *synaptopores*, which are assumed to direct the synaptic vesicles to the membrane when they are about to dis-charge the neurotransmitter. However, there are other mechanisms of neurotransmitter release. The *synaptic gap* separates two interconnected neurons. Normally, the neuro-transmitter, released into the synaptic gap, diffuses to the *postsynaptic membrane* and its receptors, which are really parts of the next neuron.

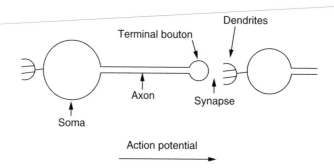

Figure 4.2 The neuron consists of several parts: dendrite, soma (body), axon, and terminal bouton (or synaptic bouton). A gap, or synapse, separates one neuron from another. A chemical messenger, or neurotransmitter, must passively diffuse across the synapse in order to transmit information from one neuron to the next. Information travels from the soma, along the axon, and to the terminal bouton, stimulating the Ca^{2+}-mediated release of a neurotransmitter.

The neuron is the fundamental structural unit of the nervous system. Each neuron is also a functioning "bioelectric unit", capable of generating and transmitting electrical information (sections 4.1.2 and 4.1.3). When a neuron is damaged by some pathological process (e.g., trauma, infection), it may neurochemically respond in one of two ways: either the neuron can quit functioning (leading to negative—loss of function—symptoms, such as the limb paralysis associated with stroke; section 4.9.3) or it can hyperfunction (leading to positive—excessive function—symptoms, such as the limb shaking associated with epilepsy).

Neurons are not the only cells contained within the central nervous system. The nervous system is also rich in *glial cells*, which function as support cells. Glial cells are biosynthetically active cells possessing protein-laden membranes that offer a wealth of potential "druggable targets" for the medicinal chemist. There are four types of glial cells: *astrocytes*, *oligodendrocytes*, *microglial* cells and *ependymal* cells. Astrocytes are workhorse cells. Most importantly, they are neurochemically active and involved in the exchange of metabolites between neurons and the blood and also in the uptake of neurotransmitter molecules from the synaptic cleft. Astrocytes are also an essential structural component of the blood–brain barrier (chapter 3), the most important pharmacokinetic hurdle to drug design for the central nervous system. Oligodendrocytes are responsible for producing and maintaining the myelin sheaths (fatty insulation layer) surrounding neuronal axons in the central nervous system. Not surprisingly, oligodendrocytes may play a role in multiple sclerosis (chapter 6). Microglia function as neural macrophages, responsible for phagocytosis (a defense mechanism which involves ingesting and removing particles or substances foreign to the brain; from the Greek, *phagein*, to eat). Ependymal cells line the fluid-filled cavities within the brain.

In addition to providing druggable targets for the drug designer, glial cells are also important in medicinal chemistry because they are the primary source of brain tumors. The majority of brain tumors arise from glial cells, not neurons. This is not surprising, given the observation that glial cells are much more active in cellular division than neurons; brain cells, unlike other cells (e.g., liver cells) do not tend to proliferate after injury. *Gliomas* are common brain tumors. Astrocytes are a frequent source of brain tumor, giving rise to *astrocytomas* and the extremely deadly *glioblastoma multiforme*. The development of anticancer agents for brain tumors is a technically challenging activity within medicinal chemistry.

4.1.2 Nerve Conduction

Nerve conduction is the process whereby electrical information is passed along the length of a given brain cell. As a process, nerve conduction is vulnerable to (and useful for) a properly designed drug. All cells show transmembrane electric potential. A microelectrode placed into a cell will indicate a potential that is 50–80 mV more negative than the potential recorded by an electrode outside the cell. This condition is a result of ion imbalance. Inside the cell, there is a high K^+ ion concentration (about 120 mM) and low Na^+ concentration (about 20 mM); the reverse is true outside the cell. In addition, there is a negative charge inside the cell because the protein anions of the cytosol are not counterbalanced by cations. The buildup of this negative charge eventually prevents the loss of more K$^+$ ions, and an equilibrium is reached; the cell becomes polarized and the transmembrane potential (*resting potential*) stabilizes.

The difference between an ordinary cell and an excitable cell becomes evident when a depolarizing current is applied. In an ordinary cell, such as an erythrocyte, the transmembrane potential is equal to zero; in a neuron, however, an explosive, self-limiting process allows the potential to overshoot zero and become about 30 mV more positive within the cell than outside it. This depolarization is called an *action potential*, and is carried first by sodium ions and then by potassium ions (see figure 4.3). Spread of the action potential along a neuron is the means by which information is transmitted in the CNS. The neuron is the fundamental anatomical unit of the brain; the action potential is the fundamental physiological (functional) unit of the brain. An action potential lasts only about a millisecond, during which time sodium rushes in and potassium rushes out through ion channel proteins opened by conformational change. The original ionic disequilibrium is then re-established through the rapid elimination of Na$^+$ ions. In myelinated nerves, such ion exchange can occur only at the nodes of Ranvier, and the action potential jumps very rapidly from node to node without a loss of potential. This wave of depolarization passes along the axon to the nerve ending and can be repeated several hundred times per second.

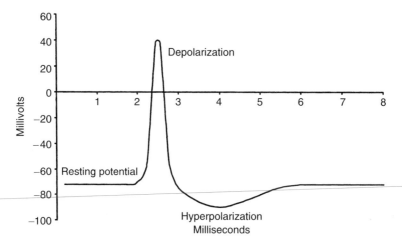

Figure 4.3 The neuron is the fundamental structural (anatomical) unit of the brain. The action potential is the fundamental functional (physiological) unit of the brain and is the means of transmitting information within the nervous system. An action potential is generated by changes in the transmembrane voltage gradient across the neuronal membrane. The action potential is initiated by the opening of voltage-gated Na$^+$ channels. The resulting wave of depolarization travels along the neuron as an electrical signal, transmitting information.

4.1.3 Synaptic Transmission

Synaptic transmission is the process whereby neurons communicate with each other and with the target organs whose physiology they are influencing; synaptic transmission permits the action potential to jump from one neuron to the next. It is imperative that the medicinal chemist understands synaptic transmission when designing neuroactive drugs. Synaptic transmission is not electrical but chemical, and is triggered by the arrival of the action potential at the nerve ending. This causes a Ca^{2+} ion influx across the membrane and into the neuron, resulting in the release of an interneuronal chemical messenger (neurotransmitter) characteristic for that particular neuron. There seem to be several different neurotransmitter release mechanisms, although none is well understood. When released, the neurotransmitter crosses the synaptic gap by passive diffusion and binds transiently to a receptor on the membrane of the postsynaptic neuron. This receptor occupation initiates the electrical axonal wave of depolarization of the next (postsynaptic) neuron; alternatively, it can trigger the activation of an enzyme such as adenylate cyclase and the formation of cAMP as a second messenger. The released neurotransmitter is then either destroyed enzymatically or taken back into the synapse and recycled. *Inhibitory neurotransmitters*, on the other hand, activate Cl^- ion uptake through the postsynaptic neuronal membrane. This effect makes the intracellular potential more negative than the original resting potential and thus *hyperpolarizes* the neuronal membrane. Naturally, a greater than normal impulse will be necessary to fire such a hyperpolarized neuron, since the threshold value of the action potential remains the same. Both excitatory and inhibitory impulses summate and trigger an all-or-none response of a particular neuron, on which hundreds of other neurons may synapse.

Besides binding to postsynaptic receptors, a released neurotransmitter also "back-diffuses" to *presynaptic receptors* or autoreceptors on the neuron from which it was just released, fulfilling an important feedback regulatory function by facilitating or inhibiting transmitter release. It has been suggested that these presynaptic receptors are also *heteroreceptors*—that is, they respond to cotransmitters as well as neurotransmitters produced by the same neuron. For instance, it is known that neurotensin regulates the release of norepinephrine, its cotransmitter. Now that the fallacy of the "one neuron—one transmitter" dogma has been revealed, it is logical to assume that multiple transmitters (neurotransmitter plus a cotransmitter) may regulate each other's release and metabolism in a given synapse and that there may be considerable overlap among presynaptic auto- and heteroreceptor functions.

4.1.4 Neurotransmitters and Neuromodulators

A *neurotransmitter* is a chemical messenger that mediates the passage of electrical information from one neuron to an adjacent neuron. To be defined as a classical neurotransmitter, a molecule must be synthesized and stored in a neuron, released from that neuron in a Ca^{2+} dependent process, diffuse to an adjacent neuron, specifically dock with a receptor on that adjacent neuron, and have its binding to this receptor blocked by a competitive antagonist. A *neuromodulator*, on the other hand, is a molecule which is present in the synaptic cleft and which modifies either the frequency or the efficiency of the neurotransmitter molecule, thereby either amplifying or attenuating the neurotransmitter action.

The traditional neurotransmitters have been recognized for a number of decades and include acetylcholine, norepinephrine, and glutamate. The number of neurotransmitters has increased rapidly in the past 10–20 years as the methodology for their detection has become more sophisticated. At this point it is well to consider that the classical definitions and concepts in this field have been undergoing considerable change, and that the distinctions between neurotransmitters, cotransmitters, neuromodulators, and *neurohormones* often become blurred. Many peptide hormones of the hypothalamus and hypophysis, for instance, have been recognized as having neurotransmitter activity at other sites, and neurohormones and the discipline of neuroendocrinology have become increasingly important in the biosciences.

In recent decades, an explosive development in the discovery of cotransmitters has greatly expanded our understanding of neurotransmission, and of the homeostatic equilibrium that is regulated by aminergic and peptidergic cotransmitters even in systems as simple as that of *Hydra*. Postsynaptically, cotransmitters can influence the same receptor on the target, bind to two different receptors on the same target, or bind to two different receptors on two different targets. This multipotential reactivity may explain the fact that some drugs and endogenous substances are partial agonists only: they may miss the help of a cotransmitter that the full agonist receives. Cross-reactivity of cotransmitter combinations may also explain the many side effects and shortcomings of neuroactive drugs that have been designed without the benefit of knowing the complete story of *in vivo* processes at the target.

It should be kept in mind that a single synapse may operate with as many as four transmitters simultaneously, in any combination of amine and peptide, or even peptide and peptide, within the groupings shown. The peptide neurotransmitters are stored separately, always in large synaptic vesicles; are synthesized in the cell body of the neuron; and are transported to the synapse after post-translational processing by fast (ATP-driven) transport systems. Amine neurotransmitters are synthesized in the synapse and are stored in small or large vesicles. Different populations of the same type of neurons may differ in their content of cotransmitters.

4.1.5 Neuronal Systems: Brain Structures Relevant to Drug Design

The neuronal systems of vertebrates are divided into the central nervous system (CNS), comprising the brain and the spinal cord, and the peripheral nervous system (PNS), comprising the autonomic nervous system and sensorimotor nervous system that serve the rest of the body.

The *brain* is really a collection of highly specialized components of enormous anatomical complexity. The brains of different mammals are very different, and the evolutionary changes in the brain are seen primarily as an increase in relative size and in the complexity of cortical folding, thus increasing the area devoted to *association* (i.e., learning and decision making). A basic schematic illustration of the human brain is shown in figure 4.4.

The central nervous system consists of the brain and *spinal cord*. The average adult brain weighs 1250–1380 grams. The brain is divided into three gross parts: the *brainstem*, the *cerebrum*, and the *cerebellum*. Structurally, the brain may be likened to a bouquet of flowers with the cerebrum (as two cerebral hemispheres) "blossoming" outwards above the brainstem; the cerebellum is attached at the back of the brainstem.

The brainstem consists of the following structures: *medulla oblongata* (at the lower end where the brainstem meets the spinal cord), *pons*, *mesencephalon* (midbrain), and

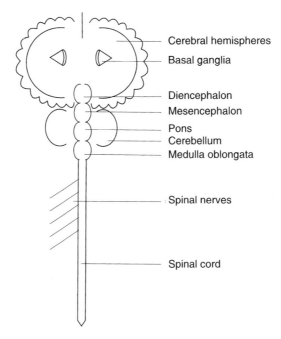

Figure 4.4 General structure of the brain: the central nervous system consists of the spinal cord and the brain. The brain consists of the brain stem (medulla oblongata, cerebellum, pons, mesencephalon, diencephalon) and the cerebrum (cerebral hemispheres, subcortical white matter, basal ganglia).

diencephalon (upper end, where the brainstem meets the cerebrum). Twelve pairs of nerves, collectively referred to as the *cranial nerves*, originate in the brainstem and subserve sensory and motor function in the head and neck. One of these nerves, the *vagus nerve* (so called because it wanders throughout the thorax and abdomen), is important to the autonomic nervous system (section 4.3.1). Within the medulla are a number of neurons actively involved in the biosynthesis of epinephrine (section 4.3.2). Lying partly in the pons and partly in the mesencephalon is the *locus ceruleus* (or *nucleus pigmentosus*), which is rich in norepinephrine-containing neurons and thus plays a role in the adrenergic neurotransmitter systems (see section 4.3.1). The dorsal (back) portion of the pons and mesencephalon is referred to as the *tegmentum* and contains a variety of nerve fibre tracts. The mesencephalon contains the *substantia nigra*, a region of the brain that is intimately involved with the dopamine neurotransmitter and is thus involved in the medicinal chemistry of Parkinson's disease (section 4.4.4). The diencephalon is divided into the following regions: *thalamus, hypothalamus, subthalamus,* and *epithalamus*. The thalamus acts as a relay station that transmits, correlates, and integrates all ascending sensory information from the body on its way to the cerebrum. The hypothalamus has a regulatory influence over the autonomic nervous system and is the point at which the nervous system (using neurotransmitters as messengers) and the endocrine system (using hormones as messengers) interface. The hypothalamus also produces responses to emotional changes and to needs signaled by hunger and thirst;

therefore, it is of interest to the design of agents for appetite control. The *pituitary gland* is attached to the hypothalamus and is crucial to the synthesis of many neurohormonal messenger molecules (chapter 5). The *pineal gland* is part of the epithalamus and is important to the timing of the onset of puberty. The pineal gland contains melatonin, biosynthetically derived from serotonin (section 4.5.1).

Running up the centre of the brainstem is a region called the *reticular activating system*. This region is influenced by a diversity of neurotransmitters (e.g., serotonin, norepinephrine, GABA) and is central to human consciousness and sleep, being involved in pharmacologically induced sleep. The *raphe nuclei* are in the middle of the brainstem (in the midline of the medulla, pons, and mesencephalon); they are associated with the reticular formation and are actively involved with the biosynthesis of serotonin (section 4.5.1). The brainstem (particularly the medulla and pons) is crucial to life, with many injuries to the brainstem being rapidly lethal; the drug designer who targets receptors in the brainstem must keep in mind a deep appreciation for the fundamental role of the brainstem in life.

The cerebrum is made up of *gray matter* structures and *white matter* structures. Gray matter is composed of neuronal cell bodies (i.e., soma); white matter is composed of nerve fibres (e.g., axons) coated in fatty insulation called myelin. The cerebrum is composed of two gray matter areas: an inner region called the *basal ganglia* (located adjacent to the diencephalon at the top of the brainstem) and an outer region called the *cerebral cortex* that lines the outer surface of the brain. Between these two gray matter regions lies an extensive zone of white matter. This zone contains the insulated wiring that carries information from the body to the brain via ascending vertical tracts, from the brain to the body via descending vertical tracts, and from one part of the brain to another part of the brain via horizontal tracts. Since multiple sclerosis (MS) is exclusively a disease of white matter (in which discrete regions of white matter become "demyelinated," giving rise to short circuiting within the brain's electrical network), this area is of importance to drug design for demyelinating diseases; interferons (chapter 6) are now used in the treatment of MS.

The innermost gray matter region of the cerebrum is the basal ganglia region. The basal ganglia consists of three major components: *caudate nucleus, putamen, and globus pallidus.* Collectively, the caudate nucleus and putamen are referred to as the *striatum*; collectively, the putamen and globus pallidus are referred to as the *lenticular nucleus*. The basal ganglia are heavily involved in motor activity. The term *extrapyramidal system* is used clinically to denote components of the basal ganglia that influence motor activity. Motor activity is intricately controlled by the interaction of three major systems: the cerebral cortex (controlling voluntary movements via the *pyramidal system* and causing spasticity and loss of volitional movement when injured), the basal ganglia (modulating static postural activities via the extrapyramidal system and causing rigidity, involuntary movements, and tremor when injured) and the cerebellum (facilitating coordination of movements and causing ataxia [staggering] when injured). Because of its central role in movement, the basal ganglia is an area in which many neurotransmitters deliver either excitatory (+) or inhibitory (−) messages (e.g., dopamine (−), acetylcholine (+), glutamate (+), γ-aminobutyric acid [GABA] (−), and substance P (+)). By manipulating these neurotransmitters, the medicinal chemist can aid in the treatment of a number of diseases, including Parkinson's disease (targeting the striatum; see section 4.4.4) and Huntington's chorea (targeting the caudate nucleus; see section 4.7.8).

The outermost layer of the cerebrum is the cerebral cortex. The 2–5 mm thick mantle of gray matter that covers the expansive surface of the cerebrum is what makes you *you*. The cortex provides the final integration of all neural mechanisms and is a place where neurotransmitter-influenced bioelectric events are of paramount importance. The neuroscientist Sherrington vividly described the cortex as "an enchanted loom where millions of flashing shuttles weave a dissolving pattern." The cortex is spread over two cerebral hemispheres, separated from each other by the deep *medial longitudinal fissure* but connected to each other by the *corpus callosum*—a broad band of white matter providing information relay between the two hemispheres. The cortex is divided into various lobes that subserve varying functions: *frontal* (motor), *parietal* (sensory), *occipital* (vision), and *temporal* (speech, memory). The *hippocampus* is an expansion of the temporal lobe and is involved in memory and epilepsy; owing to its importance to memory, it may be a future design target for anatomically targeted neurologic drugs. Discrete damage to the cortex can give rise to negative symptoms (i.e., loss of functional abilities produced by a loss of neuronal function) such as apraxia (loss of ability to carry out purposeful, skilled motor acts despite intact motor systems), alexia (inability to read), aphasia (inability to speak), and agraphia (inability to write). Focal damage to the cortex can also give rise to positive symptoms (unwanted, uncontrolled activities produced by excessive neuronal electrical discharges) in the form of seizures. Epilepsy (state of recurrent seizures) is a disorder of gray matter, as distinguished from MS, a disorder of white matter. Global damage to the cortex by neurotoxic substances such as β-amyloid may lead to neurodegenerative disorders such as Alzheimer's disease.

The outside of the cortex (i.e., the surface of the brain) is covered by membrane layers collectively referred to as the *meninges*. The meninges are trilaminar with the tough *dura mater* externally, the delicate *arachnoid* lining the dura, and the thin *pia mater* adhering to the brain. Serotonin receptors within the blood vessels of the meninges are involved with the mechanism of migraine (section 4.5.5). Infection of the meninges gives rise to *meningitis*, which is distinct from an infection involving the brain, *encephalitis* (section 9.1). The meninges also extend downwards to encase the spinal cord in a fluid-filled tube called the *thecal sac*.

The final part of the brain is the cerebellum. The cerebellum lies attached to the medulla, pons, and mesencephalon by bands of tissue referred to as *cerebellar peduncles*. It has an outer layer of gray matter and two hemispheres. As mentioned, the cerebellum is involved in the coordination of movement. It is exquisitely sensitive to its chemical environment. Inebriation with alcohol leads to a staggering gait (ataxia), reflecting the cerebellum's response to the excessive amount of ethanol. Many neuroactive agents produce cerebellar signs as a first indication of toxicity. For instance, many anticonvulsant drugs (used for epilepsy, chapter 7) produce the cerebellar sign called *nystagmus*—a jerky back-and-forth movement of the ocular pupils. The use of a *rotorod ataxia test* (ability of a rodent to stay on a slowly turning rod) is a measure of cerebellar intactness and can be used as a crude measure of neurotoxicity when evaluating new chemical entities as putative neurologic therapeutics.

The center of the central nervous system is occupied by a system of cavities containing the fluid referred to as *cerebral spinal fluid* (CSF). The average adult brain contains 150 mL of CSF, with a CSF secretion rate of 0.4 mL/min; thus, the total CSF in the brain is completely replaced 3–4 times per day. CSF is a dilute aqueous solution of Na^+ and Cl^-. CSF fulfills a number of functions, including protecting the brain from trauma by its buoyancy, functioning as a "sink" to remove certain substances from the brain, and influencing message transduction by facilitating hormonal and molecular transport within the brain. The central cavity system, which contains the CSF, is composed of the *central canal* (in the spinal cord), the *fourth ventricle* (between the pons and the cerebellum), the *third ventricle* (within the diencephalon), and the *lateral ventricles* (within the cerebral hemispheres). This fluid-filled pathway may be used to administer drugs directly (albeit with a limited distribution) into the CNS via *intrathecal* administration by delivery into the thecal sac that lies outside of the spinal cord. When there is an excess of CSF, the condition is referred to as *hydrocephalus.* Although frequently treated surgically, hydrocephalus may also be treated with enzyme inhibitors such as *carbonic* anhydrase enzyme inhibitors (e.g., acetazolamide); this observation is of interest to the medicinal chemist working on diuretics and the carbonic anhydrase system.

Since the brain is so extremely active in the electrical control of short-term homeostasis within the body, it is an ideal target for drug design. However, this high degree of activity also gives the brain a voracious appetite for glucose and oxygen as provided by the bloodstream. Indeed, the brain has the highest consumption of blood of any organ system in the body. As a generalization, blood supply to the back of the brain (brainstem, cerebellum, occipital cortex) is from the vertebrobasilar (VB) artery and the associated posterior cerebral artery (PCA); blood supply to the front of the brain (frontal and parietal cortex) is from the internal carotid (IC) and its anterior cerebral artery (ACA) and middle cerebral artery (MCA) branches. Either blockage (via atherosclerosis) or rupture (secondary to arterial hypertension) of any of these arteries will lead to a stroke, which in turn triggers a cascade of neurotransmitter events which may (or may not be) amenable to molecular manipulation by the medicinal chemist (section 4.9.3).

The other part of the central nervous system is the spinal cord. As a reversal of the trend in the brain, the spinal cord has white matter on the outside and gray matter on the inside. The spinal cord is divided into several divisions: cervical, thoracic, lumbar, and sacral. It lies protected in the bony spinal column constructed from individual vertebral bodies. Damage to the spinal cord is common and leads to severe disabilities (i.e., diplegia [paralysis of both legs] or quadriplegia [paralysis of all four extremities]—to be distinguished from hemiplegia [paralysis of one leg and arm on the same body side] which is a symptom of brain injury or damage). Currently, there are virtually no drugs available for spinal cord problems. Glycine receptors, especially those in the cervical region of the spinal cord, may have some utility in the development of therapies for the treatment of muscle spasticity following CNS injury (section 4.8). The spinal cord is extremely important because it is a conduit for all ascending information traveling up to the brain and descending information traveling down from the brain.

Commands from the CNS to the organs of the body (e.g., heart, lungs, bowel, bladder) are conveyed by the *autonomic nervous system*, whereas commands to the skeletal muscles are transmitted by the *sensorimotor system*. Likewise, information from the heart, lungs, and other viscera are conveyed back to the CNS via the autonomic nervous system and information from the skin (i.e., pain, pressure, touch) is sent to the CNS via the sensorimotor system. The autonomic nervous system is further divided into two parts: *sympathetic* and *parasympathetic*. The sympathetic portion deals with the "fight or flight" response, speeding up the heart and increasing breathing rate during times of stress; the parasympathetic portion allows us to slow down during times of relaxation.

There is considerable structural difference between the neurons of the autonomic and sensorimotor systems. In the sensorimotor system, a motoneuron may originate from a ventral horn of the spinal cord and continue without interruption, through a myelinated A-fiber, to the muscle. The neuron usually branches in the muscle and forms neuromuscular endplates on each muscle fiber, creating a single motor unit. By contrast, the autonomic nervous system interposes a peripheral *ganglionic synapse* between the CNS and an organ, acting as a kind of switching station. The neurons in the sympathetic nervous system originate in the upper and middle part of the spinal cord and form myelinated B-fibers. Each such fiber makes synaptic connection with the ganglion cell, which continues in a postganglionic, nonmyelinated C-fiber that then synapses on a smooth-muscle cell, a gland, or another neuron. In the sympathetic system, the ganglia are usually in the paravertebral chain, or within some other specialized ganglia. In the parasympathetic nervous system, the ganglia are buried in the effector organs and therefore have only short postganglionic fibers.

As will be shown in this chapter, designing drugs to treat specific diseases of the CNS is challenging and fraught with frequent failure. The diagnostic approach to neurological disease involves localization of the lesion followed by determination of the nature of the lesion. The disease is localized by examining the individual to ascertain which vertical pathways (e.g., descending tracts such as the corticospinal tract carrying information from the brain to the body, or ascending tracts such as the spinothalamic tract carrying sensory information to the brain) and horizontal pathways (e.g., spinal nerves carrying information to a particular level of the spinal cord, or visual pathways taking information from the eyes to the occipital cortex) are involved; by determining where the vertical and horizontal tracts meet, it is possible to localize the lesion. Next, it is necessary to determine the nature of the pathology that is causing problems at this location: developmental (cerebral palsy), degenerative (Alzheimer's, Parkinson's diseases), infectious (meningitis, encephalitis), inflammatory (multiple sclerosis), neoplastic (brain tumor), nutritional (thiamine deficiency causing Wernicke's encephalopathy), trauma (concussion), toxic (alcoholism, environmental poison, drug side effect), or vascular (stroke).

Designing drugs to treat such problems is not a trivial task. Although the disease may be localized, site-specific delivery of the drug is usually not possible. A focal cortical injury may be causing seizures, but the anticonvulsant drug will reach all areas of the brain, not just the focal area. In fact, it may not even be possible to get the drug into the brain. The blood–brain barrier (chapter 3) will preclude many molecules with desirable therapeutic properties. Also, various areas of the brain are highly interconnected; blocking a neurotransmitter receptor may have far-reaching consequences that extend beyond the area of interest. Despite these difficulties, medicinal chemistry of neuroactive substances is a rapidly expanding area. Furthermore, drugs targeting neurotransmitters are not exclusively used for the treatment of CNS disorders. Since the brain controls numerous functions throughout the body, modification of neurotransmitters enables the treatment of many non-neurologic problems such as high blood pressure, cardiac arrhythmias, pulmonary bronchospasm, and irritable bowel syndrome (section 4.10).

4.2 ACETYLCHOLINE AND THE CHOLINERGIC RECEPTORS

From the perspective of drug design, acetylcholine is one of the most important neuro-transmitters in the human brain; neurons that use acetylcholine as a neurotransmitter are referred to as *cholinergic* neurons. The cholinergic neuronal system can be found in the CNS (especially in the cortex and caudate nucleus), in the autonomic nervous system, and in the sensorimotor system. Acetylcholine (ACh) is the neurotransmitter in all ganglia, the neuromuscular junction, and the postganglionic synapses of the parasympathetic nervous system. However, the autonomic innervation of most organs utilizes both the parasympathetic (mediated by cholinergic neurotransmitters) and sympathetic (mediated by adrenergic neurotransmitters, section 4.3) systems, with the effects of the two usually being opposed. Table 4.1 shows the varying effects of cholinergic and adrenergic stimulation on a diversity of organ systems throughout the body. Thus, if one system causes an increase in some physiological action, the other will cause a decrease, and vice versa.

Table 4.1 Responses of Effector Organs to Autonomic Nerve Impulses and Circulating Catecholamines

Effector organs	Cholinergic impulses Response	Adrenergic impulses Receptor type	Adrenergic impulses Response
Heart			
S-A node	Decrease in heart rate; Vagal arrest	β	Increase in heart rate
Atria	Decrease in contractility and (usually) increase in conduction velocity	β	Increase in contractility and conduction velocity
A-V node and conduction system	Decrease in conduction velocity; A-V block	β	Increase in conduction velocity
Ventricles	–	β	Increase in contractility, conduction velocity, automaticity, and rate of idiopathic pacemakers
Blood vessels			
Coronary	Dilatation	α	Constriction
		β	Dilatation
Skin and mucosa	–	α	Constriction (slight)
Skeletal muscle	Dilatation	α	Constriction
		β	Dilatation
Cerebral	–	α	Constriction (slight)
Pulmonary	–	α	Constriction
Abdominal viscera	–	α	Constriction
		β	Dilatation
Renal	–	α	Constriction

(Continued)

Table 4.1 Continued

Effector organs	Cholinergic impulses Response	Adrenergic impulses Receptor type	Adrenergic impulses Response
Lung			
Bronchial muscle	Contraction	β	Relaxation
Bronchial glands	Stimulation		Inhibition
Stomach			
Motility and tone	Increase	β	Decrease (usually)
Sphincters	Relaxation (usually)	α	Contraction (usually)
Secretion	Stimulation		Inhibition
Intestine			
Motility and tone	Increase	α,β	Decrease
Sphincters	Relaxation (usually)	α	Contraction (usually)
Secretion	Stimulation		Inhibition
Liver	–	β	Glycogenolysis
Pancreas			
Acini	Secretion		–
Islets	Insulin secretion	α	Inhibition of insulin secretion
		β	Insulin secretion
Gallbladder and ducts	Contraction		Relaxation
Urinary bladder			
Detrusor	Contraction	β	Relaxation (usually)
Trigone and sphincter	Relaxation	α	Contraction
Ureter			
Motility and tone	Increase		Increase (usually)
Skin			
Pilomotor muscles	–	α	Contraction
Sweat glands	Generalized secretion	α	Slight, localized secretions
Adrenal medulla	Secretion of epinephrine and norepinephrine		–
Eye			
Radial muscle of iris	–	α	Contraction (mydriasis)
Sphincter muscle of iris	Contraction (miosis)		–
Ciliary muscle	Contraction for near vision	β	Relaxation for far vision

As do most neuronal systems, cholinergic receptors show multiplicity, and we distinguish between *nicotinic* and *muscarinic* receptors, which differ in many respects. Whereas acetylcholine (**4.1**) binds to both types of receptors, the plant alkaloids nicotine (**4.2**) and muscarine (**4.3**) trigger a response only from nicotinic or muscarinic cholinergic receptors, respectively. Nicotinic receptors are found in all autonomic ganglia (i.e., in the sympathetic system as well as the parasympathetic) and at the neuromuscular endplate of striated muscle. Muscarinic receptors occur at postganglionic

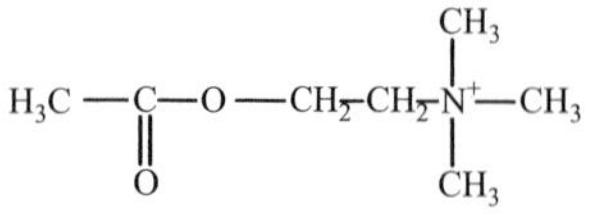

Acetylcholin (4.1)

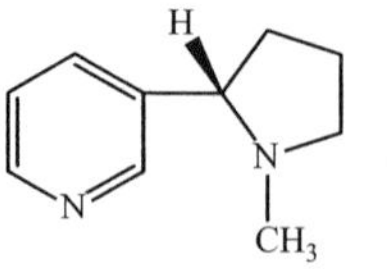

Nicotine (4.2)

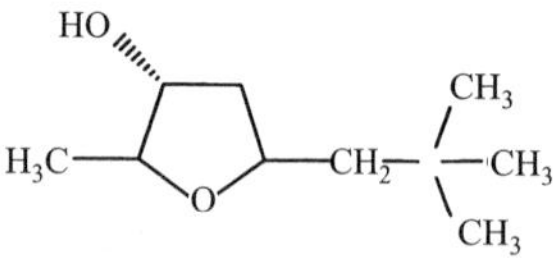

cis-L-(+)-
Muscarine (4.3)

parasympathetic terminals involved in gastrointestinal and ureteral peristalsis, glandular secretion, pupillary constriction, peripheral vasodilation, and heart rate reduction. Acetylcholine is normally an excitatory neurotransmitter, although it can occasionally show an inhibitory action in cardiac muscle. There, hyperpolarization rather than depolarization occurs because only K^+ can cross the muscle membrane. In the CNS, cholinergic inhibition is seen in the thalamus and brainstem.

4.2.1 Acetylcholine Metabolism

Acetylcholine is synthesized by the reaction

$$Choline + Acetyl\text{-}CoA \rightarrow Acetylcholine + CoA\text{---}SH$$

which is catalyzed by choline acetyltransferase. Acetyl-coenzyme A (CoA) is ubiquitous; choline is obtained from phosphatidylcholine (lecithin) and free choline. Some of this choline is recycled after ACh is hydrolyzed by acetylcholine esterase (AChE), terminating the neuronal impulse (see chapter 8). There is a high-affinity transport system (K_m = 1–5 M) for choline reuptake in the nerve endings, which can be inhibited by hemicholinium (**4.4**). Unlike most other neurotransmitters, ACh itself is not taken up by active transport into synapses.

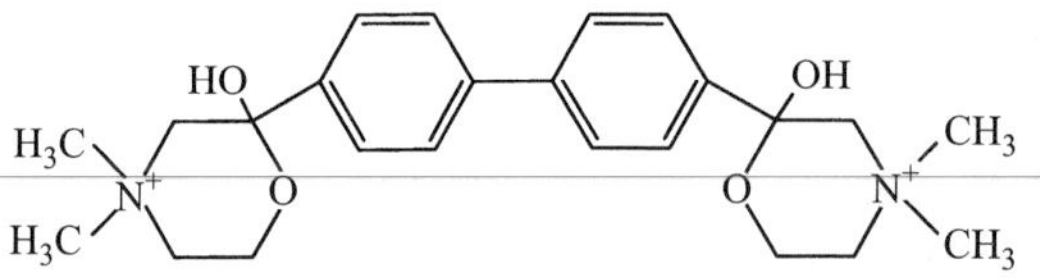

Hemicholinium (4.4)

As ACh is synthesized, it is stored in the neuron or ganglion in at least three different locations. Eighty-five percent of all ACh is stored in a depot and can be released by neuronal stimulation; it is always the newly synthesized neurotransmitter that is released preferentially. The surplus ACh can be released by K^+ depolarization only. Finally, there is stationary ACh, which cannot be released at all. It has been assumed that the neurotransmitter in cholinergic and some other neurons is released through the exocytosis of small transmitter-filled synaptic vesicles.

Acetylcholine release is inhibited by one of the most potent toxins, the *botulinus toxin* produced by the anaerobic bacterium *Clostridium botulinum*. The toxin, lethal at 1 ng/kg in humans, enters the synapse by endocytosis at nonmyelinated synaptic membranes and produces muscle paralysis by blocking the active zone of the presynaptic membrane where Ca^{2+}-mediated vesicle fusion occurs. Historically, botulinum toxin was of interest because of its role in botulism food poisoning. However, botulinum toxin ("botox") is now used as a "therapeutic." Disabling and possibly painful muscle contractions associated with neurologic disorders, such as *dystonia*, can be relieved by injecting the affected muscle with botulinum toxin. Botox injections are now also used cosmetically. Injections of the toxin into the muscles over the forehead or around the eyes will remove "age creases and skin wrinkles," "restoring the youthful appearance" much prized in our image-conscious society.

Acetylcholine is also found in non-neuronal tissues. The mammalian respiratory tract is regulated by ACh; this neurotransmitter also has a direct effect on intestinal smooth muscle and on the heart. It is therefore reasonable to consider ACh a hormone as much as a neurotransmitter, as mentioned previously.

4.2.2 The Nicotinic Acetylcholine Receptor

4.2.2.1 Isolation of Acetylcholine Receptors

The isolation of the nicotinic acetylcholine receptor glycoprotein was achieved almost simultaneously in several laboratories (those of Changeux, O'Brien, Brady, and Eldefrawi) and was helped tremendously by the discovery that the electric organ (*electroplax*) of the electric eel (*Electrophorus electricus*, an inhabitant of the Amazon River) and related species, as well as the electroplax of the electric ray (*Torpedo marmorata*) of the Atlantic Ocean and the Mediterranean Sea, contains acetylcholine receptors (AChR) in a much higher concentration than, for instance, in human neuromuscular endplates or brain tissue.

The discovery that the toxins of *Elapid* snakes bind almost irreversibly to the AChR also facilitated the isolation and study of this receptor. The structure of these venoms has been elucidated; those most widely used experimentally are the α-bungarotoxin (BTX) of the Indian cobra and the toxin of the Siamese cobra. These compounds are peptides containing from 61 to 74 amino acids, five disulfide bridges, and a high proportion of basic arginine and lysine residues, often in close proximity. Venoms are toxic because they block cholinergic neurotransmission by binding to the receptor.

The AChR is an integral membrane protein, deeply embedded into the postsynaptic membrane. It can be solubilized by nonionic detergents such as Triton X-100, Tween 80, and others, or anionic detergents such as deoxycholate, a bile acid derivative. Functionally, the regulation of ion permeability is lost when the receptor is removed from the membrane; however, the ACh and BTX binding capacity is retained and can be used for following the course of purification. In a typical isolation procedure, the electric organ is homogenized in 1 M NaCl with Na_2HPO_4 and EDTA, which solubilizes the acetylcholinesterase. The suspension is then centrifuged and the resulting pellet extracted with detergent, solubilizing the AChR. This receptor solution can then be purified further by polyacrylamide gel electrophoresis, by affinity partitioning, or, most efficiently, by affinity chromatography either on an immobilized quaternary ligand or on Siamese cobra toxin bound to an agarose bead matrix. The specific activities of the purified preparations range from 8 to 12 µmol of binding sites per gram of protein, and about 100–150 mg of receptor protein can be obtained from 1 kg of

Torpedo electric organ. Compared to normal concentration standards, this yield is exceptionally high.

4.2.2.2 Physicochemical Properties and Subunit Structure

The physical and chemical properties of the AChR have been elucidated. Optical rotatory dispersion measurements indicate that the receptor consists of about 34% helix and 28–30% β-sheet structure—a high proportion of ordered secondary structure. Some carbohydrates are part of the molecule. The DNA encoding the receptor has been cloned and sequenced, revealing the complete amino acid sequence of the subunits.

The *subunit structure* of the AChR varies according to its origin. There are four peptide chains, referred to as α (mass ~ 40 kD), β (~ 48 kD), γ (~ 58 kD), and δ (~ 64 kD), which can be separated by electrophoresis. The receptor of *Torpedo californica* has an $\alpha_2\beta\gamma\delta$ chain composition, giving it a monomeric molecular mass of 250 kD. The α chain is affinity-labeled specifically by [³H] bromoacetylcholine and by [³H] 4-N-maleimidobenzyl-trimethylammonium iodide on Cys-192 and Cys-193 and therefore must be the ACh binding subunit. The other chains are integral parts of the receptor and do not dissociate, even in 8 M urea. The different chains have different amino acid sequences but similar compositions. 4-*N*-Maleimidobenzyl-trimethylammonium iodide, a specific affinity reagent, indicates the important fact that the quaternary-ammonium-ion binding site (the -COO⁻ of glutamate) and an —SH group binding to maleimide are in close proximity. [³H] Bungarotoxin ([³H] BTX) crosslinks the α and β chains and probably also obstructs the ion channel. The β and γ chains are preferentially labeled by a nitrene obtained from pyrene-sulfonylazide, a hydrophobic reagent believed to attach itself to proteins within the core of the membrane (see figure 4.5).

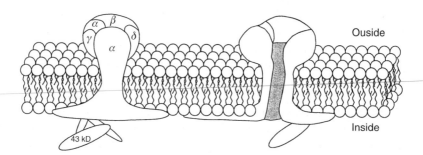

Figure 4.5 This model of the nicotinic acetylcholine receptor shows two pentameric units covalently linked through the δ subunit. One of the units is shown in cross-section, indicating the selectivity gate of the ion channel in the closed state. The 43 kD protein is shown, associated with the receptor on the cytoplasmatic side.

The *ligand binding sites* of the AChR have been explored by a number of techniques. The ACh binding site was first investigated using the MBTA affinity label. MBTA binds to the native receptor with a $K_D = 8 \times 10^{-5}$ M but binds much more strongly after reduction of the membrane preparation with dithioerythritol, a mild reducing agent. In binding to the AChR, MBTA occupies half the sites that bind [¹²⁵I] BTX.

Various pharmacologic and molecular biology studies have identified at least two distinct types of nicotinic cholinergic receptors (*cholinoceptors*). The N_N receptor is found in postganglionic neurons and in some presynaptic cholinergic terminals; the N_M receptor is found in skeletal muscle, at the neuromuscular endplate. Ligand binding to either of these receptors leads to opening of Na^+ and K^+ channels with subsequent cellular depolarization. These two receptors have differing structural features in terms of their subunit construction. The N_M receptor has a pentameric structure consisting of α, β, γ, and δ units; the N_N receptor has α and β subunits only.

4.2.3 The Muscarinic Acetylcholine Receptor

Even though the muscarinic receptor, which is present in postganglionic parasympathetic synapses, is much more stereospecific and structure-specific than its nicotinic counterpart, only since the 1980s have any molecular studies been undertaken to explore similarities and differences between the two classes of AChR. It has been labeled with the affinity label [³H] propyl-benzilylcholine-mustard (**4.5**).

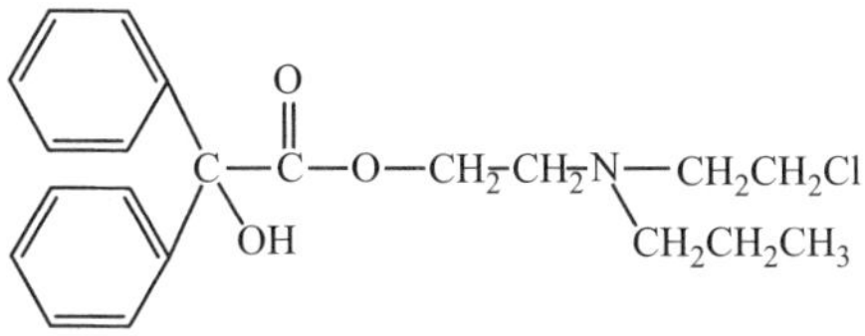

Propyl-benzilylcholine-mustard (4.5)

Muscarinic receptors may be classified into subtypes based upon their molecular structure, signal transduction properties, and various ligand affinities. To date, three types have been identified: M_1, M_2, and M_3. All three subtypes are present in the central nervous system, subserving functions such as memory, learning, pain perception, and cortical excitability. The correlation of specific mentation processes to particular receptor subtypes has not yet been achieved. Elsewhere, these three receptor subtypes participate in various processes. M_1 receptors are present on nerve cells where they facilitate impulse transmission from preganglionic axon terminals to ganglion cells. M_2 receptors mediate acetylcholine effects on the heart; these receptors open K^- channels in cardiac tissue, affecting sinoatrial pacemaker cells and thus slowing heart rate. M_3 receptors regulate smooth muscle tone in the gastrointestinal tract and in bronchi, causing stimulation of phospholipase C and increase in muscle tone. By an analogous mechanism, M_3 receptors in various glands mediate increased glandular secretion. In blood vessels, the relaxant action of ACh on muscle tone is indirectly mediated via M_3 receptors that facilitate the release of nitric oxide.

4.2.4 Cholinergic Receptor Agonists

Increased stimulation of the AChR can be achieved in two ways: (1) by binding of the directly acting cholinergic agonists to the AChR, triggering nicotinic or muscarinic effects, or both; and (2) by binding of the *indirect agonists*, which are drugs that inhibit the hydrolysis of ACh by AChE, thus prolonging the action of available ACh.

The principal directly acting cholinergic agonists include methacholine, carbachol, and betanechol—agents which, unlike acetylcholine, are used clinically.

Acetylcholine (**4.1**) has, of course, both nicotinic and muscarinic action. Because it is very rapidly hydrolyzed by AChE and even by aqueous solution, it is not used therapeutically.

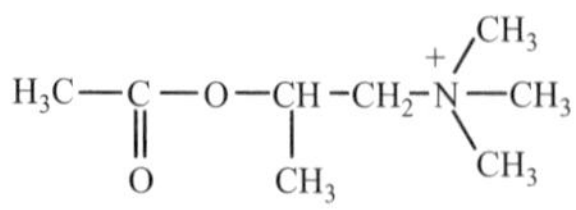

Methacholine (4.6) 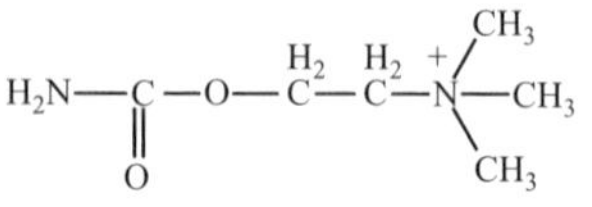   Carbachol (4.7) 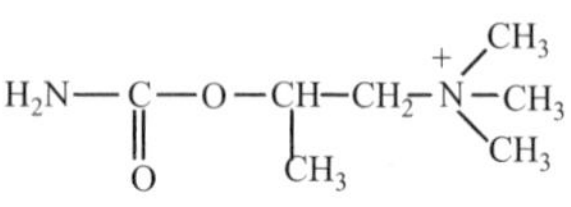   Bethanechol (4.8)

Methacholine (**4.6**) is hydrolyzed somewhat more slowly than acetylcholine because of steric hindrance of the ester by the α-methyl group. Its activity is mainly muscarinic, but it is used infrequently.

Carbachol (**4.7**) is a very potent agent because it is not an ester but a carbamate, and is hydrolyzed slowly. It is used in glaucoma to reduce intraocular pressure.

Bethanechol (**4.8**) also has a prolonged effect, and finds application in stimulation of the gastrointestinal tract and urinary bladder (both muscarinic effects) to relieve post-operative atony.

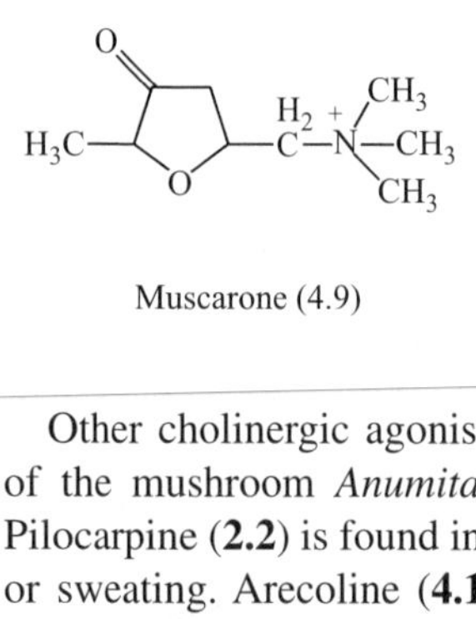

Muscarone (4.9) 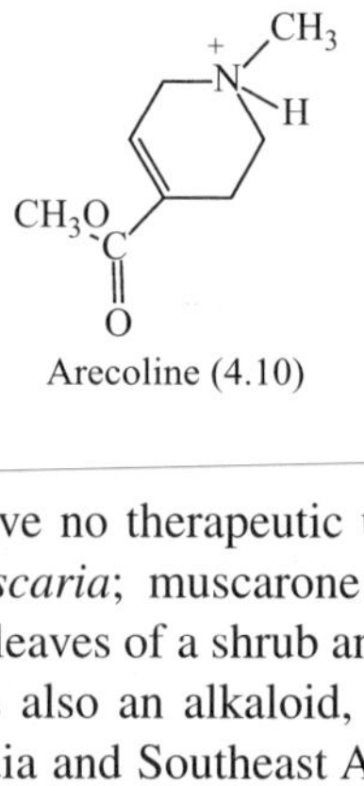   Arecoline (4.10) 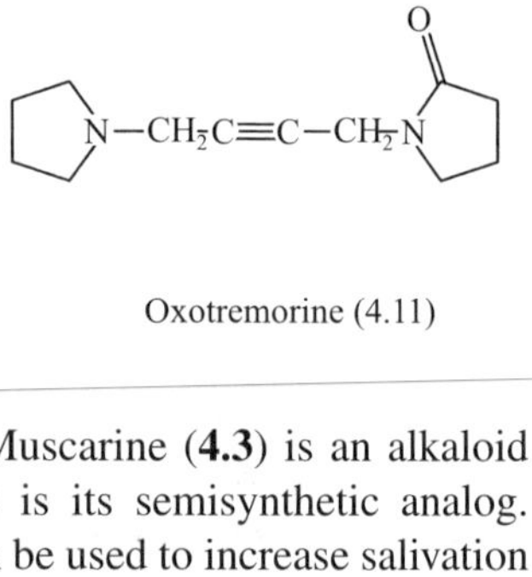   Oxotremorine (4.11)

Other cholinergic agonists have no therapeutic use. Muscarine (**4.3**) is an alkaloid of the mushroom *Anumita muscaria*; muscarone (**4.9**) is its semisynthetic analog. Pilocarpine (**2.2**) is found in the leaves of a shrub and can be used to increase salivation or sweating. Arecoline (**4.10**) is also an alkaloid, and occurs in the betel nut that is used as a mild euphoriant in India and Southeast Asia. Finally, oxotremorine (**4.11**) is a synthetic experimental agent that produces tremors and is helpful in the study of antiparkinsonian drugs.

4.2.4.1 Cholinergic Agonists: Structural Modifications of Acetylcholine

Structural modifications of acetylcholine influence the ability of analogs to function as cholinergic agonists. These modifications fall into four categories: (1) changes in the quaternary ammonium group; (2) changes in the ethylene chain; (3) changes in the ester group; and (4) the creation of cyclic analogs of the neurotransmitter.

Ammonium Group. The ammonium group of ACh can be replaced by other "-onium" compounds (phosphonium, arsonium, or sulfonium), but only with the loss of 90% of the activity. One of the methyl groups on the ammonium can be exchanged for larger alkyl residues: for instance, the dimethylethyl derivative is about 25% active. However, the insertion of larger groups or the replacement of more than one methyl leads to an almost complete loss of activity. This finding implies that the size of the quaternary ammonium group and its charge distribution are important to the activity of ACh, since the hydrophobic auxiliary binding site next to the anionic site of the receptor is optimized for two methyl groups, strengthening the ionic interaction. The uncharged carbon analog 3,3–dimethyl-butyl acetate has only 0.003% activity. It is interesting to note that many muscarinic agonists are *tertiary amines*—for example, pilocarpine (**2.2**), arecoline (**4.10**), and oxotremorine (**4.11**). At physiological pH, however, these amines are likely to be protonated and to occur in rigid ring structures. In this way, hydrogen bonding between the protonated amino group and the -COO- group of the anionic site of the receptor is not prevented by the intramolecular interaction of the -C=O and $-NH(CH_3)_2$ groups in the ligand. Such a bond would completely distort the ligand conformation and prevent normal binding to the AChR.

Ethylene Chain. The ethylene bridge of ACh ensures the proper distance between the ammonium group and the ester group, and is therefore critical in binding to the receptor. Although it is rather dangerous to assign a definite distance between the -onium and ester groups (estimated at about 0.6 nm), the "rule of five" states that there should be no more than four atoms between the N^+ and the terminal methyl group. Lengthening of the chain results in rapidly decreasing activity; interestingly, however, the 2-butyne analog has 50% ACh activity. If the ethylene is branched, only methyl groups are allowed, as shown in the muscarinic agonist methacholine (**4.6**). The α-methyl analog of ACh has more nicotinic activity.

Ester Group. The ester group does not lend itself to much modification either. Large aromatic acid moieties in the ester produce ACh antagonists rather than agonists, some of which are useful as anticholinergic agents. If ethers and ketones replace the ester, some activity is retained. The only useful replacement for the acetate has been a carbamate group, resulting in carbachol (**4.7**), which is highly active because of its slow hydrolysis.

Cyclic Analogues of ACh. Cyclic ACh analogues include the naturally occurring agonists muscarine (**4.3**), pilocarpine (**2.2**), and arecoline (**4.10**), all of which are muscarinic compounds. Dioxolanes such as (**4.12**) are muscarinic analogs of very high potency. Cyclization is a good drug design strategy in that in constrains conformational flexibility, thereby increasing receptor specificity.

2-Methyl-4 trimethyl-ammonium-
methyl-1,3-dioxolane (4.12)

4.2.4.2 Cholinergic Agonists: Mode of Binding of Acetylcholine

The acetylcholine molecule is highly flexible, and its preferred conformation is therefore hard to define. X-ray crystallographic studies suggest that ACh acts as the *gauche* conformer, and it is possible to distinguish a "methyl side" and a "carbonyl side," corresponding, respectively, to the muscarinic and nicotinic actions of the molecule. However, this is merely a general approximation, for although the muscarinic activity is quite specific, steric parameters are rather irrelevant to the action of nicotinic agonists.

It is generally accepted that the ammonium group of ACh binds ionically to a carboxylate anion of glutamate or aspartate on the receptor, aided by van der Waals interaction of the methyl groups with the adjacent hydrophobic accessory binding site. About 0.59 nm removed from this, the carbonyl group forms a hydrogen bond with an acceptor, perhaps with histidine, as in acetylcholinesterase. In muscarinic agonists a third binding point, involving the methyl group of the acetate, may assume increased significance. Whereas the primary structural requirements for nicotinic agonists are a quaternary ammonium and a carbonyl group, the muscarinic agonists are characterized by an ammonium and a methyl group. The carbonyl group is the primary hydrogen-binding site in both nicotinic and muscarinic receptors.

Once the drug binds to its receptor it can exert its therapeutic effect. Given the importance of the cholinergic system both inside and outside of the CNS, it is not surprising that therapeutic exploitation of cholinergic messenger molecules can be targeted at systemic problems. Cholinergic agonists (*cholinomimetics*) enjoy widespread use in the treatment of gastrointestinal and urinary tract problems. In clinical problems involving reduced smooth muscle activity without obstruction, cholinomimetics with muscarinic effects may be of use. These clinical problems include postoperative *ileus* (bowel paralysis following its surgical manipulation) and urinary retention (bladder atony, either postoperatively or secondary to spinal cord injury [the so-called *neurogenic bladder*]). Cholinomimetics, such as bethanechol (**4.8**), may be used for such disorders. For urinary tract problems with some degree of obstruction (e.g., *benign prostatic hypertrophy*), α-adrenergic blockade is more effective.

4.2.5 Cholinergic Receptor Antagonists

The peripheral cholinergic synapses (other than neuromuscular endplates) are muscarinic. Drugs that inhibit the interaction of ACh with the AChR are cholinergic blocking agents (or *parasympatholytics*) and must not be confused with the ganglionic and neuromuscular blocking agents, which act on nicotinic receptors.

Antimuscarinic agents have numerous clinical uses. A heart attack (especially one affecting the inferior wall of the heart) may depress the electrical system of the heart, impairing cardiac output. The judicious parenteral use of atropine or some other antimuscarinic agent may be of value in increasing the heart rate. Antimuscarinics are widely used for gastrointestinal and genitourinary indications. In the treatment of simple traveler's diarrhoea (or other mild, self-limited gastrointestinal hypermotility conditions) antimuscarinics provide rapid relief; frequently they are combined with an opioid drug (see chapter 5), which has an additive antiperistaltic effect on bowel motility. Selective M1 antimuscarinics have some value in the treatment of peptic ulcer disease (section 4.6.5). Atropine has been used to treat urinary urgency associated with bladder inflammation; oxybutynin (**4.13**), another tertiary amine antimuscarinic, has been used to relieve bladder spasm following urologic surgery. Tolterodine (**4.14**), an M3-selective antimuscarinic, has been used for the suppression of adult urinary incontinence. Scopolamine (**4.15**) is a time-honored remedy for motion sickness. A variety of antimuscarinics (e.g., homatropine (**4.16**), cyclopentolate (**4.17**), tropicamide (**4.18**)) have been used in ophthalmologic disorders (uveitis, iritis) and to enable dilation of the pupils (mydriasis) during ophthalmoscopic examination of the retina. Finally, anticholinergics have been used in the treatment of Parkinson's disease (section 4.4.4), but their use has been limited since the widespread adoption of dopaminergic therapies.

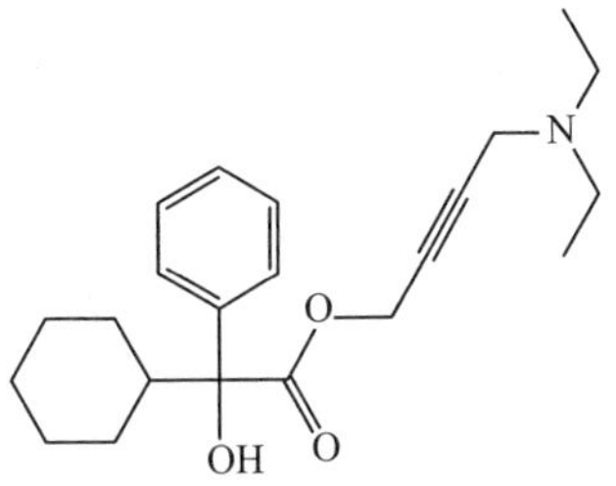

Oxybutynin (4.13)

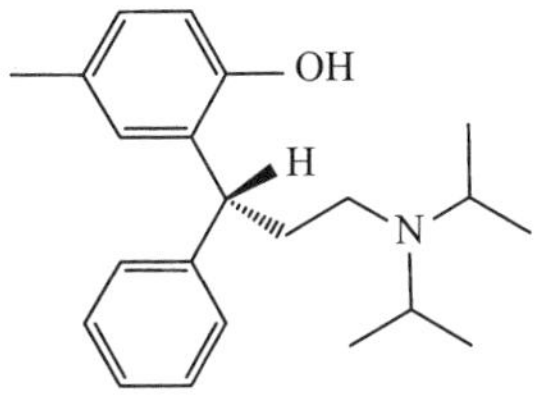

Tolterodine (4.14)

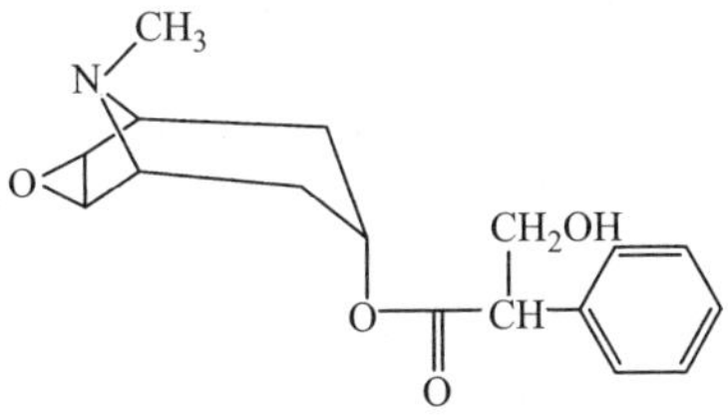

Scopolamine (4.15)

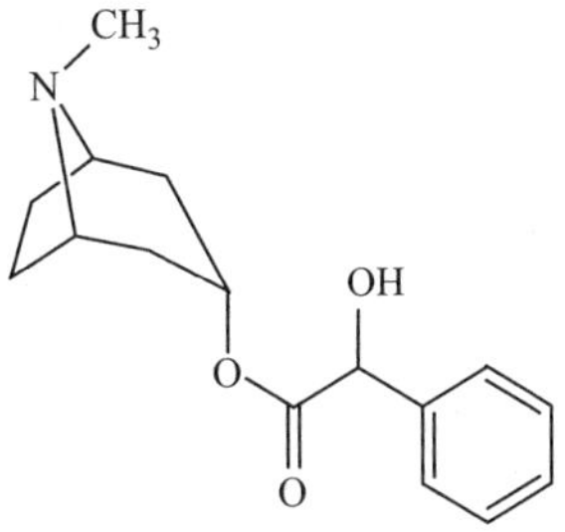

Homatropine (4.16)

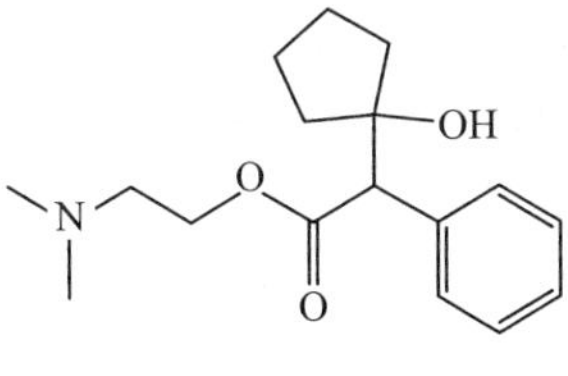

Cyclopentolate (4.17)

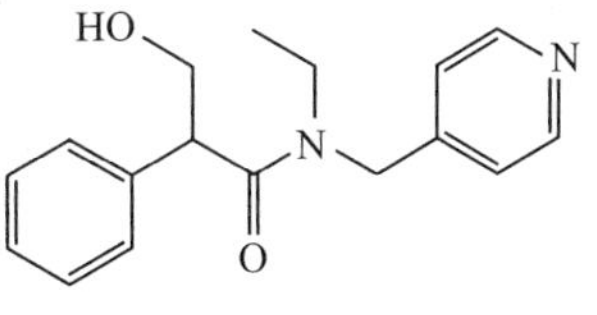

Tropicamide (4.18)

The oldest anticholinergics are the tropane alkaloids of *Atropa belladonna* (nightshade). Atropine (**2.1**) and scopolamine (**4.15**) are derivatives of tropine, a fused piperidino–pyrrolidine ring system, esterified by atropic acid. Atropine is the racemate of (−)hyosciamine, whereas scopolamine has an epoxide ring. In large doses, all of these anticholinergic agents have central excitatory and hallucinogenic effects and were prominent in medieval "witches' brews." A synthetic homolog, homatropine (**4.16**), has a shorter duration of action. These agents are mixed M_1–M_2 antagonists.

These tropine derivatives are esters of tertiary bases with a bulky acid component (atropic acid, mandelic acid). In general, a number of cholinergic blocking agents have been developed by substituting a larger acid for the acetyl group of ACh and increasing the size of the N-substituents. Among the quaternary compounds, tridihexethyl bromide (**4.19**) and propantheline bromide (**4.20**) are notable; among the tertiary amines oxyphencyclimine (**4.21**) shows high activity. Numerous analogs are known and are used therapeutically (e.g., methylatropine and methylscopolamine).

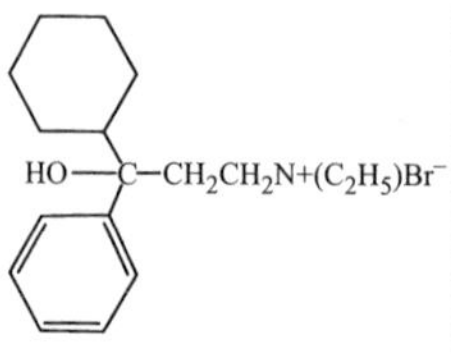

Tridihexethyl
bromide (4.19)

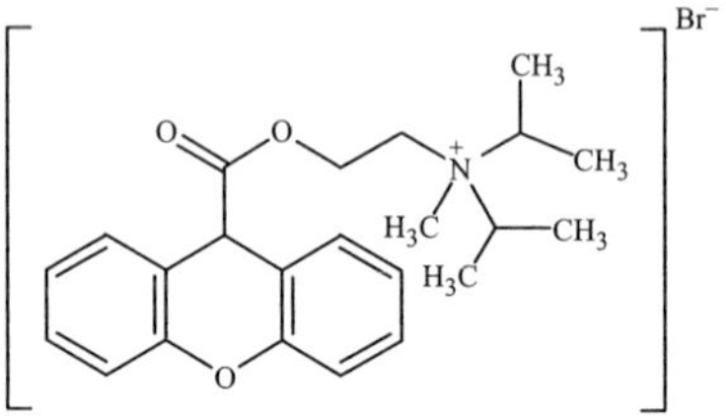

Propantheline
bromide (4.20)

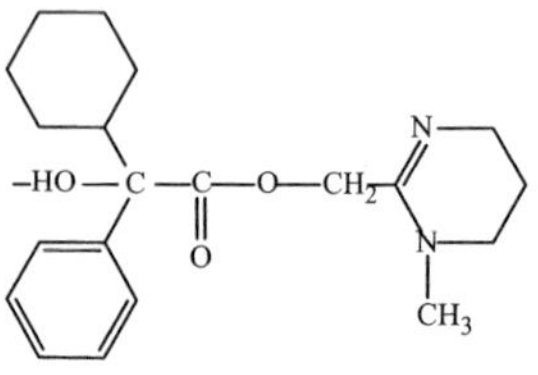

Oxyphencyclimine (4.21)

4.2.5.1 Cholinergic Antagonists: Ganglionic Blocking Agents

Ganglionic blocking agents interfere with the nicotinic ACh receptors in the ganglia. Although ganglia are, functionally, normal receptors, they probably differ structurally from the receptors at the neuromuscular endplate, and show different accessibility. Ganglionic and neuromuscular blocking agents are therefore two structurally different groups of anticholinergic drugs.

Ganglionic blockade by tetraethylammonium salts (**4.22**) has been known for a long time; however, the prototype blocking agent is hexamethonium (**1.1**), a bisquaternary compound with six methylene groups separating the two cationic groups. Some secondary and tertiary amines such as trimetaphan (**4.23**) and mecamylamine (**4.24**) were in use at one time, since they had longer durations of action in controlling hypertension by decreasing vasoconstriction. However, because none of these compounds can distinguish sympathetic from parasympathetic ganglia, they have numerous side effects. Consequently, they have largely been replaced by the more selective β-adrenergic blocking agents.

Tetraethylammonium
salts (4.22)

Hexamethonium (1.1)

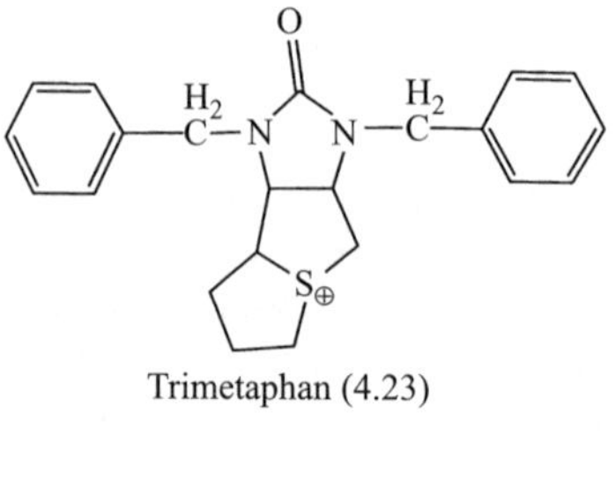

Trimetaphan (4.23)

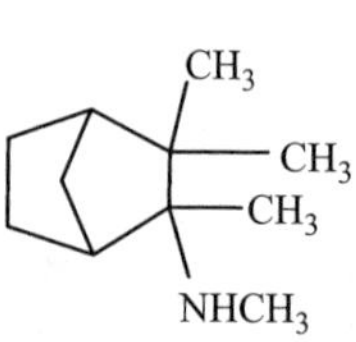

Mecamylamine (4.24)

4.2.5.2 Cholinergic Antagonists: Neuromuscular Blocking Agents

Neuromuscular blocking agents are widely used in surgery. They are capable of relaxing the abdominal muscles without the use of deep anesthesia, and make surgery much easier for both the surgeon and the patient. There are two major categories of such agents: (1) *competitive neuromuscular agents*, which occupy the same site as ACh; and (2) *depolarizing neuromuscular blocking agents*, which mimic the action of ACh but persist at the receptor.

Competitive neuromuscular agents. These agents were developed through the study of curare, the arrow poison of South American Indians. Crude curare contains a number of isoquinoline and indole alkaloids, the best known of which is tubocurarine (**3.10**), a tertiary–quaternary amine in which the distance between the two cations is rigidly fixed at about 1.4 nm (the "curarizing distance"). A similarly rigid, large molecule is the synthetic steroid derivative pancuronium (**4.25**), a bisquaternary derivative with an N^+–N^+ distance of 1.1 nm. Two acetylcholine molecules built into a rigid framework are clearly discernible. With curarization, the neuromuscular junction becomes insensitive to ACh and the motor nerve impulse, and the endplate potential falls dramatically. Numerous bulky analogs have been developed.

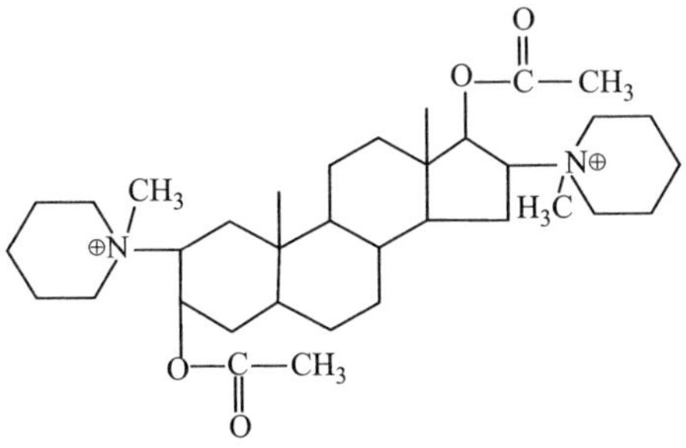

Pancuronium (4.25)

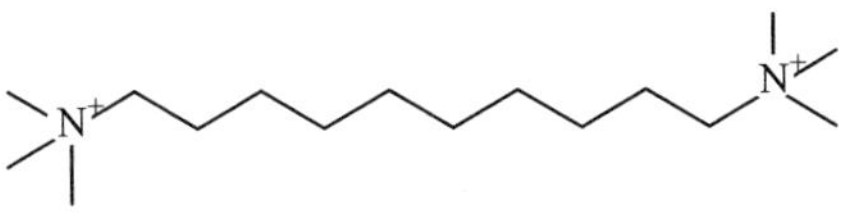

Decamethonium (1.2)

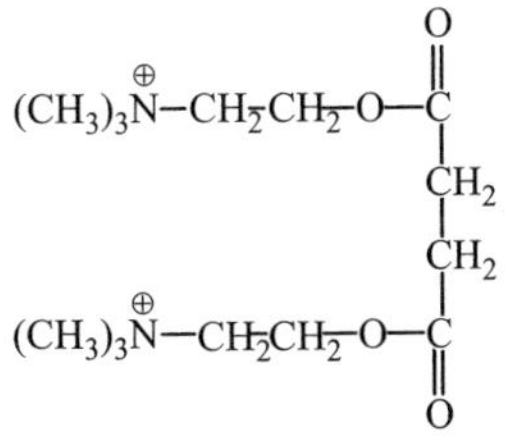

Succinylcholine (3.25)

Depolarizing neuromuscular blocking agents. These were discovered through mimicking the N^+–N^+ distance described above with aliphatic compounds. Decamethonium (**1.2**), in an extended conformation, approximates this distance, and is the prototype of the depolarizing blocking agents. This drug binds normally to the AChR and triggers the same response as does ACh—a brief contraction of the muscle—which, however, is followed by a prolonged period of transmission blockage accompanied by muscular paralysis. A related compound, succinylcholine (**3.25**, succamethonium), has the same N^+–N^+ distance, even though the ten intervening atoms are not all carbon. It has a short, self-limiting action since it is easily hydrolyzed by serum cholinesterase. Besides depolarizing muscle, both compounds depolarize autonomic ganglia.

The structure–activity relationships of neuromuscular blocking agents are instructive. The most interesting aspect of these correlations is that between the N^+–N^+ distance and the receptor structure. As the number of atoms between the -onium groups is increased beyond 10, the activity decreases until a second peak is reached at around the 16-atom distance (hexacarbacholine (**4.26**) and related compounds), which corresponds to a distance of about 2 nm. It is not necessarily the N^+–N^+ distance that is essential; any induced positive charge will be appropriate. The *p*-nitrobenzyl-hexamethonium chloride derivative (**4.27**), for instance, carries the positive charge on its two phenyl rings rather than on formal cationic ammonium ions; this is due to the electron-attracting nitro groups of this compound, which, together with the ammonium ions, dramatically decrease the π-electron density of the rings. The induced charge distance increases to about 2 nm, and the hexamethonium derivative is therefore inactive as a ganglionic blocker but becomes a very effective curarizing agent. It is interesting to note that lower invertebrates (cladocerans, annelid worms, rotifers) are more sensitive to compounds with an N^+–N^+ distance of 16 than those with an N^+–N^+ distance of 10, whereas in animals of phylogenetically higher taxa, such as mammals, this sensitivity is reversed.

Hexacarbacholine (4.26)

p-Nitrobenzyl-hexamethonium (4.27)

The nicotinic AChRs in the ganglion cell and in the neuromuscular endplate are different. The difference probably consists of dissimilar accessory sites. In addition, the neuromuscular site can accommodate not only compounds with an N^+–N^+ distance of 10 atoms but also those with an N^-–N^+ distance of 16 atoms.

4.2.6 The Clinical–Molecular Interface: Alzheimer's Dementia as a Cholinergic Disorder

Alzheimer's disease is a common neurological disorder affecting approximately 20% of persons over the age of 80 years. Clinically, Alzheimer's is characterized by loss of short-term memory, impaired cognition, decline in intellectual function, and decreased ability to carry out the activities of daily life. At the gross anatomical level, Alzheimer's disease is associated with brain atrophy and loss of the outer gray matter of the cerebral cortex. At the cellular level, the histopathology of Alzheimer's reveals "plaques and tangles," where plaques are composed of β-amyloid peptide and tangles are composed primarily of phosphorylated tau protein. It is believed that the accumulation of neurotoxic aggregates of β-amyloid peptide is concomitantly linked with degeneration of cholinergic pathways in the CNS. This degeneration subsequently leads to progressive regression of memory and learned functions and to the other symptoms of Alzheimer's disease.

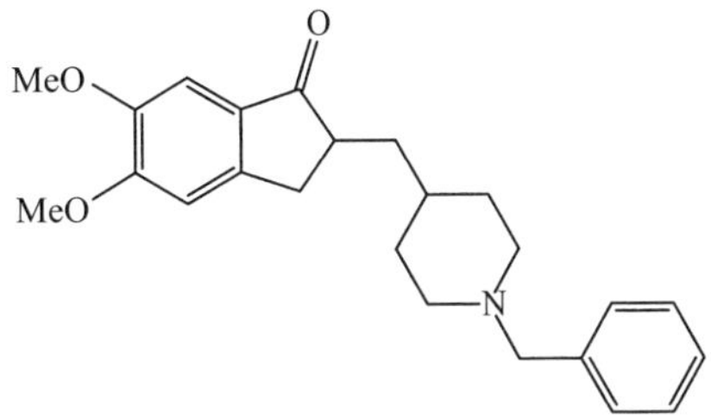

Donepezil (4.28)

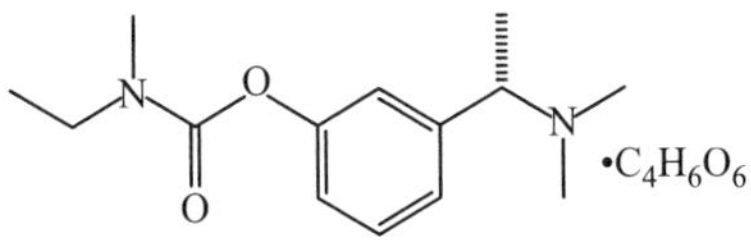

Rivastigmine (4.29)

Therapeutic approaches to Alzheimer's disease initially targeted the cholinergic systems. In early work, drug treatment with either choline replacement or cholinergic agonists was of negligible value. However, cholinesterase enzyme inhibitors such as donepezil (**4.28**), rivastigmine (**4.29**), and galantamine (**4.30**) are of definite symptomatic value in the treatment of Alzheimer's disease. By inhibiting the catabolic breakdown of acetylcholine, these agents prolong the effective half-life of acetylcholine as a neurotransmitter, thereby alleviating some of the symptoms of the disease. About 70% of people show time-limited improvement in their memory when on these agents. Unfortunately, these agents are only symptomatic and do not treat the underlying cause, namely the accumulation of neurotoxic β-amyloid aggregates. Future research in Alzheimer's disease is targeting other potential receptors, such as by blocking the synthesis of β-amyloid by inhibiting the secretase enzyme system involved in the conversion of amyloid precuror protein to β-amyloid, or by binding to β-amyloid and inhibiting its aggregation into a neurotoxic form. Since the average age of the population is on the increase, the frequency of Alzheimer's disease is increasing rapidly and requires urgent attention.

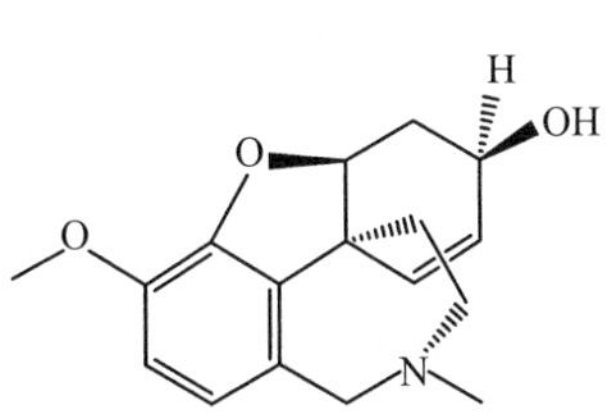

Galantamine (4.30)

4.3 NOREPINEPHRINE AND THE ADRENERGIC RECEPTORS

4.3.1 The Adrenergic Neuronal System

Norepinephrine is another important neurotransmitter, and the receptors that respond to norepinephrine and its analogs are referred to as *adrenergic* receptors. The adrenergic system, also sometimes known as the sympathetic nervous system, is found both peripherally (PNS) and centrally (CNS). Myelinated B-fibers originate in the spinal cord and meet ganglion cells remote from the effector organ. Long, unmyelinated C-fibers then transmit the impulse along the adrenergic axon from the ganglion to the synapses.

Peripherally, all organs are innervated sympathetically (as well as parasympathetically), and in most cases the adrenergic action of this system is opposite to the cholinergic effects. The neurotransmitter secreted by the nerve endings is norepinephrine and, to a lesser extent, epinephrine.

Centrally, two systems can be distinguished:

1. The noradrenergic pathways, primarily situated in the *locus ceruleus*—a deeply pigmented (hence the name, alluding to its blue color) small cell group involved in behaviour, mood, and sleep. The cortex, some thalamic and hypothalamic centers, and the cerebellar cortex are innervated from here. The noradrenergic pathways of the tegmentum are less well known.
2. The adrenergic pathways that use epinephrine as a neurotransmitter, which have been explored only recently. One of these systems is also tegmental and is mixed with noradrenergic cells. The other is thalamic–hypothalamic, involved with the vagus nerve. Some adrenergic fibers are also found in the fourth ventricle and the spinal cord.

4.3.2 Adrenergic Neurotransmitter Metabolism

The adrenergic system produces neurotransmitters belonging to the chemical class of substances known as *catecholamines.* These are derivatives of catechol (**4.31**, *o*-dihydroxybenzene), with a β-aminoethyl side chain. The biogenetically related catecholamines and the pathways leading to their biosynthesis are well-studied biochemical pathways. Starting with tyrosine (**4.32**), the main pathway goes through dihydroxyphenylalanine (**4.33**) (DOPA), dopamine (**4.34**) (DA), norepinephrine (**2.4**) (NE, also called noradrenaline in the European literature), and finally **epinephrine (4.35)** (E, or adrenaline). The pathway is shown in figure 4.6.

While dopamine is an intermediate for NE and E, it is also a neurotransmitter in its own right. Dopamine and the dopaminergic receptor, as well as drugs that act on it, are discussed below.

4.3.2.1 Key Enzymes in Catecholamine Biosynthesis

The enzymes involved in catecholamine biosynthesis have been studied intensively and are the targets of many drugs. The key enzyme is *tyrosine hydroxylase*, which requires a tetrahydrofolate coenzyme, O_2, and Fe^{2+}, and is quite specific. As usual for the first enzymes in a biosynthetic pathway, tyrosine hydroxylase is rate limiting, and

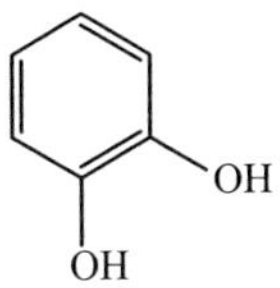

Catechol (4.31)

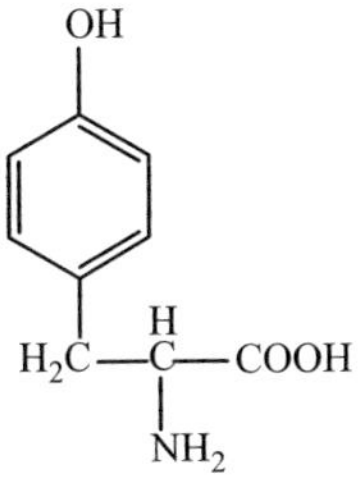

Tyrosine (4.32)

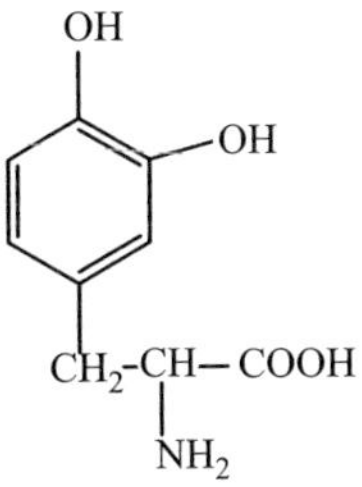

Dihydroxyphenylalanine (DOPA) (4.33)

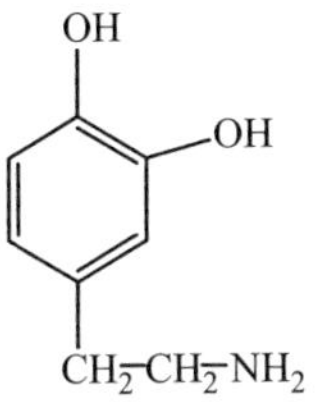

Dopamine (4.34)

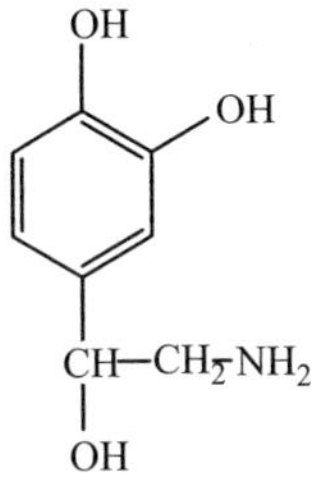

Norepinephrine (NE) (2.4)

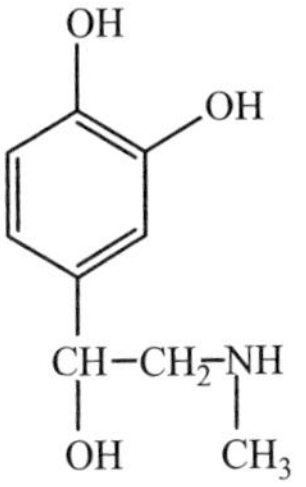

Epinephrine (E) (4.35)

is therefore the logical point for the inhibition of NE synthesis. *DOPA decarboxylase* acts on all aromatic amino acids and requires pyridoxal phosphate (vitamin B_6) as a cofactor. *Dopamine β-hydroxylase*, located in the membranes of storage vesicles, is a copper-containing protein—a mixed-function oxygenase that uses O_2 and ascorbic acid. Finally, *phenylethanolamine N-methyltransferase,* located in the adrenal medulla (the main site of epinephrine synthesis) and in the brain, uses S-adenosyl-methionine as a methyl donor.

4.3.2.2 Catecholamine Storage

Catecholamine storage utilizes synaptic vesicles of different sizes in different organs. The largest ones (up to 120 nm) are found in the adrenal medulla (part of the adrenal glands which are found just above the kidneys in the abdomen) and are called chromaffin granules. Catecholamines are stored as their ATP complexes, in a proportion of 4:1, in association with the acidic protein *chromogranin*. This keeps the neurotransmitter in a hypo-osmotic form even though its concentration is very high (up to 2.5 M) and also protects it from enzymatic oxidation by monoamine oxidase. The vesicles also contain the enzyme dopamine β-hydroxylase, proof that NE is synthesized in the vesicle. The vesicles themselves are formed in the cell body and are transported along the axon to the terminal region.

4.3.2.3 Catecholamine Release

The release of catecholamines has been studied mainly in the adrenal medulla, which is analogous to the nerve cell. In the medulla, the neuronal impulse releases ACh

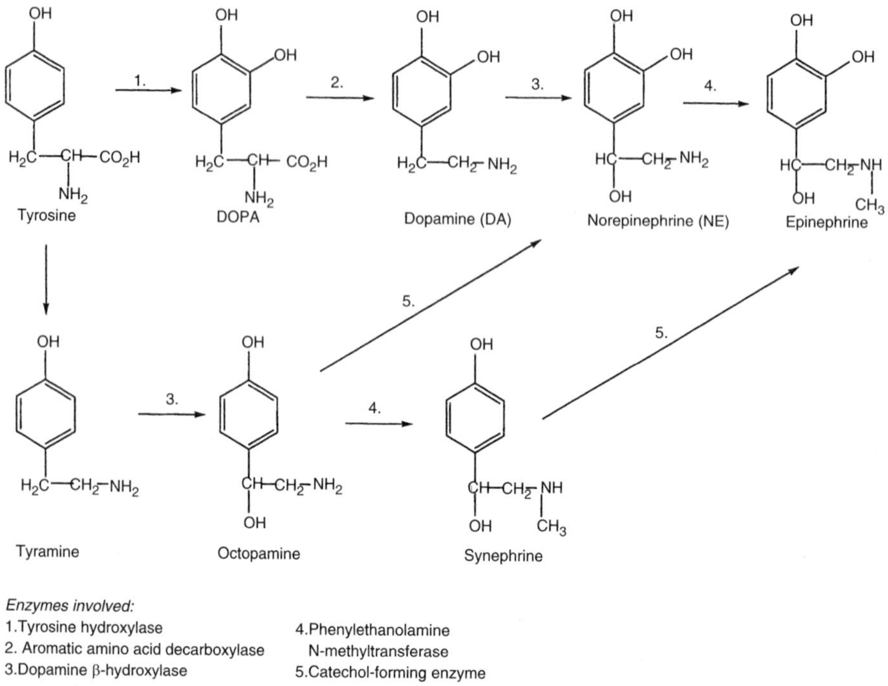

Figure 4.6 Biosynthesis of catecholamines.

(embryologically, the medulla is a modified ganglion and therefore uses ACh as a transmitter). This allows the inflow of Ca^{2+}, which triggers fusion of the chromaffin cell membrane with the secretory vesicle, resulting in exocytosis of the entire vesicle contents including all of the vesicle proteins. The release and turnover of catecholamines is subject to complex regulation, the most important type of which is modulation by presynaptic receptors. Adrenergic agonists acting on these receptors will decrease— whereas antagonists will increase—neurotransmitter release, and also seem to have an effect on regulating neurotransmitter synthesis. In addition, prostaglandins of the E (PGE) series are potent inhibitors of neural NE release through a feedback loop involving Ca^{2+} ions. These presynaptic receptors also respond to neuropeptide Y (NPY) (section 4.10.2), enkephalin (chapter 5), dopamine (section 4.4), muscarinic agonists, and angiotensin (chapter 5), in addition to adrenergic α and β agonists (section 4.3.6). Acetylcholine and cAMP also seem to regulate catecholamine release. These presynaptic heteroreceptors are more likely to have a regulatory role in adrenergic synapses than are the autoreceptors.

4.3.2.4 Catecholamine Metabolism and Reuptake

The metabolism of catecholamines is much slower and more complex than that of ACh. The degradative pathways are shown in figure 4.7. The principal, although nonspecific, enzyme in the degradation is *monoamine oxidase* (MAO), which dehydrogenates

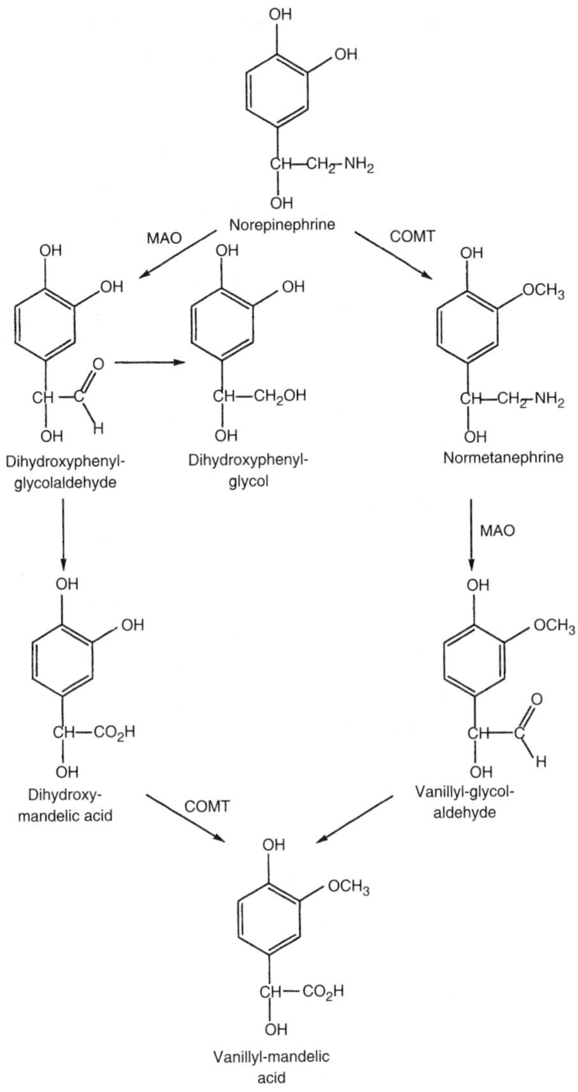

Figure 4.7 Degradative metabolism of norepinephrine.

aliphatic amines. MAO is, in itself, an important target in drug design and is discussed separately in chapter 8 since it represents a non-messenger target in drug design. The intermediate aldehyde is then oxidized to the corresponding carboxylic acid or, occasionally, is reduced to the alcohol. Monoamine oxidase is found mainly in mitochondrial membranes, and occurs in multiple isozyme forms. It is a flavoenzyme in that it contains a riboflavin coenzyme. It seems to act only on certain forms of neurotransmitter. It does not, for example, affect the bound transmitter stored in vesicles, nor, curiously, the transmitter just released. That MAO inhibitors do not increase the intensity of nerve stimulation implies that there is no enzymatic destruction of freshly released transmitter.

The other enzyme in catecholamine catabolism is *catecholamine O-methyltransferase* (COMT), a cytoplasmic enzyme that uses *S*-adenosyl-methionine to methylate the 3–OH of catecholamines and render them inactive. The methylated compounds are not taken up into the synapse.

The principal mechanism for the deactivation of released catecholamines is, however, not enzymatic destruction but *reuptake* into the nerve ending. The presynaptic membrane contains an *amine pump*—a saturable, high-affinity, Na^+-dependent active-transport system that requires energy for its function. The recycled neurotransmitter is capable of being released again, as experiments with radiolabelled [^{3}H]NE have shown, and can be incorporated into chromaffin granules as well. Many drugs interfere with neurotransmitter reuptake and metabolism, as discussed in subsequent sections.

4.3.3 Adrenergic Receptors: The α Receptor Family

The adrenergic receptors have been studied extensively and thoroughly by pharmacological methods. There are two major groups of receptors, designated as α and β, which are in turn subdivided into α_1, α_2, β_1, β_2, and β_3 receptors on the basis of their apparent drug sensitivity. The existence of receptor multiplicity was first suggested by Sir Henry Dale in the mid-1920s, but was formalized and proven by Ahlquist in 1948. Multiple receptors such as these were termed *isoreceptors,* in analogy to isoenzymes (or isozymes).

The α receptors are generally excitatory, as shown in table 4.2, and mediate a constricting effect on vascular, uterine, and intestinal muscle when stimulated by an agonist. They respond to different adrenergic agonists in the following order: epinephrine >

Table 4.2 Norepinephrine and Epinephrine Receptors

β_1 **receptors**

- Epinephrine and norepinephrine are equally potent agonists
- Found in high density in the heart and cerebral cortex
- Linked to adenylate cyclase
- Marked regional variations in brain
- Practolol is a selective antagonist

β_2 **receptors**

- Epinephrine is more potent than norepinephrine
- Found in high density in the lung and cerebellum
- Linked to adenylate cyclase
- Terbutaline and salbutamol are selective agonists

α_1 **receptors**

- Located postsynaptically on blood vessels and in peripheral tissues
- Prazosin is a selective antagonist of receptors localized in the heart

α_2 **receptors**

- Clonidine, epinephrine, and norepinephrine are selective ligands
- Clonidine is a selective agonist
- Effects mediated through stimulation of phospholipase activity and mobilization of intracellular Ca^{2+}
- Located on presynaptic nerve terminals in the periphery
- Localized in the pancreas

norepinephrine > isoproterenol. The β receptors are usually inhibitory on smooth muscle but stimulate the myocardium. Their drug sensitivity is: isoproterenol > epinephrine > norepinephrine. None of these receptors is truly tissue specific, and many organs contain both α and β adrenoceptors, although usually one type predominates.

Studies using radiolabeled agonists and antagonists have identified α receptors in both brain and peripheral tissues and have demonstrated that the binding properties are essentially the same in both of these locations. Early pharmacological and physiological studies supported the existence of two types of α receptor (α_1, α_2). These two basic types have varying anatomical and histological localizations. Some tissues possess only α_1 receptors, some possess only α_2 receptors and some possess mixtures of both. The brain, for example, contains proportions of both α_1 and α_2 receptors with highly variable distributions in different brain regions. The primary amino acid sequences of both the α_1 and α_2 receptors have been determined. The sequences of these two α receptors are not more closely related to each other than either is to any of the three proteins that make up the β-adrenergic receptor family. Not surprisingly, the α adrenergic receptors share marked structural similarities with dopamine receptors (discussed in detail in section 4.4.1) and with other members of the G-protein-linked receptor family (section 2.9).

Recent cloning and sequence analysis studies suggest that there are three subtypes of α_1 receptors and three subtypes of α_2 receptors. The three subtypes of the α_1 receptor have been designated α_{1A}, α_{1B}, and α_{1D}, and tend to be differentially distributed in the kidney, liver, and aorta, respectively. Cloning studies reveal that each arises from a different chromosome and each contains a different number of amino acids: α_{1A} [466 amino acids], α_{1B} [515], α_{1D} [560]. Similarly, there are three subtypes of α_2 receptors, designated α_{2A}, α_{2B}, and α_{2C}. As with the α_1-receptor subtypes, each α_2 receptor is encoded on a different chromosome and contains a varying number of amino acids: α_{2A} [450 amino acids], α_{2B} [450], α_{2C} [461]. The α_{1B} receptor has been shown to activate phosphoinositide-specific phospholipase C (PI-PLC), resulting in liberation of diacylglycerol (DAG) and inositol triphosphate (IP_3), while the α_{1D} receptor has been linked to activation of Ca^{2+} channels. All three of the known subtypes of the α_2 receptor are linked to inhibition of adenylyl cyclase activity. As with other receptors linked to inhibition of adenylyl cyclase, these receptors have relatively short C-terminal tails.

4.3.4 Adrenergic Receptors: The β Receptor Family

Hydropathicity analyses of the β-adrenergic receptor suggests that it possesses seven membrane-spanning hydrophobic regions (I–VII), each 20–24 amino acids in length. In addition, there is a long intracellular C-terminal hydrophilic sequence, a shorter extracellular N-terminal hydrophilic sequence, and a long cytoplasmic loop between transmembrane segments V and VI. Numerous sites accessible to phosphorylation are located on the C-terminal portion of the protein, while sites for N-glycosylation are on the N-terminal extracellular segment. The transmembrane hydrophobic helices are involved in the formation of the catecholamine binding site, and residues in the C-terminal sequence seemingly play a role in the interaction between the receptor and GTP-binding proteins. Finally, an aspartate in transmembrane segment III and two serines in segment V are believed to interact with the amino and catechol hydroxyl groups,

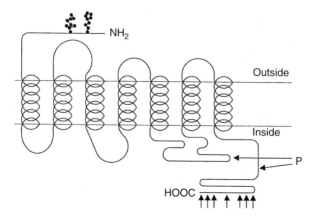

Figure 4.8 Schematic structure of the mammalian β_2-adrenergic receptor. There are seven membrane-spanning helical regions composed of hydrophobic amino acid sequences, and at least two glutamine-linked glycosylation sites near the N-terminal. P shows potential sites of phosphorylation by cAMP-linked protein kinase; arrows indicate serine and threonine molecules that can be the site of regulatory phosphorylation by receptor kinase.

respectively. This structure for the β adrenergic receptor is similar to other G-protein-linked receptors whose cDNAs have been cloned (chapter 2), and is shown in figure 4.8.

Three distinct and pharmacologically important β receptor subtypes exist: β_1, β_2, and β_3. The genomic organization of the genes encoding the biosynthesis of these three receptor proteins is somewhat unusual. The β_1 and β_2 receptor proteins are encoded by genes lacking introns. (An intron is a region of a gene that tends to be a non-coding sequence; introns range in size from fewer than 100 nucleotides to over 10,000 nucleotides. Introns differ from coding sequences in that frequently they can be experimentally altered without changing the gene function. Moreover, introns seem to accumulate mutations rapidly during evolution, leading to hypotheses that introns are composed mainly of "genetic junk".) The β_3 receptor protein, on the other hand, is encoded by an intron-containing gene, which provides an opportunity for alternative splicing as a means of introducing functional heterogeneity into the receptor. Although each of these receptors is structurally distinct, having varying numbers of amino acids (β_1 [477 amino acids], β_2 [410], and β_3 [402]), they all exert their final effect by means of a cAMP secondary messenger.

The three β-adrenoreceptor subtypes have varying localizations and functional properties. The brain contains both β_1 and β_2 receptors; the density of β_1 receptors varies in different brain areas to a much greater extent than does that of β_2 receptors. β_1 receptors predominate in the cerebral cortex; β_2 receptors are more common in the cerebellum. Likewise, there is a coexistence of β_1 and β_2 receptors in the heart, with both receptor subtypes being coupled to the electrophysiological effects of catecholamines upon the myocardium. β_2 receptors tend to predominate in the lung.

The β_3 receptor is distinct from the β_1 or β_2 receptor. In humans, the β_3 receptor is linked to obesity, diabetes, and control of lipid metabolism. mRNA for β_3 receptors is selectively expressed in brown adipose tissue in newborn humans. Polymorphism is common within the structure of the β_3 receptor. A Trp64Arg point mutation in the β_3

gene was first described in the Pima Indians, a population with a high incidence of obesity. In the human receptor, this substituted amino acid at position 64 lies at the junction of the first transmembrane spanning domain and the first intracellular loop. A large number of studies suggest associations between the Trp64Arg β_3 receptor variant and an increased capacity to gain weight, resistance to weight loss, increased blood pressure, and coronary heart disease.

The β receptor is highly stereospecific, preferentially binding only to certain stereoisomers of drugs. The conformational preference is a phenyl/NH$_3$ *trans* arrangement, meaning that the agonist molecule is extended, with the *m*-OH and β-OH coincident on the same face of the molecule. The agonist molecule therefore has a polar and a nonpolar side.

4.3.5 Adrenergic Drugs: Presynaptic and Synaptic Effects

Presynaptic adrenergic drug effects may be classified as follows:

1. Drugs acting on catecholamine synthesis
2. Drugs acting on catecholamine metabolism
3. Drugs acting on catecholamine storage
4. Drugs acting on catecholamine reuptake
5. Drugs acting on presynaptic receptors

These classes afford a logical, mechanistic approach to adrenergic drugs and each class will be discussed individually.

4.3.5.1 Drugs Interfering with Catecholamine Synthesis

These drugs include various enzyme inhibitors; mechanisms of enzyme inhibition are discussed in chapter 8. However, some of these agents have other, nonenzymatic points of attack. The most widely used of these compounds is α-methyldopa (**4.36**). Like many methyl analogs of enzyme substrates, this drug is a competitive inhibitor of DOPA decarboxylase, and was believed to decrease blood pressure by decreasing available NE through inhibition of its synthesis. Other findings, however, indicated that α-methyldopa is metabolized to α-methyl-NE, which then stimulates the central presynaptic α_2 receptors, thus decreasing NE release. The analogous α-methyltyrosine inhibits tyrosine hydroxylase, but is not used as a drug. (Other DOPA decarboxylase inhibitors will be discussed in connection with dopamine.)

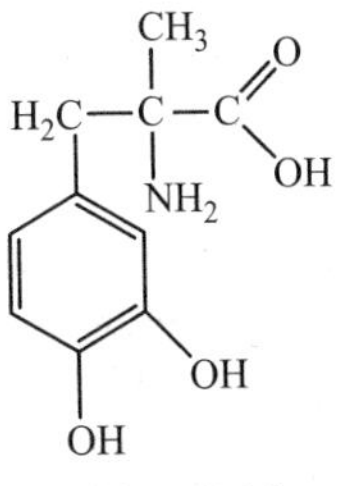

α-Methyldopa (4.36)

4.3.5.2 Drugs Interfering with Catecholamine Metabolism

This group of drugs consists primarily of compounds that block the enzyme monoamine oxidase (MAO). While useful as hypotensive and antidepressant drugs, their side effects can be serious. We discuss them briefly as enzyme inhibitors in chapter 8.

4.3.5.3 Drugs Interfering with Catecholamine Storage and Reuptake

These drugs can act in two different ways. The Rauwolfia alkaloid reserpine (**3.1**) and related natural or semisynthetic compounds interfere with the membranes of synaptic vesicles and deplete nerve endings of NE and dopamine (and, incidentally, of serotonin in serotonergic neurons). The resulting decrease in available neurotransmitter results in hypotension as well as in sedation. It seems that NE reuptake into the vesicles is also impaired. Because more effective drugs are available, reserpine is seldom if ever used, and only as a hypotensive agent. Interestingly, however, reserpine had been used for centuries in India and is one of the few examples of an "ethnopharmacologic" agent successfully introduced into Western therapeutics.

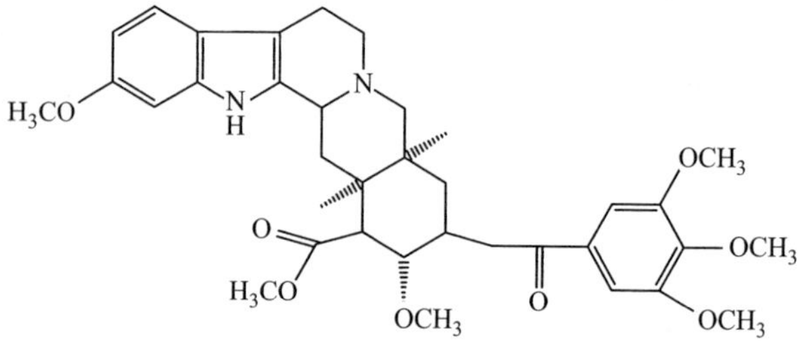

Reserpine (3.1)

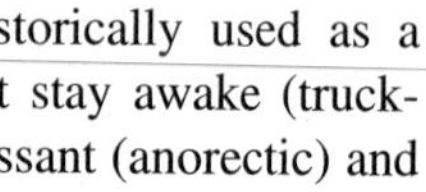

Amphetamine (4.37)

(+)-Amphetamine (**4.37**, phenylisopropylamine) has been historically used as a mood elevator and psychomotor stimulant by persons who must stay awake (truck-drivers, students), and it is still illicitly used as an appetite suppressant (anorectic) and as a mood-altering drug. A specific amphetamine-binding site related to anorectic activity has been described in the hypothalamus. In addition, amphetamines have multiple neuronal effects: they inhibit neurotransmitter reuptake, increase transmitter release, are direct α agonists, and may also inhibit the enzyme monoamine oxidase. Amphetamine is, however, a dangerous drug. In high doses, or when given intravenously (as "speed," the street drug), it can cause symptoms of paranoid psychosis by releasing dopamine in the CNS. It also has cardiovascular effects, and its use is followed by a depressive "letdown" period.

The second mode of interference with neurotransmitter storage is the prevention of neurotransmitter release from storage vesicles. Compounds acting in this way are known as adrenergic neuronal blocking agents. Among these are guanidine compounds such as guanethidine (**4.38**), and quaternary ammonium compounds such as bretylium (**4.39**).

Guanethidine, rarely used as a hypotensive drug, also causes some catecholamine depletion, but unlike reserpine it does not cross the blood–brain barrier and thus has no central sedative effects. It acts selectively because it is taken up into the neuron by the same amine pump that transports the neurotransmitter.

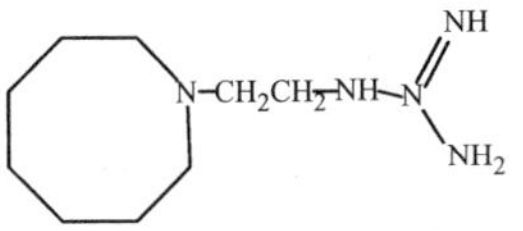

Guanethidine (4.38) 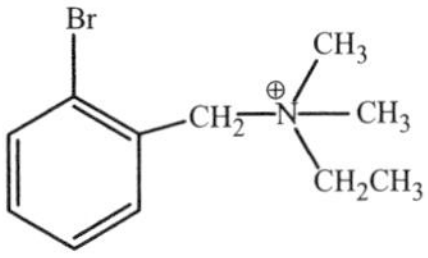   Bretylium (4.39)

4.3.5.4 Drugs Interfering with Catecholamine Uptake

This group of drugs can also be divided into two categories. The first of these consists of the *false neurotransmitters*. Tyramine (**4.40**), produced by the decarboxylation of tyrosine (and especially the β-hydroxy derivative of tyramine, octopamine, **4.41**), can be taken up through the presynaptic membrane by the rather unselective uptake-1 mechanism. Tyramine then enters the storage granules to a certain extent (even though the vesicular uptake mechanism is more specific than the presynaptic pump) and displaces NE which, when released, causes postsynaptic effects. In addition, tyramine competes with NE for monoamine oxidase and protects the neurotransmitter from destruction, thus elevating its actual concentration.

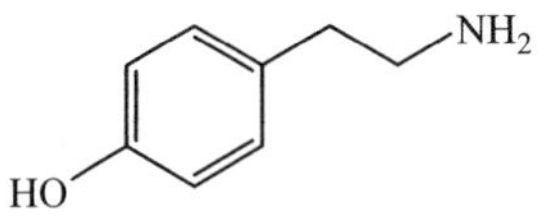

Tyramine (4.40) 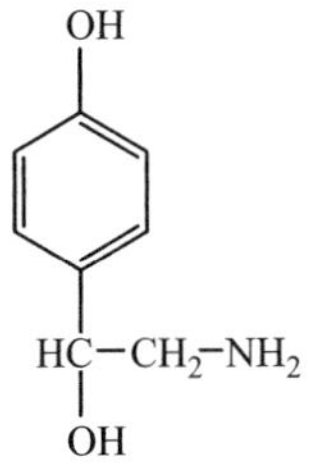   Octopamine (4.41)

Octopamine (**4.41**), which carries a β-hydroxyl group, is taken up even more readily into storage vesicles and is, in turn, released when the neuron fires. As an adrenergic agonist, octopamine is, however, only about one-tenth as active as NE; therefore, it acts as a very weak neurotransmitter. Compounds such as this behave like neurotransmitters of low potency, and are called false transmitters. On the other hand, octopamine may be a true transmitter in some invertebrates, with receptors that cannot be occupied either by other catecholamines or by serotonin.

The other group of drugs acting on catecholamine recycling are the *true reuptake inhibitors*, which block the amine pump of the reuptake-1 mechanism in central adrenergic, dopaminergic, and serotonergic neurons.

4.3.5.5 Drugs Acting on Presynaptic Adrenergic Receptors

As mentioned in section 4.3.3, there are two kinds of α receptor in brain and peripheral tissues. The crucial experiments have shown that brain tissue prelabeled with [³H]NE will release neurotransmitter upon electrical stimulation or exposure to K⁺. The release is reduced by the α agonist clonidine (**4.42**) and stimulated by the α antagonist yohimbine (**4.43**). Since the adrenoreceptor involved in this latter experiment plays a vital role in modulating neurotransmitter release, it must be presynaptic and located on the nerve-ending membrane. A similar selectivity has also been shown by peripheral tissues (heart, uterus), leading to the distinction of α₁ (postsynaptic) and α₂ (presynaptic) adrenergic receptors. There are also presynaptic β receptors, which show a feedback regulation opposite to that of the α₂ receptors; that is, their excitation by a neurotransmitter increases NE release.

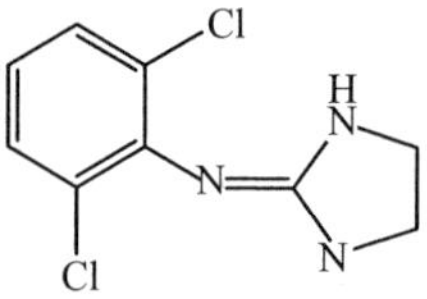

Clonidine (4.42)

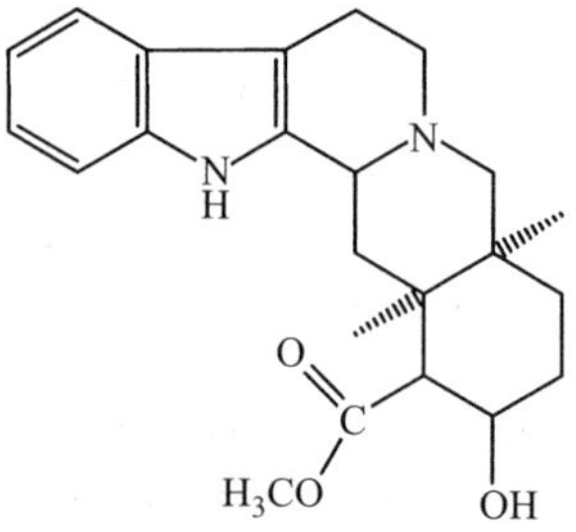

Yohimbine (4.43)

Epinephrine and norepinephrine show the same affinity for both α₁ and α₂ receptors as do some antagonists such as phentolamine (**4.44**). Sometimes receptor selectivity depends upon the drug concentration: dihydroergocryptine (**4.45**), a partial α-blocking agent, binds at a low concentration to α₁ receptors; at higher concentrations, however, α₂ binding takes over, at the point where the Scatchard plot indicates a positive cooperativity of sites. This concentration dependence is logical, considering the NE-release stimulation at a high dose of the blocking agent but not at a low dose, where the blocking action is not severe.

Other imidazolines related to clonidine, like naphazoline (**4.46**), are also α₂ agonists. In general, α-methyl substituents on phenethylamines increase their α₂ affinity, as does loss of the 3–OH group. Loss of the 4–OH group of the catechol nucleus promotes α₁ activity.

4.3.6 Adrenergic Drugs: Postsynaptic Effects

Adrenergic drugs may also exert postsynaptic effects. There is a considerable body of classical structure–activity correlation studies in the adrenergic field for these effects. It may be summarized as follows:

1. *Phenolic hydroxyls* are important for adrenergic agonist activity. Removal of the 4–OH group leaves intact only α-agonist activity, whereas removal of the 3–OH group abolishes *both* α- and β-agonist activity. The 3–OH group can, however, be replaced by a sulfonamide (soterenol) or a hydroxymethyl (salbutamol) group. 3-Amino compounds can be extremely potent. Replacement of the 4–OH group by

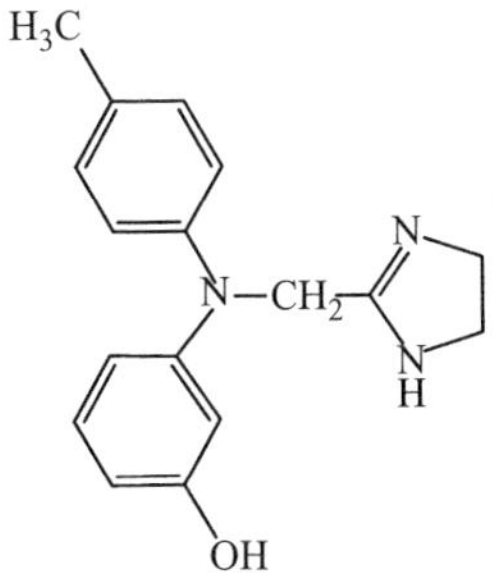

Phentolamine (4.44)

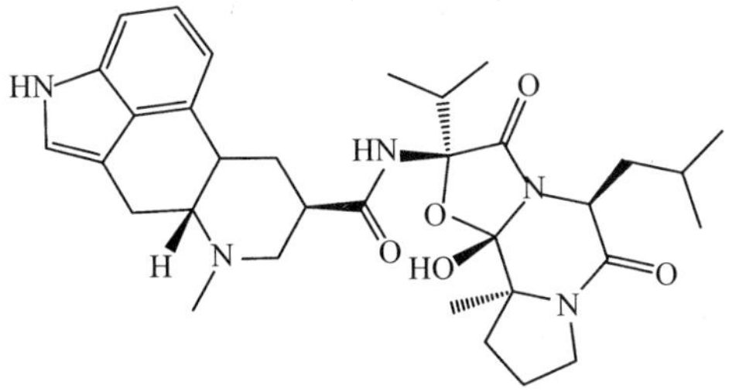

Dihydroergocryptine (4.45)

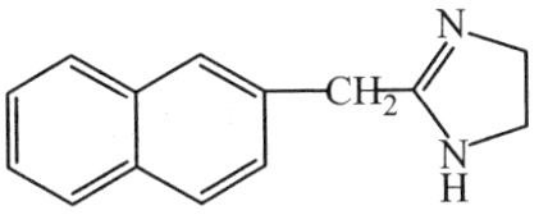

Naphazoline (4.46)

any such group leads to an almost total loss of pharmacological action; alternatively, the resultant compound may become an antagonist.

2. The *two-carbon side chain* is essential for activity, although some exceptions are known. The benzylic carbon (next to the ring) must have the *R* absolute configuration.

3. The *alcoholic hydroxyl* can be replaced only by an amino or hydroxymethyl group.

4. Small (-H, -CH₃) *N-substituents* produce α activity; larger ones (-CH(CH₃)₂, aryl) lead to β activity.

4.3.6.1 α-Adrenergic Agonists

These compounds include NE, which acts on both α and β receptors, and epinephrine, which is more active on β receptors. As mentioned previously, catecholamines lacking a 4–OH group, such as phenylephrine (**4.47**) and methoxamine (**4.48**), show almost pure α_1 activity. They are both vasoconstrictors, used in treating hypotension (low blood pressure) and nasal congestion. These drugs may also inhibit insulin release.

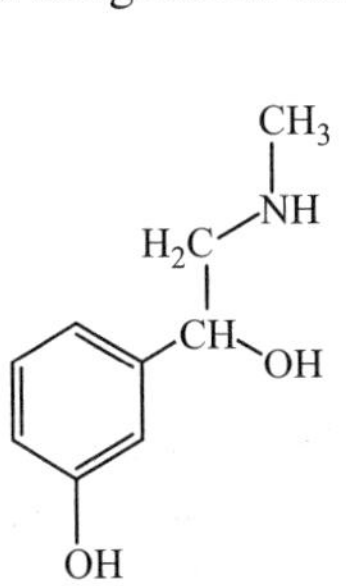

Phenylephrine (4.47)

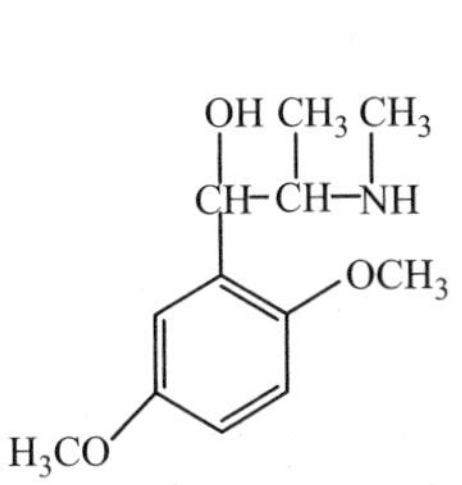

Methoxamine (4.48)

Clonidine (**4.42**) is an α_2 agonist. Therapeutically, clonidine is a central antihypertensive agent, which may perhaps act on the baroreceptor (blood pressure sensor) reflex pathway, on cardiovascular centers in the medulla, and also peripherally. As is evident from the discussion above, clonidine and α-methyldopa act in the same way. Clonidine also abolishes symptoms of opiate withdrawal and stimulates histamine H_2 receptors (section 4.6.3). It seems to have an interesting psychopharmacological activity as well, acting as an antianxiety agent that stimulates α_2 adrenoceptors and therefore decreases NE levels.

4.3.6.2 α-Adrenergic Antagonists

Because of their peripheral vasodilator effect, these drugs are used in the treatment of hypertension. They act beneficially in shock and frostbite by increasing peripheral circulation. Some, like phenoxybenzamine, also have cholinergic effects, indicating that these antagonists cross-react with the AChR.

Chemically, adrenergic blocking agents are a varied group, bearing little resemblance to the adrenergic agonists, since they use accessory binding sites of the receptor. Benzodioxanes such as piperoxan (**4.49**), and quinazolines like prazosin (**4.50**), all carry bulky, basic side chains. Phentolamine (**4.44**) is a rather nonselective older drug of imidazoline structure. There are a few "irreversible" alkylating agents such as phenoxybenzamine (**4.51**) and its congeners, which carry a β-chloroethylamine side chain capable of reacting covalently with nucleophilic -OH or $-NH_2$ groups. Although these compounds are useful drugs and experimental tools, they are slowly removed from the receptor and are therefore not truly irreversible. All of them act through IP_3 as second messenger.

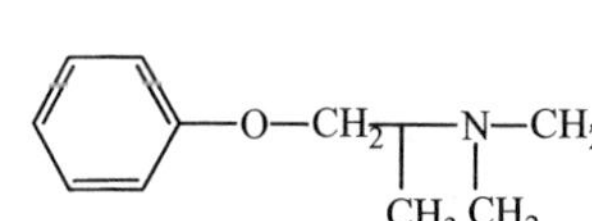

Piperoxan (4.49)

Prazosin (4.50)

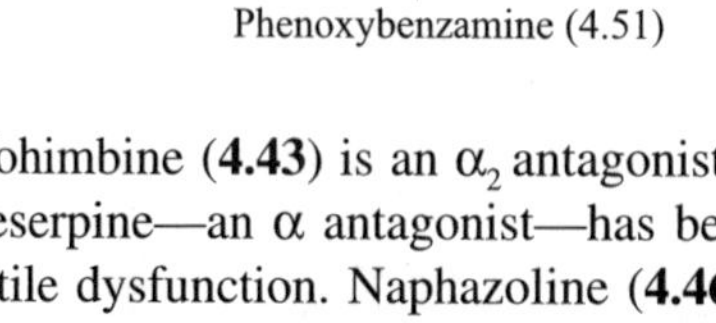

Phenoxybenzamine (4.51)

Yohimbine (**4.43**) is an α_2 antagonist. Yohimbine, an indole alkaloid closely related to reserpine—an α antagonist—has been evaluated as a potential treatment for male erectile dysfunction. Naphazoline (**4.46**) and other α-agonist imidazoline compounds are nasal decongestants, used by inhalation to decrease swelling of the nasal mucosa. Overdependence on and overuse of these drugs can lead to rebound swelling.

α-receptor blocking agents have a number of limited clinical applications. The most specific use of an α-adrenergic antagonist is in the management of *pheochromocytoma*. A pheochromocytoma is a tumor of the adrenal gland which excretes a mixture of epinephrine and norepinephrine, producing hypertension, accelerated heart rate, and cardiac arrhythmias in the patient. α-receptor antagonists may be of value in people with this tumor to prevent acute cardiac and hypertensive episodes; α-receptor blockade may also have use in the treatment of *benign prostatic hyperplasia* (BPH). This disorder of older men involves progressive urinary symptoms as the enlarging prostate slowly pinches the urethra closed. Multiple well-controlled clinical studies have shown the efficacy of α_1-receptor antagonists (e.g., prazosin (**4.50**), doxazosin (**4.52**), terazosin (**4.53**)) in patients with BPH.

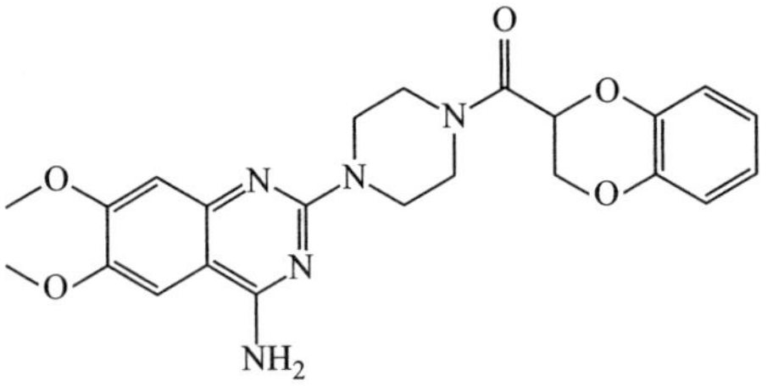

Doxazosin (4.52)

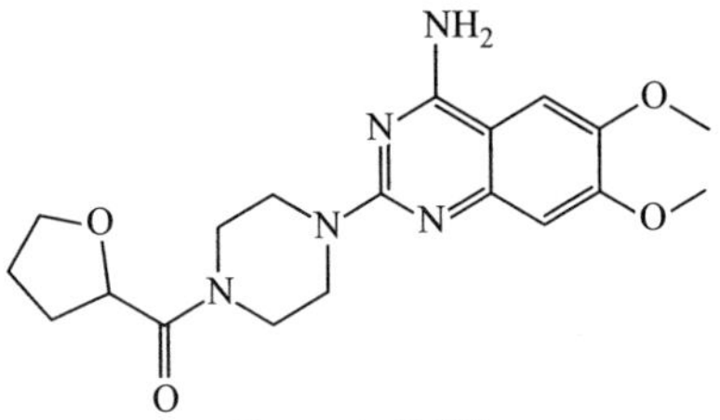

Terazosin (4.53)

4.3.6.3 β-Adrenergic Agonists

Isoproterenol (**4.54**) is a pure β agonist, and was the compound that first demonstrated the existence of adrenergic isoreceptors. It acts on both β_1 and β_2 receptors, and therefore produces a number of side effects in addition to its primary use as a bronchodilator. Another specific β agonist is methoxyphenamine (**4.55**). Studies on compounds such as these and related congeners have led to the identification of several structure–activity rules concerning β agonists with regard to β_1 and β_2 selectivities:

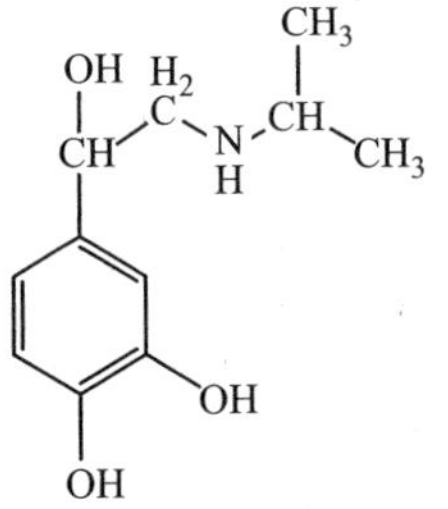

Isoproterenol (4.54)

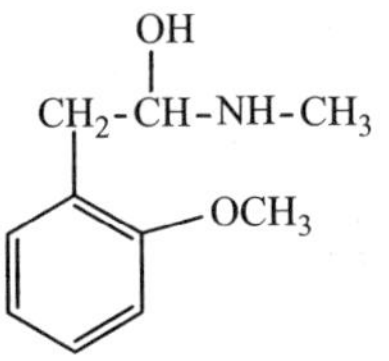

Methoxyphenamine (4.55)

1. Modification of the *catechol ring* can dramatically increase β_2 activity, such as bronchodilation. The β_2/β_1 index increases when a 3–OH group is substituted for a sulfonamide (soterenol, **4.56**), hydroxymethyl (albuterol, **4.57**), or methylamino group. Inclusion of the nitrogen into a carbostyryl ring (an α-dihydroquinolone) leads to a compound (**4.58**) that is 23,000 times more active than isoproterenol and also extremely selective. This compound carries a somewhat different N-substituent, a *tert*-butyl group, like albuterol.

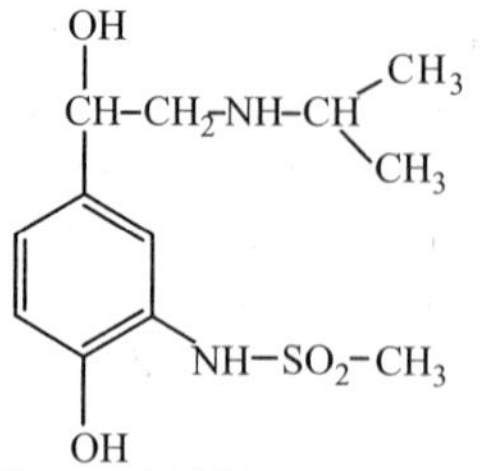

OH
|
CH-CH₂-NH-CH⟨CH₃ / CH₃

NH-SO₂-CH₃

OH
Soterenol (4.56)

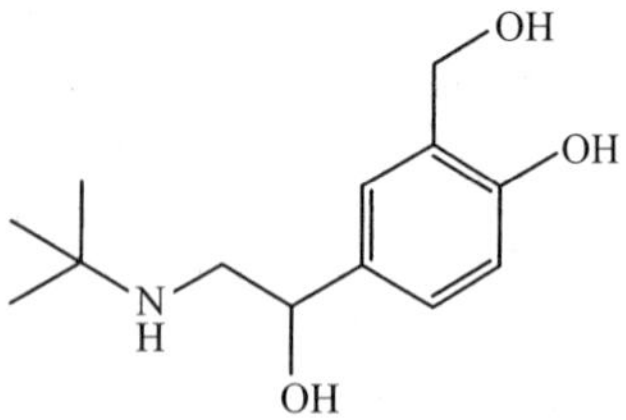

OH

OH

OH

OH
Albuterol (4.57)

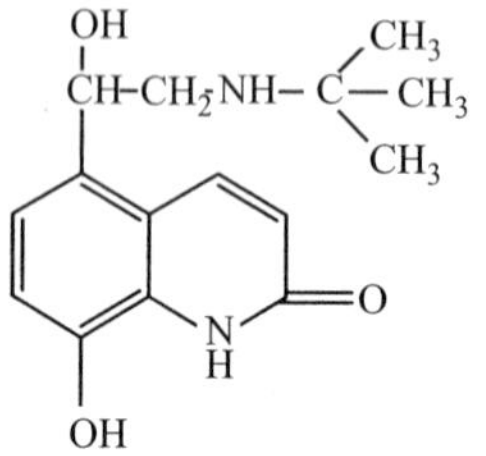

OH
|
CH-CH₂NH-C⟨CH₃ / CH₃ / CH₃

N
H
=O

OH
(4.58)

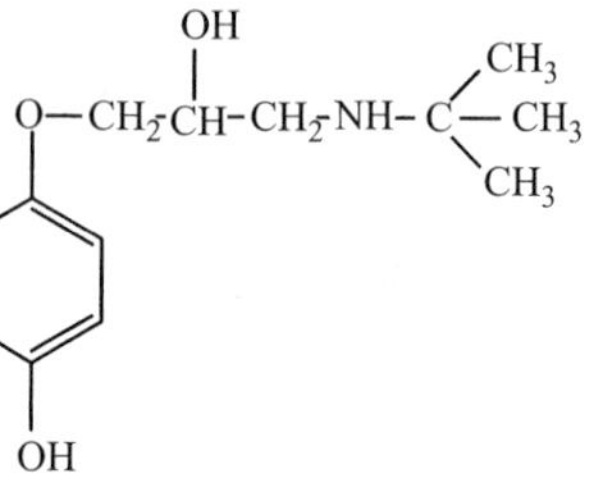

OH
|
O-CH₂CH-CH₂NH-C⟨CH₃ / CH₃ / CH₃

OH
Prenalterol (4.59)

2. Modification of the intermediate *aminoethanol side chain* between the catechol and the terminal amine can produce surprising effects, as exemplified by prenalterol (**4.59**), in which the insertion of an oxygen and a carbon atom between the alcohol and phenyl groups changes the parent compound (β_2/β_1 = 11.5) into a very selective β_1 agonist (β_2/β_1 = 0.075). Although many of these β agonists are useful in the management of heart failure, their apparently "cardioselective" (β_1) activity does not necessarily reflect true receptor selectivity.
3. Tertiary amines are not active; the β_2 activity of secondary amines is increased by branched arylalkyl chains.

These basic design principles have been of value in the design and synthesis of varying β agonists with varying β receptor selectivities.

Although β agonists (as well as β antagonists) have been thoroughly investigated for many years, active compounds have continued to emerge over the past decade, reflecting the immense clinical importance of these classes of drugs. β agonists have many potential therapeutic benefits. By stimulating β_1 receptors, hence activation of adenylate cyclase and cAMP, β_1 agonists augment all heart functions including force of contraction (systolic force, inotropism), rate of contraction (sinoatrial rate, chronotropism), and electrical conduction velocity within the heart (conduction velocity, dromotropism). Although, in theory, β_1 agonists would be valuable in heart failure, their use does carry the risk of cardiac arrhythmias. Nevertheless, the cardiostimulatory effects of β agonists such as epinephrine is exploited in the treatment of cardiac arrest. β_2 agonists produce a uterine relaxant effect and may be used in obstetrics to prevent premature labour. However, β_2 agonists enjoy a much wider application in the treatment of lung disease.

Treatment of Asthma. Broadly speaking, pulmonary diseases can be categorized into two broad groups: *restrictive lung diseases* (in which lung compliance is reduced, resulting in inadequate pulmonary inflation) and *obstructive lung diseases* (in which the tubes transporting air to and from the lungs [bronchi, bronchioles] are reduced in diameter, hindering air entry and promoting air trapping within the lungs); of these two groups, obstructive diseases are much more common. The restrictive diseases are usually caused by infiltrative diseases of the lung, such as silicosis, farmer's lung or coal miner's pneumoconiosis. The obstructive diseases may be either acute (e.g., *asthma*) or chronic (e.g., *emphysema, chronic bronchitis*; collectively these latter two disorders are called *chronic obstructive pulmonary disease* [COPD]). Asthma is the prototypic obstructive lung disease and is a medical disorder in which therapeutic manipulation of adrenergic messengers has been of crucial importance; accordingly, β_2 agonists play a central role in the day-to-day management of obstructive pulmonary diseases.

Asthma is characterized by recurrent episodic shortness of breath caused by bronchoconstriction arising from airway hyperreactivity and inflammation. Clinically, the patient with asthma wheezes and may even become cyanotic as the breathing problem worsens. Allergic inflammation of the bronchial lining is an important causative factor in asthma. Leukotrienes are formed during this inflammatory process, and as the inflammation develops the bronchi become hypersensitive to a wide range of spasmogenic stimuli, including exercise, cold air, or even cyclooxygenase inhibitor drugs (see chapter 8). The first-line treatment of choice for an acute asthma attack is the use of a short-acting aerosolized β_2 sympathomimetic. β_2 agonists (e.g., terbutaline (**4.60**), salbutamol (**4.61**)) mediate rapid bronchodilation in the lungs, and are thus a crucial component in the initial pharmacological treatment of asthma.

Terbutaline (4.60)   Salbutamol (4.61)

If β_2 mimetics have to be used more frequently than three times per week, then the pharmacological management should also attack the inflammatory component of the disease. This may be achieved using either anti-inflammatory steroids (see section 5.1) or long-acting, selective leukotriene receptor antagonists (e.g., zafirlukast, see chapter 8). Since β agonists are therapeutic for asthma, it stands to reason that β antagonists are not; in fact, the use of β antagonists can precipitate catastrophic worsening in asthmatic patients.

Following on the clinical successes of β_2 agonists, continuing work endeavors to identify therapeutic indications for other β agonists. β_3 agonists may have value in the treatment of obesity, type II diabetes, and irritable bowel syndrome. Most recent work has focused on the development of β_3 agonists for the treatment of obesity. The preliminary stages of this work have been reviewed by Weyer (1999). Various aryloxypropanolamines and arylethanolamines have been explored as molecular platforms for the development of β_3 agonists. However, development of several β_3 agonist compounds has been discontinued as a result of their lack of efficacy.

Although β agonists also mediate metabolic effects, such as the conversion of glycogen to glucose (glycogenolysis) in both liver and skeletal muscle, these effects have not yet been amenable to meaningful therapeutic exploitation.

4.3.6.4 β-Adrenergic Antagonists

β antagonists (or β-blockers) are perhaps the most important adrenergic drugs since they are used not only for neurological indications but also for a wide variety of non-neurologic indications in organ systems throughout the body, reflecting the far-reaching and pervasive influence of adrenergic neurotransmitter messenger molecules. Beta-blockers have been used extensively in the management of systemic arterial hypertension, a disease very prevalent in the Western world. Arterial hypertension ("high blood pressure"), sometimes called "the silent killer," predisposes to stroke, heart attack, and peripheral vascular disease. Hypertension may be either systolic (pressure against arterial wall during heart contraction) or diastolic (pressure against arterial wall at rest) as defined by the blood pressure recording (systolic/diastolic). If the pressure is high for prolonged periods of time, it leads to damage of the arterial wall, which in turn predisposes to atherosclerosis with thickening of the arterial wall and narrowing of the arterial diameter. Hypertension may be treated with a number of agents, including β-blockers, diuretics, and angiotensin converting enzyme (ACE) inhibitors.

In addition, β-blockers may be used for other cardiovascular indications. For instance, β antagonists protect the heart by blocking cardiac workload above basal levels; this effect is used prophylactically in the treatment of angina pectoris (a tight, squeezing retrosternal chest pain arising from decreased blood supply to the muscles of the heart as a result of partial blockage of a coronary artery). Beta-blockers also slow the heart rate, and thus may be employed to treat tachyarrhythmias (also called tachycardia)—a disorder characterized by too high a heart rate. As a side effect, β-blockers can cause bradycardia (too slow a heart rate), and can worsen an underlying asthmatic propensity.

Certain β-blockers are also used to treat neurologic disorders, such as migraine headache and benign essential tremor. Tremor may be defined as a more or less regular, rhythmic oscillation of a body part around a fixed point, usually in one plane. Benign essential tremor is a common familial disorder affecting 415 out of 100,000 adults over the age of 40 years. The tremor has a frequency of 6–8 Hz and may affect the head, larynx (and thus voice), or upper extremities. Beta-blockers may also be exploited for their anxiolytic actions whereby they reduce hand trembling and chest palpitations in people undergoing emotional stress. Musicians competing in classical music competitions, public orators, and even championship snooker players have all been known to take β-blockers to settle the "shakes" prior to major competitions—representing another aspect of "drug doping" in competitive sports. Finally, when applied topically to the eye, β-blockers can be used to treat glaucoma (increased pressure within the orb of the eye).

Structurally, β antagonists are much closer to β agonists than to either their α counterparts or anticholinergic agents. The first useful β antagonist, discovered in 1948, was dichloroisoproterenol (**4.62**, DCI), obtained by simple replacement of the catechol hydroxyls by chlorine atoms. However, DCI is also a partial β agonist, and therefore cannot be used as a hypotensive drug. Propranolol (**4.63**) was the first major β-blocker to be commercialized. Structure–activity studies using compounds such as DCI and propranolol have led to the identification of some rules and regularities:

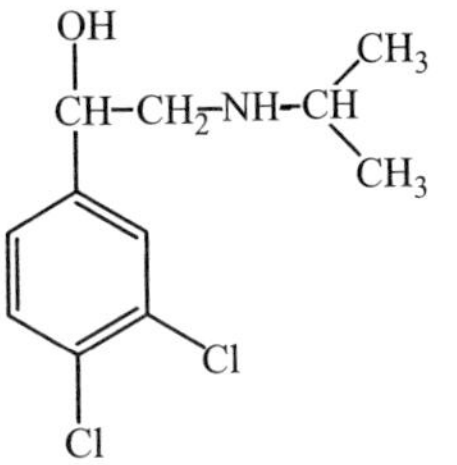

Dichloroisoproterenol, DCI (4.62)

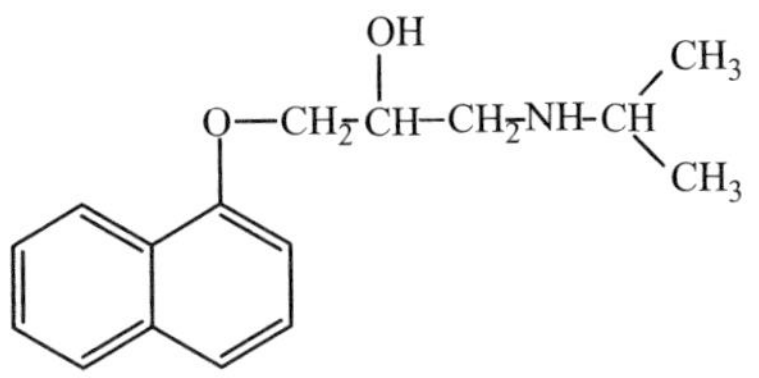

Propranolol (4.63)

1. The catechol ring system can be replaced by a great variety of other ring systems, varying from phenylether (oxprenolol, (**4.64**)) and sulfonamides (sotalol (**4.65**)) to amides (labetalol, (**4.66**)), indoles (pindolol, (**4.67**); benzpindolol, (**4.68**)), and naphthalene (propranolol, (**4.63**)).
2. The *side chain* is either the unchanged isopropylaminoethanol seen in isoproterenol or an aryloxy-aminopropanol. The side-chain hydroxyl groups are essential to activity.
3. *N-substituents* must be bulky to ensure affinity to the β receptors; isopropyl is the smallest effective substituent.

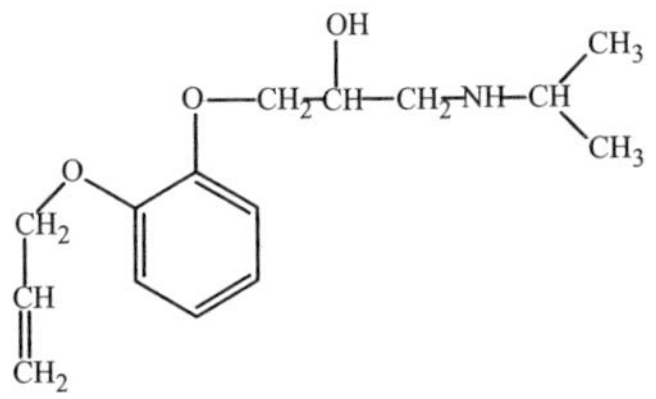

Oxprenolol (4.64) 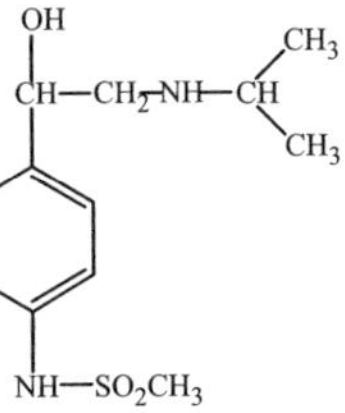   Sotalol (4.65) 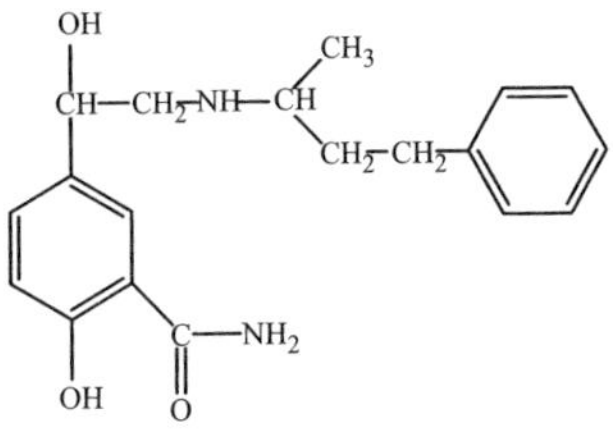   Labetalol (4.66)

It is advantageous to have selective β₁ or β₂ blockers, but this goal has been difficult to achieve since most organs have both types of β receptors in different proportions. While β-blockers—primarily propranolol (**4.63**) and the mixed α–β-blocker labetalol (**4.66**)— are antihypertensive agents, the β₁ activity of most of these compounds makes them useful in the management of some forms of angina pectoris and in cardiac arrhythmia, and they show promise in preventing second heart attacks. As mentioned, labetalol is a phenylethanolamine derivative that is a competitive inhibitor of β₁, β₂, and α₁ adrenergic receptors. It is more potent as a β antagonist than an α antagonist. Labetolol has two stereogenic carbon atoms and thus exists as a mixture of four stereoisomers; this

Pindolol (4.67)   Benzpindolol (4.68)

mixture is used clinically. The different stereoisomers have varying activities as α and β antagonists. All of the β-antagonist properties reside in the (1R,1'R)-isomer; α_1-antagonist properties are in the remaining three isomers, with the (1S,1'R)-isomer having the greatest activity. Clinically, labetolol has some advantages in the management of hypertension since the α-receptor blockade produces hypotensive vasodilation, while the β-blockade component prevents the reflex tachycardia that may be associated with the vasodilation. Unfortunately, at the pharmacokinetic level, labetolol is hindered by an extensive first-pass effect.

Because adrenergic agents such as β-blockers find such extensive use as hypotensive drugs, the etiology and drug combination treatment of hypertension are of considerable interest. A discussion in any detail of this complex and confusing field goes beyond the scope of this book, however. Other aspects of hypertension will be discussed in connection with the renin and vasopressin systems and calcium channel blockers.

Clinically, there now exists an abundance of β-blocker congeners. Propranolol was the first β-blocker to be introduced into therapy, in 1965. Now, four decades later, more than 20 different analogs have been marketed in different countries: alprenolol, bupranolol, pindolol, oxprenolol, talinolol, sotalol, timolol, metoprolol, metipranol, atenolol, bunitrolol, acebutolol, nadolol, carazolol, penbutolol, mepindolol, carteolol, befunolol, betaxolol, celiprolol, bisoprolol, bopindolol, esmolol, carvedilol, and tertatolol. (Of these compounds, acebutolol, atenolol, betaxol, bisoprolol, esmolol, and metoprolol are β_1 selective agents.) This avalanche-like increase in commercially available β-sympatholytics is driven by market forces rather than by medicinal chemistry. Variation of the basic molecular structure will often create a new patentable chemical entity, but not necessarily a drug with a novel or improved action.

4.3.7 The Clinical–Molecular Interface: Depression as an Adrenergic Disorder

Depression is one of the most common disorders of mood (affective disorders). It is characterized by a specific alteration in mood (sadness, apathy), a negative self-concept (self-reproach, self-blame), regressive and self-punitive wishes (desire to hide or die), vegetative changes (insomnia, anorexia, loss of libido), and change in activity level (agitation or listlessness). Traditionally, depression was classified as either reactive or endogenous. A reactive depression was a response to a psychosocial precipitating factor, such as death of a spouse; an endogenous depression occurred in the absence of a precipitating factor and arose from a biological predisposition. Depressions have also been categorized as *bipolar* or *unipolar*. A bipolar depression (*manic-depressive illness*) involves both manic and depressive episodes in the same person and is a biologically different disorder from a unipolar illness, which involves just the depression. Since depression is a common and potentially life-threatening disorder, drug design of antidepressants has been an ongoing activity for many decades.

According to the classical *amine hypothesis of antidepressant action*, tricyclic antidepressant drugs (TCAs) elevate the mood of patients suffering from depression and decrease the probability of suicide by interfering with the reuptake of NE or serotonin (section 4.5). Secondary amines such as desipramine (**4.69**) or nortriptyline (**4.70**) are potent inhibitors of NE uptake, whereas the tertiary amines imipramine (**4.71**), amitriptyline (**1.4**), and doxepin (**4.72**) are more effective as serotonin uptake inhibitors. According to this hypothesis, reuptake inhibition (i.e., blocking of the amine pump) increases the concentration of the neurotransmitter in the synaptic gap and thus the central adrenergic (or serotonergic) tone, resulting in mood elevation. Unfortunately, this simple and attractive hypothesis cannot explain a number of facts:

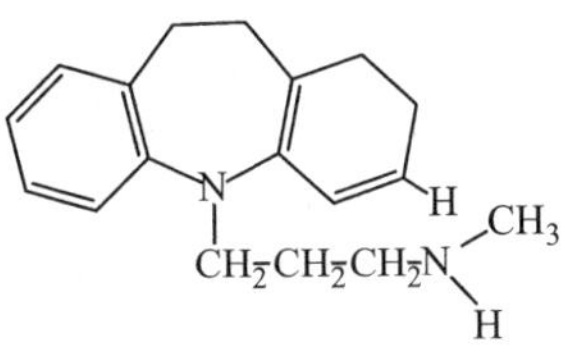

Desipramine (4.69) 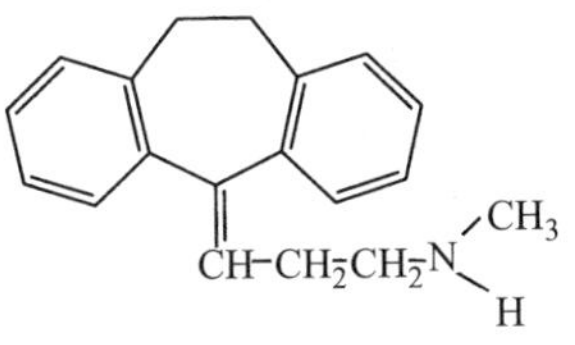   Nortriptyline (4.70) 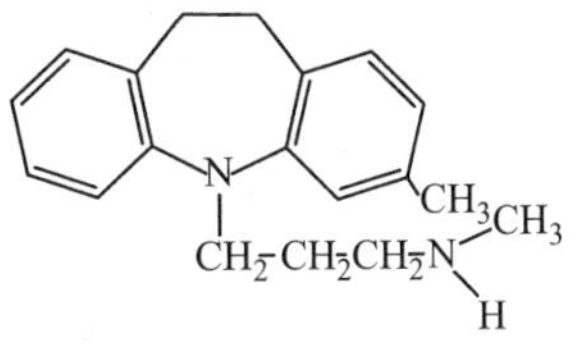   Imipramine (4.71)

1. The latency period of weeks or even months that occurs between the initiation of therapy and the antidepressant effect when tertiary tricyclics are used. Secondary amines act faster, but in both cases, although elevated neurotransmitter levels become rapidly apparent, the clinical improvement lags far behind.
2. Cocaine (**4.73**), the local anesthetic tropane alkaloid of coca leaves, is a potent NE reuptake inhibitor but has no antidepressant activity.

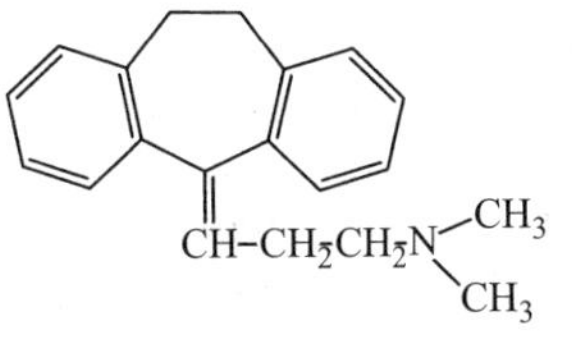

Amitriptyline (1.4) 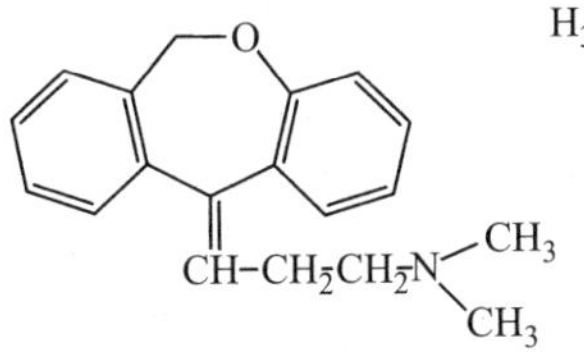   Doxepin (4.72) 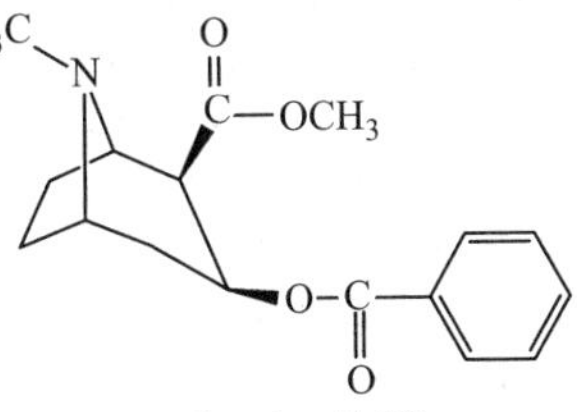   Cocaine (4.73)

The problem with the tricyclic-based neurotransmitter–receptor hypothesis of antidepressant activity is that it was based on observations in normal rat brain. Unfortunately, there are few techniques suitable for *in vivo* work on human CNS receptors of depressed patients. In addition, compounds related to dopaminergic and serotonergic functions also act as antidepressants, as will be shown later. Thus, it is likely that endogenous depression is a biologically heterogeneous syndrome; a single neurotransmitter hypothesis explaining the mode of action of all antidepressant drugs is not currently feasible.

In the 1970s and 1980s, the tricyclic antidepressants dominated the clinical management of depression. Given the somewhat distasteful and unpopular (in the lay press and movies) nature of frontal lobotomies in the 1950s and ECT (electroconvulsive therapy) in the 1960s, the molecular approach of tricyclic antidepressants seemed downright civilized. Given their widespread use (and misuse), they were sometimes simply referred to as the "tricyclics." However, it is important to observe that, like the benzodiazepine molecule (section 4.7.5), the tricyclic moiety is a *privileged structure*, with a number of successful molecules having this tricyclic structure. Antihistamines, antidopaminergics (e.g., phenothiazines, chapter 4), anticonvulsants (e.g., carbamazepine, chapter 7), and perhaps even anti-prion drugs (chapter 9) may also have tricyclic structures. Various tricyclic drugs have anticholinergic properties, and people who attempt overdoses with tricyclic drugs experience life-threatening anticholinergic side effects. In addition, tricyclic molecules may also influence brain Zn^{2+} metabolism, especially at the level of the glutamatergic NMDA receptor (section 4.9.3). Recognizing that the interaction between a drug and its receptor is a precise dance between atoms and molecular fragments, we see that the three-dimensional interplanar orientation of the various rings is crucial to an appreciation of their varying modes of action. Quantum pharmacology calculations emphasize this point.

In more recent years the tricyclic antidepressants have lost their position as the mainstay of therapy for depression and have been gradually supplanted by the *selective serotonin reuptake inhibitors* (SSRIs), such as fluoxetine. The tricyclics, however, do continue to be valuable in the management of chronic pain states such as headache or neuropathy.

An effect different from that of tricyclic antidepressants has been shown by the simple salt Li_2CO_3, commonly referred to "licarb" or simply as "lithium." It has been used since about 1970 in the long-term management of manic-depressive disorder, but it is not useful in acute mania, since its mood-stabilizing effects are seen only after 8–10 days. The mode of action of Li salts remains to be fully elucidated. Apart from interferences with transmembrane ion fluxes (via ion channels and pumps), Li^+ appears to facilitate membrane depletion of phosphatidylinositol bisphosphates, the principal lipid substrate used by various receptors in transmembrane signaling; blockade of this signal transduction pathway impairs the ability of neurons to respond to the activation of neurotransmitter receptors. Lithium may also affect the GTP-binding proteins responsible for signal transduction initiated by the formation of the agonist–receptor complex. Based upon mechanisms such as these, various reports have claimed that Li^+ accelerates catecholamine reuptake, stimulates NE turnover, and inhibits NE release—all of which are in direct opposition to TCA activities.

4.4 DOPAMINE AND THE DOPAMINERGIC RECEPTORS

Dopamine (3,4–dihydroxyphenyl-β-ethylamine, DA) (**4.34**) is a catecholamine intermediate in the biosynthesis of NE and epinephrine. There are several very important central brain structures showing specific DA receptors. Approximately 80% of DA receptors in the brain are localized in the *corpus striatum*, which receives major input from the *substantia nigra* and participates in coordinating motor movements. Additional DA receptors are found diffusely throughout the brain cortex. DA has therefore been identified as a fully-fledged neurotransmitter. Because of DA's crucial involvement in psychosis and its proven role in neurological movement disorders such as Parkinson's disease, dopaminergic drugs are the subject of very active research. Furthermore, in large doses, DA can also act on peripheral vascular α_1 adrenoceptors and cardiac β_1 receptors, but it has its own receptors in several vascular (arterial) beds, where its effect is not inhibited by the β-blocker propanolol (**4.63**).

4.4.1 Dopamine Metabolism and Receptors

Dopamine metabolism was covered in the discussion of general catecholamine biochemistry. Dopamine is stored in synaptic vesicles, and this storage can be manipulated. Although the reuptake of released DA is the major deactivating mechanism, MAO and COMT act enzymatically on DA in the same way as on NE. However, following the degradative pathway of NE, DA will finally be metabolized to homovanillic acid (3-methoxy-4-hydroxy-phenylacetic acid), since it lacks the β-hydroxyl group.

The dopamine receptors have been studied extensively by classical pharmacological methods, receptor labeling techniques, and gene cloning experiments. These experiments have revealed that, as with most neurotransmitters, several DA receptor populations exist. In the brain, DA receptors are located both pre- and postsynaptically. Five subtypes of DA receptors can be grouped into two main classes: D1-like and D2-like. There are a number of distinctions between these two classes of receptors. D1-like receptors activate adenylate cyclase; D2-like receptors inhibit adenylate cyclase. The D1-like receptors, like the β-adrenergic receptors, are transcribed from intronless genes. Conversely, the D2-like receptors contain introns, thus providing an opportunity for alternatively spliced products. In terms of molecular mass, D1-like receptors are slightly larger than D2-like receptors. D1-like subtypes include the D1 (encoded on chromosome 5) and D5 (chromosome 4) receptors; D2-like subtypes include the D2 (chromosome 11), D3 (chromosome 3), and D4 (chromosome 11) receptors. The primary amino acid sequence for the entire DA receptor class ranges from 387 residues for D4 to 477 for D5. D2-like receptors have smaller C-terminal intracellular segments but a larger intracellular loop between the sixth and seventh transmembrane segments. Two D2 receptor isoforms have been identified: D2 long and D2 short—D2 long has a 29 amino acid insert between the fifth and sixth membrane-spanning segments.

A strong correlation exists between the clinical doses of antipsychotic drugs and their affinity for brain D2 receptors. This observation led to the hypothesis that psychotic disorders resulted from overactivity of the D2 receptor subpopulation. The relative affinities of D2, D3, and D4 receptors for typical (haloperidol, chlorpromazine) and atypical (clozapine) antipsychotic molecules, together with the selective expression of D3 receptor mRNA in brain limbic areas, has led to the additional hypothesis that successful agents for psychiatric illness should also have the ability to antagonize stimulation of D3 or D4 receptors. Finally, long-term administration of antipsychotic agents leads to an increased density of D2 receptors in the basal ganglia region of the brain, causing

Parkinson's disease-like movement disorders. Therefore, the motor dysfunctions observed in patients chronically treated with antipsychotics are seemingly due to alterations in D2 receptor density.

4.4.2 Presynaptic Dopaminergic Drug Effects

Presynaptic dopamineregic drug effects may be subdivided in the same way as that in which adrenergic drugs were classified in section 4.3.5:

1. Dopamine synthesis inhibitors
2. Dopamine metabolism inhibitors
3. Dopamine storage inhibitors
4. Dopamine reuptake inhibitors
5. Presynaptic dopaminergic agonists

These enable a logical, mechanistic understanding and will be discussed individually.

4.4.2.1 Dopamine Synthesis Inhibitors

Dopamine synthesis inhibitors interfere with the enzymes involved, and are identical to those discussed in section 4.3.5 (e.g., α-methyltyrosine (**4.74**), a tyrosine hydroxylase inhibitor). In this case, α_2-adrenergic receptor effects are irrelevant, and only the classical competitive inhibitory effect is of any consequence.

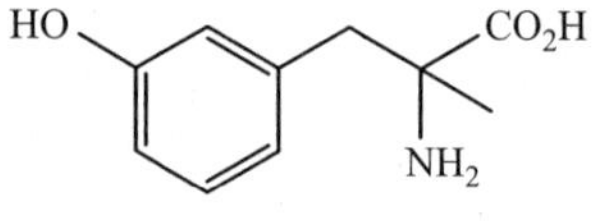

α-Methyltyrosine (4.74)

Carbidopa (**4.75**), a hydrazine analog of α-methyldopa, is an important DOPA decarboxylase inhibitor. It is used to protect the DOPA that is administered in large doses in Parkinson's disease (section 4.4.4) from peripheral decarboxylation. DOPA concentrations in the CNS will therefore increase without requiring the administration of extremely high, toxic doses of DOPA. The exclusive peripheral mode of action of carbidopa is due to its ionic character and inability to cross the blood–brain barrier. Because of this effect, carbidopa is co-administered with DOPA in a single tablet formulation as a first-line therapy for Parkinson's disease. Benserazide (**4.76**) has similar activity.

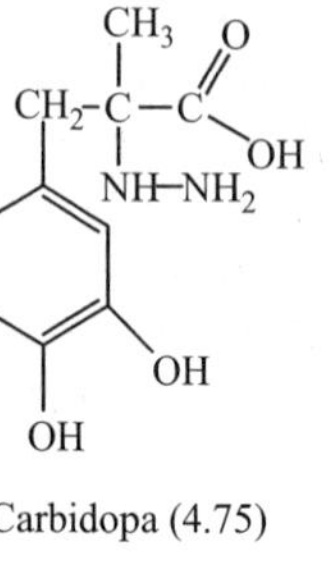

Carbidopa (4.75)

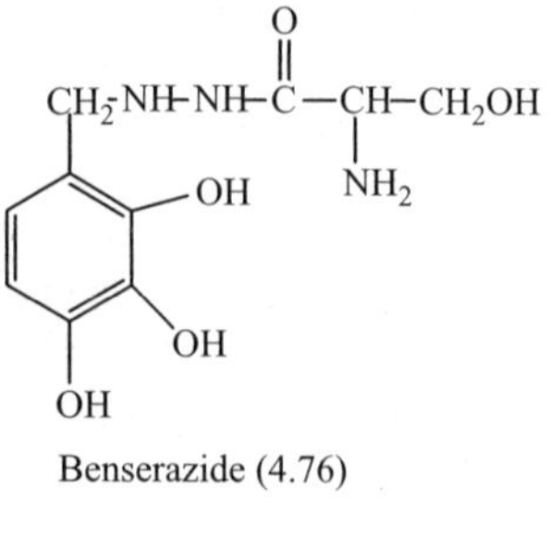

Benserazide (4.76)

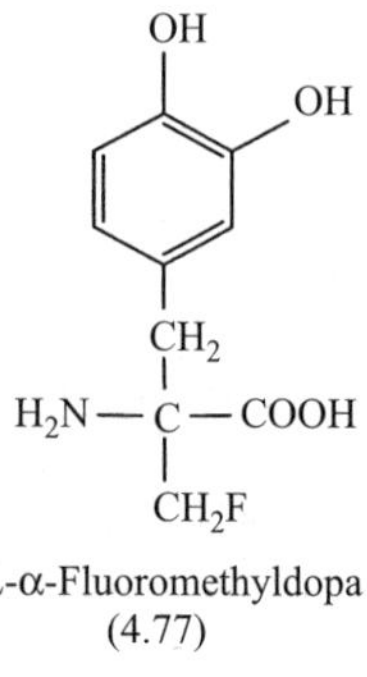

DL-α-Fluoromethyldopa (4.77)

The arylamino acid decarboxylase inhibitory action of DL-α-fluoromethyldopa (**4.77**) has also been described. By affecting the enzyme through covalent binding, this compound completely inhibits both catecholamine and serotonin synthesis. Unlike 6-hydroxydopamine, α-fluoromethyldopa does not destroy the neurons, and unlike reserpine it does not deplete chromaffin tissue in the adrenal gland.

4.4.2.2 Dopamine Metabolism Inhibitors

Dopamine metabolism inhibitors interfere with monoamine oxidase and catecholamine-O-methyltransferase. Monoamine oxidase will be discussed separately in chapter 8.

4.4.2.3 Dopamine Storage Inhibitors

The storage and release of DA can be modified irreversibly by reserpine (**3.1**), just as in vesicles containing other catecholamines and serotonin. Dopamine release can be blocked specifically by γ-hydroxybutyrate (**4.78**) or its precursor, butyrolactone, which can cross the blood–brain barrier. High doses of amphetamines do deplete the storage vesicles, but this is not their principal mode of action. Apparently, amantadine (**4.79**), an antiviral drug that is likewise beneficial in parkinsonism (and also perhaps to relieve fatigue in multiple sclerosis), may also act by releasing DA.

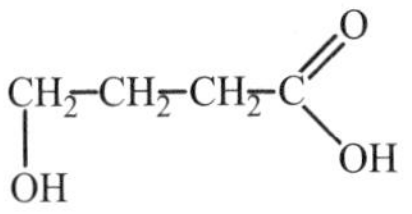

γ-Hydroxybutyrate (4.78)

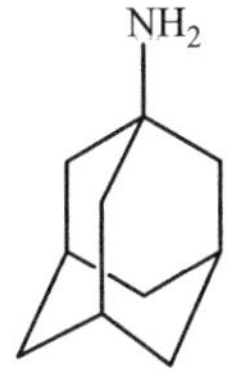

Amantadine (4.79)

4.4.2.4 Dopamine Reuptake Inhibitors

Dopamine reuptake can be inhibited specifically by benztropine (**4.80**), an anticholinergic drug, as well as by amphetamines. Several specific DA reuptake inhibitors have been discovered, such as tandamine (**4.81**), bupropion (**4.82**), and nomifensine (**4.83**), which are all potent antidepressants. Tandamine also inhibits NE uptake.

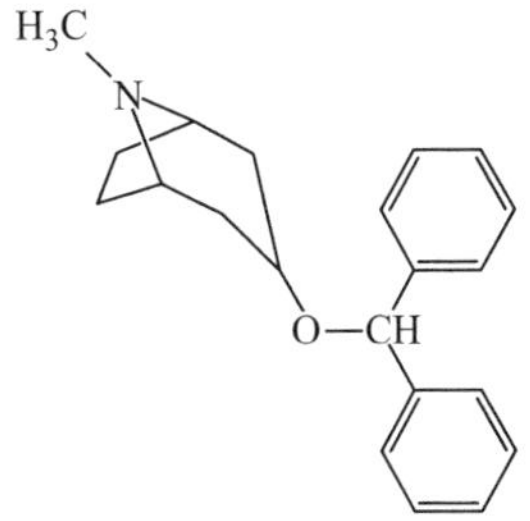

Benztropine (4.80) 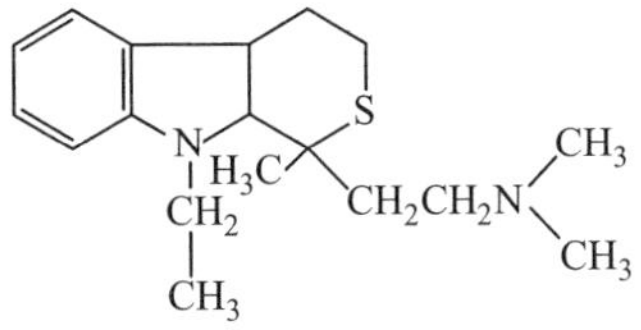   Tandamine (4.81) 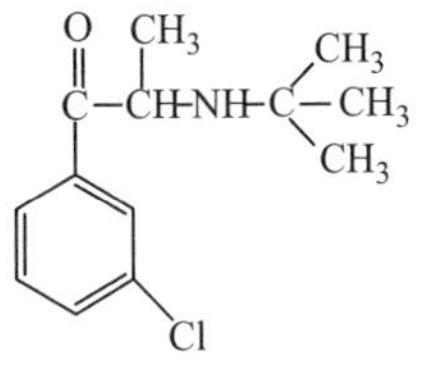   Bupropion (4.82)

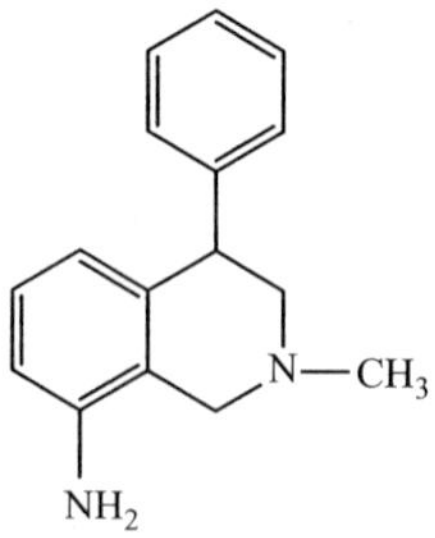

Nomifensine (4.83)

4.4.3 Postsynaptic Dopaminergic Drug Effects

4.4.3.1 Dopamine Agonists

Besides dopamine itself, several highly active DA agonists are known, all exhibiting the extended β-phenethylamine structure corresponding to a *trans* conformation. (−)-Apomorphine (**4.84**), known for its emetic (i.e., vomit inducing) effect, is both a pre- and a postsynaptic DA agonist or partial agonist, depending on the system; both hydroxyl groups are necessary for activity. Interestingly, apomorphine is now also being evaluated as an oral agent for the treatment of erectile dysfunction. N-alkylation, β-hydroxylation (of DA), and α-methyl substitution all reduce central DA activity but increase interaction with the peripheral adenylate cyclase.

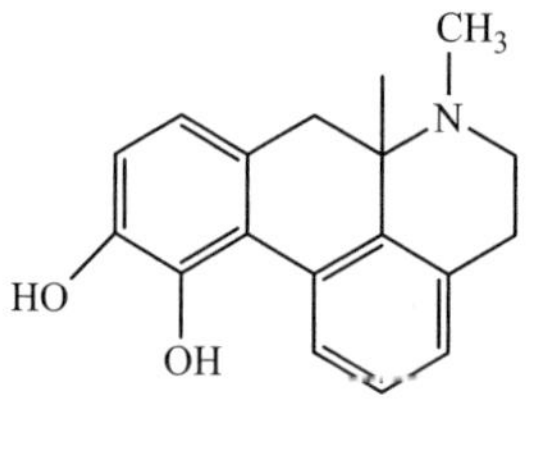

Apomorphine (4.84)

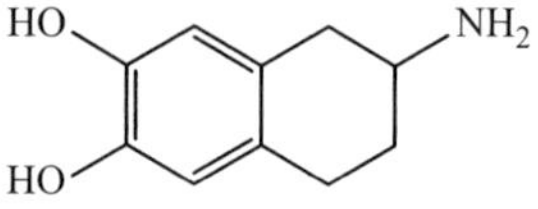

ADTN (4.85)

Extremely active compounds are found among 2-aminotetralines. 6,7-Dihydroxy-2-aminotetraline (**4.85**, ADTN) and its *N*-(*n*-propyl) derivative are well-studied 2-aminotetraline derivatives. Nomifensine (**4.83**) is related to these aminotetralines and is used as an antidepressant drug. The catechol analog of nomifensine (with two hydroxyls on the 4-phenyl ring) is also a potent inhibitor of NE and DA uptake.

The *ergot alkaloids* and their derivatives are a rich source of catecholaminergic drugs. Ergot (*Claviceps purpurea*) is a parasitic fungus found on grasses and cereals (rye). The long black *sclerotium* ("ergot") of the fungus is cultivated. Because the fungus is more valuable than the cereal crop, fields are artificially infected and the mixture of indole alkaloids is extracted from the ripe sclerotia. One of these indole peptide alkaloids, ergocryptine (**4.86**), is an α-adrenergic antagonist; however, its dihydro derivative (on the double bond of the pyridine ring) is a potent D2 agonist that is used as a vasodilator with central effects and has been advocated as a physical and mental geriatric performance enhancer. Other ergot alkaloids have a hypotensive effect and also cause smooth-muscle contraction, specifically in the uterus. This property is utilized in obstetrics to stop postpartum bleeding. There is a structural correlation between DA and ergot alkaloids: the parent tetracyclic indole acid, lysergic acid, can be considered as containing an extended phenylethylamine moiety. Amides of lysergic acid are hallucinogens and will be discussed among serotonergic drugs.

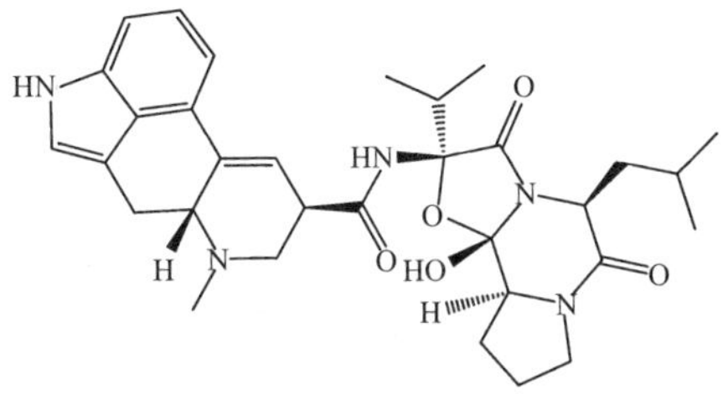

Ergocryptine (4.86) 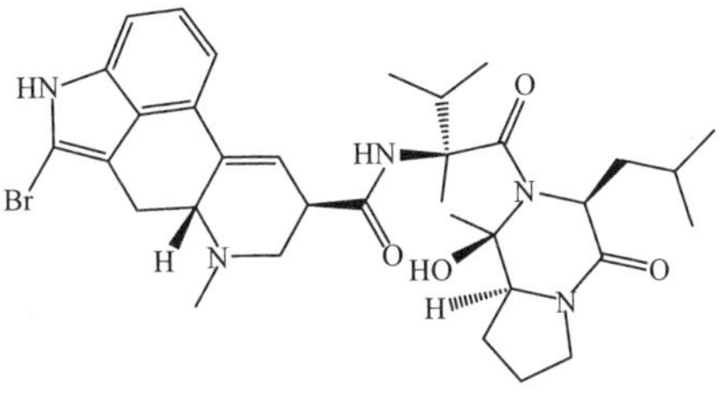   Bromocriptine (4.87)

A novel and very interesting group of compounds is represented by bromocriptine (**4.87**). It is used in parkinsonism and for inhibiting the excessive excretion of *prolactin*, a peptide hormone of the pituitary that regulates lactation. It also inhibits the secretion of *growth hormone*, another product of the anterior pituitary, and has been evaluated in the treatment of acromegaly, a form of gigantism arising from excessive growth hormone production.

4.4.3.2 Dopamine Antagonists (Neuroleptics)

The medicinal chemistry of dopamine antagonists (*antidopaminergics*) is dominated by endeavors to design therapeutic molecules for major psychiatric disorders such as psychosis. Psychosis is a severe psychiatric disorder in which mental functioning is sufficiently impaired to interfere with the patient's capacity to meet the normal demands of everyday life. This impairment involves a marked inability to correctly interpret reality, and is often accompanied by severe distortions of perception, intellectual functioning, mood, motivation, and behavior, together with personality decompensation and regression. The individual has little, if any, insight into either the nature or severity of his or her disturbance. There are many types of psychosis, including *schizophrenia, paranoid psychosis* and *affective psychosis*.

Schizophrenia is a complex group of psychotic disorders manifested by typical disturbances of thinking, mood, and behavior. The associated disintegration of mental status is attributable to a thought disorder, accompanied by misinterpretations of reality and frequently by delusions and hallucinations (i.e., they may believe that some external force is controlling their mind and actions). The term schizophrenia has generally been misinterpreted as meaning "splitting of the mind"; there is a popular misconception that this involves multiple, split personalities. (This Dr Jekyll/Mr Hyde view of schizophrenia as a state characterized by alternating, conflicting personalities is incorrect.) In schizophrenia, the split consists of incongruity between the individual's various mental functions, such as a fragmentation between thought content and emotional response. Paranoid psychosis, on the other hand, is a psychotic disorder in which delusions, generally persecutory but sometimes grandiose, are the dominant abnormality (e.g., "the government has planted listening devices in my teeth and are monitoring everything that I do; they are against me, everyone is against me"). Finally, affective psychoses are a group of psychoses characterized by a single disorder of mood, typically extreme depression, which dominates the mental life of the patient, incapacitating the patient and sometimes culminating in suicide. Clearly, the psychoses are an exceedingly complex and heterogeneous array of mental state abnormalities. It is daunting to think that a single receptor (or family of receptors) can subserve such a complicated assortment of psychiatric illnesses.

Nevertheless, dopamine antagonists are successful *antipsychotic drugs* (*neuroleptics*) and are very widely used in the symptomatic management (not cure) of all forms of psychosis. The antidopaminergics were discovered in 1952 by Delay and Daniker who, when working for the French pharmaceutical company Rhône-Poulenc, became the first to synthesize chlorpromazine (**1.3**) while searching for a drug with improved anti-histaminic properties. Instead, they recognized the major sedative action of the drug in agitated schizophrenics, and a new era in the management of affective disorders began. The tricyclic thymoleptics were derived from chlorpromazine a few years later.

Chlopromazine (1.3)

Phenothiazine (4.88)

The first and original ring system used in neuroleptic drugs is phenothiazine (**4.88**). In order for neuroleptic activity to occur, the distance between the ring nitrogen and side-chain nitrogen must be three carbon atoms. Shorter chains (like promethazine with an ethylamine side chain) are merely antihistamincs with a strong sedative action. For optimal activity, the ring substituent in position 2 must be electron attracting.

Thioxanthenes lack the ring nitrogen of phenothiazine, and the side chain is attached by a double bond. In all cases, the *cis* isomer (relative to the substituted phenyl ring) is more active. Electron-attracting substituents seem to have a cumulative effect. For instance, pifluthixol (**4.89**), with a fluorine and a trifluoromethyl substituent, is 5–10 times more potent than its parent flupenthixol (**4.90**), and has an inhibitory effect ($IC_{50} = 9.7 \times 10^{-10}$ M) on the DA-sensitive adenylate cyclase of the striatum.

The *butyrophenones* are chemically unrelated to the phenothiazines, but show a similar antipsychotic action. They were developed by P. A. Jansen and derived from fentanyl-type analgesics (see chapter 5). More than 4000 derivatives have been synthesized, of which the three most widely used antipsychotics are shown. Pimozide (**4.91**) is clearly derived from benperidol (**4.92**), even though it is no longer a butyrophenone.

Recently a number of novel, non-classical chemical structures, referred to as "atypical neuroleptics," have been described: clozapine (**4.93**), risperidone (**4.94**), olanzapine (**4.95**), and sertindole (**4.96**). These atypical neuroleptics have two distinguishing

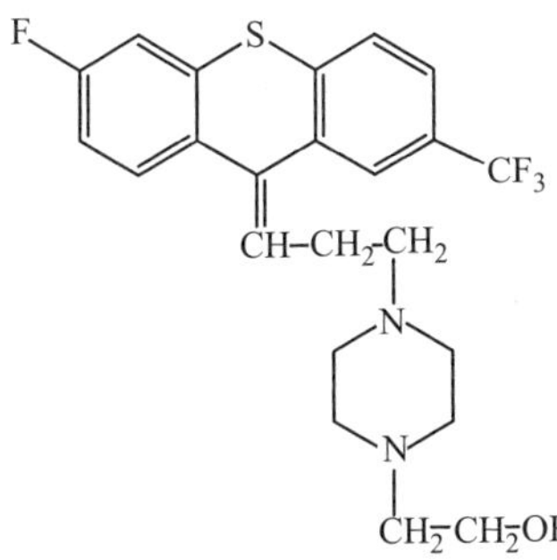

Pifluthixol (4.89)

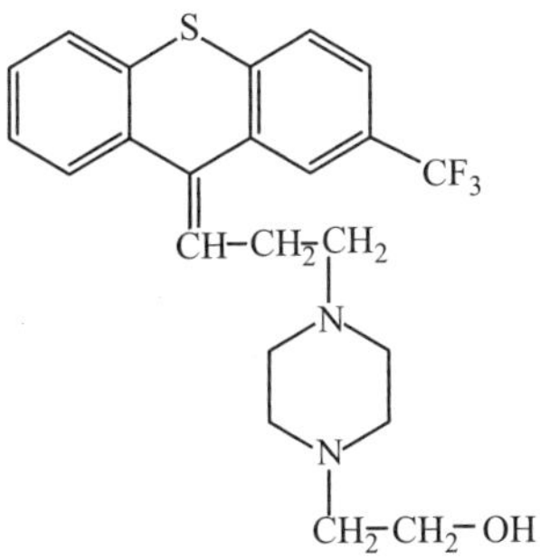

Flupenthixol (4.90)

features: an affinity for 5-HT$_2$ serotonin receptors as well as for D2 receptors, and relative anatomical selectivity for limbic structures rather than for basal ganglia structures within the brain. Clozapine also exhibits high affinity for dopamine receptors of the D4 subtypes in addition to H$_1$ histamine and muscarinic acetylcholine receptors. Clozapine may also lead to bone marrow failure (agranulocytosis), mandating the need for close blood monitoring for people on this agent.

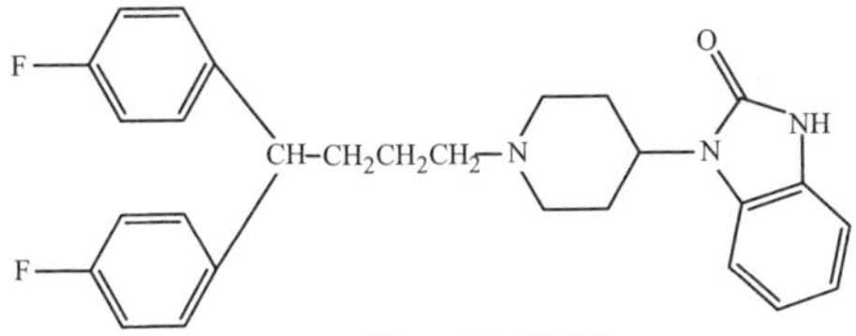

Pimozide (4.91)

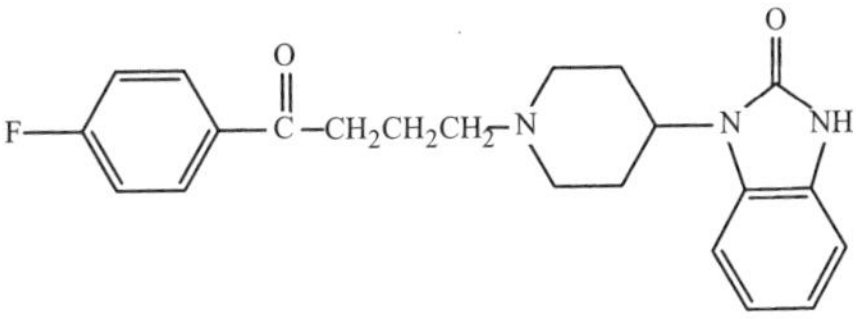

Benperidol (4.92)

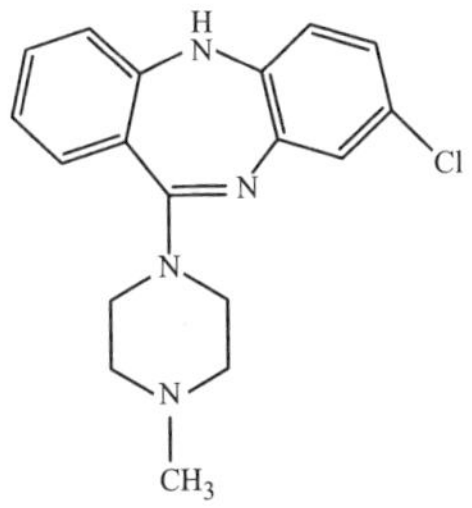

Clozapine (4.93)

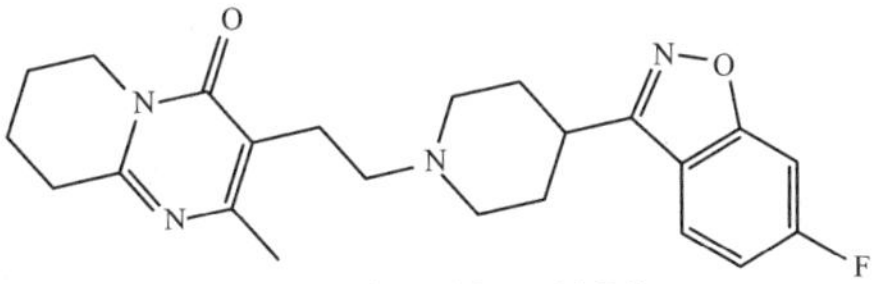

Risperidone (4.94)

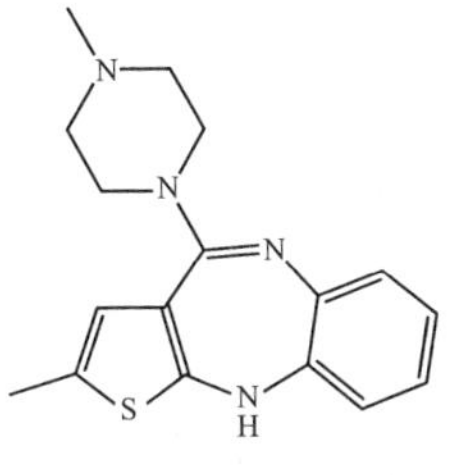

Olanzapine (4.95)

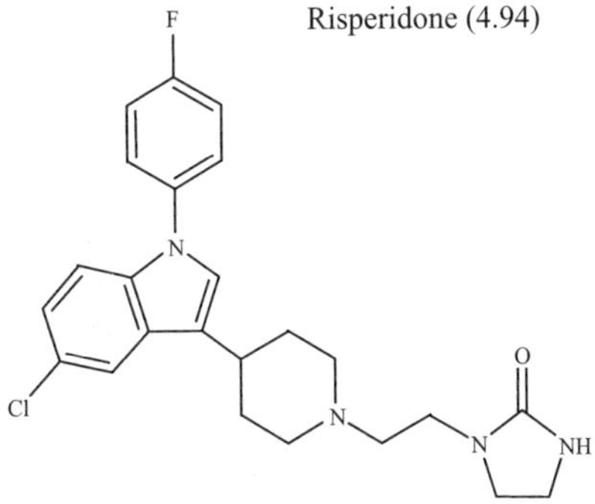

Sertindole (4.96)

The pharmacology of all these neuroleptics is extremely complex. Briefly, phenothiazines and related drugs have a calming effect on psychotic patients, without producing excessive sedation. Other central effects include the important *antiemetic effect* in disease-, drug-, or radiation-induced nausea, but not so much in motion sickness. Butyrophenones are more effective antiemetics than phenothiazines and also potentiate the activity of anesthetics.

Much of current medicinal chemistry concerning antipsychotics is exploring the various subtypes of DA receptors, and how specific or mixed antagonists at the various receptors can influence the natural history and course of major psychiatric illnesses.

The mode of action of antipsychotic neuroleptics is that of postsynaptic dopamine, especially D2, receptor blockage. The inhibition of [^{3}H] haloperidol (**4.97**) binding by neuroleptics versus the inhibition of apomorphine (**4.84**) effects shows an excellent correlation ($r = 0.94$), and even average clinical doses correlate well ($r = 0.87$) with drug binding. Although such a correlation does not prove causality, it is a strong indication of a uniform mechanism of action, especially the correlation with an *in vivo* measure of daily clinical dosage.

Haloperidol (4.97)

The most common side effects of many antipsychotics are the so-called *extrapyramidal* symptoms: rigidity and tremor (that is, parkinsonian symptoms), continuous restless walking, and facial grimacing. The final, even more severe side effect of many neuroleptics is *tardive dyskinesia,* which is manifested by stereotypic involuntary movements of the face and extremities. This syndrome, which is more prevalent in older patients after prolonged use of neuroleptics, does not respond well to antiparkinsonian drugs. Tricyclic dopamine antagonists also have complex cardiovascular side effects and antimuscarinic activity. Sedation and hypotension are also common problems. The hypotensive effect is due to α-adrenergic activity but wears off with prolonged administration, just as the sedative activity tends to disappear, even though the latter is quite useful in the management of agitated paranoid schizophrenics.

In addition to the antidopaminergic neuroleptics, there are some new and different developments probing into the origin of schizophrenia. The neuropeptide neurotensin (NT), a cotransmitter in dopaminergic neurons, may have an antipsychotic effect through modulation of DA release. Thus, drugs acting on NT receptors could be neuroleptics. Similarly, the sulfated octapeptide form of the neuropeptide cholecystokinin (CCK) inhibits DA release by presynaptic depolarization. Other studies indicate reduced CCK and somatostatin concentration in brains of schizophrenics. Other neurotransmitters, such as glutamate, adenosine, and serotonin, are also being evaluated. All these developments are potential avenues for improved control of schizophrenia, especially in patients who do not respond to conventional neuroleptic treatment or show severe tardive dyskinesia.

It should be emphasized that, whereas neuroleptic control of schizophrenic symptoms has been spectacularly successful since the 1960s, neuroleptic treatment does not cure the psychotic patient, who will almost certainly relapse if medication is discontinued. Nor does our molecular insight answer any questions about the nature, etiology, or possible biochemistry of psychiatric disorders.

4.4.4 The Clinical–Molecular Interface: Parkinson's Disease as a Dopaminergic Disorder

Deterioration of the dopaminergic neuronal pathways, known under the name of Parkinson's disease, is manifested in a collection of neurological movement disorders of unknown etiology. The symptoms include a resting tremor (sometimes referred to as a "pill rolling tremor"), difficulty in initiating movement (akinesia), rigidity, stooped posture, shuffling gait (referred to as a festinating gait), and speech and swallowing difficulties. This is an incurable and slowly progressing disease, sometimes leading to total invalidism.

The mechanism of the neurological symptoms in Parkinson's disease was discovered from the ability of reserpine to cause akinesia in humans by the depletion of central catecholamine stores. The dopamine levels in patients who died from parkinsonism were found to be extremely low because of deterioration of the dopaminergic neuronal cell bodies and the pathways connecting the substantia nigra with the corpus striatum.

Some new light was shed on the molecular cause of Parkinson's disease by an accident. In 1982, drug addicts used a "designer" drug (a noncontrolled analog of a known and illegal narcotic) contaminated with 1-methyl-4-phenyl-1,2,3,6-tetrahydropyridine (**4.98**, MPTP). Its major quaternary metabolite, MPP$^+$ (**4.99**), which is a dopaminergic neurotoxin, produced a severe and tragically permanent parkinsonism. It did so by killing dopamine-producing cells in the brain. The effect serves as a model of Parkinson's disease, although rats do not seem to be sensitive to MPTP.

MPTP (4.98)   MPP$^+$ (4.99)

Since Parkinson's disease arises from a deficiency of DA in the brain, the logical treatment is to replace the DA. Unfortunately, dopamine replacement therapy cannot be done with DA because it does not cross the blood–brain barrier. However, high doses (3–8 g/day, orally) of L(−)-DOPA (levodopa), a prodrug of DA, have a remarkable effect on the akinesia and rigidity. The side effects of such enormous doses are numerous and unpleasant, consisting initially of nausea and vomiting and later of uncontrolled movements (limb dyskinesias). The simultaneous administration of carbidopa (**4.75**) or benserazide (**4.76**)—peripheral DOPA decarboxylase inhibitors—allows the administration of smaller doses, and also prevents the metabolic formation of peripheral DA, which can act as an emetic at the vomiting center in the brainstem where the blood–brain barrier is not very effective and can be penetrated by peripheral DA.

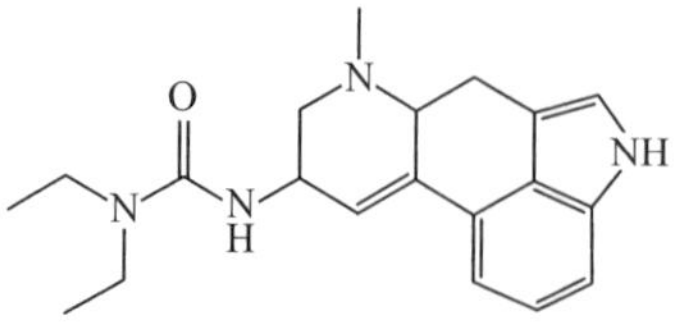

Lisuride (4.100)

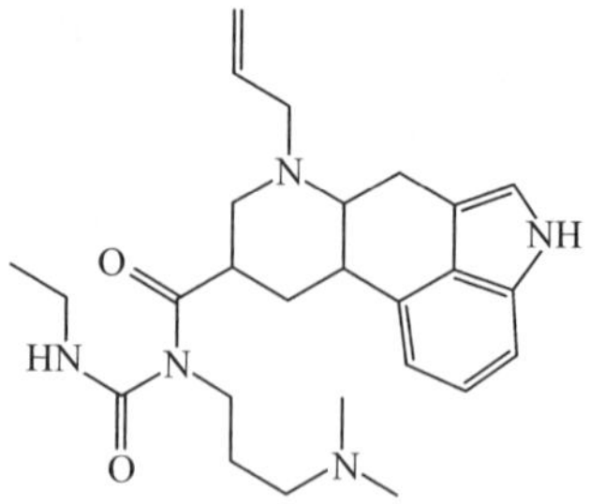

Cabergoline (4.101) 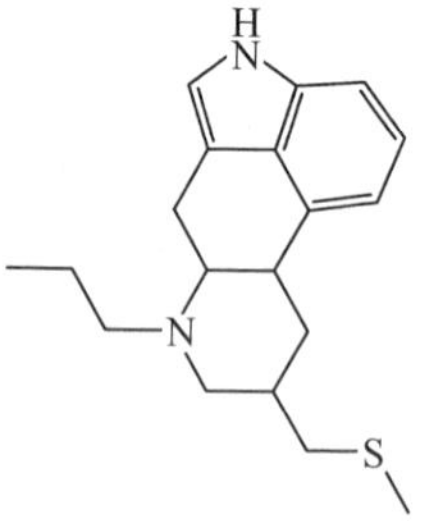   Pergolide (4.102)

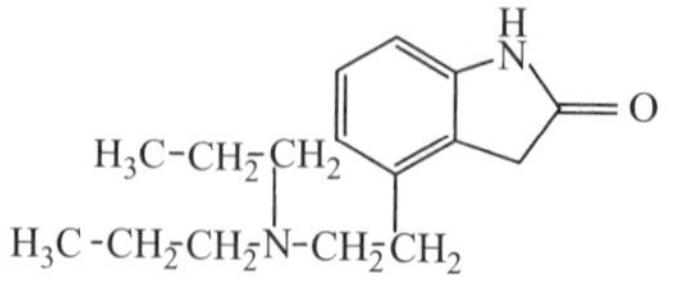

Ropinirole (4.103) 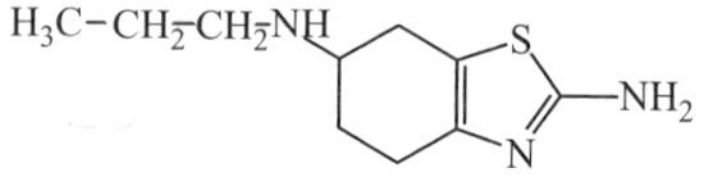   Pramipexole (4.104)

With prolonged administration, L-DOPA causes undesirable involuntary movements. Moreover, after 4–5 years, Parkinson's patients tend to become resistant to L-DOPA; this resistance occurs because the neurons required to convert the L-DOPA to dopamine are gradually dying off during the progression of the disease. This deficiency of dopamine transmission in the brain may be compensated by using ergot derivatives (e.g., bromocriptine (**4.87**), lisuride (**4.100**), cabergoline (**4.101**), pergolide (**4.102**)) or nonergot compounds (e.g., ropinirole (**4.103**), pramipexole (**4.104**)). These compounds are direct agonists for dopamine receptors and do not require conversion of L-DOPA to dopamine. These direct agonists stimulate various dopamine receptors (D1, D2 , D3) and share similar adverse effects to L-DOPA. Drugs that possibly mobilize DA stores, like amantadine (**4.79**), are sometimes of some use in treating certain forms of Parkinson's disease. Another potential mechanism of action for amantadine may involve blockade of ligand-gated ion channels of the NMDA type, ultimately leading to diminished release of acetylcholine. Since Parkinson's disease is not only a deficiency of dopamine but also an alteration in the dopamine/acetylcholine concentration ratio within the brain, anticholinergic agents may be of value. Clinically, anticholinergics are better for the suppression of the tremor of Parkinson's disease, whereas the dopaminergics are better for the bradykinesia (slow movements).

Another approach to the therapy of Parkinson's disease involves the use of enzyme inhibitors. For example, inhibition of the enzyme monoamine oxidase B (MAO-B) by selegiline (**4.105**) improves the duration of L-DOPA therapy because it inhibits the breakdown of dopamine but not of NE. Likewise, inhibitors of catechol-O-methyl-transferase (COMT) can also be exploited as agents for the treatment of Parkinson's disease. L-DOPA and dopamine become inactivated by methylation; the COMT enzyme responsible for this metabolic transformation can be clocked by agents such as entacapone (**4.106**) or tolcapone (**4.107**), allowing higher levels of L-DOPA and dopamine to be achieved in the corpus striatum of the brain.

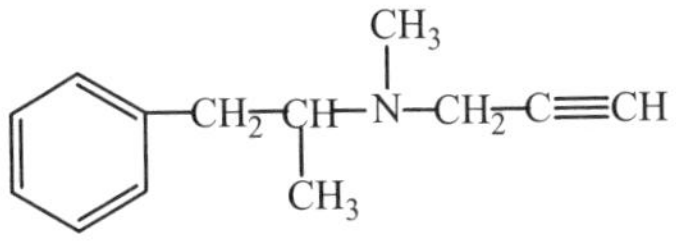

Selegiline (4.105)

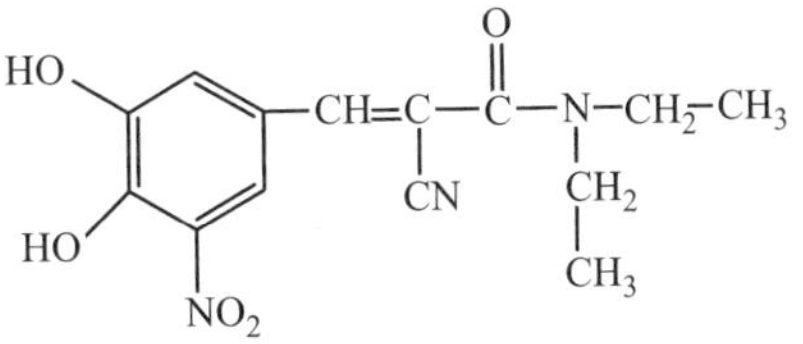

Entacapone (4.106)

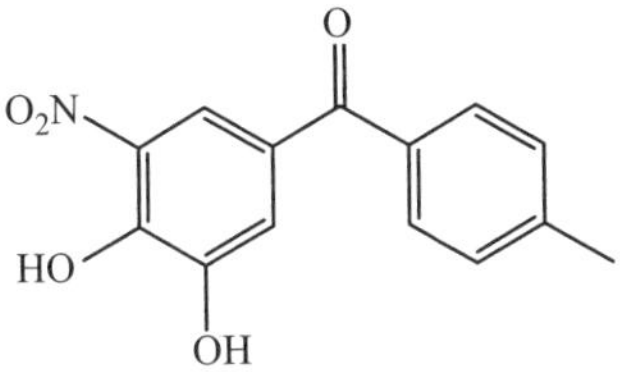

Tolcapone (4.107)

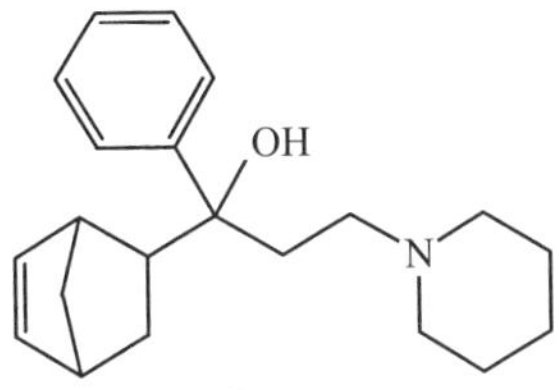

Biperiden (4.108)

In addition to successful dopaminergic-based therapies, antimuscarinic anticholinergic agents (see section 4.2.5) are also used as antiparkinsonism drugs, because the removal of inhibiting dopaminergic effects exaggerates the excitatory cholinergic functions in the striatum. Antagonists at muscarinic cholinoceptors, such as benztropine (**4.80**) or biperiden (**4.108**), suppress striatal cholinergic overactivity in the brain, thereby improving tremor and to a lesser extent rigidity; akinesia, however, is not helped by such agents. (On the other hand, the dopaminergic agents work well for akinesia, but do poorly against the tremor symptoms.)

There are many complications in the manipulation of dopamine and acetylcholine receptors for the treatment of Parkinson's disease. As discussed, anticholinergics are of benefit for the tremor of Parkinson's. However, a dementia can sometimes accompany Parkinson's disease and, as discussed in section 4.2.6, anticholinergics may in fact worsen symptoms of dementia. Also, there is an inverse relationship between Parkinson's disease and psychosis. Parkinson's disease is a deficiency of dopamine and is treated with dopaminergics; psychosis is symptomatic of an excess of dopamine and is treated with antidopaminergics. A side effect of treating Parkinson's disease with dopaminergics is confusion and psychotic delusions; a side effect of treating psychosis with antidopaminergics is the development of parkinsonian features (e.g., resting tremor, stumbling gait).

4.5 SEROTONIN AND THE SEROTONERGIC RECEPTORS

Serotonin (**4.109**, 5-hydroxytryptamine, 5-HT) is a central neurotransmitter that is also found peripherally in the intestinal mucosa and in blood platelets, where its role is incompletely elucidated; it even occurs in plants such as bananas. Although there is an enormous literature on the biochemistry and pharmacology of serotonin, our knowledge of its biological role remains somewhat fragmented. The diverse physiological effects of 5-HT influence the cardiovascular system, the cerebrovascular system, the digestive

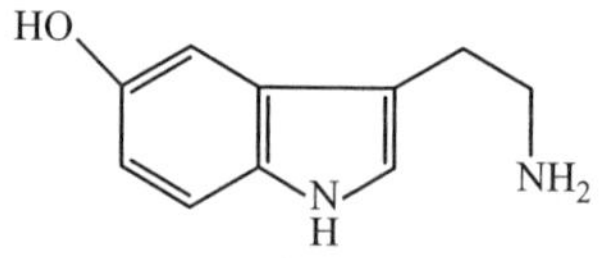

Serotonin (4.109)

system, the hematological system, and the central nervous system. Because of this diverse participation in many physiological processes, 5-HT active agents exert pharmacological effects in the plethora of disease states, including sleep disorders, modulation of circadian rhythms, eating disorders, depression, stimulation of bowel activity, migraine headache, and platelet aggregation (including the role of platelets in stroke and heart attack).

The serotonergic neuronal system in the CNS is rather restricted, localized by fluorescence histochemistry and autoradiography to the raphe region of the pons and brainstem, and projecting inferiorly to the medulla and spinal cord. The functional correlations of serotonergic neurons are equally difficult to elucidate, but work in this area has been helped by neurotoxins such as 5,6- and 5,7-dihydroxytryptamine, which destroy serotonergic neurons in the same way that 6-hydroxydopamine atrophies adrenergic networks.

4.5.1 Serotonin Metabolism

Serotonin metabolism, shown in figure 4.9, bears considerable similarities to that of the catecholamines. Serotonin itself is transformed in the pineal gland into melatonin (**4.110**), a hormone active in lightening skin pigmentation and suppressing the function of the female gonads. The β-adrenergic innervation of the pineal gland is governed by light: darkness increases cAMP formation and activation of the acetyltransferase enzyme, resulting in increased melatonin synthesis. Although situated near the thalamus, the pineal gland is not part of the CNS; it is a peripheral organ as far as the blood–brain barrier is concerned.

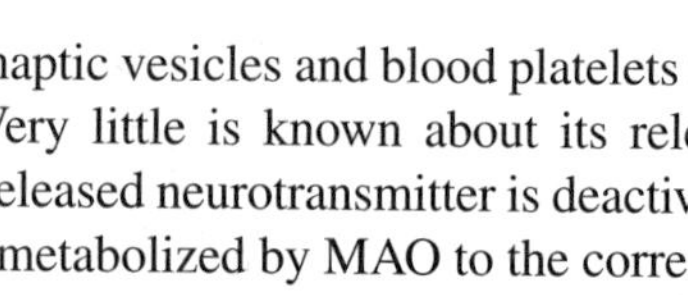

Melatonin (4.110)

Serotonin is stored in synaptic vesicles and blood platelets in the form of an ATP complex in the ratio of 2:1. Very little is known about its release, but exocytosis is the assumed mechanism. The released neurotransmitter is deactivated primarily by reuptake, but a significant amount is metabolized by MAO to the corresponding indoleacetic acid.

4.5.2 Serotonin Receptors

The characterization and classification of serotonergic receptors is continually undergoing rapid and controversial development, and has to be viewed as "in progress." Over

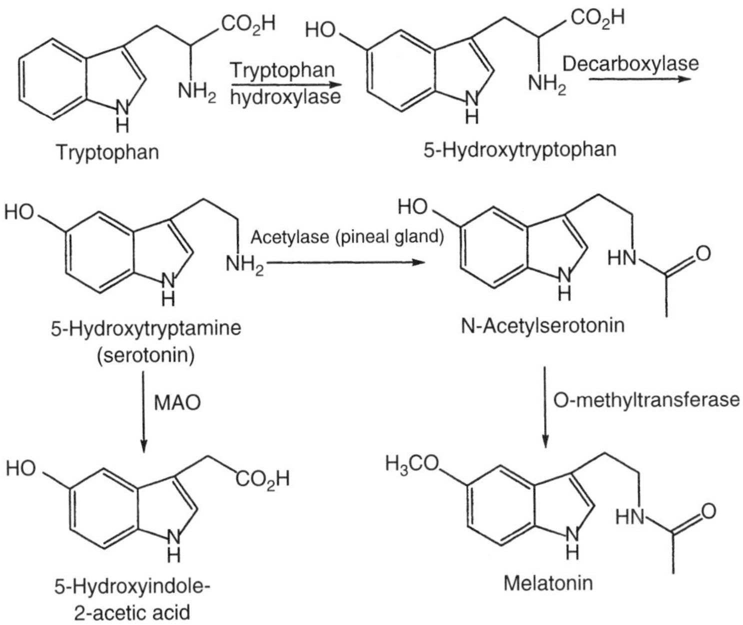

Figure 4.9 Biosynthesis and degradation of serotonin.

the past decade, receptor cloning and sequencing has greatly facilitated clarification of this confusing picture. The large number of 5-HT receptors may be divided into five subtype families. The *5-HT$_1$ family* contains receptors that are negatively coupled to adenyl cyclase through a G-protein and includes the 5-HT$_{1A}$, 5-HT$_{1B}$, 5-HT$_{1D}$, 5-HT$_{1E}$, and 5-HT$_{1F}$ receptors. Biochemical and pharmacological data suggest that the 5-HT$_{1B}$ receptor, found in rats and mice, and the 5-HT$_{1D}$ receptor, found in humans and other species, are functionally equivalent species homologs. This story is made even more confusing by the discovery of two genes that encode the human 5-HT$_{1D}$ receptor, designated 5-HT$_{1D\alpha}$ and 5-HT$_{1D\beta}$. The *5-HT$_2$ family* stimulates phosphoinositide-specific phospholipase C (PI-PLC) and includes the 5-HT$_{2A}$, 5-HT$_{2B}$, and 5-HT$_{2C}$ (formerly designated the 5-HT$_{1C}$) receptors. Activation of 5-HT$_{2A}$ receptors also mediates neuronal depolarization, a result of the closing of potassium channels. To date, the *5-HT$_3$ family* is homomeric, consisting of only one subtype protein. This receptor belongs to a ligand-gated ion channel superfamily; subunits of the 5-HT$_3$ receptor exhibit sequence similarity to the nicotinic acetylcholine and GABA$_A$ receptors. The 5-HT$_3$ receptor is a serotonin-gated cation channel that causes rapid depolarization of neurons by a transient inward ion current that is mediated by the opening of a transmembrane ion channel protein for cations. The 5-HT$_3$ receptor, like other proteins in the ligand-gated ion channel superfamily, possesses pharmacological binding sites for alcohols and anesthetic agents. The *5-HT$_4$, 5-HT$_6$, and 5-HT$_7$ family* includes receptors coupled to the stimulation of adenylyl cyclase—an effect opposite to that of the 5-HT$_1$ family. Two rat 5-HT$_7$ receptor clones, differing only in the C-terminus amino acid sequence, have been identified. The *5-HT₅ family* is a group of receptors that is coupled to neither adenylyl cyclase nor PI-PLC; the family has two subtypes: 5-HT_{5A} and 5-HT_{5B}.

This array of receptor families and subtypes may be grossly divided into two groups on the basis of their protein structure. The 5-HT_3 receptor has a tube-like structure and is a transmembrane ion channel protein. The other 5-HT receptor families tend to exhibit the typical seven transmembrane-spanning segments common to receptors coupled with G-proteins. The 5-HT receptors can also be distributed into two groups on the basis of their gene structures. The 5-HT_2 receptors are derived from genes that contain multiple introns. Other 5-HT receptors, such as the 5-HT_1 family, are encoded by genes lacking introns.

The various receptors also subserve different potential clinical applications. 5-HT_{1A} receptors are involved in psychosis, depression, and anxiety. 5-HT_{1D} and 5-HT_{1F} receptors seem to be involved with migraine headache. 5-HT_{2A} receptors are useful therapeutic targets for schizophrenia and depression. 5-HT_{2B} receptors, on the other hand, influence gastric motility and migraine headaches. The potential uses for 5-HT_{2C} ligands include anxiety, depression, obesity and cognitive disorders; 5-HT_{2C} knockout mice have cognitive disorders, epilepsy and obesity. Because of the widespread presence of neurons in the gastrointestinal tract, 5-HT_3 receptors have potential for the treatment of irritable bowel syndrome and chemotherapy-induced vomiting. Other potential therapeutic indications for 5-HT_3 drugs include anxiety, anorexia, and drug abuse. 5-HT_4 receptors are located centrally in the brain and peripherally in a variety of organs; accordingly, centrally active 5-HT_4 ligands may play a role in the treatment of schizophrenia or Parkinson's disease, while peripherally active ligands may be useful for urinary incontinence or irritable bowel syndrome. 5-HT_6 receptor ligands are effective antipsychotics and antidepressants. 5-HT_7 receptor active agents may have use in the treatment of pain associated with migraine.

4.5.3 Serotonin Receptors: Presynaptic Drug Effects

4.5.3.1 Serotonin Synthesis Inhibitors

Synthesis inhibitors block *tryptophan hydroxylase*, the first rate-determining enzyme in serotonin synthesis. Although *p*-chlorophenylalanine (**4.111**) can decrease serotonin levels by more than 90%, this agent does not cause the sedation that is seen after catecholamine depletion with reserpine. Therefore, reserpine, although capable of depleting 5-HT vesicles, causes sedation by a catecholaminergic mechanism that inhibits uptake-2. It acts on the membrane of the synaptic vesicle and seems to prevent 5-HT and catecholamine uptake into the granule.

p-Chlorophenylalanine (4.111)

4.5.3.2 Serotonin Reuptake Inhibitors

Reuptake through the presynaptic membrane is the major deactivation mechanism for serotonin. It is prevented by the tricyclic antidepressants (see section 4.3.7), among which the tertiary amines are more potent at serotonergic terminals than are the secondary bases, whereas the reverse is true for catecholaminergic synapses.

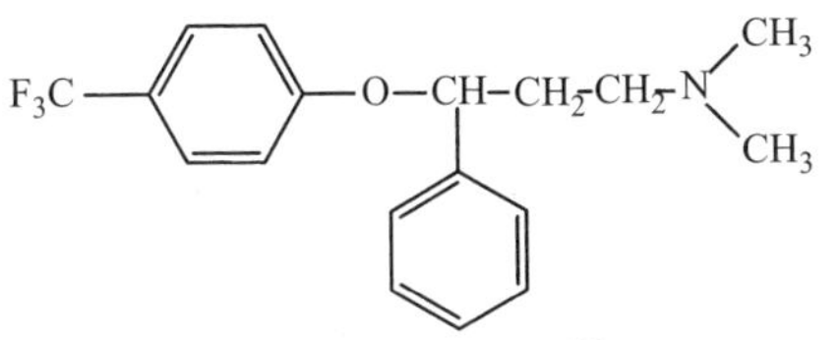

Fluoxetine (4.112)

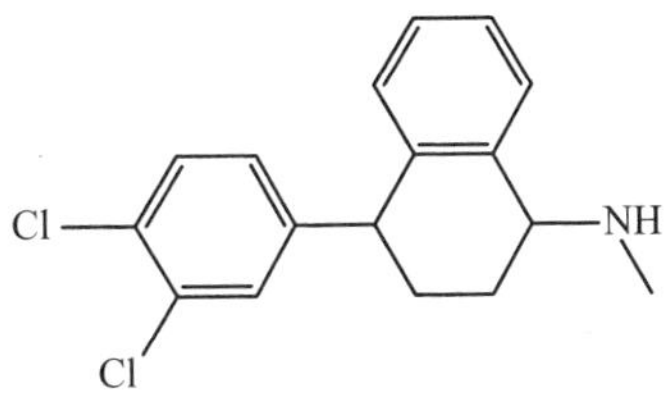

Sertraline (4.113)

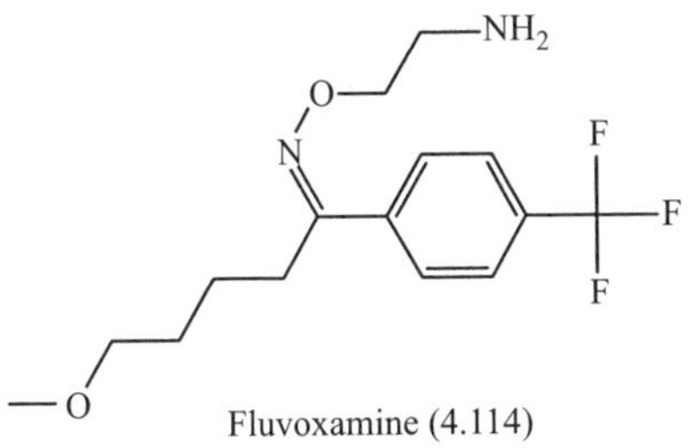

Fluvoxamine (4.114)

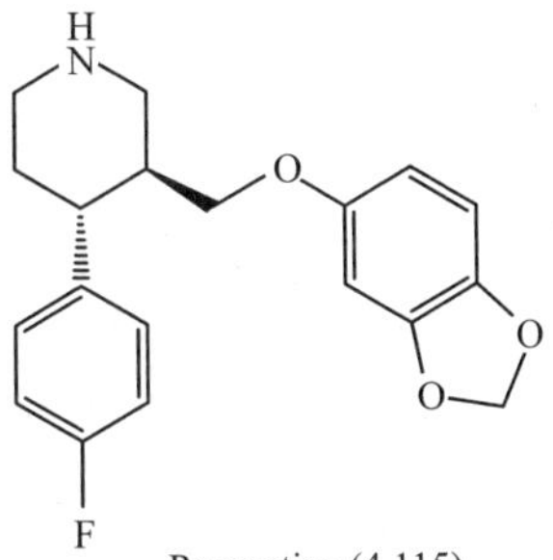

Paroxetine (4.115)

Recently, a number of selective 5-HT reuptake inhibitors have enjoyed widespread clinical success in the treatment of depression. Fluoxetine (**4.112**), sertraline (**4.113**), fluvoxamine (**4.114**), and paroxetine (**4.115**) belong to this more recently developed group of SSRIs (selective serotonin reuptake inhibitors). They are dimethylaminoethyl or dimethylaminopropyl derivatives of ring systems usually carrying an electron-attracting substituent (-CF$_3$ or -CN). All of these drugs are potent antidepressants, with a lower cardiotoxicity than the classical tricyclic agents, which strongly suggests that endogenous depression is a function of the availability of catecholamines as well as of serotonin. This does not contradict the previous discussion of the involvement of adrenergic systems (section 4.3.7). In fact, both mechanisms probably contribute to antidepressant action. A number of compounds exploit this twofold mechanism. Venlafaxine (**4.116**), for example, is an inhibitor of both serotonin reuptake and norepinephrine reuptake. At low therapeutic doses, venlafaxine behaves like an SSRI; at higher doses (e.g., more than 225 mg/day) it produces effects attributable to norepinephrine reuptake inhibition. Therefore, when administered in doses greater than 300 mg/day, it may confer a broader range of therapeutic effects than SSRIs alone.

When an SSRI agent is used with a MAO inhibitor, a dangerous pharmacodynamic interaction may occur. The combination of increased stores of monoamine together with reuptake inhibition leads to a phenomenon termed *serotonin syndrome*. This syndrome, which arises from a marked increase in synaptic serotonin, is clinically

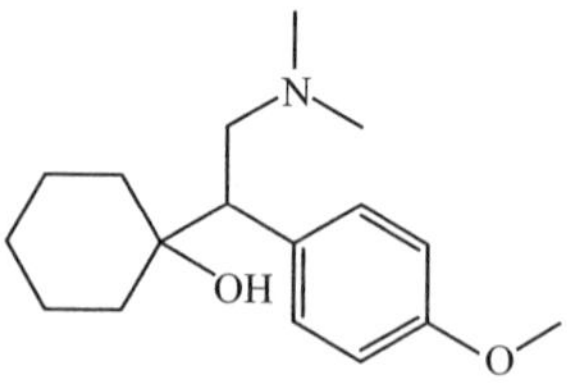

Venlafaxine (4.116)

characterized by muscle rigidity, hyperthermia (fever), myoclonus (brief lightning-like muscle twitches), and rapid changes in the patient's mental status.

4.5.4 Serotonin Receptors: Postsynaptic Drug Effects

4.5.4.1 Physiological Effects of Serotonin

The postsynaptic physiological effects of serotonin are varied and widespread. The administration of serotonin leads to powerful *smooth-muscle effects* in the cardiovascular and gastrointestinal systems. Vasodilation and hypotension may result, partly through central effects, if the serotonin concentration in the CNS is increased by administration of the serotonin precursor 5-hydroxytryptophan. Unlike serotonin, this precursor can cross the blood–brain barrier. Intestinal mobility is also influenced by serotonin.

Serotonin has an effect on the hypothalamic control of pituitary function (see chapter 5), in central thermoregulation (attributed to the 5-HT$_{1A}$ receptor), and in pain perception (probably the 5-HT$_3$ receptor), where increased serotonergic function potentiates opiate analgesia. The administration of 5-HT reuptake inhibitors like fluoxetine increases the anorectic effect of 5-hydroxytryptamine and induces a selective suppression of non-protein caloric intake in rats. The involvement of serotonin in endogenous psychiatric depression has been mentioned.

Another controversial but exciting area of research is the potential role of serotonin in sleep. 5-Hydroxytryptamine may trigger slow-wave sleep (non-REM sleep), whereas the muscarinic AChR and NE are involved in REM sleep (rapid-eye-movement sleep, paradoxical sleep, dream sleep). In addition to the aminergic regulation of sleep, recent research has identified several other presumed sleep factors: delta-sleep-inducing peptide, sleep-promoting substance, interleukin-1, and muramyl peptides.

4.5.4.2 Serotonergic Agonists

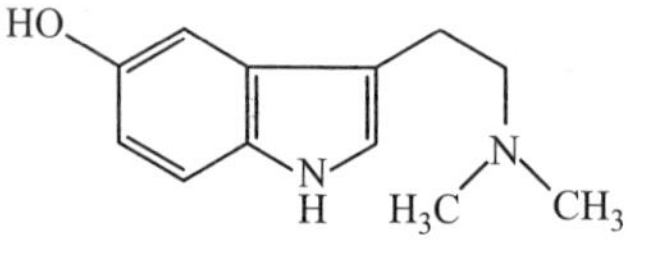

Bufotenin (4.117)

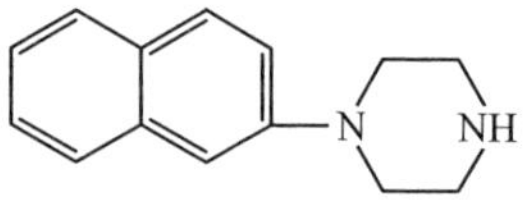

Quipazine (4.118)

The serotonergic agonists constitute a limited group of compounds, initially including serotonin (**4.109**) itself, bufotenin (**4.117**), a natural product found in plants as well as toad-skin secretion (hence the name; *Bufo* = toad), bufotenin methyl ether, and a piperazine derivative called quipazine (**4.118**). Based upon these molecular prototypes and other molecular platforms, an increasing number of analogs have been prepared and tested over the years. The majority of work has focused on 5-HT$_{1A}$ agonists. In the early 1990s, a structure–activity relationship study of substituted 8-OH-DPAT analogs demonstrated the highest 5-HT$_{1A}$ agonist activity for the *cis* C-1 substituted derivatives, with allyl being the optimal substituent whereas *cis* C-3 substitution destroyed activity. By the mid 1990s, quantum pharmacology calculations had been used to generate a binding site model of the 5-HT$_{1A}$ receptor based on 20 agonists of two chemotypes. Now, a decade later, 5-HT$_{1A}$ agonists are still being studied as potential anxiolytic/antidepressant molecules; agents that have been evaluated include alnespirone, sunepitron, and ebalzotan. Using quantum pharmacology calculations as a guide, progress has also been made in the identification of potential agonists for the 5-HT$_{1B}$ receptor. The past decade has witnessed the development of multiple 5-HT$_{1B/1D}$ receptor agonists, termed *triptans*, as potential therapeutics for migraine. A potent 5-HT$_{1F}$ agonist has been described and developed as a potential therapy for migraine. Unfortunately, its clinical development has been sidelined by nonmechanism-based liver toxicity. For the remainder of the 5-HT receptor subtypes, antagonists are, in general, more therapeutically valuable than agonists.

Another interesting serotonergic agonist is fenfluramine (**4.119**), used in the mid 1990s to control appetite. Although structurally an amphetamine, it acts by a serotonergic rather than a catecholaminergic mechanism. Fenfluramine caused serotonin release, inhibited reuptake, and was even a 5-HT agonist. Chronic use of fenfluramine caused failure of heart valves and pulmonary hypertension, a pathology perhaps related to involvement of 5-HT$_{2B}$ receptors. A number of young women taking fenfluramine as a diet aid died. Assigning the causality of these deaths to the fenfluramine molecule was delayed by the fact that fenfluramine was frequently co-administered with a much older amphetamine-like anorexiant called phentermine (**4.120**), in a combination agent called "Fen-Phen."

Fenfluramine (4.119)

Phentermine (4.120)

4.5.4.3 Serotonin Antagonists

Cinanserin (**4.121**) is not only an antiserotonin drug but also an analgesic and immunosuppressant. Cyproheptadine (**4.122**) is an antihistamine, in addition to being a 5-HT blocking agent. The other compounds shown are all semisynthetic derivatives of lysergic acid, obtained from ergot alkaloids. Methysergide (**4.123**) is related to the ergot alkaloid ergonovine, an oxytocic (uterus-contracting) drug. It is one of the most potent

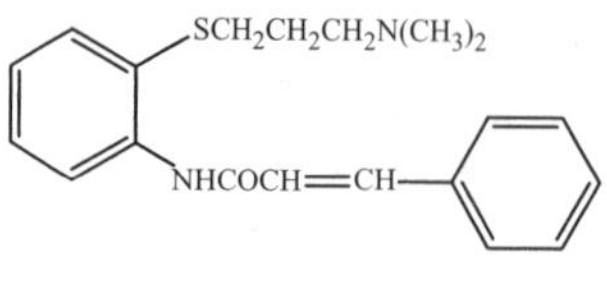

Cinanserin (4.121)

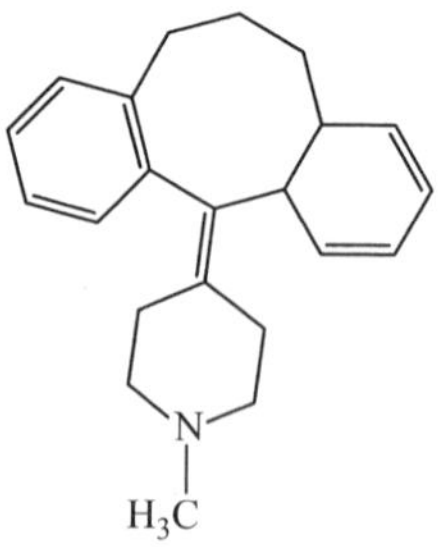

Cyproheptadine (4.122)

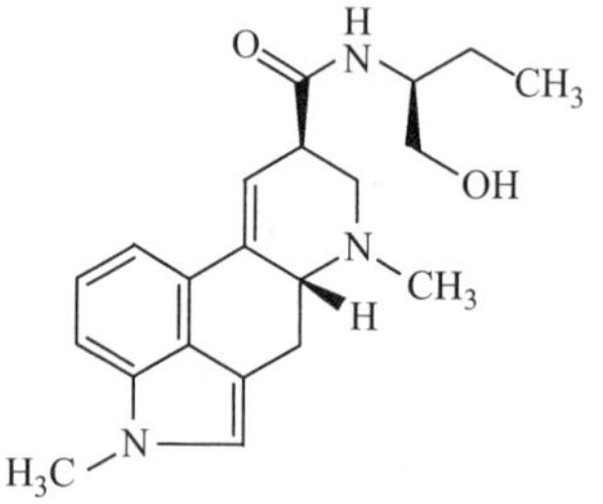

Methysergide (4.123)

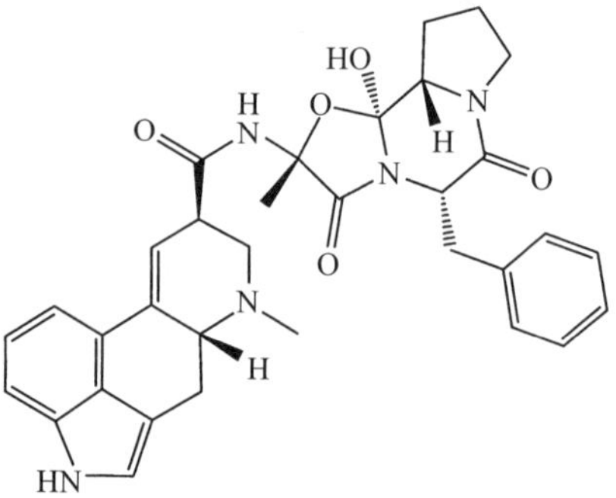

Ergotamine (4.124)

5-HT antagonists and is used for the prevention, but not the treatment, of migraine headaches. Ergotamine (**4.124**), another ergot alkaloid, can be used for treating existing migraines, since it probably acts as a vasoconstrictor. Ketanserin (**4.125**), a 5-HT$_2$ antagonist, is a clinically effective hypotensive agent that acts through a unique mechanism. Related compounds show anxiolytic activity. The diethylamide of lysergic acid, LSD (**4.126**), is a widely studied and abused hallucinogen. (The acronym originates from the German name: Lyserg-Säure-Diethylamid.)

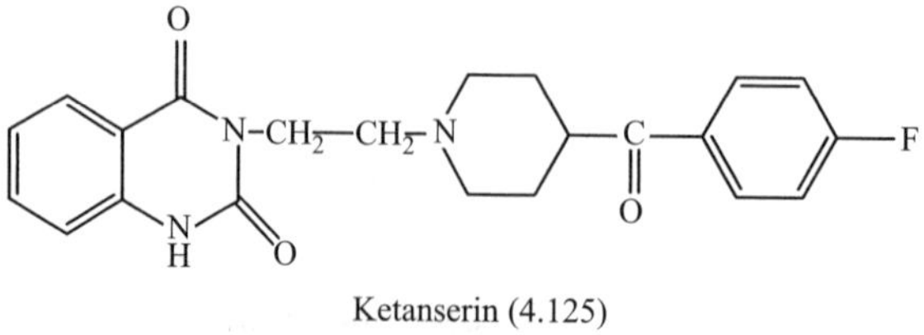

Ketanserin (4.125)

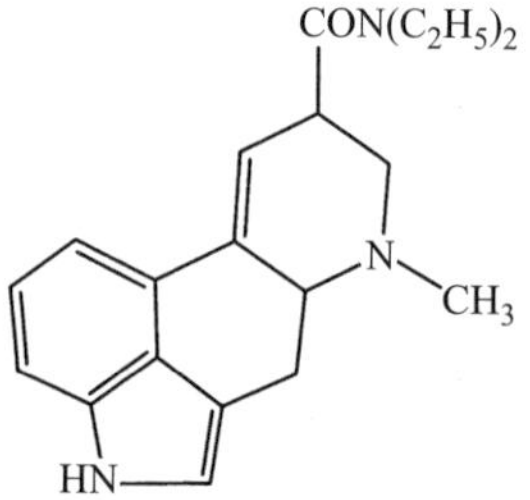

LSD (4.126)

Because of the widespread importance of serotonin, the development of serotonin antagonists is a continuing area of research endeavor. 5-HT$_{2C}$ antagonists have been clinically evaluated as antidepressants; 5-HT$_6$ antagonists may have clinical utility as antipsychotics.

4.5.4.4 Serotonergic Drugs as Hallucinogens and Psychotomimetic Agents

These drugs seem to act on central 5-HT neurons in a manner that is not clear. They decrease the turnover of serotonin, possibly through a presynaptic receptor in the raphe cells. Since 5-HT is an inhibitory neurotransmitter in many of its actions, the removal of this inhibition could lead to behavioral changes. However, to discredit this simple hypothesis, there are a number of LSD derivatives that are not hallucinogens (e.g., the 2-bromo derivative). The effects of LSD are also seen in animals with raphe lesions that have destroyed the serotonergic neurons.

There are indications that other psychotomimetic agents also act through a central 5-HT mechanism. Four groups can be distinguished:

1. Lysergic acid diethylamide and related indolalkylamines
2. Phenylethylamines (e.g., mescaline)
3. Cannabis derivatives
4. Anticholinergics

LSD (**4.126**) is a rather structure-specific compound. Only the (+)-isomer is active, and alkylamides other than the diethyl derivative, including some cyclic analogs (the pyrrolidide and morpholide analogs), have very low activity. Lysergic acid does contain the 3-indolylethylamine moiety, and it is therefore not surprising that other such structures are also hallucinogens. Among natural products, psilocin (**4.127**) and its phosphate ester, psilocybin, occur in a Mexican mushroom (*Psilocybe*). Harmaline (**4.128**), an alkaloid (from *Peganum harmala* and some other plants), and some related compounds are also effective hallucinogens.

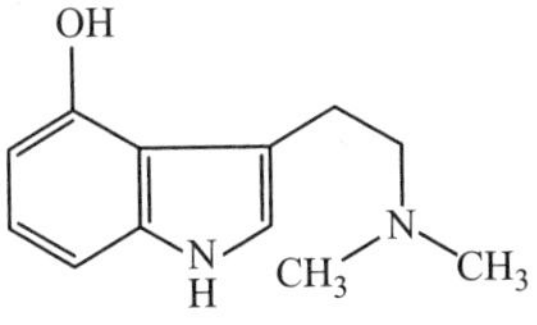

Psilocin (4.127) 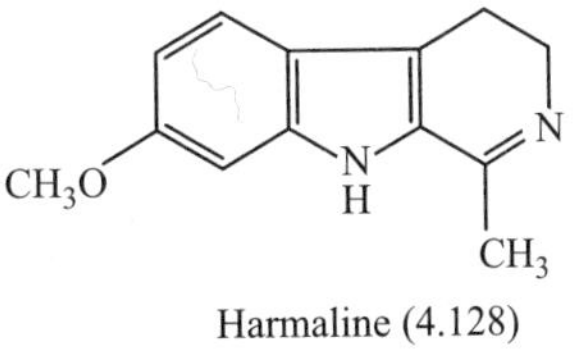   Harmaline (4.128)

After the ingestion of hallucinogens, a great variety of symptoms may occur, including dizziness; perceptual changes of size, time, and distance; visual hallucinations; mood changes; and potential panic. These effects may last for about 12 hours. Tolerance develops quickly, and there is cross-tolerance with phenylethylamines but not with amphetamines. The psychological hazards ("bad trip") of LSD use are very real; the physiologically harmful effects, if any, are not clear cut.

Phenylethylamine, a fragment of LSD, and a number of alkoxyphenylethylamines have been used by various aboriginal societies as hallucinogens. The best known is mescaline (**4.129**), which occurs in a number of cacti (*Lophophora, Trichocereus*) native to Mexico. In large doses (300–500 mg) it causes vivid and colorful hallucinations, perception of the environment as unusually beautiful, and increased insight ("mind-expanding experience"). The effect is increased by attaching a methoxy group to the ortho position and using alkyl substituents.

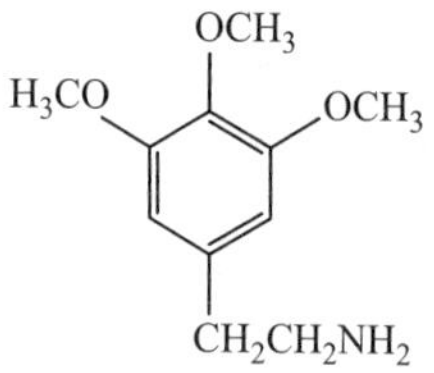

Mescaline (4.129) 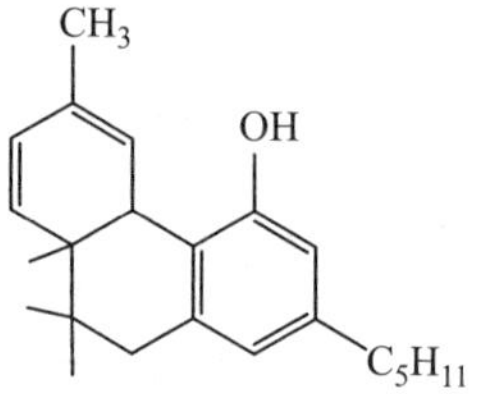 Δ^9-Tetrahydrocannabiol (4.130) 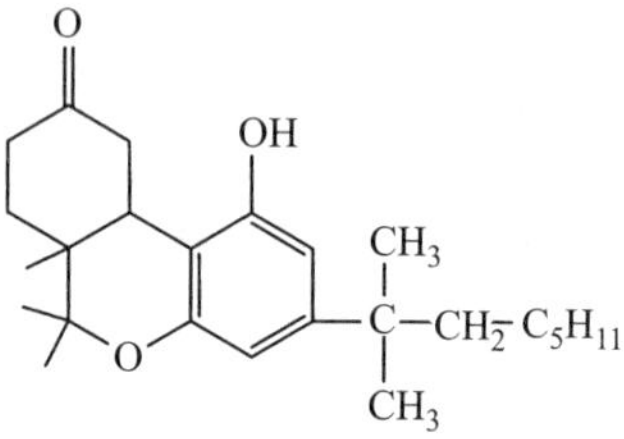   Nabilone (4.131)

The third group of psychotomimetic compounds comprises the *cannabinoids*, represented by Δ^9-tetrahydrocannabinol (THC) (**4.130**), the principal active ingredient of marijuana and hashish, produced by the hemp *Cannabis sativa*. It is not considered a hallucinogen, is not habit forming, and seems to have no adverse physiological effects except in habitual consumers of large quantities. Extremely high doses may lead to paranoid depersonalization.

The emotional propaganda (both pro and con) surrounding cannabinoids notwithstanding, cannabinoid derivatives—if not THC itself—have interesting therapeutic possibilities. One of the cannabinoid derivatives in clinical use, nabilone (**4.131**), has a very selective antiemetic activity in patients suffering from toxic side effects of cancer chemotherapy. Other derivatives show anticonvulsant and analgesic activity, and also decrease ocular pressure in glaucoma. The structure–activity correlations of these compounds have been explored quite thoroughly.

4.5.5 The Clinical–Molecular Interface: Migraine Headache as a Serotonergic Disorder

Although migraines produce severe headaches, not all severe headaches are migraines. As a headache disorder, migraines have distinctive diagnostic criteria. A *classical migraine* onsets with an "aura," which may be brightly colored lights or bright lightening displays in the visual fields. The pain then occurs as a pounding, pulsatile, throbbing headache localized to one side of the head and associated with photophobia (dislike of light), phonophobia (dislike of noise), nausea, and perhaps vomiting. A *common migraine* lacks the aura phase. Migraine is a familial problem in 70% of patients and occurs more frequently in women than in men, with 14–15% of women experiencing migraine headaches. A *complicated migraine headache* is a migraine associated with neurologic problems, such as weakness over half of the body. In younger women

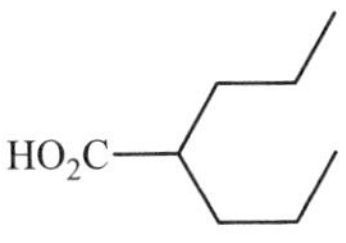

Valproic acid
(4.132)

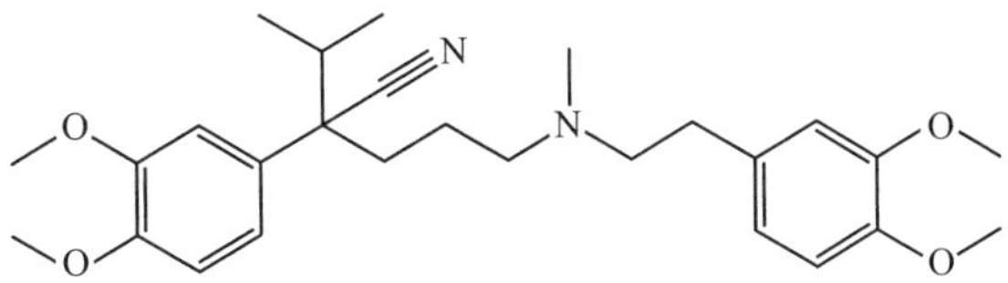

Verapamil
(4.133)

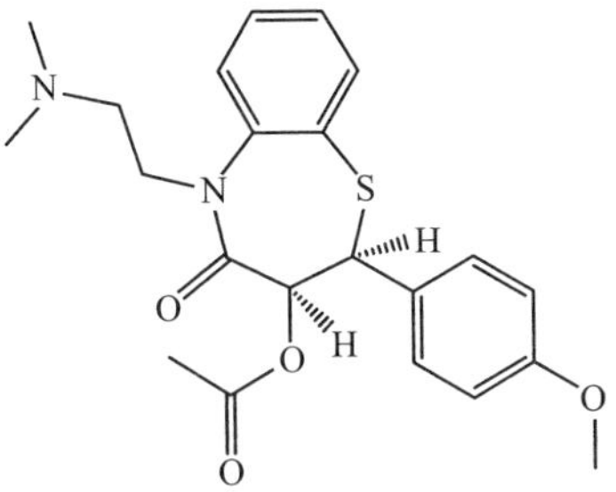

Diltiazem
(4.134)

(especially those who both smoke and are receiving the birth control pill), migraine can be a risk factor for stroke.

Since migraine is so common and so painful, numerous therapies have been put forth over the years. The simplest therapies are the analgesics such as acetaminophen, acetylsalicylic acid, or codeine. Such analgesics are acceptable if the headaches are infrequent. However, daily use of such analgesics can, paradoxically, make the headaches worse in a phenomenon referred to as *analgesic rebound headaches.* For people who experience several migraines per week, the use of a prophylactic agent to prevent migraine occurrence is indicated. A wide variety of drugs have enjoyed success as migraine prophylactic agents, including β-adrenergic blockers (propranolol), tricyclic antidepressants (amitriptyline), Na^+ channel active anticonvulsants (valproic acid, **4.132**), and Ca^{2+} channel active agents (verapamil, **4.133**, diltiazem, **4.134**).

However, the class of agents with the greatest success against migraine has been based on a serotonergic approach. The success of serotonergic drugs reflects the fact that the mechanism of migraine involves a vascular ("vasomotor") component. Current studies on migraine have revealed the involvement of numerous serotonergic nerve endings in blood vessels within the meninges (coverings over the brain); during a migraine attack these cranial blood vessels become swollen and inflamed. Over the past decade, the acute treatment of migraine has been revolutionized by the "triptans," including sumatriptan (**4.135**), naratriptan (**4.136**), rizatriptan (**4.137**), and zolmitriptan (**4.138**). These compounds are selective agonists for $5\text{-}HT_{1D}$ and $5\text{-}HT_{1B}$ receptors, which are found on blood vessels in the meninges and mediate vasoconstriction in these vessels. In addition to the triptans, the ergot alkaloids are also effective against migraine. The efficacy of ergot derivatives in migraine is so specific that it almost constitutes a diagnostic test. Traditionally, ergotamine, when given during the aura, is particularly effective. Dihydroergotamine (**4.139**) is commonly employed intravenously

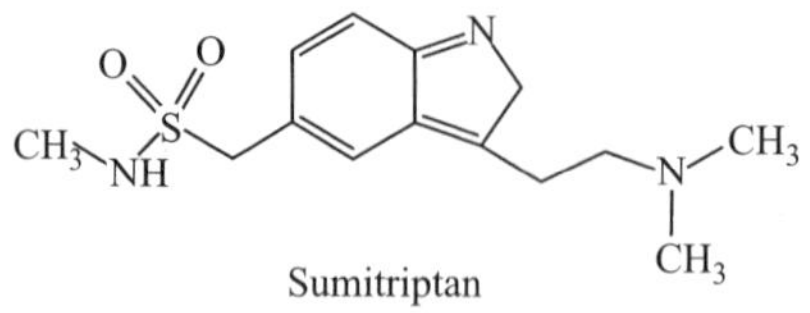

Sumitriptan
(4.135)

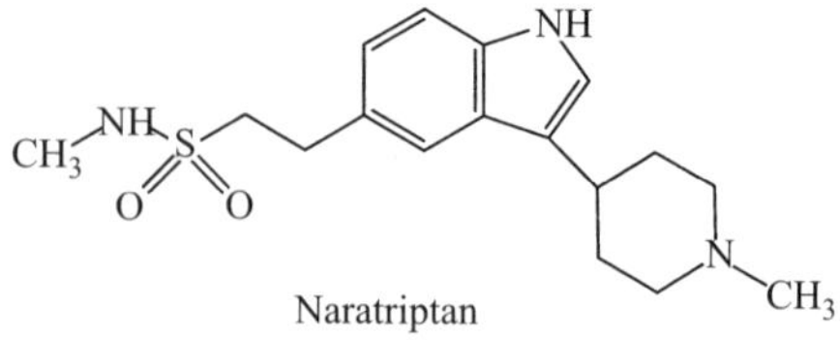

Naratriptan
(4.136)

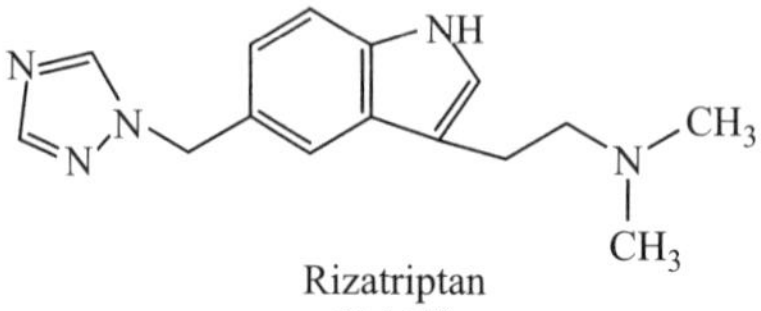

Rizatriptan
(4.137)

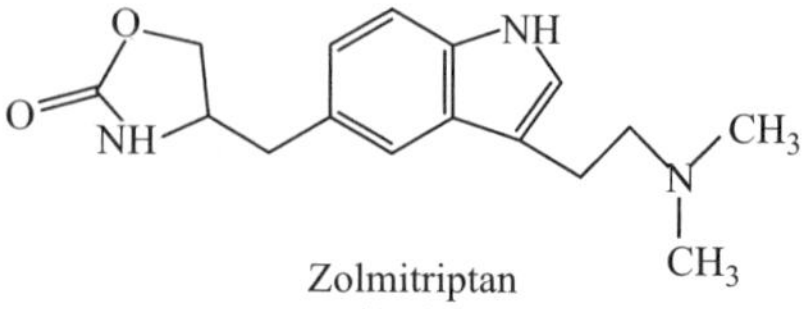

Zolmitriptan
(4.138)

for the treatment of a severe intractable migraine. Another ergot derivative, methysergide (**4.123**), is a safer compound for chronic prophylactic use. The interactions of ergots with serotonin receptors is complex, since these agents can be agonists, antagonists, or partial agonists at one or more of the various serotonin receptors. Ergotamine and methysergide are serotonin antagonists as well as partial agonists effective at 5-HT$_2$ receptors.

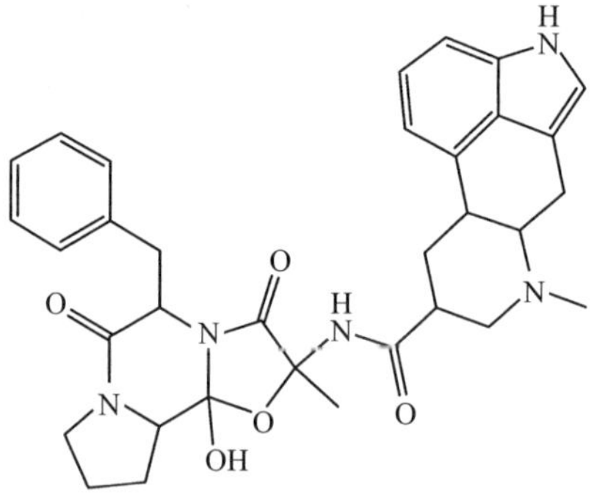

Dihydroergotamine
(4.139)

4.6 HISTAMINE AND THE HISTAMINE RECEPTORS

The role of histamine as a central neurotransmitter has long been recognized, and a considerable amount of research has been directed toward elucidating its central effects and receptors. The additional discovery of the duality of the histamine receptor (i.e., both inside and outside the brain) has added another dimension to this complex field, leading to new and successful therapeutic as well as theoretical investigations.

4.6.1 Structure, Conformation, and Equilibria of Histamine

Protonation has been an important aspect in the design of some histamine antagonists. Figure 4.10 shows the tautomeric equilibria between different histamine species and the respective mole percentages of these species. The most important among them is the N^t—H (tele-) tautomer, which also appears to be the active form of the agonist on both receptors. Tautomerism does not appear to be important in H_1 receptor binding (in the intestine); however, it does seem to be important to gastric H_2–receptor activity. Histamine may play the role of a proton-transfer agent, in a fashion similar to the charge-relay role of the imidazole ring in serine esterases. The percentage of the monocation tautomers is greatly influenced by substituents in position 4, which alter the electron density on the N^π atom, an important consideration in modifying the receptor-binding properties of histamine.

Histamine metabolism differs from that of classical neurotransmitters because histamine is so widely distributed in the body. The highest concentrations in human tissues are found in the lung, stomach, and skin (upto 33 μg/g tissue). Histamine metabolic pathways are simple; histamine is produced from histidine in just one step (see figure 4.11). The principal production takes place in the *mast cells* of the peritoneal cavity and connective tissues. The gastric mucosa is another major storage tissue. Histamine can be found in the brain as well.

Histamine is released from mast cells in antigen–antibody reactions, as in anaphylaxis and allergy, which are the most widely known physiological reactions to histamine. However, these potentially fatal reactions are not caused by histamine alone. Other agents present in mast cells, such as serotonin, acetylcholine, bradykinin (a nonapeptide), and a "slow-reacting substance" or leukotriene (see chapter 8) also contribute. In the stomach, where histamine induces acid secretion, its release seems to be regulated by the peptide hormone pentagastrin.

4.6.2 Histamine Receptors

Classically, these receptors have also been divided into three groups. The first of these, the H_1 receptors, were described by Schild in 1966. The H_2 receptors were discovered in 1972 by Black et al. The H_3 receptor subtype was described by Arrang in 1983. The H_1 receptor is found in the smooth muscle of the intestines, bronchi, and blood vessels and is blocked by the "classical" antihistamines. The H_2 receptor, present in gastric parietal cells, in guinea pig atria, and in the uterus, does not react to H_1 blockers but only to specific H_2 antagonists. H_2 receptors also appear to be involved in the immunoregulatory system and may be present in T lymphocytes, basophil cells, and mast cells. H_3 receptors are found predominantly in brain but are also localized in stomach, lung, and cardiac tissue.

H_1, H_2, and H_3 receptors are present in the CNS. Tricyclic antidepressant drugs seem to interact with histamine receptors in the CNS. Histamine receptor subtypes in the CNS and the central neurotransmitter role of histamine have been the subject of many recent investigations. Currently there are three central histamine receptors:

1. H_1 receptors are widely distributed, especially in the cerebellum, thalamus, and hippocampus, and are located on neurons, astrocytes, and blood vessels. H_1 receptors in brain vary widely from species to species. Since histamine does not easily cross

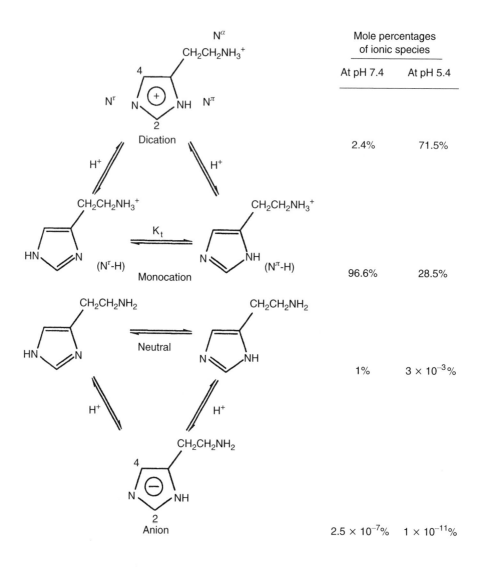

Figure 4.10 Top: Ionic and tautomeric equilibria between histamine species and their respective mole percentages. Bottom: 4-substituted histamine derivatives act as proton-transfer agents in their tele-tautomeric form. This may be important for H2 receptor ligands.

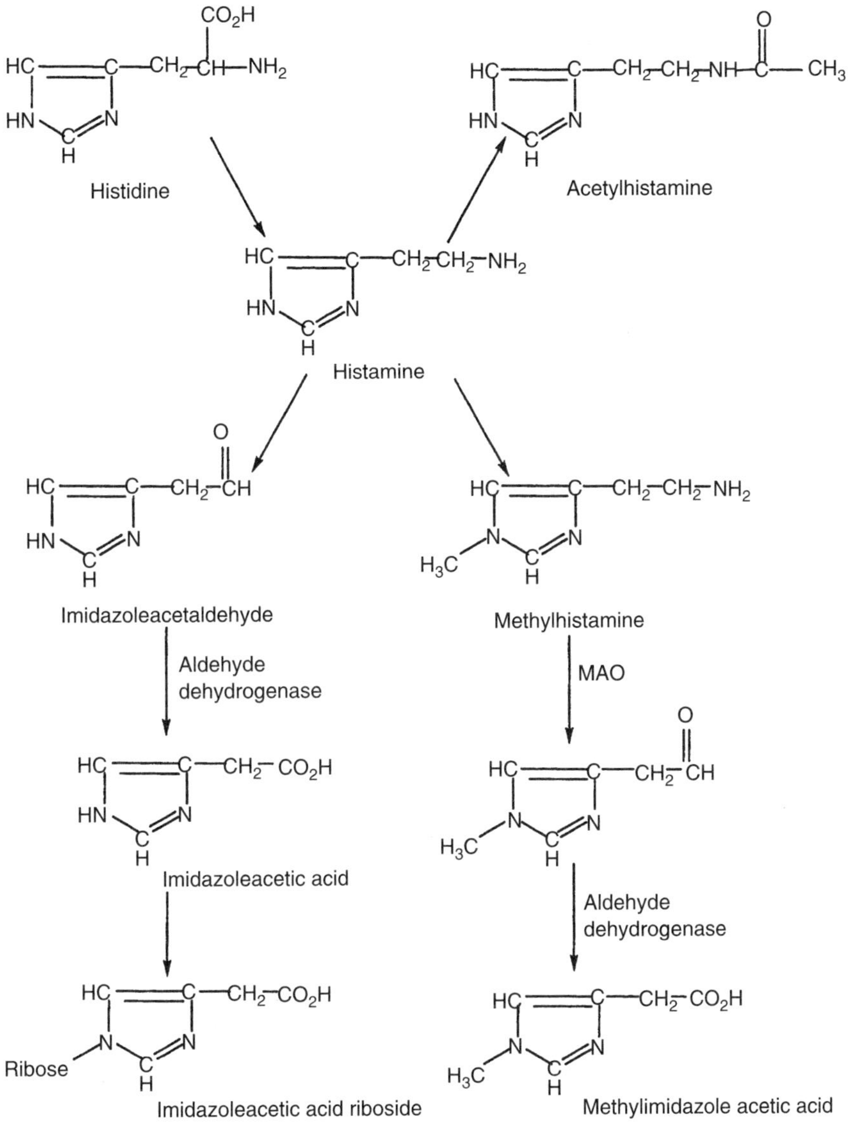

Figure 4.11 Histamine metabolism.

the blood–brain barrier, central receptors can use only the locally synthesized histamine. Occupation of the H_1 receptor by an agonist does not activate adenylate cyclase (AC); it appears to use phosphoinositol as second messenger. H_1 receptors are not very specific and are occupied by antidepressants and neuroleptics as well. This explains the sedative effect of all three classes of drugs. The H_1 receptors are easily solubilized and have been purified on lectin affinity columns, indicating their glycoprotein nature.

2. H_2 receptors are localized to the cortex and striatum and are found in neurons, glial cells (astrocytes), and blood vessels. They are coupled to AC; their stimulation has a central disinhibitory effect, due to the decrease in Ca^{2+}-activated K^+ conductance. Thus the role of the central histamine receptor may not be information transmission, but sensitization of brain areas to excitatory signals from "waking amines."

3. H_3 receptors have been described and seem to be localized in cortex and substantia nigra; these seem to be presynaptic autoreceptors, controlling histamine release and synthesis. They are activated by histamine concentrations that are two orders of magnitude lower than those necessary for triggering postsynaptic receptors. Their blockade may potentially lead to increased blood flow and metabolism combined with a central arousal, whereas their stimulation (or inhibition of central H_2 receptors) could have a sedative effect.

Knowledge of the physiological role of histamine in the CNS and evidence for the existence of discrete neuronal networks that could be called histaminergic are still evolving. Histamine-mediated hypothermia, emesis, and hypertension have been shown to exist, and the well-known sedative effects of H_1 antihistamines are centrally mediated.

4.6.3 Histamine Agonists

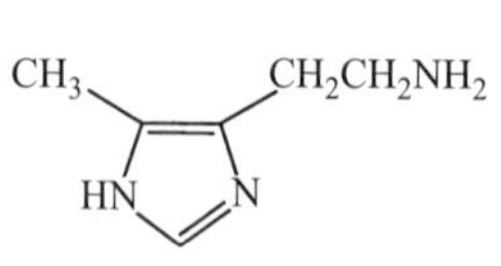

4-Methylhistamine (4.140)   2-Methylhistamine (4.141)   2-Pyridylhistamine (4.142)   2-Thiazolylhistamine (4.143)

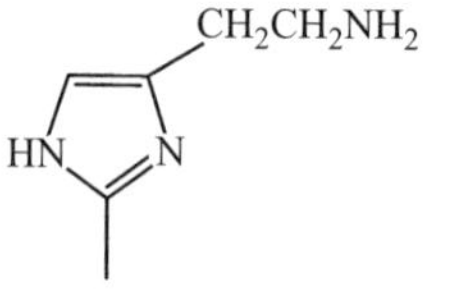

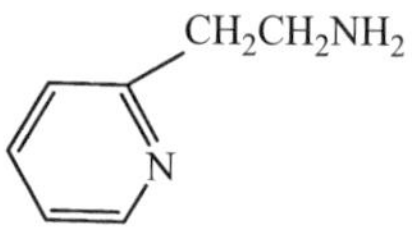

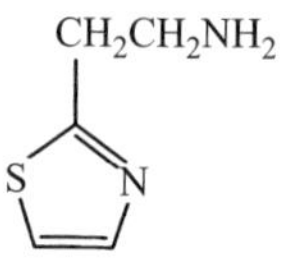

A selection of agonists is shown. 4-Methylhistamine (4.140) is selective for the H_2 receptor, whereas the 2-methyl derivative (4.141) is a weak but usable H_1 agonist. The fact that the 2-pyridyl (4.142) and 2-thiazolyl rings (4.143) also lend H_1 activity to histamine derivatives shows that tautomerism is not an issue in H_1 activity. Large alkyl

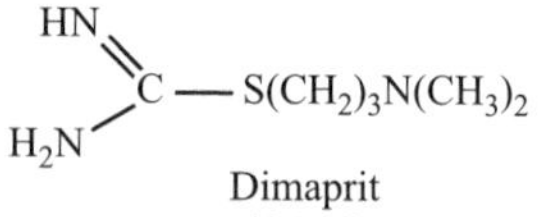

Dimaprit
(4.144)

groups on C-4 decrease activity and lead to partial agonists, whereas side-chain N-substitution enhances the antagonistic properties of the molecule.

An interesting histamine agonist is dimaprit (4.144), which was described in the late 1970s. It is a selective H_2 agonist, having between 19% and 70% H_2 activity, with no effect on the H_1 receptor. The isothiourea system in dimaprit has a planar electron sextet, like that of the imidazole ring in histamine, and is capable of tautomerism as well as of donating and accepting hydrogen. It produces a higher maximum gastric acid secretion in dogs than does histamine.

The pharmacological effects of histamine may be summarized as follows:

1. The circulatory effects are manifested as arteriolar dilation and increased capillary permeability, causing plasma loss. The localized redness, edema (hives, wheal), and diffuse redness seen in allergic urticaria (rash) or physical skin injury result from these circulatory changes. Vasodilation also causes a decrease in blood pressure.
2. The effects on the heart (H_2 response) are minor, but the heart rate increases.
3. Humans and guinea pigs are very prone to bronchoconstriction by histamine (an H_1 effect), and severe asthmatic attacks can be triggered by small doses, provided the person suffers from asthma and is therefore very sensitive to histamine.
4. Stimulation of gastric acid secretion is the most important H_2 response; it is blocked only by H_2 antagonists. As mentioned before, the hormone gastrin may be involved in histamine release, because H_2 antagonists block gastrin-induced acid secretion.
5. H_3 receptors are involved in mediating the neuroregulatory influence of the brain on stomach, lung, and heart. Structural alterations of histamine result in (R)-α-methyl-histamine (**4.145**), a potent and selective H_3 agonist. Replacement of the amino group with bioisosteric polar cationic groups yields imetit (**4.146**) and immepip (**4.147**), other potent and selective H_3 agonists.

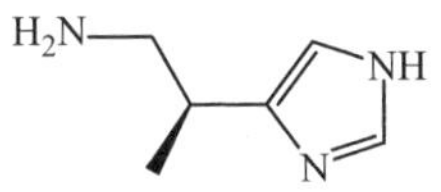

(R)-α-Methylhistamine (4.145)　　　Imetit (4.146)　　　Immepip (4.147)

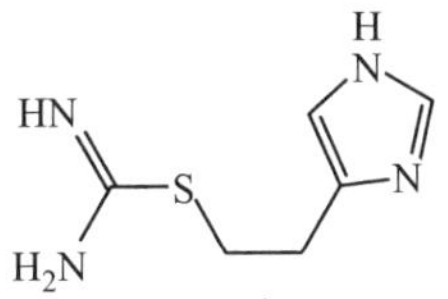

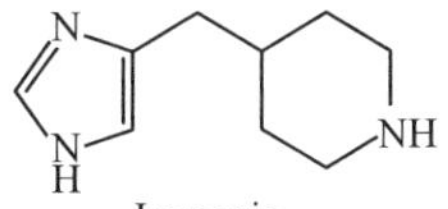

4.6.4 Histamine Antagonists

4.6.4.1 *H₁ Antagonists*

Antagonists of the H_1 receptor were first discovered by Bovet in 1933. They do not bear any close resemblance to the agonist since their binding involves accessory binding sites.

Ethylenediamines, aminoalkyl ethers, and aminopropyl compounds, for which X is nitrogen, oxygen, and carbon, respectively, show a general H_1 antagonist structure. Cyproheptadine (**4.122**), a serotonin antagonist, is also a potent antihistamine (about 150 times more active than diphenhydramine, **2.5**), and so is promethazine (**3.5**) and its derivatives, which, formally at least, can be considered a result of ring closure connecting the two aryl rings in a diphenyl-ethylenediamine.

The unpleasant sedative CNS effect of most antihistamines, combined with their slight anticholinergic activity, is exploited for the prevention of *motion sickness.* Diphenhydramine (**2.5**), in the form of an 8-chlorotheophylline salt (dimenhydrinate, **4.148**), is widely used for this purpose. The theophylline derivative was originally added to counteract the drowsiness produced by diphenhydramine, since it is a central excitant related to caffeine.

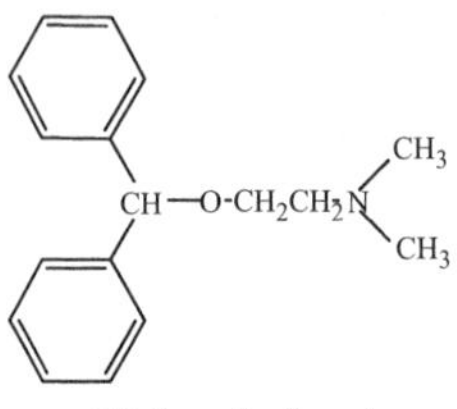

Diphenhydramine	Promethazine	Dimenhydrinate
(2.5)	(3.5)	(4.148)

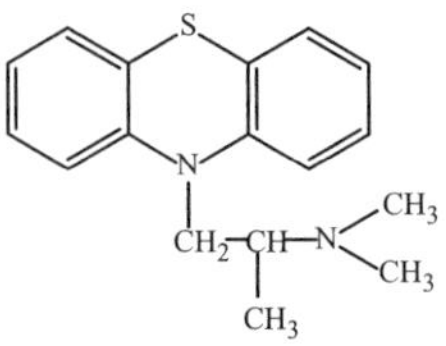

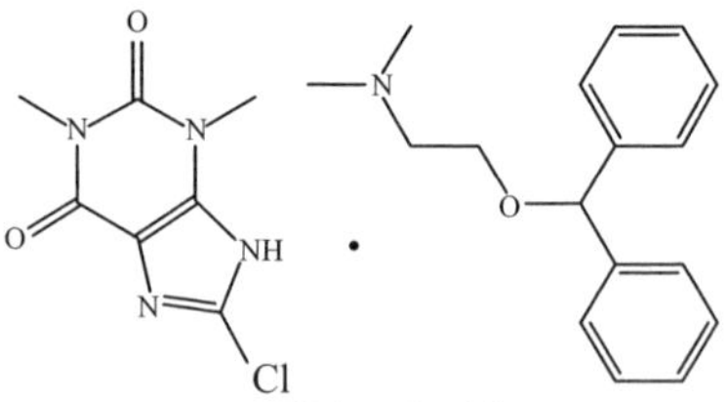

Several nonsedative H_1 inhibitors have been marketed—for example, astemizole (**4.149**) and terfenadine (**4.150**). They are quite polar molecules and therefore cannot cross the blood–brain barrier to reach central histamine receptors. This is a good example of drug design exploiting knowledge of the pharmacokinetic processes to preclude undesirable CNS side effects.

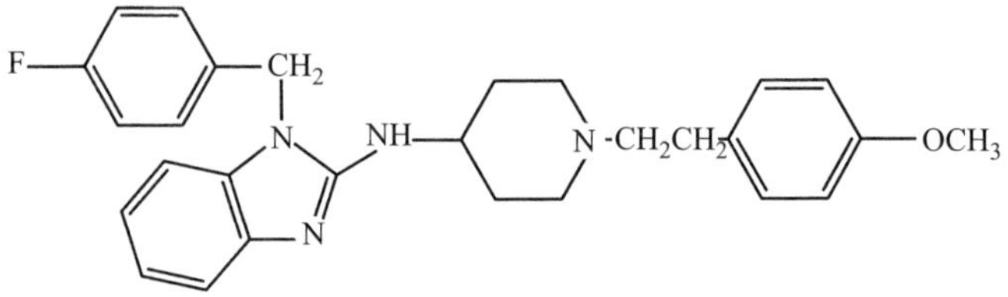

Astemizole
(4.149)

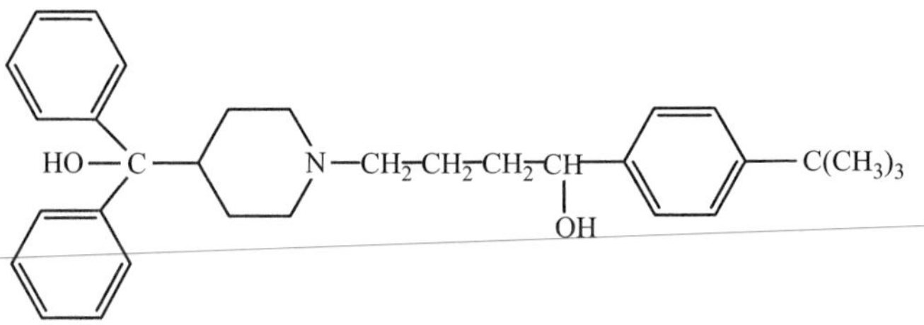

Terfenadine
(4.150)

4.6.4.2 H_2 Antagonists

Antagonists of the H_2 receptor were first reported in 1972 by Black and co-workers, and work in this area was successively continued by the same group in an elegant series of investigations based on considerations that were guided by molecular pharmacological principles. One of the compounds that showed weak H_2-antagonist activity, guanyl-histamine, was the point of departure in the development of these drugs. Extension of the side chain was found to increase the H_2-antagonist activity, but some agonist effects were retained. When the very basic guanidino group was replaced by the neutral thiourea, burimamide (**4.151**) was obtained. Although an effective drug, it cannot be absorbed orally. The addition of a 4-methyl group further improved binding to the H_2 receptor. Introduction of the electron-withdrawing sulfur atom into the side chain reduced the ring pK_a. The proportion of the cationic form was also decreased, and the tele tautomer became predominant. Reduced ionization improved the membrane permeability of the molecule; the oral absorption of the resulting compound, metiamide (**4.152**), was excellent, and the compound had an activity 10 times higher than that of burimamide. However, metiamide still showed some side effects in the form of hematological and kidney damage, which were attributed to the thiourea group.

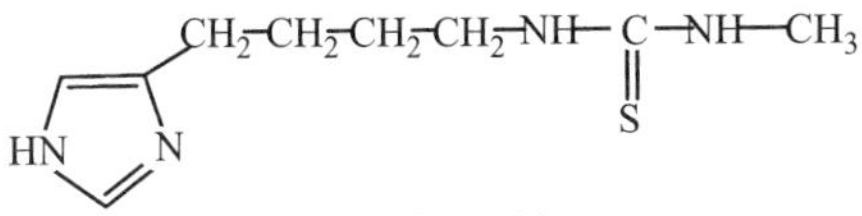

Burimamide
(4.151)

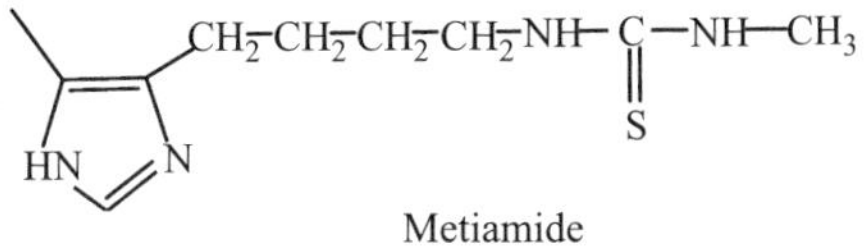

Metiamide
(4.152)

A satisfactory replacement was found by substituting another electron-withdrawing group on guanidine while retaining the appropriate pK_a. A cyano group proved suitable, and the safe and effective cimetidine (**4.153**) resulted, which became a drug of choice in treating peptic ulcer. It then became clear that an imidazole nucleus was not absolutely necessary for H_2-antagonist activity. The furan derivative ranitidine (**4.154**) is even more active than cimetidine, and famotidine (**4.155**) is seven times more active still. Since none of these compounds is lipid soluble (their average partition coefficient is only 2, compared with coefficients of up to 1000 for typical H_1 antagonists), they do not produce any sedative CNS action since they cannot cross the blood–brain barrier.

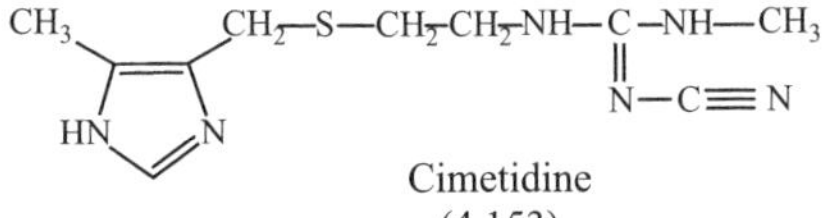

Cimetidine
(4.153)

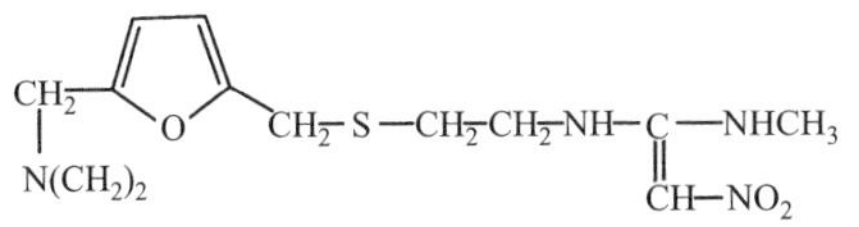

Ranitidine
(4.154)

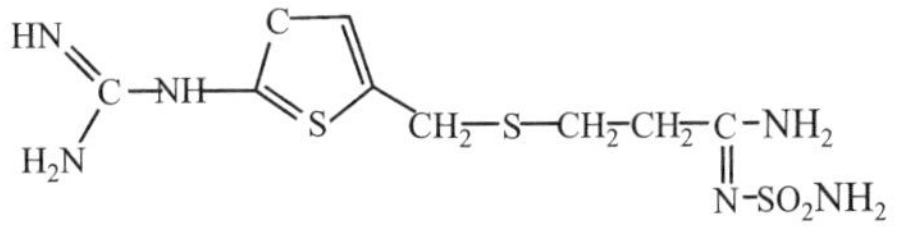

Famotidine
(4.155)

H_2 Antagonists and the Treatment of Peptic Ulcers. Treatment of peptic ulcers is a complicated and multilevel therapy in which H_2 antagonists are very successful and widely used (and abused). Peptic ulcers may affect either the stomach (*gastric ulcers*, less common overall but more common in people with iatrogenic [i.e., physician-induced] ulcers from the use of nonsteroidal anti-inflammatory drugs [NSAIDs]) or the duodenum (*duodenal ulcers*). The lining of the stomach or duodenum is attacked by the digestive juices to such an extent that the protective mucous layer on the surface has been "eaten through," exposing the inner submucosal connective tissue layer. This mucosal damage is promoted by *Helicobacter pylori* bacteria that colonize the gastric lining. The penetrating damage of the digestive acids may erode into blood vessels, causing life-threatening gastrointestinal hemorrhages ("upper GI bleeds"), or it may perforate through the stomach, causing either peritonitis (inflammation of the peritoneal cavity within the abdomen) or pancreatitis. To facilitate healing, prevent ulcer recurrence, and relieve pain, the medicinal chemistry approach is multipronged and involves lowering aggressive acid output, augmenting the mucous-based protection, and/or eradicating the *Helicobacter pylori*.

The concentration of acid in the stomach may be reduced either by neutralizing the acid or by inhibiting acid production. Acid neutralization is inexpensively achieved by using a nonabsorbable antacid such as $CaCO_3$, $Mg(OH)_2$, or $Al(OH)_3$. Inhibition of acid production may be realized by one of three approaches. First, pirenzepine (**4.156**) is an anticholinergic agent (relative M1 receptor specificity) that does not cross the blood–brain barrier, but which binds to acid-secreting cells in the gut to downregulate their production. Secondly, omeprazole (**4.157**) is a *proton pump* (H$^+$, K$^+$-ATPase) *inhibitor* that blocks the transport of H$^+$ into the gut. Thirdly, H$_2$ histamine antagonists (e.g., cimetidine, ranitidine) can be used to prevent acid secretion from parietal cells that are contained within the stomach.

Pirenzepine
(4.156)

Omeprazole
(4.157)

The next major class of drugs for peptic ulcer disease is the mucoprotectants and other protective agents. Sucralfate (**4.158**) is a carbohydrate-based drug (chapter 8) which forms an impenetrable paste that adheres to the stomach lining defect, providing a protective barrier. Misoprostol (**4.159**) is a semisynthetic prostaglandin derivative that promotes mucus production. Carbenoxolone (**4.160**) has a mineralocorticoid-type action that also promotes mucus production.

Finally, since the microorganism *Helicobacter pylori* plays an important role in the pathogenesis of ulcers, antibacterial agents such as amoxicillin (**4.161**), metronidazole (**4.162**), or even colloidal bismuth compounds (e.g., **4.163**) may also be used.

The nonsurgical treatment of peptic ulcer is a superb example of how multiple molecular approaches can be used to therapeutically attack a single clinical problem from multiple directions. Also, the medical management of peptic ulcer disease demonstrates how antagonists of neurotransmitter messenger molecules (acetylcholine, histamine) can be used to treat nonneurological disorders. The role of β-adrenergic agonists and antagonists in the treatment of cardiopulmonary diseases is a similar example. Since these messenger molecules are useful both inside and outside the CNS, the drug designer must

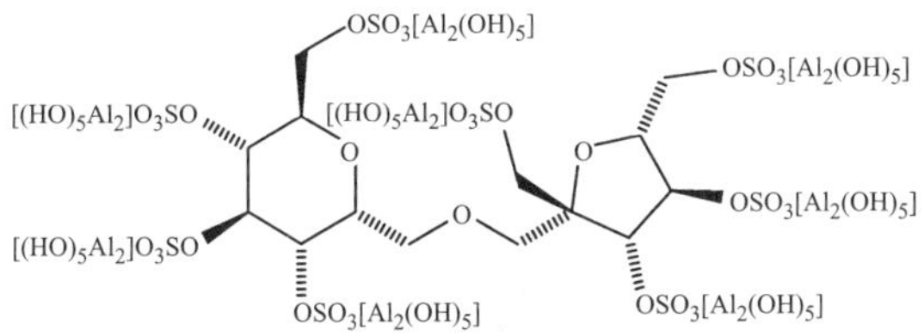

Sucralfate
(4.158)

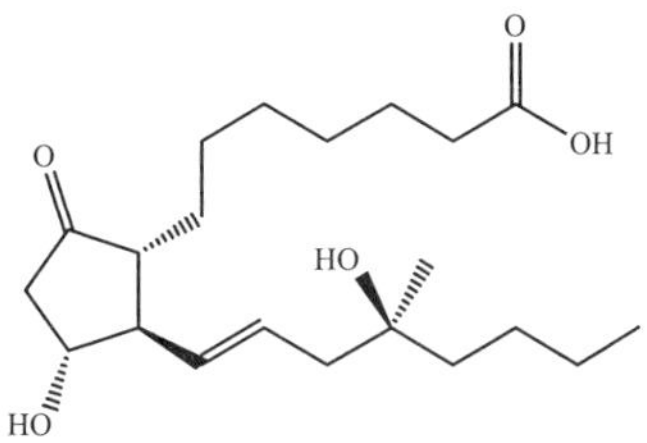

Misoprostol
(4.159)

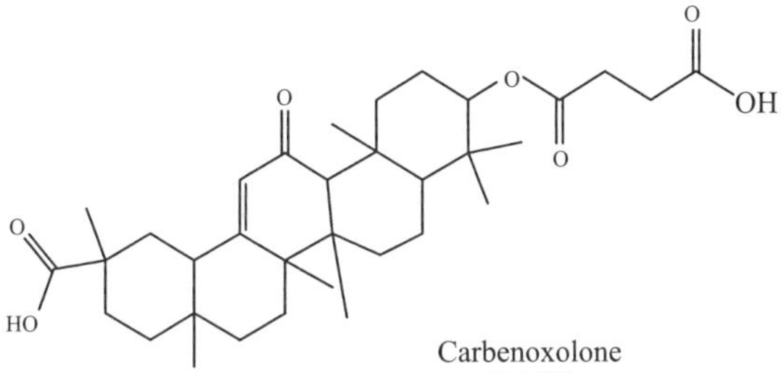

Carbenoxolone
(4.160)

also remember that drugs designed to target CNS receptors can have systemic side effects and that drugs targeting systemic problems can have CNS side effects.

4.6.4.3 H₃ Antagonists

Several lines of evidence suggest a role for H_3 receptors in cognitive processes. The use of H_3 antagonists in learning and memory disorders has been suggested. Thioperamide, a prototypic H_3 antagonist, enhances arousal patterns in cats, an observation which has been confirmed using other nonthiourea H_3 blockers.

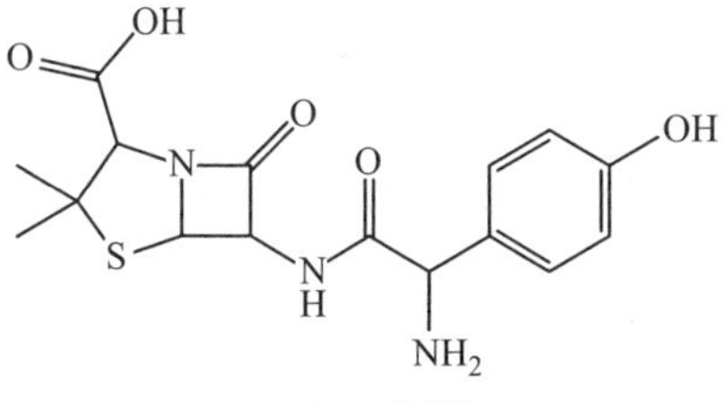

Amoxicillin
(4.161)

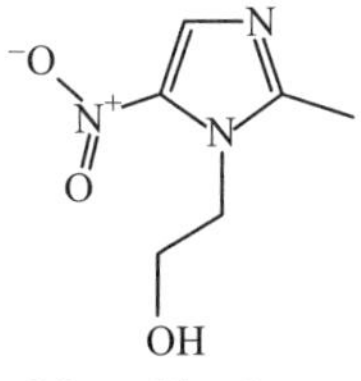

Metronidazole
(4.162)

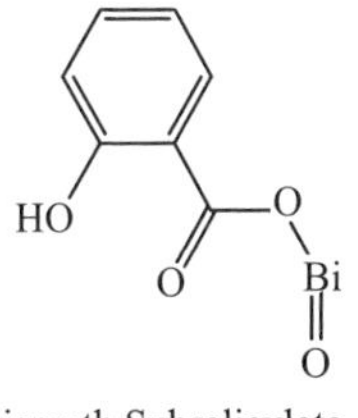

Bismuth Subsalicylate
(4.163)

4.6.5 The Clinical–Molecular Interface: The Many Uses and Abuses of H₁ Receptor Antagonists

H_1 receptor antagonists (antihistamines) are among the most widely used therapeutic agents. Indeed, sometimes the exact same molecule (e.g., diphenhydramine) is marketed in different aisles in the same pharmacy under many brand names for different indications. Antihistamines are usually the first drugs used to treat allergic reactions. They are very effective at reducing the "runny nose and itchy eyes" (allergic rhinitis and allergic conjunctivitis) of allergies. However, since they are effective at reducing the runny nose of allergies, antihistamines are frequently used in "cold remedies." The nasal congestion of a cold (viral upper respiratory tract infection) arises via a different mechanism from the nasal congestion of allergies; thus, the use of antihistamines in over-the-counter cold remedies is questionable at best. Regrettably, the antihistamines are usually listed as one of several active agents within a "shotgun" cold remedy. A person taking such a medication may be taking the antihistamine unknowingly. This can cause problems. Since H_1 receptors are diffusely located throughout the brain, cold remedies can cause drowsiness (a danger to those operating moving equipment) and may, rarely, even trigger seizures in a person with a predisposition to epileptic seizures.

The same molecules used to treat allergies and "cold symptoms" have many other uses. Antihistamines are particularly effective as antiemetics in suppressing nausea associated with gastrointestinal illnesses. They can also be used to treat the symptoms of motion sickness or even vestibular disturbances (vertigo). Because of their ability to induce sedation, antihistamines are widely used in over-the-counter sleep aids.

4.7 INHIBITORY AMINO ACID NEUROTRANSMITTERS: γ-AMINOBUTYRIC ACID (GABA)

GABA is the most comprehensively studied inhibitory neurotransmitter, and there are many reviews of its biochemistry and pharmacology. The reason for this great interest is the discovery that the most popular drugs of the 1970–1980s, the benzodiazepine tranquilizers or "anxiolytics," as well as the previously popular barbiturates, act on the GABAergic neuronal system.

4.7.1 Neuronal Systems and GABA Metabolism

There are numerous GABAergic neuronal pathways in the CNS. γ-Aminobutyric acid is found in high concentrations in the cerebellum, is also found in the hypothalamus, thalamus, and hippocampus, and occurs in low concentrations in practically all brain structures as well as in the spinal cord. The amounts present are relatively high—on a μmol/g order of magnitude—rather than the nanomolar quantities seen with most major neurotransmitters. γ-Aminobutyric acid also occurs in glial cells, where its role is less well defined.

The biosynthesis of GABA occurs only in the neurons, since it cannot penetrate the blood–brain barrier, and no peripheral precursor is known. The synthesis is tied to the Krebs cycle through α-ketoglutarate. γ-Aminobutyric acid is formed by the decarboxylation of L-glutamate, catalyzed by *glutamic acid decarboxylase* (GAD), an enzyme found only in the mammalian CNS and in the retina. This reaction is irreversible. The cofactor of GAD is pyridoxal phosphate (vitamin B_6). Since GAD is the rate-determining enzyme, GABA metabolism can be regulated by the manipulation of this enzyme, the manipulation of pyridoxal, or both.

γ-Aminobutyric acid can be deactivated and recycled by the transamination reaction with α-ketoglutarate to yield glutamate. This reaction circumvents the usual oxidative route, insofar as glutamate can be decarboxylated to yield GABA once again. This transamination is catalyzed by the enzyme *GABA transaminase* (GABA-T), which is widely distributed. Therefore, free GABA cannot be found anywhere except in the brain. The transaminase enzyme also depends on pyridoxal phosphate as a cofactor.

4.7.2 Characterization of GABAergic Receptors

GABAergic receptors have been investigated extensively over the past 20 years. The great increase in research activity in this area was largely due to the recognition that the extremely widely used benzodiazepine tranquilizers (e.g., diazepam, Valium) act through the GABA receptor. Currently three major GABA receptors are recognized: GABA$_A$, GABA$_B$, and GABA$_C$.

The GABA$_A$ receptor was first cloned using partial protein sequence followed by cDNA expression of GABA-activated channels in *Xenopus* oocytes. The GABA$_A$ receptor is an ionophore complex; it is a 275 kDa heteropentameric glycoprotein composed of five different subunit peptides selected from a group of at least 19 different but closely related polypeptides. These subunit peptides are divided into six classes: α, β, γ, δ, ε, and ρ. There are 6 subtypes of the α subunit, 4 subtypes of the β subunit, 4 subtypes of the γ subunit, and 3 subtypes of the ρ subunit. There is 30% sequence homology between subunit peptides and 70% sequence homology within a subunit class. A typical GABA$_A$ receptor could consist of two α subunits, two β subunits, and a γ subunit to yield the pentameric structure. However, an array of other compositions is possible, depending upon which subunit proteins are used. Different combinations of subunits with differing pharmacologies and conductances are expressed in different areas of the brain. Each GABA$_A$ subunit contains four α-helical membrane-spanning domains (M1–M4). A membrane-spanning region from each of the M2 domains from each of the pentameric subunits forms the walls of a central ion channel pore. The segment between M3 and M4 within each subunit is a long variable intracellular domain that contributes to receptor specificity in regulating intracellular mechanisms. The GABA$_A$ receptor is a member of a superfamily of ligand-gated ion channel receptors, which also includes the acetylcholine receptor and the 5-HT$_3$ serotonin receptor.

Less is known about GABA$_B$ receptor structure. GABA$_B$ receptors are coupled indirectly to K$^+$ channels. These receptors, which are always inhibitory, are coupled to G-proteins. When activated, GABA$_B$ receptors decrease Ca^{2+} conductance and inhibit cAMP production. The GABA$_C$ receptor is probably little more than a subtype of the GABA$_A$ receptor. It contains the ρ subunit peptide and is located primarily, if not exclusively, in the retina.

The neuronal activity of GABA shows different inhibitory mechanisms reflecting the GABA$_A$ and GABA$_B$ receptors. The first mechanism is the conventional hyperpolarization of an excitatory neuron by increased Cl$^-$ ion flux, which makes the neuron unable to fire when it receives a normal impulse. The second is the partial (presynaptic) depolarization of an excitatory neuron, which causes a decrease in neurotransmitter release when this neuron receives an electrical impulse.

4.7.3 GABA Receptors: Presynaptic Drug Effects

Presynaptic drug effects can interfere with the metabolism, storage, release, and reuptake of GABA, as they can with the functioning of other neurotransmitters.

4.7.3.1 GABA Synthesis Inhibitors

GABA synthesis inhibitors act on the enzymes involved in the decarboxylation and transamination of GABA. Glutamic acid decarboxylase (GAD), the first enzyme in GABA biosynthesis, is inhibited easily by carbonyl reagents such as hydrazines [e.g., hydrazinopropionic acid (**4.164**) or isonicotinic acid hydrazide (**4.165**)], which trap pyridoxal, the essential cofactor of the enzyme. A more specific inhibitor is allylglycine (**4.166**). All of these compounds cause seizures and convulsions because they decrease the concentration of GABA.

Hydrazinopropionic acid
(4.164)

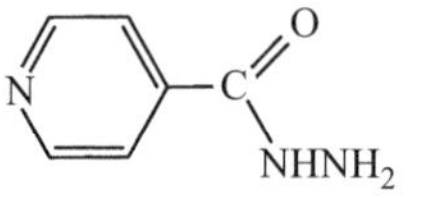

Isonicotinic acid hydrazide
(4.165)

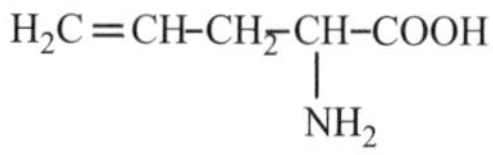

Allylglycine
(4.166)

4.7.3.2 GABA Metabolism Inhibitors

In contrast to GABA synthesis inhibitors, inhibitors of GABA-T, the transaminase active in eliminating GABA, increase the concentration of this neurotransmitter. The most potent of these agents are gabaculine (**4.167**) and vigabatrin (**4.168**), both of which protect against drug-induced seizures.

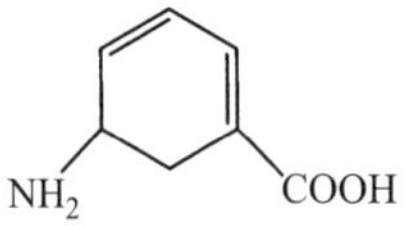

Gabaculine
(4.167)

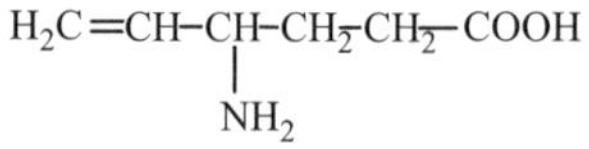

Vigabatrin
(4.168)

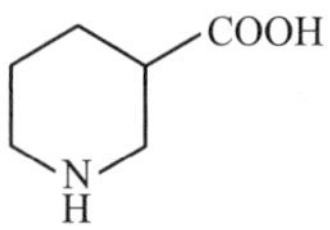

Nipecotic acid
(4.169)

4.7.3.3 GABA Reuptake Inhibitors

Another mechanism involves several GABA reuptake inhibitors, such as nipecotic acid (**4.169**) and other related compounds such as tiagabine (**4.170**). Mechanistically, these may be thought of as glial GABA uptake blockers. By this mechanism, they block uptake of GABA into adjacent glial cells and thus block GABA breakdown.

4.7.3.4 Agents Affecting GABA Release

High doses of imipramine, haloperidol, and chlorpromazine (at 1 μM concentrations) are known to inhibit GABA release *in vitro*. Baclofen, [β-(*p*-chlorophenyl)-GABA] (**4.171**), is a valuable compound which enhances GABA release and is therefore an indirect agonist.

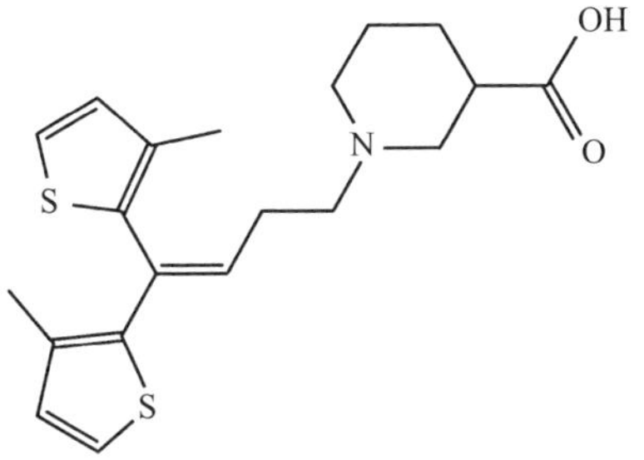

Tiagabine
(4.170)

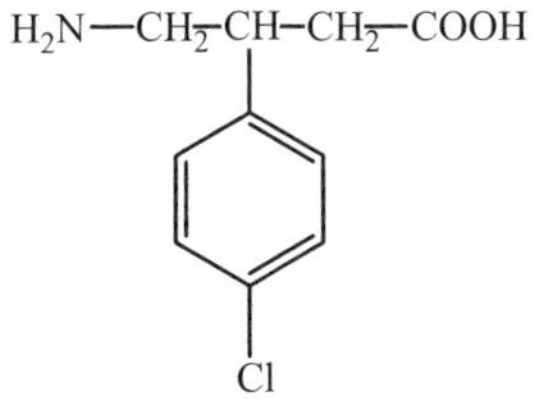

Baclofen
(4.171)

It is an orally active muscle relaxant and is used in treating the spasticity and muscle rigidity of spinal cord injuries, cerebral palsy, and other related neurologic disorders.

4.7.4 GABA Receptors: Postsynaptic Drug Effects

4.7.4.1 GABA Agonists

Directly acting GABA agonists usually bear some resemblance to the neurotransmitter. Muscimol (**4.172**), an isoxazole isolated from the mushroom *Amanita muscaria* ("deadly fly agaric"), is a hallucinogen with a receptor affinity greater than that of GABA (0.9 nM versus 9.4 nM on Triton-treated membranes) on the $GABA_B$ receptor. A number of related compounds have been synthesized and have also proved to be active, among them THIP (gaboxadol) (**4.173**). Progabide (**4.174**) is another novel GABA agonist that bears a resemblance to the benzodiazepines.

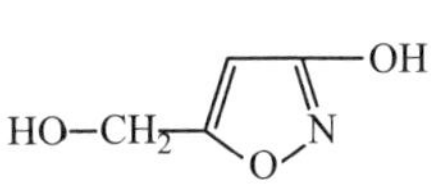

Muscimol
(4.172)

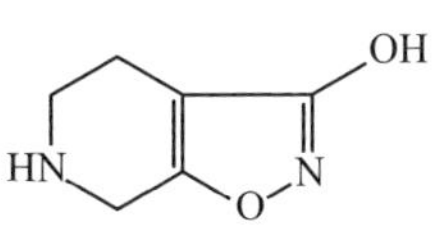

THIP
(4.173)

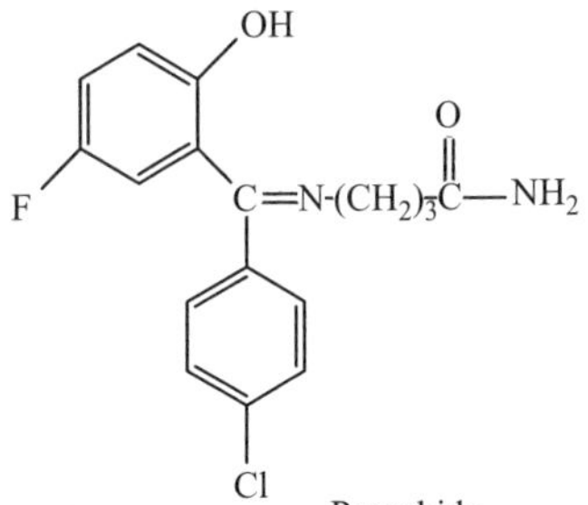

Progabide
(4.174)

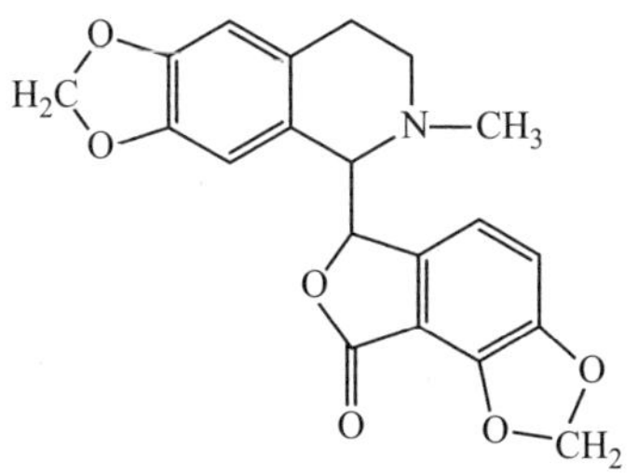

(+)-Bicuculline
(4.175)

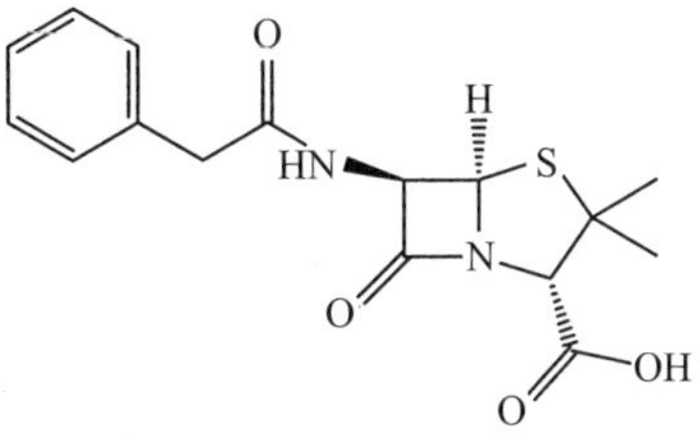

Benzylpenicillin
(4.176)

4.7.4.2 GABA Antagonists

An important direct antagonist is the alkaloid (+)-bicuculline (**4.175**), which binds to all synaptic GABA sites. Being a lactone, it is sensitive to hydrolysis. Its binding is influenced by salts; that is, [^{3}H] bicuculline binding in the presence of $50 \mu M$ NaSCN or $200 \mu M$ NaClO$_4$ is more "specific" than in the absence of salts, and only 30–50% of it can be displaced by GABA or muscimol. This is a further indication of GABA receptor multiplicity. Interestingly, benzylpenicillin (**4.176**) can antagonize GABA in doses below $2 \mu M$, and can thus be epileptogenic. Under most circumstances this is not clinically relevant since benzylpenicillin does not cross the blood–brain barrier. However, when used to treat diseases such as meningitis (in which the structural integrity of the blood–brain barrier is jeopardized), benzylpenicillin can contribute to the development of seizures.

4.7.5 GABAergic Drugs: Benzodiazepines

The benzodiazepines are probably the most clinically important class of GABA-active compounds. Benzodiazepines modify affective responses to sensory perceptions; specifically, they render individuals less responsive to anxiety-producing stimuli and therefore exert a strong *anxiolytic* action. In addition, benzodiazepines exert sedating, anticonvulsant, and muscle relaxant effects.

The benzodiazepines were discovered by Leo Sternbach at the Hoffman–La Roche laboratories, and their pharmacology was elucidated by Randall of the same company. An enormous variety of these compounds exist. Since about 3500 benzodiazepine compounds have been investigated, the neurologic structure–activity relationships of these drugs have been well established and the central features can be generalized as follows:

1. R$_1$ should be an electron-attracting group. Other substituents should not be attached to any of the carbons on that ring.
2. R$_2$ and R$_3$ can be varied. Replacement of the lactam oxygen by sulfur decreases activity.
3. The phenyl group is necessary for activity; halogen substituents are preferred in the *ortho* position.

Despite these GABAergic structure–activity properties, the benzodiazepine moiety is an extremely versatile building block or molecular platform upon which to design non-GABAergic drugs. In fact, benzodiazepines have been referred to as *privileged structures*. The term "privileged structure" was introduced by Evans et al. in describing the development of benzodiazepine-like cholecystokinin antagonists based on asperlicin, a natural product. By their definition, Evans and co-workers concluded that the benzodiazepine ring system was a privileged structure since it was "a single molecular framework able to provide ligands for diverse receptors" and that "judicious modification of the benzodiazepine structure could be a viable alternative in the search for new receptor agonists and antagonists." For instance, benzodiazepines are found in multiple types of CNS agents and are ligands for both ion channels and G-protein coupled receptors. Derivatives of benzodiazepines which are analgesics, cholecystokinin antagonists, angiotensin II antagonists, vasopressin antagonists, and bradykinin agonists have been described. Because of this versatility, libraries of benzodiazepines have been created to provide leads in drug discovery. Quantum pharmacology calculations have been used to design *in silico* libraries of benzodiazepine analogs for use as potential building blocks in the design of bioactive molecules.

Nevertheless, the GABAergic properties of benzodiazepines remain their most important clinical application. Over the past 30 years, the most widely used benzodiazepine drug has been diazepam (**1.6**). It is an anxiolytic, sedative, and muscle relaxant; the anxious, depressed person becomes more outgoing and relaxed. There have been many diazepam analogs. Oxazepam (**4.177**) and lorazepam (**4.178**) have similar effects. Temazepam (**4.179**), flunitrazepam (**4.180**), and flurazepam (**4.181**) are useful sedative-hypnotics. Clonazepam (**4.182**) is a clinically useful anticonvulsant. Brotizolam (**4.183**), a novel benzodiazepine analog, seems to be an effective sedative-hypnotic. Midazolam (**4.184**) is an imidazolo-benzodiazepine that is water soluble and thus easily injectable. It is a hypnotic sedative with marked amnestic (i.e., memory loss) properties and is used in dentistry, endoscopic procedures, and induction to anesthetics in the elderly and in

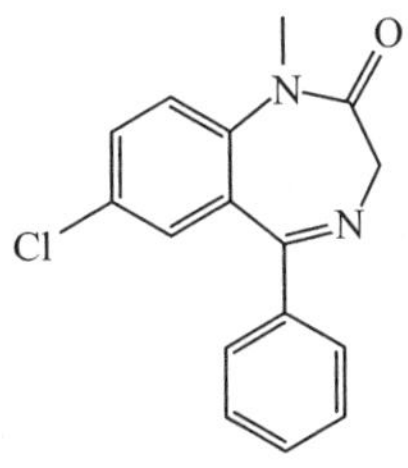

Diazepam
(1.6)

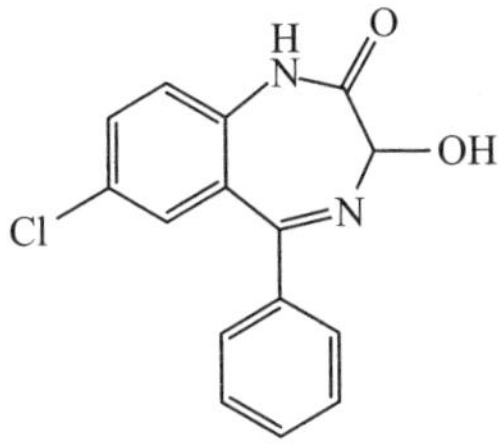

Oxazepam
(4.177)

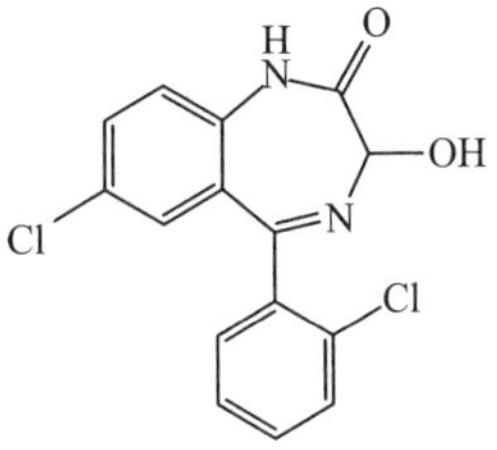

Lorazepam
(4.178)

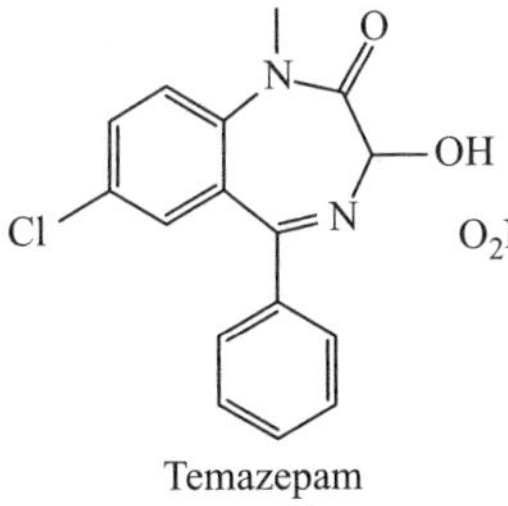

Temazepam
(4.179)

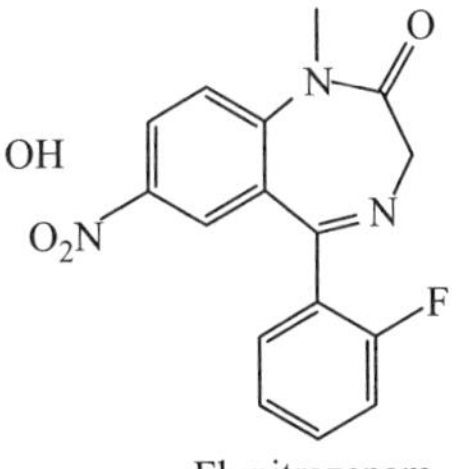

Flunitrazepam
(4.180)

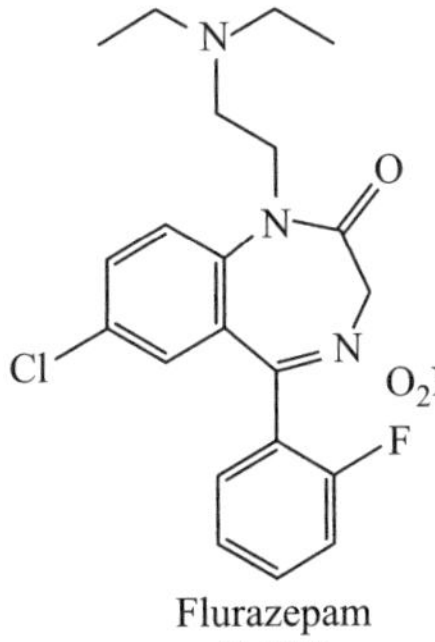

Flurazepam
(4.181)

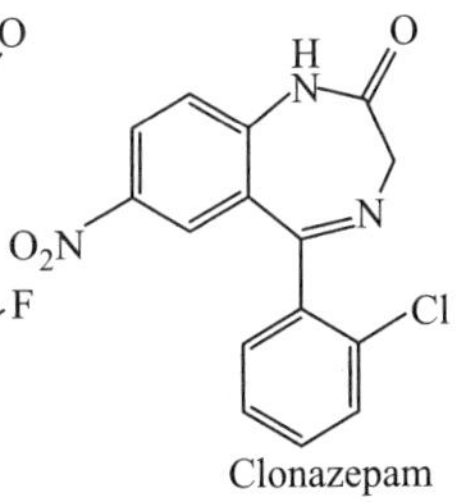

Clonazepam
(4.182)

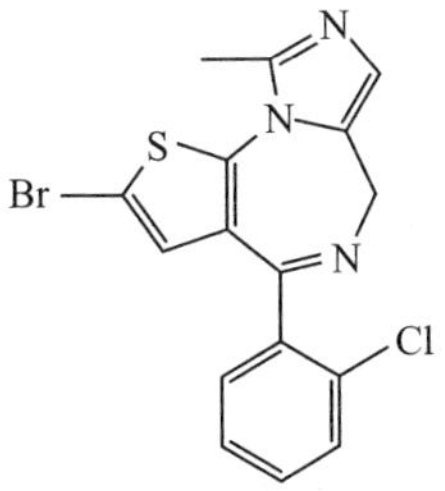

Brotizolam
(4.183)

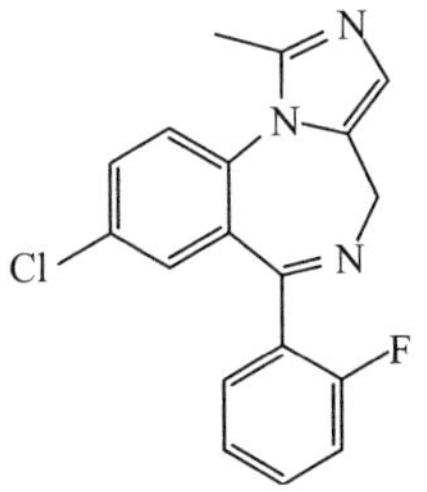

Midazolam
(4.184)

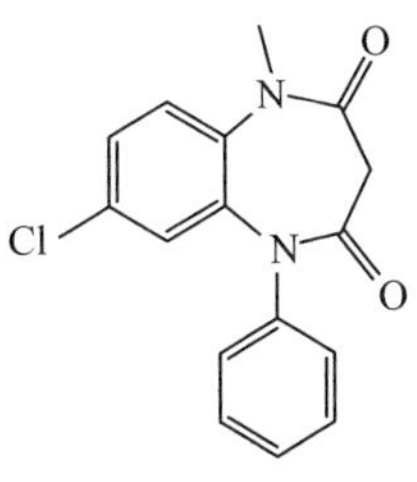

Clobazam
(4.185)

cardiac patients. These previous analogs are termed 1,4-benzodiazepines since the two nitrogen atoms within the diazepine ring are situated in a 1–4 arrangement with two carbon atoms between them; clobazam (**4.185**) is a 1,5-benzodiazepine, having its nitrogens in a 1–5 arrangement with three carbons between them. Clobazam is now a benzodiazepine of choice for epilepsy, with improved anticonvulsant activity in the presence of decreased likelihood of tolerance developing to the anticonvulsant effects.

4.7.5.1 Mode of Benzodiazepine Action

The mode of action of benzodiazepines is apparently based on augmentation of inhibitory neurons. The benzodiazepine receptor is an integral part of the GABA$_A$ receptor, which is itself a chloride ion channel. From electrophysiological studies it is known that these benzodiazepines increase the frequency of channel opening in response to GABA. The resulting increased chloride conductance of the neuronal membrane effectively short-circuits responses to depolarizing inputs. At the molecular level, benzodiazepine receptor agonists increase the affinity of GABA to its receptor. Thus, at a given concentration of GABA, binding to the receptors will be increased, causing an augmented response and a corresponding diminution in excitability.

This mode of action imparts a variety of clinical uses upon the benzodiazepine class of molecules. Probably first and foremost is their use as anxiolytics in the treatment of anxiety disorders. In the context of psychiatric disorders, anxiety consists of apprehension, tension, and excessive concern over danger that is either minor in degree or largely unrecognized; it is accompanied by signs of increased activity of the sympathetic nervous system. *Free-floating anxiety* occurs when there is no conscious recognition of a specific external danger. *Generalized anxiety disorder* is characterized by the frequent presence of excessive free-floating anxiety and overconcern, to the level that it interferes with emotional comfort and effectiveness in living. Benzodiazepines are particularly effective in the management of such disorders. Benzodiazepines are also quite useful in the treatment of panic disorders (anxiety attacks associated with episodic fearfulness) and phobic disorders (anxiety attacks associated with intense fear of a situation that the person consciously recognizes as harmless). Thus, while the antidopaminergics ("major tranquilizers") are useful for the treatment of major psychiatric disorders, the benzodiazepines ("minor tranquilizers") are the drugs of choice for minor psychiatric disorders. Benzodiazepines have also found utility as anticonvulsants, sleep-inducing agents, and general anesthetics.

4.7.5.2 Receptor Characterization and Drug Classification

Several receptors have been described for benzodiazepine agonists. The benzodiazepine receptor is part of the GABA$_A$ protein. As stated above, this protein is a pentamer composed of combinations of structurally related subunit families (α, β, γ, δ, ρ), some of which exist in multiple isoforms. The α-subunit bears the benzodiazepine (BDZ) binding site. The Type I BDZ receptor, which displays high affinity for a wide range of benzodiazepine analogs, contains an α_1 subunit, whereas the Type II BDZ receptor, which has a lower affinity for such agents, contains α_2 or α_3 subunits. Since an increasing number

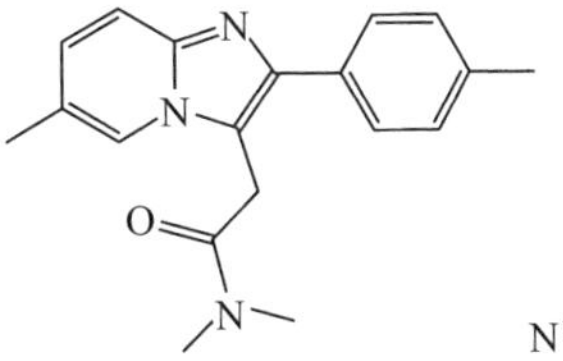

Zolpidem
(4.186)

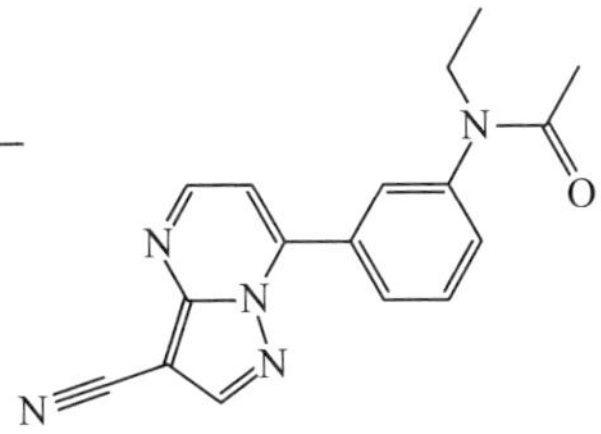

Zaleplon
(4.187)

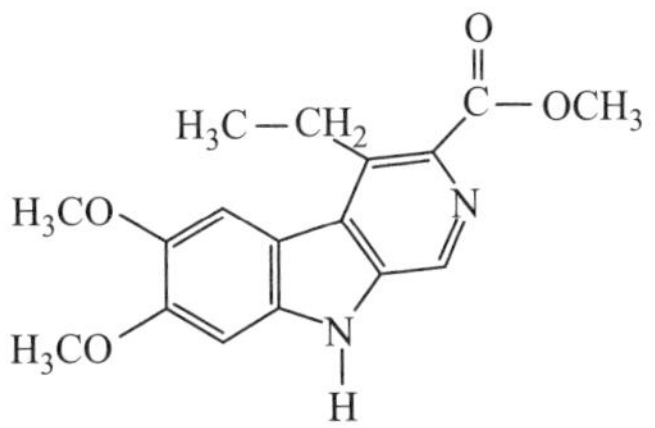

DMCM
(4.188)

of compounds with affinity for these receptors are not benzodiazepines, it may be more acceptable to employ the alternative nomenclature: ω_1 for Type I BDZ and ω_2 for Type II BDZ. The ω_1 receptor is located in brain areas involved with sedation; ω_2 receptors are highly concentrated in areas responsible for cognition, memory, and psychomotor functioning. Zolpidem (**4.186**) was the first non-benzodiazepine ω_1 agonist marketed; it is a hypnotic agent with minimal anticonvulsant and anxiolytic effects. Zaleplon (**4.187**) is another ω_1 agonist used in the treatment of insomnia.

Inverse benzodiazepine agonists such as DMCM (**4.188**) are anxiogenic and convulsive: they are called inverse agonists because they bind to agonist sites but have effects opposite to those of GABA. *Competitive benzodiazepine antagonists* (e.g., Ro 15-1788, **4.189**) also bind to this site; they are inactive by themselves, but prevent agonist and inverse agonist binding.

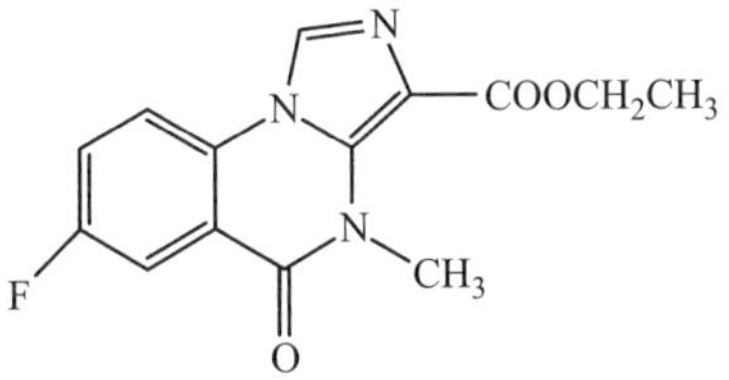

Ro 15-1788
(4.189)

4.7.6 GABAergic Drugs: Barbiturates

A large and still used group of these drugs is the barbiturates—sedative-hypnotic compounds that are used in anesthesia and as "sleeping pills." Barbiturates are acidic because of tautomerism with the enolate. Their pK_a is about 7.3, and therefore even slight changes in body pH will influence their ionization and, consequently, absorption and distribution. The replacement of oxygen by sulfur in position 2 leads to increased lipophilicity and very rapid penetration of the blood–brain barrier. Therefore, compounds like thiopental (**4.190**) are ultrashort-acting intravenous anesthetics, used in surgery for short operations or for inducing anesthesia prior to use of inhalation anesthetic. Methylation on the N-1 atom has a similar effect, as in hexobarbital (**4.191**). Branched side chains on C-5 lead to longer activity (pentobarbital, **4.192**; and amobarbital, **4.193**); short side chains, like ethyl, lead to the longest duration of action because the slow entry of the resultant molecule into the CNS affects the onset of action. Aromatic substituents produce anticonvulsant activity. Barbiturate enhancement of GABA binding is proportional to the anesthetic activity of the barbiturate.

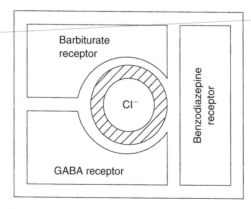

Thiopental
(4.190)

Hexobarbital
(4.191)

Pentobarbital
(4.192)

Amobarbital
(4.193)

The principal disadvantages of barbiturates as hypnotics include the development of physical dependence, a relatively low therapeutic index (and the potential of poisoning, as in suicide), suppression of REM sleep, and possible hangover effects. As mentioned above, benzodiazepines (e.g., flurazepam or brotizolam) are hypnotics as effective as barbiturates and are much safer in terms of their therapeutic index, addiction potential, and REM sleep-deprivation effects. Thus benzodiazepines have displaced barbiturates as sedative hypnotics.

Like benzodiazepines, barbiturates bind to the $GABA_A$ receptor—however, at a different site from the benzodiazepines. Measurements of mean ion channel open times show that barbiturates act by increasing the proportion of channels opening to the longest open state (i.e., 9 msec), resulting in an overall increase in Cl^- flux into the neuron.

Figure 4.12 shows a schematic model of the $GABA_A$ channel with representations of benzodiazepine and barbiturate binding sites. There is also a binding site for neuro-steroids, and analogs of such agents may emerge as useful anticonvulsants or general anesthetics in future years.

Figure 4.12 The GABA-A channel functions as a receptor for many multiple different drug classes, including benzodiazepines (e.g., diazepam) and barbiturates (e.g., phenobarbital). There is also a steroid-binding site on the GABA-A channel which may be useful in the future design of general anesthetics.

4.7.7 GABAergic Drugs: Anticonvulsants

Benzodiazepines and barbiturates are used as anticonvulsant drugs in the treatment of epilepsy. Epilepsy, a medical disorder characterized by recurrent seizures, has many different forms. The four most common seizure types are generalized tonic–clonic seizures (old name: grand mal seizures), generalized absence seizures (petit mal seizures), complex partial seizures (psychomotor or temporal lobe seizures), and simple partial seizures (focal seizures).

The drug of choice for generalized tonic–clonic seizures is an iminostilbene derivative called carbamazepine (1.5), with hydantoins such as phenytoin (4.194) being a close second. Other drugs useful in the treatment of simple partial, complex partial, and generalized tonic–clonic seizures include valproic acid (4.132), topiramate (4.195), lamotrigine (4.196), vigabatrin (4.168), gabapentin (4.197), tiagabine (4.170), phenobarbital (4.198), and primidone (4.199). The drug of choice for absence epilepsy is ethosuximide (4.200), with valproic acid (4.132), clobazam (4.185), lamotrigine (4.196), and topiramate (4.195) being acceptable alternatives. Vigabatrin can actually worsen absence seizures.

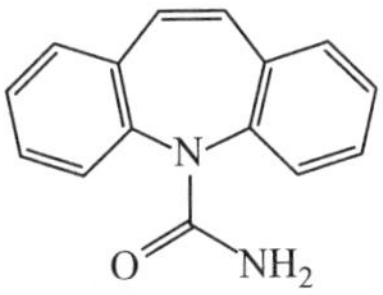

Carbamazepine
(1.5)

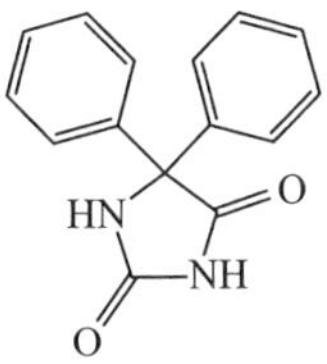

Phenytoin
(4.194)

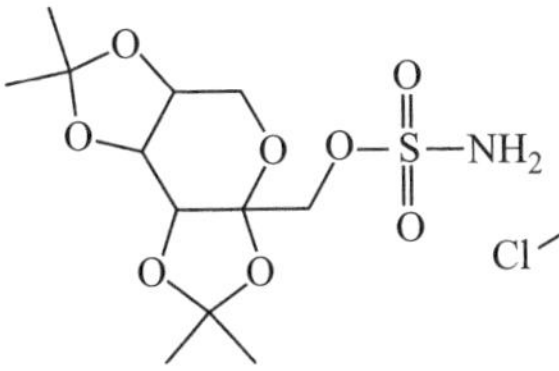

Topiramate
(4.195)

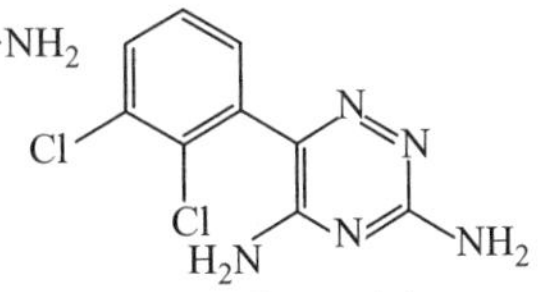

Lamotrigine
(4.196)

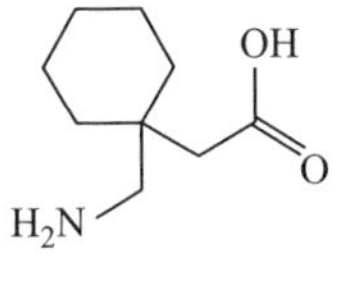

Gabapentin
(4.197)

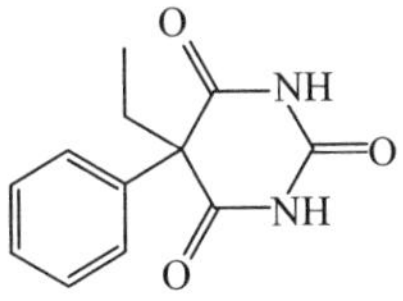

Phenobarbital
(4.198)

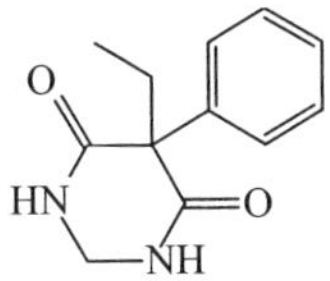

Primidone
(4.199)

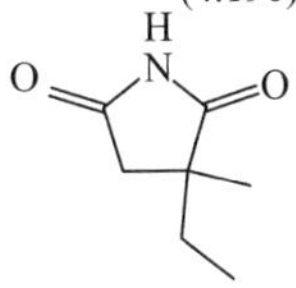

Ethosuximide
(4.200)

These various drugs have differing mechanisms of action. Carbamazepine and phenytoin block the voltage-gated sodium channel. In addition, carbamazepine may increase available adenosine A_1 receptors; it has been proposed that adenosine is a natural anticonvulsant. Lamotrigine also blocks the voltage-gated sodium channel, but in addition decreases brain levels of glutamate. Topiramate likewise blocks the sodium channel, as well as being a $GABA_A$ agonist, a glutamate antagonist (at the AMPA site), and a carbonic anhydrase inhibitor. Phenobarbital binds to the $GABA_A$ receptor and, at high concentrations, also blocks the voltage-gated sodium channel; primidone is metabolically converted to phenobarbital in the liver following its administration. Valproic acid blocks the voltage-gated sodium channel and the T-type Ca^{2+} channel at routine doses; at high doses, it also inhibits *succinic semialdehyde dehydrogenase*, the enzyme oxidizing the semialdehyde. As this metabolite accumulates, GABA-T activity is decreased by end-product inhibition, and the neurotransmitter concentration increases, thus inhibiting seizures. Ethosuximide, like valproate, blocks the T-type Ca^{2+} channel. Vigabatrin is a GABA transaminase inhibitor that causes dramatic increases in brain GABA levels. It produces side effects in the form of psychosis and visual field constriction (due to retinal damage); both of these side effects may be GABA-mediated. Finally, as stated above, clobazam is a $GABA_A$ agonist.

γ-Aminobutyric acid also seems to be involved in a number of other physiological and pathological functions, including feeding, sleep, hormonal secretion, cardiovascular functions, and, most importantly, general anesthesia.

4.7.8 GABAergic Drugs: General Anesthetics

Richard Feynman, the Nobel-prize winning physicist, once wrote:

> I wonder why. I wonder why.
> I wonder why I wonder.
> I wonder *why* I wonder why
> I wonder why I wonder!

This simple quatrain wonderfully epitomizes the essence of the human mind. Awareness of one's inner and outer environments is central to the functioning of the human mind. Consciousness is the enigmatic phenomenon that is crucial to the interface between brain and mind. General anesthetics are drugs that temporarily rob us of our consciousness; GABA is a key molecular target for general anesthesia.

Many general anesthetic agents work at the level of the $GABA_A$ receptor. This GABAergic mechanism for general anesthesia accounts for the observation that many benzodiazepines (midazolam, **4.184**), barbiturates (thiopental, **4.190**), and other anticonvulsants are both anticonvulsants and general anesthetics. In addition, a number of general anesthetics, such as propofol (**4.201**) and etomidate (**4.202**), bind to the $GABA_A$ receptor. This recent implication of GABA receptors in general anesthesia provides intriguing evidence for the role of GABAergic mechanisms in a molecular-level understanding of human consciousness—that most enigmatic of physiological processes. Indeed, the riddle of human consciousness has been described by Francis Crick, the Nobel-prize-winning biologist, as the major unsolved problem in biology.

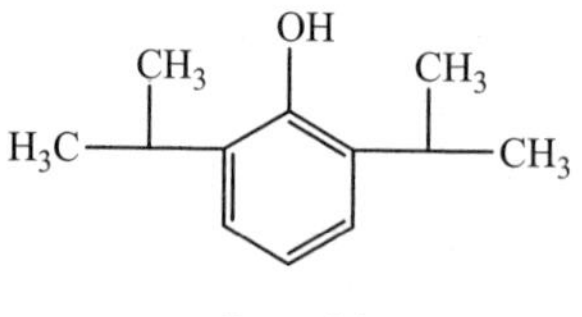

Propofol
(4.201)

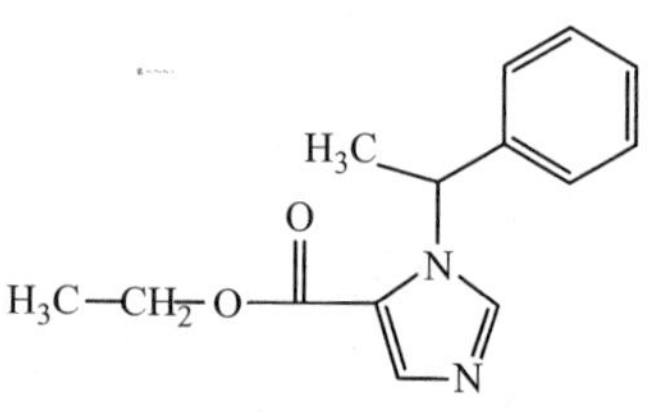

Etomidate
(4.202)

4.7.9 The Clinical–Molecular Interface: Huntington's Disease as a GABAergic Disorder

Huntington's disease or Huntington's chorea, a neurodegenerative disorder, has its symptomatic basis in defective central GABA metabolism. Huntington's is a hereditary disease that manifests itself in disabling involuntary movements. These movements are called chorea (from the Greek for dance). Choreiform movements are frequent, brief, sudden, random twitches that can affect any part of the body. Other movement symptoms such as athetosis (slower, writhing movements) and dystonia (abnormal posturing of body parts) may be superimposed upon the choreiform movements. As the disease progresses, muscle rigidity appears and eventually a progressive dementia.

This progressive autosomal dominant disorder tends to become clinically apparent in early adulthood. The gene abnormality is on the short arm of chromosome 4. This gene abnormality, called a trinucleotide repeat disorder, results in too many glutamine amino acids being incorporated into a particular protein. The gene product ("huntingtin") is cytotoxic to neurons producing neurotransmitter abnormalities that culminate in the disease manifestations. Although many neurotransmitter abnormalities are implicated, GABA is a central player.

Both GABA and glutamic acid decarboxylase (GAD), the enzyme central to GABA's biosynthesis, are markedly reduced in the brains of people with Huntington's chorea. Furthermore, the GABAergic neurons projecting from the caudate nucleus to the substantia nigra within the brain show lesions in patients with Huntington's. The caudate nucleus is a brain region implicated in the coordination of movement. Concomitant with the reduction in GABAergic inhibition are imbalances in dopamine and acetylcholine. The clinical manifestations of the disease therefore seem to evolve from functional underactivity of GABAergic pathways coupled to functional overactivity of dopaminergic pathways, leading to a symptom complex characterized by concomitant movement and mentation disorders.

Attempts to treat Huntington's disease (or at least to alleviate the symptoms) by enhancing central GABAergic processes have been disappointing at best. Since direct replacement of the lost inhibitory functions is not possible, the usual treatment of Huntington's disease consists of inhibiting excessive dopaminergic activity by the DA-antagonist neuroleptics haloperidol or chlorpromazine, thus restoring the balance of GABAergic and dopaminergic functions. Extended use of the DA antagonists is associated with iatrogenic (i.e., physician-induced) parkinsonism. Moreover, these pharmacological manipulations are symptomatic at best, sometimes yielding inconsistent, minimally beneficial responses.

4.8 INHIBITORY AMINO ACID NEUROTRANSMITTERS: GLYCINE

Glycine (**4.203**) is the simplest amino acid and was first identified as a neurotransmitter in 1965. It is found predominantly in the brainstem and spinal cord, but also diffusely throughout the CNS. Like GABA, it is a predominantly inhibitory transmitter, but acts thus through a vastly different pharmacology. The biosynthetic precursor of glycine is serine, which is metabolically converted to glycine through a process catalyzed by the enzyme serine hydroxymethyltransferase. The action of glycine is terminated by cellular re-uptake, mediated through a high-affinity transporter system. There are two glycine receptors: the strychnine-sensitive Cl⁻ channel glycine receptor, and the strychnine-insensitive glycine subsite on the NMDA receptor complex.

The first receptor is the strychnine-sensitive Cl⁻ channel glycine receptor. The plant alkaloid strychnine (**4.204**) is an antagonist for this receptor, binding with nanomolar affinity. The glycine receptor, like other members of the family, is formed by a pen-tameric arrangement surrounding a central ion pore; there is significant homology with the GABA$_A$ receptor. This receptor exists as a macromolecular complex composed of two homologous polypeptides: α(48 kDa) and β(58 kDa). Binding sites for glycine and strychnine are found in the subunit. Mutations of the α subunit, particularly a leucine for arginine substitution at position 271, lead to a very rare neurological disorder termed *hyperekplexia*. Patients with this disease demonstrate an exaggerated startle response to environmental stimuli, jumping dramatically or even collapsing in response to minor situational stimuli such as a car door shutting. The strychnine-sensitive Cl⁻ channel glycine receptor functions as an anion channel. Permitting chloride anion to enter neu-rons causes neuronal hyperpolarization (increased negative charge within the cell), which in turn decreases neuronal excitability. Therefore, this glycine receptor is inhibitory and, when binding to this receptor, glycine is acting as an inhibitory neuro-transmitter. Autoradiography studies with [³H]strychnine show that these receptors are clustered mainly in the spinal cord and brainstem.

Glycine
(4.203)

Strychnine
(4.204)

The strychnine-insensitive glycine subsite on the NMDA receptor complex is the other major glycine receptor. On this receptor, unlike the strychnine-sensitive receptor, glycine is linked to neuronal excitation. Glycine is a necessary cofactor for the func-tioning of the excitatory NMDA receptor; when binding to this receptor, glycine is acting as an excitatory neuromodulator.

Both the strychnine-sensitive and insensitive receptors are involved in pathological processes and thus are targets for drug design. The strychnine-sensitive site may be a useful target when designing drugs (as agonists) to treat spasticity. Spasticity is a symp-tom arising from damage to the spinal cord or to the descending corticospinal tract from the brain that is characterized by increased tone, sometimes painful, in the muscles of the arms and legs. The strychnine-insensitive receptor may be a useful target when designing drugs (as antagonists) to treat either epilepsy or stroke.

In medicinal chemistry, the strychnine-insensitive receptor has received more attention. For example, the 5-nitro derivative of 6,7-dichloroquinoxalinedione has been identified as a highly potent antagonist of the glycine subsite on the NMDA receptor.

This compound has displayed significant activity in animal models of epilepsy (maximal electroshock assay) and stroke (focal ischemia model). In addition, a number of 3-substituted indole-2-carboxylates were shown to be powerful glycine site antagonists. From these studies, and from related molecular modeling investigations, it became apparent that the NMDA glycine recognition site possessed a large lipophilic pocket adjacent to the 3-position of the indole nucleus; reduction of the indole completely abolished compound affinity for the receptor site.

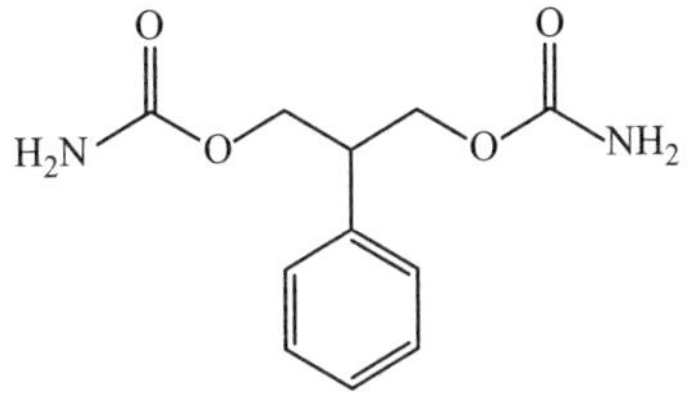

Felbamate
(4.205)

Compounds targeting the glycine receptors have had mixed success as drug candidates. In terms of drugs developed for epilepsy, felbamate (**4.205**), active at the NMDA glycine subsite, is a potent anticonvulsant but has run into toxicity problems, causing aplastic anemia (suppression of the bone marrow). Drugs targeting the NMDA glycine site and designed for stroke have had even greater developmental problems. Although many of these compounds work very well against animal models of stroke, they have uniformly failed to be of significant value when used in Phase III human trials of the same disease. This observation calls into question the utility of some of the animal models of stroke that are currently employed.

4.9 EXCITATORY AMINO ACID
NEUROTRANSMITTERS: GLUTAMATE

Glutamate (**4.206**) and aspartate (**4.207**) have long been known as excitatory transmitters, first in crustacean muscle and later in the vertebrate CNS. As amino acids, they have many other important biochemical roles; thus their concentration is uniformly high throughout the nervous system. Certain areas in the spinal cord, interneurons of the reflex arc, and a pathway from the cortex to the striatum are presumed sites of activity. Because of the uniform glutamate and aspartate distribution, mapping of the receptors was accomplished only recently. For the same reason, specific nonendogenous natural-product agonists had to be used for receptor characterization, in the same manner as in the differentiation of the nicotinic and muscarinic cholinoceptors.

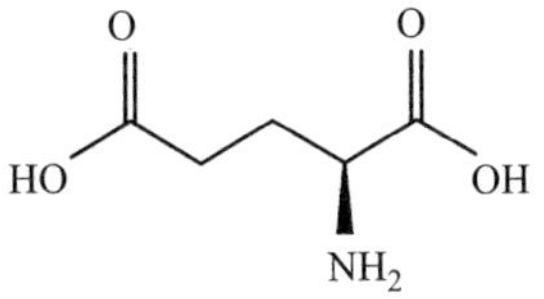

Glutamate
(4.206)

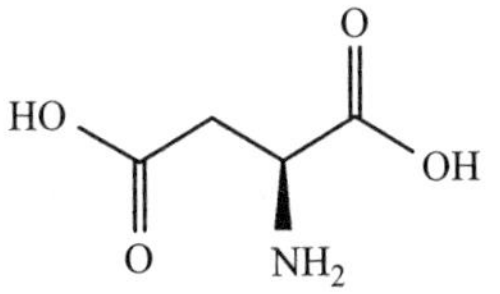

Aspartate
(4.207)

4.9.1 Glutamate Receptors

The array of receptors that mediate the activities of glutamate is somewhat confusing. Broadly speaking, the glutamate receptors can be categorized in two major groups: ionotropic and metabotropic. (Ionotropic receptors exert their effect by influencing trans-membrane ionic fluxes via ion channels; metabotropic receptors exert their influence by controlling intracellular processes via G-proteins.) The ionotropic receptors may be sub-divided into three types: N-methyl-D-aspartate (NMDA; **4.208**), α-amino-3-hydroxy-4-isoxazole propionic acid (AMPA; **4.209**), and kainate (KA; **4.210**). These three receptor types are both functionally distinct and defined by distinct molecular families of receptor genes. The three ionotropic receptors increase transmembrane cation conductance, especially of Ca^{2+} (in the case of the NMDA receptor). The metabotropic receptor family is even more confusing. In 1989, the notion of metabotropic receptors was barely emerging; by 1995, there were three metabotropic receptors; by 1999 there were reportedly eight of them. As stated above, the metabotropic receptors mediate their effects via G-proteins. These ionotropic and metabotropic receptors influence a variety of neuro-chemical and neurophysiological events. The ionotropic receptors, especially the NMDA receptor, have received the greatest amount of study. They have been implicated in the mechanism of information processing, memory, and learning, through long-term poten-tiation of neuronal pathways. They have also been involved in pathological processes such as epilepsy and stroke.

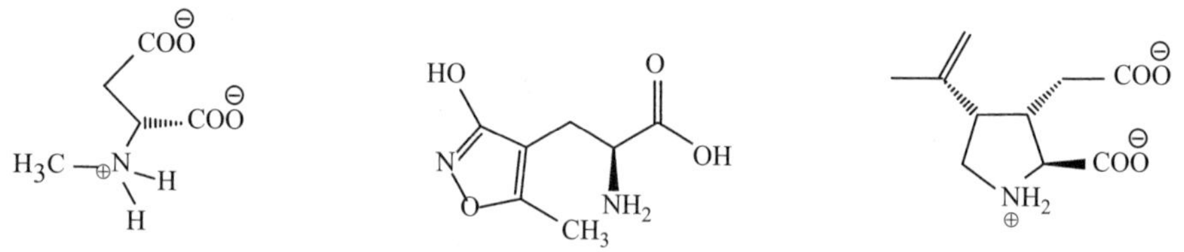

N-Methyl-D-aspartate
(4.208)

α-Amino-3-hydroxy-4-isoxazole propionic acid
(4.209)

Kainate
(4.210)

4.9.2 Glutamate Receptor Agonists and Antagonists

An immense literature regarding drug design and excitatory amino acids has emerged over the past 15 years. Initial work focused strongly on the NMDA receptor. The design of NMDA agonists and antagonists has been both helped and hindered by the complex-ity of the NMDA receptor complex. In addition to the anticipated agonist and competi-tive antagonist binding sites, there are a number of other functional subsites on the receptor complex: glycine, polyamine, Zn^{2+} and Mg^{2+}. Each one of these additional bind-ing sites responds functionally to either endogenous or exogenous ligands and is thus also a reasonable target for drug design. NMDA-related drug design is an area in which contributions from molecular modeling and physical chemistry have been invaluable.

The majority of NMDA agonists are closely related to the structure of glutamic acid. Thus, 4-methylene-L-glutamic acid (**4.211**) is a potent NMDA agonist. Bioisosteric replacement of a carboxylate group also produces agonists; D,L-(tetrazol-5-yl)glycine (**4.212**), in which a tetrazole bioisostere replaces a carboxylate, is a potent NMDA agonist. Likewise, with appropriate substitutions and replacement, glutamic acid analogs can also be competitive antagonists. In these antagonists, bioisosteric replacement of carboxylates with phosphonates is a frequent design strategy. Incorporation of a ketone functional group into the carbon backbone of 2-amino-5-phosphonovaleric acid (termed "AP5" in the NMDA literature) affords a potent, orally active NMDA antagonist called (R)-4-oxo-5-phosphononorvaline (**4.213**). This latter compound may be conformationally constrained through the creation of a *cis*-2,3-disubstituted piperidine nucleus to yield another agent with potent NMDA receptor affinity. Introduction of *trans*-4-methyl substituents into this piperidine derivative yields even greater receptor affinity.

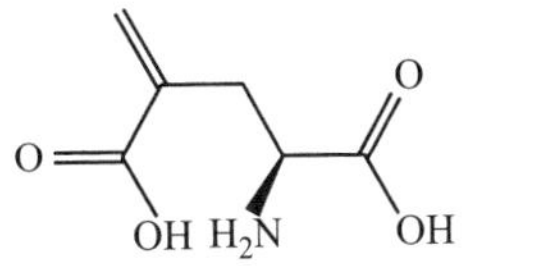

4-Methylene-L-glutamic acid
(4.211)

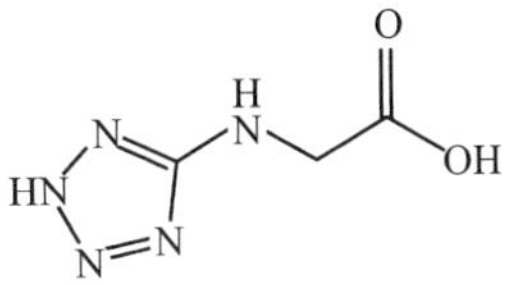

D,L-(tetrazol-5-yl)glycine
(4.212)

A variety of non-competitive NMDA antagonists have also been identified over the years. Probably the most famous of these molecules is dizocilopine or MK-801 (**4.214**). MK-801 has been massively studied as both a potential drug and a molecular probe for studying and understanding neurodegeneration. Although it never successfully emerged as a drug, for toxicology considerations, the structure–activity relationships of MK-801 and related iminomethanobenzocycloheptenes have received considerable study. NMDA antagonists have also emerged from studies targeting the subsites on the NMDA receptor complex. Work on glycine subsite antagonists is discussed in the section on glycine (section 4.8). Research on the polyamine site has also been fruitful. The polyamine antagonist eliprodil (**4.215**) was shown to protect cultured hippocampal brain cells from NMDA-mediated toxicity and demonstrated some activity in rat models of brain trauma and brain ischemia; oxindole variants of eliprodil have also been prepared and have shown enhanced activity and improved selectivity.

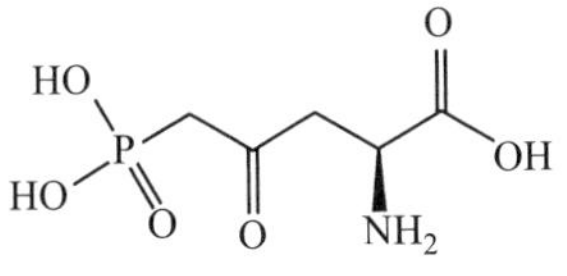

(R)-4-oxo-5-phosphononorvaline
(4.213)

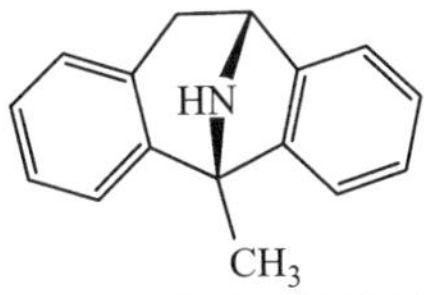

Dizocilopine (MK-801)
(4.214)

The zinc subsite has proved to be more challenging. Designing drugs to uniquely interfere with the biochemistry of a metallic anion in the central nervous system is non-trivial. It can be appreciated, from studies that attempt to design ion-specific chelating agents, that developing drugs to uniquely target one metal ion (e.g., zinc) over another (e.g., copper) can be difficult. Nevertheless, tricyclic antidepressants and phenothiazines, including desmethylimipramine and ethopropazine, have been suggested to act as zinc site ligands. The relationship between the zinc site and the magnesium site has also been considered. Magnesium itself has been studied as a drug. Models of stroke and heart attack have shown that magnesium sulphate infusions have neuroprotective effects equivalent in magnitude to those of noncompetitive NMDA antagonists.

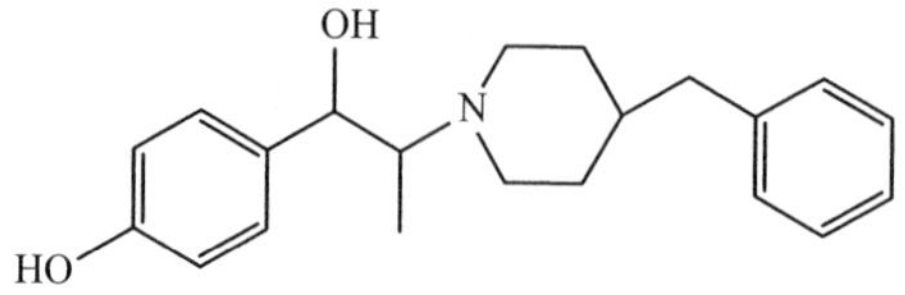

Eliprodil
(4.215)

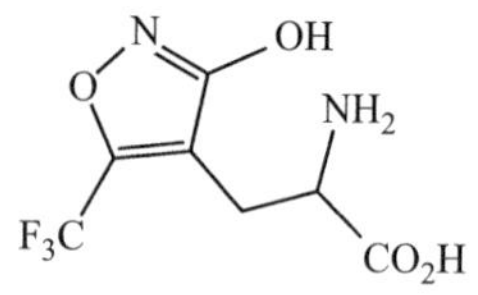

Trifluoro-AMPA
(4.216)

Glutamatergic and antiglutamatergic agents can also be prepared by targeting the non-NMDA receptors of glutamate. Trifluoro-AMPA (**4.216**) is a potent AMPA agonist in *in vitro* studies of rat cortex. Interestingly, replacement of the hydroxyl functional group present on the isoxazole ring of AMPA with a chlorine atom results in complete loss of binding affinity. Competitive AMPA antagonists have greater clinical use than AMPA agonists. Structure–activity studies have been performed on 2-phosphonoethylphenylalanine (**4.217**); analogs from this class, particularly the 5-methyl analog, are neuroprotective and effective in blocking seizures. In studies using rat spinal cords, acromelic acid and stizolobic acid were kainate agonists. 6,7-Dinitroquinoxalinedione (DNQX; **4.218**) and 6-cyano-7-nitroquinoxalinedione (CNQX; **4.219**) were the first examples of kainate antagonists. S-4-Carboxyphenylglycine (**4.220**) and R,S-α-methyl-4-carboxyphenylglycine (**4.221**) are metabotropic receptor antagonists.

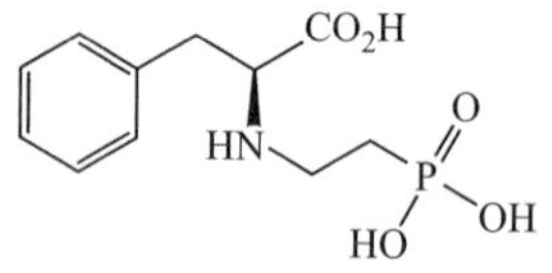

2-Phosphonoethylphenylalanine
(4.217)

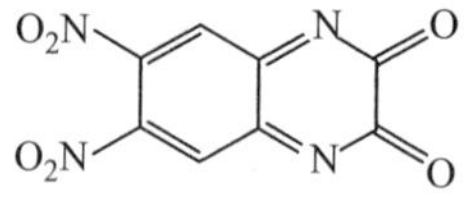

DNQX
(4.218)

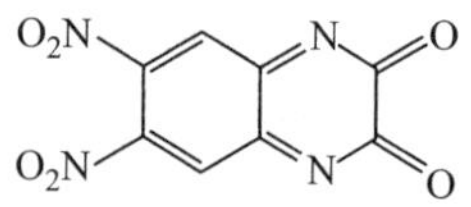

CNQX
(4.219)

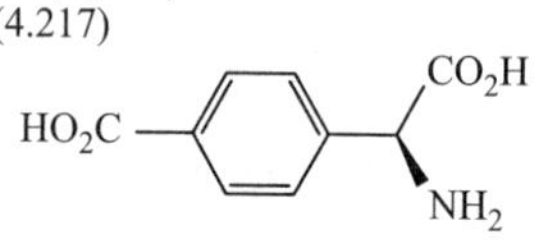

S-4-Carboxyphenylglycine
(4.220)

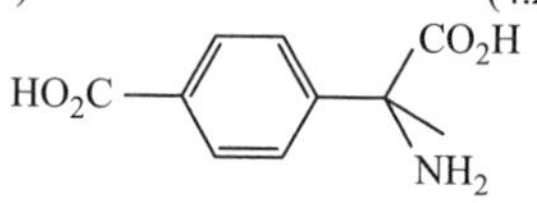

R,S-α-Methyl-4-Carboxyphenylglycine
(4.221)

In 2005, Rothstein et al. devised a novel antiglutamatergic approach based upon the identification of molecules that increase brain expression of the glutamate transporter protein (GLT1), thereby accelerating inactivation of glutamate within the synapse. Using a screening strategy, they identified multiple β-lactam antibiotics (e.g., penicillin, ceftriaxone) as potent stimulators of GLT1 expression – a result that demonstrates the utility of the β-lactam moiety as a privileged platform for drug design. Ceftriaxone was neuroprotectant in vitro when used in models of stroke and amyotrophic lateral sclerosis (ALS).

4.9.3 The Clinical–Molecular Interface: Stroke as a Glutamatergic Disorder

Stroke is a leading cause of death and disability and is the most common neurological disorder of the elderly. A stroke is defined as the acute onset of a neurologic deficit (e.g., paralysis of motor movement in the arm and leg on the same side of the body—hemiplegia) associated with an abrupt alteration in blood supply to a discrete region of the brain. Strokes may be of two types: hemorrhagic (in which the artery actually ruptures or leaks) or ischemic (in which the artery is blocked by debris or progressive narrowing and no blood can go through it). Ischemic strokes are by far the more common and are a byproduct of atherosclerosis and other disorders of blood vessels.

The medical treatment of stroke has been revolutionized in the past five years by the introduction of therapy with tPA (tissue plasminogen activator). This is a "clot busting drug" which is used to clean out the debris blocking the artery. Although this agent may be effective if used early (i.e., within three hours of the onset of the stroke), its delayed use can be associated with catastrophic brain hemorrhages. Moreover, the three-hour time window is too tight for many patients with stroke. Thus the need to develop anti-stroke agents is a continuing challenge.

A conceptual approach which has been pursued in the development of stroke therapies is based upon rescue of the ischemic brain using *neuroprotectant drugs*. This approach has been based upon the notion that the reduced supply of blood (and hence the supply of oxygen and glucose) to the brain leads to depolarization of cell membranes and subsequently to the release of glutamate; this stimulates the NMDA and AMPA receptors, causing an increase in intracellular calcium which in turn promotes lipolysis and proteolysis, thereby producing irreversible brain cell damage. The role of a neuroprotectant drug would be to interrupt this neurotoxic biochemical cascade. The majority of drugs developed to fulfill this role as a neuroprotectant have targeted glutamate and glutamate receptors: competitive NMDA antagonists (e.g., selfotel, **4.222**), NMDA polyamine site antagonists (e.g., eliprodil (**4.215**), ifenprodil (**4.223**), NMDA glycine site antagonists (e.g., gavestinel, **4.224**), NMDA Mg site antagonists (e.g., magnesium sulphate), other NMDA antagonists (e.g., cerestat, **4.225**), inhibition of glutamate release (e.g., lamotrigine), and AMPA antagonists (e.g., NBQX). The majority of these drugs worked (and worked impressively at times) in rodent models of stroke in which an artery was tied off by a ligature, thus producing a stroke. However, these successes in animal models did not translate into successful drugs for human stroke. These various neuroprotectant agents, targeting glutamatergic processes, have failed in human trials due to insufficient efficacy. Clearly human stroke has a complicated biochemistry which is poorly understood and which does not respond to a straightforward

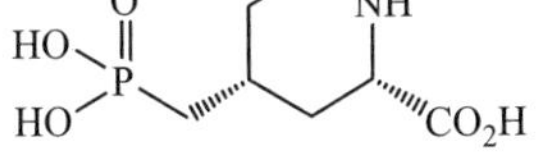

Selfotel
(4.222)

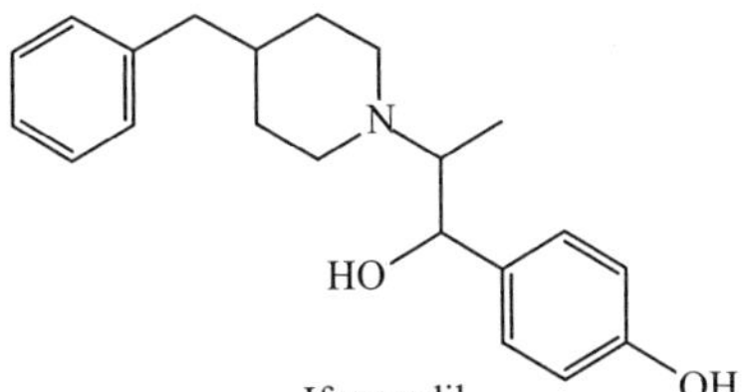

Ifenprodil
(4.223)

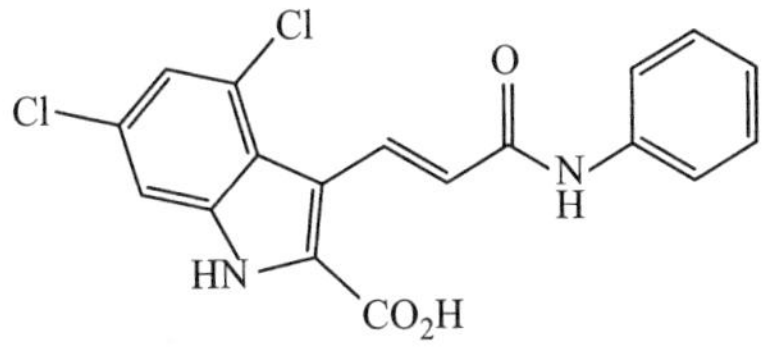

Gavestinel
(4.224)

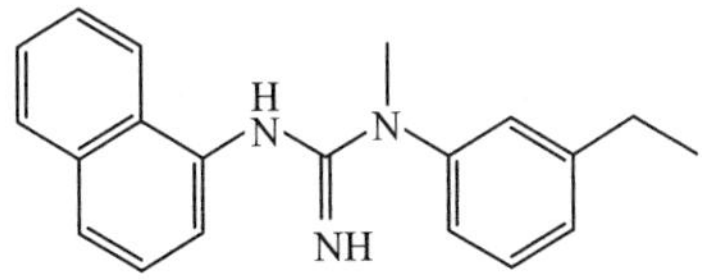

Cerestat
(4.225)

approach using antiglutamatergic therapies. The rodent model of stroke does not fully capture these complexities in a realistic manner.

In addition to failed trials with antiglutamatergic neuroprotectants, other approaches to neuroprotectant design for stroke have likewise failed: antagonism of voltage-gated Ca^{2+} channels, antagonism of voltage-gated Na^+ channels, inhibition of nitric oxide synthase, and scavenging of free radicals. Several recent high-profile failures in designing therapies for stroke suggest that a deeper understanding of stroke as a biological and pathological phenomenon may be required as a prerequisite to the improved design of stroke therapies.

4.10 LARGE-MOLECULE NEUROTRANSMITTERS: PEPTIDES

Although the majority of neurotransmitters are relatively small molecules, there is also a variety of neurotransmitters that are peptides. Peptide neurotransmitters present unique challenges for the drug designer. Whereas the design of agonists and competitive antagonists for small-molecule neurotransmitters (e.g., glutamate) may be conceptually straightforward, the same is not the case with peptidic neurotransmitters. Designing a small molecule to either mimic or block a peptide is a nontrivial task. The peptide neurotransmitter is a large, floppy molecule, which may interact with a receptor via many points of contact. Determining the bioactive conformation of a peptide is difficult; as described in chapter 1, computational drug design around peptides is hindered by a multiple minima problem. As described in chapter 3, peptidomimetic chemistry describes an approach to producing small-molecule drugs based on larger peptides. Regardless of these design challenges, a number of important peptide neurotransmitters are being actively studied in the search for new therapeutics. Currently, more than 20 peptide neurotransmitters that could be useful platforms for drug design have been identified (see table 4.3). Although all of these are interesting, we will only discuss several representative examples.

4.10.1 Corticotropin Releasing Factor

Corticotropin releasing factor (CRF) is a neuropeptide that falls into the broad spectrum of having neurotransmitter/neurohormonal/neuromodulator activities. CRF prepares the host organism for response to a variety of stressors including impending physical trauma, insults to the endocrine or immune systems, and difficult social interactions. In the CNS, CRF influences neurons located in higher cortical centers. Not surprisingly, hyper- or hypo-activity of the CRF system can participate in the mechanisms of a variety of human disorders: depression, stress-induced gastrointestinal disorders, drug addiction, pain, and eating disorders. The CRF system has also been implicated in Alzheimer's disease and Parkinson's disease. CRF exerts its actions through one of two different types of G-protein-coupled receptors (CRF-1, CRF-2), each being encoded by separate genes. Distinct subtypes of these two receptors (CRF-1α, CRF-1β, CRF-1γ, CRF-1σ; CRF-2α, CRF-2β, CRF-2γ) arise from different sequence modifications and differ in their anatomical location and in their responsiveness to various exogenous ligands.

The design and synthesis of low molecular weight nonpeptide ligands for the CRF receptors has been, and continues to be, a very active area of research. The design of CRF-1 antagonists has been more successful than that of CRF-2 agents; this design process has been greatly facilitated by molecular modeling and quantum pharmacology

Table 4.3 Neuroactive Brain Peptides

Class	Peptide
Gastrointestinal peptides	Cholecystokinin
	Bombesin
	Gastrin
	Glucagon
	Insulin
	Leucine-enkephalin
	Methionine-enkephalin
	Motilin
	Neurotensin
	Secretin
	Somatostatin
	Substance P
	Thyrotropin-releasing hormone
	Vasoconstrictive intestinal polypeptide
Hypothalamic releasing hormone	Corticotropin-releasing hormone
	Gonadotropin-releasing hormone
	Growth hormone-releasing hormone
	Somatostatin
	Thyrotropin-releasing hormone
Neurohypophyseal hormones	Oxytocin
	Vasopressin
Pituitary peptides	Adrenocorticotropic hormone
	α-Melanocyte-stimulating hormone
	β-Endorphin
	Growth hormone
	Luteinizing hormone
	Prolactin
	Thyrotropin
Other	Angiotensin II
	Bradykinin
	Calcitonin
	Galanin
	Neuropeptide Y
	Substance K

calculations. Theoretical pharmacophore analyses using such calculations indicate that CRF-1 ligands should possess a hydrogen bond acceptor in a core aromatic heterocycle that is flanked by a branched alkyl group and a 2,4-disubstituted aromatic substituent. The core aromatic heterocycle may be six-membered or even five-membered, such as thiazoles, pyrazoles, or imidazoles.

As mentioned, CRF antagonists can be employed in the treatment of a wide variety of disorders. One particularly important area is in the treatment of *irritable bowel syndrome* [IBS], a disorder characterized by abdominal discomfort or pain associated with a change in the frequency of stool passage and a change in the form of the stool. CRF-1 mediates the increase in colonic motility, whilst CRF-2 mediates the associated delay in colonic emptying; accordingly, CRF-1 and CRF-2 antagonists may provide novel therapeutic approaches for IBS. (It is highly significant that neurologically active agents are used in the treatment of gastrointestinal disorders. The gut has an extensive innervation and is controlled by a complex array of neurons, making it an ideal target for neuroactive agents.)

4.10.2 Neuropeptide Y

Neuropeptide Y (NPY) is another messenger molecule that demonstrates the close interplay between the nervous and gastrointestinal systems. Neuropeptide Y, a member of the pancreatic polypeptide family, is a 36-amino-acid neuropeptide that is widely distributed in both the central and peripheral nervous systems. NPY elicits its physiological effects via interaction with at least six different G-protein-coupled receptors (Y1–Y6), all of which demonstrate varying anatomical distributions: Y1 (vascular smooth muscle of blood vessels and brain); Y2 (peripheral nervous system and several brain regions); Y3 (heart, colon, adrenals, brainstem); Y4 (small intestine, pancreas); Y5 (brain, testis, spleen); and Y6 (hypothalamus, kidney). Arising from this plethora of receptors throughout the body, NPY plays central roles in multiple conditions: obesity (via Y1 and Y5); anxiety (via Y1); heart rate and blood pressure (via Y3); and ethanol abuse. Molecular modeling-assisted rational drug design has been used to develop antagonists to many of these receptor subtypes.

4.10.2.1 Neuropeptides and Obesity

Neuropeptides such as NPY may represent novel approaches to the treatment of obesity. Obesity is a disorder of epidemic proportions in developed countries, especially North America. To make matters worse, obesity spawns a host of other health problems: diabetes, high blood pressure, osteoarthritis, heart attack, and stroke. It is easy to say that the solution is simple ("eat less"); in reality, this is not happening, and the general public is inundated with a library of self-help diet books and a litany of drastic (and sometimes dangerous) diets. An appetite suppressant pill is a superficial fix that fails to address the complexities of obesity, hence the high rate of failure. Arguably, there may be a need for two separate diet pills: one that helps take weight off and one that helps to keep weight off (many people find it "easy to lose the weight and even easier to put it back on"). These two processes may be biochemically distinct, affording separate targets for drug design. While the gratification that comes from eating may seem to be centered in the upper abdomen, in reality it comes from the brain. Neuropeptides, such as NPY, may offer novel approaches to the pharmacological management of this ever-growing problem.

4.10.3 Galanin

Gly-Trp-Thr-Leu-Asn-Ser-Ala-Gly-Tyr-Leu-Leu-Gly-Pro-His-Ala-
Val-Gly-Asn-His-Arg-Ser-Phe-Ser-Asp-Lys-Asn-Gly-Leu-Thr-Ser
Galanin
(4.226)

Galanin is a neuropeptide with widespread distribution in both the central and peripheral nervous systems. Human galanin (**4.226**) is a 30-amino-acid peptide that inhibits the release of other neurotransmitters and in doing so plays a role in memory acquisition, sexual behaviour modulation, gastrointestinal mobility, and the appreciation of pain.

Throughout the 1990s, molecular cloning enabled the identification of at least three novel galanin receptors: GALR1–3. These receptors operate via G-protein effectors. Work on developing nonpeptidic small-molecule antagonists for these receptors is currently under way. This design work is being greatly facilitated by a combination of experimental and molecular modeling methodologies. The conformation of the galanin peptide has been modeled using ^{1}H-NMR spectroscopic techniques; a three-dimensional model of the GALR1 receptor has been deduced from molecular modeling studies based on the transmembrane helices of the bacteriorhodopsin protein. These diverse structural studies may permit the development of antagonists which may be useful in the treatment of obesity, dementia-associated cognitive decline, and peripheral nerve injury.

4.10.4 Neurokinins

Neurokinins (or tachykinins) constitute a family of neuropeptides that includes substance P (SP), neurokinin A (NKA) and neurokinin B (NKB). Neurokinins are widely distributed throughout the central and peripheral nervous systems. The first neurokinin to receive extensive study was substance P. Substance P (for "powder"), an undecapeptide of the sequence Arg–Pro–Lys–Pro–Gln–Gln–Phe–Phe–Gly–Leu–Met–NH$_2$, is an extremely active excitatory peptide neurotransmitter. The mechanisms of its biosynthesis and inactivation remain incompletely elucidated. Nonetheless, the localized neuronal presence of SP and its release in the salivary gland and in several brain regions suggest that it has a neurotransmitter or neurohormonal role. Its coexistence in some serotonergic and histaminergic neurons in the CNS is an interesting phenomenon. Substance P may be involved in pain mediation, as suggested by the fact that its injection into the brain has produced analgesia and it may regulate catecholamine turnover. SP is involved in mediating pain responses peripherally as well as centrally. Small fibers of the peripheral pain-sensitive neurons use SP as excitatory transmitter, and neurotransmitter release is inhibited by opiates at the dorsal horn of the spinal cord. In the CNS, the effect of SP is inhibitory, analgesic, and is stimulated by the endogenous opiate neuropeptide met-enkephalin. In stressed animals, SP can block the analgesic effect of endogenous opiates. The peripheral sensory effects of SP are mediated by the C-terminal fragment (6–8 amino acids) of SP, whereas the central analgesic effects are due to the N-terminal fragment, which also stimulates learning and memory. Thus post-translational peptide-modifying enzymes may decide which effect will prevail, in addition to the receptors at the site of action.

Over the past decade, considerable effort has been expended on the design and synthesis of nonpeptidic antagonists for the three receptors (NK1–3) to which SP, NKA, and NKB bind. These antagonists could have therapeutic use in conditions such as migraine, arthritis, inflammatory diseases such as cystitis, psoriasis, asthma, anxiety, depression, and emesis (vomiting). A variety of NK1–3 receptor antagonists have been designed and synthesized. The majority of these compounds are polyheterocyclics.

4.11 SMALL-MOLECULE NEUROTRANSMITTERS: GASES (NITRIC OXIDE, CARBON MONOXIDE)

Peptide neurotransmitters present problems for the drug designer because they are too large. Gaseous neurotransmitters likewise present problems for the drug designer, but because they are too small. It is not feasible to synthesize analogs of a substance like NO—it only has two atoms. Rather, it becomes necessary to appreciate the metabolism of the neurotransmitter and to endeavor to modify the activities of enzymes and other cofactors. Given the emerging importance of NO, this activity is under serious pursuit by many academic and industrial groups. A growing body of literature also supports the observation that carbon monoxide may also be a gaseous neurotransmitter. Together, NO and CO are recognized as *unconventional neurotransmitters.*

4.11.1 Discovery of NO as a Messenger Substance

Nitric oxide was first reported by Joseph Priestly (the discoverer of oxygen) in 1790. Until fairly recently, nitric oxide (NO) was considered primarily for its role as a toxic smog pollutant. As a chemical substance, it was only of interest to inorganic and organometallic chemists. However, by the early 1990s it was realized that NO was also a truly unique messenger molecule involved in a wide range of physiological processes. Perhaps most importantly, NO was recognized as a neurotransmitter within the central nervous system—an amazing observation given the simplicity of the molecule. In recognition of this significance, the editors of the research journal *Science* dubbed NO with the honour "Molecule of the Year" in 1992.

The discovery of NO as a messenger occurred over a fairly brief timespan—less than 10 years. In 1980, Furchgott and Zawadzki, while studying isolated smooth muscle preparations, discovered that, following stimulation with acetylcholine, a short-lived vasodilating substance was released into blood vessels. They called this substance *endothelium derived relaxing factor* (EDRF). In 1987, three independent research groups reported that EDRF and NO were one and the same molecule. Subsequent research revealed that NO was generated by many cells throughout the body and was even a neurotransmitter in the CNS. The 1998 Nobel Prize for Physiology and Medicine was bestowed upon Furchgott, Ignarro, and Murad for their pioneering contributions to the identification and elaboration of NO as a messenger substance.

4.11.2 Metabolism of NO

The synthesis of NO requires merely one step: the conversion of L-arginine into NO and citrulline. This conversion is catalyzed by the nitric oxide synthase (NOS) enzyme. Three distinct isoforms of the NOS enzyme have been cloned: Isoform I (nNOS; chromosome 12) is a Ca^{2+}-dependent neuronal form of the enzyme; Isoform II (mNOS or iNOS; chromosome 17) is a Ca^{2+}-independent macrophage inducible form of the enzyme found in microglia; Isoform III (eNOS, chromosome 7) is a Ca^{2+}-dependent form found in the endothelial cells that line blood vessels. Since NO is an extremely important messenger substance, the NOS enzyme is exquisitely regulated by processes such as phosphorylation and hormonal control.

Once synthesized, NO behaves somewhat differently from classical neurotransmitters. NO is not released from neurons in a Ca^{2+}-dependent exocytotic process; rather, it diffuses freely out of the neuron and to the next neuron. Once it reaches its target enzyme, NO does not interact with specific membrane-associated receptor proteins; instead, it interacts with second-messenger molecules in the receiving neuron (by interacting with the heme moiety of soluble guanylyl cyclase to increase cGMP levels), precipitating a cascade of intracellular biochemical processes. Once in the interneuronal region, NO may even function as a retrograde messenger, influencing the metabolism and release of neurotransmitters from presynaptic terminals. Also, unlike conventional neurotransmitters, NO is not inactivated by enzymatic processes but decays to nitrite after 30 seconds.

It is possible that biosynthesis from L-arginine is not the only synthetic route for the generation of NO. NO may also be generated nonenzymatically by reduction of nitrate ion to NO. This process occurs under the reducing conditions that evolve in the brain during the ischemia of stroke. Furthermore, there is an excess release of NO associated with NMDA receptor stimulation during a stroke. Coupled to this realization that NO plays a role in the neurotoxic cascade of stroke has been an attempt to pharmacologically manipulate NO as a stroke therapy. Regrettably, as with the glutamate story (discussed above), NO-based drugs have failed to provide a "magic bullet" for the treatment of stroke.

4.11.3 Role of NO in the Central and Peripheral Nervous Systems

NO plays a significant role in the CNS, where it functions as both a neurotransmitter and a neuromodulator. As a neuromodulator, NO can influence ligand-gated ion channels within the brain. In the brain, NO plays a role in memory and learning. Moreover, as discussed above, NO is a participant in the molecular cascade that leads to brain damage during stroke.

NO exerts an equally important role in the PNS. Nonadrenergic, noncholinergic (NANC) neurons are distributed throughout both the gastrointestinal and reproductive tracts. NO is a neurotransmitter/neuromodulator within many of these NANC neurons. Penile erection, for example, is caused by NO release from NANC neurons. Accordingly, impotence is a clinical indication for the use of an NO donor such as nitroglycerin ointment. NO enables penile engorgement by activating guanylyl cyclase, which increases the concentration of cGMP, which in turn stimulates the dephosphorylation of myosin light chains in the smooth muscles within the penile arterial walls. In fact, any drug that increases cGMP might be of value in the treatment of erectile dysfunction. Sildenafil (**4.227**) (more famously known by its trade name Viagra) is an enzyme inhibitor that increases cGMP levels by blocking cGMP's catabolic breakdown as catalyzed by the phosphodiesterase (PDE isoform 5) enzyme located in the corpora cavernosa of the penis.

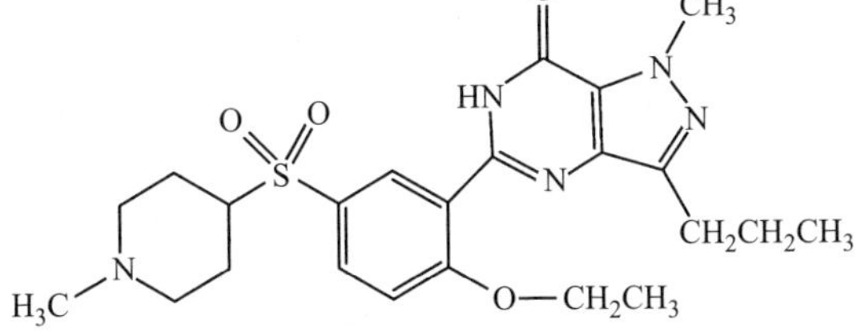

Sildenafil (Viagra)
(4.227)

4.11.4 Non-Neurologic Roles of NO

Nitric oxide has many roles as a messenger outside of the nervous system. In organ systems, ranging from the brain to the gut, NO exhibits significant vascular effects by influencing vascular smooth muscle tone and blood pressure. A reduction in NO synthesis (achieved using NOS inhibitors) increases vascular tone and elevates the mean arterial blood pressure. Conversely, an enhancement of NO-mediated effects tends to decrease vascular tone, causing dilation of blood vessels and reduction of blood pressure. These vascular effects of NO are diffusely distributed throughout the body.

Not surprisingly, NO is involved in diseases in which altered vascular tone and blood pressure are important mechanistic participants. For instance, NO may play a major role in arterial hypertension associated with pregnancy. *Pre-eclampsia* and *eclampsia*, medical conditions for which no explanation currently exists, are major causes of maternal and perinatal morbidity throughout the world. Clinically, these conditions are characterized by increased blood pressure, edema (swelling of the extremities), and proteinuria (protein in the urine, arising from kidney injury) in the mother and by intrauterine growth retardation in the fetus. As the condition progresses, the mother experiences severe elevation in blood pressure, resulting in organ damage and failure (e.g., liver and kidney dysfunction, seizures). Some workers speculate that NO may play a central role in pre-eclampsia. One hypothesized mechanism invokes reduced production of NO that is associated with enhanced formation of thromboxanes and free radicals. Indeed, the use of L-arginine as a nutritional supplement to augment NO levels has been proposed as a therapeutic approach to pre-eclampsia.

NO also seems to play a role in the hypotension of *septic shock*. Septic shock is a life-threatening clinical condition occurring as a complication of bacterial infections and characterized by hypotension, shock, organ failure, and death. In severe Gram-negative bacterial infections, increased urinary excretion of nitrates, an oxidative byproduct of NO, has been described; bacterial wall lipopolysaccharides activate the NOS enzyme.

NO exerts other effects on blood vessels. For example, atherosclerotic plaques arising from hypercholesterolemia produce narrowed blood vessels and reduced formation of NO. In animal models, the thickening of blood vessel walls after surgical procedures such as angioplasty can be blocked using NO donors, NO inhalation, or NOS gene transfer. One instance in which NO participation in atherosclerosis is particularly evident concerns organ transplantation. Accelerated atherosclerosis in the blood vessels within the transplanted organ is a common chronic condition that may lead to transplant failure and death. By reducing free radical toxicity, NO may act as a cytoprotective agent under such transplantation conditions; in fact, careful (and not excessive) L-arginine supplementation has been demonstrated to reduce transplant organ atherosclerosis.

Of equal significance is the role played by NO on platelet function. Platelets are small bodies (2–4 μm in diameter) found in the circulating blood at a concentration of 300,000 per μL of blood; when a blood vessel is transected or injured, platelets clump together, stopping the leakage of blood. However, excessive platelet "stickiness" can lead to unwanted blood vessel blockage, enhancing the likelihood of occluded blood vessels in the heart (precipitating a heart attack) or brain (leading to a stroke). Nitric oxide is a potent inhibitor of platelet adhesion and aggregation.

In the lungs, NO affects not only blood vessels but also the bronchi and bronchioles as well. In newborns with defective gas exchange, NO inhalation decreases pulmonary arterial blood pressure, enabling more blood to be oxygenated. In adults with obstructive lung diseases, NO inhalation seems to relax airway smooth muscle, thus acting as a bronchodilator.

Finally, NO has significant involvement in both acute and chronic inflammation. In acute inflammation, NO promotes swelling and increased vascular permeability. In animal models of acute inflammation, NOS inhibitors have a dose-dependent protective effect. In models of chronic inflammation (arthritis), NO is detrimental and L-arginine supplementation causes inflammatory exacerbation. At a molecular level, NO stimulates inflammation by activating the cyclooxygenase enzyme. Further supportive data are provided by the observation that fluid drained from the swollen joints of people with arthritis contains peroxynitrate and other oxidation products of NO.

4.11.5 Drug Design Exploitation of NO

Although the discovery of NO as a neurotransmitter is relatively recent, the use of NO-related drugs has a much longer history. Patients have been "popping nitros" for years to alleviate the chest pain associated with angina. Organic nitrates, nitrites, nitroso compounds, and various other nitrogen-containing therapeutics, including sodium nitroprusside, exert their pharmacological effect (i.e., vasodilation) through the release and/or generation of NO. NO achieves vasodilation by activating guanylate cyclase, which in turn activates cGMP-dependent protein kinases which then phosphorylate the myosin light chain kinase enzyme, causing its inactivation, which leads to an associated muscle relaxation in the wall of the artery. As this muscle relaxes, the vessel dilates, heralding improved blood flow and a reduction in blood pressure. Contraction of smooth muscle in arterial walls is regulated by reversible phosphorylation of myosin. Smooth muscle myosin consists of two heavy chain proteins (MW = 200,000) associated with two pairs of light chains; when combined with actin, they participate in a protein kinase-dependent biochemical cascade that ultimately culminates in either muscle contraction or relaxation.

Since these nitrate-based compounds cause blood vessels to dilate, they are called nitrovasodilators. Nitrovasodilators have clinical utility in the treatment of angina and in the treatment of malignant hypertension (high blood pressure that is severely out of control). Some of the available nitrovasodilator products include amyl nitrite (**4.228**), nitroglycerin (**4.229**), isosorbide dinitrate (**4.230**), erythrityl tetranitrate (**4.231**), pentaerythritol tetranitrate (**4.232**) and sodium nitroprusside (**4.233**).

The recent expanded appreciation of NO's ubiquitous presence as a messenger substance in a multitude of organ systems (especially the brain) has motivated the search for additional NO-related therapeutic molecules. As discussed above, NO is a simple two-atom molecule not amenable to "analoging." Rather, the drug designer must either upregulate or downregulate the enzyme (NOS) central to the biosynthesis of NO. Traditionally, arginine analogs have represented the largest class of compounds to be exploited as NOS inhibitors. N-methyl-L-arginine (**4.234**) is one of the most exhaustively studied arginine analogs. Owing to the importance of NO, ongoing research continues to design NOS inhibitors, with an emphasis on compounds that are not arginine analogs.

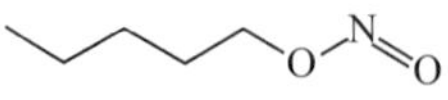

Amyl nitrite
(4.228)

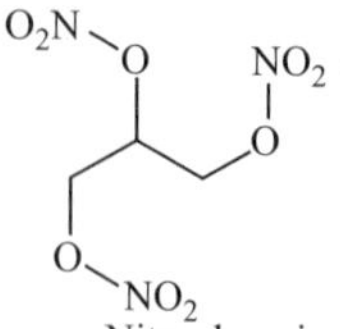

Nitroglycerin
(4.229)

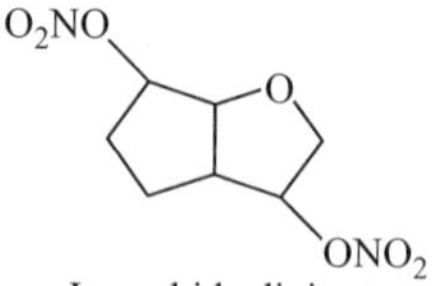

Isosorbide dinitrate
(4.230)

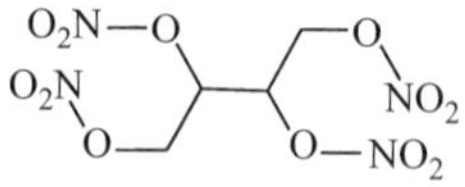

Erythrityl tetranitrate
(4.231)

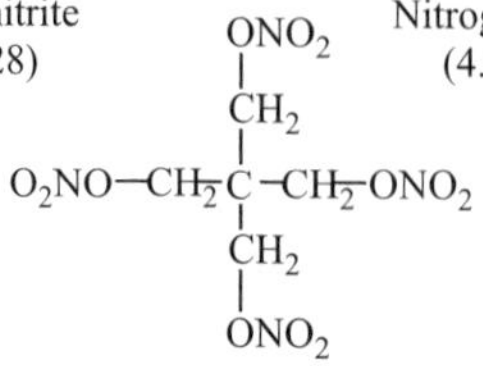

Pentaerythritol tetranitrate
(4.232)

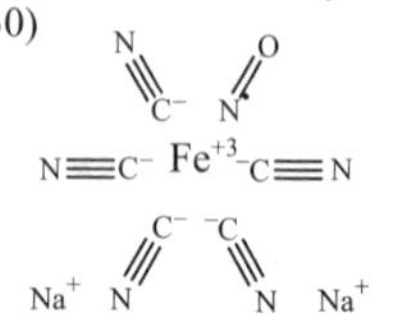

Sodium nitroprusside
(4.233)

4.12 NEUROMODULATORS: TAURINE AND β-ALANINE

Taurine (2-aminoethanesulfonic acid; **4.235**) is an inhibitory neurochemical that probably acts primarily as a neuromodulator rather than a neurotransmitter. It is formed from cysteine, and its accumulation can be prevented by the cardiac glycoside ouabain. Although receptor sites and specific actions cannot be elucidated without an antagonist, taurine has been implicated in epilepsy and, potentially, in heart disease. There are a large number of physiological effects attributed to taurine, among them cardiovascular (antiarrythmic), central (anticonvulsant, excitability modulation), muscle (membrane stabilizer), and reproductive (sperm motility factor) activity. Analogs of taurine, phthalimino-taurinamide (**4.236**) and its N-alkyl derivatives, are less polar than taurine and are potent anticonvulsant molecules.

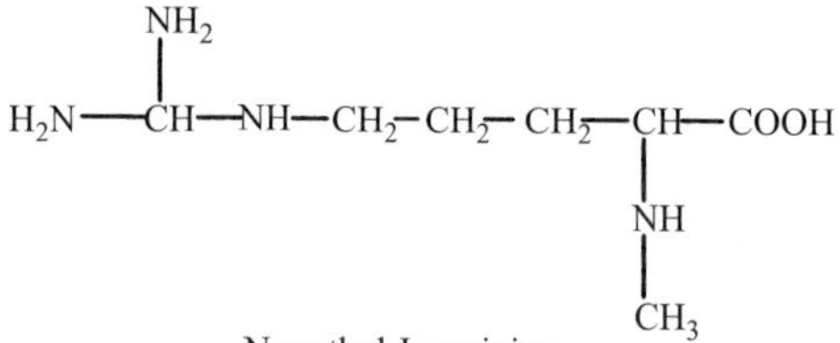

N-methyl-L-arginine
(4.234)

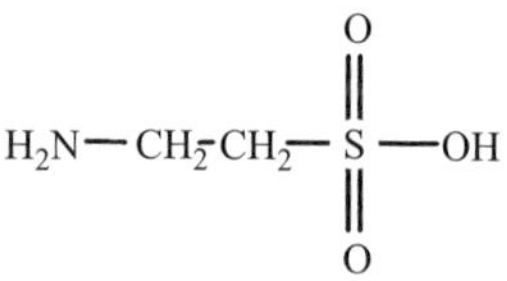

Taurine
(4.235)

Like taurine, β-alanine (**4.237**) is an inhibitory neuromodulator in the human CNS. Numerous studies support this observation: β-alanine occurs naturally in the CNS, is released by electrical stimulation, has binding sites, and inhibits neuronal excitability. Structurally, β-amino acids (such as β-alanine) are intermediate between α-amino acids (e.g., glutamate, glycine) and γ-amino acids (e.g., GABA). The existence of unique β-alanine receptors is controversial, and no β-alanine antagonists have yet been identified. However, studies have verified a dual action of β-alanine on both glutamatergic and GABAergic processes. β-Alanine has five receptor sites: glycine co-agonist site on the NMDA cation channel complex (strychnine insensitive); glycine receptor site (strychnine sensitive); GABA$_A$ receptor; GABA$_C$ (GABA-ρ) receptor; and blockade of "glial GABA

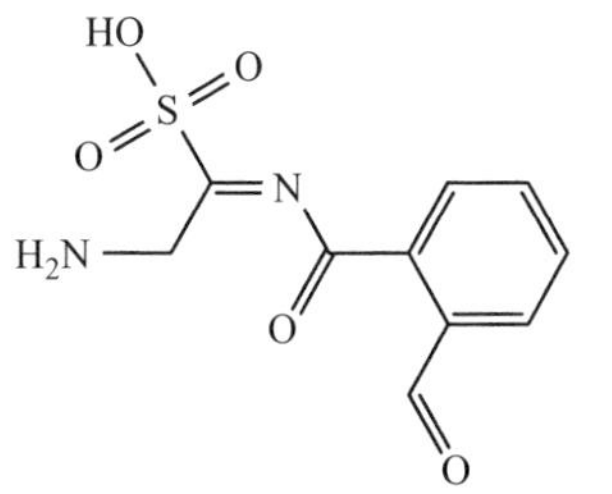

Phthalimino-taurinamide
(4.236)

$H_2N-CH_2CH_2-COOH$

β-Alanine
(4.237)

uptake" (via GAT proteins). β-Alanine binding sites have been identified throughout the hippocampus, other limbic structures, and the neocortex. Finally (of great importance), an active transport shuttle capable of transporting β-alanine and related analogs across the blood–brain barrier has been identified. Such a "beta-shuttle" may permit β-amino acid analogs to gain access to the CNS via active uptake rather than passive diffusion.

4.13 PURINERGIC NEUROMODULATION AND THE ADENOSINE RECEPTORS

Purines have numerous biochemical functions: they are the building blocks of nucleic acids, the energy transducers ATP and GTP, and the second messenger cAMP. They have also been known to have other direct physiological functions in bronchial constriction, in vasodilatation, and in inhibition of platelet aggregation and of central neuronal firing. Owing to this widespread importance, the existence of purinergic receptors ("P" receptors) has been speculated upon for many years. Originally, it was thought that P receptors could bind only purines; however, it was later realized that a subpopulation of purine receptors could also bind pyrimidines. The major purine bases found in nucleotides are adenine, guanine, hypoxanthine, and xanthine; the major pyrimidine bases found in nucleotides are cytosine, thymine, and uracil. Of the compounds active at purinergic receptors, adenosine is the most noteworthy. Adenosine has been found in numerous synapses and, in view of its bewildering multitude of actions, has been called the "most mysterious neuromodulator." Intensive research has shed some light on purinergic activity that is defined as neuromodulation rather than neurotransmission, even though adenosine receptors are reasonably well elucidated. Adenosine (**4.238**) is not a classical neurotransmitter because it is not stored in neuronal synaptic granules or released in quanta.

Purinergic receptors are subdivided into two main groups, the P_1 and the P_2 receptors. The P_1 purinergic receptor is further subdivided into four types of adenosine receptors: A_1, A_{2A}, A_{2B}, and A_3. All four of these belong to the superfamily of receptors that signal via G-proteins and contain seven hydrophobic transmembrane-spanning segments. To date, eleven subtypes of P_2 receptors have been identified: P_{2x1-7}, P_{2y1}, P_{2y2}, P_{2y4}, P_{2y6}. The P_{2x} receptors are coupled to ligand-gated channels; the P_{2y} receptors are G-protein coupled. These various receptors have been cloned from mammalian species.

Purines act both pre- and postsynaptically. Adenosine inhibits the release of NE and ACh in autonomic neuronal terminals, and both adenosine and ATP function as pre- and

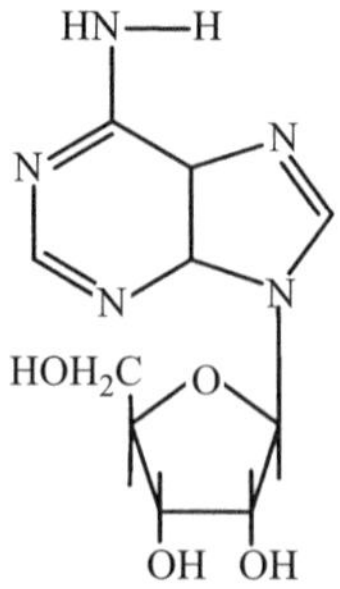

HN—H

Adenosine
(4.238)

postjunctional membrane-potential modulators in ganglia. It appears that NE and ATP are cotransmitters in sympathetic synapses, where ATP mediates the fast phasic contraction of smooth muscle through P_2 receptors. Because of this diverse biochemical involvement, purinergic receptors, especially those binding adenosine, are implicated in various pathologies and therapeutics. Adenosine is probably best known for its ability to protect organs, including the heart and brain, from ischemic injury during an acute heart attack or stroke. Multiple adenosine receptors are implicated in asthma since adenosine causes bronchoconstriction. A_1 receptor agonism influences intraocular pressure within the eye; thus, agents active at this receptor may have a therapeutic role in glaucoma. There exists strong evidence that adenosine possesses strong antinociceptive properties and may accordingly be useful in the treatment of neuropathic pain within the peripheral nervous system. Adenosine receptors are also widely distributed in the CNS, where they depress neuronal activity. Adenosine is considered to be an endogenous anticonvulsant/neuroprotectant following head injury. Not surprisingly, adenosine antagonists are CNS stimulants.

Purinergic agonists include adenosine and all adenosine phosphates, as well as a number of highly active synthetic N6-substituted adenosine derivatives (cyclohexyl, phenylisopropyl) (**4.239, 4.240**). The P_1 receptor is very sensitive to changes in the ribofuranose ring (e.g., epimerization to arabinose), whereas the P_2 receptor is not. Selective A_1 receptor agonists can be obtained simply by the addition of an N6 substituent, particularly an α-branched substituent. Almost universally, A_{2A} agonists can be achieved by substituting the C2 atom. Finally, for optimal A3 receptor agonism, structure–activity studies have shown that the methyl amide at C5' and the benzyl substituent at N6 are preferred.

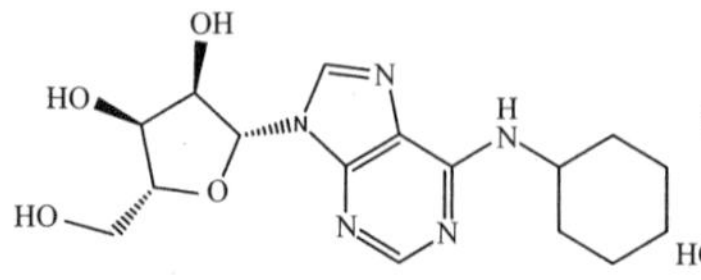

N6-cyclohexyl-adenosine
(4.239)

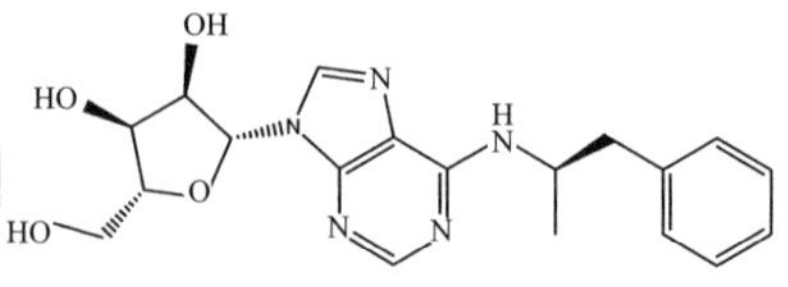

N6-(R-phenylisopropyl)-adenosine
(4.240)

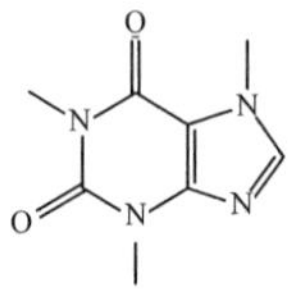

Caffeine
(4.241)

Purinergic antagonists include clonidine, a potent P_1 antagonist that is also an α_2 and H_2 agonist (see section 4.3.6), facilitating purine release. Methylxanthines, especially caffeine (**4.241**), are potent P_1 antagonists. When one considers the enormous amount of caffeine consumed in the world, this discovery is significant for understanding the symptoms of caffeine addiction. Many classes of A_1 receptor antagonists have been described; the majority of these have traditionally been xanthine analogs. Likewise, a number of A_{2A} receptor antagonists based on xanthines and related heterocyclic core structures have been described. Finally, a structurally diverse set of A_3 receptor antagonists has recently been identified.

At the time of writing, many compounds based on the adenosine receptor are in preclinical or early clinical development.

Selected References

Overview of Relevant Neuroanatomy and Neurophysiology

M. L. Barr, J. A. Kiernan (1988). *The Human Nervous System: An Anatomical Viewpoint*, 5th ed. Philadelphia: Lippincott.

M. Göthert (1985). Role of autoreceptors in the function of the peripheral and central nervous system. *Arzneimittelforschung 35*: 1909–1916.

T. Hökfelt, B. Evaritt, B. Meister, T. Melander, M. Schalling, O. Johansson, J. M. Lundberg, A. L. Hulting, S. Werner, C. Cuello, C. Hemming, C. Ouimet, J. Walaas, P. Greengard, M. Goldstein (1986). Neurons with multiple messengers, with special reference to neuroendocrine systems. *Recent Prog. Hormone Res. 42*: 1–70.

P. M. Laduron (1985). Postsynaptic heteroreceptors in the regulation of neuronal transmission. *Biochem. Pharmacol. 34*: 467–470.

S. Z. Langer (1981). Presynaptic receptors. *Pharmacol. Rev. 32*: 337–363.

C. J. Pazoles, J. L. Ives (1985). Cotransmitters in the CNS. *Annu. Rep. Med. Chem. 20*: 51–60.

L. F. Reichardt, R. B. Kelly (1983). A molecular description of nerve terminal function. *Annu. Rev. Biochem. 52*: 871–926.

Acetylcholine and Cholinergic Receptors

E. A. Accili, G. Redaelli, D. DiFrancesco (1998). Two distinct pathways of muscarinic current responses in rabbit sino-atrial node myocytes. *Pflugers Arch. 437*: 164.

B. C. Bowman (1986). Mechanisms of action of neuromuscular blocking drugs. In: G. N. Woodruff (Ed.). *Mechanisms of Drug Action*, vol. 1. London: Macmillan, pp. 65–96.

O. E. Brodde, M. C. Michel (1999). Adrenergic and muscarinic receptors in the human heart. *Pharmacol. Rev. 51*: 651.

W. H. Bunnelle, M. J. Dart, M. R. Schrimpf (2004). Design of ligands for the nicotinic acetylcholine receptors: the quest for selectivity. *Curr. Top. Med. Chem. 4*: 299–334.

M. P. Caulfield, N. J. M. Birdsall (1998). Classification of muscarinic acetylcholine receptors. *Pharmacol. Rev. 50*: 279.

B. M. Conti-Tronconi, M. A. Raftery (1982). The nicotinic cholinergic receptor: correlations of molecular structure with functional properties. *Annu. Rev. Biochem. 51*: 491–530.

Y. Dunant, M. Israël (1985). The release of acetylcholine. *Sci. Am. 252*: 58–66.

R. M. Eglen, S. S. Hedge, N. Watson (1996). Muscarinic receptor subtypes and smooth muscle function. *Pharmacol. Rev. 48*: 531.

A. D. Fryer, D. B. Jacoby (1998). Muscarinic receptors and control of airway smooth muscle. *Am. J. Respir. Crit. Care Med. 158*: S154.

K. Fukuda, T. Kubo, I. Akiba, A. Maeda, M. Mishima, S. Numa (1987). Molecular distinction between muscarinic acetylcholine receptor subtypes. *Nature 327*: 623–625.

H. C. Hartzell (1982). Physiological consequences of muscarinic receptor activation. In: J. W. Lamble (Ed.). *More About Receptors*. Amsterdam: Elsevier Biomedical Press.

R. C. Hogg, D. Bertrand (2004). Nicotinic acetylcholine receptors as drug targets. *Curr. Drug Targ.: CNS Neurol. Disorders 3*: 123–130.

V. Itier, D. Bertrand (2001). Neuronal nicotinic receptors: from protein structure to function. *FEBS Lett. 504*: 118–125.

A. Karlin, P. N. Kao, M. DiPaola (1986). Molecular pharmacology of the nicotinic acetylcholine receptor. *Trends Pharmacol. Sci. 7*: 304–308.

D. A. Kharkevich (1981). Main trends in the search for new neuromuscular blocking agents. *Trends Pharmacol. Sci. 2*: 218–220.

J. Kistler, R. M. Stroud, M. W. Klymkowsky, R. A. Lalancette, R. H. Fairclough (1982). Structure and function of an acetylcholine receptor. *Biophys. J. 37*: 371–378.

R. R. Levine, N. J. M. Birdsall, N. M. Nathanson (1999). Subtypes of muscarinic receptors. *Life Sci. 64*: 355.

J. P. Margiotta, P. C. Pugh (2004). Nicotinic acetylcholine receptors in the nervous system. *Adv. Mol. Cell Biol. 32*: 269–302.

M. P. McCarthy, J. P. Ernest, E. F. Young, J. Choe, R. M. Stroud (1986). The molecular neurobiology of the acetylcholine receptor. *Annu. Rev. Neurosci. 9*: 383–413.

M. Noda, H. Takahashi, T. Tanabe, M. Toyosato, Y. Furutani, T. Hirose, M. Asak, S. Inayama, T. Miyata, S. Numa (1982). Primary structure of α-subunit precursor of *Torpedo californica* acetylcholine receptor from cDNA sequence. *Nature 299*: 793–802.

A. Noma (1986). GTP-binding proteins couple cardiac muscarinic receptors to potassium channels. *Trends Neurol. Sci. 10*: 142–143.

K. Peper, R. J. Bradley, F. Dryer (1982). The acetylcholine receptor at the neuromuscular junction. *Physiol. Rev. 62*: 1271–1340.

J. L. Popot, J.-P. Changeux (1984). Nicotinic receptor of acetylcholine: structure of an oligomeric integral membrane protein. *Physiol. Rev. 64*: 1162–1239.

J. D. Schmitt, M. Bencherif (2000). Targeting nicotinic acetylcholine receptors: advances in molecular design and therapies. *Ann. Rep. Med. Chem. 35*: 41.

J. P. Snyder (1985). Molecular models for muscarinic receptors. *Trends Pharmacol. Sci. 6*: 464–466.

S. H. Snyder (1984). Drug and neurotransmitter receptors in the brain. *Science 224*: 22–31.

E. R. Spindel (2003). Neuronal nicotinic acetylcholine receptors: not just in brain. *Am. J. Physiol. 285*: L1201–L1202.

J. C. Venter (1983). Muscarinic receptor structure. *J. Biol. Chem. 258*: 4842–4848.

V. P. Whittaker (1986). The storage and release of acetylcholine. *Trends Pharmacol. Sci. 7*: 312–315.

Norepinephrine and Adrenergic Receptors

R. Aantaa, A. Marjamaki, M. Scheinin (1995). Molecular pharmacology of alpha 2-adrenoceptor subtypes. *Ann. Med. 27*: 439.

M. R. Bristow (1997). Mechanism of action of β-blocking agents in heart failure. *Am. J. Cardiol. 80*: 26L.

A. K. Cho, G. S. Takimoto (1985). Irreversible inhibitors of adrenergic nerve terminal function. *Trends Pharmacol. Sci. 6*: 443–447.

E. Costa, G. Racagni (Eds.) (1992). *Typical and Atypical Antidepressants: Molecular Mechanisms. Advances in Biochemical Psychopharmacology*, vol. 31. New York: Raven Press.

A. Davis (1984). Molecular aspects of the imipramine "receptor." *Experientia 40*: 782–794.

J. D. Fitzgerald (1993). Do partial agonist β-blockers have improved clinical utility? *Cardiovasc. Drugs Ther. 7*: 303.

S. Goldstein (1996). β-Blockers in hypertensive and coronary heart disease. *Arch. Int. Med. 156*: 1267.

R. M. Graham (1996). Alpha-1-adrenergic receptor subtypes: molecular structure, function and signaling. *Circ. Res. 78*: 737.

P. Guicheney, P. Meyer (1979). Biochemical approach to pre- and postsynaptic α-adrenoceptors. *Trends Pharmacol. Sci. 1*: 69–71.

J. R. Hampton (1994). Choosing the right β-blocker. *Drugs 48*: 549.

M. P. Heintzen, B. E. Strauer (1994). Peripheral vascular effects of β-blockers. *Eur. Heart J. 15*: 2.

E. J. M. Helmreich, T. Pfeuffer (1985). Regulation of signal transduction by β-adrenergic hormone receptors. *Trends Pharmacol. Sci. 6*: 438–443.

B. I. Hoffman, R. J. Lefkowitz (1980). Radioligand binding of adrenergic receptors: new insights into molecular and physiological regulation. *Annu. Rev. Pharmacol. Toxicol. 20*: 591–608.

P. A. Insel, R. D. Feldman (2004). Norepinephrine receptors. *Encyclopedia of Endocrine Diseases 3*: 375–381.

G. Kolata (1987). Manic-depression gene tied to chromosome 11. *Science 235*: 1139–1140.

W. Kostowski (1981). Brain noradrenalin, depression and antidepressant drugs: facts and hypotheses. *Trends Pharmacol. Sci. 2*: 314–317.

S. Z. Langer (1981). Presynaptic receptors. *Pharmacol. Rev. 32*: 337–363.

S. Z. Langer, C. Moret, R. Raisman, M. L. Dubocovich, M. Briley (1980). High-affinity ³H-imipramine binding in rat hypothalamus: association with uptake of serotonin but not norepinephrine. *Science 210*: 1133–1135.

R. J. Lefkowitz, J. L. Benovic, B. Kobilka, M. C. Caron (1986). β-Adrenergic receptors and rhodopsin: shedding new light on an old subject. *Trends Pharmacol. Sci. 7*: 444–448.

S. B. Liggett (2003). Polymorphisms of adrenergic receptors: variations on a theme. *Assay and Drug Dev. Tech. 1*: 317–326.

A. Lovitzki (1986). β-Adrenergic receptors and their mode of coupling to adenylate cyclase. *Physiol. Rev. 66*: 819–854.

E. Malta, G. A. McPherson, C. Raper (1985). Selective β₁-adrenoceptor agonists—fact or fiction? *Trends Pharmacol. Sci. 6*: 400–403.

C. J. Ohnmacht, J. B. Malick, W. J. Frazee (1983). Antidepressants. *Annu. Rep. Med. Chem. 18*: 41–50.

S. M. Paul, B. Hulihan-Giblin, P. Skolnik (1982). (+)-Amphetamine binding to rat hypothalamus: relation to anorectic potency of phenylethylamines. *Science 218*: 487–489.

M. T. Plascik, D. M. Perez (2001). α1-Adrenergic receptors: new insights and directions. *J. Pharmacol. Exp. Therap. 298*: 403–410.

S. R. Post, H. K. Hammond, P. A. Insel (1999). Beta-adrenergic receptors and receptor signaling in heart failure. *Ann. Rev. Pharmacol. Toxicol. 39*: 343.

R. R. Ruffolo, J. P. Hieble (1994). α-Adrenoceptors. *Pharmacol. Ther. 61*: 1.

R. Schorr, R. J. Lefkowitz, M. G. Caron (1981). Purification of the β-adrenergic receptor: identification of the hormone binding subunit. *J. Biol Chem. 256*: 5820–5826.

K. M. Small, D. W. McGraw, S. B. Liggett (2003). Pharmacology and physiology of human adrenergic receptor polymorphisms. *Ann. Rev. Pharmacol. Toxicol. 43*: 381–411.

R. D. Smith, J. R. Regan (1986). Antihypertensive agents. *Annu. Rev. Med. Chem. 21*: 63–72.

P. Sneddon, D. P. Westfall, J. S. Fedan (1982). Cotransmitters in the motor nerves of the guinea-pig vas deferens: electrophysiological evidence. *Science 218*: 693–695.

S. M. Stahl, L. Palazidou (1986). The pharmacology of depression: studies of neurotransmitter receptors lead the search for biochemical lesions and new drug therapies. *Trends Pharmacol. Sci. 7*: 349–354.

R. K. Sunhara, C. W. Dessauer, A. G. Gilman (1996). Complexity and diversity of mammalian adenylyl cyclases. *Ann. Rev. Pharmcol. Toxicol. 36*: 461.

S. Swillens, J. E. Dumont (1980). A unifying model of current concepts and data on adenyl cyclase activation by β-adrenergic agonist. *Life Sci. 27*: 1013–1028.

J. R. Teerlink, B. M. Massie (1999). β-Adrenergic blocker mortality trials in congestive heart failure. *Am. J. Cardiol. 84*: 94R.

P. A. van Zwieten, P. B. Timmermans (1979). The role of central α-adrenoceptors in the mode of action of hypotensive drugs. *Trends Pharmacol. Sci. 1*: 39–41.

P. A. van Zwieten (1993). An overview of the pharmacodynamic and therapeutic potential of combined α and β-adrenoceptor antagonists. *Drugs 45*: 509.

J. C. Venter, P. Horne, B. Eddy, R. Greguski, C. M. Fraser (1984). α_1-Adrenergic receptor structure. *Mol. Pharmacol. 26*: 196–205.

C. Weyer, J. F. Gauthier, E. Danforth (1999). Development of β-3-adrenoceptor agonists for the treatment of obesity and diabetes. *Diabetes Metab. 25*: 11.

B. Xu (2001). The importance of beta-adrenergic receptors in immune regulation: a link between neuroendocrine and immune system. *Med. Hypotheses 56*: 273–276.

H. Zhong, K. P. Minneman (1999). Alpha-1-adrenoceptor subtypes. *Eur. J. Pharmacol. 375*: 26.

Dopamine and Dopaminergic Receptors

R. M. Bilder (1997). Neurocognitive impairment in schizophrenia and how it affects treatment options. *Can. J. Psychiatry 42*: 255.

P. B. Bradley, S. H. Hirsch (1986). *The Psychopharmacology and Treatment of Schizophrenia.* New York: Oxford University Press.

T. S. Braver, D. M. Barch, J. D. Cohen (1999). Cognition and control in schizophrenia: a computational model of dopamine and prefrontal function. *Biol. Psychiatry 46*: 312.

A. Breier, P. H. Berg (1999). The psychosis of schizophrenia: prevalence, response to atypical antipsychotics, and prediction of outcome. *Biol. Psychiatry 46*: 361.

A. Carlsson, N. Waters, M. L. Carlsson (1999). Neurotransmitter interactions in schizophrenia: therapeutic implications. *Biol. Psychiatry 46*: 1388.

M. L. Carlsson, A. Carlsson, M. Nilsson (2004). Schizophrenia: from dopamine to glutamate and back. *Curr. Med. Chem. 11*: 267–277.

I. Creese, D. R. Sibley, M. W. Hamblin, S. E. Leff (1983). The classification of dopamine receptors: relationship to radioligand binding. *Annu. Rev. Neurosci. 6*: 43–71.

B. Cusack, A. Nelson, E. Richelson (1994). Binding of antidepressants to human brain receptors: focus on newer generation compounds. *Psychopharmacol. 114*: 559.

L. L. Demchyshyn, B. F. O'Dowd, S. R. George (2003). Structure of mammalian D1 and D5 dopamine receptors and their function and regulation in cells. *Neurol. Dis. Ther. 56*: 45–76.

R. S. Duman, G. Heninger, E. Nestler (1997). A molecular and cellular theory of depression. *Arch. Gen. Psychiatry 54*: 597.

S. B. Dunnett, A. Bjorklund (1999). Prospects for new restorative and neuroprotective treatments in Parkinson's disease. *Nature, 399*: A32.

R. H. Edwards (1993). Neural degeneration and the transport of neurotransmitters. *Ann. Neurol. 34*: 638.

J. R. Fozard (1982). Highly potent irreversible inhibitors of aromatic L-amino acid decarboxylase. *Trends Pharmacol. Sci. 3*: 429.

J. Gerlach, L. Peacock (1995). New antipsychotics: the present status. *Int. J. Psychopharmacol. 10*: 39.

M. Goldstein, D. B. Calne, A. Lieberman, M. O. Thorner (1980). *Ergot Compounds and Brain Function.* New York: Raven Press.

A. Hilditch, G. M. Drew (1985). Peripheral dopamine receptor subtypes—a closer look. *Trends Pharmacol. Sci. 6*: 396–400.

O. Hornykiewicz (1977). Psychopharmacological implication of dopamine antagonists: a critical evaluation of current evidence. *Annu. Rev. Pharmacol. Toxicol. 17*: 545–549.

S. D. Iversen (1986). *Psychopharmacology.* Oxford: Oxford University Press.

J. Jankovic (1999). New and emerging therapies for Parkinson's disease. *Arch. Neurol. 56*: 785.

M. D. Jibson, R. Tandon (1998). New atypical antipsychotic medications. *J. Psychiatr. Res. 32*: 215.

I. Jirkovsky, W. Lippman (1978). Antidepressants. *Annu. Rep. Med. Chem. 13*: 1–10.

J. M. Kane (1996). Schizophrenia. *New Engl. J. Med. 334*: 445.

J. M. Kane (1999). Pharmacologic treatment of schizophrenia. *Biol. Psychiatry 46*: 1396.

M. Kapsimali, S. Le Crom, P. Vernier (2003). A natural history of vertebrate dopamine receptors. *Neurol. Dis. Ther. 56*: 1–43.

J. W. Kebabian, T. Agui, J. C. van Oene, K. Shigematsu, J. M. Saavedra (1986). The D_1 dopamine receptor: new perspectives. *Trends Pharmacol. Sci. 7*: 96–99.

W. C. Koller, E. Tolosa (1998). Current and emerging drug therapies in the treatment of Parkinson's disease. *Neurology 50* (Supplement 6): 1–65.

A. E. Lang, A. M. Lozano (1998). Parkinson's disease. *New Engl. J. Med. 339*: 1044.

J. W. Langston (1985). Mechanism of MPTP toxicity: more answers, more questions. *Trends Pharmacol. Sci. 6*: 375–378.

T. A. Larsen, D. B. Calne (1985). Recent advances in the study of Parkinson's disease. In: D. Bousfield (Ed.). *Neurotransmitters in Action.* Amsterdam: Elsevier, pp. 252–256.

H. Y. Meltzer (1999). Treatment of schizophrenia and spectrum disorders: pharmacotherapy, psychosocial treatments, and neurotransmitter interactions. *Biol. Psychiatry 46*: 1321.

J. C. Nelson (1997). Treatment of refractory depression. *Depress. Anxiety 5*: 165.

C. B. Nemeroff, S. T. Cain (1985). Neurotensin–dopamine interaction in the CNS. *Trends Pharmacol. Sci. 6*: 201–205.

K. A. Neve, C. J. DuRand, M. M. Teeter (2003). Structural analysis of the mammalian D2, D3, and D4 dopamine receptors. *Neurol. Dis. Ther. 56*: 77–144.

A. G. Phillips, R. F. Lane, C. D. Blaha (1986). Inhibition of dopamine release by cholecystokinin: relevance to schizophrenia. *Trends Pharmacol. 7*: 126–127.

N. P. Quinn (1984). Anti-Parkinson drugs today. *Drugs 28*: 236–262.

J. M. Schaus, J. A. Clemens (1985). Dopamine receptors and dopaminergic agents. *Annu. Rep. Med. Chem. 20*: 41–50.

P. G. Strange (1986). Isolation and characterization of D_1 dopamine receptors. *Trends Pharmacol. Sci. 7*: 253–254.

C. A. Tamminga (2002). Partial dopamine agonists in the treatment of psychosis. *J. Neural Transmission 109*: 411–420.

E.-K. Tan (2003). Dopamine agonists and their role in Parkinson's disease treatment. *Exp. Rev. Neurotherap. 3*: 805–810.

R. Tintner, J. Jankovic (2003). Dopamine agonists in Parkinson's disease. *Exp. Op. Inv. Drugs 12*: 1803–1820.

P. Tyrer, A. Mackay (1986). Schizophrenia: no longer a functional psychosis. *Trends Neurosci. 9*: 537–538.

H. M. van Praag (1978). Amine hypothesis of affective disorders. In: L. L. Iversen, S. D. Iversen, S. H. Snyder (Eds.). *Handbook* of *Psychopharmacology*, vol. 13. New York: Plenum Press, pp. 187–297.

Serotonin and Serotonin Receptors

D. N. Bateman (1993). Sumatriptan. *Lancet. 341*: 221.

D. Deleu, Y. Hanssens, E. A. Worthing (1998). Symptomatic and prophylactic treatment of migraine. *Clin. Neuropharmacol. 21*: 267.

M. Dukat (2004). 5-HT3 serotonin receptor agonists: a pharmacophoric journey. *Curr. Med. Chem.: CNS Agents 4*: 77–94.

W. B. Essman (Ed.) (1978). *Serotonin in Health and Disease*, 5 vols. New York: Spectrum.

J. R. Fozard (1987). 5-HT$_3$ receptors and cytotoxic drug-induced vomiting. *Trends Pharmacol. Sci. 8*: 44–45.

J. R. Fozard, H. O. Kalkman (1994). 5-Hydroxytryptamine and the initiation of migraine: new perspectives. *Naunyn-Schmiedebergs Arch. Pharmacol. 350*: 225.

R. W. Fuller (1990). Pharmacology of central serotoninergic neurons. *Annu. Rev. Pharmacol. Toxicol. 20*: 111–127.

J. C. Gillin, W. B. Mendelson, N. Sitaram, R. J. Wyatt (1978). The neuropharmacology of sleep and wakefulness. *Annu. Rev. Pharmacol. Toxicol. 18*: 563–579.

R. A. Glennon (1987). Central serotonin receptors as targets for drug research. *J. Med. Chem. 30*: 1–12.

R. A. Glennon (2003). Higher-end serotonin receptors: 5-HT5, 5-HT6, and 5-HT7. *J. Med. Chem. 46*: 2795–2812.

R. A. Glennon, M. Dukat (1995). Serotonin receptor subtypes. In: F. Bloom, D. Kupfer (Eds.). *Psychopharmacology: The Fourth Generation of Progress*. New York: Raven Press.

M. Gothert (1982). Modulation of serotonin release in the brain via presynaptic receptors. *Trends Pharmacol. Sci. 3*: 437–440.

A. R. Green (1985). *Neuropharmacology of Serotonin*. New York: Oxford University Press.

L. Gyermek (1995). 5-HT3 receptors: pharmacologic and therapeutic aspects. *J. Clin. Pharmacol. 35*: 845.

S. Inoué, A. A. Borbély (1985). *Endogenous Sleep Substances and Sleep Regulation*. Utrecht: VNU Science Press.

D. Julius (1991). Molecular biology of serotonin receptors. *Ann. Rev. Neurosci. 14*: 335.

J. M. Krueger (1985). Somnogenic activity of muramyl peptides. *Trends Pharmacol. Sci. 6*: 218–221.

M. Lejoyeux, J. Ades, F. Rouillon (1994). Serotonin syndrome: incidence, symptoms and treatment. *CNS Drugs 2*: 132.

D. N. Middlemiss, M. Hibert, J. R. Fozard (1986). Drugs acting at central 5-hydroxytryptamine receptors. *Annu. Rep. Med. Chem. 21*: 41–50.

T. Nogrady, P. D. Hrdina, G. M. Ling (1972). Investigation of serotonin–ATP association *in vitro* by NMR and UV spectroscopy. *Mol. Pharmacol. 8*: 565–577.

S. J. Peroutka (1995). Serotonin receptor subtypes: Their evolution and clinical significance. *CNS Drugs 4*: 19.

S. J. Peroutka, R. M. Lebovitz, S. H. Snyder (1981). Two distinct central serotonin receptors with different physiological functions. *Science 212*: 827–829.

B. P. Richardson, G. Engel (1986). The pharmacology and function of 5-HT$_3$ receptors. *Trends Neurosci. 9*: 424–428.

A. J. Robichaud, B. L. Largent (2000). Recent advances in selective serotonin receptor modulation. *Ann. Rep. Med. Chem. 35*: 11.

D. V. S. Sankar (1975). *LSD: A Total Study.* Westbury, New York: PJD Publishing.

P. R. Saxena (1995). Serotonin receptors: subtypes, functional responses and therapeutic relevance. *Pharmacol. Ther. 66*: 339.

H. Sternbach (1991). The serotonin syndrome. *Am. J. Psychiatry 148*: 705.

L. H. Tecott, D. Julius (1993). A new wave of serotonin receptors. *Curr. Opin. Neurobiol. 3*: 310.

S. C. Veasey (2003). Serotonin agonists and antagonists in obstructive sleep apnea: therapeutic potential. *Am. J. Resp. Med. 2*: 21–29.

S. P. Vickers, M. J. Bickerdike, C. T. Dourish (1999). Serotonin receptors and obesity. *Neurosci. News 2*: 22–28.

H. Wang (2003). Serotonin receptor as a potential therapeutic target for pulmonary vascular remodeling. *Drug Dev. Res. 58*: 69–73.

Histamine and Histamine Receptors

L. F. Alguacil, C. Perez-Garcia (2003). Histamine H3 receptor: A potential drug target for the treatment of central nervous system disorders. *Curr. Drug Targets: CNS Neurol. Dis. 2*: 303–313.

R. A. Bakker, H. Timmerman, R. Leurs (2002). Histamine receptors: specific ligands, receptor biochemistry, and signal transduction. *Clin. All. Immunol. 17*: 27–64.

B. S. Bochner, L. M. Lichtenstein (1991). Anaphylaxis. *New Engl. J. Med. 324*: 1785.

T. H. Brown, R. C. Young (1985). Antagonists of histamine at its H_2-receptor. *Drugs of the Future 10*: 51–69.

F. M. Cuss (1999). Beyond the histamine receptor: effects of antihistamines on mast cells. *Clin. Exper. Allergy 29*: 54.

J. Del Valle, I. Gantz (1997). Novel insights into histamine H2 receptor biology. *Am. J. Physiol. 273*: G987.

J. P. Desager, Y. Horsmans (1995). Pharmacokinetic–pharmacodynamic relationships of H1-antihistamines. *Clin. Pharmacokinet. 28*: 418.

G. J. Durant (1979). Chemical aspects of histamine H_2-receptor agonists and antagonists. In: F. Gualtieri, M. Gianella, C. Melchiorre (Eds.). *Recent Advances in Receptor Chemistry.* New York: Elsevier/North Holland, pp. 245–266.

M. Feldman, M. E. Burton (1990). Drug therapy: H2 receptor antagonists. *New Engl. J. Med. 323*: 1672.

C. R. Ganellin, M. E. Parsons (Eds.) (1982). *Pharmacology of Histamine Receptors.* Bristol: Wright PSG.

C. R. Ganellin, J.-C. Schwartz (Eds.) (1985). *Frontiers in Histamine Research.* New York: Pergamon.

G. L. Garay, J. M. Muchowski (1985). Agents for the treatment of peptic ulcer disease. *Annu. Rep. Med. Chem. 20*: 93–105.

J. G. Gleason, C. D. Perchonock, T. J. Torphy (1986). Pulmonary and antiallergy agents. *Annu. Rep. Med. Chem. 21*: 73–83.

E. E. J. Haaksma, R. Leurs, H. Timmerman (1990). Histamine receptors: subclasses and specific ligands. *Pharmacol. Ther. 47*: 73.

S. T. Holgate (1999). Antihistamines: back to the future. *Clin. Exper. Allergy 29*: Suppl. 3.

D. MacGlashan, Jr (2003). Histamine: a mediator of inflammation. *J. All. Clin. Immunol. 112*: S53–S59.

H. J. Nielsen (1996). Histamine-2 receptor antagonists as immunomodulators: new therapeutic views. *Ann. Med. 28*: 107.

D. M. Richards, R. N. Brogden, R. C. Heel, T. M. Speight, G. S. Avery (1984). Astemizole: a review of its pharmacological properties and therapeutic efficacy. *Drugs 28*: 38–61.

J.-C. Schwartz, J.-M. Arrang, M. Garbarg (1986a). Three classes of histamine receptors in the brain. *Trends Pharmacol. Sci. 7*: 24–28.

J.-C. Schwartz, J.-M. Arrang, M. Garbarg, M. Komer (1986b). Properties and roles of the three subclasses of histamine receptors in the brain. *J. Exp. Biol. 124*: 203–224.

F. E. Simons, K. J. Simons (1999). Clinical pharmacology of new histamine H1 receptor antagonists. *Clin. Pharmacokinet. 36*: 329.

K. Sugano, Y. Fukushima (2000). Molecular biology of histamine H2 receptor. *Therap. Res. 21*: S245–S250.

H. Timmerman (1999). Why are non-sedating antihistamines non-sedating? *Clin. Exp. Allergy 29*: 13–18.

A. Togias (2003). H1-receptors: localization and role in airway physiology and in immune functions. *J. All. Clin. Immunol. 112*: S60–S68.

GABA and GABA Receptors

P. R. Andrews, G. A. R. Johnston (1979). GABA agonists and antagonists. *Biochem. Pharmacol. 28*: 2697–2702.

E. Ben-Menachem (1995). Vigabatrin. *Epilepsia 36* (Suppl. 2): S95.

N. Bouche, B. Lacombe, H. Fromm (2003). GABA signaling: a conserved and ubiquitous mechanism. *Trends Cell Biol. 13*: 607–610.

A. C. Bowling, R. J. Delorenzo (1982). Micromolar activity BZD receptors: identification and characterization in the CNS. *Science 216*: 1247–1249.

M. J. Brodie, R. J. Porter (1990). New and potential anticonvulsants. *Lancet 336*: 425.

H. M. Bryson, B. R. Fulton, D. Faulds (1995). Propofol: an update of its use in anaesthesia and conscious sedation. *Drugs 50*: 513.

M. Chebib, G. A. R. Johnston (1999). The "ABC" of GABA receptors: a brief review. *Clin. Exp. Pharmacol. Physiol. 26*: 937–940.

E. Costa (1982). Do benzodiazepines act through a GABAergic mechanism? In: A. M. Creighton, and S. Turner (Eds.). *Chemical Regulation of Biological Mechanisms*. London: Royal Society of Chemistry.

E. Costa, G. di Chiara, G. L. Gessa (Eds.) (1980). *GABA and Benzodiazepine Receptors. Advances in Biochemical Pharmacology*, vol. 26. New York: Raven Press.

E. Costa, A. Guidotti (1996). Benzodiazpines on trial: a research strategy for their rehabilitation. *Trends Pharmacol. Sci. 17*: 192.

F. V. DeFeudis (1982). γ-Aminobutyric acid and analgesia. *Trends Pharmacol. Sci. 3*: 444–446.

R. C. Effland, M. F. Forsch (1992). Anti-anxiety agents, anticonvulsants and sedative-hypnotics. *Ann. Rep. Med. Client. 17*: 11–19.

F. J. Ehlert (1986). "Inverse agonists," cooperativity and drug action at benzodiazepine receptors. *Trends Pharmacol. Sci. 7*: 28–32.

S. L. Erdö (1985). Peripheral GABAergic mechanisms. *Trends Pharmacol. Sci. 6*: 205–208.

U. Grunert (1999). Distribution of GABAA and glycine receptors in the mammalian retina. *Clin. Exp. Pharmacol. Physiol. 26*: 941–944.

W. Haefely, E. Kyburz, M. Gerecke, H. Möhler (1985). Recent advances in the molecular pharmacology of benzodiazepine receptors and the structure–activity relationship of their agonists and antagonists. In: B. Testa (Ed.). *Advanced Drug Research*, vol. 14. New York: Academic Press, pp. 165–322.

W. Hevers, H. Luddens (1998). The diversity of GABA$_A$ receptors: pharmacological and electrophysiological properties of GABA$_A$ channel subtypes. *Mol. Neurobiol. 18*: 35.

J. K. Ho, R. A. Harris (1981). Mechanism of action of barbiturates. *Annu. Rev. Pharmacol. Toxicol. 21*: 83–111.

R. Hoehn-Saric (1998). Generalized anxiety disorder: Guidelines to diagnosis and treatment. *CNS Drugs 9*: 85.

G. A. R. Johnson, M. Willow (1982). GABA and barbiturate receptors. In: J. W. Lamble (Ed.). *More About Receptors*. Amsterdam: Elsevier Biochemical Press.

G. A. R. Johnson, M. Chebib, J. R. Hanrahan, K. N. Mewett (2003). GABAC receptors as drug targets. *Curr. Drug Targets: CNS Neurol. Dis. 2*: 260–268.

E. R. Korpi, G. Grunder, H. Luddens (2002). Drug interactions at GABAA receptors. *Prog. Neurobiol. 67*: 113–159.

M. D. Krasowski, N. L. Harrison (1999). General anaesthetic action on ligand-gated ion channels. *Cell. Mol. Life Sci. 55*: 1278.

P. Krogsgaard-Larsen, P. Jacobsen, E. Falch, H. Hjeds (1982). GABA-agonists: chemical, molecular, pharmacologic and therapeutic aspects. In: A. M. Creighton, S. Turner (Eds.). *The Chemical Regulation of Biological Mechanisms*. London: Royal Society of Chemistry.

M. Lancel (1999). Role of GABA$_A$ receptors in the regulation of sleep: Initial sleep responses to peripherally administered modulators and agonists. *Sleep 22*: 33.

H. Luddens, E. R. Korpi (1995). Biological function of GABA-A/benzodiazepine receptor heterogeneity. *J. Psychiatr. Res. 29*: 77.

A. Manocha (1998). Pharmacology of GABAc receptors. *Ind. J. Pharmacol. 30*: 218–226.

S. P. Nordt, R. F. Clark (1997). Midazolam: a review of therapeutic uses and toxicity. *J. Emerg. Med. 15*: 357.

E. J. Peck, Jr (1980). Receptors for amino acids. *Annu. Rev. Physiol. 42*: 615–627.

G. Racagni, A. O. Donoso (Eds.) (1986). *GABA and Endocrine Functions*. New York: Raven Press.

J. G. Richards, P. Schoch, H. Möhler, W. Haefely (1986). Benzodiazepine receptors resolved. *Experientia 42*: 121–126.

R. Sankar, D. F. Weaver (1997). Antiepileptic drugs: basic principles of medicinal chemistry. In: J. Engel, T. Pedley (Eds.). *Epilepsy: A Comprehensive Textbook*. Philadelphia: Lippincott-Raven, vol 2, pp. 1393–1403.

P. Skolnick, S. M. Paul (1981). Benzodiazepine receptors. *Annu. Rep. Med. Chem. 16*: 21–29.

M. Steriade (2004). Local gating of information processing through the thalamus. *Neuron 41*: 493–494.

G. Toffamo, A. Leon, M. Massotti, A. Guidotti, E. Costa (1980). GABA-modulin: a regulating protein for GABA receptors. In: G. Pepeu, M. J. Kuhar, S. J. Enna (Eds.). *Receptors for Neurotransmitters and Peptide Hormones*. New York: Raven Press, pp. 132–142.

S. Vicini (1999). New perspectives in the functional role of GABA$_A$ channel heterogeneity. *Mol. Neurobiol. 19*: 97.

A. Wassef, J. Baker, L. D. Kochan (2003). GABA and schizophrenia: a review of basic science and clinical studies. *J. Clin. Psychopharmacol. 23*: 601–640.

P. J. Whiting (1999). The GABA$_A$ receptor gene family: new targets for therapeutic intervention. *Neurochem. Int. 34*: 387.

P. J. Whiting (2003). GABA-A receptor subtypes in the brain: a paradigm for CNS drug discovery? *Drug Discov. Today 8*: 445–450.

M. Williams, N. Yokoyama (1986). Anxiolytics, anticonvulsants and sedative hypnotics. *Annu. Rep. Med. Chem. 21*: 11–20.

C. F. Zorumsky, K. E. Isenberg (1991). Insights into the structure and function of GABA-benzodiazepine receptors: ion channels and psychiatry. *Am. J. Psychiatry 148*: 162.

Glycine and Glutamate

B. Bettler, C. Mulle (1995). Neurotransmitter receptors: AMPA and Kainate. *Neuropharmacology 34*: 123.

H. Betz (1987). Biology and structure of the mammalian glycine receptor. *Trends Neurosci. 10*: 113–117.

P. Bregestovski (2002). Function, modulation and pharmacol. of inhibitory glycine receptor channels. *Neurophysiol. 34*: 85–90.

M. Cascio (2002). Glycine receptors: lessons on topology and structural effects of the lipid bilayer. *Biopolymers 66*: 359–368.

G. L. Collingridge (1985). Long term potentiation in the hippocampus: mechanisms of initiation and modulation by neurotransmitters. *Trends Pharmacol. Sci. 6*: 407–411.

P. J. Conn, J. P. Pin (1997). Pharmacology and functions of metabotropic glutamate receptors. *Ann. Rev. Pharmacol. 37*: 205.

J. R. Cooper, F. E. Bloom, R. H. Roth (1996). *The Biochemical Basis of Neuropharmacology*, 7th ed. New York: Oxford University Press.

S. Daniels (2003). General anesthetic effects on glycine receptors. In: J. Antognini, E. Carstens, D. Raines (Eds.). *Neural Mechanisms of Anesthesia*. Totowa, New Jersey: Humana Press, pp. 333–344.

R. Dingledine (1986). NMDA receptors: what do they do? *Trends Neurosci. 9*: 47–49.

M. Hollmann, S. Heinemann (1994). Cloned glutamate receptors. *Ann. Rev. Neurosci. 17*: 31.

M. Jansen, G. Dannhardt (2003). Antagonists and agonists at the glycine site of the NMDA receptor for therapeutic interventions. *Eur. J. Med. Chem. 38*: 661–670.

P. Krogsgaard-Larsen, T. Honore (1983). Glutamate receptors and new glutamate agonists. *Trends Pharmacol. Sci. 4*: 31–33.

M. Mayer (1987). Two channels reduced to one. *Nature 325*: 480–481.

D. T. Monaghan, V. R. Holets, D. W. Toy, C. W. Cotman (1983). Anatomical distribution of four pharmacologically distinct ^{3}H-L-glutamate binding sites. *Nature 306*: 176–179.

S. Nakanishi (1994). Metabotropic glutamate receptors: synaptic transmission, modulation and plasticity. *Neuron. 13*: 1031.

E. J. Peck (1980). Receptors for amino acids. *Annu. Rev. Physiol. 42*: 615–627.

J. D. Rothstein, S. Patel, M. R. Regan, et al. (2005). β-Lactam antibiotics offer neuroprotection by increasing glutamate transporter expression. *Nature 433*: 73–77.

N. A. Sharif (1985). Multiple synaptic receptors for neuroactive amino acid transmitters—new vistas. *Int. Rev. Neurobiol.*, vol. 26. New York: Academic Press, pp. 85–150.

Neuropeptide Neurotransmitters

S. H. Buck, E. Burcher (1986). The tachykinins: a family of peptides with a brood of 'receptors.' *Trends Pharmacol. Sci. 7*: 65–68.

Z. Fathi (1998). Galanin receptors: recent developments and potential use as therapeutic targets. *Ann. Rep. Med. Chem. 33*: 41–49.

P. J. Gilligan, P. R. Hartig, D. W. Robertson, R. Zaczek. Corticotropin-releasing hormone receptors and discovery of selective non-peptide CRH-1 antagonists. *Ann. Rep. Med. Chem. 32*: 41–50.

M. R. Hanley (1985). Substance P antagonists. In: D. Bousfield (Ed.). *Neurotransmitters in Action*. Amsterdam: Elsevier, pp. 170–172.

S. G. Mills (1997). Recent advances in neurokinin receptor antagonists. *Ann. Rep. Med. Chem. 32*: 51–60.

R. A. Nicoll, C. Schenker, S. E. Leeman (1980). Substance P as a transmitter candidate. *Annu. Rev. Neurosci. 3*: 227–268.

T. Pedrazzin, F. Pralong, E. Grouzmann (2003). Neuropeptide Y: the universal soldier. *Cell. Mol. Life Sci. 60*: 350–377.

L. Protas, J. Qu, R. B. Robinson (2003). Neuropeptide Y: neurotransmitter or trophic factor in the heart? *News Physiol. Sci. 18*: 181–185.

B. F. B. Sandberg, L. L. Iversen (1982). Substance P. *J. Med. Chem. 25*: 1009–1015.

J. Saunders, J. P. Williams (2001). New developments in the study of corticotropin releasing factor. *Ann. Rep. Med. Chem. 36*: 21–29.

A. W. Stamford, E. M. Parker (1999). Recent advances in the development of neuropeptide Y receptor antagonists. *Ann. Rep. Med. Chem. 34*: 31–39.

A. W. Stamford, J. Hwa, M. van Heek (2003). Recent developments in neuropeptide Y receptor modulators. *Ann. Rep. Med. Chem. 38*: 61.

C. Swain, N. M. J. Rupniak (1999). Progress in the development of neurokinin antagonists. *Ann. Rep. Med. Chem. 34*: 51–60.

Neuromodulators: Taurine

M. Galarreta, J. Bustamante, R. M. Del Rio, J. M. Solis (1996). A new neuromodulatory action of taurine: long-lasting increase of synaptic potentials. *Adv. Exp. Med. Biol. 403*: 463–471.

R. I. Huxtable, L. A. Sebring (1986). Towards a unifying theory for the action of taurine. *Trends Pharmacol. Sci. 7*: 481–485.

S. S. Oja, P. Saransaari (1996). Taurine as osmoregulator and neuromodulator in the brain. *Metabol. Brain Dis. 11*: 153–164.

M. F. Olive (2002). Interactions between taurine and ethanol in the central nervous system. *Amino Acids 23*: 345–357.

P. Saransaari, S. S. Oja (2000). Taurine and neural cell damage. *Amino Acids 19*: 509–526.

C. E. Wright, H. H. Tallan, Y. Y. Lin, G. E. Gault (1986). Taurine: biological update. *Annu. Rev. Biochem. 55*: 427–453.

Neuromodulators: Purines

G. Burnstock (1983). *Purinergic Receptors.* London: Chapman and Hall.

J. R. Cooper, F. E. Bloom, R. R. Roth (1986). *The Biochemical Basis of Neuropharmacology,* 5th ed. New York: Oxford University Press.

J. W. Daly (1985). A_3 Adenosine receptors. *Adv. Cyclic Nucleotide Res. 18*: 29–46.

A. Dobolyi, A. Reichart, T. Szikra, G. Juhasz (1998). Purine and pyrimidine nucleoside content of the neuronal extracellular space in rat: An *in vivo* microdialysis study. *Adv. Exp. Med. Biol. 431*: 83–87.

B. B. Fredholm, B. Johansson, I. van der Ploeg, P. S. Hu, S. Jin (1993). Neuromodulatory roles of purines. *Drug Dev. Res. 28*: 349–353.

K. A. Jacobson (2002). Purine and pyrimidine neucleotide (P2) receptors. *Ann. Rep. Med. Chem. 37*: 75.

K. A. Jacobson, S. Tchilibon, B. V. Joshi, Z.-G. Gao (2003). *Ann. Rep. Med. Chem. 38*: 121.

T. Katsuragi, T. Furukawa (1985). Novel, selective purinoceptor antagonists: investigation of ATP as a neurotransmitter. *Trends Pharmacol. Sci. 6*: 337–339.

D. Satchell (1984). Purine receptors: classification and properties. *Trends Pharmacol. Sci. 5*: 340–343.

S. H. Snyder (1985). Adenosine as a neuromodulator. *Annu. Rev. Neurosci. 8*: 103–124.

G. L. Stiles (1986). Adenosine receptors: structure, function and regulation. *Trends Pharmacol. Sci. 7*: 486–490.

5

Messenger Targets for Drug Action II

Hormones and their receptors

5.1 OVERVIEW OF RELEVANT ANATOMY AND PHYSIOLOGY OF HORMONES

In 1902, Bayliss and Starling demonstrated that the effect exerted by the duodenum on pancreatic secretion was due to a bloodborne factor that seemingly acted as a "chemical messenger." Their research led to the identification of this chemical messenger, which they called *secretin*. They went on to suggest that secretin was not unique, hypothesizing that many chemical agents are secreted by various cells throughout the body and that these agents, upon distribution by the bloodstream and circulation, influence the function of organs that are located some distance away. They coined the term *hormone* to describe such chemical messengers that are synthesized in one organ system and distributed via the circulation to distant organ systems to elicit an altered biochemical response. Insulin is the prototypic example of a hormone; it is biosynthesized in the pancreas and then distributed via the bloodstream to organs and tissues throughout the body where it influences carbohydrate, protein, and fat metabolism. The medical discipline of *endocrinology* deals with the diagnosis and treatment of diseases related to hormones and hormonally responsive systems.

Hormones are central to homeostasis since they facilitate chemical control over metabolic and biochemical processes throughout the entire body. The endocrine system, which exerts control over *chemical processes* via hormones, is crucial to homeostasis over an intermediate time scale. The endocrine system is distinct from the nervous system, which employs neurotransmitters to control *electrical and electrochemical processes* and to influence homeostasis over a shorter time scale. The endocrine system is also distinct from the immune system, which employs immunomodulators to control *cellular processes* and to influence homeostasis over a longer-term time scale. There is, of course, a large amount of overlap among these three control systems.

The endocrine system is composed of hormone-producing organs within the body. Probably the most important endocrine organ is the *pituitary gland*, located at the base of the skull and intimately associated with the *hypothalamus*; both the pituitary and the hypothalamus secrete a wide variety of peptidic hormones and are crucial to the

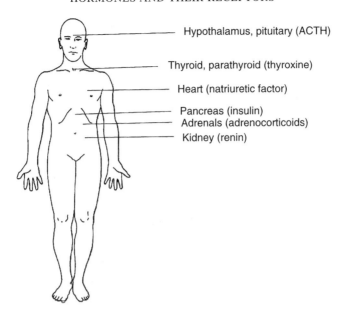

Figure 5.1 Anatomy of hormone producing organs. The endocrine system is formed by the hormone-secreting organs. These organs are diffusely distributed throughout the body and affect the metabolic control of a wide variety of biochemical processes. Accordingly, hormones and hormone receptors are important messenger targets for drug design.

nervous system–endocrine system linkage. The *thyroid gland*, located in the neck, secretes *thyroxine* as its messenger hormone; by means of this messenger molecule, the thyroid gland stimulates oxygen consumption at a cellular level, helps to regulate lipid and carbohydrate metabolism, maintains metabolic levels, and influences growth and maturation. The *parathyroid glands*, located on either side of the thyroid, secrete *parathyroid hormone*, which is involved in the control of calcium metabolism and bone physiology. The *pancreas* is located in the upper abdomen near the duodenum; unique clusters of cells within the pancreas, called islets of Langerhans, secrete the polypeptide hormones, *glucagon* and *insulin*, which regulate metabolism of proteins, lipids, and carbohydrates. The *adrenal glands*, located above each kidney, secrete a variety of steroid hormones (e.g., *glucocorticoids, mineralocorticoids, sex hormones*) and are important to the endocrine system–immune system linkage. The *kidneys* are themselves endocrine organs, producing *renin*, a hormonal messenger involved in the regulation of sodium and fluid volumes within the body. The *gonads* secrete gender-specific steroidal sex hormones and are, of course, involved in the development and functioning of the reproductive system. Even the heart has an endocrine function, producing *natriuretic peptides* as hormones. Superimposed upon this classification of endocrine organs and classical hormones are the neuropeptides, which may have neurotransmitter and/or neurohormonal effects. The existence of these peptide neurohormones makes the brain a partial endocrine organ as well as a neural organ. The various endocrine organs involved in hormonal metabolism are shown in figure 5.1.

Hormones are ideal targets for rational drug design. Since hormones are precise chemical messengers influencing specific metabolic function throughout the body, their pharmacological manipulation by the administration of either agonists or antagonists permits therapeutic modulation of a wide range of biochemical events. Moreover, since many hormones are small molecules (i.e., steroids or short peptides with molecular weights less than 1000) they are readily studied using molecular modeling calculations to facilitate rational drug design. When broadly categorized on a molecular basis, hormones may be either steroid-based or peptide-based. Both categories have been and continue to be widely exploited for purposes of drug design. This chapter examines the role of both steroid hormones (sections 5.2–5.13) and peptide hormones (sections 5.14–5.23) in drug design.

5.2 STEROID HORMONES: INTRODUCTION

The steroid hormones include the sex hormones (*estrogens, progestins, androgens*) and the *adrenocorticoids* (glucocorticoids, mineralocorticoids), as well as their biosynthetic precursor, cholesterol. Steroid hormones are based on the four-ring steran (**5.1**) carbon skeleton. Other pharmacologically interesting steroids, like the heart-active cardenolides, are compounds of plant origin.

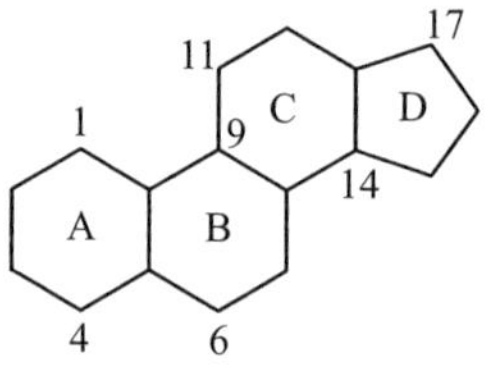

Steran (5.1)

Minor changes in the stereochemistry and substitution pattern of the steran skeleton result in vastly different yet specific physiological and pharmacological effects, which in turn influence developmental, metabolic, and behavioral phenomena. The organic chemistry and biochemistry of steroids is the subject of many excellent books and an enormous amount of research and patent literature. This chapter compares and contrasts the structure and mode of action of various steroids, their role in regulating hormonal secretion, and the timing of this regulatory action.

One of the most unique and powerful features of steroid hormones is the nature of the steroid receptor. Unlike most other hormones or drugs, which target protein receptors usually embedded in membranes, steroids target the genes themselves, buried deep within the nucleus of the cell.

5.3 STEROID HORMONES: RECEPTOR BIOCHEMISTRY

Steroid receptors are highly specific macromolecules found in central regulatory organs (e.g., pituitary, hypothalamus), in various end-point target tissues (e.g., uterus, vagina, prostate), and in lower concentrations in the brain, liver, kidney, ovary, and many other organs. Steroid hormones exhibit remarkable tissue selectivity when binding to these receptors. Within the target organs in which the steroids bind, the steroid molecules exert their influence directly on protein synthesis, at the level of transcription of the genetic message. Steroid hormones work primarily by regulating tissue-specific gene expression; the hormones enter the nucleus of a cell, bind to target genes in the DNA of that cell, and subsequently influence protein synthesis. Although steroids may also affect other cellular processes by influencing various enzyme systems through cAMP-dependent protein kinases, their effects on protein synthesis are of primary importance.

While the macromolecules and target tissues involved show extreme specificity for the appropriate steroid hormones and their congeners, the general scheme of the steroid–receptor mechanism is remarkably uniform. We can therefore deal with this receptor model in a general way, mentioning specific details as appropriate in the subsections of this chapter.

The general steroid–receptor hypothesis is based mainly on estrogen and progesterone receptors. The currently accepted mechanism is unique and consists of several steps at different subcellular structures:

1. Cytoplasmic receptor activation
2. Translocation of the hormone–receptor complex to the nucleus
3. Binding of the complex to DNA acceptor sites
4. Activation of transcription and influencing protein synthesis

The steroid hormones are transported to their target cells via the bloodstream in a protein-bound form, but diffuse into the cell as free steroids. At this point, they encounter a cytoplasmic steroid receptor protein. The complement DNAs (cDNAs) of all the major steroid receptors have been cloned, providing the complete amino acid sequence of each. These receptors are large proteins; the estrogen receptor, for example, has a molecular weight of approximately 75,000. The general model of the steroid receptor protein consists of several functional domains. The "E" domain (ligand binding region) is composed of the C-terminal 250 amino acids; this section has the steroid binding site and additional sites for binding to chaperone proteins (discussed in the next paragraph). The "D" domain (hinge region), located adjacent to the E domain, is involved with the translocation of the steroid–receptor complex into the nucleus of the target cell. Next, the "C" domain (DNA binding region) is made up of 70 amino acids clumped into two finger-shaped regions, each coordinated with a Zn ion; these so-called zinc fingers are crucial to the process whereby the steroid recognizes and binds to the DNA once the nucleus has been penetrated. Finally, at the N-terminal of the steroid receptor protein is the "A/B" domain (DNA modulator region), which enables the steroid–receptor complex that has bound to the DNA to activate genes and initiate transcription.

Upon entering the cell, the steroid molecule initially binds to the steroid receptor protein (E domain) to form the steroid-hormone–receptor complex. This complex concomitantly binds to an additional eight or more other peptides (also via the E domain); these peptides are termed *chaperone peptides* and consist of macromolecules such as *heat shock proteins* (e.g., hsp70, hsp90). The chaperone peptides help to twist and turn the steroid receptor protein into an improved three-dimensional shape for final and optimal binding of the steroid molecule. Following binding of the chaperone peptides, the steroid-hormone–receptor complex becomes a "mature steroid-hormone–receptor complex." Following optimal binding of the steroid to its receptor, the mature complex dissociates, releasing the chaperones and converting the steroid hormone receptor complex into an activated form. The activated receptor is then phosphorylated, dimerized, and transported into the nucleus with the aid of the D domain. Once in the nucleus, the zinc fingers of the C domain bind to the DNA and the A/B domain effects gene activation. The synthetic response in the cell is very rapid: within 15 minutes a considerable increase in the concentration of RNA polymerase can be detected, and within 30 minutes induced protein synthesis is measurable. These early responses can even be triggered by hormones that have a lower than optimal affinity for the receptor (such as estriol in uterus, which binds much more weakly than estradiol).

The steroid–receptor complex remains in the nucleus for a limited time only, and eventually dissociates from the chromatin. About 40% of the receptors released in this dissociation are recycled and used again; the rest are destroyed and resynthesized. Steroid hormones can even regulate the level of synthesis of their own receptors, and sometimes the synthesis of other steroid receptors as well.

5.4 STEROID HORMONES: STRUCTURE AND CONFORMATION OF AGONISTS AND ANTAGONISTS

All steroids are based on the steran (**5.1**) skeleton, a fully hydrogenated cyclo-pentano-phenanthrene. Traditionally, the rings of this skeleton are labeled A, B, C, and D. All four rings are in the chair conformation in naturally occurring steroids; additionally, rings B, C, and D are always *trans* with respect to each other, whereas rings A and B can be *trans* (as in cholestane, **5.2**) or *cis* (as in coprostane, **5.3**). It is simple to conceptualize this ring anellation (fusion) if one observes the relation of substituents (including hydrogen) on the carbon atoms common to the rings in question. For rings A and B, the relative positions of the 19-methyl group (attached to C-10) and the hydrogen on C-5 determine the structure, and their *trans* or *cis* configuration is easily visualized. In general, neighboring substituents are *trans* if they are diaxial or diequatorial, and *cis* if they are axial–equatorial. The two methyl groups on C-10 and C-13 are always axial relative to rings B and D, with the C-10 substituent (which is not necessarily methyl) being the conformational reference point.

A somewhat obsolete but still valid nomenclature determines substituent conformations relative to the plane of a cyclohexane-type ring; thus, the 19-methyl group in the steroid ring system is designated β and is *above* the plane of the molecule, while H-5 in cholestane (**5.2**) is α and *below* the plane. The α–β convention for steroids must not be compared or confused with the usual axial–equatorial convention since, with the latter convention, the flipping of a cyclohexane ring (for example) from one chair form to another changes the position of a β substituent from axial to equatorial, and vice versa. Since a substituent designated α (or β) will remain α (or β) but can be either axial or equatorial, confusion can arise. The stability, reactivity, and spectroscopy of a substituent will, however, change, depending on its axial or equatorial position. Equatorial substituents are normally more reactive and less stable than their epimers and show slightly different absorption spectra. The physiological and pharmacological properties of the different molecules are also different, as might be expected.

These considerations apply to all cycloalkane derivatives, including steroids. However, the chair form of a ring is inherently more stable than the boat form. Moreover, the fused-ring nature of the system lends it a very considerable rigidity, and *cis–trans* isomerization would necessitate the breaking and formation of covalent bonds. Therefore, steroid substituents maintain their conformation at room temperature, whereas cyclohexane substituents usually do not. Steroids are classified according to their substituents in addition to their occurrence.

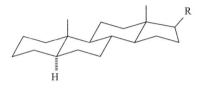

Cholestane (5.2), A-B trans

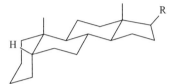

Coprostane (5.3), A-B *cis*

5.5 STEROID HORMONES: STEROID BIOSYNTHESIS

The biosynthesis of steroids is complex, as one would expect since all the compounds in this group must be derived from a single precursor (cholesterol (**5.4**)). The primary source of all the compounds involved in steroid synthesis is acetate, in the form of acetyl-coenzyme A. Cholesterol, besides being ingested in food, is synthesized in large amounts, and an adult human contains about 250 g of cholesterol. In contrast, the steroid hormones are produced at the milligram level or lower.

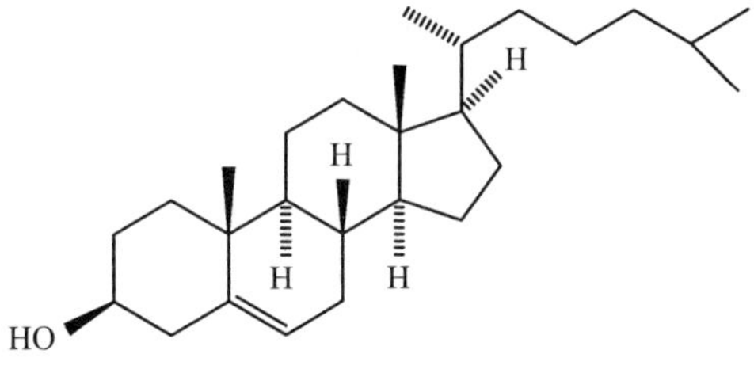

Cholesterol (5.4)

The regulation of steroid biosynthesis is achieved by an intricate network of peptide hormones. The pituitary-based production of these hormones is under neuroendocrine influence, starting with the hypothalamus in the CNS, which itself is under dopaminergic control. The hypothalamus produces a number of small peptide hormones that act as the releasing factors for a second series of peptide hormones synthesized in the anterior pituitary gland. Among these hormones, *adrenocorticotropic hormone* (ACTH) regulates corticosteroid synthesis in the adrenal cortex, whereas *luteinizing hormone*

(LH) and *follicle-stimulating hormone* (FSH) (the *gonadotropins*) act on the ovaries and testes. Gonadotropins induce the production of estrogens and progestins in the female, which, in turn, produce appropriate changes in the reproductive tract. In the male, LH and FSH regulate androgen formation. At the same time, the steroids have a feedback regulatory effect on the hypothalamo-pituitary axis, setting up an exquisitely tuned regulatory loop. In addition, gonads produce the protein *inhibin*, which suppresses production of follitropin (FSH) and *gonadoliberin* in the pituitary and hypothalamus, respectively, and therefore, indirectly, steroid hormone production. There are many opportunities to exercise direct or indirect drug control over a large number of metabolic and reproductive phenomena in such a sensitive multicomponent feedback system. Understanding the biochemistry, regulation, and molecular mechanism by which the steroids act is therefore important in rational drug design and therapy.

5.6 STEROID HORMONES: CHOLESTEROL AS A BIOSYNTHETIC PRECURSOR

Since cholesterol is the biosynthetic precursor of all the steroid hormones, a consideration of cholesterol biochemistry is central to medicinal chemistry and rational drug design as applied to steroids. Although cholesterol plays a role in a variety of processes, its participation in gallstones and atherosclerosis are probably its two most important aspects.

5.6.1 Cholesterol and Gall Bladder Disease

Cholesterol is metabolized in the liver to bile acids, which are necessary for digestion since they act as natural detergents and solubilize dietary fats. The bile acids, once synthesized in the liver, are stored in the gallbladder and then delivered via the common bile duct to the duodenum to aid with digestion. All of the hydroxyl groups of the various bile acids (a maximum of three OH groups) are axial, the A–B ring anellation is *cis*, and the polarity of the side-chain is increased by conjugation with glycine or taurine. Consequently, there are both hydrophilic and hydrophobic portions of the molecule, which can thus act as a detergent and form inclusion compounds with fatty acids, promoting their absorption through the intestinal wall. The excessive excretion of cholesterol can lead to its crystallization and the formation of *gallstones* through a process called *cholelithiasis*. Gallstones are more frequent in middle-aged women and present with upper abdominal pain. Most stones are formed within the gall bladder and then migrate into the common bile duct. The opening of the duct into the duodenum is only 2–3 mm in diameter and thus the duct becomes occluded, leading to pain. A possible complication is the development of an infection in the blocked bile duct, leading to a potentially life-threatening disorder called *ascending cholangitis*.

Cholesterol-containing gallstones can be dissolved slowly using long-term administration of chenodeoxycholic acid (**5.5**, CDCA) or ursodeoxycholic acid (**5.6**, UDCA). These two molecules are physiologically occurring stereoisomeric bile acids (7-OH group, being α in CDCA and β in UCDA). UCDA is better tolerated and is effective at doses of 8–10 mg daily; CDCA may cause liver toxicity and diarrhea, and requires a dose of 15 mg/day. Provided that the gallstone is small (<15 mm), the stone may be dissolved with 1–2 years of treatment. Stone formation may recur after cessation of the therapy. Not surprisingly, surgical therapy is still preferred in the treatment of gallstones, rather than the slow and somewhat inefficient use of these pharmacological agents.

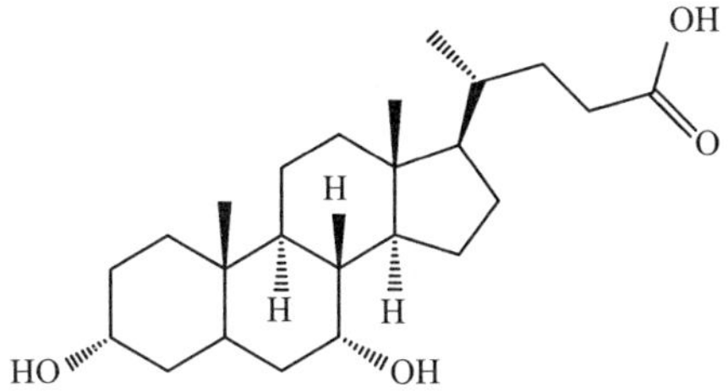

Chenodeoxylcholic acid (5.5) 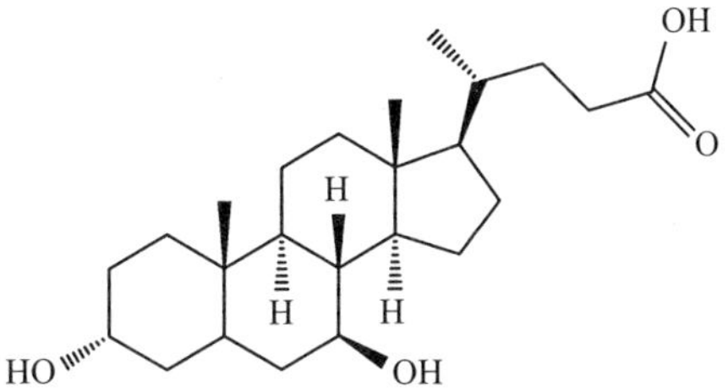   Ursodeoxycholic acid (5.6)

5.6.2 Cholesterol and Vascular Disease

More important than its involvement in gallstone formation, cholesterol's role as a central culprit in the dreaded pathological process of *atherosclerosis* ("hardening of the arteries") has attracted extensive attention. Cholesterol metabolism and its involvement in atherosclerosis play a fundamental role in research efforts directed at decreasing the incidence of cardiovascular disease, the number-one killer in Western civilization, and cerebrovascular disease, the number-three killer. There is a statistical correlation between serum cholesterol levels and the incidence of atherosclerosis. In the latter condition, damaging lesions (called *atheromas* or *atherosclerotic plaques*), which contain cholesterol and other lipids, form in arterial walls. Initial atheroma formation is followed by fibrosis of these plaques, with narrowing of the arterial diameter, causing a decrease in blood flow, changes in the blood-clotting mechanism, and eventual total arterial occlusion. When arteries become blocked and cannot supply blood, the pathological process of *ischemia* occurs (ischemia is death or injury to cells arising from an interruption of the blood supply to those cells). This, in turn, may lead to *myocardial ischemia* (MI, "heart attack") or *cerebral ischemia* ("stroke"), arising from reduced blood supply to the cells of the heart or brain, respectively. (Some people, with extensive atherosclerosis throughout their entire body, are called "vasculopaths" and are susceptible to heart attacks, strokes, and a variety of other vascular problems throughout their body.) Regrettably, the effects of cholesterol metabolism, dietary factors (e.g., excessive consumption of saturated fatty acids), exercise, and life-style in the etiology of cardiovascular disease is a highly controversial topic, fraught with half-truths and misinformation, especially in popular press reports and from the so-called "health gurus" who promote their products on the world-wide web.

Encouragingly, there is increasing, robust scientific evidence that atherosclerosis can be arrested and even decreased by pharmacologic approaches. However, the evaluation of drug therapy for this condition is complicated by the fact that the regulation of cholesterol metabolism alone does not necessarily have a beneficial effect on atherosclerosis.

In addition to high levels of serum cholesterol and cholesterol esters, a combination of other molecular factors appears to be necessary for the development of atherosclerosis. Among these additional molecules, *lipoproteins* are crucial. Lipoproteins are macromolecular assemblies consisting of lipids (cholesterol, triglycerides) noncovalently bound with protein and, to a lesser extent, carbohydrate. These collections of molecules have a spherical shape and consist of a nonpolar core surrounded by a monolayer of phospholipids whose polar groups are oriented out; within the phospholipid monolayer are a small number of cholesterol molecules and proteins, called *apolipoproteins*. Lipoprotein particles solubilize lipids, preventing insoluble aggregates in the plasma and enabling cholesterol transport in the aqueous plasma. These solubilized collections of cholesterol and other lipids can more readily penetrate damaged arterial walls, thereby promoting atherosclerosis.

The various types of lipoproteins that carry cholesterol can be separated and categorized by ultracentrifugation. From a molecular viewpoint, the ratio of these various serum lipoproteins is of major clinical importance. There are three main types of lipoproteins: very low-density (VLDL), low-density (LDL), and high-density (HDL) lipoproteins. VLDL is 60% triglyceride, 18% phospholipid, and 12% cholesterol; LDL consists of 50% cholesterol and 10% triglycerides; HDL is 25% cholesterol and 50% protein. The first two types (VLDL, LDL) seem to increase atherosclerosis, whereas the high-density lipoproteins seem to decrease the incidence of atheroma formation. High-density lipoproteins may even facilitate the removal of cholesterol from the arterial wall. This is probably accomplished in two ways: by the increased esterification of cholesterol, and by inhibition of the LDL–cholesterol complex uptake by the cells of the arterial wall. Since atherosclerosis is a fundamental pathology, leading to stroke and heart attack, manipulations of cholesterol and lipoproteins are important in drug design.

According to our present knowledge, there are several ways to treat disorders (referred to as *hypercholesterolemias* or *hyperlipoproteinemias*) that enhance cholesterol-containing atheroma formation. The class of compounds referred to as *statins* dominates the treatment of these disorders.

The statins (lovastatin (**5.7**), simvastatin (**5.8**), pravastatin (**5.9**), fluvastatin (**5.10**), cerivastatin (**5.11**), atorvastatin (**5.12**)) target liver biochemistry. As discussed above, cholesterol is an essential constituent in human metabolism, and the liver obtains 60% of its required cholesterol from *de novo* synthesis, starting with acetylcoenzyme-A. A crucial step in this synthesis is the conversion of hydroxymethylglutaryl CoA (HMG CoA) to mevalonic acid, a conversion catalyzed by the HMG CoA reductase enzyme. The pharmacophore of these statin drugs resembles the substrate of this enzyme. In the presence of statin-type drugs, liver cells increase the synthesis of LDL receptor proteins and uptake and removal of LDL from the systemic circulation. The end result of this effect is an overall reduction in harmful atherosclerosis-promoting lipids. Lovastatin and simvastatin are lactones that undergo significant first-pass effect in the liver, being hydrolyzed to bioactive metabolites; pravastatin and fluvastatin, on the other hand, are organic acids that are administered in a bioactive form. A potentially dangerous side effect of the statins is skeletal muscle damage, a risk that is augmented if the statin is co-administered with a fibric acid analog. The action of statins is intensified if they are co-administered with ion exchange resins.

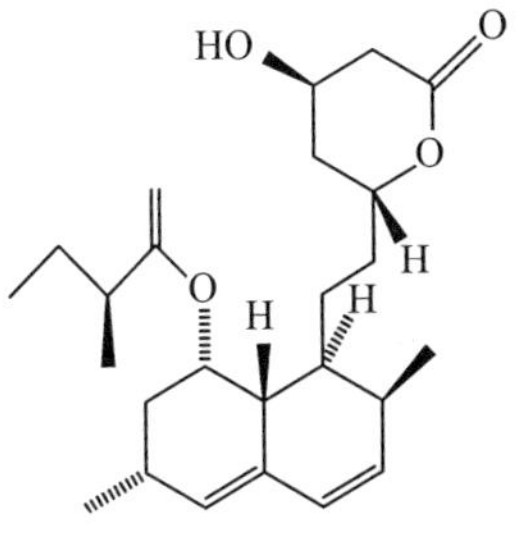

Lovastatin (5.7)

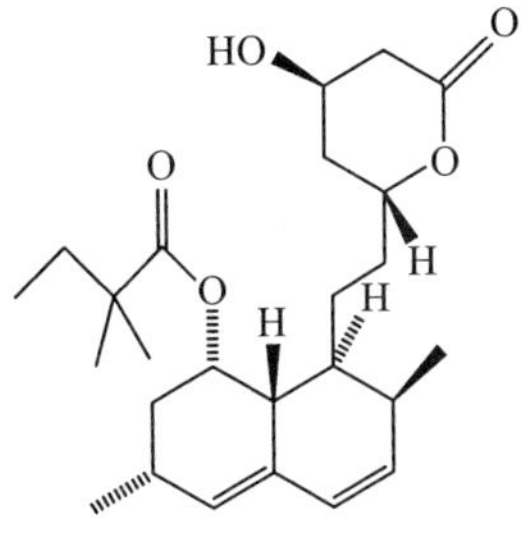

Simvastatin (5.8)

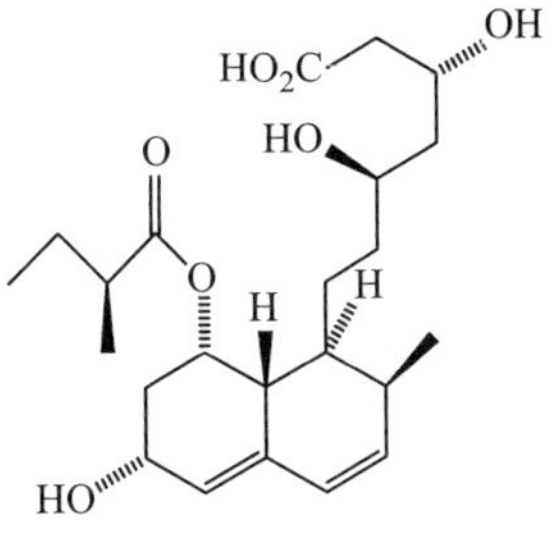

Pravastatin (5.9)

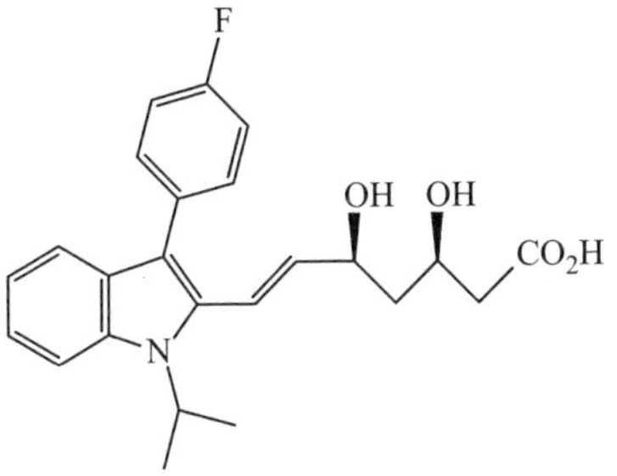

Fluvastatin (5.10)

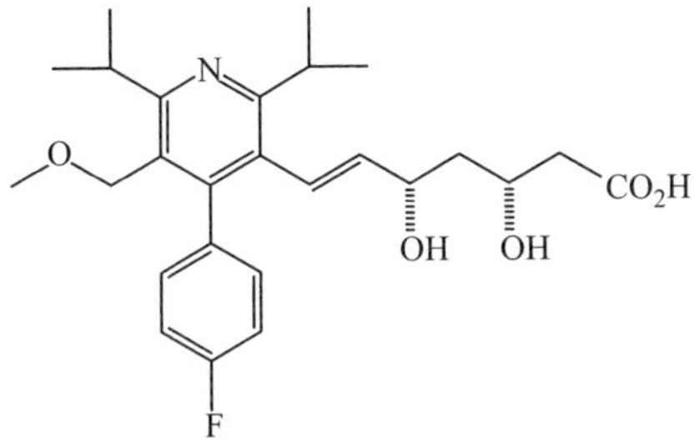

Cerivastatin (5.11)

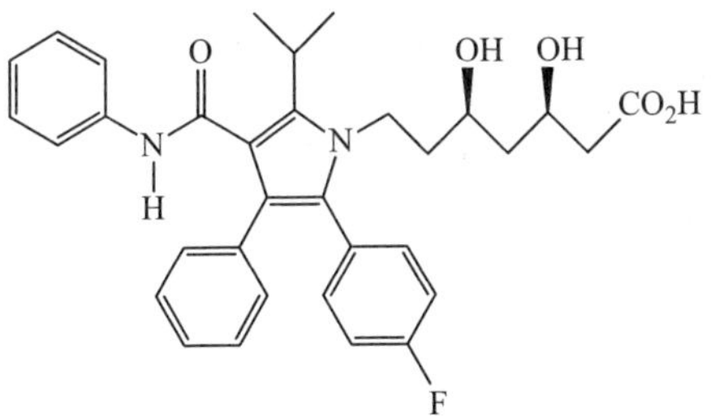

Atorvastatin (5.12)

In addition to treatment with the statins, hypercholesterolemia is sometimes treated with the use of nonabsorbable anion-exchange resins like cholestyramine (**5.13**) and colestipol, which sequester bile acid in the intestine, excrete them, and thus increase their synthesis in the liver by a feedback mechanism. Increased bile acid synthesis increases cholesterol metabolism and also decreases LDL concentrations. Unfortunately, these resins interfere with the absorption of other fats and fat-soluble vitamins (A, D, E, and K). They

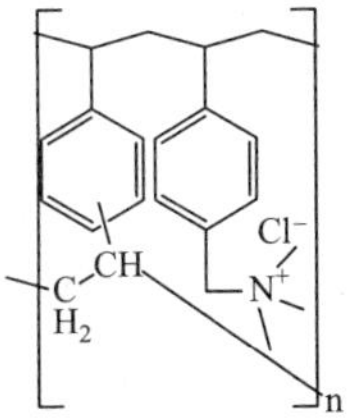

Cholestyramine
(5.13)

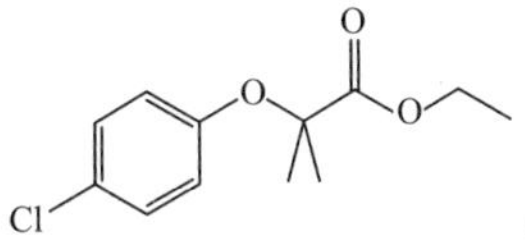

Clofibrate
(5.14)

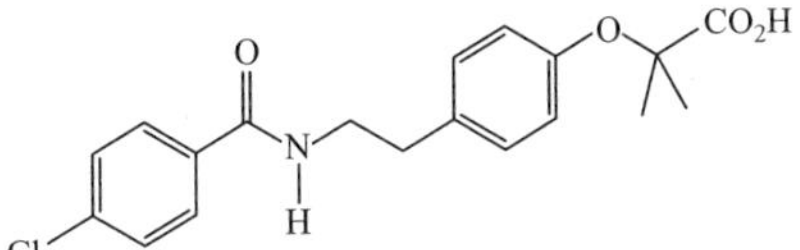

Bezafibrate (5.15)

also adsorb co-administered drugs such as digitoxin and diuretics, thereby decreasing their bioavailability.

Another drug that is still sometimes used to inhibit cholesterol biosynthesis is clofibrate (**5.14**), an isobutyrate derivative. Bezafibrate (**5.15**), etofibrate (**5.16**), and gemfibrozil (**5.17**) are other bioactive analogs in this class of drugs. These agents seem to block cholesterol synthesis at the point at which cholesterol exerts feedback inhibition on its own synthesis: at the formation of mevalonate. They also block acetyl-CoA carboxylase, the enzyme producing malonyl-CoA. However, clofibrate is far from being an ideal drug as it has numerous side effects. It is being largely replaced by the statins.

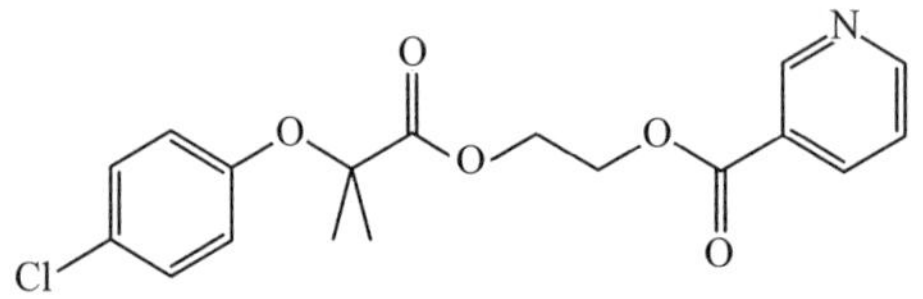

Etofibrate (5.16) 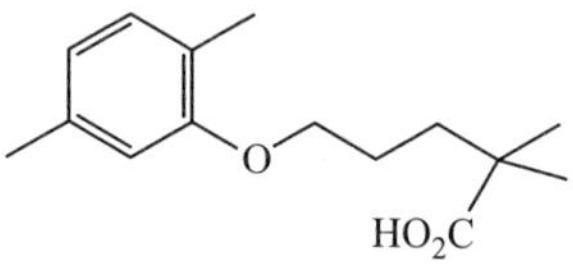 Gemfibrozil (5.17)

Nicotinic acid (**5.18**), and related derivatives such as pyridylcarbinol (**5.19**), xanthinol nicotinate (**5.20**), acipimox (**5.21**), given in large doses, influence the lipoprotein ratio, decreasing the concentrations of very low and low-density lipoprotein, but have no effect on HDL–cholesterol complexes. Acipimox (**5.21**) is a new pyrazine derivative that is 20 times more active than nicotinic acid. When first administered, the use of these agents is associated with flushing and hypotension.

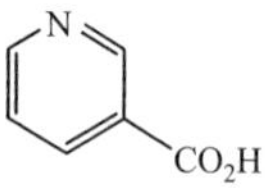

Nicotinic acid (5.18) 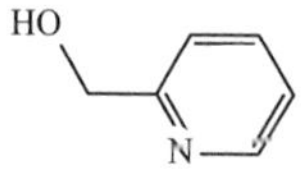 Pyridylcarbinol (5.19) 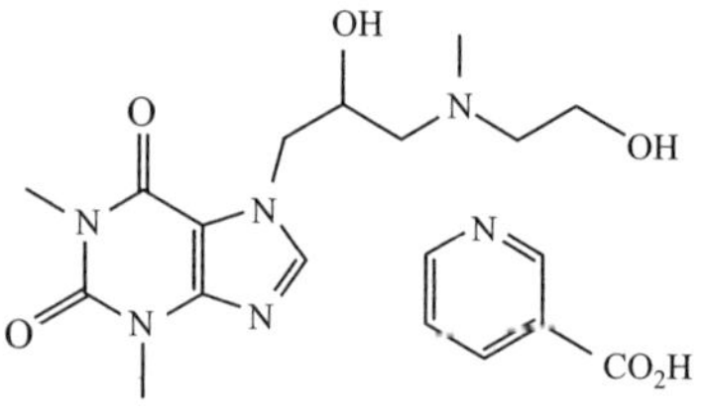 Xanthinol nicotinate (5.20) 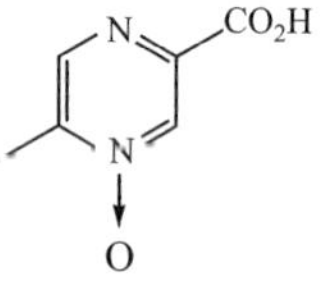 Acipimox (5.21)

Since atherosclerosis is an important cause of pathology in our aging population, drug design will continue to focus on this challenging but fruitful area of research.

5.7 STEROID HORMONES: SEX HORMONES—INTRODUCTION

Of the various steroid hormones produced from cholesterol, the sex hormones are extremely important. The reproductive steroid hormones are divided into three classes:

1. Estrogens, which regulate ovulation and the development of the secondary female sex characteristics
2. Progestins, or gestagens, which maintain pregnancy
3. Androgens, the male sex hormone

These three classes of sex hormones are of fundamental importance in medicinal chemistry, not only for their role in influencing reproductive susceptibilities, but also as potential therapeutics in a range of other conditions, including cancer.

5.8 STEROID HORMONES: SEX HORMONES—ESTROGENS

Of these three classes of sex steroid hormones, the estrogens are the best studied. Estrogens are produced mainly in the ovaries when the latter are stimulated by follicle-stimulating hormone (FSH). Under such stimulation, the estrogen levels rise until the middle of the menstrual cycle (when ovulation takes place), remain at a fairly constant concentration, and then decline if fertilization does not take place. The final result is menses, the shedding of the uterine endometrium (lining). Estrogens also regulate uterine growth in immature animals and, as noted above, are responsible for all female secondary sex characteristics.

5.8.1 Naturally Occurring and Synthetic Steroidal Estrogens

Estrone (5.22)   Estradiol (5.23)   Estriol (5.24)

The major estrogens produced by women are estrone (**5.22**, E_1), estradiol (**5.23**, 17β-estradiol, E_2), and estriol (**5.24**, E_3). Estradiol is the major secretory product of the ovary. Estrone and estriol are formed either in the liver, from estradiol, or in peripheral tissues, from androgens such as androstenedione. During the first part of the menstrual cycle, estrogens are produced in the ovarian follicles (80–350 µg/day estradiol); after ovulation, estrogens are synthesized (via a different biosynthetic pathway) in the *corpus luteum*, which is a proliferation of cells that replace clotted blood within the ruptured ovarian follicle. During pregnancy, large amounts of estrogen are synthesized by the fetoplacental unit (1000-fold increase when compared to the nonpregnant state). Estriol synthesized by the fetoplacental unit enters the maternal circulation and is excreted in the urine.

Although estrogens are easily isolated from the urine of pregnant women, their most abundant source has historically been from the urine of horses, especially pregnant mares. Somewhat surprisingly, however, the most prolific source of estrogen is stallions, in which urinary estrogens are produced as metabolites of androgens. Horses excrete a variety of equine estrogens, including equilenin (**5.25**), a steroid containing the naphthalene ring system, and equilin (**5.26**); these equine estrogens are not synthesized in humans but readily bind to human estrogen receptors. All of these equine steroids have an aromatic A-ring and therefore lack a 19-methyl group. These estrogenic substances are excreted in very large quantities in equine urine and can be recovered and used for medicinal applications in human patients.

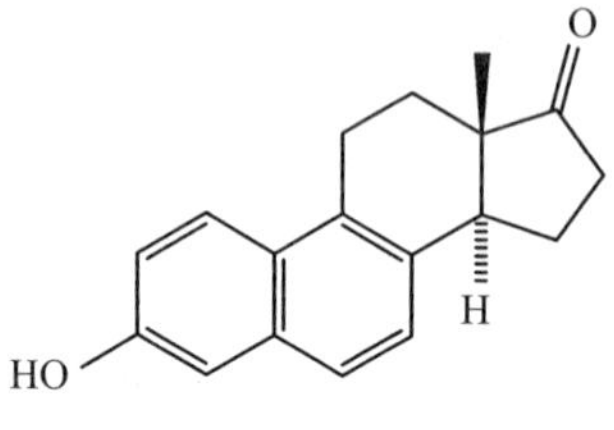

Equilenin (5.25) 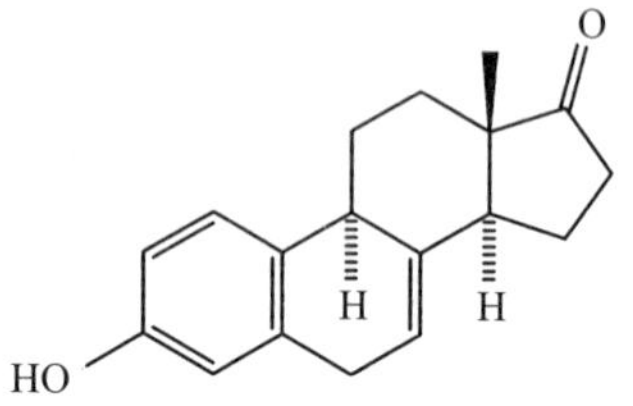   Equilin (5.26)

Today, a variety of therapeutic estrogens are produced semisynthetically from estrogen intermediates synthesized from diosgenin and other natural precursors. Two semisynthetic, orally active estrogens are ethinyl estradiol (**5.27**) and its 3-methyl ether (**5.28**, mestranol). Both of these are used in oral contraceptives (see section 5.8.3). Quinestrol (**5.29**) is another semisynthetic estrogen. The most important property of the semisynthetic estrogens is their increased oral effectiveness.

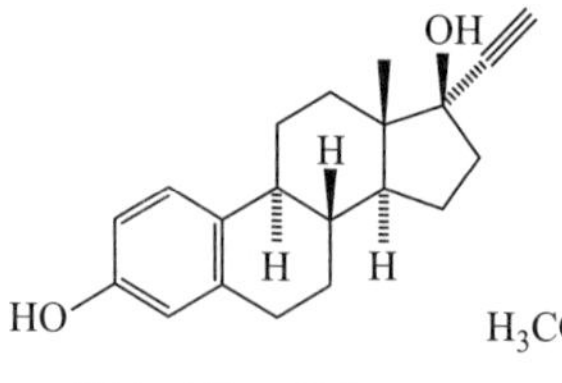

Ethinyl Estradiol (5.27) 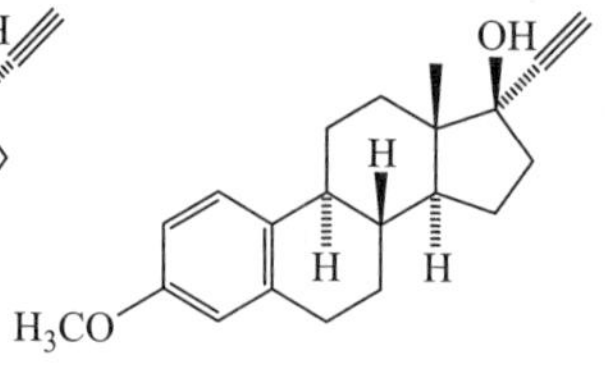   Mestranol (5.28) 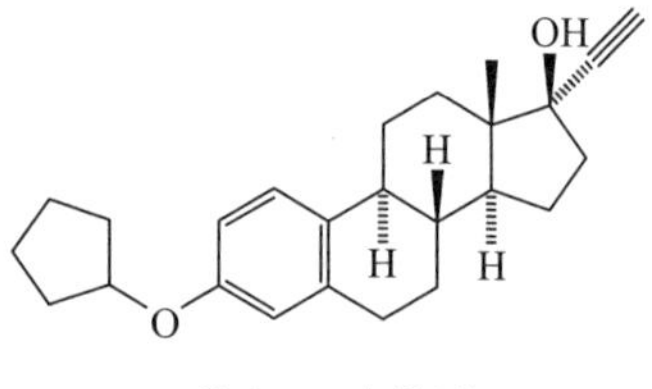   Quinestrol (5.29)

5.8.2 Nonsteroidal Estrogens

In addition to estrogen agonists based on steroid structures, a variety of nonsteroidal estrogens have also been synthesized and used clinically: dienestrol (**5.30**), diethylstilbestrol (**5.31**), benzestrol (**5.32**), hexestrol (**5.33**), methestrol (**5.34**), methallenestril (**5.35**), and chlorotrianisene (**5.36**).

Of these various nonsteroidal compounds, the first were *trans*-diethylstilbestrol (DES) (**5.31**) and its reduced derivative hexestrol (**5.33**). The way in which these nonsteroidal stilbene (diphenylethylene) derivatives are usually drawn suggests a resemblance to the steroid skeleton. However, this resemblance is purely incidental, since the two ethyl groups are not indispensable for estrogenic activity. For example, four methyl groups will give comparable pharmacological activity. It seems, however, that the

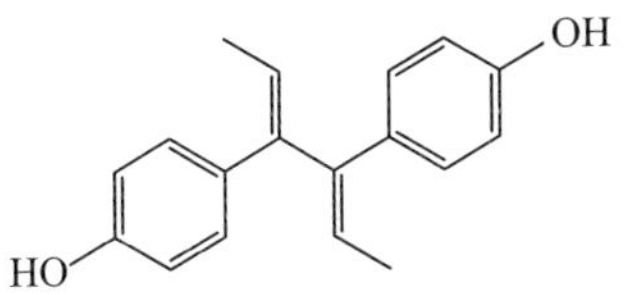

Dienestrol(5.30)

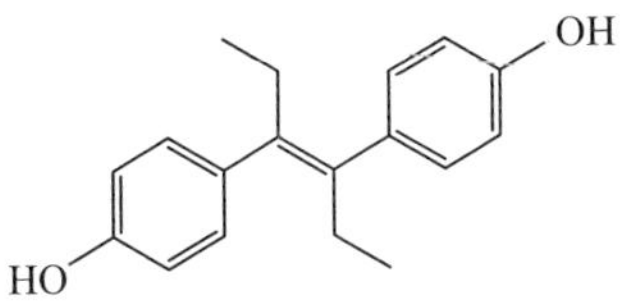

Diethylstilbestrol (5.31)

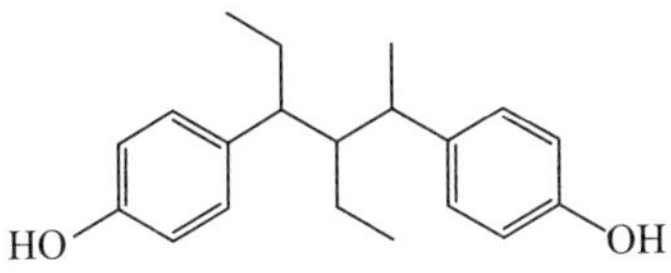

Benzestrol (5.32)

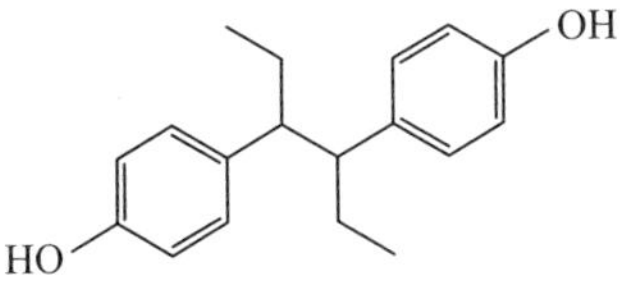

Hexestrol (5.33)

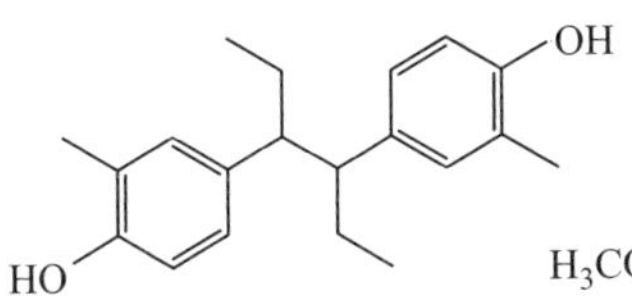

Methestrol (5.34)

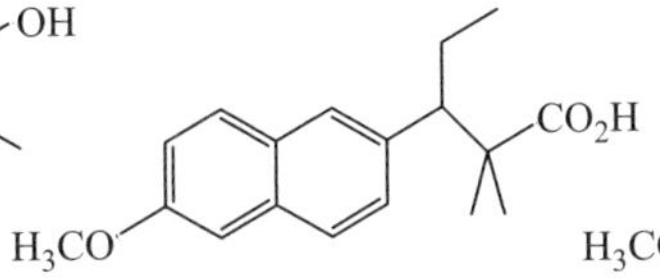

Methallenestril (5.35)

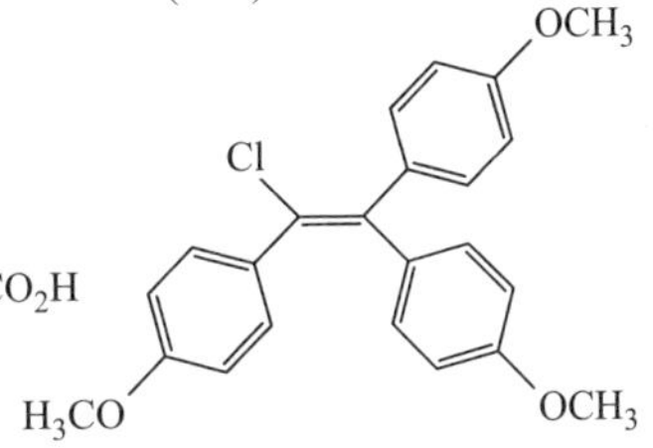

Chlorotrianisene (5.36)

"thickness" of the molecule is important to its activity: the planes of the two phenyl groups must have a 60° torsion angle, and four alkyl methyl groups will provide the same steric hindrance as two ethyls; the dimethyl derivative is therefore inactive. In addition to the geometric orientation of the two phenyl groups, the interatomic distance between the two phenolic OH groups in DES is also important. Indeed, as long ago as 1946, Schuler suggested that the distance between the two OH groups in DES matched the 3-OH to 17-OH distance in estradiol. Modern physical chemistry structural studies, however, have shown that the OH–OH distances are 12.1 Å in DES but only 10.9 Å in estradiol; this discrepancy can be eliminated by hydrogen bonding two water molecules to the 17-OH moiety of estradiol and by considering the water -OH group rather than the estradiol -OH group in the distance calculation. Recent quantum pharmacology calculations by Wiese et al. support these observations and suggest that the estrogen receptor can recognize as many as ten low-energy conformations of DES, provided the OH–OH geometries are within an acceptable range. Further support of this conclusion comes from the thousands of DES analogues that have been synthesized and biologically evaluated.

   In terms of toxicology, DES has been a fascinating drug. There have been a number of reports of vaginal adenocarcinoma (vaginal cancer) in young women (so-called "DES-babies") whose mothers were treated with DES during pregnancy in the mistaken belief that DES would reduce the possibility of miscarriage. Although the incidence of cancer was low—less than 1 case per 1000 mothers exposed—it nonetheless quite appropriately led to the discontinued use of DES in pregnancy. This agent should probably only be used for the treatment of cancer (e.g., of the prostate.)

5.8.3 Therapeutic Uses of Estrogens

The most common use of estrogens is in birth control pills. Since estrogen is combined with progestins in oral contraceptives, these agents are discussed in section 5.9.1. Estrogens are also used therapeutically to replace or augment hormones whose natural production is insufficient during menopause, in menstrual disorders, or as a result of insufficient development of the female reproductive tract. Estrogens are very useful in treating the "hot flashes" of early menopause and in helping menopausal conditions such as atrophic vaginitis. When used postmenopausally, estrogens may theoretically decrease the risk of heart disease by lowering the incidence of atherosclerosis; however, this is an extremely controversial area around which there is much debate. For example, in 2004 researchers evaluating the use of estrogen replacement in postmenopausal women reported that it was not achieving heart disease prevention, and may be increasing the risk of stroke; this resulted in the National Institutes of Health (NIH) instructing participants in the estrogen-alone study of the Women's Health Initiative (WHI) to stop taking their study medication.

Estrogens taken after menopause have also been unequivocally shown to reduce the severity of osteoporosis ("bone thinning"), especially in combination with exercise and adequate nutrition.

Another therapeutic role for estrogens is in the treatment of cancer. In androgen-dependent prostate carcinoma, estrogens are used therapeutically to suppress androgen formation and thus tumor growth. Estrogens are also used to treat inoperable breast cancer in men and postmenopausal women. However, antiestrogens, such as tamoxifen, have fewer side effects and are usually preferred (see section 5.8.4).

Although estrogens can be used to treat cancer, the relationship between estrogens and cancer is a complicated one. For example, estrogens can actually stimulate existing breast cancers in certain premenopausal women. While no increased incidence of breast cancer has been demonstrated for short-term estrogen therapy, a small increase in the incidence of this tumor may occur with prolonged therapy. Several studies have also shown an increased risk of endometrial carcinoma in certain patients taking estrogens alone. However, the concomitant use of a progestin may prevent this increased risk and may even reduce the likelihood of endometrial cancer. Other non-cancer side effects associated with the administration of estrogens include migraine headaches, gallstone formation, breast tenderness, and nausea.

5.8.4 Antiestrogens

Antiestrogens (estrogen receptor antagonists) are used for two purposes: fertility drugs and antitumor agents. The first application is based on the fact that estradiol, the natural hormone, inhibits the secretion of the gonadotrophic hormones LH and FSH by feedback inhibition. The result of this inhibition is the production of a single ovum in every menstrual period, thus preventing overlapping pregnancies. Antiestrogens, such as clomiphene (5.37), block this inhibition in women who are infertile because of anovulation resulting from excessive estradiol production. Antiestrogens are therefore useful in helping infertile women to become pregnant. As a complication of their use, multiple pregnancies are rather common in women treated with antiestrogens. This incidence of multiple births can rise up to 10% in women treated with antiestrogens for infertility, which is much higher than the normal rate. Also, antiestrogen use is associated with birth defects in 2–3% of live newborns.

Antiestrogens (e.g., tamoxifen (**5.38**)) are also active as antitumor agents in estrogen-dependent mammary carcinoma (breast cancer), a neoplasm that has estrogen receptors. Hormonally responsive estrogen receptors are found in about two-thirds of all breast tumors. Paradoxically, the use of tamoxifen as an antiestrogen to treat breast cancer is associated with an increased risk of endometrial cancer as a delayed side effect. Alternatively, breast tumors can sometimes be treated with androgens, preferably the nonvirilizing derivatives (see section 5.10), in addition to removal of the ovaries.

Both clomiphene (**5.37**) and tamoxifen (**5.38**) are aminoether derivatives of stilbene. They are both structurally related to the nonsteroidal estrogens such as diethylstilbestrol (DES) and chlorotrianisene. However, the latter two compounds are estrogen agonists, whereas the former two are estrogen antagonists. When a nonsteroidal estrogen, such as DES, enters a cell's nucleus, it first binds to DNA and then triggers two activation factors (TAF1, TAF2) which in turn displace a repressor molecule, allowing transcription and protein synthesis; in this manner, DES functions as an estrogenic agonist. With antiestrogens, on the other hand, the aromatic ring structure permits DNA binding, but the dialkyl aminoethyl side-group precludes triggering of the TAF1 activation factor; accordingly, these agents are antagonists.

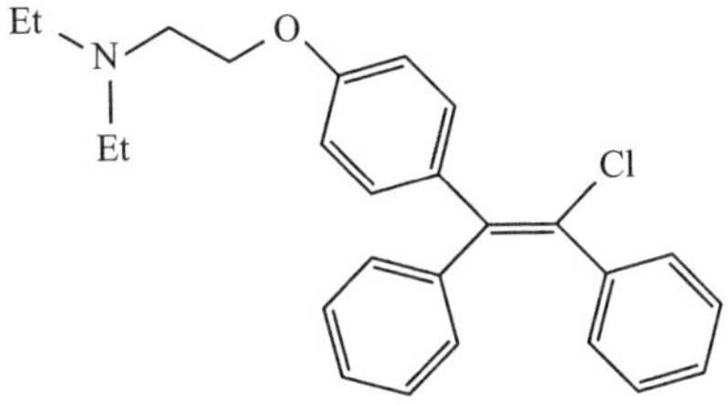

Clomiphene (5.37)

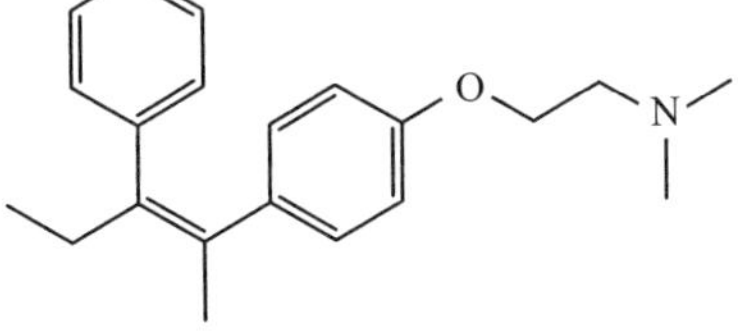

Tamoxifen (5.38)

5.9 STEROID HORMONES: SEX HORMONES—PROGESTINS (GESTAGENS)

These hormones are essential for the maintenance of pregnancy. The only natural progestin hormone, progesterone (**3.38**), is produced by the corpus luteum, an endocrine tissue formed in the ovary by the ruptured ovarian follicle after the level of luteinizing hormone peaks. If pregnancy occurs, the corpus luteum persists for the first three months of the pregnancy; after that, its role is taken over by the placenta as the major source of progesterone as well as estrogens. Through its feedback effect on the hypothalamus, progesterone prevents ovulation and also stops uterine contraction, thus avoiding dislodgement of the fertilized egg or embryo. In the absence of pregnancy, the progesterone level, together with the estrogen concentration, declines, resulting in estrus or menses—the shedding of uterine endometrium along with the unfertilized egg. At the same time, low steroid levels disinhibit hypothalamo-pituitary endocrine secretion, the peptide hormone levels rise, and the cycle starts again.

Progesterone itself has been used in the past in the treatment for threatened spontaneous abortion. However, the clinical trials that supported this therapeutic application have been called into question and, presently, progestins do not have any place in the treatment of threatened or habitual abortion; indeed, in the United States, the Food and Drug Administration (FDA) has strongly warned against the use of progestins during pregnancy. Likewise, there have been claims that progestins may reduce the unpleasant effects of PMS (premenstrual syndrome). As with the use of progestins for threatened abortion, there are no medical data to support this claim. Progestins have been used successfully as palliative treatments in inoperable endometrial cancer.

The main use of progestins is based on their antifertility effect. Since progesterone is poorly absorbed, it cannot be given orally; furthermore, it is not particularly potent and has an unacceptably short serum half-life of about 5 minutes. In early work, ethisterone (5.39), an acetylenic compound prepared from androsterone, was an orally active progestin, but also had male hormone action. In subsequent work, a large number of

Ethisterone (5.39)

semisynthetic progesterone derivatives (*progestogens*) were synthesized (levonorgestrel (5.40), desogestrel (5.41), norethindrone (5.42), norgestrel (5.43), and ethynodiol (5.44)) and found their principal use in oral contraceptive formulations. In preparing these synthetic progestogens, maintaining the progesterone 4-en-3-one A ring is essential for preserving receptor binding. Adding 17 α-alkyl groups slows metabolism; incorporating either 6-methyl or 6-chloro substituents enhances bioactivity while also slowing metabolism.

5.9.1 Combinations of Progestins and Estrogens (Oral Contraceptives)

The concept of oral contraception was pioneered by Pincus in the early 1950s. Modifications of the ethisterone molecule—removal of the 19-methyl group or the introduction of additional methyl groups in position C-6 of 17α-acetoxyprogesterone—led to a series of highly active progestins. Acetylation increases the lipid solubility and therefore extends the duration of activity of such derivatives, whereas the introduction of a methyl group on C-6 (and C-11) interferes with metabolic destruction of the drug

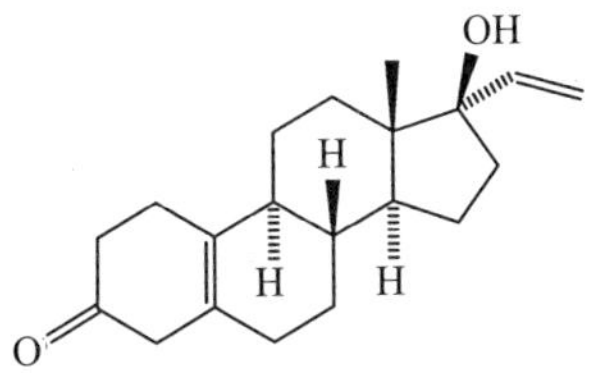

Levonorgestrel (5.40) 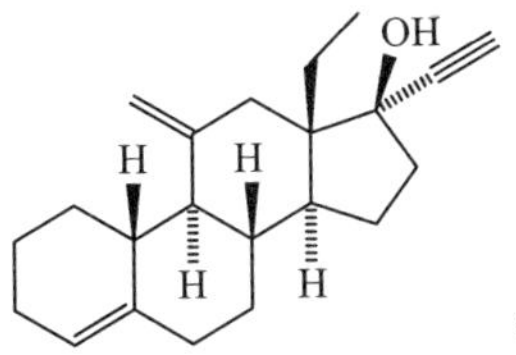　　　Desogestrel (5.41) 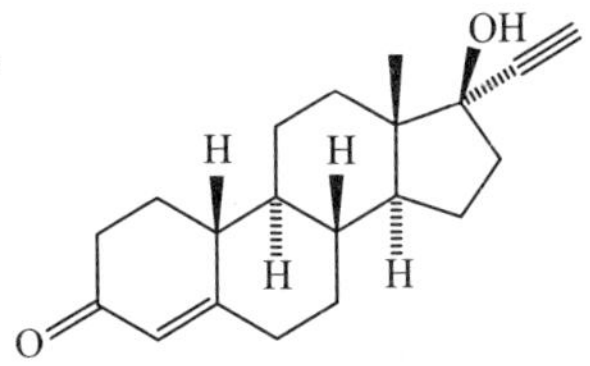　　　Norethindrone (5.42)

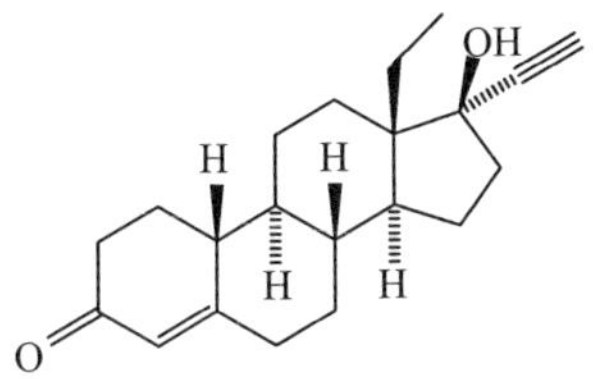

Norgestrel (5.43) 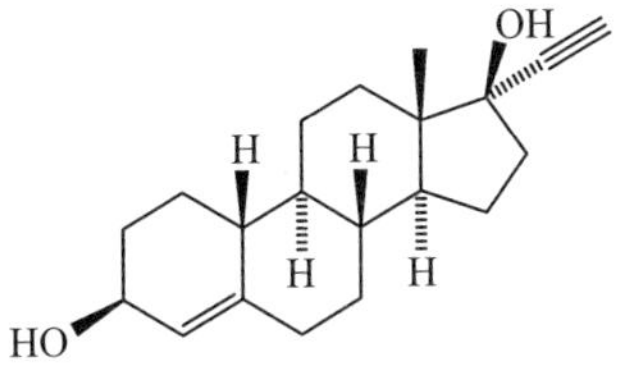　　　Ethynodiol (5.44)

and increases its oral potency. These medicinal chemistry advances were crucial to the design of oral contraceptive agents.

Oral contraception based on hormonal manipulation may be achieved in three ways, as follows.

5.9.1.1 Ovulation Inhibitors

To achieve inhibition of ovulation, a progestogen (e.g., levonorgestrel, norethindrone, norgestrel, norgestimate) and an oral estrogen (e.g., ethinyl estradiol, mestranol) are combined in varying amounts and/or at varying times during the menstrual cycle. The desired contraceptive effect is due to inhibition of ovulation through the hypothalamo-pituitary mechanism; administration of exogenous steroids during the first half of the menstrual cycle suppresses FSH production, thereby inhibiting maturation of ovarian follicles and preventing ovulation. Adding a progestogen allows the secretory phase of the endometrium to be elicited, and when the exogenous steroids are temporarily withdrawn, menstruation occurs. These combination estrogen/progestogen agents exert a secondary contraceptive effect via alteration of the viscosity of the cervical mucus to physically impede the sperm.

These combination oral contraceptives are formulated in different "phasic" preparations. Phasic contraceptives vary the progestin dose during the cycle, mimicking its variation under physiological conditions. Monophasic and triphasic formulations are among the most popular. A typical monophasic oral contraceptive contains the same dose of estrogen and progestogen (e.g., ethinyl estradiol 30 µg and levonorgestrel 150 µg) given on 21 days of the menstrual cycle with 7 days off. A typical triphasic combination oral contraceptive varies the doses of the steroid hormones over the course of the cycle (e.g., ethinyl estradiol 30 µg and levonorgestrel 50 µg [days 1–6]; ethinyl estradiol 40 µg and levonorgestrel 75 µg [days 7–11]; and ethinyl estradiol 30 µg and levonorgestrel 125 µg [days 12–21]; inert tablets [days 22–28]).

Of the various pharmacologic approaches to contraception, the combination pills are by far the most convenient and the most effective. In addition, several noncontraceptive benefits are recognized for combination oral contraceptives:

1. Reduced incidence of cancer of the endometrium
2. Reduced incidence of benign breast disease
3. Reduced likelihood of functional ovarian cysts
4. Less menstrual blood loss and more regular cycles
5. Decreased severity of premenstrual syndrome and improvement in acne and hirsutism (facial hair)

These potential benefits are balanced by a slightly increased risk of gallbladder disease, hypertension, myocardial infarction, cerebral infarction, and pulmonary embolism. The increased risk of stroke and heart attack associated with the "pill" is accentuated when compounded by other risk factors, including smoking, migraine headaches, and advancing age. Indeed, convincing data support an upper age limit of 35 years for oral contraceptive use by women who smoke.

5.9.1.2 Mini-Pill

A low dosage of progestin ("mini-pill") is used, in the form of medroxyprogesterone acetate, which is active at a very low dose. The mini-pill does not inhibit ovulation, but rather interferes with the endometrium and the cervical mucus. The use of this pill prevents most of the side effects of oral contraception, specifically nausea, water retention, and in some cases thrombophlebitis. However, a lower success rate and other frequent side effects have reduced the widespread acceptance of this preparation. Nevertheless, the mini-pill has a role to play in certain specific situations. For example, in an uncommon form of epilepsy called *catamenial epilepsy*, female patients will experience seizures at particular times during their menstrual cycle, reflecting the fact that seizure focus is stimulated by estrogens but inhibited by progestins. In such women, the mini-pill may afford not only birth control but also improved seizure control.

5.9.1.3 "Morning After" Pill

This agent refers to the administration of a high dose of both an estrogen and a progestin. Preferably, this is given 12–24 hr after coitus. The morning after pill induces menstrual bleeding which in turn prevents the fertilized ovum from implanting. Alternatively, implantation of the ovum can also be blocked with mifepristone, which is a progesterone and glucocorticoid receptor antagonist; this agent induces a noninvasive abortion in early pregnancy.

It is to be expected that these extremely widely used oral methods of contraception will eventually be replaced by luteinizing hormone analogs as antiovulatory agents, or possibly by immunological methods directed against human chorionic gonadotrophin (hCG), which has both LH and FSH activity. Prostaglandins also show promise, and reliable male contraception is also a future possibility.

5.10 STEROID HORMONES: SEX HORMONES—ANDROGENS

These steroids control the development of male characteristics: sperm production and growth of the sex organs (penis, testes), prostate, and seminal vesicles (their *androgenic*

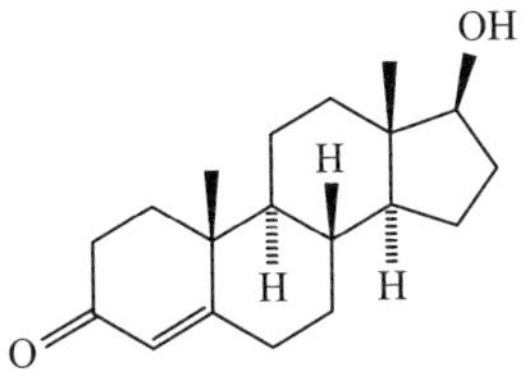

Testosterone (5.45)

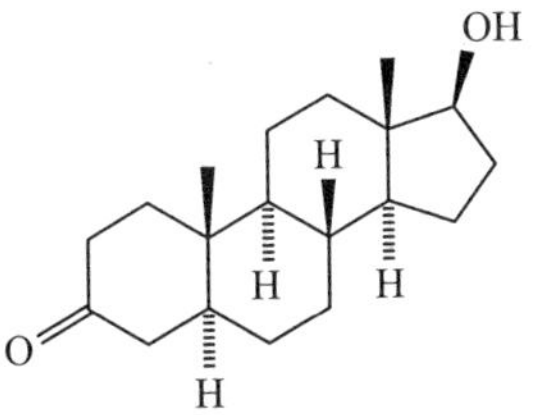

Dihydrotestosterone
(5.46)

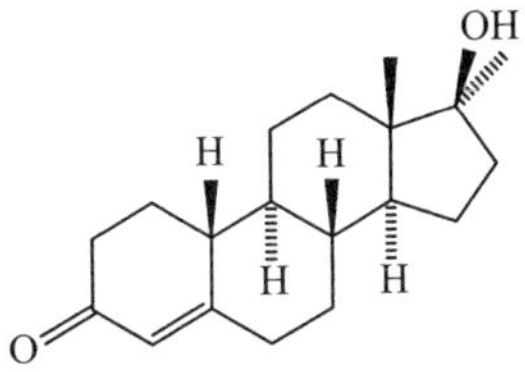

7α-methyl-19-nortestosterone
(5.47)

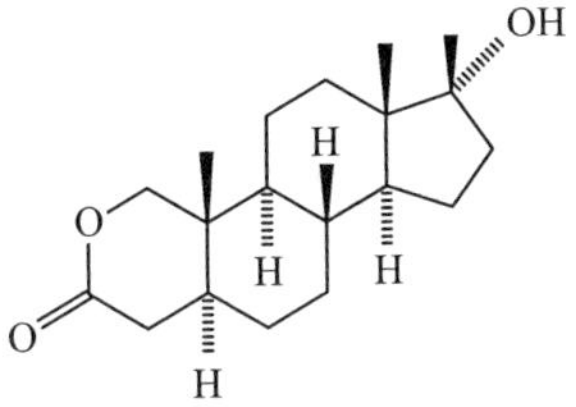

Oxandrolone (5.48)

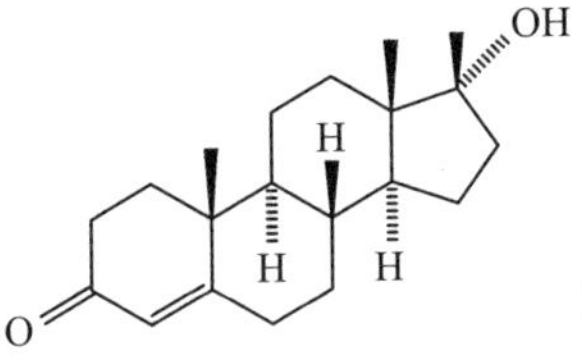

17-α-Methyltestosterone
(5.49)

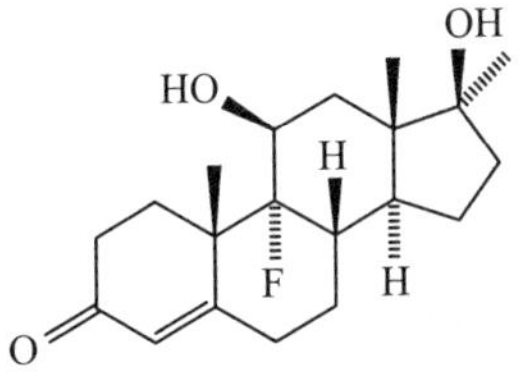

Fluoxymesterone
(5.50)

effect). Androgens also control metabolic effects during growth in adolescence and stimulate protein synthesis (causing "nitrogen retention") (their *anabolic effect*).

Testosterone (**5.45**) is the prototypic androgen sex hormone. Testosterone is synthesized by the testes, and must be reduced to dihydrotestosterone (**5.46**, DHT) before it will bind to the receptor. Among highly active synthetic testosterone analogs, 7α-methyl-19-nortestosterone (**5.47**) and oxandrolone (**5.48**) have about 100 times greater activity than testosterone as androgens. 17α-Methyltestosterone (**5.49**) is orally active.

The distinction of anabolic from androgenic action is important in the anabolic therapy of such wasting conditions as cancer, trauma, osteoporosis, and the effects of immobilization. These conditions necessitate nitrogen and mineral retention; however, the masculinizing effects of androgenic agents would be undesirable in female patients. Fluoxymesterone (**5.50**), the 9α-fluoro-11-hydroxy-17α-methyl derivative of methyltestosterone, has 10 times the androgenic and 20 times the anabolic activity of the parent compound, but is used mainly as an androgen in hormone-replacement therapy, hypogonadism, and some forms of breast cancer. Some other anabolic agents are methandrostenolone (**5.51**), and stanozolol (**5.52**), clostebol, and nandrolone. Use of anabolic steroids by athletes is a widespread but dangerous and banned misuse of these drugs. Nevertheless, their use does elicit protein anabolic effects and is thus associated with enhanced muscle bulk and more rapid recovery after injury. A nonvirilizing

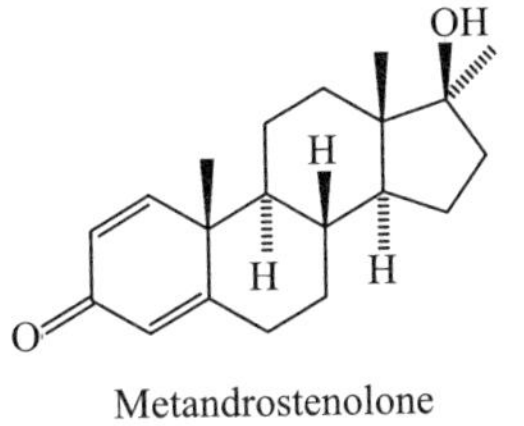

Metandrostenolone
(5.51)

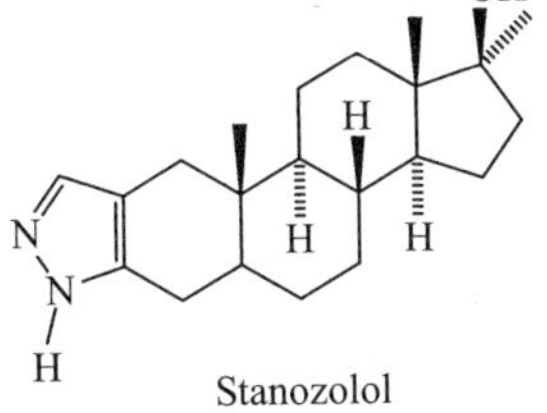

Stanozolol
(5.52)

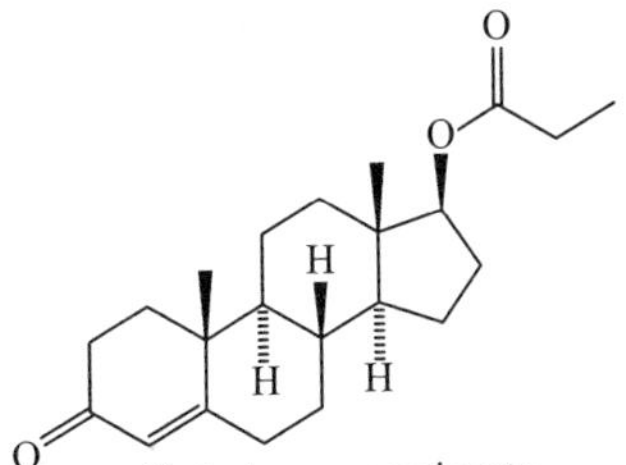

Testosterone propionate
(5.53)

androgen that has been useful in the treatment of breast cancer in young women is the 2-methyl derivative of testosterone propionate (**5.53**).

5.10.1 Androgens: Molecular Action

The receptor-mediated molecular action of androgens differs only slightly from the general steroid hormone model described previously, which is based principally on estrogens and progesterone. In the prostate, testosterone is reduced enzymatically to DHT, which then binds with high affinity ($K_D = 10^{-11}$ M) to a cytosol receptor. The binding of androgens to the cytosol receptor is a process requiring minutes to several hours. Molecular "flatness," especially in the A–B ring area, is a prerequisite for effective binding, and A–B *cis* compounds are therefore inactive; accessory binding sites for a 7α-methyl group seem to be present, which strengthens the binding; the 17β-hydroxyl group is essential for activity. The intracellular cytosol steroid receptor is bound to stabilizing proteins such as heat shock protein 90 (Hsp90); when the cytosol receptor binds to a molecule of steroid, the resulting steroid–receptor cytosol complex becomes unstable and releases the stabilizing proteins (Hsp90). The steroid–receptor complex then enters the nucleus, binding to a specific sequence of nucleotides (androgen response element) on a gene, thus regulating transcription by RNA polymerase II and associated transcription factors. The resulting mRNA is exported to the cytoplasm for the production of protein that enables the final androgen response. The genomic effects of androgens arise principally from proteins synthesized in response to RNA transcribed by a responsive gene; however, some of the genomic effects are indirect and are mediated by the autocrine and paracrine effects of autocoids such as growth factors, lipids, glycolipids, and cytokines produced by neighboring cells, likewise induced by hormonal action.

5.10.2 Antiandrogens

Antiandrogens such as cyproterone acetate (**5.54**) or the nonsteroidal flutamide (**5.55**, a substituted anilide) are competitive antagonists on the cytosol receptor. They do not prevent DHT formation; rather, they inhibit the nuclear retention of DHT in the prostate. They cause feminization in male fetuses and decrease libido in males. Cyproterone is also an active progestogen. In men, antiandrogens are used commonly in the treatment of prostatic cancer and uncommonly to inhibit sex drive in "hypersexuality"; in women, antiandrogens are used to treat virilization. Bicalutamide (**5.56**) and nilutamide (**5.57**) are potent, orally active antiandrogens that may be used in the treatment of metastatic prostate carcinoma.

Finasteride (**5.58**), a steroid-like enzyme inhibitor, represents an alternative approach to antiandrogens. Finasteride inhibits the 5α-reductase-catalyzed conversion of testosterone to DHT. In those tissues in which DHT is the active androgenic species (e.g., prostate), the androgenic stimulus is reduced; in those tissues in which testosterone is the active androgenic species (e.g., muscle), the androgenic influence is minimally affected.

5.10.3 Androgens and Male Contraception

Male contraception, involving the suppression of spermatogenesis, is an intensely investigated area, but there have been few positive results that could be used on a practical

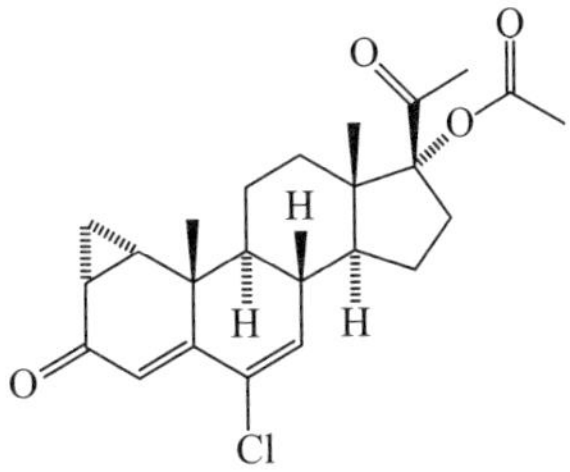

Cyproterone acetate (5.54)

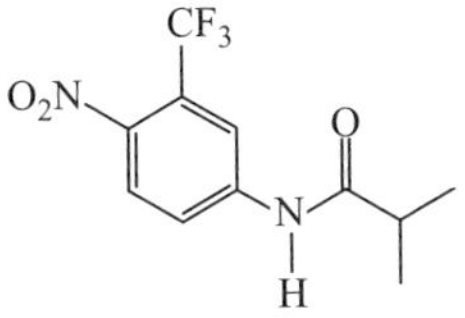

Flutamide (5.55)

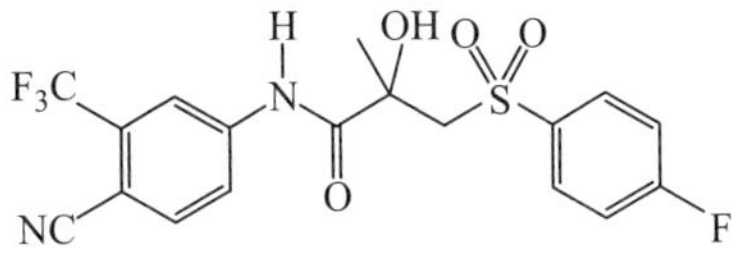

Bicalutamide (5.56)

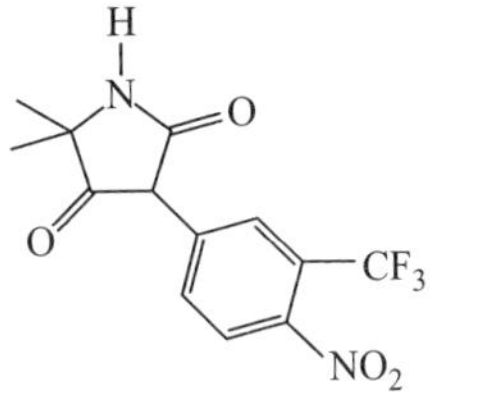

Nilutamide (5.57)

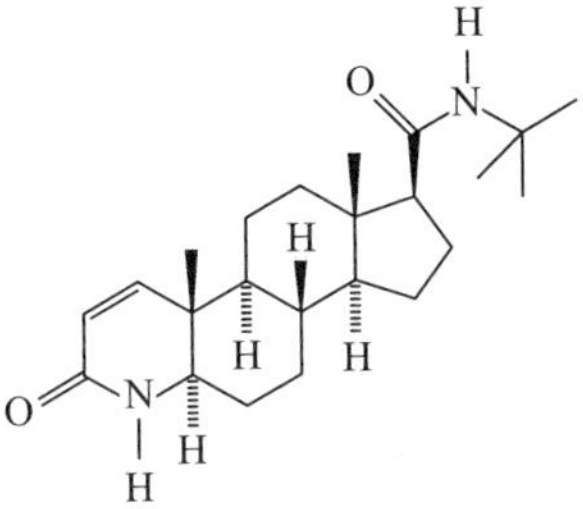

Finasteride (5.58)

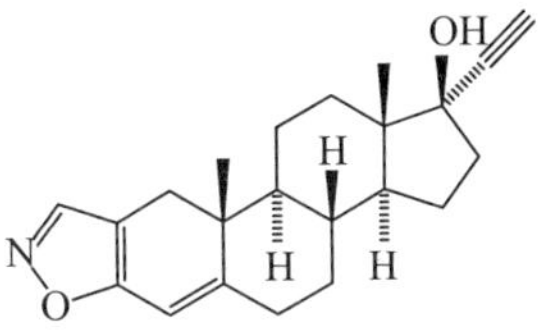

Danazol (5.59)

scale. Testosterone enantate injections produced azoospermia in only 65% of males in a pivotal multicentre study. Danazol (**5.59**), a gonadotropin inhibitor in females, is also only partially active in males, even if it is taken together with long-acting testosterone derivatives. Cyproterone acetate, the potent antiandrogen, likewise failed to yield reliable contraception. To date, no complete spermatozoan elimination has been achieved with male contraceptive agents, although arguably a total absence of sperm is probably not necessary for sterility.

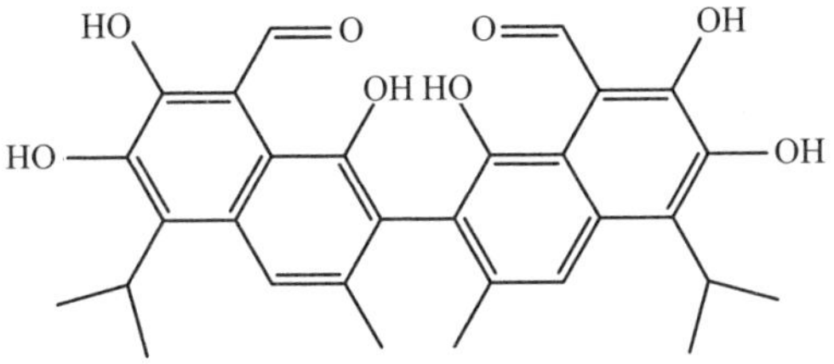

Gossypol (5.60)

Gossypol (**5.60**), a phenolic compound isolated from cottonseed oil, has direct antispermatogenic activity. Gossypol does not affect endocrine function but seems to act on the spermatid phase of spermatogenesis. In some studies it is 99% effective in reducing sperm counts. It has been most extensively studied in China. Unfortunately, it has considerable and only partly reversible side effects. The most noteworthy of these is the production of transient paralysis. Due to its toxicity, gossypol will never succeed as a drug molecule. Nevertheless, it may be an important "lead compound" around which to design improved agents. Gossypol does have interesting structural chemistry, existing as two different optical isomers. Despite the fact that the (+) and (−) enantiomers have differing bioactivities, most clinical studies have been done on racemic mixtures. The optical isomers arise from the restricted conformational rotation around the single bond connecting the two naphthalene ring systems.

5.11 STEROID HORMONES: ADRENOCORTICOIDS (ADRENAL STEROIDS)—OVERVIEW

The adrenal gland is a cap-like organ sitting at the upper pole of the kidney. Histologically, the gland consists of the inner *adrenal medulla* and the outer *adrenal cortex* (shell). In accord with this histological differentiation, there are two endocrine organs within the adrenal gland, one surrounding the other. The inner adrenal medulla secretes the catecholamines: epinephrine and norepinephrine. The outer adrenal cortex secretes corticosteroid hormones: glucocorticoids and mineralocorticoids. The adrenal cortex is divided into three zones: *zona glomerulosa* (outer), *zona fasciculata* (middle), and *zona reticularis* (inner). Although these three zones are variable in their anatomical distinctness, they are definite in their biochemical distinctness. All three cortical zones secrete corticosterone: the zona glomerulosa biosynthesizes aldosterone; the zona fasciculate and zona reticularis biosynthesize cortisol (and sex hormones as well).

Like most endocrine glands, the adrenal cortex is regulated by hypothalamo-pituitary peptides. The hypothalamus secretes *corticotropin releasing factor* (CRF) (see section 5.15.2.6), which controls the release of *adrenocorticotropin* (ACTH), a peptide consisting of 39 amino acids. Corticotropin secretion is under feedback regulation by the adrenal steroids, as well as being under the control of higher CNS centers; stress or epinephrine can also increase corticosteroid production. Adrenocortical secretion is controlled primarily by ACTH from the anterior pituitary, but mineralocorticoid secretion is also subject to independent control by circulating factors, of which the most important is *angiotensin II*.

On the basis of biochemical effects, the two groups of corticosteroids can be readily distinguished:

1. Glucocorticoids, which act on carbohydrate, fat, and protein metabolism
2. Mineralocorticoids, which regulate electrolyte balance through Na^+ retention

Since these two classes of corticosteroids are both important in medicinal chemistry and drug design, they will be discussed separately.

5.12 STEROID HORMONES: ADRENOCORTICOIDS—GLUCOCORTICOIDS

Glucocorticoids have important metabolic effects on carbohydrates, proteins, and lipids; they stimulate phosphoenolpyruvate carboxykinase, glucose-6-phosphatase, and glycogen synthase. Glucocorticoids promote gluconeogenesis by increasing the pyruvate carboxylase concentration, and thus increase the concentration of oxaloacetate in the mitochondrial pathway of pyruvate–phosphoenolpyruvate synthesis. Another effect appears to be the inhibition of pyruvate oxidation. Glucocorticoids stimulate the release of amino acids during the course of muscle catabolism. Glucocorticoids stimulate hormone-sensitive lipase and thus lipolysis. By promoting the supply of glucose from gluconeogenesis, inhibiting peripheral glucose uptake, stimulating lipolysis, and releasing amino acids from muscle catabolism, glucocorticoids ensure the supply of glucose to the brain, especially in times of fasting or starvation. As do all steroids, glucocorticoids increase the rate of enzyme synthesis in the nucleus of target cells and achieve their effect on overall protein/enzyme synthesis in this manner. The principal target of glucocorticoids is the liver, although other organs—notably the muscles and brain—are also rich in glucocorticoid receptors.

Most glucocorticoids have some mineralocorticoid effect, which is usually considered an undesirable activity. Through structural chemistry and structure–activity relationship (SAR) studies, molecular modifications can separate the two activities.

Glucocorticoids provide a valuable lesson in drug design. Since they influence so many enzymes in so many cell types, the pharmacological effects of glucocorticoids are likewise many and far reaching. However, so too are their side effects. If the drug designer is targeting a receptor that has widespread distribution and is not localized to a single tissue or cell type, the likelihood of unwanted side effects is concomitantly increased.

5.12.1 Glucocorticoid Receptor

The glucocorticoid receptor protein belongs to a superfamily of nuclear receptor proteins that includes steroid, Vitamin D, thyroid, retinoic acid, and other receptors that interact with promoters that regulate the transcription of target genes. Although the glucocorticoid receptor differs from the estrogen or progesterone receptors, the basic principles of its action seem to be the same. The glucocorticoid receptor is 800 amino acids in length and has three functional domains: the C-terminal has the glucocorticoid binding region, the middle portion has the DNA binding region (containing nine cysteine residues folded into a "two finger" structure stabilized by Zn^{2+} ions), and the N-terminal, which has the receptor-specific region. A gene for the classic glucocorticoid receptor has been identified. Binding to the glucocorticoid receptor is somewhat dependent on temperature, being optimal at 37°C.

Within the blood, the glucocorticoid steroid is bound to corticosteroid binding globulin (CBG); however, the steroid molecule enters the cell in an unbound state. Upon entering the cell, it binds to the glucocorticoid receptor, which is itself bound to two stabilizing proteins, including two molecules of heat shock protein (Hsp90). When the steroid binds, the complex becomes unstable and the Hsp90 molecules are released. The steroid–receptor complex is now able to enter the nucleus within the cell as an activated dimer. The complex then binds to a glucocorticoid response element (GRE) on a gene within the nucleus, thereby permitting regulation of transcription by the RNA polymerase II enzyme.

5.12.2 Glucocorticoids: Structure–Activity Correlations

The structure–activity relationships of glucocorticoids are based on two natural hormones, cortisol (**3.8**) and corticosterone (**5.61**). The characteristic structural features of these hormones are the conjugated 3-ketone, the 11-OH group, and the 17β-ketol side chain. Molecular modifications have been aimed at deriving compounds with glucocorticoid (and anti-inflammatory) actions but lacking in mineralocorticoid effects and side effects. Substituents added to the cortisol molecule may alter bioactivity and receptor binding affinity independently of other functional groups present on the molecule. While a high receptor binding ability does not necessarily reflect overall pharmacological activity, it is an important factor in glucocorticoid drug design. Two of these important modifications are as follows:

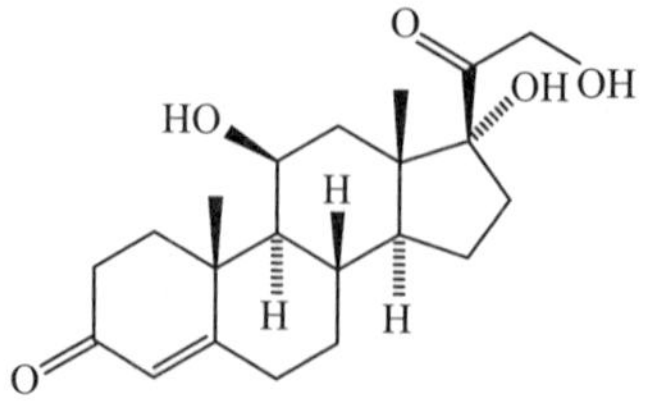

Cortisol (3.8) 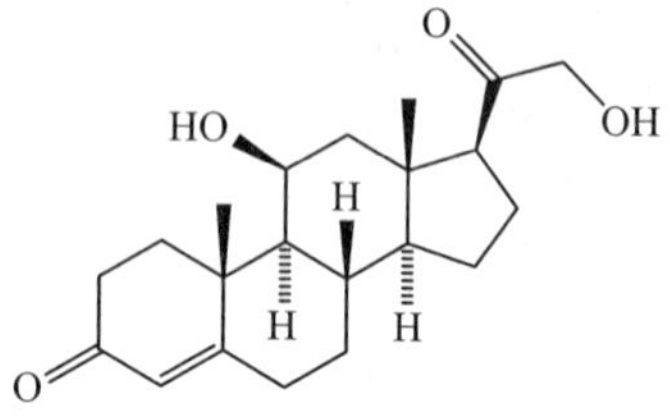   Corticosterone (5.61)

1. *Halogenation.* 9α-Fluorocortisone is only about 10 times more active than its parent compound, but its mineralocorticoid activity is 300–600 times greater. This is undesirable since it leads to edema; thus, the compound fludrocortisone (**5.62**) is used only topically, in ointments.

2. *Additional double bonds.* Δ¹-compounds (where Δ^1 indicates the position of a double bond) were introduced, like prednisone (**3.7**), a Δ^1-11-ketone, and prednisolone, its 11-hydroxy analogue. Changing the geometry of the A ring increased the potency without augmenting mineralocorticoid activity.

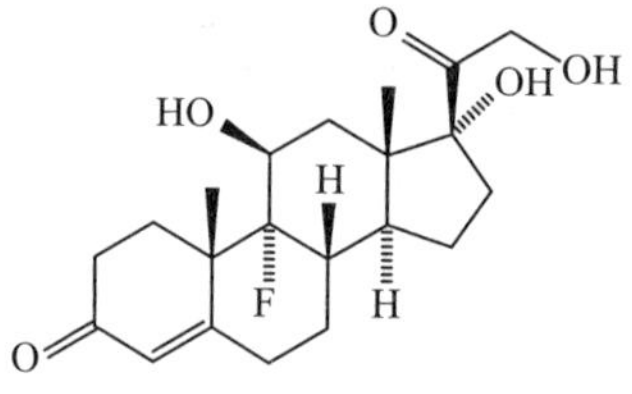

Fludrocortisone (5.62) 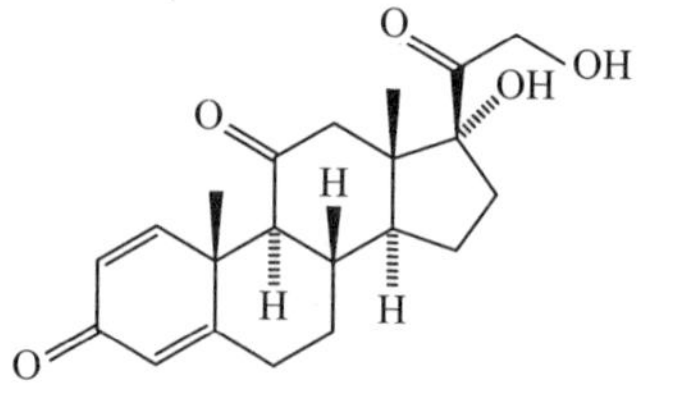   Prednisone (3.7)

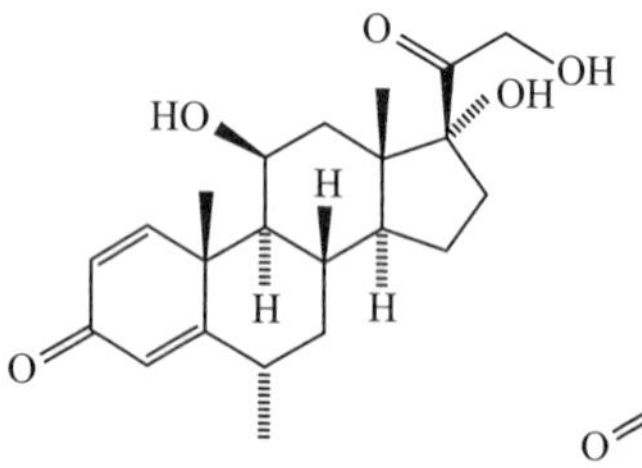

Methylprednisolone (5.63) 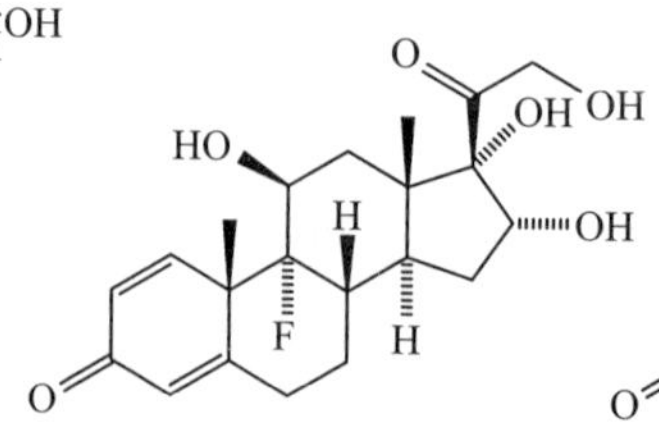   Triamcinolone (5.64) 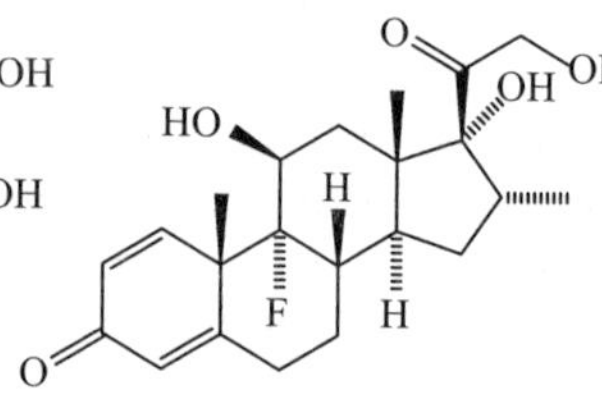   Dexamethasone (5.65)

The introduction of a methyl group in methylprednisolone (**5.63**) resulted in a slight increase in activity; however, the greatest improvements in activity came from the combination of a double bond, halogen, and methyl substituents. Triamcinolone (**5.64**), in the form of its acetomide (an acetone ketal), shows a 9α-fluoro group in addition to Δ^1 unsaturation, and a 16α-OH. It is used in treating psoriasis and other dermatological problems. Dexamethasone (**5.65**), perhaps the most active and highly stable glucocorticoid, is the 16α-methyl analog of triamcinolone. It is interesting to note that progesterone binds well to the glucocorticoid receptor despite a missing 11-oxygen functional group, but it fails to elicit gene activation in glucocorticoid target cells, thus shedding light on the role of the 11-OH group.

The optimum glucocorticoid structure shows a 1α, 2β-half-chair conformation for ring A, with ring D as a 13-envelope (C-13 is bent up) or a half-chair. Halogenation is most effective in positions 6, 7, 9, and 12. The compounds bind on their β face by hydrophobic binding forces.

5.12.3 Glucocorticoids: Pharmacological Activity

The most important clinical application of glucocorticoids and their semisynthetic analogs is their *anti-inflammatory* activity, discovered in 1949 by Hench and co-workers. The profound anti-inflammatory effects of glucocorticoids arise from the combined effects of these steroids on both the cellular and molecular mediators of inflammation; these effects are separate from the metabolic effects described above and further indication of the widespread diversity of macromolecules to which steroids can bind. Glucocorticoids suppress inflammation at the cellular level by downregulating the concentration, distribution, and function of *leukocytes* (white blood cells) that profoundly influence inflammation and response to infection within the body (In this way, steroids help to mediate the overlap between the endocrine systems [chapter 5] and the immune systems [chapter 6]). Glucocorticoids also suppress inflammation at the molecule level by suppressing inflammatory cytokines, chemokines, and other molecular mediators of inflammation.

From a clinical perspective, the anti-inflammatory effects of glucocorticoids are extremely important and in some cases life saving. While initially revolutionizing the symptomatic treatment of osteoarthritis and rheumatoid arthritis, these drugs did not provide a lasting cure for these crippling, painful, and widespread diseases. Moreover, any benefits in disorders such as arthritis were more than negated by the potential for severe long-term side effects that are inevitable with prolonged use of steroids. Despite these limitations, anti-inflammatory glucocorticoids have found clinical use in many other systemic inflammatory diseases (e.g., systemic lupus erythmatosus). Anti-inflammatory glucocorticoids have also been used to suppress pathological inflammation in various organ systems, including the heart (myocarditis, pericarditis), lungs (pneumonitis, allergic bronchitis, asthma), bowel (regional ileitis, Crohn's disease), and skin (dermatitis). In the case of skin symptoms, the external application of steroid drugs is possible, eliminating many systemic side effects. Patients suffering from acute leukemia and lymphoma (especially Hodgkin's disease) may also benefit from glucocorticoids. Anti-inflammatory steroids have been used to treat exacerbations of multiple sclerosis and to prevent strokes and blindness in patients with temporal arteritis (inflammation of the temporal artery blood vessel).

Glucocorticoids are also used as hormonal replacement therapy in patients in whom the adrenal cortex has been destroyed by some pathological process. The resulting syndrome of hypoadrenalism, called *Addison's disease*, is lethal if not treated. The symptoms of this disease are weakness, anemia, nausea, hypotension, depression, and abnormal skin pigmentation. Hyperadrenalism is the opposite of Addison's disease and is produced by an excessive circulating amount of glucocorticoids. Hyperadrenalism, or *Cushing's syndrome,* may result from adrenal tumors, which secrete glucocorticoid steroids, or from increased ACTH levels secreted by a tumor (microadenoma) of the pituitary gland that produces a specific clinical subtype of Cushing's syndrome, called *Cushing's disease.* Patients who have been taking therapeutic glucocorticoid treatment for inflammatory diseases for a prolonged period of time develop a similar medical problem, called *Cushingoid syndrome.* The symptoms of Cushing's syndrome, Cushing's disease, and Cushingoid syndrome are all essentially identical and consist of facial fattening ("mooning"), facial cheeks becoming red, acne, osteoporosis (with susceptibility to deterioration ["aseptic necrosis"] of the head of the femur), hypertension, weight gain, body mass redistribution (thin legs with a fat, pendulous abdomen), fat collection over the upper back ("buffalo hump"), reddish-purple striae over the skin ("stretch marks"), thinning of the skin with poor wound healing, and decreased resistance to infection.

As has been emphasized, the therapeutic administration of glucocorticoids is associated with side effects and complications. In addition to the Cushingoid syndrome described above, there are many other side effects of steroid use. These include loss of blood glucose control in a known diabetic, initiation of diabetes in a previously well person, worsening of infections (especially bacterial or fungal infections), psychiatric problems (psychosis, hypomania, or depression), peptic ulcers, *hypokalemia* (i.e., decreased concentration of K^+ in the blood, which could predispose to cardiac arrhythmias), and cataracts in the posterior subcapsular region of the lens within the eye. If used for less than two weeks, the likelihood of these side effects is uncommon. However, if used for a prolonged period of time, the possibility of side effects becomes increasingly common. Another problem that can occur if corticosteroids are used for more than two weeks is *adrenal suppression.* If a person is receiving steroids for an anti-inflammatory indication, the administration of the exogenous steroids may suppress the body's capacity to produce its own endogenous steroids, resulting in *iatrogenic* (i.e., "treatment induced") *hypoadrenalism.* Therefore, if corticosteroids have been administered for a prolonged time, they must be tapered off (using a gradually decreasing dose) rather than abruptly stopped.

5.12.4 Glucocorticoids: Inflammation versus Infection

The use of anti-inflammatory steroids also demands an appreciation of the differences between the processes of *infection* and *inflammation.* An infection is a pathological process whereby an exogenous agent (fungus, bacterium, virus, prion) invades the body, either locally or systemically, causing some form of injurious dysfunction. Inflammation, on the other hand, is the body's physiological endogenous response to an injury, be it caused by an infective agent or some other process (trauma, chemical). Although inflammation is a normal physiological response, it may become pathological under certain circumstances; for example, inflammation of the bronchi in the lungs due to inhalation of a toxic chemical may cause so much swelling in the bronchial lining that the flow of air into the lung is blocked, resulting in a pathological state. Infections are usually accompanied by inflammation. The reddened area around a skin abscess reflects the body's inflammatory response to the infection. If one puts topical steroids on such an infection, the redness disappears, reflecting suppression of the inflammation and giving the erroneous impression that the infection has been successfully treated; in truth, the infection is now spreading much more aggressively since the body's inflammatory defense system has been suppressed. Nevertheless, sometimes it is necessary to cautiously use anti-inflammatory steroids in the presence of an infection. For example, pneumonia may be associated with a dangerous inflammation of the airways, causing decreased air entry into the lungs. Under such circumstances, the combined use of an anti-inflammatory steroid with the appropriate antibiotic is justified.

5.13 STEROID HORMONES: ADRENOCORTICOIDS—MINERALOCORTICOIDS

Mineralocorticoids such as the natural hormone aldosterone (**5.66**) regulate electrolyte concentration by stimulating Na^+ retention in kidney cells. 11-Desoxycorticosterone (**5.67**) and fludrocortisone (**5.62**) are much less active, but are used for maintaining electrolyte balance in adrenal insufficiency. Aldosterone synthesis is probably regulated by ACTH and angiotensin (see section 5.21), a peptide hormone. This hormone has its own receptors in kidney cells. Hyperaldosteronism can play a role in high blood pressure. Aldosterone antagonists such as spironolactone (**5.68**) are therefore useful as hypotensive diuretic agents, because the increase in Na^+ excretion that they promote is always paralleled by an increased urine volume.

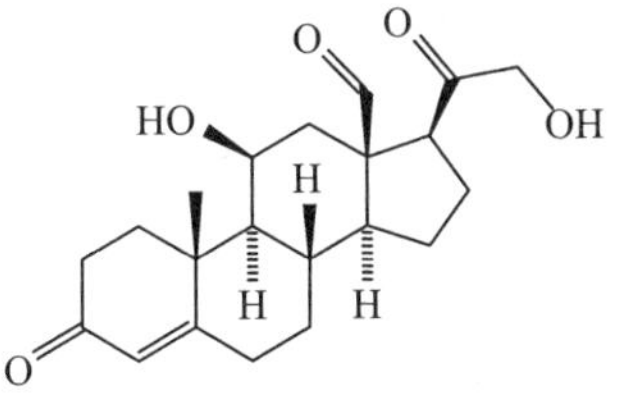

Aldosterone (5.66)

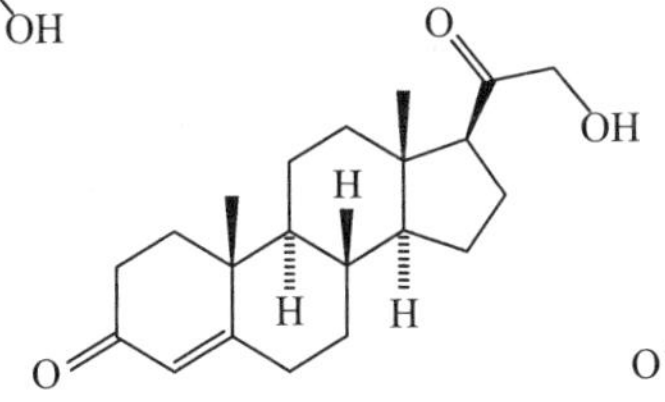

11-Desoxycorticosterone (5.67)

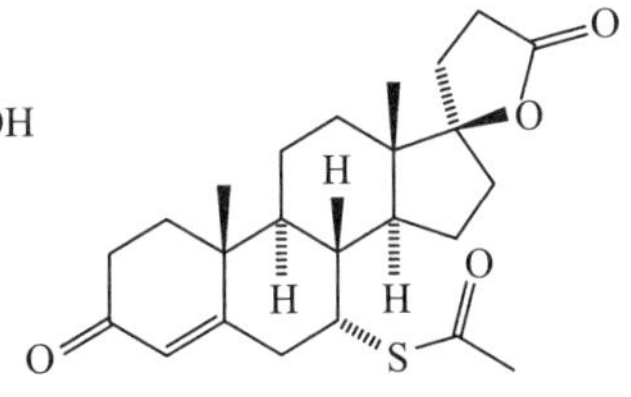

Spironolactone (5.68)

Structurally, aldosterone differs from glucocorticoids in having an 18-aldehyde group and lacking the 17α-OH. The aldehyde participates in a tautomeric equilibrium, forming a cyclic hemiacetal ring. Spironolactone, a synthetic compound and antiandrogen, has a lactone ring attached to C-17 through one common carbon (a "spiro" compound) and the 7-thiolester group. The structure of the mineralocorticoid receptor has been deduced through cloning of its cDNA. It shows considerable structural and functional similarity to the glucocorticoid receptor.

5.14 PEPTIDE HORMONES: INTRODUCTION

Peptides are polymers of amino acids. Each amino acid contributes three atoms (-N(-H)-Cα(-H,-R')-C(=O)-) to the *backbone* of the peptide. The side-chains of the individual amino acids then extend outwards from the core backbone. A *dipeptide* is composed of two amino acids; a *tripeptide* is composed of three amino acids; and so on. An *oligopeptide* is composed of a small number of amino acids; a protein is composed of a larger number of amino acids. Most peptide hormones are oligopeptides. Except for glycine, each of the twenty naturally occurring amino acids contributes *chirality* to the peptide by having *stereogenic centers* at each Cα atom. This enables stereospecific interactions between the peptide hormone and its receptor. This latter feature is invaluable in drug design since it permits analogs of peptide hormones to be designed with the capability of a unique stereospecific interaction with their receptors.

5.15 PEPTIDE HORMONES OF THE BRAIN

Although it is the dominant organ of the neural system, the brain also has an endocrine function, enabling the all-important overlap between neural and endocrine control systems. The most obvious and classically recognized hormonal function of the brain arises from the peptide hormones of the hypothalamus. The hypothalamus is intimately connected with the pituitary, producing the *hypothalamic–pituitary axis*. The hypothalamus is part of the brain; the pituitary, although located within the skull, is not part of the brain but is part of the endocrine system. Peptide hormones from the hypothalamus influence pituitary function and thus endocrine function throughout the body.

In addition to these classical hypothalamic peptide hormones, the brain also produces a number of other peptides that can be considered as peptide hormones. These peptides, or neuropeptides, are functionally difficult to categorize and have properties of neurotransmitters, neurohormones, and neuromodulators. One given peptide can exhibit one or more of these properties, depending upon its location in the nervous system. A neuropeptide may diffuse to an adjacent neuron to elicit a response, thus acting as a neurotransmitter; conversely, the same neuropeptide may also be transported to cells that are further away by the bloodstream before eliciting a response, thus acting as a hormone. These "multi-tasking" neuropeptides are found in the central nervous system, peripheral nervous system, and gastrointestinal tract.

The functions of CNS neuropeptide neurohormones seem to be multiple, because the possible variations and fine-tuning of signals offer nearly inexhaustible possibilities that may even satisfy the requirements of the enormous complexity of the homeostatic integration and control of feeding, temperature regulation, circulation, pain, learning, and many other behavioral and developmental challenges that continuously confront organisms. Our modest initial recognition of the immense complexity of neuroendocrine functions, combined with our rudimentary understanding of ion channel function, raises the hope that therein lies the key to future understanding of higher mental functions such as learning and memory.

5.15.1 Peptide Hormones of the Brain: Non-Hypothalamic

Since the mid-1970s, a major revolution has occurred in our understanding of neurotransmission and endocrinology, combining these two previously distinct disciplines into the single unified field of *neuroendocrinology*. Neuropeptides can mediate communication between neurons, either directly or indirectly. Thus neurons are capable of synthesizing peptides and using them for communication either over a short range (synaptically, i.e., neural) or over a long range (via the circulation, i.e., endocrine); conversely, endocrine glands produce compounds that can act as neurotransmitters as well as hormones. About 100 neuropeptides have been described, and the end is not in sight. Neuropeptides share a number of functional characteristics with small amine neurotransmitters: their release is normally Ca^{2+} dependent, and they operate through ion channels or second messengers. There are, however, a number of complicating factors peculiar to this all-important group of messengers.

We have already discussed the co-occurrence of small amine and peptide neurotransmitters: their release is normally Ca^{2+} dependent, and they operate through signal transmission. They are also capable of regulating each other's release and even the synthesis, clustering, and affinity of receptors. Neuroendocrine cells are capable of producing more than one peptide, and thus an amine–peptide as well as a peptide–peptide combination is possible. It is known, for instance, that the vagus nerve contains substance P, vasointestinal peptide, enkephalin, cholecystokinin, and somatostatin— peptides with a hybrid combination of neural and hormonal communication properties.

Our ideas about the selectivity of peptide neurohormones have also undergone profound development. Since most of these neurohormones act both centrally and peripherally, one has to surmise that the receptors in different organs are *isoreceptors,* in the sense of, say, the β_1- and β_2-adrenoceptors. Although the neuropeptide binds to both, the "command" executed will be appropriate to the receptor and the organ. In addition, different parts of the peptide may carry a different message, as in the peripheral pain mediation of substance P by the C-terminus and the central analgesia by the N-terminus. This hierarchy of selectivities is necessary in a neurohormone that is distributed by the blood circulation because all cells are equally exposed to its message. There is also a difference in the onset and duration of action between the ultrafast, small amine neurotransmitters and the slow but durable and persistent peptides. It seems that the two classes also differ in the average amount present in tissue (1–10 µmol/mg tissue for amines; fmol to pmol/mg tissue for peptides); this difference, however, need not involve the rate of turnover. Peptide concentrations can also vary by several orders of magnitude in different organs.

The metabolism of neuropeptides is not noticeably different from that of other proteins. Neuropeptides are unusual, however, in that the majority are synthesized in the form of *prohormones* that may contain several copies of smaller individual peptides, sometimes even of unrelated activity. These have to be modified (e.g., by glycosylation, disulfide bridge formation, methylation, etc.), cleaved by exo- and endopeptidases, and the fragments further modified (e.g., by C-terminal amidation, pyroglutamate formation). These processes vary from organ to organ; thus the same prohormone can undergo alternative post-translational modification, appropriate to the specific needs of the organ or species. Neuropeptides do not undergo reuptake like the majority of amine transmitters, but rather are eliminated by proteolysis. Protein synthesis must occur in the cell body of the neuron, and the protein must be packaged in the Golgi apparatus and transported by a fast transport system (3–5 µm/sec) to the synapse; thus we need to learn more about neuropeptide economies in terms of production and use. It is obvious that the complex nature of the neuropeptide and the variety of factors involved in neuroendocrine physiology provide drug designers with ample opportunities for interference with these processes.

A final interesting facet concerning neuropeptides/neurohormones concerns the existence of a family of "gut–brain" neuropeptide receptors. As numerous researchers have observed, the gut seemingly has almost as much neuronal activity going on as the brain. Peptides that function as either neurotransmitters or neurohormones within the CNS may exhibit similar properties in the gut. For example, receptors for calcitonin, vasoactive intestinal peptide, parathyroid hormone, secretin, glucagon, and growth hormone releasing factor are found both in the brain and in the gut. The representation of such receptors in both brain and gut must be considered during the drug design process, in which organ-specific delivery of the drug molecule is a desired therapeutic end-point.

5.15.2 Peptide Hormones of the Brain: Hypothalamic Neurohormones

The hypothalamus is located in the diencephalon in the upper portion of the brain stem. Parts of the hypothalamus produce a number of hormonal peptides that are distributed by the circulation (via the portal vein) and reach the pituitary gland (or hypophysis), situated immediately below the hypothalamus. The pituitary, unlike the hypothalamus, is not part of the CNS because it lies outside the blood–brain barrier. The hypothalamic peptides are hormones, but since they are secreted by neurons they can also be considered neurotransmitters, and some even fulfill the role of true neurotransmitters. Additionally, these hypothalamic hormones regulate the synthesis and release of other peptide hormones produced by the pituitary, and are thus called *releasing hormones* ("releasing factors" or "inhibitory factors", as the case may be). The release of these hypothalamic neurohormones is regulated by higher brain centers through cholinergic, dopaminergic, and GABAergic intervention; their synthesis is adjusted by feedback mechanisms from the target organs.

The correlation of the hypothalamus and its hormones with the hormones of the anterior pituitary gland is summarized in figure 5.2. There is no direct vascular connection between the hypothalamus and the posterior lobe of the pituitary that would correspond to the portal vein system for the anterior lobe of the gland. Not all hypothalamo-pituitary hormones will be discussed in the subsequent sections. Only those that are well-defined chemical entities or have a direct connection with drug action are considered.

5.15.2.1 Hypothalamic Releasing Factors

These peptides were isolated and their structure was elucidated primarily in the laboratories of Andrew Schally and Roger Guillemin, who shared the Nobel Prize in medicine for this work in 1977. Since the early 1970s neuroendocrinology has undergone an explosive development as a result. These hormones have an extremely high binding affinity, with a K_D ranging from 2 to 20×10^{-10} M. They act on the plasma membrane receptors of pituitary cells, triggering an energy-requiring process that probably involves Ca^{2+} and cAMP. Brain cell membranes also show the presence of receptors for these hormones, as would be expected in a system regulated by feedback. The hypothalamic hormones consist of *thyroliberin, corticoliberin, gonadoliberin, melanoliberin, somatocrinin* (growth hormone releasing factor), *somatostatin, prolactin*

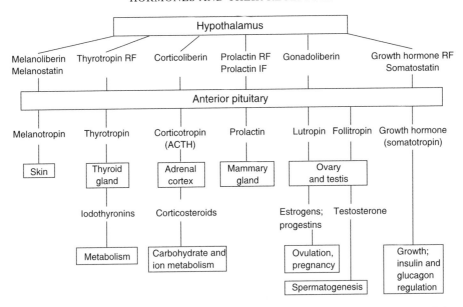

Figure 5.2 The Hypothalamic Pituitary Axis: The hypothalamus is part of the diencephalon within the brain. The pituitary, although located within the skull, is more correctly a part of the endocrine system than the nervous system. Together, the hypothalamus and pituitary form the interface between the nervous system and endocrine system and exert control over the majority of other hormone secreting organs. Releasing and inhibiting factors permit the hypothalamus to control the pituitary. Pituitary hormones are released into the general circulation, affecting metabolic function throughout the thorax and abdomen.

releasing factor, and *prolactin inhibiting factor*. A number of these hypothalamic hormones are important to the medicinal chemistry of drug design.

5.15.2.2 Thyroliberin

Thyroliberin (thyrotropin releasing hormone, TRH) was the first releasing factor to be isolated and synthesized. It has the simple tripeptide structure of pyroGlu-His-Pro-NH$_2$. The three rings on the peptide decrease the accessibility of the peptide bonds to hydrolysis by proteolytic enzymes and account for some oral activity, which is rare among peptide hormones. Nevertheless, the hormone is quickly inactivated *in vivo*. It is active in picogram amounts and liberates 200–2000 times its own amount of thyrotropin (thyroid stimulating hormone, TSH). Interestingly, it also promotes the release of prolactin, another pituitary hormone, even though there is a specific prolactin-releasing factor (PRF) in the hypothalamus; however, prolactin can also be released by numerous other substances.

There are indications that thyroliberin is neurotropic, acting as a neuromodulator or transmitter in the brain and spinal cord, and that it exhibits antidepressant activity.

Agonist analogs of TRH have been synthesized, showing that the π electron system and the basic imidazole ring are necessary for activity: the N-formyl-Pro-Met-His-Pro-NH$_2$ (**5.69**) has 40% of the full activity of TRH but is resistant to serum inactivation. Some analogs can differentiate between pituitary and CNS receptors: homo-pyroGlu-His-Pro-NH$_2$ (**5.70**) is equipotent to TRH in the pituitary but about 10 times more active in the CNS.

Antagonists of TRH have also been synthesized. For example, cyclopentylcarbonyl-thienylalanyl-pyrrolidine amide inhibits TSH release at high doses. Thyroliberin is used diagnostically only, to distinguish between hypothalamic and pituitary hypothyroidism.

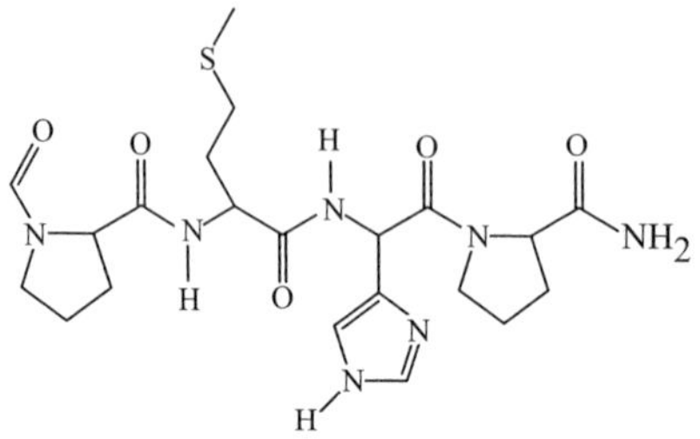

N-formyl-Pro-Met-His-Pro-NH$_2$ (5.69)

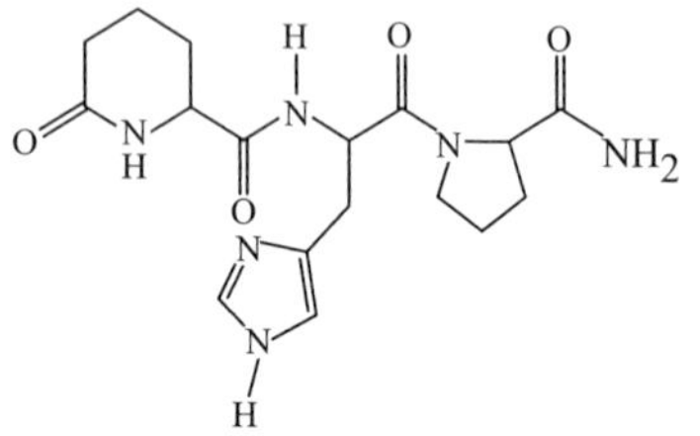

Homo-pyroGlu-His-Pro-NH$_2$
(5.70)

5-oxoPro-His-Trp-Ser-Tyr-D-Leu-Leu-Arg-ProNHC$_2$H$_5$
Leuprolide (5.71)

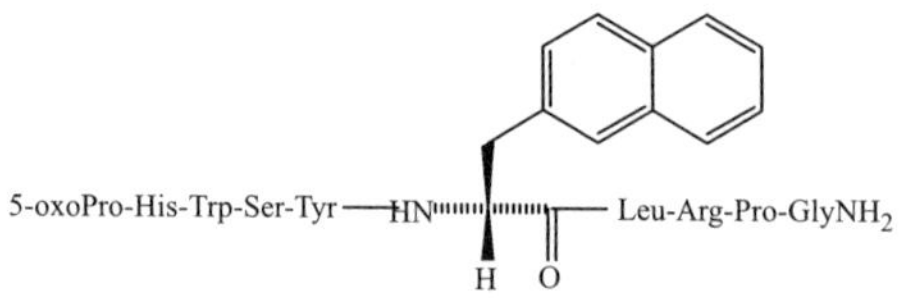

5-oxoPro-His-Trp-Ser-Tyr ——— HN‖‖‖‖‖‖‖‖‖‖‖ ——— Leu-Arg-Pro-GlyNH$_2$
H   O

Nafarelin (5.72)

5.15.2.3 *Gonadoliberin*

Gonadoliberin (gonadotropin releasing hormone, GnRH; luteinizing hormone-releasing factor, LHRH) releases gonadotropins, *lutropin* (LH) and *follitropin* (FSH). It was first isolated and synthesized by the Schally group and has an identical structure in all vertebrates: pyroGlu-His-Trp-Ser-Tyr-Gly-Leu-Arg-Pro-Gly-NH$_2$. Gonadoliberin does not release equal amounts of LH and FSH. Apparently, GnRH synthesis is under GABAergic control and acts through adenylate cyclase.

Agonist and "superagonist" analogs of GnRH have been prepared among the more than 1000 compounds investigated. The [D-Trp[6], Pro[9]-*N*-Et] and [D-Ala[6], *N*-Me-Leu[7]] analogs (where the superscript indicates the amino acid position in the original peptide) have a 150-fold greater activity than GnRH. Many of these highly active analogs have a long-lasting action because the D-amino acids in their structure prevent attack by proteolytic enzymes. Additionally, they are active not only parenterally (by injection) but also as nasal sprays or intravaginal suppositories. Some of the agonists that have been prepared include leuprolide (**5.71**), nafarelin (**5.72**), goserelin (**5.73**), and histrelin (**5.74**). An inspection of these structures suggests the structure–activity relationships. For example, D-amino acids at position 6 and ethylamide substituents for glycine at position 10 tend to increase duration of action and potency.

5-oxoPro-His-Trp-Ser-Tyr-D-Ser(t-Bu)Leu-Arg-Pro-NHNHCONH$_2$

Goserelin (5.73)

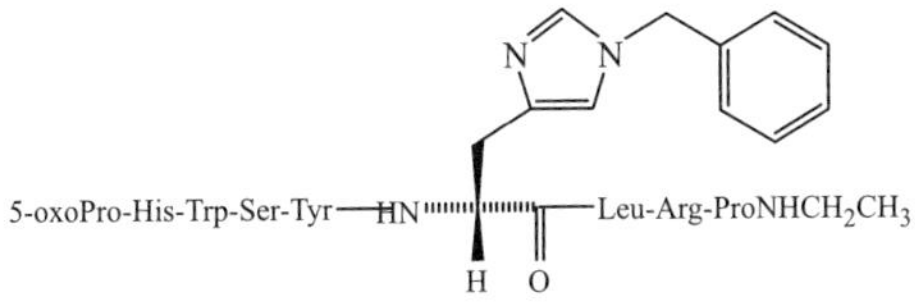

5-oxoPro-His-Trp-Ser-Tyr—HN''''''|''''''|—Leu-Arg-ProNHCH$_2$CH$_3$

H   O

Histrelin (5.74)

GnRH agonists have a number of clinical uses. They induce ovulation and spermatogenesis, increasing gonadotropin and sex-steroid levels. Therefore, GnRH agonists can be used to treat both male and female infertility. Paradoxically, they also inhibit spermatogenesis in rats after repeated administration over a prolonged period, thus allowing the potential for decreasing male fertility without decreasing libido; this means that GnRH analogs can be used either to increase or to decrease fertility in men or women. In addition to the treatment of fertility disorders, GnRH agonists can be used to treat prostate cancer, uterine fibroids, endometriosis, and polycystic ovary syndrome.

Gonadoliberin antagonists can be obtained by modifying the first three amino acids of the natural peptide. The N-terminal is considered the active center, whereas the rest of the molecule serves only in the binding process. The [D-Phe^2, Pro^3, D-Trp^6], [D-pyro-Glu^1, D-Phe^2, D-Trp^3, D-Trp^6] and similar compounds can block ovulation at doses of 200 to 750 µg.

5.15.2.4 Somatocrinin (Growth Hormone Releasing Hormone)

The *growth hormone releasing hormone* (or *factor*) (GHRH; *somatocrinin*) was isolated only in 1982. It consists of either 40 or 44 amino acids. In addition to stimulating growth hormone production it also stimulates release of *somatomedins*, which are responsible for the many anabolic effects of growth hormone. Its production by recombinant DNA techniques has made it possible to undertake physiological studies on its activity.

Tyr-Ala-Asp-Ala-Ile-Phe-Thr-Asn-Ser-Tyr-Arg-Lys-Val-Leu-Gly

NH$_2$Arg-Ser-Met-Ile-Asp-Gln-Leu-Leu-Lys-Arg-Ala-Ser-Leu-Gln

Sermorelin (5.75)

Growth hormone releasing peptides (GHRPs) are small synthetic peptide analogs of GHRH that stimulate GH secretion. Sermorelin (**5.75**, GHRH$_{1-29}$) is a clinically available acetate salt of the synthetic 29-amino acid N-terminal segment of GHRH. Even shorter peptide fragments have biological activity. GHRPs, such as sermorelin, may be used diagnostically to evaluate pituitary function in children with short stature, and may be used therapeutically to promote growth in children with short stature arising from a neuroendocrine dysfunction. Although GHRPs have not been clearly shown to cause malignancies, their capacity to induce long-term carcinogenesis is a theoretical possibility that has not been fully studied. Interestingly, GHRH antagonists have been shown to reduce the growth rate of human malignancies in nude mice.

5.15.2.5 Somatostatin

Somatostatin (SS, growth hormone release-inhibiting hormone, GH-RIH) is perhaps the best investigated and most important of the inhibitory factors produced by the hypothalamus. It is a cyclic tetradecapeptide isolated by the Guillemin group in 1973. It can be routinely produced by methods of genetic engineering that incorporate the human somatostatin gene into the genome of *Escherichia coli*, providing an inexpensive and efficient source of the hormone.

Somatostatin is very active at nanomolar levels, but is also very labile, and shows a half-life of only a few minutes when injected. It is hydrolyzed by endopeptidases between the Trp^8-Lys^9 residues, and the therapeutic applications of the native hormone are therefore rather restricted.

The principal activity of somatostatin is inhibition of the release of growth hormone (somatotropin) from the pituitary. Excessive growth hormone production leads to *acromegaly*, a form of gigantism, whereas its lack results in dwarfism. Since acromegaly is a relatively rare endocrine disease, other actions of somatostatin have received more attention—primarily its action on the pancreas. Somatostatin suppresses the release of the pancreatic hormones insulin and glucagon. According to present views, *diabetes* (more properly referred to as *diabetes mellitus* to distinguish it from *diabetes insipidus*) is a common endocrine disease caused by lack of insulin and a loss of insulin receptors— as well as a relative excess of glucagons—and causes hyperglycemia (an excessive blood glucose concentration), faulty glucose metabolism, lipolysis, and amino acid mobilization from proteins. In turn, this produces a number of chronic pathological effects including blindness, neuropathy, and accelerated atherosclerosis. The contemporary treatment of diabetes mellitus concentrates on the replacement of insulin or the administration of nonpeptidic hypoglycemic drugs. It is conceivable that administration of supplementary somatostatin (or of analogs thereof) to decrease the release of glucagon may also be used to favorably influence glucose metabolism. Since obesity is becoming an epidemic disorder, especially in North America, diabetes mellitus is likewise anticipated to achieve epidemic rates of occurrence in future years.

Somatostatin also reduces gastric acid secretion and has potential use in treating gastric ulcers. Additionally, being distributed throughout the CNS, mainly in the spinal cord, and found in nerve endings, it is assumed to be a neurotransmitter. It potentiates some of the effects of L-DOPA, induces sedation and hypothermia, and affects sleep patterns by inhibiting central epinephrine secretion. The suspected CNS targets include the extrapyramidal motor system and perhaps cognition.

In view of the many actions of somatostatin that potentially have great therapeutic importance, a very large number of synthetic analogs have been prepared, with particular attention being given to overcoming the impracticably short half-life of the native hormone.

Modifications designed to enhance the enzyme resistance and prolong the activity of SS derivatives have been quite successful. The use of D-amino acids instead of the normal L-enantiomers (e.g., in Trp^8), or replacement of the disulfide link by a nonreducible ethylene bridge, leads to an increased duration of activity, approaching 3 hours. Several analogs show a greatly increased effect, like the [D-Ala^2, D-Trp^8]somatostatin, which has 20 times the activity of SS on growth hormone release. The NH-terminal outside the cyclic dodecapeptide is not essential for activity. Selectivity of action results from

OH

OH

D-Phe-Cys-Phe-D-Trp-Lys-Thr-Cys ——NH

Octreotide (5.76)

manipulation of the cysteines in SS analogs. When Cys^3 is replaced by its D-enantiomer, insulin release is preferentially inhibited; when the L-Cys^{14} is replaced, an increase in the inhibition of glucagon secretion occurs. When the C-terminal cysteine forms a lactam with the N-terminal of Ala^1, resulting in cyclo-SS, growth-hormone repression becomes enhanced.

Of the various somatostatin analogs developed, octreotide (**5.76**) has been the most clinically effective. Octreotide is 45 times more potent than somatostatin in inhibiting GH release but decreases insulin release by only a factor of 2. It is useful in the treatment of acromegaly and in the treatment of hormone-producing tumors such as gastrinomas or glucagonomas. Radiolabeled octreotide has been used to localize neuroendocrine tumors that express somatostatin receptors. Gallstones will occur in 30% of patients receiving octreotide for more than 6 months.

5.15.2.6 Corticotropin Releasing Factor

Corticotropin releasing factor (CRF) is a neurohormone that has attracted considerable attention in the area of drug design. CRF exerts its actions by interacting with one of two subtypes of G-protein-coupled receptors (CRF_1, CRF_2), each being encoded by separate genes (cDNA encoding a third receptor has been isolated from catfish pituitary). All of these receptors possess seven putative transmembrane domains and are positively coupled to adenylate cyclase. Ongoing research is endeavoring to develop receptor-specific agents for four CRF receptors: CRF_1, $CRF_{2\alpha}$, $CRF_{2\beta}$, and $CRF_{2\gamma}$. Aiding this design process is the fact that the primary amino acid sequences of these various receptors are known. For example, the $CRF_{2\alpha}$ receptor has 411 amino acids and has 71% sequence homology with the CRF_1 receptor protein.

Over the past decade, studies with CRF peptide analogs have provided invaluable structure–activity relationship information. Deletion of residues from the N-terminal of CRF reduces potency but maintains efficacy. However, when leucine 8 is deleted, the resulting analogs start to behave as antagonists rather than agonists. Binding affinity of the peptide for its receptor(s) can be modified by alterations to residues 8–32. Such information is useful in the drug design strategy.

Since peptides have limited utility as drugs, the design and synthesis of low molecular weight non-peptidic ligands for CRF receptors is an active research area. These drug design and discovery programs employ peptidomimetic strategies for obtaining non-peptidic drugs from peptides (discussed in chapter 3) and combinatorial chemistry libraries with high throughput screening (also discussed in chapter 3). These non-peptide ligands tend to have a core aromatic heterocycle with appended hydrogen bonding acceptors and other functional groups with binding potential appended to it. Five-membered (pyrazoles, imidazoles) and six-membered (pyrimidines, pyridines) heterocycles have been explored as potential structural cores.

These various CRF-active agents may have clinical application in a variety of disorders. Selective small-molecule CRF_1 antagonists have been shown to reverse anxiogenic (anxiety-producing) effects in experimental models. Such compounds may be useful in the treatment of anxiety and agitated depression. CRF_1 and CRF_2 antagonists may be useful in the treatment of stress-induced gastrointestinal dysfunction and in other bowel motility disorders such as irritable bowel syndrome. Such agents may also be useful in treating stress-induced relapse of drug abuse as well as the anxiogenic behaviors that occur during acute drug or alcohol withdrawal. Finally, CRF antagonists may be of value in the clinical management of stress-related alterations in food intake and weight gain.

5.16 PEPTIDE HORMONES OF THE ANTERIOR PITUITARY

The pituitary is one of the most important endocrine organs. Hormones are produced by both the posterior pituitary and the anterior pituitary; the posterior pituitary produces *oxytocin* and *vasopressin*. The anterior pituitary is directly influenced by the hypothalamic hormones. The anterior pituitary hormones include: *thyrotropin* (TSH, which influences the ability of the thyroid gland to produce other hormones called iodothryonines); *corticotropin* (ACTH, which influences the ability of the adrenal cortex to produce corticosteroids); *lutropin* and *follitropin* (which influence the capacity of the ovary and testis to produce estrogen, progestins and testosterone); *somatotropin* (growth hormone, modulating insulin and glucagon and influencing growth); *prolactin* (which influences the mammary glands); and *melanotropin* (which affects skin pigmentation). A number of these pituitary hormones are important to the medicinal chemistry of drug design.

5.16.1 Gonadotropins

The gonadotropins are produced by the anterior pituitary (adenohypophysis) and the placenta. This group of glycoproteins (carbohydrate-containing proteins) includes the following hormones:

1. Lutropin (LH, luteinizing hormone, which in the male is called interstitial cell-stimulating hormone, ICSH)
2. Follitropin (FSH, follicle-stimulating hormone)
3. Human chorionic gonadotropin (hCG)
4. Human menopausal gonadotropin (hMG)

These are very complex hormones, having molecular weights of around 28,000. The α subunit consists of 89 amino acids; the β subunit consists of 115 amino acids in LH and FSH, and of 145 in hCG. While the α subunits of all gonadotropins are very similar in their amino acid sequence, the β subunits of the various hormones are quite different. The carbohydrate portions of both the α and β subunits contain oligosaccharides attached at specific amino acids (Asn), which branch at a mannose group and contain galactose, glucosamine, galactosamine, and acetylneuraminic acid residues. The carbohydrate portion of the hormones influences their biological and immunological properties as well as their stability.

The gonadotropins are released in a pulsed manner. Lutropin and follitropin act together to regulate ovarian functions, egg maturation, and follicular transformation to the *corpus luteum* in females. In the male, spermatogenesis depends on these hormones. Ovarian and testicular steroids are also produced as a result of gonadotropin action, and these in turn have a feedback regulatory effect on the hypothalamus and pituitary. Human chorionic gonadotropin, produced by the placenta, shows LH activity and is more stable than the other gonadotropins.

Therapeutically, gonadotropins are used to induce ovulation in infertile women. The antiestrogens clomiphene (**5.37**) and tamoxifen (**5.38**) are also used for this purpose since they counteract the ovulation-inhibitory effect of estrogens. Pregnancy tests depend on the presence of an increased hCG concentration in the urine after fertilization.

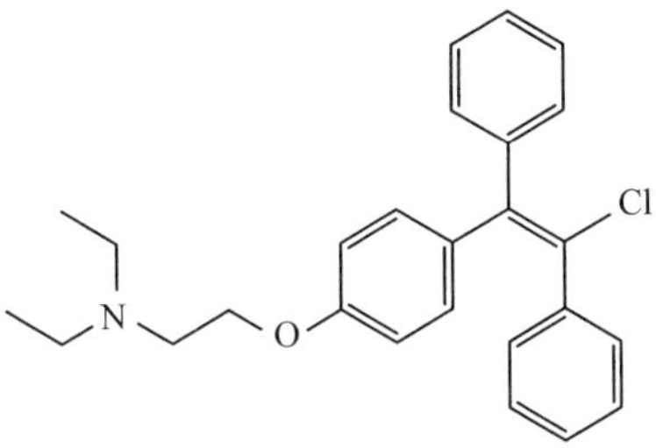

Clomiphene (5.37) 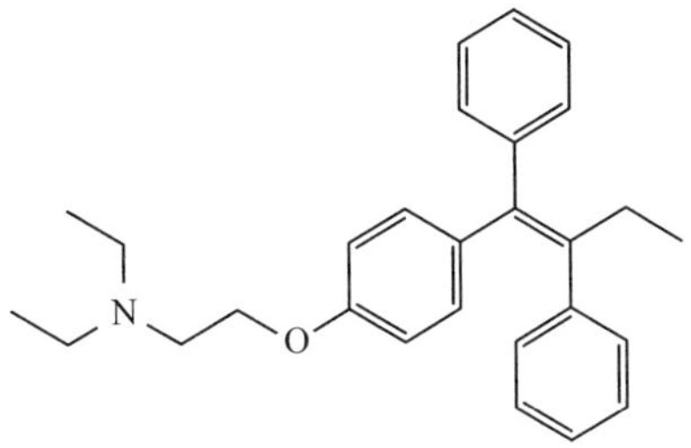   Tamoxifen (5.38)

5.16.2 Corticotropin

Corticotropin (adrenocorticotropic hormone, ACTH) regulates the function of the adrenal cortex and has numerous other effects on metabolism. It contains 39 amino acids in the form of a random coil, owing to the presence of several proline residues that prevent helix formation. Species differences are seen in amino acids 25–39; the rest of the ACTH molecule is identical in all animals and humans. The first 24 amino acids are responsible for all of the biological action of ACTH; synthetic human $ACTH_{1-24}$ is known as cosyntropin.

Since a large number of ACTH analogs have been synthesized, the information contained in the molecule has been analyzed in detail. The receptor binding core seems to reside in positions 15–18, whereas steroid synthesis is regulated by the sequence of amino acids 6–13. The N-terminal 13 residues are identical to melanocyte-stimulating hormone, found in animals but not in humans. The $ACTH_{4-10}$ fragment has shown remarkable behavioral effects in humans: it acts as a stimulant and restores optimal performance during long, monotonous tasks; in rats, it is a positive reinforcer in self-administration experiments. In humans, the N-terminal 24 residues of ACTH have the following biological activities:

1. A direct effect on the adrenal cortex: the regulation of steroid synthesis
2. Indirect effects mediated by the adrenal gland: thymus involution and an increase in glucose utilization
3. Extra-adrenal effects: melanotropic hormone and growth hormone release; lipolytic action; and influencing such neurobehavioral effects as stretching and yawning

The primary effect of ACTH seems to be mediated via cAMP production by interaction of the hormone with differing populations of receptors. The difference between the extent of steroidogenesis and cAMP formation indicates different receptor affinities in different organs.

Clinically, ACTH stimulation of the adrenals is used diagnostically to detect adrenal insufficiency; plasma cortisol levels are measured before and 60 minutes following an intravenous injection of cosyntropin. Adrenocortical insufficiency is known as Addison's disease; Addison's classic description, in 1855, namely "general debility, remarkable feebleness of the heart, irritability of the stomach, and a peculiar change of the colour of the skin", summarizes the clinical features of this disease, which is uniformly fatal if undetected and untreated. Therapeutically, corticotropin therapy has been essentially abandoned in favor of the direct administration of glucocorticoids. However, ACTH is still rarely used in the treatment of the infantile spasm seizure disorder.

5.17 PEPTIDE HORMONES OF THE POSTERIOR PITUITARY

5.17.1 Oxytocin (OT)

Oxytocin (OT) is a nonapeptide in which six amino acids form a ring closed by a disulfide bridge, while the ring itself forms an antiparallel "pleated sheet." The "tail" portion of the peptide, composed of Pro-Leu-Gly-NH$_3$, is also rigidly held in a folded conformation. Oxytocin causes the powerful contraction of some smooth muscles and plays a vital role in milk ejection (not to be confused with milk secretion, which is regulated by prolactin). It also has uterotonic action, contracting the muscles of the uterus, and is therefore used clinically to induce childbirth.

5.17.2 Vasopressin (VP)

Vasopressin occurs in two variations: arginine-vasopressin (AVP) and lysine-vasopressin (LVP), in which Arg^8 is replaced by Lys. The conformation of these hormones is almost identical to that of oxytocin, except that the terminal "tail" is conformationally free and not held by the ring. The physiological role of the vasopressins is the regulation of water reabsorption in the renal tubules (i.e., an antidiuretic action). In high doses, they promote the contraction of arterioles and capillaries and an increase in blood pressure; hence the name of these hormones. Because of their very similar structures, OT and VP overlap in a number of effects.

5.17.3 Structure–Activity Correlations of OT and VP

The elucidation of the conformation–activity relationships of OT was of the utmost importance in the design of highly active analogs of these hormones. Amino acids 3, 4, 7, and 8 are not involved in the hydrogen bonding that determines the ring conformation, and can therefore act as sites of binding to the oxytocin receptor. The Tyr^2-hydroxyl group, the intact hexapeptide ring, and the amide of Asn^5 are essential for the biological activity of this hormone. Therefore, the "corner" amino acids 3, 4, and 8 can be varied, yielding more selective compounds: the [Thr^4,Gly^7] OT has an oxytocin/antidiuretic activity ratio of 135,000:1, compared to the 200:1 ratio of the natural hormone. On the other hand, the [1-deamino, D-Val^4, D-Arg^4] VP has a 125,000:1 antidiuretic/pressor activity ratio (AVP has a 1:1 ratio). A few oxytocin inhibitors have also been prepared, such as the [3, 5-dibromo-Tyr^2] OT, which inhibits the uterotonic effect of the hormone.

5.17.4 Clinical Applications of OT and VP

The clinical applications of OT and vasopressin are widespread.

5.17.4.1 Clinical Uses of Oxytocin

Oxytocin is used to induce labor in childbirth and to promote the expulsion of the placenta, although the antidiuretic activity of the native hormone is a disadvantage. Other drugs with uterotonic activity include the *ergot alkaloids*. Ergot is from the fungus *Claviceps purpurea*, which infects cereals, mainly rye. A number of indole alkaloids have been isolated from this source, in which the indole moiety is lysergic acid. The latter forms amides with both cyclic tripeptides (e.g., ergocristine (**5.77**)) and with the amino-alcohol L-alaninol in ergonovine (**5.78**). The peptide alkaloids have a slow and cumulative action, whereas the water-soluble ergonovine and its derivatives are fast acting. The latter is used to prevent postpartum hemorrhage by the compression of uterine blood vessels through uterine muscle contraction. Some of these alkaloids are α-adrenergic blocking agents and have been used with moderate success in the treatment of migraine headaches.

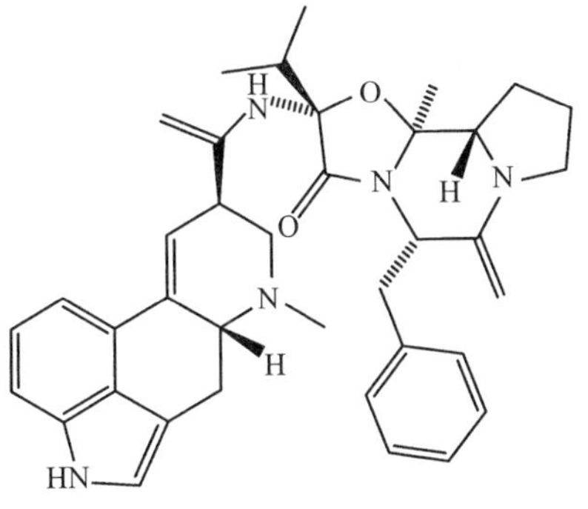

Ergocristine (5.77) 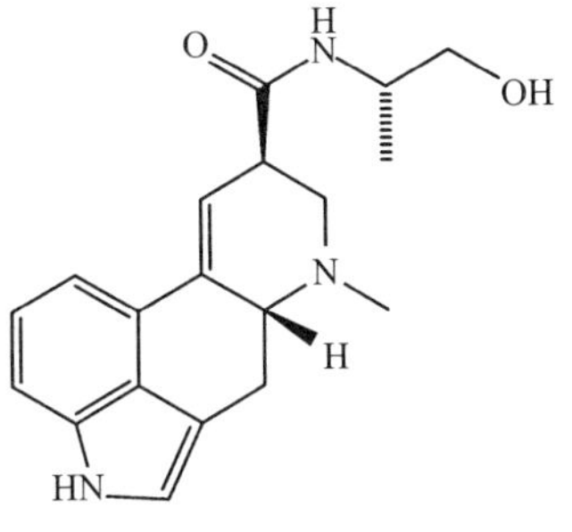   Ergonovine (5.78)

5.17.4.2 Clinical Uses of Vasopressin

The treatment of *diabetes insipidus* is the most logical indication for vasopressin use. Diabetes insipidus (not to be confused with diabetes mellitus, which arises from a deficiency of insulin—frequently from pathology of the pancreas), is the disorder arising from a deficiency of vasopressin—frequently from pathology of the posterior pituitary. From a symptom perspective, both diabetes insipidus and mellitus are associated with polyuria (passing large amounts of urine) and polydypsia (drinking large amounts of fluid).

Diabetes insipidus is treated with highly active synthetic vasopressin analogs. Desmopressin acetate (**5.79**, DDAVP) is synthetic 1-desamino-8-D-arginine vasopressin and is the synthetic vasopressin analog of choice for the treatment of diabetes insipidus. Desmopressin may be given intranasally, intramuscularly, or intravenously. Lypressin (**5.80**, 8-L-lysine vasopressin) and felypressin (**5.81**, 2-L-phenylalanine-8-L-lysine vasopressin) are two other, less clinically successful, synthetic vasopressin analogs. These synthetic agents have largely supplanted the use of pitressin (**5.82**), a sterile water extract from the posterior pituitary of healthy domestic animals slaughtered at an abattoir.

$$O$$

$''$ CH$_3$CO$_2$H

S

Tyr-Phe-Gln-Asn-Cys-Pro-D-arg-Gly-NH$_2$

Desmopressin Acetate (5.79)

Cys-Tyr-Phe-Gln-Asn-Cys-Pro-Lys-Gly-NH$_2$

Lypressin (5.80)

Cys-Phe-Phe-Gln-Asn-Cys-Pro-Lys-Gly-NH$_2$

Felypressin (5.81)

Cys-Tyr-Phe-Gln-Asn-Cys-Pro-Arg-Gly-NH$_2$

Pitressin (5.82)

In addition to the treatment of diabetes insipidus, vasopressin receptor modulators may have utility in a variety of other disorders. Vasopressin interacts with at least three G-protein-mediated receptor subtypes: V_{1a}, V_{1b}, and V_2. V_{1a} receptors are located predominantly in vascular smooth muscle and blood platelets, contributing to vasoconstriction and platelets aggregation; V_{1b} receptors are in the anterior pituitary and influence the release of ACTH and β-endorphin; V_2 receptors are localized in the kidney and regulate a subset of transmembrane water channels, called aquaporin-2 channels. In accord with this surprising diversity of receptor distribution, vasopressin modulators have been suggested as therapeutics for a variety of common disease states. V_2 and V_{1a}/V_2 mixed receptor antagonists have been targeted for the treatment of congestive heart failure and hypertension (congestive heart failure is a disorder in which the heart begins to fail as a mechanical pump, resulting in both "forwards failure", in which the heart fails to meet the blood perfusion requirements of the body, and "backwards failure", in which the heart cannot cope with the influx of fluid that it receives, causing fluid to pool in either the lungs [pulmonary edema] or the legs [peripheral edema]). V_{1a} receptor antagonists have been targeted for the treatment of dysmenorrhea (difficult and painful menstruation) and glaucoma (disease of the eye characterized by increased intraocular pressure arising either from excess production or decreased drainage of the aqueous humor within the eye). Vasopressin is a local neurotransmitter used by the hypothalamus and has also been implicated in learning and memory. Work on the development of non-peptidic receptor specific vasopressin mimics and blockers is an active area of research.

5.18 PEPTIDE HORMONES OF THE OPIATE SYSTEM

Relief from pain has been an age-old aspiration of humankind. Natural substances—opium alkaloids from the latex of the poppy (*Papaver somniferum*, the "sleep-bringing poppy")—have been used since ancient Chinese and classical Greek times to modify pain perception, but also misused for their euphoric effect. The opium alkaloids, called opiates and including morphine, are centrally acting major *analgesics* (pain relievers) that have a strong narcotic action, producing sedation and even loss of consciousness. In contrast, the over-the-counter pain relievers like aspirin or acetaminophen are non-narcotic and are better called minor analgesics.

At the time of the discovery of opiate receptors in 1973, it was assumed that only opiates of plant origin existed. Thus, the presence, in animals, of very specific receptors for a substance of plant origin was puzzling. Furthermore, these receptors were also found in organs and brain regions not implicated in pain perception. The answer to these problems started to emerge in 1975, with the discovery of the endogenous opioid peptides, the natural analgesics of animal organisms whose receptors, fortuitously, also bind opiate alkaloids. There are several types of such peptides: the large *endorphins (endogenous morphins)* isolated from the pituitary; and the small peptides, most importantly the pentapeptide *enkephalins* [*kephalos* (Gr.) = "head"]. It eventually became evident that the opioid peptides are neurohormones that are involved not only in pain perception, but in a number of other physiological activities as well. On the basis of their many neuronal effects, opioid peptides could also be considered peptide neurotransmitters. That idea, however, might be an oversimplification; hence this topic has been considered in this chapter, which deals with hormones.

5.18.1 Endogenous Opioid Peptide Hormones

In 1974, Liebeskind showed the existence of a central pain-suppressive system, and was able to produce analgesia by electrical stimulation of the periventricular gray matter within the brain. This *electroanalgesia* could be reversed by opiate antagonists, and showed a cross-tolerance with morphine-induced analgesia. These results indicated the existence of a neuronal system that uses an endogenous neuromodulator or neurotransmitter with opiate-like properties.

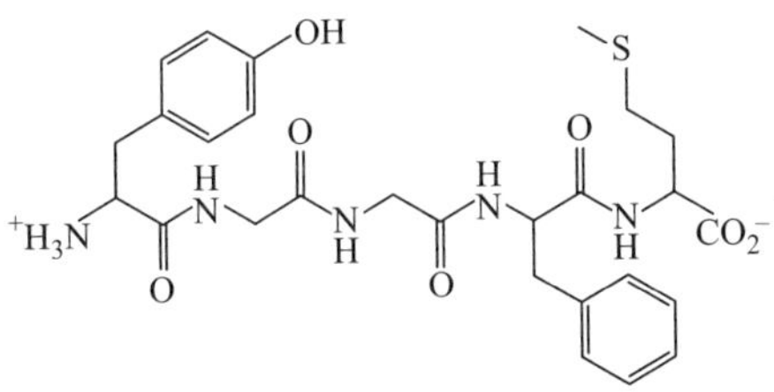

Met-enkephalin (5.83) 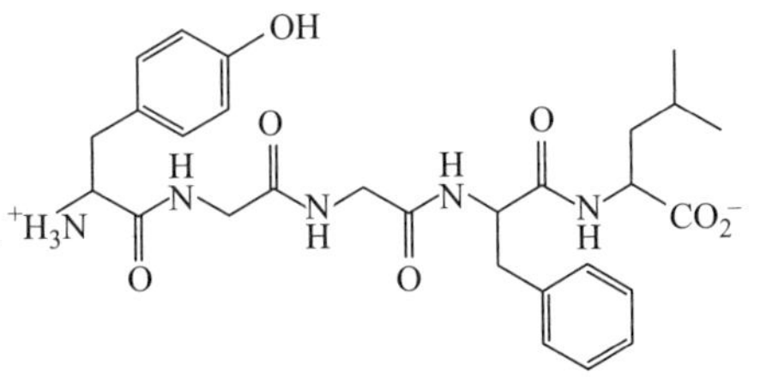   Leu-enkephalin (5.84)

The isolation of such endogenous opiates was reported simultaneously by four laboratories: those of Goldstein in Palo Alto, Hughes in Aberdeen, Snyder in Baltimore, and Terenius in Uppsala. Acetone extracts of pig, calf, and rat brains yielded, after purification, two pentapeptides, called enkephalins, with the structures NH$_2$-Tyr-Gly-Gly-Phe-Met-COOH (**5.83**, Met-enkephalin) and NH$_2$-Tyr-Gly-Gly-Phe-Leu-COOH (**5.84**, Leu-enkephalin). These are present in a 4:1 ratio in pig brain but in a 1:4 ratio in cattle brain. Since the genetic code for methionine (AUG) differs by only one base from one of the leucine *codons* (CUG), a point mutation or genetic drift may account for this difference. The ratio of the two enkephalins also varies in different brain regions.

The Goldstein group also isolated from the pituitary gland a larger peptide that proved to be identical to a fragment of β-lipotropin, a pituitary peptide hormone with a questionable physiological role. This peptide was called β-endorphin, and was found to be almost 50 times more analgetic than morphine if injected directly into the brain. Other endorphins and a smaller peptide, dynorphin, were also found in the pituitary. The heptapeptide dermorphin (**5.85**) was isolated from the skin of the frog *Phyllomedusa*.

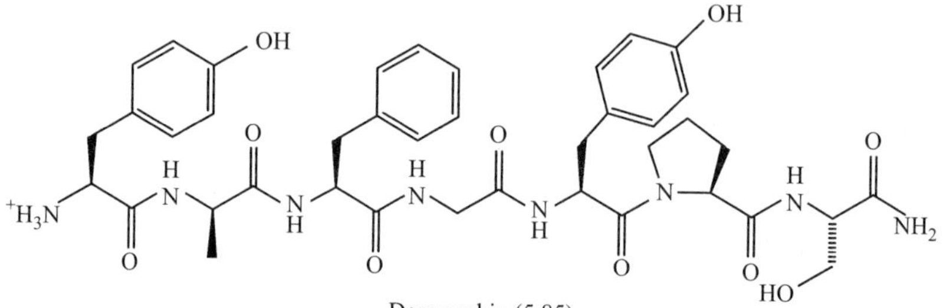

Dermorphin (5.85)

5.18.2 Structure–Activity Correlations of Opioid Peptide Hormones

Naturally, many structure–activity investigations have been reported on molecularly modified enkephalin peptide hormones. Besides increasing opiate activity, the principal goal of this work is to prevent the rapid hydrolysis between Tyr and Gly, the way in which all enkaphalins become deactivated. Removing the Tyr[1] from enkephalins or interfering with its phenolic hydroxyl or amino groups abolishes the activity of these substances. When the natural L-Tyr[1] is replaced by its D-enantiomer, activity is lost in the enkephalins as well as the endorphins. On the other hand, replacement of the Gly[2] residue with D-Ala renders the peptide resistant to hydrolysis, to the extent that some of the synthetic enkephalin analogs retain their activity when taken orally. D-Ala[2] analogs combined with modifications of Met[5] have produced potent derivatives. Met[5]-amides are also resistant to hydrolysis, and some potent compounds have been discovered among them.

5.18.3 Properties of the Opioid Hormone Receptors

There are various opioid receptors; the three major classes of opioid receptors are mu (μ), delta (δ) and kappa (κ) receptors. The μ receptor is the principal pain-modulating site in the CNS, mediating the action of morphine. There is considerable interest in the κ receptor, which mediates a sedating analgesia with decreased addiction liability and respiratory depression and which allows for some structural flexibility. Unfortunately, the κ receptor seems to be coupled to the sigma (σ) receptor, which is implicated in psychotomimetic and dysphoric side effects.

All of these receptors are G-protein-coupled receptors with a high degree of amino acid sequence homology amongst them. On pharmacological grounds, there appears to be multiple subtypes of each receptor: μ_1, μ_2, δ_1, δ_2, κ_1, κ_2, κ_3.

5.18.4 Enkephalinergic Pathways

The many actions of enkephalin opiate peptides notwithstanding, the principal interest of medicinal chemists is still the analgesic effect of opiate alkaloids and their analogs. An understanding of pain and the central pain pathways is therefore essential to the study of these agents, and the distinction between pain and pain perception must be made. There is a considerable personal and psychological component involved in this phenomenon, aptly called the "puzzle of pain." Two major pathways are involved: the first is the *neospinothalamic* path that mediates sharp localized pain; the second is the *paleospinothalamic* path involved in the dull, burning pain that responds well to opiates. The dorsal horn of the spinal cord is involved in collecting the *nociceptive* (pain) stimuli. However, these stimuli are experienced by cortical centers and interpreted emotionally by the limbic system. The distribution of enkephalin opiate receptors along these pathways has been demonstrated, and the intracerebral or epidural injection of opiates in very small doses can produce long-lasting analgesia. Electrical stimulation of the central and periventricular gray matter within the brain leads to the same effect.

There are many peripheral organs that possess enkephalin opiate receptors: the *ileum,* the most distal part of the small intestine, and the *vas deferens* are the most significant. The receptors in the ileum are responsible for the antidiarrheal activity of opiates. This is also the explanation for the severe constipation that may occur when people use opiates for pain relief.

5.18.5 Physiological Effects of Opiate Hormones

The principal opiate effect in mammals is analgesia—the reduction of pain perception. The endogenous peptides probably also modulate dopaminergic and cholinergic centers in the brain. Most opiates have a number of undesirable side effects. First and foremost are the addictive narcotic effects: euphoria and sedation in humans, apes, and dogs; excitation, fright, or convulsions in cats, cattle, and horses. It might be argued that sedation and euphoria are useful components rather than undesirable effects in alleviating the anxiety that accompanies pain. Indeed, high doses of morphine cause a deep narcosis. Another dangerous side effect is respiratory depression, which involves a decrease in the CO_2 sensitivity of the respiratory center, causing CO_2 retention and cerebral vasodilation. This is the potential cause of death when opiates are overdosed. Some humans suffer from nausea and emesis upon morphine administration.

Withdrawal symptoms are experienced when exogenous opiate agonists are abruptly discontinued in persons dependent on or addicted to opiates. Dependence is a physiological (as well as psychological) adaptive state. It varies with the particular drug, its dosage, and the duration of addiction; the symptoms of abstinence may therefore also vary in severity. These symptoms become manifest as a loss of appetite and weight, mydriasis, chills and sweating, abdominal cramps, muscle spasms, tremor, and *piloerection* ("gooseflesh"—hence the slang "cold turkey" for opiate withdrawal symptoms).

Withdrawal symptoms can be immediately precipitated in addicts by the administration of narcotic antagonists.

5.18.6 Biochemical Effects of Opiate Hormones

Since they are linked to G-proteins, opioid receptors affect intracellular Ca^{2+} and protein phosphorylation. Another principal biochemical effect of opiates is the inhibition of adenylate cyclase (AC), which decreases cAMP production.

Hypotheses concerning the mechanisms of tolerance, addiction, and withdrawal symptoms are based on adaptive syndromes. In tolerance without dependence, receptors become uncoupled from the G_i unit of the receptor complex, and higher doses are necessary to produce analgesia. When dependence develops, the chronic deficit of cAMP triggers a compensatory feedback loop to rectify the low cAMP levels, which are due to inhibition of AC (i.e., its low turnover number). Compensation can be achieved either by an increased synthesis of new AC molecules or, perhaps, by an increase in turnover number of the existing AC. The consequently increased amount of AC will, even in its inhibited state, produce sufficient cAMP to meet all of the requirements for the cells. In this condition, the organism will function normally *only* in the presence of opiate, and physiological habituation results. When the level of opiate drops, either because of drug withdrawal or through the administration of an antagonist, the previously inhibited AC produces cAMP at a normal rate; the concentration of AC increases to higher than normal levels, causing a sudden flood of cAMP; this triggers the multiple and diffuse withdrawal symptoms, which may even prove fatal.

It is proper to distinguish between habituation and addiction. The former is a biochemical–physiological process, whereas the latter has very considerable psychological and socioeconomic components as well.

5.18.7 Pharmacological Effects of Opiate Hormones

Although the pharmacological activity of the endorphins and enkephalins is analgesic, the endorphins have a wider range of effects. Among other effects, they induce catatonic rigidity in animals, which is reversible by opiate antagonists. The hope of connecting endorphins to schizophrenia was not fulfilled; nor was the expectation of finding the ideal, nonaddicting analgesic among these peptides. Strangely, and unfortunately, repeated doses of endorphin or enkephalin give rise to addiction and withdrawal symptoms. There are indications that acupuncture analgesia operates through enkephalin mobilization, since it is reversed by opiate antagonists.

5.18.8 Opium Alkaloids

Opium alkaloids are nonpeptide agonists for the opioid peptide hormone receptors. The dried latex of *Papaver somniferum* (opium), or the seed capsule of the plant itself, are the sources of almost 25 alkaloids. Some simple isoquinolines from opium, like papaverine (**5.86**), are antispasmodics. The principal alkaloid (~10% of the total) is morphine (**3.11**), which is also an isoquinoline (rings C and E) but can additionally be considered a phenanthrene derivative (rings A, B, and C).

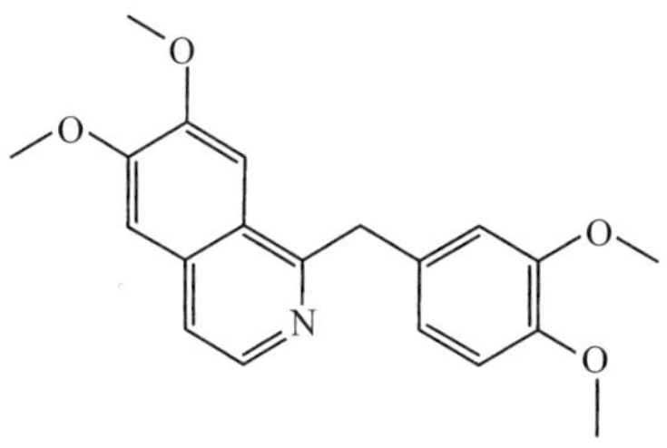

Papaverine (5.86)

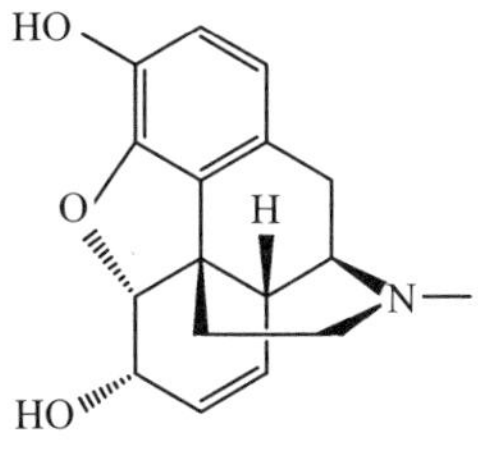

Morphine (3.11)

Even though pure morphine has been available since 1803, its structure was finally elucidated only in 1925, by Sir Robert Robinson. Detailed structure–activity relationships have been worked out during the many years of study of morphine and its derivatives.

The *phenolic hydroxyl* on C-3 is a very important functional group on an essential ring. There are very few active opiates without the aromatic ring or the phenolic hydroxyl group. The latter probably amplifies the van der Waals binding to the receptor of the aromatic ring through hydrogen bonding. Masking this hydroxyl by acetylation or methylation, as in heroin or codeine, changes the analgesic effect. Codeine, which is a natural alkaloid, has only about one-tenth the effect of morphine if given parenterally, because it must be partially demethylated in the liver and thus transformed to morphine; thus, when given intracerebrally, it is totally inactive as an analgesic. It is widely used as an antitussive (cough suppressant) drug. Weakening of the electron density at the 3-OH position by the introduction of electron-attracting substituents (e.g., $-NO_2$) in position 1 also has an inactivating effect.

Heroin (diacetylmorphine) is a highly addictive drug, being twice as active as morphine. The reason may be that its increased lipophilic character results in better transport characteristics, and the increased activity due to esterification of the 6-OH group compensates for the loss of potency due to masking of the 3-OH by the acetyl group.

The *alcoholic hydroxyl* at C-6 can be modified (as in heroin) or even omitted. For instance, heterocodeine (**5.87**, morphine-6-methylether) is about five times more active than morphine. The 6-keto derivative (**5.88**, morphinone) and the 6-methylene analog are both active analgesics.

The $\Delta^{7,8}$ *double bond* is not essential to the activity of morphine. Dihydromorphine or dihydromorphinone are active compounds with a reduced duration of action but increased activity.

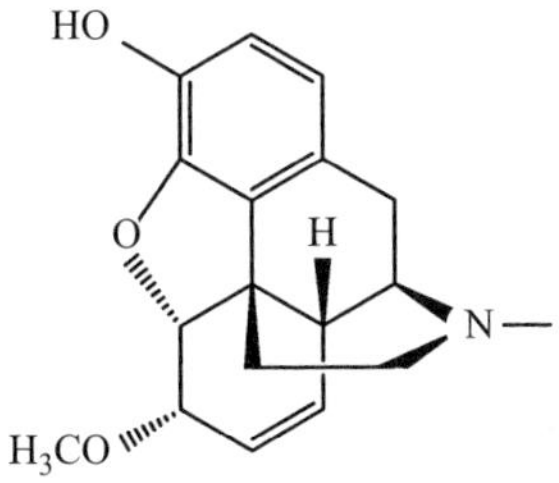

Heterocodeine (5.87)

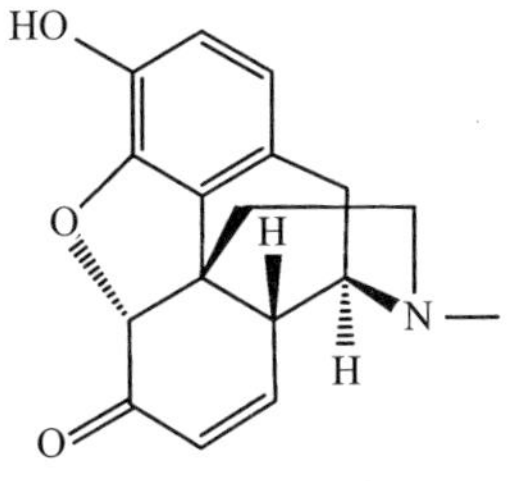

Morphinone (5.88)

The *N-methyl substituent* on morphine is not absolutely essential to its analgesic activity; such activity is instead more a question of the proper partition coefficient. Thus N-normorphine, the secondary amine, has only one-eighth of the central activity of morphine but is equiactive with the latter on the guinea pig ileum, indicating that it cannot cross the blood–brain barrier because of its polarity. Higher alkyl substituents usually render the molecule less active, although the activity rises dramatically if the side chain carries an aromatic ring. For instance, the N-furfurylmethyl- and N-phenethyl-normorphine derivatives and their analogs can be up to 50 times more active than morphine as a result of the involvement of auxiliary binding sites.

The most important and dramatic change results when the N-methyl group of morphine is replaced by an N-alkene or N-cyclopropylmethyl group. The resulting compounds show antagonist properties.

Derivatives with a C-14 *hydroxyl group* such as oxymorphone (**5.89**, 7, 8-dihydro-14-hydroxymorphine-6-one) show increased potency (up to five times that of morphine), probably as a result of the introduction of an additional hydrogen-bonding substituent. The stereochemistry of this hydroxyl is of considerable importance in terms of activity.

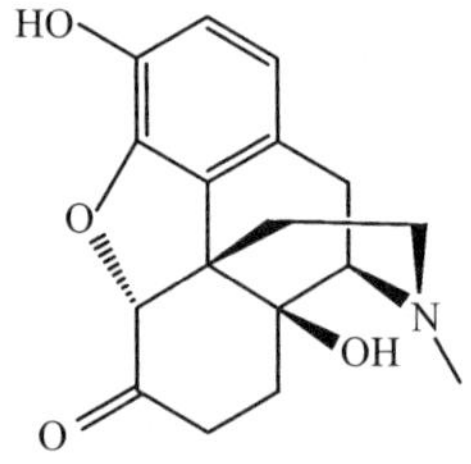

Oxymorphone
(5.89)

5.18.9 Synthetic Morphine Analogs

Molecular modifications in the form of simplifications of the morphine ring system have been undertaken for many years, and have resulted in the development of some spectacularly successful compounds devoid of addictive properties.

5.18.9.1 Morphinans

Omission of the furan ring (i.e., oxygen bridge) of morphine results in compounds known as morphinans, such as (−)levorphanol (**3.12**), which is five to six times more active than morphine. Loss of the phenolic hydroxyl decreases the activity of this compound to 20% of that of morphine. The (+) isomer, dextrorphan (**5.90**), is totally devoid of analgesic activity because it cannot bind to the receptor. However, dextrorphan has antitussive properties that are valuable, and is used in the form of its methyl ether (dextromethorphan). Shifting of the phenolic hydroxyl group to position 2 or 4 results in a total loss of activity. The N-phenethyl (**5.91**) derivative of levorphanol shows an analgesic effect about twenty times greater than that of the parent compound.

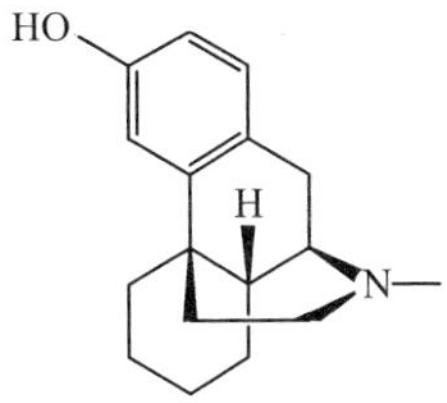

(-) Levorphanol
(3.12)

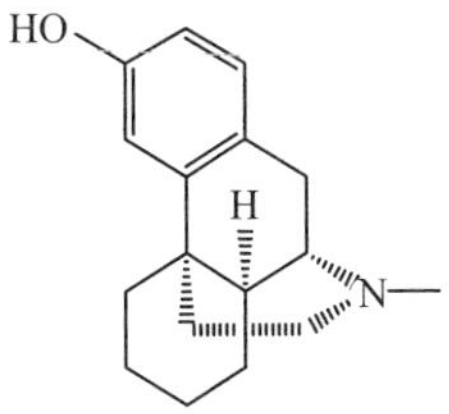

(+)- Dextrorphan
(5.90)

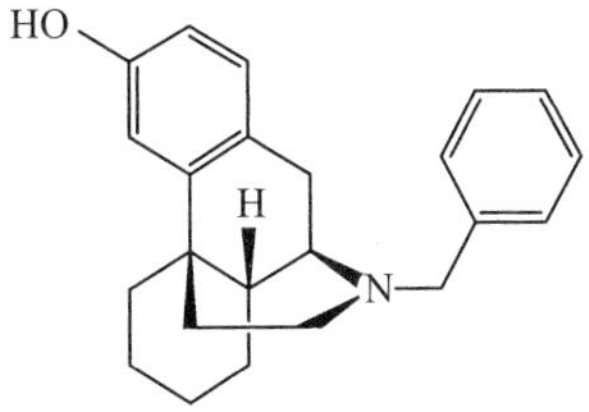

N-Phenethyl-levorphanol
(5.91)

5.18.9.2 Benzomorphan and Metazocine Series

Opening of the C ring as well as omission of the D ring of morphine results in the 6,7-benzomorphan series. Two methyl groups were retained on ring B. Study of the stereochemistry of these methyl groups has led to the major discovery of nonaddictive analgesics of the metazocine series. Compounds in which the two alkyl substituents are *cis* (as are the corresponding carbon atoms in morphine) are powerful analgesics. However, they cannot relieve withdrawal symptoms in addicted animals, which means that these drugs are not addictive. The *trans* isomers, on the other hand, while also potent analgesics, will relieve withdrawal. The (−) isomers in this series will precipitate withdrawal symptoms in addicted animals even with an N-methyl (that is, agonist) substituent, which indicates that they are mixed agonist–antagonists. Although there are complicating factors and exceptions to this rule with some of the compounds, derivatives of exceptionally low addictive capacity and satisfactory analgesic potency have been prepared, such as pentazocine (**3.13**), which can be used in chronic applications without the danger of addiction. Some of these derivatives, like the N-cyclopropylmethyl compound (**5.92**, cyclazocine), cannot be used because of unpleasant dysphoric and hallucinogenic properties. Another benzomorphan, bremazocine (**5.93**), is a powerful κ agonist of long duration, and is devoid of addictive properties and respiratory depressant activity. On the basis of receptor binding, it is about 200 times more active than morphine and has a very low sodium shift, like all of the mixed agonist– antagonist (or metagonist) benzomorphans.

5.18.9.3 Piperidine Derivatives

Piperidine derivatives represent the ultimate simplification of the morphine skeleton. They were first developed in the 1940s, and one—meperidine (**3.14**, Demerol)—is perhaps still the most widely used synthetic opiate in clinical practice despite its addictive properties. It is obviously a morphine analog, but only the A and E rings are retained. The addition of a 3-OH group results in the bemidone (**5.94**) series, while modification of the ester group to a ketone gives the ketobemidones (**5.95**), which have more than six times the activity of meperidine. The derivative carrying the N-phenethyl side chain (**5.96**, anileridine) has also proved to be potent. Remarkably, N-alkene or N-cyclopropylmethyl derivatives in this series do not show antagonist properties. It should be noted that the original C-13 must be quaternary in all of these compounds.

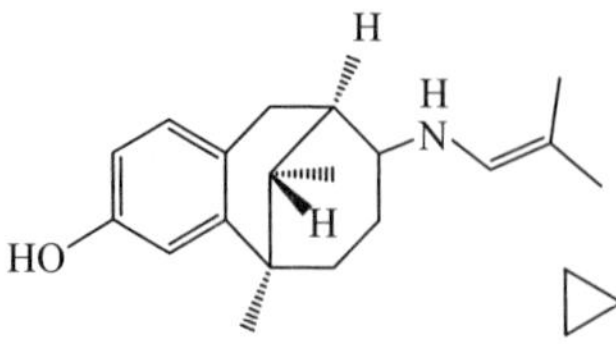

Pentazocine
(3.13)

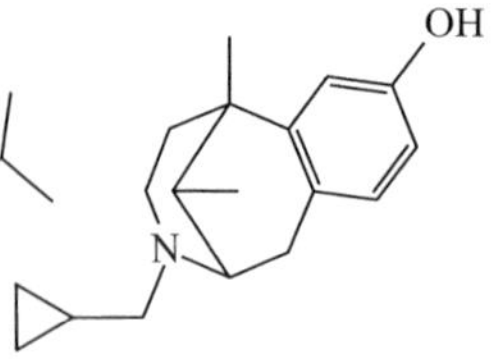

Cyclazocine
(5.92)

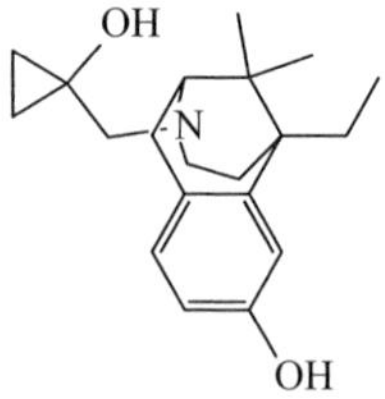

Bremazocine
(5.93)

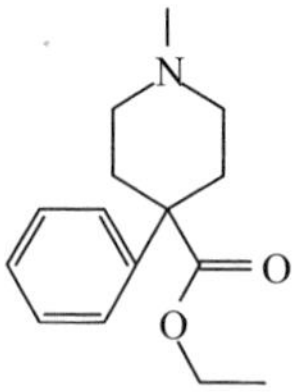

Meperidine
(3.14)

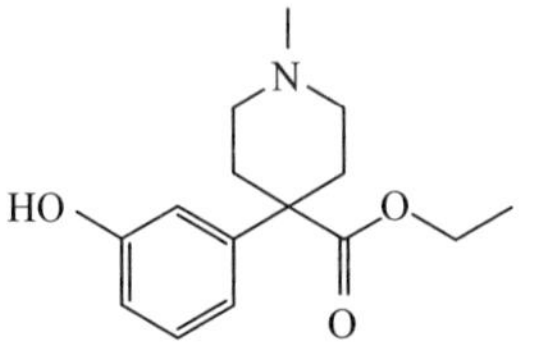

Bemidone
(5.94)

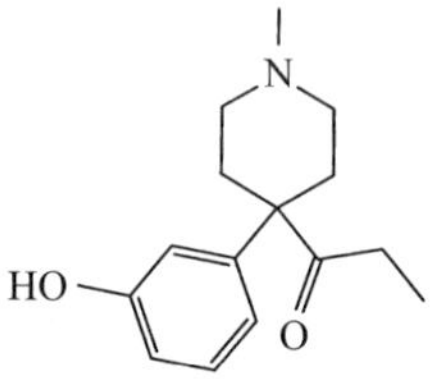

Ketobemidones
(5.95)

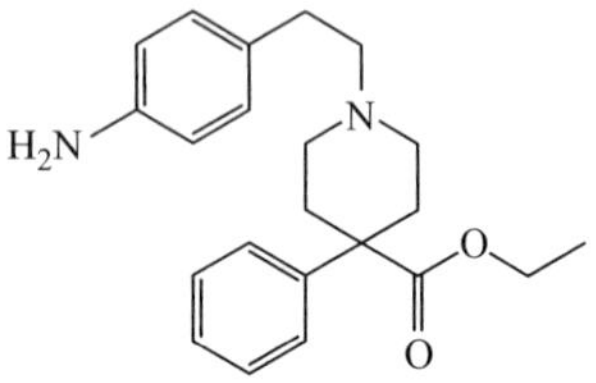

Anileridine
(5.96)

Lengthening of the N-substituent in meperidine leads to active analgesics such as the propiophenone analog, which is 200 times more active than meperidine. However, the butyrophenone derivative suddenly becomes a highly active neuroleptic without any analgesic activity.

Perhaps the most successful modification of the 4-phenylpiperidine derivatives of morphine is the 4-anilino compounds, such as fentanyl (**3.15**). This drug is 50–100 times more active than morphine, owing mainly to its excellent transport across the blood–brain barrier and into the CNS as a result of its high lipophilicity. Spectacular activities (~ 5000–6000 times that of morphine) have also been achieved by introducing ether or keto substituents (sufentanil), as in the meperidine or ketobemidone series. Fentanyl derivatives are very fast acting and of short duration. They are used in neuroleptanalgesia in surgery, in combination with neuroleptics or major tranquilizers like droperidol.

5.18.10 Opiate Antagonists

Opiate addiction affects the life of millions of humans in the Western hemisphere alone. Since addiction is a potentially serious medical problem, the identification of opiate antagonists is of clinical relevance. Although N-allyl-norcodeine was first described in 1915, the discovery of the antagonist effects produced by substituting three-carbon side chains on the morphine molecule came only in the early 1940s. Nalorphine (**5.97**) was the first clinically useful antagonist, having a dramatic reviving effect on patients on the verge of death from opiate-induced respiratory failure. It also precipitates withdrawal symptoms in addicts. However, although nalorphine is a mixed opiate agonist–antagonist and thus a potentially valuable nonaddicting drug, its unpleasant psychomimetic and hallucinogenic properties preclude its use as an analgesic.

The analogous compound derived from 14-hydroxymorphone, naloxone (**5.98**), was discovered in 1961 and is a pure antagonist. Its cyclopropylmethyl analog, naltrexone

(**5.99**), is even more useful, since it is longer acting. Levallorphan (**5.100**) and cyclorphan (**5.101**) are morphinane derivatives, whereas oxilorphan (**5.102**) is their 14-OH analog. The cyclobutyl homolog of oxilorphan, butorphanol (**5.103**), is 4–10 times more active than morphine, has 50–70 times the activity of pentazocine as an analgesic, and 30 times the activity of pentazocine as an antagonist.

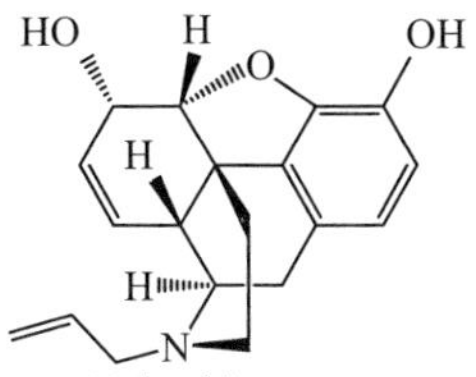

Nalorphine (5.97)

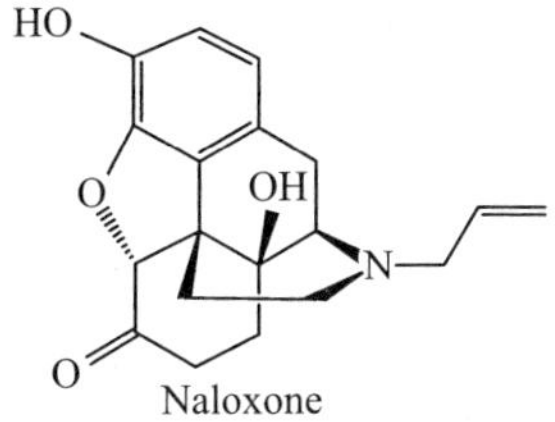

Naloxone (5.98)

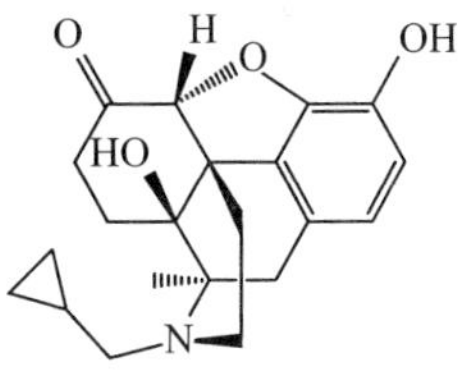

Naltrexone (5.99)

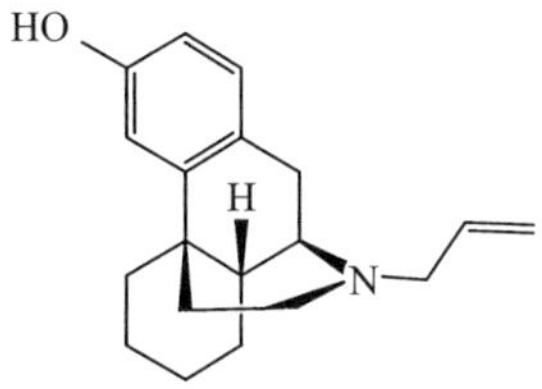

Levallorphan (5.100)

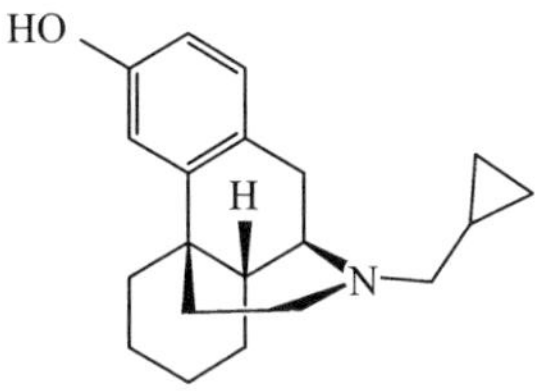

Cyclorphan (5.101)

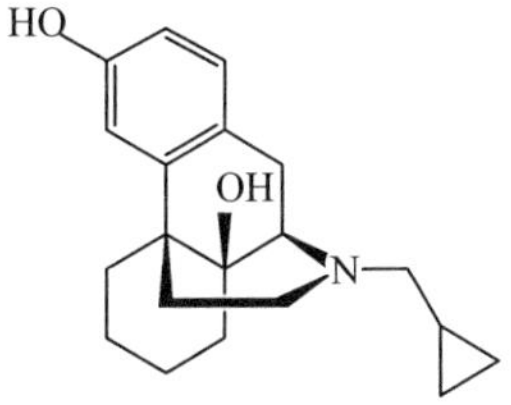

Oxilorphan (5.102)

In the benzomorphan series, cyclazocine and pentazocine are useful mixed agonist–antagonists. Unfortunately, the former has considerable hallucinogenic properties, although pentazocine is a very useful analgesic. Among the oripavines, buprenorphine (**5.104**) and diprenorphine (**5.105**) are valuable agonist–antagonists.

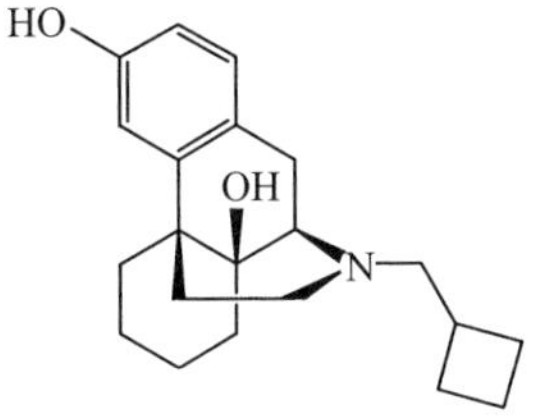

Butorphanol (5.103)

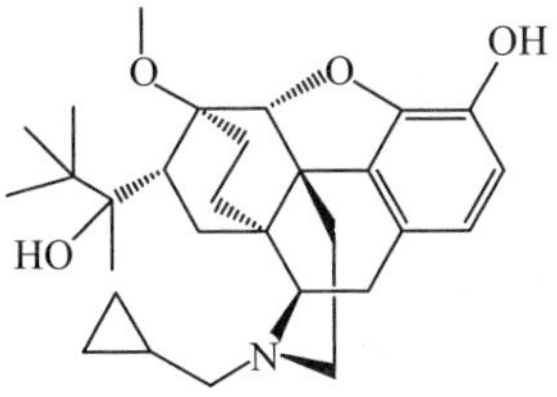

Buprenorphine (5.104)

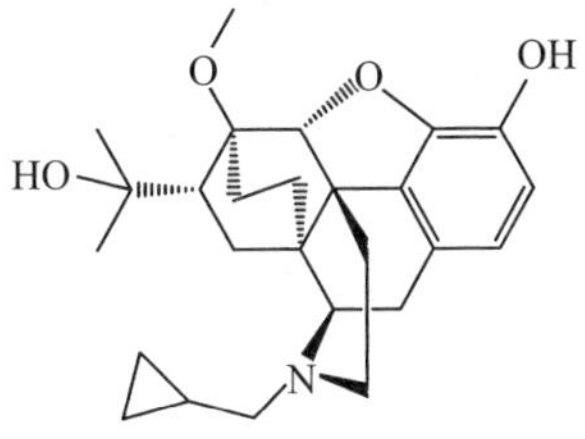

Diprenorphine (5.105)

5.19 PEPTIDE HORMONES OF THE THYROID AND PARATHYROID GLANDS

The thyroid and parathyroid glands are two important endocrine organs that are heavily committed to the biosynthesis of hormones as chemical messengers. The thyroid gland, which surrounds the larynx, has an enormous variety of metabolic functions. It is itself regulated by *thyroliberin*, which in turn regulates production of *thyrotropin* (thyroid stimulating hormone, TSH). The latter is a pituitary glycoprotein whose α subunit is identical with that of lutropin (LH) and whose specific β moiety is composed of 112 amino acids carrying a single carbohydrate side chain. The majority of the thyroid gland contains follicular cells that produce thyroid hormones (*iodothyronines*). The thyroid also contains parafollicular cells (sometimes called "C cells") which biosynthesize another hormone, *calcitonin*. Four parathyroid glands exist in two pairs, one pair embedded on the back surface of each of the two thyroid gland lobes. The parathyroid glands produce parathyroid hormone (PTH).

5.19.1 Iodothyronines

L-Thyroxine (5.106)   T₃L-Thyroxine Analog (5.107)

L-Thyroxine (**5.106**, tetraiodothyronine, T_4) is 3,5,3',5'-tetraiodo-*p*-hydroxyphenoxy-phenylalanine. The 3,5,3'-triiodo analog (**5.107**, T_3) is also a naturally occurring and active hormone. In most of its physiological effects, T_3 is more active than T_4. Both hormones are amino acid derivatives, discovered more than 60 years ago and synthesized in the thyroid gland from tyrosine by inclination and transfer of the iodophenol portion of a second iodotyrosine molecule. The synthesis takes place in the follicles (acini) of the thyroid gland, a large endocrine organ weighing about 20 g in adults. The lumen of the follicle is filled with a colloidal, viscous solution of *thyroglobulin*. This protein binds T_3 and T_4 with high affinity and is also efficient in binding circulating iodide, which enters the thyroid gland by active transport coupled to an ATPase. Iodide uptake by the gland is enhanced by thyrotropin (TSH) and inhibited by large anions such as ClO_4^-. The iodide is then oxidized and attached to tyrosine by peroxidase and a flavoprotein monooxygenase in a process that involves NADPH. Thyroglobulin (MW, 650,000; 19 S) binds the thyronines very strongly, and the hormones can enter the circulation and reach other cells only after the binding protein is lysed. This proteolysis is stimulated by TSH.

The iodothyronines are very insoluble molecules and are kept in solution by *transport proteins*. The most important of these is *thyroxine binding globulin* (TBG), which carries about 65% T_4 and 70% T_3. It is a small (MW 60,000–65,000) glycoprotein consisting of four subunits. It has a single, high-affinity binding site for T_4, with an estimated K_D of 1.2×10^{-10} M.

Thyroxine binding prealbumin (TBPA) carries about 30% T_4 but no T_3. Its affinity is only of the order of 10^{-8} M, but it is much more abundant in serum than is TBG. The amino acid sequence and the structure of this protein are known. Four identical subunits (127 amino acids each) form a prolate ellipsoid. Noncovalent interactions between the subunits form a channel of 1 nm diameter along the long axis, which has a funnel-shaped opening of 2.5 nm. The T_4 molecule is held in one arm of this channel, binding to Lys^9 and Lys^{15}. The 4'-OH is a requirement for binding, which shows negative cooperativity. It is interesting that, in addition to T$_4$, four molecules of the small retinol-binding protein, the vitamin A carrier, are also bound to TBPA. *Serum albumin* also transports T$_4$, but with a K_D of only 10^{-6} M. In addition, there is a small amount of free T$_3$ and T$_4$ in the serum (about 2 µg/L).

5.19.1.1 Biological Effects

The biological effects of iodothyronines are numerous:

1. These hormones induce *amphibian metamorphosis,* the change of a tadpole into a frog—an obviously very complex series of biochemical and developmental reactions. Human fetuses will show skeletal abnormalities as well as neuromuscular and mental retardation if born with an inadequately functioning thyroid gland.
2. Thyroid deficiency (hypothroidism) has been connected to cretinism and myxedema. Cretinism occurs when hypothyroid children are born intellectually handicapped, are small, and have coarse hair and thick skin. Myxedema, seen in older hypothyroid people, is characterized by subcutaneous semifluid deposits, causing puffiness of the hands and face. The basal metabolism of these patients is depressed to 30–40% below normal, and their body temperature and pulse rate are also reduced. Women suffering from hypothyroidism may give birth to children afflicted with cretinism.
3. In hyperthyroidism the metabolic rate is increased, resulting in vasodilatation and sweating, weight loss, cardiac arrhythmias, diarrhea, and agitation. When the thyroid gland enlarges as a result of increased activity, a goiter may develop. In certain geographical locations, this can be caused by a chronic lack of iodide; therefore iodized table salt is widely used throughout the world. In addition, hyperthyroidism may result in exophthalmos, in which the eyeballs protrude markedly. This probably occurs through an autoimmune process (Grave's disease, the most common cause of hyperthyroidism), and can be very persistent even after hyperthyroidism is cured. The increased metabolic rate in the disease manifests itself as an increase in the oxygen demand of all tissue, except those of the brain, and an increased sensitivity to β-adrenergic agonists.

5.19.1.2 Mode of Action

The mode of action of the iodothyronines is the regulation of protein synthesis in the nuclei of cells sensitive to these hormones. The T$_4$ and T$_3$ hormones dissociate from the thyroid binding proteins and enter the cell, probably by passive diffusion. Within the cell the T$_4$ is metabolized to T$_3$ by 5'-deiodinase; the T$_3$ then enters the nucleus and binds to a specific T$_3$ receptor protein. A nuclear receptor for thyroid hormones has been isolated, showing a K_D for T_3 on the order of 10^{-9}–10^{-10} M in many cell types. It is a nonbasic, nonhistone protein associated with nuclear DNA and involved in regulating the transcription of mRNA for a presumably large number of enzymes. This protein has been isolated and cloned, and the nature of the T$_3$ binding pocket has been characterized. The receptor belongs to a family of receptors that are homologous with the c-*erb* oncogene; other family members include the vitamins A and D receptors and steroid hormone receptors. The T_3 receptor exists in α and β forms; varying concentrations of these two forms in different tissues permit the varying and subtle differences of biological action of T_3 in different organ systems.

5.19.1.3 Structure–Activity Correlations

The structure–activity relationships of thyroid hormones and related structural analogs have been studied using both qualitative and quantitative methods, including the Hansch correlation. The structural requirements for receptor binding, and therefore hormone activity, are:

1. Two aromatic rings perpendicular to each other and separated by a spacer atom (O, S, or C) that holds the rings at an angle of about 120°.
2. Halogen or methyl groups on the 3 and 5 positions of the ring that bears the alanine side chain (these substituents keep the rings perpendicular to each other and participate in hydrophobic bonding to the receptor).
3. An anionic side chain two or three carbons long, para to the bridging atom, forming an ion pair with the nuclear receptor (the -NH$_2$ group decreases receptor affinity but plays a role in transport of the hormones and delays their metabolic degradation).
4. A phenolic 4'-OH group, which may be generated metabolically (e.g., by oxidation *in vivo*) if originally absent.
5. A lipophilic halogen, alkyl, or aryl substituent in the 3' position. An isopropyl group has the optimal effect.
6. A 5' substituent reduces activity in direct proportion to its size. It interferes with the binding of the 4'-OH group, increases the binding to transport proteins, and therefore reduces the concentration of available free hormone.

5.19.1.4 Thyroid Agonists and Antagonists

Hypothyroidism can occasionally be treated with a regular intake of iodide, but in the case of non-functioning of the gland thyroxine must be used. Synthetic levothyroxine is the preparation of choice.

The treatment of *hyperthyroidism* is less straightforward. Surgery, in the form of thyroidectomy, is sometimes employed, although concomitant inadvertent removal of the parathyroid glands may complicate the subsequent hormone treatments. Irradiation with ingested $^{125}I_2$, primarily a γ-emitting isotope, or $^{131}I_2$, a β-emitting isotope, is useful because it destroys thyroid follicles selectively. While these methods offer advantages over thyroidectomy, all of the dangers inherent in radioisotope treatment must be carefully considered prior to their use. Alternatively, several pharmacologic agents can be used.

Goitrogens, or compounds that produce hypothyroidism (or preferably *euthyroidism* [i.e., normal thyroid functioning] in a hyperthyroid individual), were recognized many years ago. Of historical interest, some are naturally occurring, like progoitrin (**5.108**), which is transformed into the cyclic compound goitrin (**5.109**) in rutabaga plants. Large anions such as thiocyanate (SCN$^-$), pertechnetate (TcO$_4^-$), or perchlorate (ClO$_4^-$) were once used clinically (especially potassium perchlorate) as competitive inhibitors of iodide

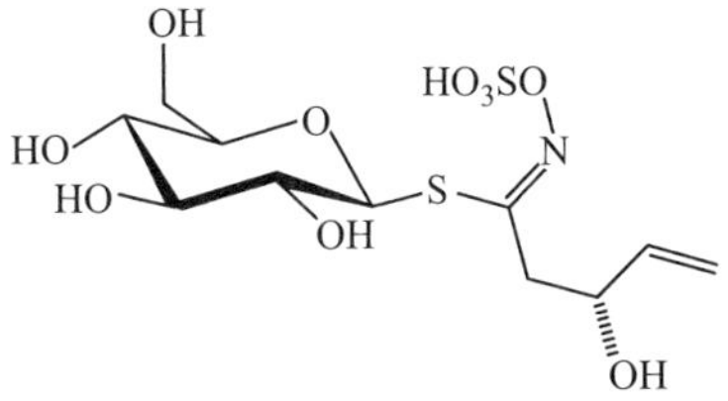

Progoitrin (5.108) 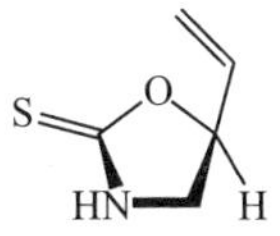   Goitrin (5.109) 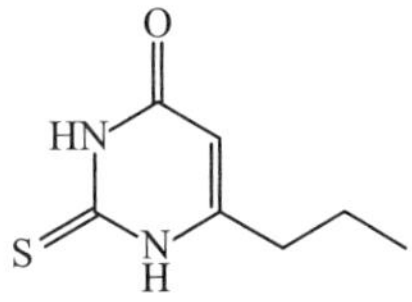   6-Propylthiouracil
                                                          (5.110)

uptake by the thyroid gland; but they have severe side effects. In recent years, the drugs of choice for controlling hyperthyroidism are thioamides such as 6-propylthiouracil (**5.110**) and methimazole (**5.111**), both of which are cyclic derivatives of thiourea, inhibiting the thyroid peroxidase that is essential to iodine organification during biosynthesis of the thyroid hormones. In some countries, carbimazole (**5.112**), which is metabolically converted to methimazole *in vivo,* is also used. Structure–activity studies of these orally active antithyroid agents reveal that the thiocarbamide group (-N-C(=S)-R) is essential for their antithyroid activity.

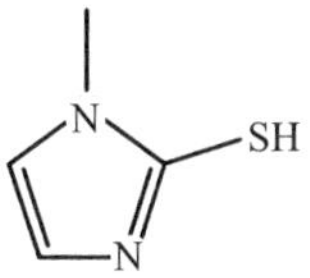

Methimazole   Carbimazole (5.112)
(5.111)

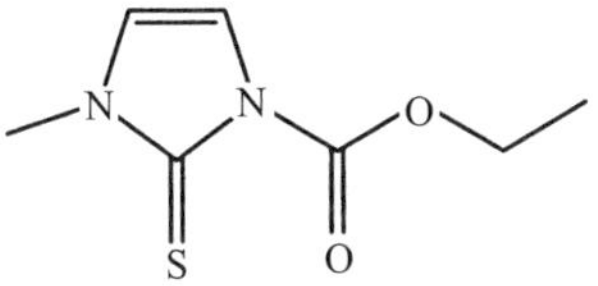

5.19.2 Calcitonin

Calcitonin (CT) is a peptide containing 32 amino acids. Calcitonin obtained from different species is essentially identical over the first nine residues, as well as containing a glycine at position 28 and a C-terminal prolylamide; this prolylamide, as well as a disulphide linkage between residues 1 and 7, is crucial for bioactivity. Residues 10–27 can be varied to influence the potency and half-life of the resulting peptide analog. Calcitonin isolated from fish exhibits potency about 20–25 times higher than that of mammalian CT because of its higher receptor affinity in bone and kidney. Calcitonin decreases the Ca^{2+} content of plasma by increasing Ca^{2+} and PO_4^{2-} excretion in the urine, as well as inhibiting the absorption of these ions in the intestine. It also decreases the activation (hydroxylation) of vitamin D, another regulator of Ca^{2+} and phosphorus metabolism. In addition, CT prevents the mobilization of Ca^{2+} from bone. CT has been used to treat postmenopausal osteoporosis and the hypercalcemia that accompanies malignancies such as metastatic breast carcinoma or multiple myeloma. Salmon calcitonin (Miacalcin) is available in a nasal spray formulation.

5.19.3 Parathyrin

Parathyrin (parathyroid hormone, parathormone, PTH) is secreted by the parathyroid glands. A peptide of 84 amino acids, it has two precursors during its biosynthesis, a pre-pro-PTH with an additional 31 amino acids and a pro-PTH with an additional 6 amino acids only. Parathyroid hormone increases the serum Ca^{2+} content and decreases the inorganic phosphate content of the body by inhibiting renal phosphate absorption and mobilizing Ca^{2+} from bone. It therefore has an action opposite to that of calcitonin. Both hormones act through the production of cAMP, but at different receptors in different zones of the bones and kidneys. Teriparatide acetate is a synthetic polypeptide constructed from the N-terminal 34 amino acid residues of PTH; this peptide possesses the full structural complement necessary for bioactivity. It is used as a diagnostic tool during the clinical evaluation of patients with hypocalcemia due to hypoparathyroidism.

5.20 PEPTIDE HORMONES OF THE PANCREAS:
INSULIN AND GLUCAGON

The homeostatic regulation of nutrient levels is a basic biochemical task that is currently not well understood. Three peptide hormones occupy a central role in this regulation of carbohydrate, lipid, and amino acid metabolism: insulin, glucagon, and somatostatin. A lack of insulin leads to *diabetes mellitus,* characterized by high blood glucose levels, excretion of glucose in the urine (hence *mellitus*: "honeyed"), and failure to utilize carbohydrates and lipids. The disease is invariably fatal if untreated. Recently, hypotheses of immunological and viral origin of diabetes are being examined.

Whereas insulin causes hypoglycemia, the other pancreatic hormone, glucagon, mobilizes glucose from its stores and causes hyperglycemia. Somatostatin, originally discovered as a hypothalamic hormone that inhibits growth-hormone release, is also found in the pancreas, where it inhibits the secretion of both insulin and glucagon. The interrelationship of these three hormones may ultimately be central to long-term treatment strategies for diabetes.

5.20.1 Insulin

5.20.1.1 Insulin: Structure, Metabolism, and Receptors

Structure. In 1916, Schafer discovered that the β cells of the "islets of Langerhans" in the pancreas secrete an antidiabetic substance, which he named insulin ("of an islet"). This was followed by the isolation of the peptide by Banting and Best in 1922, its sequencing by Sanger in 1955, and its total synthesis by several groups in the early 1970s. The cloning of the human insulin gene and its transfer into bacteria has been achieved, and human insulin is now produced by fermentation technology.

Insulin is synthesized by the pancreatic β cells in the form of proinsulin (**5.113**), in which a connecting "C-peptide" consisting of amino acids 31–60 joins the A chain at A^1 through an Arg–Lys, to connect with B^{30} through an Arg–Arg fragment. In this way

Proinsulin (5.113)

the two connecting disulphide bonds can meet properly. Proinsulin is more soluble than insulin, but has only about 35% insulin-like activity *in vitro*.

Insulin therefore consists of two peptide chains that are connected by two disulfide bonds, since the C-peptide is cleaved off. There are some species-specific differences in the amino acid sequence of the hormone. X-ray diffraction studies have shown that insulin occurs as a hexameric protein containing two Zn atoms. The dimers are first held by four hydrogen bonds and a hydrophobic bond along the β^{21-30} sequence in the form of an antiparallel β sheet. The dimers then bind by interaction of the B^{14}–Ala, B^7–Leu, and B^{18}–Val residues. The core of the hexamer contains water.

Structure–Activity Correlations. This detailed knowledge of the three-dimensional structure of insulin led to the recognition that its biological activity resides in an area of the molecule rather than in specific amino acid residues, just as dimerization and further association of the molecule also depend on an intact spatial structure. The foregoing concept is corroborated by structural modifications of the hormone. The last three amino acids of the B chain can be removed without a loss of activity, but cleavage of the C-terminal of the A chain (Asn^{21}) results in a total loss of activity. Amino acids can be replaced inside the chains only if such substitution does not change the overall geometry of the molecule. The structure–activity relationships of insulin derivatives are inconsistent and not always comparable.

Characterization of Binding. Different equilibrium binding constants for insulin have been reported by different authors. As measured with ^{125}I-labeled insulin in adipocytes (fat cells), the K_D varies from 5×10^{-11} to 3×10^{-9} M. There are about 10,000–12,000 binding sites in every fat cell. Although Scatchard plots indicate a single population of receptors in fat cells, other cell populations (liver cells, lymphocytes) show a Scatchard plot that is not linear. A possible explanation is based on the negative cooperativity of the receptor, which implies that binding is inhibited and dissociation accelerated in the presence of more insulin. This downregulation of insulin receptors provides an intrinsic mechanism that enables certain cells to limit their response to excessive or prolonged insulin concentrations.

The insulin receptor is composed of two heterodimers; each heterodimer is composed of an α unit and a β unit. The α unit is extracellular and contains the insulin recognition and binding sites; the β unit spans the cellular membrane and contains a tyrosine kinase. Although insulin can bind to a single $\alpha\beta$ dimer, it binds with higher affinity to the $\alpha\beta\alpha\beta$ tetrameric complex. When insulin binds to an α unit, the tyrosine kinase associated with the corresponding β unit is stimulated. Following this, intracellular proteins such as IRS-1 and IRS-2 (IRS=insulin receptor substrate) are phosphorylated by the β subunit tyrosine kinase, and they in turn activate a network of phosphorylations within the receptor cell.

Biochemical Effect. The biochemical effects of insulin include:

1. Facilitation of glucose transport into cells
2. Enhancement of intracellular glucokinase activity
3. Enhancement of amino acid incorporation into proteins

4. Stimulation of DNA translation
5. Increased lipid synthesis
6. Stimulation of Na^+, K^+, and P_i transport into cells

Insulin is therefore seen as a hormone-promoting anabolism rather than catabolism, since it promotes the synthesis of glycogen, proteins, and lipids. It is quite clear from the preceding list of insulin actions on metabolic functions that the hormone has a much broader biochemical role than the regulation of carbohydrate metabolism and removal of free circulating glucose. In the absence of insulin, therefore, there are profound changes in the entire metabolic pattern. Since carbohydrate, the principal nutrient in most diets, cannot be utilized properly, the energy requirements of the diabetic organism must be met in other ways. One of these is an increased gluconeogenesis from protein, with the consumption of body tissues; there is also an increased excretion of nitrogen and diminished protein synthesis (therefore, the patient's body wastes despite the high glucose in the bloodstream—"starvation in the midst of plenty"). The glucose produced in such a tedious and harmful way is also excreted and wasted. Lipolysis occurs, resulting in lipemia (an excess of circulating lipids), and ketone bodies such as acetoacetate and acetone are produced excessively.

In recent years, tremendous progress has been achieved in the elucidation of the cellular and molecular mechanisms of insulin action. The acute cellular action of insulin is initiated by rapid clustering of occupied receptors on the cell surface. Within three minutes, a redistribution of glucose transporters from the cytoplasm to the plasma membrane can be measured; lipolysis is also increased.

5.20.1.2 Insulin Preparations and Insulin-like Drugs

Insulin Preparations. Since diabetes mellitus is a defect of one or more of insulin production, secretion, or action, the administration of insulin replacement as a treatment for diabetes in the 1920s was a landmark discovery. Historically, most commercial insulin came from either bovine or porcine sources. Beef insulin differs from human insulin by three amino acid substitutions; pork insulin differs by only one residue. For many years, standard insulin preparations were 70% beef and 30% porcine. However, the biosynthesis of human insulin has now displaced the animal insulins, especially bovine insulin which was more antigenic. Mass production of human insulin by recombinant DNA methods is achieved by inserting the human proinsulin gene into either *E. coli* or yeast and treating the resulting proinsulin to yield the human insulin molecule.

Insulin preparations may be divided into four major types:

1. Ultra-short-acting insulin, very rapid onset, short duration (e.g., lispro insulin)
2. Short-acting insulin, rapid onset of action (e.g., regular insulin)
3. Intermediate-acting insulin (e.g., NPH or isophane, lente, or zinc insulin)
4. Long-acting insulin, slow onset of action (e.g., ultralente insulin)

The short-acting insulins are dispensed with the insulin molecules in solution, thereby enabling a rapid onset of action. The intermediate- and long-acting insulins are dispensed as turbid suspensions such that mobilization of the insulin molecule from the subcutaneous point of injection is retarded; these suspensions are achieved by creating poorly water-soluble complexes of the anionic insulin with either polycationic protamine in phosphate buffer (NPH insulin) or zinc in acetate buffer (lente, ultralente insulin). In addition to these "formulation tricks," a clever manipulation of peptide chemistry is used to influence the pharmacokinetics of ultra-short-acting insulin. Ultra-short-acting insulin is produced by recombinant DNA technology whereby two amino acids near the C-terminal of the B-chain are reversed: the lysine at position B29 is moved to B28, and the proline at B28 is moved to B29 (hence the name lispro insulin). This amino acid reversal does not influence the ability of the resulting insulin analog to bind to insulin receptors, but it does decrease the likelihood of the insulin molecule polymerizing to hexamers (common in human insulin); accordingly, the lispro insulin remains in a monomeric form, enabling it to be rapidly absorbed and to bind to receptors more quickly. These multiple preparations of insulin enable "fine tuning" such that insulin receptors can be stimulated over a therapeutically desirable time frame. Indeed, standard insulin therapy frequently consists of split-dose injections of mixtures of short-acting and intermediate-acting insulins.

The standard mode of insulin therapy has traditionally been by subcutaneous injection using disposable needles/syringes. However, other routes of administration, including continuous subcutaneous insulin infusion pumps and inhalation of finely powdered aerosolized insulin, are currently being explored.

Hypoglycemic Agents. Hypoglycemic agents are orally administered drugs (small organic molecules) that lower the blood glucose level and substitute for the action of insulin that is missing for any reason (insufficient insulin production, increased destruction, or the presence of anti-insulin antibodies). Oral antidiabetic agents can be divided into four groups:

1. Insulin secretagogs (e.g., sulfonylureas [1st and 2nd generations], meglitinides)
2. Biguanides (e.g., metformin, phenformin)
3. Thiazolidinediones (e.g., rosiglitazone, pioglitazone)
4. α-Glucosidase inhibitors (e.g., acarbose, miglitol)

The insulin secretagogs increase insulin release from the pancreas. Within the pancreas, sulfonylureas bind to a high-affinity sulfonylurea receptor that is associated with an ATP-sensitive potassium ion channel. When a sulfonylurea binds to its receptor, the efflux of potassium ions through the channel is inhibited; this causes cellular membrane depolarization, which leads to opening of voltage-gated calcium channels, resulting in calcium influx, which in turn produces a release of preformed insulin. The first generation sulfonylureas include tolbutamide (**5.114**), tolazamide (**5.115**), acetohexamide (**5.116**), and chlorpropamide (**3.24**); all of these have somewhat short half-lives, apart from chlorpropamide which has a half-life of 32 hours. The second generation sulfonylureas consist of glyburide (**5.117**), glipizide (**5.118**) and glimepiride (**5.119**). It is debatable whether these agents have more efficacy than chlorpropamide, but they do produce fewer side effects; there are some patients who fail on tolbutamide or tolazamide, but who respond to a more potent second-generation agent. Meglitinides, such as repaglinide (**5.120**), are a newer class of insulin secretagog. The insulin secretagogs are widely used in the treatment of adult Type II diabetes.

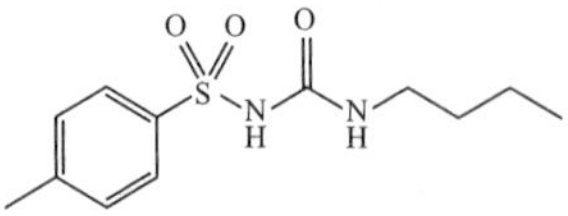

Tolbutamide (5.114)

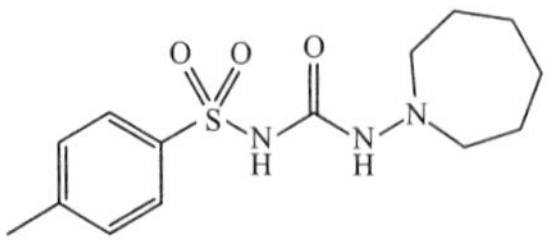

Tolazamide (5.115)

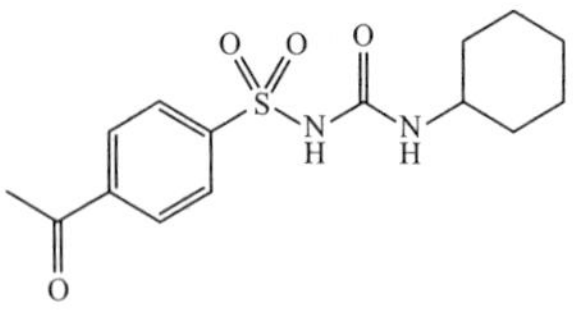

Acetohexamide (5.116)

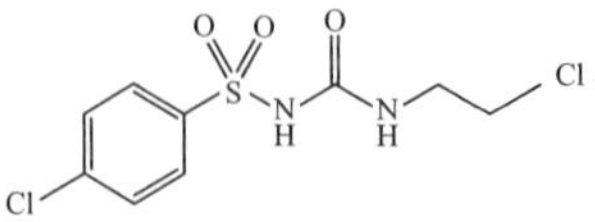

Chlorpropamide (3.24)

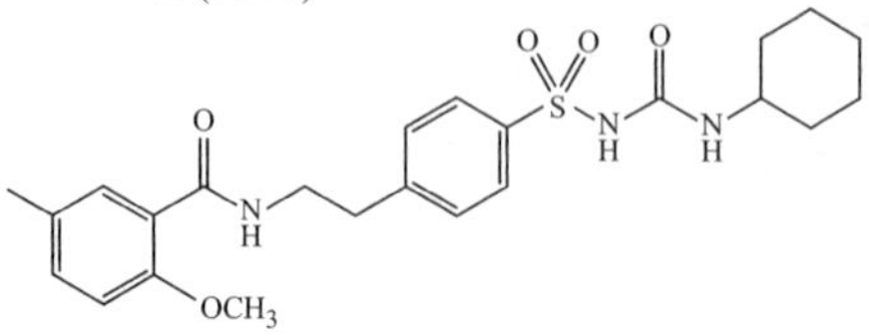

Glyburide (5.117)

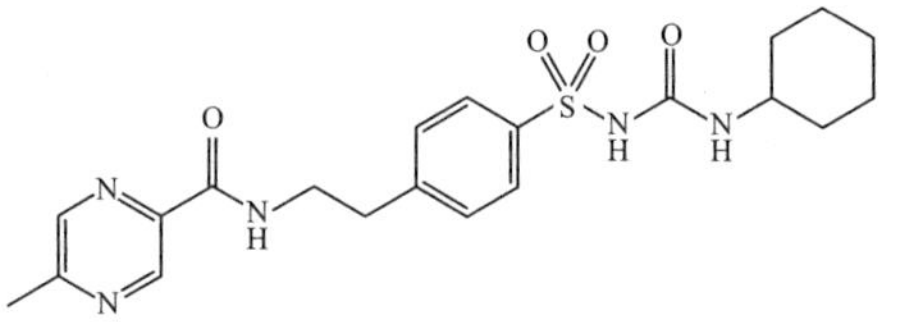

Glipizide (5.118)

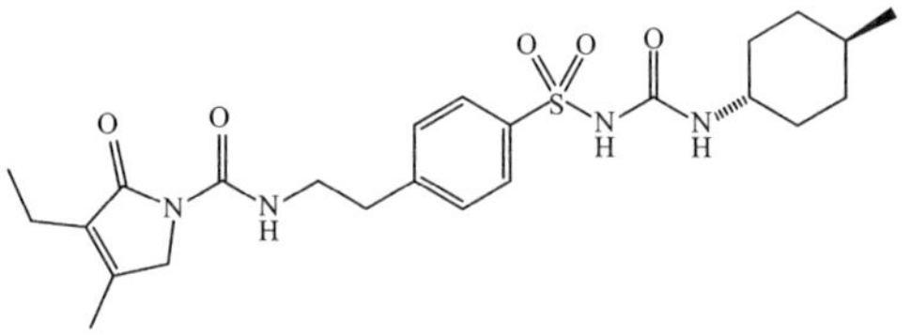

Glimepiride (5.119)

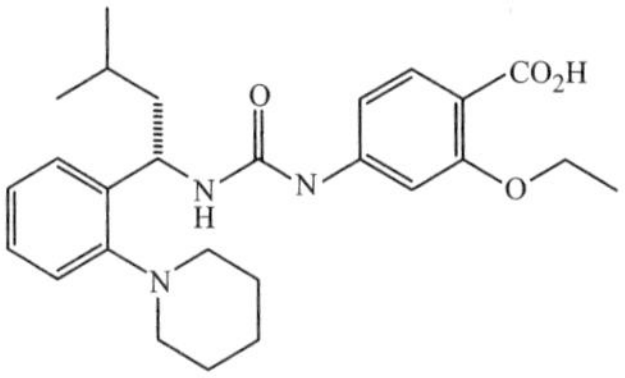

Repaglinide (5.120)

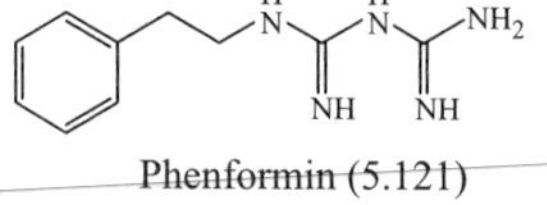

Phenformin (5.121)

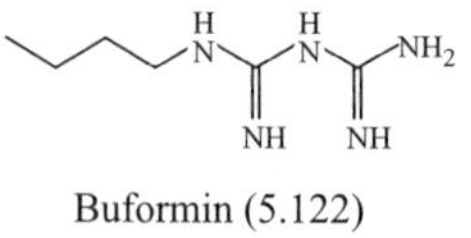

Buformin (5.122)

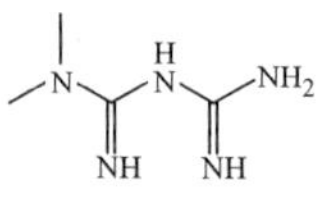

Metformin (5.123)

The biguanides consist of phenformin (**5.121**), buformin (**5.122**) and metformin (**5.123**). Phenformin was discontinued in the USA because of its implication in causing lactic acidosis. The mechanism of action of these agents is still being elucidated, but it probably involves direct stimulation of tissue glycolysis with reduced hepatic gluconeogenesis and slowed glucose absorption from the gastrointestinal tract.

These agents are most widely used in patients with refractory obesity who have ineffective insulin action.

The thiazolidinediones consist of troglitazone (**5.124**), rosiglitazone (**5.125**), and pioglitazone (**5.126**). Shortly after its initial marketing troglitazone was discontinued, initially in the USA and UK, because of more than 60 cases of liver failure or death. These agents work by enhancing target tissue insulin sensitivity by increasing glucose uptake and metabolism in muscle and adipose tissues. Thiazolidinediones are *euglycemic agents*, which when used alone ("monotherapy") can reduce glucose levels to the normal range without causing hypoglycemia. They can also be used in combination with either biguanides or sulfonylureas. They should not be used in people with liver disease. The α-glucosidase inhibitors include acarbose (**5.127**) and miglitol (**5.128**). These agents are competitive inhibitors of the α-glucosidase enzyme and thus modify the digestion and absorption of starch and disaccharides from the gut.

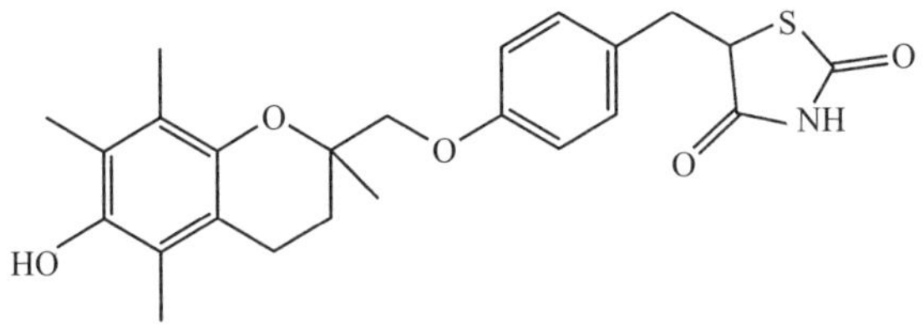

Troglitazone (5.124)

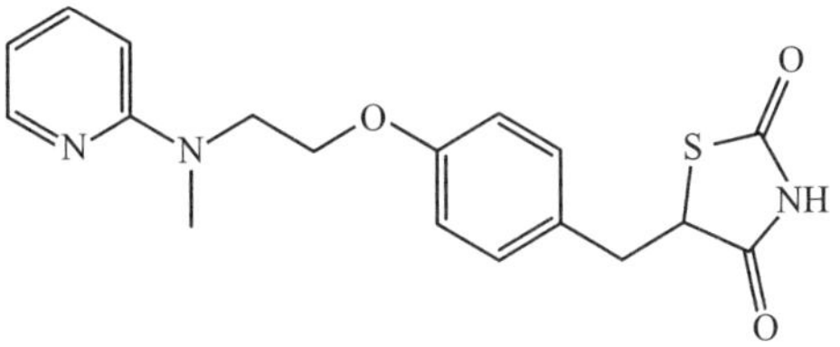

Rosiglitazone (5.125)

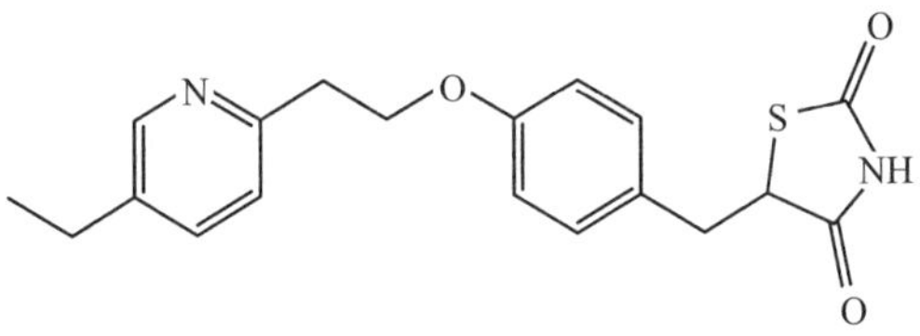

Pioglitazone (5.126)

5.20.1.3 Insulin and the Treatment of Diabetes Mellitus

Treating Diabetes Mellitus. There are two main types of diabetes. Type I diabetes (previously called insulin-dependent diabetes mellitus, IDDM) is a severe form which occurs most commonly in juveniles and young adults and which results from an absolute insulin deficiency arising from pancreatic B cell destruction, presumably via an immune-mediated mechanism. Type II diabetes (previously called non-insulin-dependent diabetes mellitus, NIDDM) is a milder, heterogeneous form of diabetes which occurs more

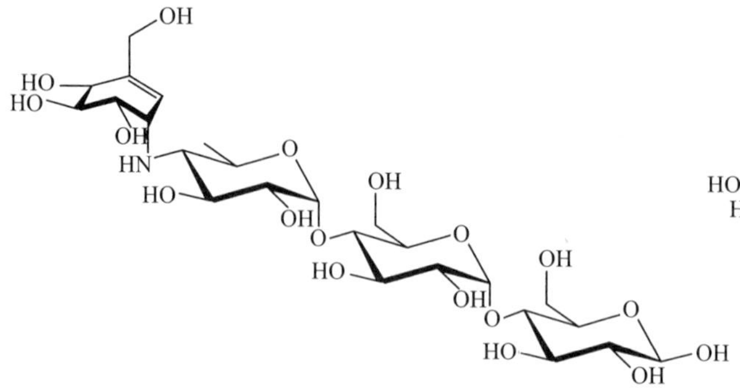

Acarbose (5.127)

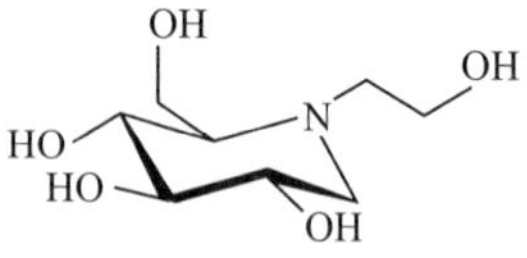

Miglitol (5.128)

frequently in older people and which results from relative insulin deficiency and insulin resistance, frequently arising from obesity which causes impaired insulin action. Insulin is the mainstay of therapy for Type I diabetes. For Type II diabetes, however, weight reduction is the first line of therapy, not the administration of drugs. Nevertheless, oral hypoglycemic agents, and sometimes even insulin, are used in Type II diabetes.

Managing the Complications of Diabetes. Diabetes is more than just a disorder of elevated glucose; it is a systemic disease that affects many organ systems. In addition to the metabolic problems there are numerous neurological, circulatory, and renal complications, even when the blood glucose level is properly controlled. The main reason is the unnatural administration of insulin by injection, instead of the constant secretion by the pancreas in response to changing blood glucose levels. Diabetics have a predisposition of atherosclerosis, with an increased risk for heart attacks and stroke.

Another complication of diabetes is blindness, which is due to blood vessel damage at the back of the eye (*proliferative retinopathy*); this accounts for about 12% of all blindness. In hyperglycemia, fructose is only slowly metabolized, and sorbitol accumulates in tissues. Because aldose reductase is found in kidneys, optic nerve, and peripheral neurons, retinopathy and painful neuropathies develop in poorly controlled or long-standing diabetes as a result of sugar alcohol (sorbitol) accumulation. *Aldose reductase inhibitors,* such as tolrestat (**5.129**) or sorbinil (**5.130**), have been evaluated as agents to ameliorate these additional symptoms of diabetes.

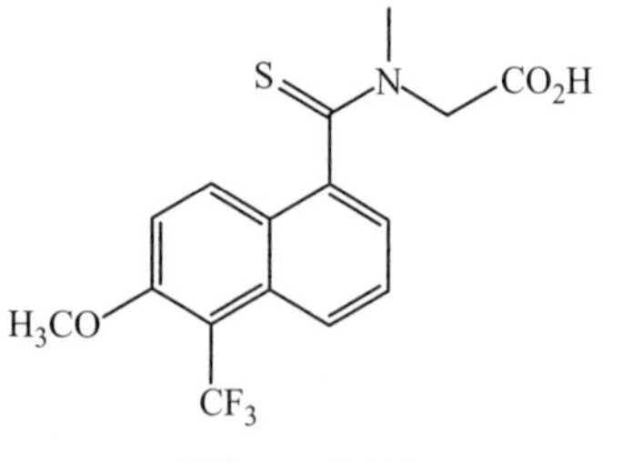

Tolrestat (5.129)

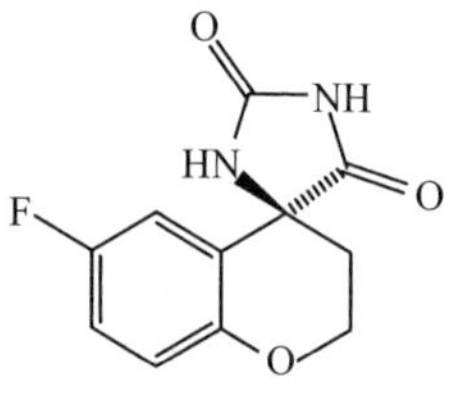

Sorbinil (5.130)

5.20.2 Glucagon

Glucagon, the second pancreatic hormone, was discovered as an impurity in early insulin preparations. It is a peptide containing 29 amino acids and is biosynthesized as proglucagon (160 residues) which is cleaved to produce glicentin, which in turn is cleaved to yield glucagon. The hormone activates receptors in liver cell membranes and acts through adenylate cyclase and cAMP. It triggers glycogenolysis and thus elevates blood glucose levels, and also activates protein phosphorylation in cell organelles. The glucagon receptor has been characterized.

The hyperglycemic action of glucagon is believed to play a role in diabetes. The hormone is produced by the α cells of the islets of Langerhans, which are not impaired in diabetes. Animal studies employing antibodies against glucagons suggest that glucagon plays a role in maintaining elevated blood glucose. It is possible that therapies that block glucagon may provide a significant advance in the management of Type II diabetes mellitus. Due to this promising potential for therapeutic utility, balanced against the shortcomings of peptides as drugs, the search for non-peptidic glucagon receptor modulators has been an active research area for the past decade. Moreover, successful cloning and expression of the glucagon receptor has enabled research groups to search for new chemical entities as novel ligands for this receptor. Since 1992, various families of quinoxalines, acyl hydrazides, and pyrimidones have been evaluated as glucagon agonists and antagonists. This work has been greatly facilitated by X-ray crystallography and molecular modeling calculations.

5.21 PEPTIDE HORMONES OF THE KIDNEY (RENIN–ANGIOTENSIN SYSTEM)

Blood pressure is regulated by a multitude of interrelated factors involving neural, hormonal, vascular, and volume-related effects. Two of these factors have been dealt with in preceding chapters in sections on the adrenergic neuronal system and the hypothalamic hormone vasopressin. Another group of peptide hormones involved in blood pressure regulation, the angiotensins, was recognized many years ago through the enzyme that activates some of them. Since this enzyme is produced in the kidneys, it was named *renin*.

The renin–angiotensin system regulates blood pressure through several feedback mechanisms. A decrease in blood pressure due to blood loss, sodium loss, or caused experimentally by clamping of the renal artery, stimulates the juxtaglomerular cells of the kidney to secrete renin, a proteolytic enzyme. This enzyme acts on a circulating protein, angiotensinogen, cleaving off angiotensin I, which is then further cleaved in the lung and kidneys by angiotensin-converting enzyme to yield angiotensin II. This hormone has a powerful constricting action on arterioles, and elevates the blood pressure instantly. Long-range effects are also achieved, either by angiotensin II or its cleavage product, angiotensin III. One or both trigger aldosterone release, causing Na^+ retention and an increase in fluid volume (i.e., antidiuresis), which results in elevation of the blood pressure. Since the angiotensins are quickly hydrolyzed, their effect is transitory and therefore suitable for continuous homeostatic regulation of the blood pressure.

5.21.1 Characterization of the Renin–Angiotensin System

Angiotensinogen (pre-proangiotensin) is produced by the liver; it is a glycoprotein with a molecular weight of 57,000. Its production is under endocrine control by adrenocorticoids, thyroid hormones, and estrogens (with the latter being prominent during pregnancy).

Renin, a highly specific endopeptidase (aspartyl protease), is a glycoprotein composed of 340 amino acids; it is biosynthesized in the kidneys. It cleaves angiotensinogen between leucine residue 10 and valine residue 11 to yield the bioinactive decapeptide angiotensin I. Renin has a precursor, prorenin, which seems to be activated by pepsin or trypsin. Prorenin has a molecular weight of about 60,000. Renin secretion is strictly regulated by (1) renal vascular baroreceptors that sense the vessel wall tension in arterioles; (2) Na^+ or Cl^- receptors; (3) an angiotensin feedback mechanism; and (4) the CNS, through catecholamines.

Angiotensin II, the octapeptide cleavage product of angiotensin I, is produced by angiotensin-converting enzyme (ACE). ACE, a widely distributed exopeptidase that is most abundant in the lungs and kidneys, is a glycoprotein containing a Zn^{2-} ion; it is a peptidyl dipeptidase that converts angiotensin I to angiotensin II by cleaving off the C-terminal His-Leu dipeptide. Angiotensin II is a strong vasoconstrictor, and its effect depends on its C-terminal Phe^8 residue. If this terminal amino acid is replaced by any aliphatic amino acid, the activity of the hormone is lost; replacement by threonine leads to antagonist action. The physiological effects of angiotensin II are widespread: in arterioles, it causes smooth muscle contraction, elevating the blood pressure with a potency 40 times that of norepinephrine; in the autonomic nervous system, it causes the release of epinephrine and norepinephrine from the adrenal medulla; in the adrenal cortex, it stimulates the biosynthesis of aldosterone; in the kidneys, it causes renal vasoconstriction, sodium reabsorption, and inhibition of the release of renin; in the central nervous system, it stimulates drinking, and increases the secretion of vasopressin and ACTH.

Angiotensin II mediates its bioactivity via interaction with angiotensin (AT) receptors. Two distinct receptor subtypes have been identified: AT_1 and AT_2. These receptors are integral membrane proteins. The relative concentrations of the two receptors differ from tissue to tissue; the AT_1 receptor dominates in vascular smooth muscle. Thus, most of the pharmacological effects of angiotensin II are mediated via the AT_1 receptor, a G-protein-coupled receptor, leading to the generation of inositol triphosphate and diacylglycerol.

5.21.2 Inhibitors of the Renin–Angiotensin System

There are several points in the renin–angiotensin system that have proven to be amenable to inhibition. Renin, angiotensin-converting enzyme, and the angiotensin receptors are the most important sites of regulation.

5.21.2.1 Renin Inhibitors

Renin inhibitors have been found among naturally occurring phospholipids and synthetic phosphatidylethanolamine derivatives. Pepstatin (5.131), isolated from *Streptomyces* strains, is a pentapeptide with an acylated N-terminus and the unusual 4-amino-3-hydroxy-6-methylheptanoic acid (AHMH) residues. It is a general protease inhibitor that leads to a significant, short-duration lowering of the blood pressure following intravenous injection. Two partially orally active renin inhibitors have been designed and developed by medicinal chemists (remikiren (**5.132**), enalkiren (**5.133**)); these were cleverly designed as peptidomimetic transition state analogs of the cleavage site of human angiotensinogen.

Pepstatin (5.131)

5.21.2.2 Angiotensin-Converting Enzyme Inhibitors

Angiotensin-converting enzyme inhibitors are among the most effective antihypertensive drugs. The first such drug to be developed was teprotide (**5.134**), a nonapeptide identical in sequence to peptides isolated from the venom of a Brazilian viper (*Bothrops jararaca*). The four prolines and the pyroglutamate in this peptide make it resistant to degradation, with the result that it has a long-lasting action but is not hypotensive in normal animals. Teprotide competitively inhibits the degradation of angiotensin I by the converting enzyme. Due to its peptide structure, it is not orally active and must be administered intravenously.

In 1977, a new drug, captopril (**5.135**), (2S)-I-(3-mercapto-2-methylpropionyl)-L-proline, was developed by Cushman, Ondetti, and coworkers. It shares many of the actions of teprotide but may be administered orally and is more than 10 times as active, with a K_D of 1.7×10^{-9} M. Cushman and colleagues studied peptide analogs to gain an improved understanding of ACE enzymatic properties and then combined the results of these studies with the fact that ACE has properties similar to those of the carboxypeptidase A enzyme, another zinc-containing exopeptidase. Assembling this knowledge, they postulated a model receptor of the enzyme active site, containing three principal binding subdomains: a positively charged arginine residue, the positively charged Zn^{2+} cation, and a hydrophobic pocket. They set out to design compounds to fit into this model receptor. An important starting point arose from the observation that D-2-benzylsuccinic acid is a potent inhibitor of carboxypeptidase A. To exploit this, a variety of analogs of 2-benzylsuccinic acid were prepared, including a family of succinyl-L-proline analogs. Although bioactive, this family needs improvement in terms of potency. To achieve this improvement, the succinyl-L-proline backbone was altered to include a sulfhydryl group whose sulphur atom would enable enhanced binding to the zinc atom within the ACE active site. This relatively simple *bioisosteric* substitution (and the addition of a 2-D-methyl group) enhanced potency 1000-fold and produced the clinical candidate molecule captopril.

Although the sulfhydryl group of captopril produced superb ACE inhibition, it also caused two side effects which are sometimes seen with sulphur-containing drugs: skin rashes and a metallic taste disturbance. In an attempt to overcome these side effects,

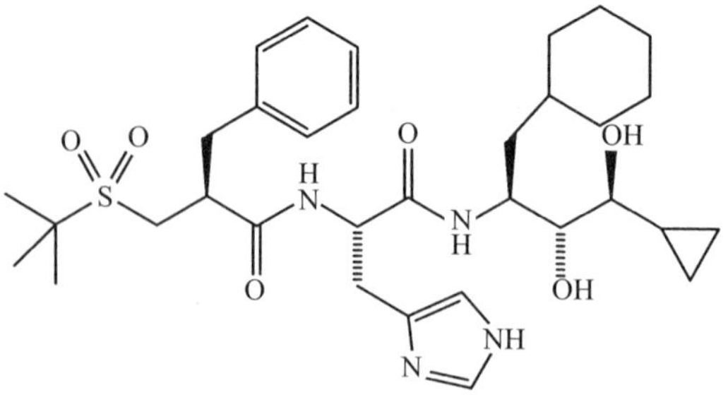

Remikiren (5.132)

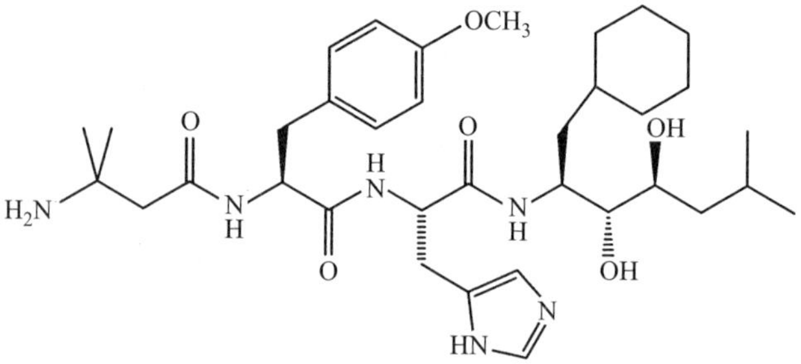

Enalkiren (5.133)

Pyr-Trp-Pro-Arg-Pro-Gln-Ile-Pro-Pro

Teprotide (5.134)

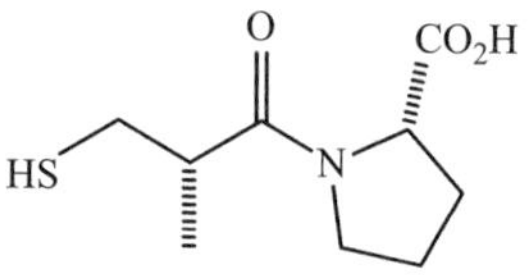

Captopril (5.135)

researchers at Merck sought to replace the sulfhydryl group with a bioisosterically equivalent carboxylate group that could bind to Zn^{2+}. Thus they sought to design dicarboxylate-containing ACE inhibitors. As a prototypic structure around which to execute this design strategy, they designed a model tripeptide in which the N-terminal residue was isosterically replaced with an N-carboxymethyl group. An analog series based upon this central pharmacophore structure yielded enalaprilat (**5.136**) as a clinical candidate. Although it exhibited excellent activity when administered intravenously, enalaprilat demonstrated unacceptably poor oral bioavailability. However, esterification of enalaprilat produced enalapril (**5.137**), an agent with superior oral bioavailability. Since enalapril needs de-esterification for bioactivation, it functions as a *prodrug*.

The success of enalapril led to a variety of other additional dicarboxylate inhibitors. The first of these was lisinopril (**5.138**). This agent was developed at the same time as

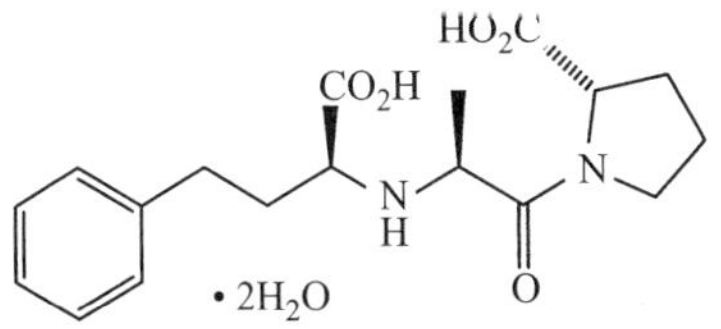

Enalaprilat (5.136)

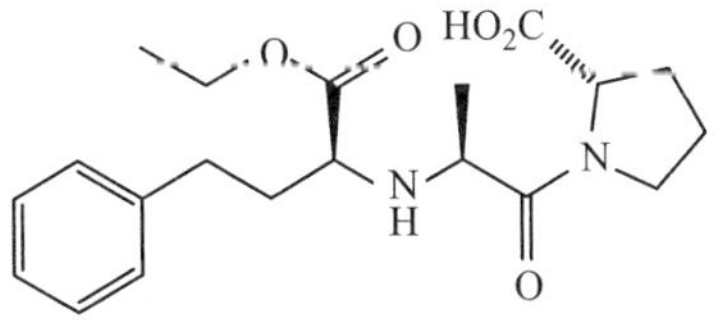

Enalapril (5.137)

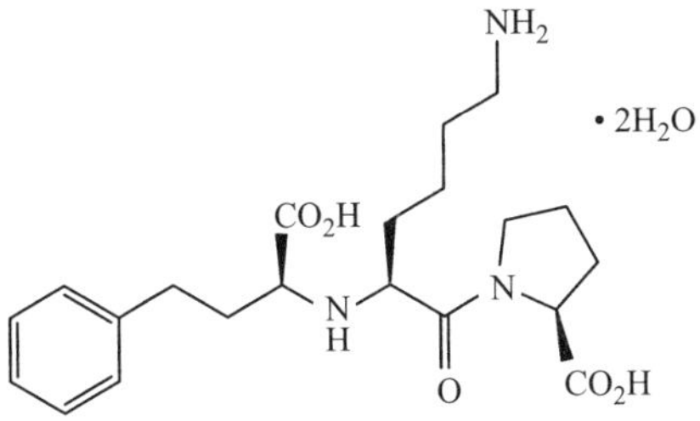

Lisinopril (5.138)

enalapril and is a lysine derivative of enalaprilat. Lisinopril and captopril are currently the only two ACE inhibitors in clinical use that are not prodrugs. The remainder of the ACE inhibitors were developed by varying the ring system of the C-terminal amino acid. The pyrrolidine ring system used in captopril, enalapril, and lisinopril was replaced with larger bicyclic or spiro ring systems. This led to compounds such as moexipril (**5.139**), perindopril (**5.140**), quinapril (**5.141**), ramipril (**5.142**), and trandolapril (**5.143**). Despite many clinical similarities, these agents differ in their absorption, dosing with other drugs, and duration of actions; for example, quinapril has a $t_{1/2}$ of 3 hours whereas ramipril has a $t_{1/2}$ of 13–17 hours. The quest for ACE inhibitors devoid of the sulfhydryl group also led to the evaluation of phosphonate-containing inhibitors, on the basis of the notion that phosphinic acid is bioisosterically equivalent to sulfhydryl and carboxylate groups in terms of Zn^{2+} chelation. This lead to the development of fosinopril (**5.144**) as a prodrug that is hydrolyzed by liver enzymes to the bioactive fosinoprilat.

Clinically, ACE inhibitors decrease vascular smooth muscle tone in arterioles and promote *natriuresis* (urinary secretion of Na^+) without increasing heart rate; accordingly, ACE inhibitors not only treat hypertension but also decrease morbidity associated with congestive heart failure. ACE inhibitors may also delay the progression of diabetes-associated kidney damage.

ACE inhibitors were designed to block the conversion of angiotensin I to angiotensin II. However, they also inhibit the degradation of other peptides including bradykinin, substance P, and enkephalins. This inhibition of bradykinin metabolism causes ACE-related side effects, including cough and angioedema.

5.21.2.3 Angiotensin Receptor Antagonists

Angiotensin receptor antagonists were developed through modification of the angiotensin molecule. The most effective compound is the [Sar^1, Val^5, Ala^8] angiotensin II, called saralasin (**5.145**) (sar = sarcosine = N-methylglycine). Unfortunately, the peptidic

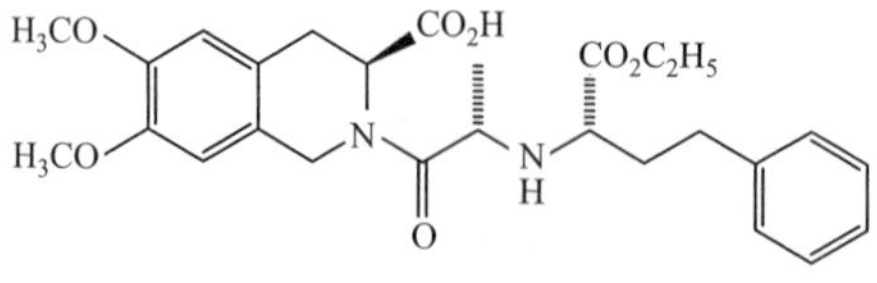

Moexipril (5.139)

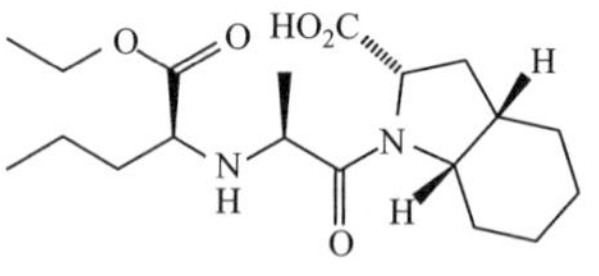

Perindopril (5.140)

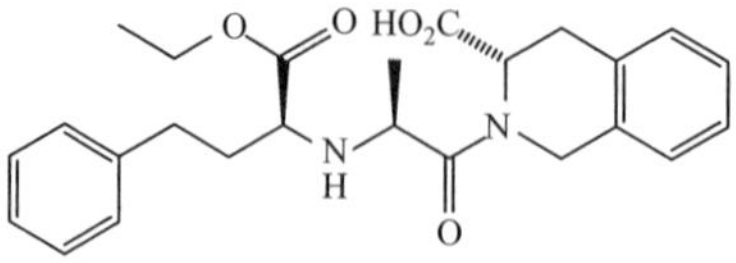

Quinapril (5.141)

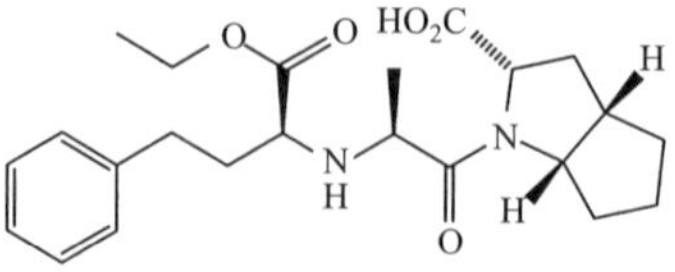

Ramipril (5.142)

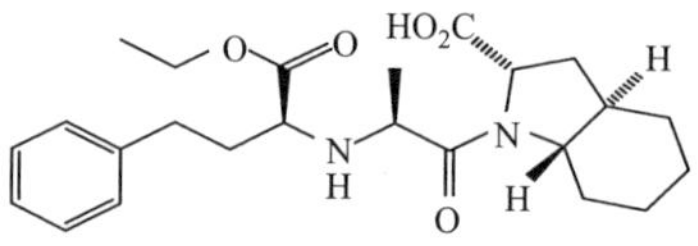

Trandolapril (5.143)

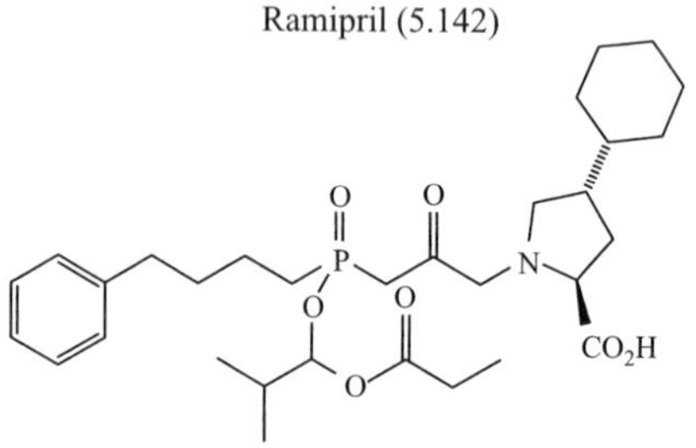

Fosinopril (5.144)

Sar-Arg-Val-Tyr-Val-His-Pro-Ala

Saralasin (5.145)

angiotensin receptor antagonists are active only during continuous intravenous infusion, and are therefore seldom if ever used.

A more recent class of nonpeptidic angiotensin II antagonists has been designed and developed. These are potent, orally active, specific, competitive inhibitors of the AT_1 receptor and exhibit blood pressure lowering efficacy analogous to that of enalapril. Representative congeners in this class of compounds include losartan (**5.146**), valsartan (**5.147**), eprosartan (**5.148**), candesartan (**5.149**), and irbesartan (**5.150**). Losartan was designed with the aid of computer modeling studies in which the structure of angiotensin II was overlapped with a series of antihypertensive imidazole-5-acetic acid analogs.

5.22 PEPTIDE HORMONES OF THE HEART (NATRIURETIC FACTORS)

In 1981, De Bold and his co-workers discovered that extracts of heart atria (but not ventricles) cause profound natriuresis, diuresis, and hypotension in rats. It was found that secretory granules in the atria contain a series of peptides responsible for these homeostatic regulatory effects, and that the heart is de facto an endocrine organ—a landmark observation. These atrial natriuretic peptides (ANP) are derived from a prohormone (preproANP) consisting of 151 amino acids that is produced in response to atrial stretch, high blood volume, and high sodium concentration. The prohormone is subsequently modified and cleaved into shorter segments. ANP circulates as a 28-amino-acid peptide with a single disulphide bridge that forms a 17-residue ring.

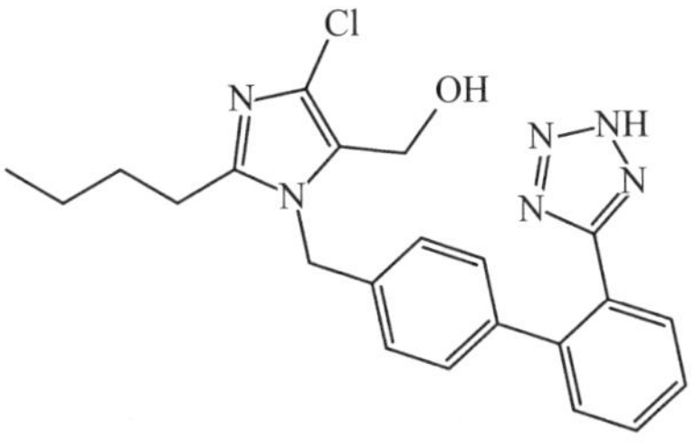

Losartan (5.146)

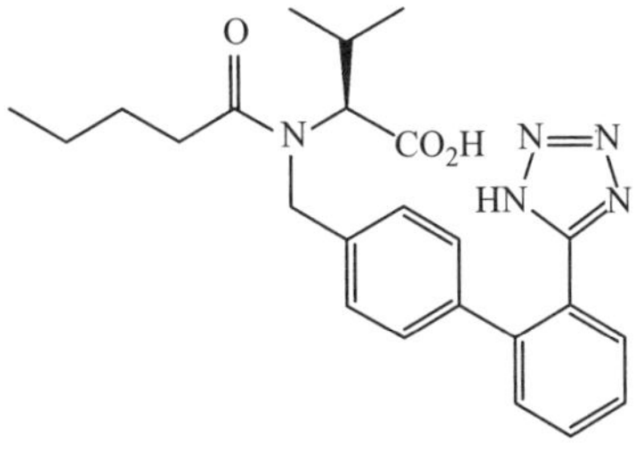

Valsartan (5.147)

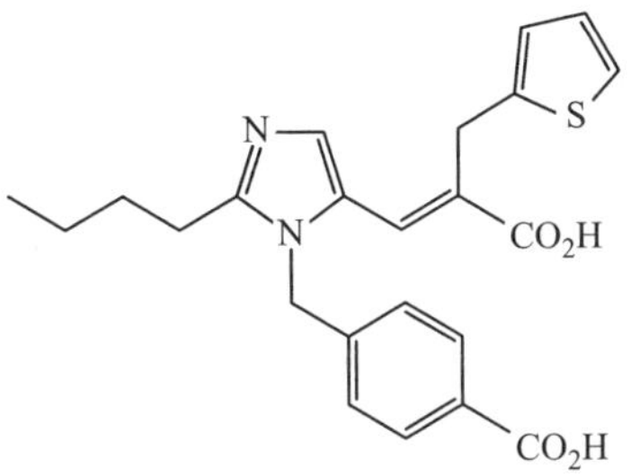

Eprosartan (5.148)

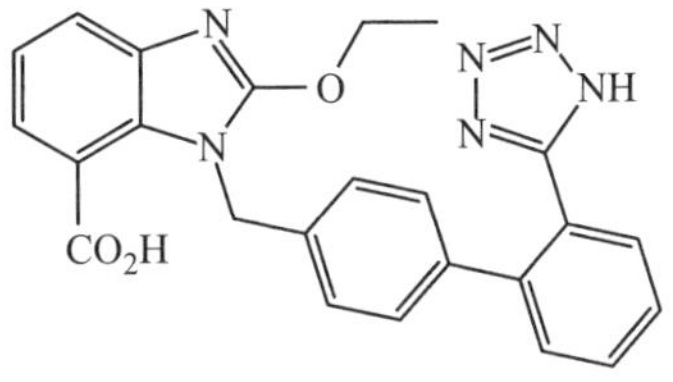

Candesartan (5.149)

ANP now belongs to a family of natriureteic peptides that also includes BNP (brain natriuretic peptide) and CNP (C-type natriuretic peptide). BNP and CNP have different N- and C-termini to ANP. CNP has a shorter N-terminal than either ANP or BNP. CNP has less natriuretic and diuretic activity than ANP or BNP.

ANP receptors have been isolated and characterized. Three receptor subtypes, termed ANP_A, ANP_B, and ANP_C, have been described. The ANP_A receptor is a 120 kDa protein with ANP and BNP as its primary ligands; the ANP_B receptor has CNP as its primary ligand. The ANP_A and ANP_B receptors are coupled to guanylyl cyclase; the ANP_C receptor is not.

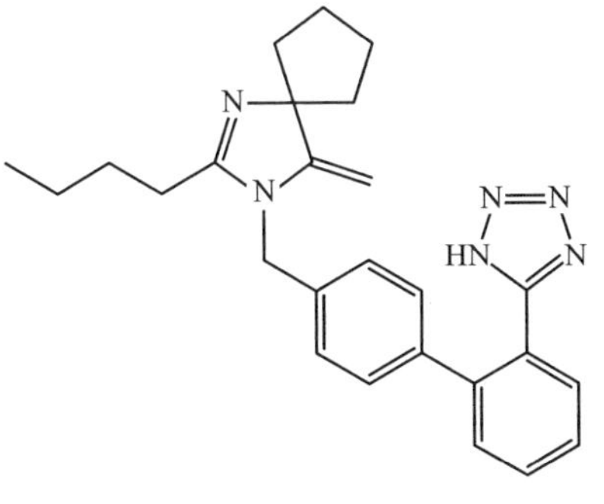

Irbesartan (5.150)

The natriuretic peptides have a short $t_{1/2}$ when administered systemically. They are rapidly metabolized by the neutral endopeptidase (NEP) enzyme. When ANP is administered, there is a prompt and profound increase in urine flow and sodium excretion, with a concomitant and dramatic hypotensive effect. Modification of these peptides may lead to the discovery of more active and stable analogs for the treatment of edema in congestive heart failure or renal insufficiency. Moreover, since natriuretic peptides may play a role in actually preventing the development of hypertension (rather than merely lowering blood pressure once hypertension has developed), this family of peptides may be of future value in drug design related to therapies for hypertension. Since hypertension is the most common cardiovascular disease, drug design and development for treating high blood pressure is an extremely important endeavor within medicinal chemistry.

5.23 PEPTIDE HORMONES AND THE DESIGN OF DRUGS FOR HYPERTENSION

The "bottom line" in medicinal chemistry is to design a molecule with the properties of a drug (see chapter 1) to fit a macromolecule with the properties of a receptor (see chapter 2) by exploiting a variety of design and optimization processes (see chapter 3). From both a philosophical and a practical perspective, the overall process of drug design may be pursued by either of two very broad approaches. In the first approach (*one target–multiple disease approach*), one selects a single *druggable target* and then designs therapeutic molecules for pathological process(es) implicated in that target. For instance, if the voltage-gated Na+ channel were selected as the target, then drugs designed to target this protein could conceivably be used as local anesthetics, anticonvulsants, or cardiac antiarrhythmics. An alternative to this approach (*multiple targets–one disease*) is to select a specific pathological process (or disease) and then design drugs to treat this process, using a diversity of different and varied druggable targets. The medicinal chemistry of treating hypertension is a superb example of this latter approach. Although hormones (mineralocorticoids, vasopressin, renin–angiotensin system, natriuretic factors) are a logical starting point, considering their influence on fluid and electrolyte homeostasis, they are not the only source of design strategies in antihypertensive drug design.

Systemic arterial hypertension ("high blood pressure") does not typically make the afflicted individual feel unwell; however, after many years, it leads to vascular damage and to the secondary complications thereof; hence, the designation of hypertension as "the silent killer." The ultimate aim of the pharmacological management of hypertension is to prevent these complications and thus to prolong not only life expectancy but also quality of life.

Hypertension is a sustained elevation of the systemic arterial pressure. The arterial pressure is determined by the cardiac output (amount of blood pumped) and the peripheral resistance in the arterial vessels (pressure = flow × resistance); the peripheral resistance is determined by the viscosity of the blood and by the caliber (and distensibility) of the resistance vessels. The systolic pressure is the pressure being exerted against the walls of the arteries during the time of peak cardiac contraction and blood ejection; the diastolic pressure is the pressure being exerted against the arterial walls while the heart vessels are refilling but not forcibly ejecting blood. Blood pressure is measured as the ratio of the systolic pressure over the diastolic pressure. Although definitions vary, hypertension reflects either a systolic pressure greater than 160 torr or a diastolic pressure greater than 96 torr. If the diastolic value exceeds 115 torr, then the afflicted individual has severe hypertension.

In hypertension, the prolonged elevation of pressure against the arterial walls either damages the lining of the artery, promoting atherosclerosis and eventual partial or total blockage of the artery at the point or zone of damage, or causes the arterial wall to rupture. Not surprisingly, there are many long-term secondary complications arising from hypertension: coronary artery atherosclerosis (producing angina pectoris which may progress to a myocardial infarction [heart attack] that may or may not be complicated by cardiac arrhythmias), cerebral artery atherosclerosis (producing transient ischemic attacks which may progress to a cerebral infarction [stroke] that may or may not be complicated by seizures), peripheral artery atherosclerosis (producing intermittent claudication—lower leg calf pain during physical exertion), renal artery atherosclerosis (producing decreased renal function and ultimately kidney failure), heart failure (mechanical failure of the heart, either because it is pumping against a high peripheral resistance or because the muscles of the heart are damaged by a myocardial infarction), or cerebral haemorrhage (from rupture of a blood vessel within the brain).

Since optimal blood pressure is crucial to the health of the organism as a whole, it is also not surprising that there are many control systems within the body to influence and adjust the blood pressure. Components of both the peripheral and central nervous systems, as well as the hormonal systems of the kidney, heart, and peripheral vascular network, all work in concert to continuously adjust blood pressure on a short and intermediate timeframe basis. Arising from this complexity is the availability of numerous druggable targets that may be exploited for antihypertensive drug design. These include one or more of the following four general mechanistic targets:

1. *Neurotransmitter receptors* as messenger targets (e.g., adrenergic drugs such as β-blockers, chapter 4)
2. *Hormone receptors* as messenger targets (e.g., renin–angiotensin system drugs, section 5.21)
3. *Endogenous cellular structures* as nonmessenger targets (e.g., membrane targets such as Ca^{2+} ion channels, section 7.1) and
4. *Endogenous macromolecules* as nonmessenger targets (e.g., carbonic anhydrase inhibitors or other agents as diuretics, chapter 8)

In keeping with these four mechanistic classes of antihypertensives, the clinically useful agents for hypertension can be grouped into four corresponding major categories:

1. *Sympathoplegic agents*: lower blood pressure by inhibiting cardiac function, increasing venous pooling in capacitance vessels (rather than in arterial resistance vessels), and reducing peripheral vascular resistance (e.g., β-adrenergic antagonists such as propranolol (**4.63**), α-agonists such as clonidine (**4.42**), and biogenic amine depletories such as reserpine (**3.1**))
2. *Hormonal agents* that inhibit the production or action of angiotensin: reduce peripheral vascular resistance and possibly blood volume (e.g., ACE inhibitors such as enalapril (**5.137**), AT_1 antagonists); other hormones that influence blood pressure could also be targeted

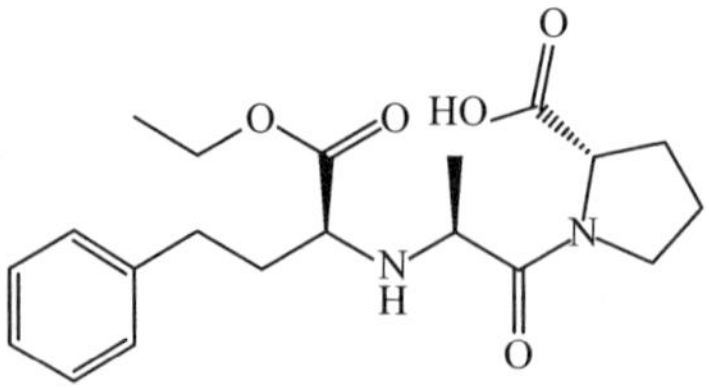

Enalapril (5.137)

3. *Direct vasodilators*: reduce pressure by dilating resistance vessels (arteries) by vascular smooth muscle relaxation (e.g., Ca^{2+} channel blockers such as verapamil (**4.133**) or nifedipine (**5.151**); vasodilators such as dihydralazine (**5.152**))
4. *Diuretics*: lower blood pressure by reducing blood volume and promoting sodium depletion (e.g., diuretics such as hydrochlorothiazide (**5.153**))

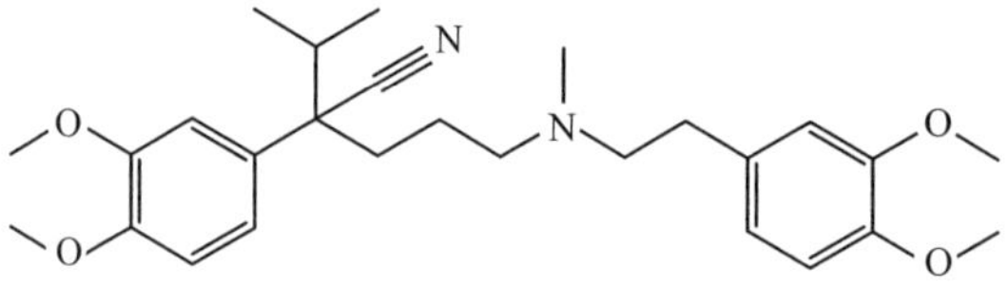

Verapamil (4.133)

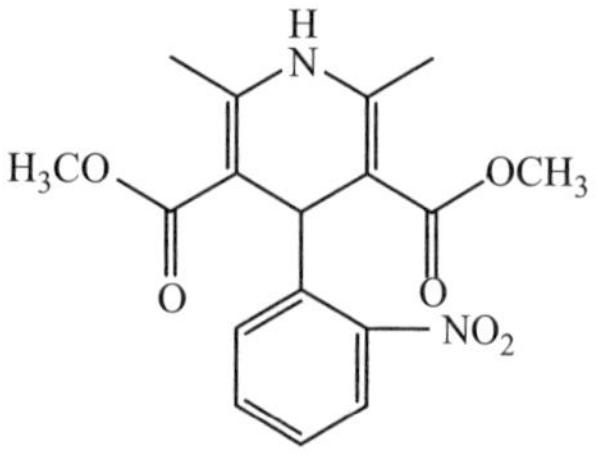

Nifedipine (5.151)

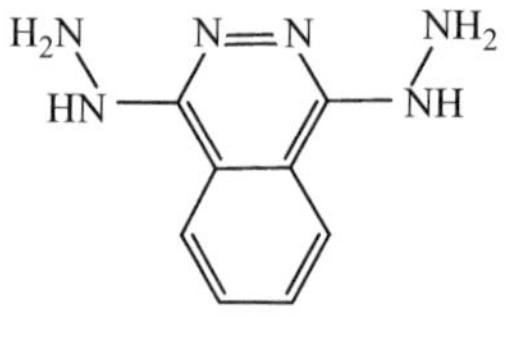

Dihydralazine
(5.152)

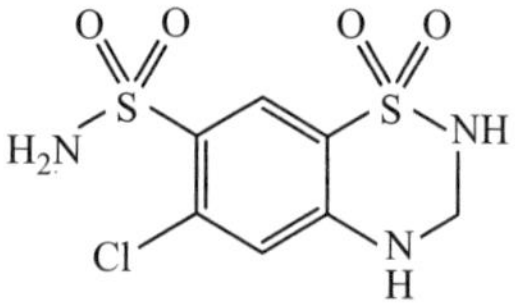

Hydrochlorothiazide
(5.153)

From a practical clinical perspective, the initial therapy of hypertension involves the selection of an agent from one of the following four starter groups:

1. Diuretics
2. Beta-blockers
3. ACE inhibitors or AT_1 antagonists
4. Ca^{2+} channel antagonists

If the therapeutic results are inadequate, the option is either to change to another drug in one of the remaining three groups (maintaining one-drug therapy or *monotherapy*) or to combine with a drug from one of the remaining three groups (initiating *polytherapy*). In severe cases of hypertension, combination therapy is achieved by combining one or more drugs from the four starter groups listed above with one or more drugs from the following four groups:

1. Central α_2 agonists (e.g., clonidine)
2. Alpha-blockers (e.g., prazosine)
3. Vasodilators (e.g., dihydralazine, minoxidil)
4. Depletors of neurotransmitter biogenic amine stores (e.g., reserpine)

When selecting an agent, hypertensive complications and/or concomitant co-morbidities must be considered. For instance, when hypertension is being treated in the presence of angina pectoris, a β-blocker or Ca^{2+} channel antagonist is preferable to a diuretic; when hypertension is being treated in the presence of prostatic hyperplasia (with associated impairment of urination), α-blockers are preferred. For uncomplicated mild hypertension, monotherapy is preferable to polytherapy. When polytherapy is being used, drugs that rationally complement each other—by acting through different mechanistic pathways—should be employed. For example, the combination of an ACE-inhibitor with a diuretic (thiazide) is a potent and rational combination. Combining two ACE inhibitors would not be a rational combination; nor would combining a β-blocker with verapamil, since both produce *bradycardia* (decreased heart rate) as a side effect. Whether pursuing monotherapy or polytherapy, non-pharmacological measures, such as weight reduction or the use of a low Na^+ diet, are useful adjuncts.

5.24 PEPTIDE AND STEROID HORMONES AS STARTING POINTS IN DRUG DESIGN

As has been demonstrated in this chapter, hormones are superb starting points in the design and discovery of new chemical entities as potential therapeutics. Hormones are molecular messengers that have the capacity to influence a variety of metabolic and chemical processes throughout the organism. Consequently, the design of agonists or antagonists to a particular hormonal receptor imparts a capacity for selective modification of hormonal function. Moreover, being small molecules, the majority of hormones can be readily "analoged" and explored as putative drugs.

Although steroids have been studied for many years, steroid hormones continue to be a rich area of medicinal chemistry study. Steroids continue to be evaluated for their therapeutic role in the treatment of cancer, especially malignancies whose growth characteristics are hormonally responsive. In recent years, the increased recognition of the role of steroids in the brain has resulted in ongoing projects to evaluate steroids as general anasthetics and anticonvulsants. Also, the search for compounds that bind to steroid receptors, but which are not steroidal in their molecular structure, is another important area of research.

Likewise, peptide hormones are also an area of continuing drug design research. The plethora of peptide hormones both within the brain and external to the brain continues to grow in complexity. Accordingly, peptidic hormones mediate a diversity of metabolic processes and are logical targets for a variety of pathological states. The geometries and conformations of short peptides can be comprehensively studied using molecular mechanics force field calculations in conjunction with molecular dynamics or Monte Carlo techniques for searching conformational space (chapter 1). In addition, peptidomimetic chemistry techniques using bioisosteric substitutions can be used to design peptide mimics which are bioactive yet "less peptidic" in molecular structure (chapter 3). Finally, if the hormone receptor has been isolated, high throughput screening in conjunction with a combinatorial library may lead to the identification of a nonpeptidic organic molecule lead compound to be optimized as a potential drug. By exploiting this multipronged strategy, drug design that targets hormones and hormonal receptors persists as a viable and useful medicinal chemistry pursuit.

Selected References

Steroid Hormones

R. Alonso, I. Lopez-Coviella (1998). Gonadal steroids and neuronal function. *Neurochemical Research 23*: 675–688.

J. L. Arriza, C. Weinberger, G. Cerelli, T. M. Glaser, B. L. Handelin, D. E. Housman, R. E. Evans (1987). Cloning of human mineralocorticoid receptor c-DNA: structural and functional kinship with the glucocorticoid receptor. *Science 227*: 268–275.

C. Bai, A. Schmidt, L. P. Freedman (2003). Steroid hormone receptors and drug discovery: therapeutic opportunities and assay designs. *Assay Drug Dev. Technol. 1*: 843–852.

S. Barker (2003). Anti-estrogens in the treatment of breast cancer: current status and future directions. *Curr. Opin. Invest. Drugs (Thomson Curr. Drugs) 4*: 652–657.

M. R. Bell, F. H. Batzold, R. C. Winneker (1986). Chemical control of fertility. *Annu. Rep. Med. Chem. 21*: 169–177.

S. Bhasin (1993). Androgen treatment of hypogonadal men. *J. Clin. Endocrinol. Metab. 81*: 757.

M. S. Brown, J. L. Goldstein (1984). How LDL receptors influence cholesterol and atherosclerosis. *Sci. Am. 255*: 58–66.

M. J. Coghlan, S. W. Elmore, P. R. Kym, M. E. Kort (2002). Selective glucocorticoid receptor modulators. *Ann. Rep. Med. Chem. 37*: 167.

H. Danielson, J. Sjövall (Eds.) (1985). *Sterols and Bile Acids. New Comprehensive Biochemistry*, vol. 12. Amsterdam: Elsevier.

J. Gorski (1986). The nature and development of steroid hormone receptors. *Experientia 42*: 744–750.

A. B. Grey, J. P. Stapleton, M. C. Evans, I. R. Reid (1995). The effect of anti-estrogen tamoxifen on cardiovascular risk factors in normal post-menopausal women. *J. Clin. Endocrinol. Metab. 80*: 8192.

O. Jenne, C. W. Bardin (1984). Androgen and antiandrogen receptor binding. *Annu. Rev. Physiol. 46*: 107–118.

V. C. Jordan (1984). Biochemical pharmacology of antiestrogen action. *Pharmacol. Rev. 36*: 245–276.

R. S. Newton, B. R. Krause (1986). Approaches to drug intervention in atherosclerotic disease. *Annu. Rep. Med. Chem. 21*: 189–200.

G. H. Rasmussen (1986). Chemical control of androgen action. *Annu. Rep. Med. Chem. 21*: 179–188.

D. R. Rowley, D. J. Tindall (1986). Androgen receptor protein: purification and molecular properties. In: G. Litwack, (Ed.) *Biochemical Action of Hormones* vol. 13. New York: Academic Press, pp. 305–324.

D. F. Smith, D. O. Toft (1993). Steroid receptors and their associated proteins. *Mol. Endocrinol. 4*: 7.

Y. Tabata, H. Osada (2003). Estrogen and progesterone receptors as molecular targets for anti-cancer drug. *Drug Deliv. Syst. 18*: 336–342.

M. R. Walters (1985). Steroid hormone receptors and the nucleus. *Endocrinol. Rev. 6*: 512–543.

Peptide Hormones

M. R. Bell, F. H. Batzold, R. C. Winekker (1986). Chemical control of fertility. *Annu. Rep. Med. Chem. 21*: 169–177.

P. M. Conn (1984). Molecular mechanism of gonadotropin releasing hormone action. In: G. Litwack (Ed.). *Biochemical Action of Hormones*, vol. 11. New York: Academic Press, pp. 67–92.

F. V. DeFeudis, J.-P. Moreau (1986). Studies on somatostatin analogs might lead to new therapies for certain types of cancer. *Trends Pharmacol. Sci. 7*: 384–386.

A. S. Dutta, B. J. A. Furr (1985). Luteinizing hormone releasing hormone (LHRH) analogs. *Annu. Rep. Med. Chem. 20*: 203–214.

A. M. Felix, E. P. Heimer, T. F. Mowles (1985). Growth hormone releasing factors. *Annu. Rep. Med. Chem. 20*: 195–192.

D. M. Gash, G. J. Thomas (1985). What is the importance of vasopressin in memory processes? In: D. Bousfield (Ed.). *Neurotransmitters in Action*. Amsterdam: Elsevier, pp. 305–308.

R. Guillermo, P. Brazeau, P. Böhlen, F. Esch, N. Ling, W. B. Wehrenberg, B. Bloch, C. Mougin, F. Zeytin, A. Baird (1984). Somatocrinin, the growth hormone releasing factor. In: R. O. Greep (Ed.). *Recent Progress in Hormone Research*, vol. 40. New York: Academic Press, pp. 233–299.

W. K. Hagmann (2002). Therapeutic applications of non-peptidic δ-opioid agonists. *Ann. Rep. Med. Chem. 37*: 159.

V. J. Hruby (1985). Design of peptide hormone and neurotransmitter analogs. *Trends Pharmacol. Sci. 6*: 259–262.

R. M. Jones, P. D. Boatman, G. Semple, Y.-J. Shin, S. Y. Tamura (2003). Clinically validated peptides as templates for de novo peptidomimetic drug design at G-protein-coupled receptors. *Curr. Opin. Pharmacol. 3*: 530–543.

D. Larhammar (1996). Structural diversity of receptors for neuropeptide Y, peptide YY and pancreatic polypeptide. *Regul. Pept. 65*: 165.

D. R. Lynch, S. H. Snyder (1986). Neuropeptides: multiple molecular forms, metabolic pathways, and receptors. *Annu. Rev. Biochem. 55*: 773–799.

J. van Nispen, R. Pinder (1986). Formation and degradation of neuropeptides. *Annu. Rep. Med. Chem. 21*: 51–62.

D. Regoli, A. Boudon, and J.-L. Fauchere (1994). *Pharmacol. Rev. 46*: 551.

A. V. Schally, D. H. Coy, C. A. Meyers (1978). Hypothalamic regulatory hormones. *Annu. Rev. Biochem. 47*: 89–118.

E. J. Trybulski (2001). Vasopressin receptor modulators: from non-peptide antagonists to agonists. *Ann. Rep. Med. Chem. 36*: 159.

P. Vermeij, D. Blok (1996). New peptide and protein drugs. *Pharm. World Sci. 18*: 87–93.

Thyroid Hormones

E. N. Cheung (1995). Thyroid hormone action: determination of hormone–receptor interaction using structural analogs and molecular modeling. *Trends Pharmacol. Sci. 6:* 31–34.

D. S. Cooper (1984). Antithyroid drugs. *N. Engl. J. Med. 311*: 1353–1357.

C. Fleck, M. Schwertfeger, P. M. Taylor (2003). Regulation of renal amino acid (AA) transport by hormones, drugs and xenobiotics—a review. *Amino Acids 24*: 347–374.

N. J. Gittoes, J. A. Franklin (1998). Hyperthyroidism: current treatment guidelines. *Drugs 55*: 543.

F. S. Greenspan (1991). Thyroid disease. *Med. Clin. North Am. 75*: 1.

R. S. Lindsay, A. D. Toft (1997). Hypothyroidism. *Lancet 349*: 413.

M. I. Surks, R. Sievert (1995). Drugs and thyroid function. *N. Engl. J. Med. 333*: 1688.

B. D. Weintraub, M. S. Szkunlinski (1999). Development and in vitro characterization of human recombinant thyrotropin. *Thyroid 9*: 447.

Insulin and Glucagon

A. D. Cherrington (1999). Banting Lecture 1997. Control of glucose uptake and release by the liver in vivo. *Diabetes 48*: 1198.

M. P. Czech (1985). The nature and regulation of the insulin receptor: structure and function. *Annu. Rev. Physiol. 47*: 357–391.

S. Gammeltoft (1984). Insulin receptors: binding kinetics and structure–function relationship of insulin. *Physiol Rev. 64*: 1321–1378.

V. J. Hruby, J.-M. Ahn, D. Trivedi (2001). The design and biological activities of glucagon agonists and antagonists, and their use in examining the mechanisms of glucose action. *Curr. Med. Chem. Immunol. Endocr. Metabol. Agents 1*: 199–215.

R. Iyengar (1986). Structural characterization of the glucagon receptor. In: N. Kraus-Friedman (Ed.). *Hormonal Control of Gluconeogenesis*, vol. 2. Boca Raton: CRC Press, pp. 21–34.

P. F. Kador, J. H. Kinoshita, N. E. Sharpless (1985). Aldose-reductase inhibitors: a potential new class of agents for the pharmacological control of certain diabetic complications. *J. Med. Chem. 28*: 841–849.

W. L. Lee, B. Zinman (1998). From insulin to insulin analogs: progress in the treatment of type 1 diabetes. *Diabetes Rev. 6*: 73.

A. Ling (2002). Small-molecule glucagon receptor antagonists. *Drugs Fut. 27*: 987–993.

C. A. Lipinski, N. J. Hutson (1984). Aldose-reductase inhibitors as a new approach to the treatment of diabetic complications. *Annu. Rep. Med. Chem. 19*: 169–177.

M. E. V. Mora, A. Scarfone, M. Calvani, A. V. Greco, G. Mingrone (2003). Insulin clearance in obesity. *J. Am. Coll. Nutr. 22*: 487–493.

E. van Obberghen, S. Gammeltoft (1986). Insulin receptors: structure and function. *Experientia 42*: 727–734.

L. Pirola, A. M. Johnston, E. van Obberghen (2003). Modulators of insulin action and their role in insulin resistance. *Int. J. Obes. 27*(suppl. 3): S61 S64.

C. R. Rasmussen, B. E. Maryanoff, G. F. Tutwiler (1981). Diabetes mellitus. *Annu. Rep. Med. Chem. 16*: 173–188.

A. R. Saltiel, J. A. Fox, P. Sherline, P. Cuatrecasas (1986). Insulin-stimulated hydrolysis of a novel glycolipid generates modulators of cAMP phosphodiesterase. *Science 233*: 967–972.

P. R. Shepherd, B. B. Kahn (1999). Glucose transporters and insulin action—implications for insulin resistance and diabetes mellitus. *N. Eng. J. Med. 341*: 248.

Hormones and Hypertension

J. M. Connell (1986). Essential hypertension: rational pharmacotherapy. *Trends Pharmacol. Sci. 7*: 412-419.

D. W. Cushman, M. A. Ondetti (1980). Control of blood pressure by angiotensin blockage. *Trends Pharmacol. Sci. 1*: 260–263.

T. Hedner, X. Sun, I. L. Junggren, A. Pettersson, L. Edvinsson (1992). Peptides as targets for antihypertensive drug development. *J. Hypertens. 10*: S121.

T. Inagami, R. C. Harris (1993). Molecular insights into angiotensin II receptor subtypes. *News Physiol. Sci. 8*: 215.

G. Lembo (2002). Hormones and signalling pathways. *Dev. Cardiovasc. Med.* 242 (Cardiovascular Genomics): 77–81.

M. A. Ondetti, D. W. Cushman (1978). Inhibitors of the renin–angiotensin system. *Annu. Rep. Med. Chem. 13*: 82–91.

Natriuretic Hormones

E. J. Benaksas, E. D. Murray, W. J. Wechter (1995). Natriuretic hormones. II. *Prog. Drug Res. 45*: 245–288.

A. J. de Bold, H. B. Borenstein, A. T. Veress, H. Sonnenberg (1981). A rapid and potent natriuretic response to intravenous injection of atrial myocardial extracts in rats. *Life Sci. 28*: 89–94.

M. Curtin, J. Genest (1986). The heart as an endocrine gland. *Sci. Am. 254*(2): 76–81.

R. W. Lappe, R. L. Wendt (1986). Atrial natriuretic factor. *Annu. Rep. Med. Chem. 21*: 273–281.

K. Nakao, Y. Ogawa, S. Suga, H. Imura (1992). Molecular biology and biochemistry of the natriuretic peptide system. II: Natriuretic peptides. *J. Hypertens. 10*: 907.

P. Needleman (1986). Atriopeptin biochemical pharmacology. *Fed. Proc. 45*: 2096–2100.

G. Thibault, F. Amiri, R. Garcia (1999). Regulation of natriuretic peptide secretion by the heart. *Annu. Rev. Physiol. 61*: 193.

R. L. Vandlen, K. E. Arcuri, L. Hupe, M. E. Keegan, M. A. Napier (1986). Molecular characteristics of receptors for atrial natriuretic factor. *Fed. Proc. 45*: 2366–2370.

6

Messenger Targets for Drug Action III

Immunomodulators and their receptors

The final messenger system that can be exploited for purposes of understanding drug action is the immune system. The function of the immune system is to preserve the integrity of the body at a cellular level. It is designed to eliminate pathogens and to protect the host. In fulfilling this goal, the immune system makes extensive use of white blood cells as cellular messengers and mediators. Like neural and endocrine messenger systems, the immune system is capable of manipulating biochemical information for the purpose of maintaining homeostasis and combating disease. The immune system stores information concerning previous infections and transmits information through the integrated use of cellular (white blood cell) and molecular (cytokine) elements.

The immune system is well integrated with the two other principal messenger systems. For example, steroid molecules exert a significant effect upon both the immune and endocrine systems, representing a molecular overlap between the hormonal and immune messenger networks. Various researchers are also investigating a possible interaction between the nervous system and the immune system; for instance, can a neurochemically mediated psychiatric disorder, such as severe depression, lead to a suppression of immune system functioning? Although somewhat controversial, this remains a fascinating area of research, rich in potential. Such opportunities must be explored, since the immune system possesses numerous valuable druggable targets for drug design.

The immune system also represents an important messenger system for drug design as applied to chronic inflammatory diseases, including rheumatoid arthritis, bronchial asthma, and inflammatory bowel disease. As discussed in this chapter, cytokines and their receptors are the key druggable target molecules within the immune system; in future, they will undoubtedly emerge as important drug design targets for the treatment of chronic inflammatory diseases such as rheumatoid arthritis.

6.1 OVERVIEW OF THE IMMUNE SYSTEM AS A SOURCE OF DRUG TARGETS

The immune system responds when the normal, healthy operation of the body (*homeostasis*) is perturbed. In response to such a perturbation, the immune system orchestrates the body's attack on the invading disease (e.g., a bacterial *pathogen*) while retaining the ability to recognize "self" (thereby preventing the body from attacking itself). Agents that augment the immune system are therefore important in the treatment of infectious diseases, such as AIDS, or in the management of disorders in which the distinction between "self" and "nonself" becomes blurred (e.g., cancer). Drugs that suppress the immune system may also be of benefit in the treatment of disorders arising from dysregulation of the immune system (e.g., *autoimmune diseases*; see section 6.4) or in helping a person retain an organ or tissue that has been transplanted from a donor. A wide variety of other diseases, including disorders as diverse as rheumatoid arthritis or multiple sclerosis, also benefit from the manipulation of the immune system. Finally, there is an increasing recognition of the potential role of the immune system in an even greater range of diseases not traditionally thought to arise from a dysfunctional immune system. For instance, recent epidemiological studies suggest that anti-inflammatory agents may play a role in delaying the onset or progression of Alzheimer's disease.

Although the potential for the development of drugs that target the immune system is obvious, the approach to be pursued in achieving this goal is not. The precise molecular-level functioning of the immune system has yet to be fully elucidated and many of the molecular details remain obscure. Accordingly, the medicinal chemistry of the immune system is in its infancy when compared to the successes of drug design for other messenger target systems such as neurotransmitters. Furthermore, since the immune system tends to function at a cellular level, therapeutics designed to address immune system targets may possibly be achieved more readily using biological approaches in preference to classical drug molecule approaches. Stem cell research and other biological therapeutics may offer a better long-term approach to disorders of the immune messenger system. Nonetheless, small organic molecules have historically represented the most viable and successful approach to therapeutics; thus it is imperative that drug design for immune targets be pursued.

6.1.1 Structures of the Immune System Relevant to Drug Action

In designing drugs to target immune-mediated messengers, it is important to appreciate the anatomy and biochemistry of the immune system. The anatomy of the immune system is not nearly as well delineated as that for the other messenger systems, such as the nervous system. The "foot soldiers" of the immune system are the *leukocytes* (white blood cells), which do the majority of the work within the immune system. Leukocytes may be subcategorized as follows:

Agranular leukocytes (Agranulocytes)
    Lymphocytes
        T cells (T_H1 cells, T_H2 cells, CDL cells)
        B cells
    Monocytes
Granular leukocytes (Granulocytes)
    Neutrophils
    Basophils
    Eosinophils

The initial distinction is between granular and agranular leukocytes, depending upon whether they contain small inclusions, called granules, as seen at the level of the light microscope.

Since the leukocytes are the "foot soldiers" of the immune system, they have a number of "training camps and bases of operation." The bone marrow appears to be the source of stem cells for all of the cellular elements of blood. A single *pluripotential stem cell* serves as a precursor for all leukocytes, both lymphocytes and granulocytes. The *thymus,* a small organ located just behind the sternum, is responsible for the maturation and development of T lymphocytes; it also exerts a measure of control and maintenance over other activities of the immune system. The thymus is sometimes removed to control an unusual immune-mediated disease called *myasthenia gravis.* The *spleen,* located in the upper left quadrant of the abdomen, has diffusely packed areas of T cells and B cells and serves as a major filter for blood-borne antigens. Since the spleen is vulnerable to rupture during abdominal trauma, it is sometimes surgically removed (*splenectomy*); people without a spleen are somewhat more susceptible to certain types of bacterial infections. *Lymph nodes* are regionalized collections of lymphocytes and macrophages. Enlarged lymph nodes may represent a response to a localized infection ("swollen glands" in the neck during a bacteria-induced "sore throat") or regional spread of a cancer (firm, enlarged lymph nodes in the armpit of a woman with breast cancer). (See figure 6.1.)

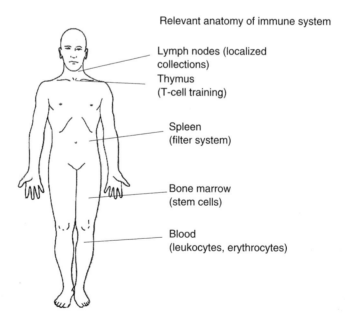

Relevant anatomy of immune system

Lymph nodes (localized collections)
Thymus (T-cell training)

Spleen (filter system)

Bone marrow (stem cells)

Blood (leukocytes, erythrocytes)

Figure 6.1 Relevant anatomy of the immune system. The immune system is distributed throughout the body, permitting it to fulfill its goal of maintaining structural homeostasis in the body at a cellular level. A number of immunomodulator molecules are the messengers of the immune system.

6.1.1.1 Agranular Leukocytes

Lymphocytes are the most common agranular leukocyte. They are small cells, about 8–10 μm in diameter, which are almost entirely filled with nucleus. The bloodstream circulates lymphocytes to every tissue in the body. The task of the lymphocyte is to specifically recognize unwanted molecular targets and then to set in motion the complex machinery of the immune system. This unwanted molecular target, usually a protein, is called an *antigen*, this term being applied to any molecule that is viewed as being foreign or unwanted by the host. There are two main types of lymphocytes: B lymphocytes are the principal mediators for the process of *humoral immunity*; T lymphocytes are the principal mediators for the process of *cell-mediated immunity*.

B Lymphocytes. B lymphocytes are responsible for humoral immunity. Humoral immunity refers to the production of *antibodies* (also called *immunoglobulins*), which are glycoprotein molecules capable of binding to the antigen. There are five classes of immunoglobulins, called IgG, IgM, IgA, IgE, and IgD. The basic unit of every immunoglobulin macromolecule is composed of four peptide chains: two light chains, and two heavy chains. This basic structure is bifunctional, possessing one fragment (the F(ab)$_2$ fragment) that binds to the antigen, and a second fragment (the Fc fragment) that activates other leukocytes, such as neutrophils, to "zero in" on the targeted F(ab)/antigen complex, thereby attacking and, hopefully, killing the unwanted cell. Figure 6.2 shows the structure of an antibody protein.

T Lymphocytes. T lymphocytes are the principal mediators of cell-mediated immunity. Cell-mediated immunity refers to an immune process that does not directly involve antibodies. In cell-mediated immunity the lymphocyte binds to the unwanted cell and is directly able to destroy it. Rejection of a kidney transplant is an example of cell-mediated immunity.

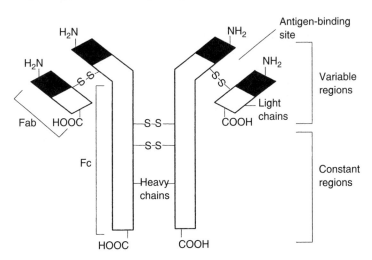

Figure 6.2 Immunoglobulins or antibodies are produced by B lymphocytes and facilitate the destruction of foreign invaders via the humoral component of the adaptive immune system. They consist of four peptide chains: two heavy, two light. The variable regions of the immunoglobulin molecule permit binding to the foreign invader.

Monocytes. After the lymphocyte-initiated killing is over, the resulting cellular debris is cleaned up by another important phagocytic cell called a *macrophage.* The macrophage is derived from the circulating agranular leukocyte called a *monocyte.* In recent years, the immunological importance of monocytes and macrophages has been more greatly appreciated.

6.1.1.2 Granular Leukocytes

Granular leukocytes are the dominant leukocytes. They are heavily involved with the killing of unwanted intruder cells, and are of many types.

Neutrophils. *Neutrophils* are the main type of granulocyte and are the predominant circulating white blood cell of the human body. Neutrophils work in close harmony with lymphocytes; in fact, T lymphocytes, upon being activated through an interaction with an antigen, release chemical messengers (*chemotactic factors*) to attract both B lympho-cytes and neutrophils to the site of the immune response. This process is referred to as *chemotaxis.* Immunoglobulins, produced by B lymphocytes, bind to the "unwanted" antigen via the F(ab)$_2$ end of the antibody molecule, leaving the Fc end of the molecule exposed and free; the neutrophil possesses a surface receptor to recognize this Fc por-tion of the antibody molecule, to which it then binds. In turn, this leads to the creation of a neutrophil–antibody–antigen–unwanted cell complex. The neutrophil then kills the unwanted cell by the process of *phagocytosis,* in which the foreign substance (for example, a bacterium) is literally engulfed by the neutrophil—completely "swallowed" via a membrane-bound sphere called a *vacuole.* Once the vacuole is engulfed, enzyme-containing bodies (called *phagolysosomes*) contained within the neutrophil fuse with the vacuole and release their destructive enzymes, leading to the death of the unwanted cell.

Basophils and Eosinophils. These two types of granulocyte represent only a small proportion of the circulating leukocytes. The basophil possesses receptors specific for the Fc portion of IgE molecules. It is involved in allergic reactions and can play a role in the clinical state of *anaphylaxis,* which is a life-threatening allergic response. Eosinophils are involved with the immune reactivity of drug allergies and with the body's immune response to parasitic infestations such as worms (see chapter 9).

These diverse cellular constituents of the immune system provide protection and repair within the body by two principal components: the innate and the adaptive immune systems. From a functional perspective, these two immunological systems are crucial to the process of future drug design.

6.1.2 The Innate Immune System

The innate immune system constitutes the front line of defense against disease and has three structural components:

1. Histological (at the tissue level) defense systems (e.g., skin)
2. Cytological (at the cellular level) defense systems (e.g., macrophages, neutrophils)
3. Biochemical (at the molecular level) defense systems (e.g., complement proteins)

The tissue-level (histological) defense systems are composed primarily of the skin and the mucous membranes (*mucosa*) that line the airways, mouth, and other portions of the gastrointestinal system. Healthy, intact skin is probably the most important barrier against chemical or biological insults.

When the barriers of the tissue-level defences are breached, the cytological and biochemical components of the innate immune system take over. Within these latter two components, the *complement protein cascade* plays a dominant role. Complement is a system of approximately 20 separate molecules (nine "classical" complement proteins and ten or more cofactors) that continuously circulate within the bloodstream in an inactive form. Frequently, they are stabilized in their inactive form through association with cell membranes. However, when a "nonself" molecule becomes present within the body, the complement system is activated. The nine proteins of complement are split into fragments during the activation process. The complement protein C3b attaches itself to the invading cell through a process called *opsonization*, thereby tagging the invader so that it can be recognized and destroyed by neutrophils and macrophages. Through this process, C3b acts as an *opsonin*. Complement proteins C3a and C5a then attract the neutrophils and macrophages to the tagged (opsonized) invader. Complement proteins C5b, C6, C7, C8, and C9 associate with one another to form a *membrane attack complex* (MAC) that punches holes into the membrane of the invading cell. In carrying out this function, complement is assisted by other molecules called *cytokines* or *chemokines* (e.g., interleukin-8, IL-8) that facilitate the process whereby neutrophils and monocytes leave the bloodstream to enter the tissue in which the invader is located.

The molecular by-products produced by complement activation and by the activities of the innate immune system serve as molecular cues that, in turn, activate the adaptive immune system.

6.1.3 The Adaptive Immune System

If the innate system fails, the adaptive immune system is mobilized to continue the fight. This system constitutes the second line of defense against disease and has two structural components:

1. The humoral immune system
2. The cell-mediated immune system

The adaptive immune system is much more sophisticated than the innate immune system. Three characteristics reflect this degree of sophistication:

1. Unlike the innate system, which is relatively nonspecific, the adaptive system responds to unwanted cells or molecules in a *specific* manner.
2. The adaptive system can discriminate between nonself (i.e., "foreign") and self, thereby preventing an undesirable attack upon its own molecules or cells.
3. The adaptive immune system has the capacity to learn and to establish memory, thereby enabling it to respond more efficiently to a previously encountered unwanted antigen—an important property central to the therapeutic effectiveness of vaccines.

The two components of the adaptive immune system (humoral and cell-mediated) are not competitive with each other; rather, they work together in harmony. This harmonious attack is facilitated by the cooperative efforts of both B-type and T-type lymphocytes and is heavily dependent upon the actions of cytokines—molecular messengers that coordinate the activities of the various types of lymphocytes.

T cells and B cells are the centerpiece of adaptive immunity. When an unwanted protein or molecule has successfully evaded the innate immune system, a macrophage, acting as an *antigen-presenting cell* (APC), enzymatically digests part of the unwanted protein and then attaches the resulting peptide fragments to a surface protein, called a Class II MHC (major histocompatibility complex) protein; this attachment process is facilitated by *adhesion molecules* (e.g., ICAM-1). The resulting processed "antigenic peptide fragment–class-II MHC surface protein" complex forms a macromolecular assembly that is then "presented" to a circulating subtype of T cell called a *T helper cell* or a T_H cell. These activated T_H cells secrete a cytokine molecule (interleukin-2, IL-2), which triggers activation and proliferation of two new classes of lymphocytes, called T_H1 and T_H2 cells. The T_H1 cells then produce two additional cytokines (interferon-γ [IFN-γ]; tumor necrosis factor-β [TNF-β]), which activate two additional classes of leukocytes (activated natural killer cells [NKs]; activated cytotoxic T cells [CTLs]). The NK and CTL cells then attack and kill the unwanted intruder cell; in addition, the NK cells produce another messenger molecule, called a *lymphokine*, which activates a group of "promiscuous killer cells," called *lymphokine-activated killer* (LAK) cells. While the T_H1 cells are initiating this complex cascade, the T_H2 cells are concomitantly producing two different cytokine molecules (interleukin-4 [IL-4] and interleukin-5 [IL-5]). IL-4 and IL-5 then establish a molecular interface with the B-cell lymphocyte system. IL-4 and IL-5 induce the proliferation and differentiation of B-cell lymphocytes, resulting in the increased production of antibody immunoglobulins. As discussed above, the immunoglobulins "set up" the unwanted cellular intruder for attack and removal by neutrophils and phagocytic macrophages. A further level of fine-tuning is imposed on this immune cascade by one additional cytokine molecule (interleukin-10, IL-10), which mediates a feedback control loop between the T_H1 and T_H2 subsets of cells.

Since the immune system exerts a major influence upon inflammation in the body, drug design that targets immunocompetent systems would appear to have many useful applications. Logically, drug design that targets immune messengers and their receptors can be divided into three broad groupings:

1. Immunosuppressive agents
2. Immunostimulating agents
3. Immunomodulating agents

A survey of the relatively small number of drugs available for influencing the immune system reveals that the majority of them are immunosuppressive agents. Clearly, this is an area of medicinal chemistry with immense potential for growth and innovation. As the various cytokine messengers and their receptors become better understood, important opportunities for drug design will emerge.

6.2 DESIGN OF IMMUNOSUPPRESSIVE DRUGS

Logically, immunosuppressive drug design can be divided into five main categories:

1. Agents ("small" molecules) that inhibit lymphocyte proliferation
   "Nonspecific" cytotoxic agents

Azathioprine, cyclophosphamide, methotrexate, leflunomide
"Lymphocyte-specific" cytotoxic agents
Mycophenolate mofetil
2. Agents ("small" molecules) that inhibit cytokine production or action
Corticosteroids
Prednisone, dexamethasone
Natural product derivatives
Cyclosporine, tacrolimus, sirolimus
Thalidomide
3. Agents ("small" molecules) that target cytokine receptors
4. Cytokines
Interferons (IFN-β)
5. Antibodies that target lymphocytes or cytokine receptors

Of these five categories, the nonspecific cytotoxic agents and corticosteroids are the most widely used clinically.

Although commonly employed as immunosuppressants, the cytotoxic agents are traditionally used to treat cancer. These drugs inhibit the growth of rapidly dividing cells. Since blood cells, such as leukocytes or lymphocytes, tend to proliferate rapidly, they are preferentially susceptible to cytotoxic agents. However, the majority of these agents (azathioprine (**6.1**), cyclophosphamide (**6.2**), methotrexate (**3.31**)) are not specific for immune cells and simply block the proliferation of all rapidly reproducing cells. Azathioprine is an imidazolyl derivative of mercaptopurine and thus functions as an antimetabolite; cyclophosphamide is an alkylating agent; methotrexate is a derivative of folic acid and functions as an antimetabolite; and leflunomide (**6.3**) is a prodrug of an inhibitor of pyrimidine biosynthesis.

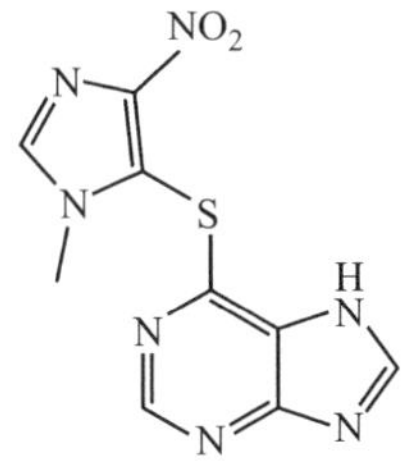

Azathioprine (6.1)

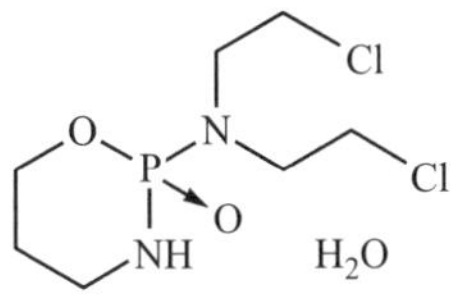

Cyclophosphamide (6.2)

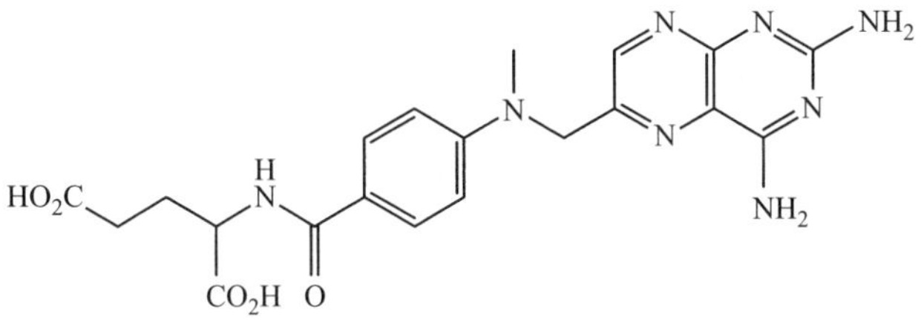

Methotrexate (3.31)

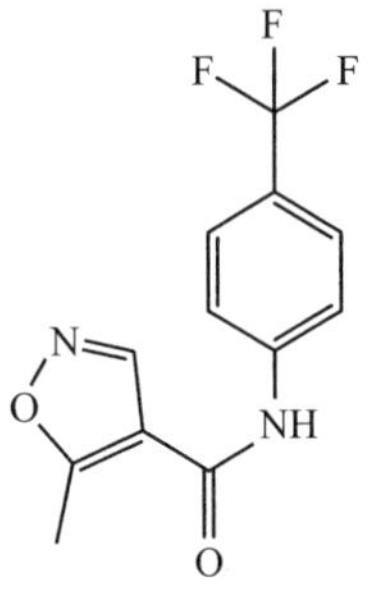

Leflunomide (6.3)

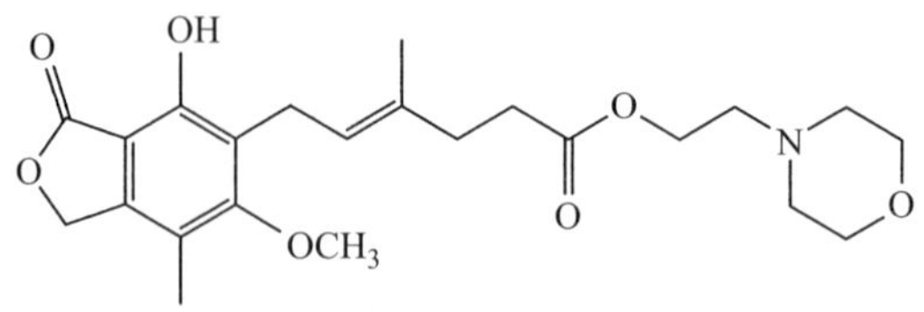

Mycophenolate Mofetil (6.4)

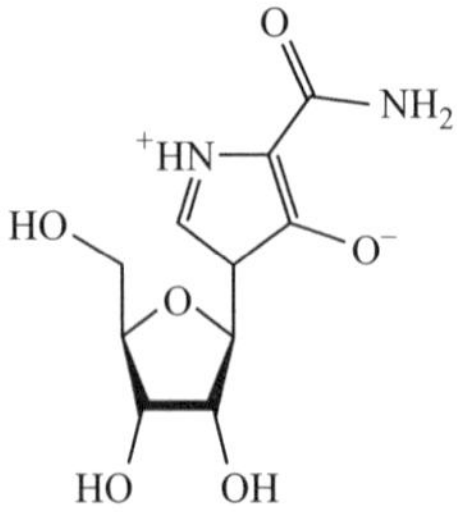

Mizoribine (6.5) 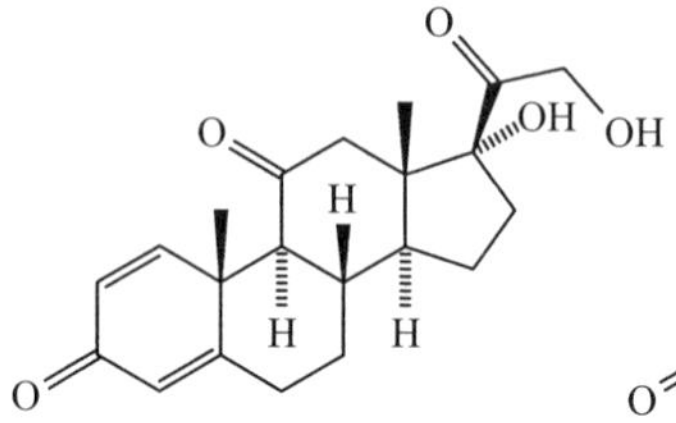   Prednisone (3.7) 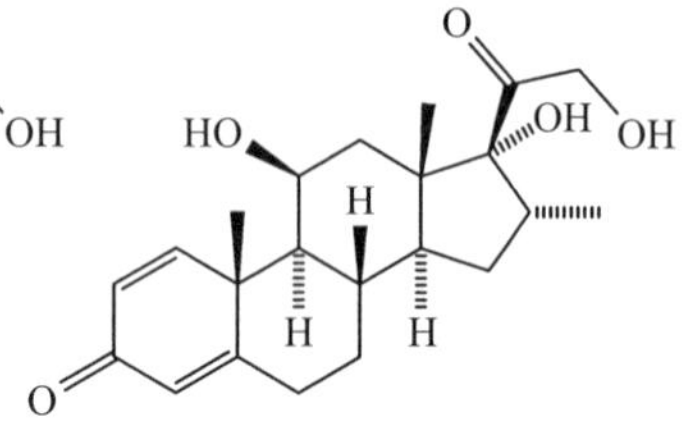   Dexamethasone (5.65)

Unlike these nonspecific agents, mycophenolate mofetil (**6.4**) tends to be a lymphocyte-specific cytotoxic agent. Mycophenolate mofetil is a semisynthetic derivative of mycophenolic acid, isolated from the mold *Penicillium glaucum*. It inhibits both T and B lymphocyte action. Since it inhibits the enzyme inosine monophosphate dehydrogenase, which catalyses purine synthesis in lymphocytes, this agent has a more specific effect on lymphocytes than on other cell types. Mizoribine (**6.5**) is a closely related drug which inhibits nucleotide synthesis, preferentially in lymphocytes.

The next major class of immunosuppressive agents is the corticosteroids, which includes drugs such as prednisone (**3.7**) and dexamethasone (**5.65**). Corticosteroids affect immune function through many different mechanisms. They interfere with the cell cycle of lymphoid cells and are even toxic to specific subsets of T cells. Corticosteroids impair the process of chemotaxis, decrease the number of circulating lymphocytes, and partially inhibit IL-1, IL-2, and INF-γ production. Additionally, corticosteroids lower the concentrations of specific immunoglobulins. Because of this broad spectrum of activity, corticosteroids are the most widely used immunosuppressive agents. Regrettably, corticosteroids are also hormones and thus their use is associated with significant and potentially severe side effects. Long-term use of corticosteroids results in a distinctive cluster of side effects known as *Cushing's syndrome*. To avoid these side effects, the cytotoxic agents, such as azathioprine and cyclophosphamide, are occasionally used; accordingly, the cytotoxic agents are sometimes referred to as *steroid sparing agents*.

In addition to steroids, a structurally diverse group of natural products is used to inhibit cytokine production and action. The first of these is cyclosporine (**6.6**). Cyclosporine is an antibiotic polypeptide isolated from fungi. It influences immune function through several routes. Following administration, cyclosporine is taken up by lymphocytes and binds to a receptor protein, called *cyclophilin*, located in the cytoplasm of the lymphocyte. By interacting with this cytoplasmic protein, cyclosporine inhibits the enzyme phosphatase calcineurin, which in turn plays an integral role in T-cell signal transduction. Cyclosporine also inhibits the gene transcription of IL-2, IL-3, IFN-γ, and other cytokines produced by activated T cells.

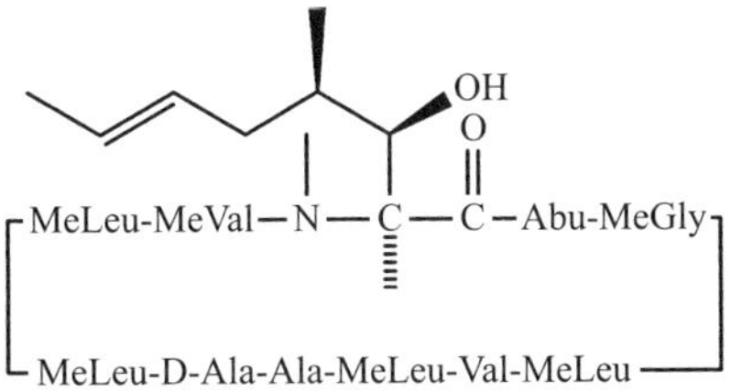

Cyclosporin A (6.6)

Although it is not chemically related to cyclosporine, tacrolimus (**6.7**) has a similar mechanism of action. Tacrolimus is an immunosuppressant macrolide antibiotic derived from *Streptomyces tsukubaenis*. Like cyclosporine, tacrolimus inhibits the same cytoplasmic phosphatase, calcineurin, which catalyzes the activation of a T-cell-specific transcription factor (NF-AT) involved in the biosyntheses of interleukins such as IL-2. Sirolimus (**6.8**) is a natural product produced by *Streptomyces hydroscopicus*; it blocks the ability of T cells to respond to cytokines.

Thalidomide (**6.9**) is another agent that alters cytokine action. Thalidomide is the infamous drug that produced disastrous birth defects (such as *phocomelia*, in which the hands or feet are attached close to the body, resembling flippers, because of very short limbs) when administered to pregnant women. Thalidomide stimulates responsiveness to IL-4 and IL-5 cytokine stimulation, thus favoring the T_H2 subtype of cell and distorting the balance between T_H1 and T_H2 effects. Despite its unfortunate history, thalidomide is an effective agent, now widely used for a variety of immune-related disorders, including *leprosy* and *systemic lupus erythmatosus*.

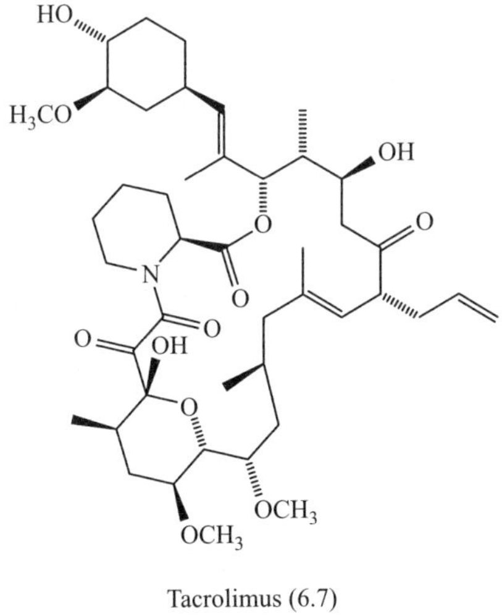

Tacrolimus (6.7)

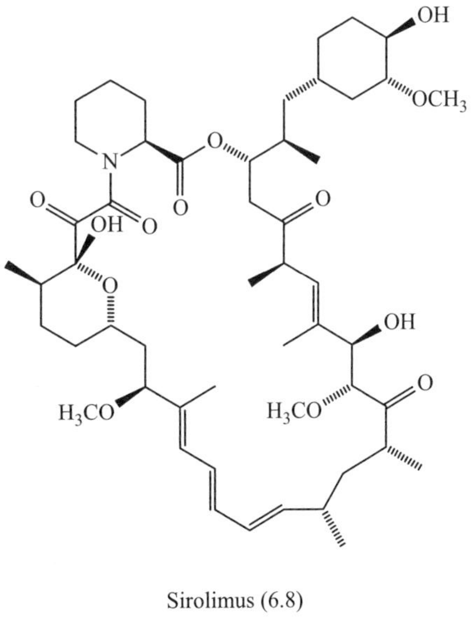

Sirolimus (6.8)

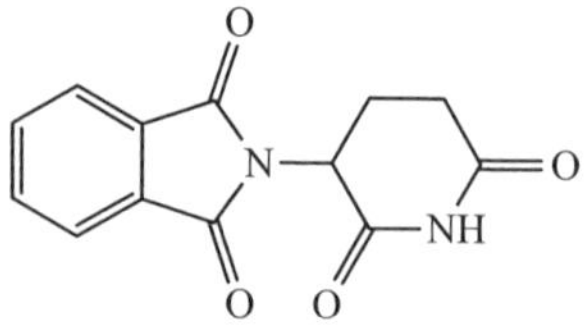

Thalidomide (6.9)

In addition to these small-molecule therapeutics, there are also several macromolecular or biological therapeutics. For example, muromonab is a monoclonal antibody that blocks antigen recognition by T lymphocytes. Infliximab is a chimereic IgG monoclonal antibody possessing a human Fc immunoglobulin region; infliximab binds TNF-α cytokine, thus suppressing the synthesis of IL-1 and IL-6 and therefore decreasing leukocyte activation. However, amongst these various biological macromolecules, the interferons are those with the most established efficacy as potential immunosuppressive therapeutics.

6.2.1 Interferons

The interferons (IFN) are a family of cytokine molecules; cytokines are described in more detail in section 6.3.1. Interferons are glycoproteins that can be divided into three families: IFN-α, IFN-β, and IFN-γ. IFN-α and IFN-β are sometimes collectively referred to as Type I IFNs because they share a 30% sequence homology in their primary amino acid structure. IFN-γ is structurally separate and is a Type II IFN. Type I IFNs are acid stable; Type II IFNs are acid labile. IFN-α is biosynthesized in lymphocytes and macrophages; IFN-β is produced in fibroblasts, epithelial cells, and macrophages; and IFN-γ is biosynthesized in various types of lymphocyte, including CD4, CD8, and NK cells.

The effects of interferons on the human immune system are highly variable. IFN-β tends to suppress certain aspects of immune function, whereas IFN-α can inhibit immune cell proliferation; IFN-γ, on the other hand, displays immune-enhancing properties. All three types of interferon have been studied preclinically and even clinically.

6.2.1.1 IFN-α

More than 24 human genes code for 16 variants of IFN-α molecules. Human IFN-α proteins contain either 165 or 166 amino acid residues, with an overall molecular weight of 19,000 Da. The two primary human subtypes are IFN-α-2a and IFN-α-2b, which differ only at position 23 (a lysine residue in IFN-α-2a and arginine in IFN-α-2b). IFN-α type proteins exhibit complex antineoplastic and antiviral properties. Recombinant variants have been studied in the treatment of hepatitis C.

6.2.1.2 IFN-β

This protein tends to be somewhat unstable for routine clinical use. Accordingly, a 165-amino acid, stable, recombinant analog of IFN-β-1b (in which a serine replaces cysteine at position 17) is used. The IFN-β proteins have a molecular weight of 18,500 Da. Various IFN-β proteins have been studied for the treatment of multiple sclerosis.

6.2.1.3 IFN-γ

This glycoprotein has a molecular weight of 15,500 Da. One member of this family, IFN-γ-1b, is 140 amino acids long and has been studied clinically as a treatment for lung cancer and kidney cancer.

6.2.2 Clinical–Molecular Interface: Multiple Sclerosis and IFN-β

CB is a 32-year-old female who presented to the emergency department with the sudden onset of visual loss in her right eye. She was completely blind in that eye. She had previously been well, but on detailed questioning other symptoms became apparent. Three years earlier, she had experienced numbness of her entire left leg. The problem had lasted 12 weeks, but had completely resolved. Over the past year, she had noticed that sometimes, when she flexed her neck forward to look downwards, she would experience an "electric shock" sensation down her back. Over the past several months, she had noticed transient problems with double vision whenever she took a hot bath; upon getting out of the bathtub, the double vision would promptly resolve. A clinical diagnosis of multiple sclerosis was put forth. Laboratory support for this diagnosis was obtained when a magnetic resonance imaging (MRI) scan revealed multiple areas of demyelination within the brain. Her current problem with visual loss was treated acutely with intravenous methylprednisolone. She was then started on a regimen of IFN-β to decrease the likelihood of future exacerbations.

Multiple sclerosis is a fascinating disease that is typically treated by pharmacological manipulation of the immune system. It most frequently affects people who live in temperate climates (rather than from hotter, equatorial countries), or at least people who spent their first 15 years of life in a temperate climate. Multiple sclerosis is referred to as a *demyelinating disease*, because it episodically involves the patchy loss of myelin from the brain and spinal cord. The discrete area of myelin loss is called a *plaque* and can occur in any white-matter area of the central nervous system. Since the plaques can occur in various places in the brain and sometimes years apart, multiple sclerosis is said to be defined by "demyelinating lesions separated in space and time." If one regards myelin as an insulating fat substance wrapped around axons in the brain, then demyelination causes localized "short circuits" within the brain, thereby giving rise to a wide variety of different symptoms. Transient (8–12 weeks) loss of vision in an eye (*optic neuritis*) is a frequent early symptom of multiple sclerosis. The disease is now frequently treated with interferon-β or sometimes with large doses of corticosteroids to reduce the severity of a particular exacerbation of the disease.

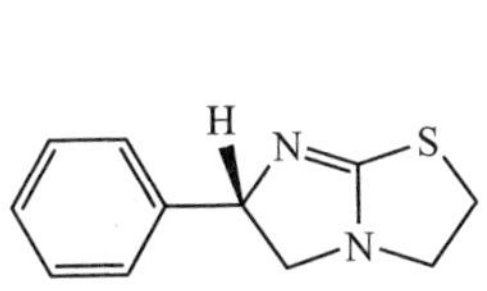

Levamisole (6.10) 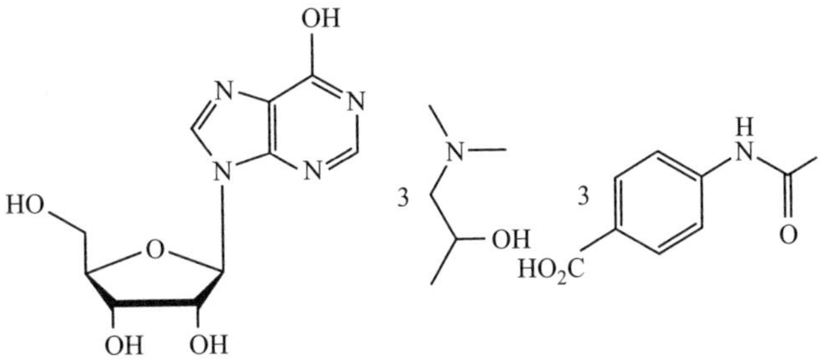   Inosiplex (6.11)

6.3 DESIGN OF IMMUNOMODULATING DRUGS

The design of new chemical entities to either stimulate or somehow modulate the immune system represents one of the new frontiers of medicinal chemistry. The clinical indications for such agents would be considerable, including people with immunodeficiency states arising from cancer or chronic infections. However, this is a drug design area that is still very much in its infancy. Serendipity has led to the identification of several compounds with limited immunomodulating activity. For example, levamisole (**6.10**) is a compound that activates macrophages, while inosiplex (**6.11**) augments T cell function.

Given the importance of immunomodulators, this is a drug design area that is too important to be left to the vagaries of serendipity. The cytokines represent the best hope for rational drug design in the area of immunomodulation.

6.3.1 Cytokines

"Cytokine" is a general term used for a diverse assortment of water-soluble protein molecules that mediate interactions between the various cells of the innate and adaptive immune systems. Over 100 human cytokines are currently under study and more than 300,000 research papers have been published in this area over the past 10–15 years. The interferons, discussed in section 6.2.1, are a subset of the cytokines, as are the interleukins, discussed in section 6.1.3. The nomenclature used to describe cytokines is somewhat confusing and varying subclasses of cytokines are recognized:

1. Lymphokines—cytokines derived from lymphocytes
2. Monokines—cytokines derived from monocytes
3. Chemokines—cytokines that regulate leukocyte movement
4. Interferons (IFN)—cytokines that modulate ("interfere" with) leukocyte function
5. Interleukins (IL)—cytokines that mediate communication between leukocytes
6. Tumour necrosis factors (TNF)—cytokines that are immunotoxic to tumors and bacteria
7. Hematopoietic growth factors—cytokines that regulate proliferation of blood cells
8. Granulocyte colony-stimulating factors (G-CSF)—cytokines that stimulate the growth of granulocytic leukocytes
9. Macrophage colony-stimulating factors (M-CSF)—cytokines that stimulate the growth of macrophages
10. Derived growth factors—cytokines that regulate proliferation of other blood elements such as platelets (e.g., platelet-derived growth factor [PDGF])

Cytokines will act only on cells that express cytokine receptors. The classification of cytokine receptors is also somewhat confusing. Cytokine receptors can be divided into five broad groupings:

1. Class I cytokine receptors (IL-2, IL-7, IL-9, IL-13, G-CSF)
2. Class II cytokine receptors (INF-α, INF-β, INF-γ)
3. Chemokine receptors (IL-8)
4. TNF receptors (TNF-α, TNF-β)
5. Immunoglobulin superfamily receptors (IL-1, M-CSF)

Regrettably, the receptor classification system is much more complex than this. For example, the chemokine class of cytokine receptors is further subdivided into at least 16–20 different receptor subtypes, including CCR1, CCR2 ... CCR11, CXCR1, CXCR2 ... CXCR5. These receptors interact with more than 25–30 different types of chemokines.

Complicated though the classes of cytokines and their receptors may be, this explosion of data does offer significant hope that druggable targets can and will be identified for purposes of drug design. To date, most of the work to develop therapeutics has been done through a more biological approach; i.e., developing antibodies against cytokine targets. However, small organic molecules capable of targeting cytokines and cytokine receptors are beginning to emerge:

1. By screening compound libraries, selective antagonists for the chemokine CCR3 receptor have been identified; these contain arylpiperidine motifs linked by a 3–6 carbon tether to a second aromatic group. Such compounds may have utility in treating chronic inflammatory disorders such as bronchial asthma.
2. Screening assays are also being used to optimize a series of benzimidazole derivatives and a series of 4-pentadienamide derivatives as antagonists for the chemokine CCR2 receptor. Such compounds may be useful in treating the inflammation that occurs in the walls of blood vessels undergoing degenerative atherosclerotic processes.
3. The cytokine TNF-α is synthesized through the actions of the TNF-α-converting enzyme (TACE), which cleaves the 76-amino acid TNF-α from its 233-amino acid precursor. Based upon computer-aided drug design (sequence homology studies) and X-ray crystallographic studies, rational drug design is leading to TACE inhibitors as potential therapeutics for inflammatory bowel diseases such as ulcerative colitis or Crohn's disease.
4. The biosynthesis of the IL-12 cytokine is dependent upon various enzymes, including phosphodiesterase 4 (PDE4). Thus, PDE4 enzyme inhibitors act as functional IL-12 antagonists, and may have clinical application in the treatment of rheumatoid arthritis.
5. Cytokine-suppressing anti-inflammatory drugs (CSAIDs) are a new class of agents that inhibit the production of cytokines such as TNF-α. Many prototypic CSAIDs are built around a bicyclic imidazole framework and are being optimized through the application of quantitative structure–activity relationship studies of analogous series.

An exploration and exploitation of cytokines represents the future of drug design for the purposes of manipulating the immune messenger systems. Table 6.1 lists a number of cytokines being pursued as drug design targets. Drugs based upon the biochemistry of cytokines will have widespread utility in many chronic inflammatory disorders, especially the collagen-vascular diseases (see section 6.4).

6.3.2 Growth Factors

Just as interferons and interleukins are a subset of the class of molecules known as cytokines, in turn, cytokines are a subset of a larger family of proteins known as growth factors (see figure 6.3). Although not all growth factors are involved in regulation of the immune system, they are nevertheless protein messenger molecules that influence cell

Table 6.1 Cytokines Being Exploited as Therapeutic Targets

Cytokine	Properties
Granulocyte colony stimulating factor (G-CSF)	Granulocyte production
Granulocyte-macrophage colony stimulating factor (GM-CSF)	Granulocyte, monocyte production
Interferon-α (IFN-α)	Activates NK cells
Interferon-β (IFN-β)	Activates NK cells
Interferon-γ (IFN-γ)	Activates TH1 cells, NK cells, CTLs, macrophages
Interleukin-1 (IL-1)	T cell activation, B cell proliferation and differentiation
Interleukin-2 (IL-2)	T cell proliferation, TH1, NK, LAK cell activation
Interleukin-3 (IL-3)	Hematopoietic stem cell proliferation and differentiation
Interleukin-4 (IL-4)	TH2 and CTL activation, B cell proliferation
Interleukin-5 (IL-5)	Eosinophil proliferation, B cell proliferation
Interleukin-6 (IL-6)	HCF, TH2, CTL, and B cell proliferation
Interleukin-7 (IL-7)	CTL, NK, LAK and B cell proliferation
Interleukin-8 (IL-8)	Neutrophil chemotaxis
Interleukin-9 (IL-9)	T cell proliferation
Interleukin-10 (IL-10)	TH2 suppression, CTL activation, B cell proliferation
Interleukin-11 (IL-11)	B cell differentiation
Interleukin-12 (IL-12)	TH1 and CTL activation
Interleukin-13(IL-13)	B cell proliferation
Interleukin-14 (IL-14)	B cell proliferation
Interleukin-15 (IL-15)	TH1, CTL, NK/LAK activation
Interleukin-16 (IL-16)	Chemotaxis
Interleukin-17 (IL-17)	Cytokine production, enhanced cytokine biosynthesis
Macrophage colony stimulating factor (M-CSF)	Monocyte production, macrophage activation
Tumor necrosis factor-α (TNF-α)	Macrophage activation
Tumor necrosis factor-β (TNF-β)	Promotion of chemotaxis

division and proliferation. Growth factors are gene products that exert important roles in the regulation of cell division and tissue proliferation. They are proteins or glycoproteins that, upon binding to cellular receptors, regulate the proliferation and differentiation of cells and tissues. The growth factors include:

1. Cytokines
2. Keratocyte growth factor
3. Insulin-like growth factor
4. Fibroblast growth factor
5. Epidermal growth factor
6. Vascular endothelial growth factor
7. Connective tissue growth factor

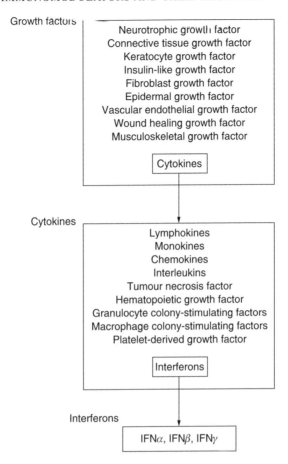

Figure 6.3 A vast array of chemical mediators influence immunomodulation and cellular growth. These various mediators are linked to one another. For example, interferons are a subset of cytokines, which in turn are a subset of growth factors.

  8. Neurotrophic growth factor
  9. Wound-healing growth factor
10. Musculoskeletal growth factor

Whereas the cytokines influence the development of blood cells and other cellular elements of the immune system, the other growth factors regulate the development of a wide range of other cell types. Within each of these categories are additional subdivisions into further types of growth factors. For example, neurotrophic growth factors include many different types of *neurotrophins,* including nerve growth factor (NGF) and brain-derived neurotrophic factor (BDNF). These various factors influence the development of neurons.

   Not surprisingly, growth factors are messenger molecules that are being exploited to treat a wide variety of diseases at the cellular level of attack. These diseases include

Figure 6.4 Drugs targeting vascular endothelial growth factor.

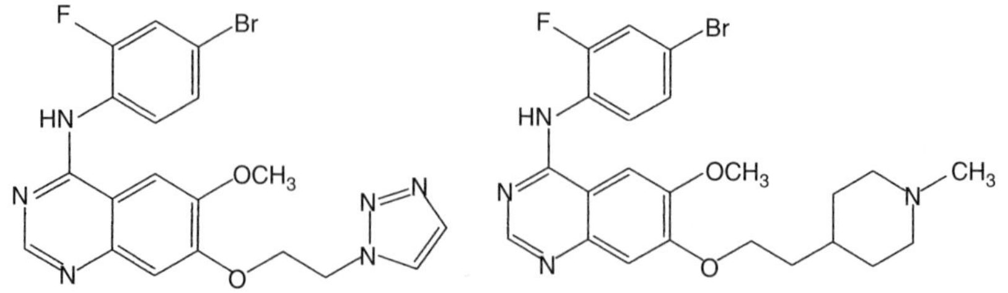

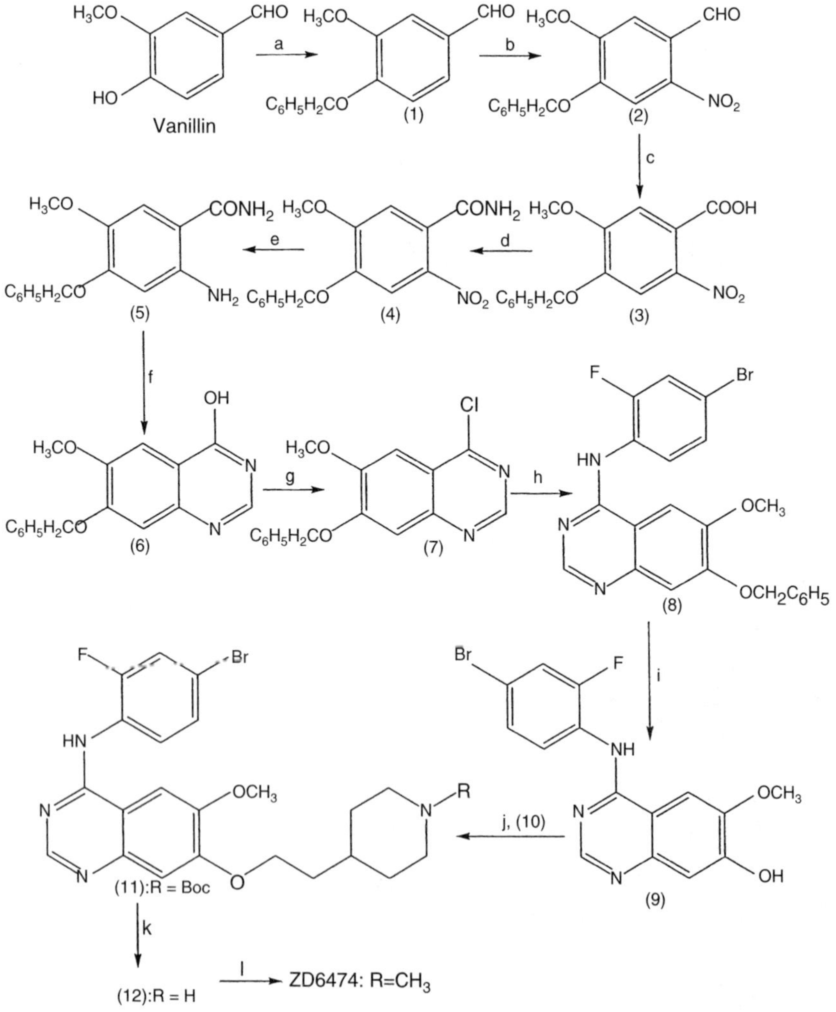

(Continued)

Scheme b

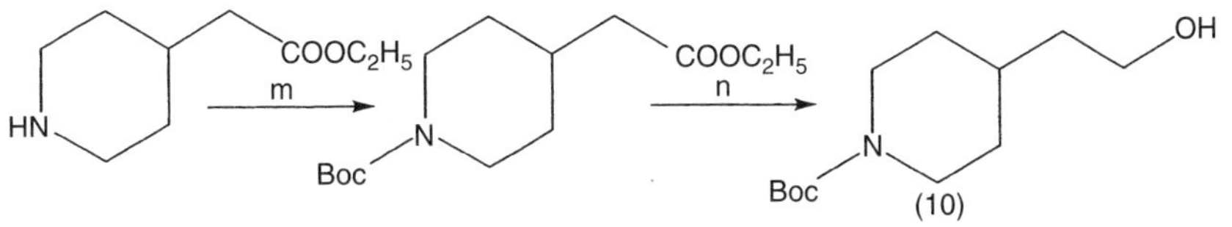

ZD6474 is one of the many immunomodulating molecules currently in clinical trials for the treatment of various types of cancers. Its aqueous solubility (330 μM at pH 7.4) and anti-tumor activity at a dosage as low as 12.5 mg/kg per day in mice made this compound a good candidate for clinical trials in replacement of the more potent but less bioavailable analog ZD4190.

Unlike conventional anticancer drugs whose ability to inhibit tumor growth is attributable to a direct cytotoxic or cytostatic effect on tumor cells, ZD6474 exerts its anticancer activity by blocking tumor angiogenesis *via* the inhibition of the vascular endothelial growth factor (VEGF) signaling pathway. The total synthesis of the drug candidate ZD6474 (Hennequin et al., 1999) has been achieved in a twelve-step procedure starting from the readily available vanillic aldehyde, a naturally occurring catechol derivative that is directly extracted from vanilla beans (Hocking, 1997) (reaction scheme a).

The alcohol 10 introduced at step j was prepared following a literature procedure from ethyl-2-(piperidin- 4-yl)acetate after N-Boc protection and selective reduction of the ester group using lithium aluminum hydride (Villalobos et al., 1994) (reaction scheme b).

References for Figure 6.4

L. F. Hennequin, A. P. Thomas, C. Johnstone, E. S. E. Stokes, A. P. Ple, J.-J. M. Lohmann, D. J. Ogilvie, M. Dukes, R. S. Wedge, J. O. Curwen, J. Kendrew, C. Lambert-van der Brempt (1999), *J. Med. Chem. 42*: 5369–5389.

M. B. Hocking (1997), *J. Chem. Educ. 74*: 1055–1059.

A. Villalobos, J. F. Blake, C. K. Biggers, T. W. Butler, D. S. Chapin, Y. L. Chen, J. L. Ives, S. B. Jones, D. R. Liston, A. A. Nagel, D. M. Nason, J. A. Nielsen, I. A. Shalaby, W. F. White (1994), *J. Med. Chem. 37*: 2721–2734.

cancer, hepatitis C, osteoporosis, diabetic foot ulcers, wound care, and muscle wasting. Drugs based on the neurotrophic growth factors are being developed to treat chronic pain disorders such as diabetic neuropathy. Small-molecule drugs based on vascular endothelial growth factors are being developed as anticancer agents, because they inhibit the ability of the tumor to grow the new blood vessels (*angiogenesis*) that are crucial to supporting its rapid rate of growth. The synthesis of such an anti-angiogenesis agent is given in figure 6.4. Clearly, there are many diseases that can be targeted using small molecules based on molecular messengers such as interleukins, cytokines, or growth factors.

6.4 THE CLINICAL–MOLECULAR INTERFACE: COLLAGEN DISEASES

The collagen diseases (also known as collagen-vascular diseases, connective tissue diseases, or rheumatoid diseases) are a group of clinicopathological entities presenting with common and overlapping clinical, cellular, and molecular features. Some of the disorders that are collectively referred to as collagen diseases include:

1. Rheumatoid arthritis
2. Systemic lupus erythmatosus
3. Scleroderma
4. Dermatomyositis
5. Polyarteritis nodosa

Each of these major diseases exhibits prominent constitutional symptoms coupled with varying patterns of organ involvement. The common cellular features of this group of diseases include widespread immune-mediated inflammatory damage to connective tissue and blood vessels. Connective tissues are composed of varying combinations of collagen, elastin, proteoglycans, and other less well-characterized glycoproteins.

Although the precise molecular mechanisms underlying collagen diseases have yet to be fully elucidated, they all appear to be autoimmune diseases. Autoimmunity occurs when the body turns the immune system against itself, failing to distinguish "self" tissues from "nonself" (foreign) tissues. When the body mounts an immune response against itself, there is an activation of self-reactive T and B lymphocytes that generates cell-mediated or humoral immune-mediated actions directed against self proteins. Rheumatoid arthritis and systemic lupus erythmatosus are both well-characterized examples of autoimmune diseases. In rheumatoid arthritis, self-directed antibodies (*rheumatoid factors*) form immune complexes, leading to inflammation of the joints and various other organs, including the kidneys. In systemic lupus erythmatosus, autoantibodies are made against self-DNA, red blood cells, white blood cells, platelets, mitochondria, ribosomes, histone proteins, and many other proteins.

Although abundant evidence supports the existence of such an autoimmune phenomenon, the causative event that heralds this self-directed immune-mediated attack remains uncertain. A number of mechanisms have been proposed to afford a molecular-level explanation of autoimmunity. One such explanation is *molecular mimicry*. Molecular mimicry occurs when a protein associated with a foreign substance bears structural similarities to a protein found in the host. For example, if a person experiences an infection from bacteria, there is a possibility that a protein in the bacterium shares certain similar geometrical and conformational features with a protein already existing in the person. Thus, an immune response directed against the bacteria will cross-react with organs in the host organism.

Because of this self-directed immune attack, the symptoms and signs of such diseases are typically diffuse, arising from many different organ systems. Systemic lupus erythmatosus is a superb example of this. The clinical symptoms include weight loss, arthritis, skin rashes ("butterfly rash" of the face), hair loss (alopecia), cold and blue finger tips (Raynaud's phenomenon), abdominal pain, lung inflammation, heart murmurs, enlarged spleen, enlarged liver, psychosis, and epilepsy. Rheumatoid arthritis may also involve several organs. Since rheumatoid arthritis can extend well beyond joint disease, some clinicians regard this disorder as *rheumatoid disease* rather than rheumatoid arthritis.

The treatment of collagen disease is based on immunosuppressive therapies. Immunosuppressive agents, such as corticosteroids, are widely used. In addition, cytotoxic agents (azathioprine, cyclophosphamide, and methotrexate) have also been administered.

Selected References

Immunomodulators

Y. Azuma, K. Ohura (2004). Immunological modulation by lidocaine–epinephrine and prilocaine–felypressin on the functions related to natural immunity in neutrophils and macrophages. *Curr. Drug Targ.: Immune, Endocr. Metabol. Disord. 4*: 29–36.

S. Chada, R. Ramesh, A. M. Mhashilkar (2003). Cytokine- and chemokine-based gene therapy for cancer. *Curr. Op. Mol. Ther. 5*: 463–474.

W. J. Fantl, S. Rosenberg (2000). Anti-angiogenesis as a therapeutic strategy for cancer. *Ann. Rep. Med. Chem. 35*: 123.

M. Fischereder, M. Kretzler (2004). New immunosuppressive strategies in renal transplant recipients. *J. Nephrol. 17*: 9–18.

I. Krause, G. Valesini, R. Scrivo, Y. Shoenfeld (2003). Autoimmune aspects of cytokine and anticytokine therapies. *Am. J. Med. 115*: 390–397.

E. C. Lavelle, P. McGuirk, K. H. G. Mills (2004). Molecules of infectious agents as immunomodulatory drugs. *Curr. Top. Med. Chem. 4*: 499–508.

S. H. Powis (1998). Lessons from an age-old war. *Nature Medicine 4*: 887–888.

M. J. Schultz, J. Kesecioglu, T. Van Der Poll (2004). Immunomodulating properties of macrolides: animal and human studies. *Curr. Med. Chem.: Anti-Inf. Ag. 3*: 101–107.

J. S. Skotnicki, J. I. Levin (2003). TNF—a converting enzyme (TACE) as a therapeutic target. *Ann. Rep. Med. Chem. 38*: 153.

B. K. Trivedi, J. E. Low, K. Carson, G. J. LaRosa (2000). Chemokines: Targets for novel therapeutics. *Ann. Rep. Med. Chem. 35*: 191.

A. O. Tzianabos (2000). Polysaccharide immunomodulators as therapeutic agents: structural aspects and biologic function. *Clin. Microbiol. Rev. 13*: 523–533.

7

Nonmessenger Targets for Drug Action I

Endogenous cellular structures

7.1 CELLULAR STRUCTURES: RELEVANT ANATOMY AND PHYSIOLOGY

The goal of medicinal chemistry is to discover and develop novel chemical compounds (new chemical entities) that will influence the function of the host organism in some beneficial manner. As discussed in chapters 4–6, the most obvious approach is either to mimic or to block endogenous messengers used by the organism itself to control its own biochemistry. These endogenous messengers may be neurotransmitters, hormones, or immunomodulators working at the electrical, molecular, or cellular level, respectively. However, not all pathologies afflicting the human organism can be addressed by manipulating these messengers. Accordingly, it becomes necessary to target other cellular components (this chapter) and/or endogenous macromolecules (chapter 8) that are not normally directly controlled through binding to endogenous messengers. To identify such cellular targets for drug design requires an appreciation of cellular structure. The study of microscopic cellular structure is termed *cytology*. Cytology should be distinguished from *histology* (the microscopic study of tissues; i.e., functional aggregates of similar cells, such as neural tissue) and *gross anatomy* (the study of organs and body parts, such as the brain).

Cells are the fundamental building blocks of the human body. On average the human body contains 10^{14} cells, ranging in size from nerve cells with a length of 0.5–1.0 m to red blood cells with a diameter of 7 μm. From a structural perspective, the cell can be subdivided into three major components:

1. Cell membrane
2. Cytoplasm
3. Nucleus

Each one of these components is composed of a complicated array of substituent macromolecules and offers targets suitable for drug design (see figure 7.1).

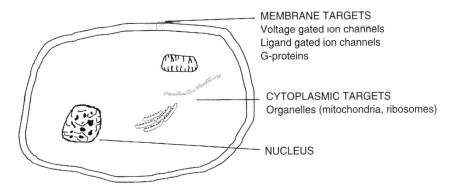

MEMBRANE TARGETS
Voltage gated ion channels
Ligand gated ion channels
G-proteins

CYTOPLASMIC TARGETS
Organelles (mitochondria, ribosomes)

NUCLEUS

Figure 7.1 Drug targets at the level of cellular structure. The mammalian cell presents a variety of druggable targets. The most important ones are located at the level of the cell membrane. Within the cell, cytoplasmic organelles, such as mitochondria, are beginning to be exploited as potential drug targets. The nucleus, at the center of the cell, is an important target for the development of antineoplastic agents for the treatment of cancer.

The cell membrane envelops the entire cell and is semipermeable. The aqueous interior of the cell has an ionic composition markedly different from that of the extracellular fluid. Traditionally, the cell membrane is described as a lipid bilayer composed of two layers of lipids, usually phospholipids, with their anionic head groups oriented either outwards to the extracellular environment or inwards to the intracellular environment. The long alkyl chain tails of the lipids face each other within the interior of the cell membrane. The membrane is 10 nm in thickness. Within this lipid bilayer, a wide variety of proteins are interspersed between various lipid molecules: "protein icebergs floating in a sea of fat." Singer and Nicolson envisioned the cell membrane as a lipid–protein *fluid mosaic* made up of a discontinuous bimolecular array of phospholipids in which globular proteins are erratically embedded. Some of these proteins extend across the entire expanse of the lipid (*integral* or *transmembrane* proteins), whilst others extend only partly into the membrane lipids and are associated with only one face of the membrane (*peripheral* proteins) (see figure 7.2).

OUTSIDE

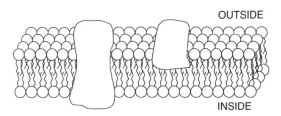

INSIDE

Figure 7.2 The cell membrane consists of protein "icebergs" floating in a sea of lipid. The proteins are a collection of ion channels, ion pumps, and G-proteins. These afford a rich diversity of potential drug targets.

The composition of the cell membrane varies from one anatomical location to another. For example, the cell membranes of neurons are 75% lipid and 25% protein, while those of epithelial cells in the intestinal villi are 75% protein and 25% lipid. The mixture of lipids that constitutes the cell membranes of neurons is different from that found in other organ systems; the same applies to the membrane proteins of neurons. Cell membranes are dynamic structures: the phospholipids and fatty acids are in a state of constant turnover; the protein components tend to turnover with a cycle ranging from 9 days to 6 months.

The *cytoplasm* is the semi-fluid polyphasic colloid that comprises the bulk of the cell's interior between the cell membrane and the nucleus; it contains enzymes responsible for catalyzing the biosynthetic machinery of the cell and *organelles* responsible for specific tasks within the cell. Cytoplasm must be differentiated from *protoplasm*; protoplasm is the whole material contained within the cell membrane and is further differentiated into the material found within the nucleus (*nucleoplasm*) and material external to the nucleus (*cytoplasm*). Organelles are important functional structures within the cytoplasm. Various structures visible by light microscopy are classified as organelles, including mitochondria, endoplasmic reticulum, and lysosomes.

Mitochondria are rod-like organelles, 400 μm in length, with a smooth outer membrane and an inner membrane folded into projections called *cristae*. Mitochondria are enzyme-rich organelles which produce adenosine triphosphate (ATP) and function as the energy source of the cell. The *endoplasmic reticulum* (ER) is a complex system of fluid-filled tubes with membranes similar to the cell membrane. The ER permeates all parts of the cytoplasm. There are two types of endoplasmic reticulum: *rough endoplasmic reticulum* (RER) and *smooth endoplasmic reticulum* (SER). RER is made "rough" by being studded with *ribosomes*, which are small granular particles that are the site of protein synthesis. SER has many functions, including being the site of lipid and steroid biosynthesis. The *Golgi apparatus* is a specialized portion of the ER composed of flattened sacs and used in the packaging of substances for secretion. *Lysosomes* are membrane-bound sacs of hydrolytic enzymes whose function within the cytoplasm is to engulf and digest foreign substances entering the cell. Under certain circumstances the lysosome can be suicidal for the cell, causing autodigestion of the cell following release of its hydrolytic enzymes. In fact, certain venoms, such as cobra venom from the *African Rhingal*, destroy cells by "turning on" lysosomes to release their enzymes. Proteinaceous rods, 4–5 nm thick, are found in the cytoplasm of most cells, especially muscle cells, providing rigidity, support, and participating in contractility processes; these rods are called *fibrils*. In certain places these fibrils come together to form tubular assemblies, called *microtubules*, which are involved in cellular transport and motion.

The final principal component of the cell is the *nucleus*. This is located in the center of the cell and is surrounded by a double membrane, the outer layer being derived from the ER of the cytoplasm and the inner layer coming from the nucleus itself. The two leaflets of the double membrane are fused in places, producing nuclear pores that enable the transfer of macromolecules from the cytoplasm to the nucleus. Two important components of the nucleus are *chromatin* and the *nucleolus*. Chromatin represents polymers of DNA complexed with protein. The nucleolus is a complex substructure, composed of ribonucleoprotein granules, that controls the synthesis of RNA destined to form the ribosomes of the cytoplasm. Cells engaged heavily in protein synthesis have large nucleoli and may even have more than one nucleolus. The cell nucleus is the site of the storage and replication of most of the cell's hereditary information.

The three components of cellular structure offer a plethora of *druggable targets* for the drug designer. The most important of these components is the cell membrane. Drugs are designed to interfere with the functioning properties of the cell. The cell membrane is ideally suited to enabling such an opportunity. The normal communication of signals from the extracellular environment to the intracellular environment within a given cell (thereby regulating biochemistry within that cell), or the communication of signals from one cell to the next cell (thereby regulating the biochemistry of adjacent cells), is all mediated by macromolecular proteins (e.g., ion channel proteins, G proteins) within the cell membrane. The macromolecular constituents within the membrane structure are all important targets for drug design.

Although the cytoplasm contains many extremely important organelles, it is somewhat less important as a source of cellular components as druggable targets. This may arise from the importance of these organelles to the life of the cell. Mitochondria, for example, are the energy sources for the cell. Any interference with mitochondrial function could easily result in rapid cell death; cyanide (CN-), for example, kills by poisoning the electron transport chain within mitochondria. Accordingly, targeting such an important organelle for drug design is fraught with the associated problems of toxicity. The cytoplasm does contain a number of smaller molecules that are involved in cellular biochemistry. For instance, the cytoplasm is a rich source of chemical catalysts (e.g., protein enzymes) and chemical mediators (e.g., lipid-derived eicosanoids). These chemicals are an important source of drug targets; however, since they are not necessarily associated with a particular organelle or cellular structure, they will be discussed separately in chapter 8.

The cell nucleus is another important source of druggable targets. Surprisingly, the nucleus is not as important to the survival of an individual cell as are many of the cytoplasmic organelles. A cell can live without its nucleus, it just cannot reproduce. (Mature adult human red blood cells, for example, do not have nuclei.) On the other hand, a cell cannot live without its mitochondria. Therefore, the cell nucleus is an important structure to target when designing drugs for diseases in which one wishes to stop cellular reproduction (e.g., cancer, viral or bacterial infections).

This chapter discusses potential drug targets in each of the main components of cellular structure, starting with the cell membrane and working inwards to the cell nucleus.

7.2 TARGETING CELL MEMBRANE STRUCTURES: LIPID COMPONENT

Our present ideas about the nature of biological membranes, which are so fundamental to all biochemical processes, are based on the Singer–Nicholson mosaic model. This model of the membrane is based on a phospholipid bilayer that is, however, asymmetrical. In the outside monolayer, *phosphatidylcholine* (lecithin) predominates, whereas the inner monolayer on the cytoplasmic side is rich in a mixture of phosphatidylethanolamine, phosphatidylserine, and phosphatidylinositol. *Cholesterol* molecules are also inserted into the bilayer, with their 3-hydroxyl group pointed toward the aqueous side. The hydrophobic fatty acid tails and the steran skeleton of cholesterol form the inner, hydrophobic part of the lipid bilayer, which behaves as a liquid. Cholesterol serves to make the bilayer more rigid and less permeable. There are also *glycolipids* in the outer monolayer.

7.2.1 Targeting Specific Membrane Lipids as Putative Receptors

Since the membrane is so vital to cellular functioning, it would seem reasonable that the lipid component of the cell membrane might be a viable drug target. However, as discussed in section 2.2, lipids are not ideal molecules to function as putative receptors. Lipid molecules, unlike peptides, are somewhat structurally bland and do not offer a complex and unique array of heteroatoms from which to form potential receptors. Also, insufficient lipid variability from cell to cell may render it difficult to uniquely target lipids contained within the cells of a particular organ. Accordingly, attempts to target endogenous lipids within cell membranes as possible drug receptors have been frequently unsuccessful.

In addition, the notion that "lipid insertion," or some other lipid-mediated event, is the mechanism of action for general anesthetics has not withstood the test of time. For many years it was hypothesized that general anesthetics inserted themselves into membranes, thereby modifying membrane fluidity, which in turn modified membrane function, in turn producing anesthesia. However, since the mid-1990s this postulate has been largely rejected and replaced with the observation that most general anesthetics interact with ligand-gated ion channel proteins such as the GABA-A receptor or the NMDA receptor. Likewise, the notion that ethanol exerts its CNS-depressant influence exclusively through ethanol–membrane interactions has similarly been downplayed. Nevertheless, a variety of data do support the possibility of ethanol–membrane interactions. Ethanol is a small lipid-soluble molecule and does dissolve in the lipid bilayer; nuclear magnetic resonance spectroscopic studies have shown that ethanol is located near the membrane surface, with the hydroxyl group anchored to the polar headgroup of the phospholipids. Other studies support the hypothesis that ethanol specifically interacts with the polar headgroup of phosphatidylcholine, and that ethanol, at physiologically attainable levels, dose-dependently decreases the order in brain membranes. Also, an abnormal lipid, identified as phosphatidylethanol, has been found in the cell membranes of ethanol-intoxicated rats. Finally, cells chronically exposed to ethanol demonstrate an increased cholesterol content and thus decreased fluidization. Despite these observations, the role of ethanol in binding to membrane proteins is probably more pharmacologically relevant than these lipid interactions.

There are a few unique circumstances under which membrane lipids can successfully be targeted as drug receptors. Since membrane integrity is essential to cell survival, targeting endogenous cell membrane lipids is difficult but targeting exogenous cell membrane lipids is desirable. If there is an unwanted "foreign" cell type within the body, and if the membrane of this unwanted cell can be uniquely targeted, then designing drugs to specifically interact with membrane lipids becomes a viable design strategy. Antifungal agents, such as Amphotericin B, offer a superb example of this strategy. Amphotericin B has a selective fungicidal effect by exploiting the differences in lipid composition between fungal and human cell membranes. The predominant sterol in human cell membranes is cholesterol, whereas ergosterol is the principal cell membrane sterol in fungi. Amphotericin B binds to cell membrane ergosterol in the fungi, and the resulting drug–lipid complex produces leaks in the membrane that ultimately culminate in death of the fungal cell. Approaches to the design of antifungal agents are described in greater detail in section 9.5.

7.2.2 Targeting Biochemical Injuries to Cell Membrane Lipids

Although it is difficult to target drugs against endogenous lipid molecules, it is much easier to target pathological processes involving membrane lipids. Acute ischemic or traumatic injury provides a good working example of such a design strategy. Ischemic and traumatic injuries to the central nervous system are common causes of irreversible damage to the brain and spinal cord. One of the leading molecular mechanisms for such irreversible damage is via membrane lipid peroxidation.

Lipid peroxidation of cellular membrane components is a chain reaction that ultimately destroys the polyunsaturated chains of membrane phospholipids. This process occurs when a reactive oxygen species (e.g., superoxide anion, hydroxyl radical, peroxyl radical, hydrogen peroxide with iron) or a carbon radical attacks the fatty acid chains of membrane phospholipids. A radical (sometimes called a "free radical") is a neutral chemical species that contains an odd number of electrons and thus has a single unpaired electron in one of its orbitals. Radicals are highly reactive because they contain an atom with an odd number of electrons (usually seven) in its valence shell, rather than a stable noble gas octet. A radical can achieve a stable valence shell octet of electrons through an abstraction process with another molecule, leaving behind a new radical. Such processes lead to reactions such as radical substitutions or radical additions. Radical reactions normally require three steps: initiation, propagation, and termination. During propagation, the radical reaction becomes a self-sustaining cycle, making the overall process a chain reaction. The unsaturated fatty acids of the phospholipids of the cell membrane are particularly susceptible to attack by radicals because of their allylic hydrogens. Hydrogen abstraction initiates the chain reaction; the resulting allyl radical product then abstracts hydrogens from adjacent chains or reacts with oxygen to form a lipid peroxyl radical. This initial radical attack can be catalyzed by iron, which can decompose lipid hydroperoxides to peroxyl and alkoxy radicals that in turn initiate new lipid radical chain reactions. This cascade of chain reactions effectively attacks the cell membrane.

The structural integrity of the cell membrane is irreversibly damaged by the process of membrane lipid peroxidation. The damaged membrane becomes leaky and extracellular calcium enters the cell. This in turn activates calcium-dependent phospholipases and protein kinases, subsequently leading to fatty acid cleavage and other biochemical alterations within the cell. Ultimately this leads to damage or death of the cell.

The importance of cell death mediated by oxidative damage has led to the popularity of antioxidants as potential therapeutics. A variety of naturally occurring (vitamin C, vitamin E) and synthetic (lazaroids) antioxidants have been studied as possible remedies for a wide variety of ailments. Large doses of vitamin E have been studied as a putative therapy in Alzheimer's disease, functioning through the inhibition of amyloid-induced oxidative destruction of neuronal membranes within the brain.

7.3 TARGETING CELL MEMBRANE STRUCTURES:
PROTEIN COMPONENT

The other major structural features of cell membranes are the proteins. A large number of protein molecules are embedded in the lipid bilayer to a greater or lesser extent. Some are anchored only superficially onto an outer or inner monolayer, but there are also some spanning the entire width of the membrane (*transmembrane proteins*). Structurally, the membrane proteins include (1) simple *helical proteins,* like *glycophorin* of red blood cells; and (2) *globulins,* like the complex multisubunit ionophores (e.g., the cholinoceptors; see section 4.2). They are anchored in the hydrophobic interior of the bilayer by stretches of apolar amino acids that form hydrophobic bonds with the lipid hydrocarbon chains, but their hydrophilic parts protrude into the outer and inner aqueous phase and serve as the major communication link between cells. Most proteins carry oligosaccharides on their outer surfaces and are therefore called glycoproteins; the oligosaccharide "antennae" often serve as a recognition structure—for example, erythrocyte blood type factors.

Functionally, the membrane proteins can be divided into three broad categories:

1. Transmembrane ion channel proteins
   Voltage-gated ion channel proteins (e.g., Na^+ channel, K^+ channel, Ca^{2+} channel)
   Ligand-gated ion channel proteins (e.g., GABA-A receptor)
   Gap-junction channels
2. Signal transduction proteins
   G-protein coupled receptors (e.g., adenylate cyclase system, phospholipase C system)
3. Transport proteins
   Energy-dependent pumps (e.g., Na^+, K^+-ATPase)
   Carrier proteins (e.g., large neutral amino acid transporter, Na^+/glucose-cotransport protein)

The voltage-gated ion channel proteins (VGICs) are protein pores that regulate and permit the passage of ions from the extracellular environment to the intracellular milieu or vice versa. Conformational changes in the shape of the protein permit the ion transport process to be regulated. In VGICs, these conformational changes are driven by alterations in the transmembrane voltage gradient. In ligand-gated ion channels (LGICs), the conformational changes occur following binding of a small molecule to the LGIC protein. Neurotransmitters are typical ligands that bind to such a protein. The GABA-A receptor is a classic example of such a protein.

Gap junction channels represent protein assemblies that are central to the process of electrical transmission between neurons. Classically, information exchange between neurons was described as occurring via chemical transmission, involving the diffusion of a neurotransmitter molecule from one neuron to the next, a process described in detail in chapter 4. In comparison to chemical transmission, electrical transmission is much simpler. In response to electrochemical gradients, ions simply flow from a stimulated presynaptic cell through the gap junction channels that directly connect the cytoplasm of one cell to the cytoplasm of a second adjacent cell, thus changing the electrical activity within the postsynaptic cell. A *gap junction* is an aggregate of gap junction channels. A gap junction channel is made from two mirror-image symmetrical components called *connexons* (one in each of the two adjoining cells); each connexon is composed of six homologous subunits, the *connexin* proteins. Each connexin protein traverses the membrane four times (via segments M1–M4); the portion of the protein between M2 and M3 forms an intracytoplasmic "hinge region" which characterizes the connexins in either α or β families; the M3 transmembrane segment is amphipathic and lines the lumen of the channel pore; the C1 (between M1 and M2) and C2 (between M3 and M4) extracellular loops contain three cysteine residues; the amino and carboxy termini are both intracytoplasmic. Connexin proteins are encoded by a single gene family; in humans, these genes map to at least four chromosomes, including chromosome X. Gap junctions form electrotonic synapses between neurons that carry electrical current from cell to cell predominantly through the transport of K^+ ions. Experimentally, gap channels may be blocked with lipophilic agents such as heptanol, octanol, glycyrhetinic acid, or oleic acid; no gap junction active agent is currently in human clinical use, although this is an active area of ongoing research. Therapeutically, gap channel agents could be used to treat neurological disorders such as epilepsy or migraine. Also, since gap junctions mediate intercellular communication that influences cellular growth rate, gap junction agents may provide a novel therapeutic strategy to the treatment of tumorigenesis.

G-coupled proteins transduce a message from the external environment to the interior of the cell through a series of coupled molecular events. These have been described in chapter 2.

Another transport structure, used for the passage of large molecules such as hormone–receptor complexes, is the *coated pit*. The coat is formed by a network of *clathrin* molecules in a regular pattern. After binding of the ligand—and, frequently, clustering of ligand–receptor complexes—the coated pit deepens, invaginates, and undergoes endocytosis, to become a *coated vesicle*. The clathrin coat is shed and the vesicle fuses with a preformed endosome, which releases the enclosed molecules, often to a lysosome. The receptors, still embedded in the internalized membrane, can then be recycled to the plasma membrane via vesicles pinched off the endosome. Alternately, internalized ligand–receptor complexes can be transported intact to the other end of the cell and the ligand released by exocytosis.

There are also a number of specialized proteins involved in functions unique to given cellular populations. Cells that have to withstand severe deformation—such as red blood cells, which must squeeze through narrow capillaries—contain a scaffold directly below the plasma membrane. Large glycoprotein molecules hold small *ankyrin* molecules on the inside. Long filaments of *spectrin* connect the ankyrins, and short *actin* chains secure the crossover points of spectrin molecules like braces. This subsurface mesh reinforces the delicate plasma membrane, providing the toughness and flexibility required during the long life of erythrocytes. The mesh can also anchor enzymes such as protein kinases.

7.4 TARGETING CELL MEMBRANE STRUCTURES: VOLTAGE-GATED ION CHANNELS

7.4.1 Channel Structure

The elucidation of the amino acid sequence of many channels now allows structural and phylogenetic comparisons. The nicotinic acetylcholine receptor (AchR) Na^+-channel, the GABA-A ionophore channel, and the glycine receptor—the latter two both being Cl⁻ channels—show about 50% homology, have closely related subunits, and are all organized with four transmembrane domains. These three proteins constitute one important family of channel receptors.

Another major channel family is the voltage-gated ion channel family. The sodium and calcium channels have four structurally similar regions with each region consisting of six transmembrane domains; within each of the four regions exists an Arg-X-X motif, where X is a hydrophobic amino acid. This Arg-rich zone is thought to be the voltage sensor that transforms electric fields into conformational changes. The potassium channel is much smaller, but part of it is 50% homologous with the voltage sensor of the Na⁺ and Ca²⁺ channels. The K⁺ channel is believed to be the most ancient one and may have developed into the more modern structures by gene duplication.

7.4.1.1 Voltage-Gated Ion Channels in Neuronal Membranes

Na⁺ and K⁺ ions are transported independently across neuronal membranes. The discovery of the highly specific effects of some channel-blocking neurotoxins has helped to clarify our concept of these ion channels. Tetrodotoxin (**7.1**, TTX) is found in the liver and ovaries of the puffer fishes and in the eggs of some amphibians (e.g., the newt *Taricha*). The potential dangers notwithstanding, the flesh of the puffer fish (Fugu fish) is a delicacy in Japan, prepared by licensed chefs. Yet, despite their skills, there are quite a few losers every year in this gastronomic game of Russian roulette. Saxitoxin (**7.2**, STX) is produced by marine dinoflagellates of the genera *Gonyaulax* and *Gymmodinium*, important members of the phytoplankton. Under certain environmental conditions they multiply explosively, causing "red tides." These algae are consumed by shellfish, which remain unaffected by the poison but can produce extremely toxic effects known as paralytic shellfish poisoning in humans who consume as little as 1 mg of STX in the shellfish.

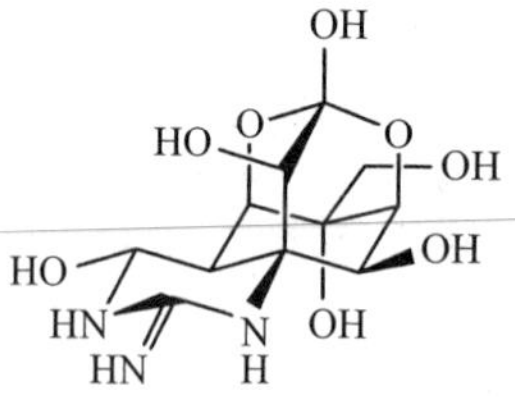

Tetrodotoxin (7.1)

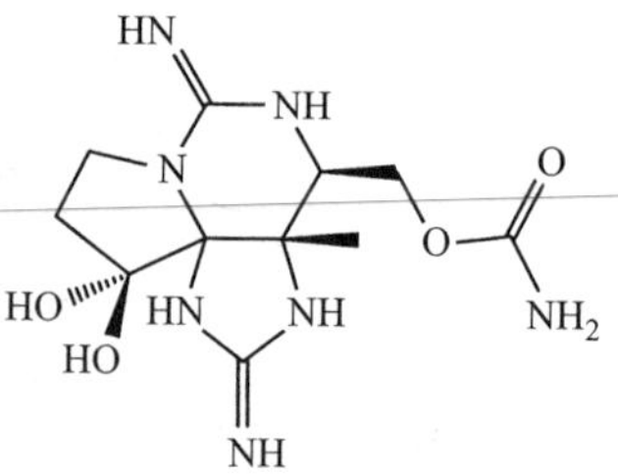

Saxitoxin (7.2)

Both TTX and STX exert their effect by blocking the inward Na⁺ current during neuronal depolarization while not affecting the outward K⁺ current. The toxins are effective only if applied from the outside; they are ineffective if perfused into the axon. Both toxins seem to be Na⁺-channel specific by virtue of their guanidinium groups, since guanidinium ions can pass through Na⁺ channels and interact with the open ion channel only. The affinity of both toxins is very high, ranging from 2 to 8 nM.

The potassium channel, on the other hand, is blocked by both tetraethylammonium salts (**7.3**, TEA) and nonyl-triethylammonium (**7.4**) salts, indicating the presence of a hydrophobic binding site that accommodates the nonyl group. Both blocking agents must be applied intra-axonally, which is understandable if one considers that the K^+ current is always directed outward.

Tetraethylammonium (7.3)   Nonyl-triethylammonium (7.4)

7.4.1.2 Characterization of the Voltage-Gated Ion Channels

MacKinnon, who was awarded the 2003 Nobel Prize in Chemistry, has recently provided the first in-depth experimental structure of an ion-selective channel. He studied a transmembrane K^+ channel composed of four identical subunits clustered symmetrically around a central pore. Each subunit contributes two membrane-traversing α-helices, which are connected by a peptide loop, called the P-region, that constitutes the selectivity filter of the channel. These two α-helices are tilted away from the central axis towards the extracellular side of the channel. Four inner α-helices from each of the subunits line the pore toward the cytoplasmic end. The inner and outer mouths of the pore are lined with amino acid side chains. The selectivity filter is lined by three main-chain carbonyl groups from the protein backbone of each of the four subunits. The channel is occupied by three K^+ atoms.

MacKinnon's landmark studies have gone a long way to providing a rigorous understanding of the K^+ channel. These results can be extrapolated to a more fundamental understanding of both the Na^+ and Ca^{2+} channels.

7.4.2 Targeting the Sodium Channel Protein: Local Anesthetics

Local anesthetics are molecules that are capable of reversibly blocking impulse conduction along nerve fibres (and other excitable membranes) that exploit the sequential opening of sodium channels as the primary biochemical process for the transmission of electrical information; by means of this action, these molecules can inhibit perception of pain sensation from specific regions of the body. Blocking Na^+ channels is an ideal method for achieving local anesthesia. The pain-producing stimulus is still present, but the nerves that carry the pain-conveying information to the brain have been blocked; thus, the brain never receives the information, which renders it incapable of perceiving the pain. Local anesthetics must be differentiated from general anesthetics. Local anesthetics block perception of pain from a discrete body region while leaving the individual conscious; they function by blocking voltage-gated ion channels, such as the Na^+ channel, within the peripheral nervous system (i.e., the peripheral spinal nerves). General anesthetics block perception of pain within the entire body by inducing a state of unconsciousness; they function by blocking ligand-gated ion channels, such as the

GABA-A or NMDA channels, within the central nervous system (i.e., the brain). The term anesthesia may therefore be defined as a reversible suppression of sensation (particularly to pain) with or without loss of consciousness, depending whether it is local or general anesthesia.

Historically, local anesthetics have been known for many years. Cocaine, the first such agent, was isolated in 1860 and introduced for clinical application in 1884. Procaine was developed as a synthetic analog of cocaine in 1905 and lidocaine was synthesized in 1943. The development of new chemical entities as putative local anesthetics remains an ongoing activity in medicinal chemistry.

From a structural perspective, local anesthetic molecules are composed of three building blocks: a lipophilic (usually aromatic) group connected via an intermediate alkyl/alkylene chain group (that typically incorporates either an amide or an ester linkage) to a terminal ionizable group (usually a tertiary amine). This is shown in figure 7.3. By carefully balancing these three chemical building blocks, a wide variety of different local anesthetics with varying properties have been designed. The lipophilic portion is essential for bioactivity. Aromatic groups are strongly preferable to lipophilic nonaromatic groups. Structural modifications to the aromatic group profoundly affect the chemical and pharmacological properties of the local anesthetic molecule. For instance, an electron-donating substituent in the *ortho* or *para* (or both) positions tends to increase local anesthetic potency. Insertion of methylene or ethylene linkers between the aromatic group and the carbonyl moiety of either the amide or the ester tends to decrease bioactivity. The intermediate chain is typically linked directly to the aromatic ring via the amide or ester and then, more distally, to the terminal ionizable group by a one–three carbon linker. The insertion of small branching alkyl groups adjacent to either the amide or ester sterically hinders catabolic cleavage of these moieties, thereby prolonging the biological half-life. In the lidocaine series, lengthening the chain between the amide and the tertiary amine from one, to two, to three carbons increases the pK_a of the amino group from 7.7, to 9.0, to 9.5. The terminal hydrophilic ionizable group is preferentially a secondary or tertiary alkyl amine. However, nitrogen heterocycles such as pyrrolidine or morpholine can also be used.

The local anesthetics can be broadly categorized on the basis of the chemical nature of the linkage contained within the intermediate alkyl chain group. The amide local anesthetics include lidocaine (**7.5**), mepivacaine (**7.6**), bupivacaine (**7.7**), etidocaine (**7.8**), prilocaine (**7.9**), and ropivacaine (**7.10**); the ester local anesthetics include cocaine (**7.11**), procaine (**7.12**), benzocaine (**7.13**), and tetracaine (**7.14**). Since the pharmacodynamic interaction of both amide and ester local anesthetics with the same Na$^+$ channel receptor is essentially identical, the amide and ester functional groups are bioisosterically equivalent. However, amide and ester local anesthetics are not equal from a pharmacokinetic perspective. Since ester links are more susceptible to hydrolysis than amide links,

Figure 7.3 Generalized structure of a local anesthetic. Local anesthetics consist of three fundamental structural units: a lipophilic part, a hydrogen bonding part, and a terminal amine. The majority of local anesthetic drugs possess these three structural units.

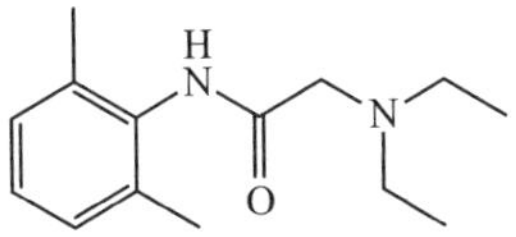

Lidocaine (7.5)

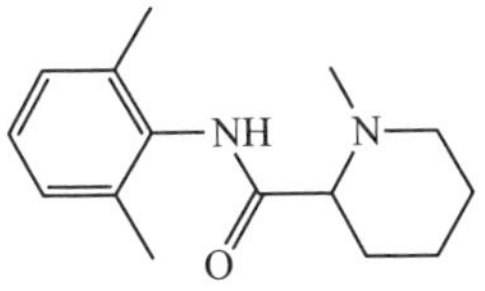

Mepivacaine (7.6)

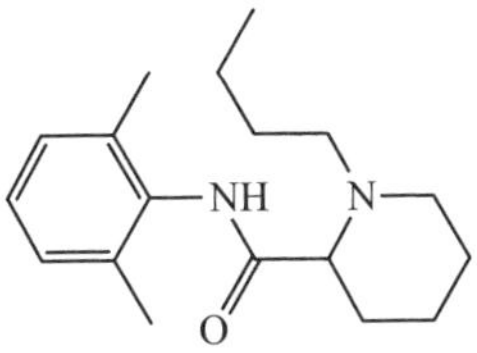

Bupivacaine (7.7)

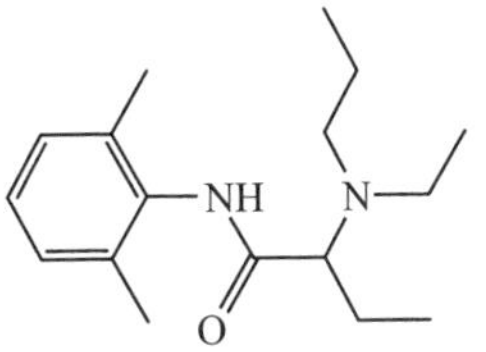

Etidocaine (7.8)

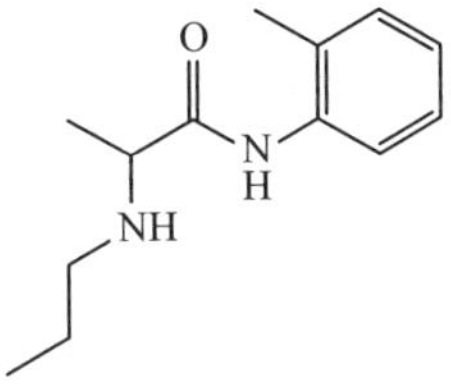

Prilocaine (7.9)

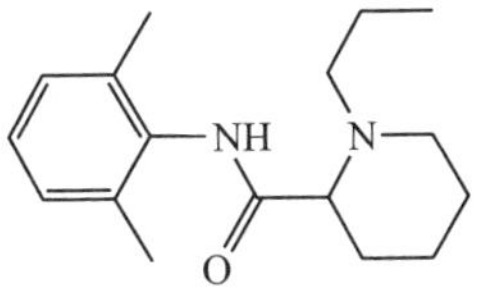

Ropivacaine (7.10)

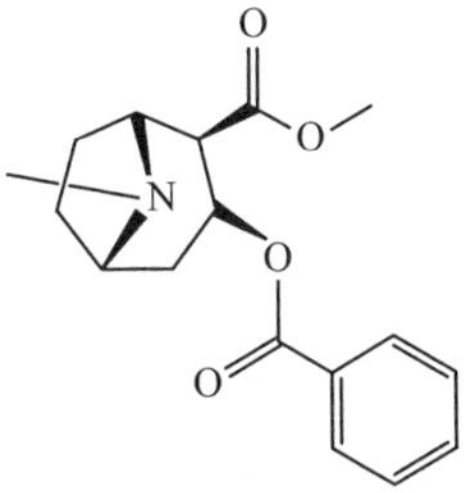

Cocaine (7.11)

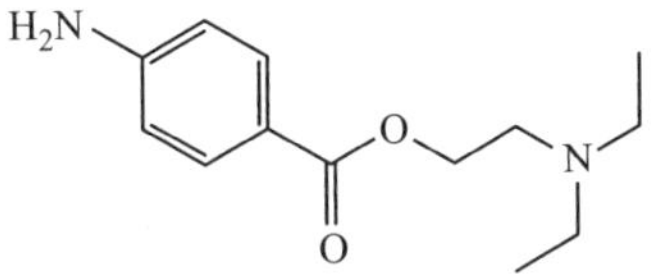

Procaine (7.12)

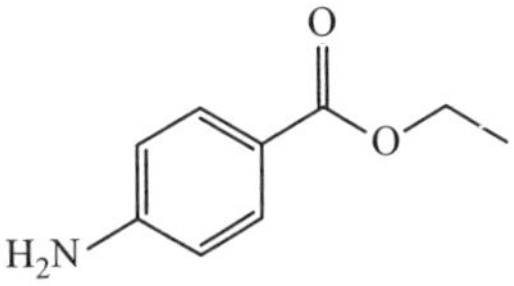

Benzocaine (7.13)

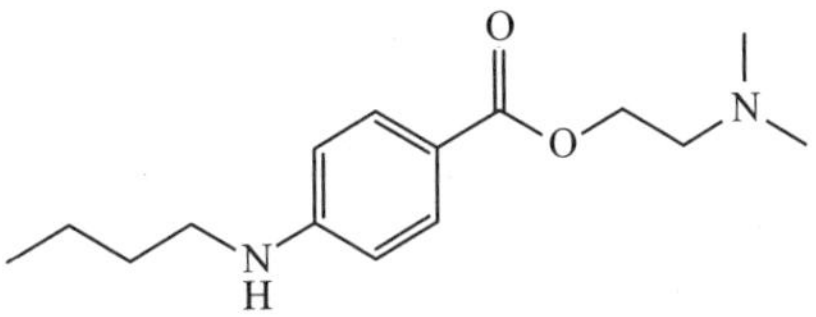

Tetracaine (7.14)

esters typically have a shorter duration of action. This observation enables the drug designer to engineer, with insight, a compound that will be either a short-duration or long-duration local anesthetic.

The receptor for local anesthetic molecules is the voltage-gated Na^+ channel protein. Local anesthetics act as blockers of the Na^+ channel. This blockade is described as *voltage dependent*. The Na^+ channel protein undergoes significant changes in its shape (conformation) as it opens and closes during the process of temporarily permitting Na^+ ions to enter a cell (a process driven by changes in the transmembrane voltage gradient). Three dominant conformations have been described: activated (open), inactivated (closed and temporarily unable to open), and resting (closed but waiting to open). Local anesthetics tend to bind with greater affinity to the activated conformation in preference to the resting conformation of the channel protein; they therefore more effectively block actively firing nerve axons than resting fibres. The local anesthetic receptor is located near the intracellular end of the Na^+ channel protein (this binding site is distinct from the extracellular binding site for such biological toxins as tetrodotoxin (**7.1**) and saxitoxin (**7.2**)). The interaction of the local anesthetic with the receptor protein is seemingly via a three-point binding interaction between the local anesthetic pharmacophore and the Na$^+$ channel protein. The three points of contact include a lipophilic/pi electron stacking pocket for the aromatic ring, a hydrogen-bonding surface for the amide or ester group, and an anionic zone for coulombic electrostatic interaction with the cationic terminus of the local anesthetic.

Since local anesthetics contain a tertiary amine ionizable group, they are weak bases. *In vivo* local anesthetics can thus coexist either as an uncharged base or as a cation. The ratio of molecular forms is important to bioactivity and may be calculated using the Henderson–Hasselbalch equation. Since the pK_a of most local anesthetic molecules is in the 8.0–9.0 range, they tend to exist predominantly in the cationic form. This charged form is optimal for the pharmacodynamic interaction of the local anesthetic with its receptor, since this interaction involves an electrostatic interaction. However, the charged form is less than optimal for the pharmacokinetic delivery of the local anesthetic molecule through the lipid membrane to its intracellular binding site. Therefore, having local anesthetics exist in a mixture of charged and uncharged forms is best suited for the combined task of drug delivery (favored by the neutral form) and subsequent drug binding (favored by the charged form). By increasing the proportion of the drug that exists in the uncharged form at a physiological pH of 7.3–7.4, it is possible to design a local anesthetic with a faster time to onset of action. This is because the neutral form can more rapidly penetrate biological membranes and thus be more quickly bioavailable to its intracellular receptor site, although it will bind to the receptor with lower affinity. Infected tissues have a more acidic pH, and thus local anesthetics that are injected into such regions will have a lower fraction in the neutral form; accordingly, local anesthetics are less effective under such circumstances because they are less likely to reach their intracellular receptor site.

Since local anesthetics possess an ionizable tertiary amine group, they tend to exist at least in part, in a highly polar charged ionic form at physiological pH. This prevents them from diffusing across apolar lipid barriers such as the blood–brain barrier. This is an immense benefit when attempting to design a drug with reduced CNS toxicity—an important consideration when one recalls the large number of Na$^+$ channel proteins which exist within the brain and which could, in principle, be blocked by local anesthetic molecules! However, the inability to cross the blood–brain barrier precludes the use of local anesthetics in the therapeutic treatment of abnormalities of electrical transmission within the CNS (i.e., seizures). The heart, on the other hand, is an electrically active organ that is not protected by the blood–brain barrier. Since the heart is a systemic organ external to the brain, local anesthetic molecules (or analogs thereof) can be used in the therapeutic treatment of abnormalities of electrical transmission within the heart's conduction system (i.e., arrhythmias). Conversely, local anesthetic molecules can also produce cardiac side effects by virtue of binding to Na$^+$ channels within the heart.

Clinically, local anesthetics may be used in a variety of pharmaceutical forms and administered in many ways, tailored to the desired clinical indication:

1. Topical anesthesia—direct application to the skin, or a mucous membrane, of the local anesthetic in the form of a spray, cream, or gel
2. Infiltration anesthesia—relatively nonspecific injection of the local anesthetic into the skin and deeper tissues of the area to be anesthetized

3. Regional nerve block anesthesia—injection of a small amount of local anesthetic into the tissue immediately surrounding a nerve supplying the region to be anesthetized

4. Intravenous regional anesthesia—injection of local anesthetic into a suitable vein supplying the limb to be anesthetized; the blood flow from this limb is then restricted by a tourniquet

5. Epidural anesthesia—injection of local anesthetic into the space external to the dural sac that encloses the spinal cord, enabling nerves to the pelvis to be selectively anesthetized during obstetrical labor and delivery

This multitude of applications has enabled local anesthetics to be used with ever-increasing effectiveness and at times to replace the considerably more dangerous use of general anesthetics.

Local anesthetics are frequently coadministered with vasoconstrictor molecules such as epinephrine. Normally, they are applied or injected locally and then taken up by local blood vessels into the systemic circulation, ultimately leading to their metabolic breakdown. The co-administration of a vasoconstrictor decreases the systemic absorption of the local anesthetic, thereby increasing its effective half-life in the area of administration and decreasing the probability of systemic toxicity (i.e., cardiac toxicity) secondary to systemic distribution.

Certain local anesthetic molecules are "dirty" or "not clean in terms of receptor interactions." This means that their molecular geometry enables them to bind to a variety of other receptors, and not just specifically to the voltage-gated sodium channel. In this regard, cocaine is a prototypic molecule. It is a local anesthetic molecule which, however, also inhibits neurotransmitter reuptake at noradrenergic synapses, thereby imparting sympathomimetic properties. Cocaine also readily enters the CNS, producing a short, intense amphetamine-like effect. In the CNS, cocaine inhibits dopamine reuptake in the "pleasure centres." Since it combines local anesthetic and sympathomimetic pharmacophores in the same molecule, cocaine was once widely used for "packing noses" during *epistaxis* (nosebleeds); the sympathomimetic vasoconstrictor properties helped to decrease blood loss during the packing process. This vasoconstrictor property accounts for the necrosis of the nasal septum seen in people who "snort" cocaine.

7.4.3 Targeting the Sodium Channel Protein: Antiarrhythmic Drugs

Historically and romantically, the heartbeat is recognized as the quintessential hallmark of life. Normally, the heart beats at 60–100 beats per minute (bpm), with each beat yielding a ventricular contraction that ejects blood out to the body. Each heartbeat is an electrical event that originates from a collection of electrically excitable cells within the heart called the *sinoatrial node* (SA), anatomically located at the upper pole of the heart. The sinoatrial node is the primary pacemaker of the heart. The electrical impulse generated in the sinoatrial node spreads rapidly downward from the atria chambers of the heart and reaches the *atrioventricular node* (AV), a collection of electrically excitable cells that constitutes the electrical interface between the atria and ventricles of the heart. From the AV node, the impulse propagates throughout the ventricles via an electrical conduction system referred to as the *His–Purkinje system*. The electrical transmission through the His–Purkinje system is coupled to mechanical contraction of the heart's ventricular muscles. The electrical generation and conduction system of the heart is analogous to excitable membranes elsewhere in the body and is focused around the sequential opening of voltage-gated Na^+ channels, producing an action potential, with additional ionic contributions from K^+ and Ca^{2+} channels.

Any disturbance in the generation or conduction of impulses within the heart's electrical system leads to abnormalities of heartbeat called *arrhythmias*. Arrhythmias do not themselves constitute a disease, but rather are symptomatically indicative of some underlying pathology within the heart. Over 80% of people with myocardial infarctions experience arrhythmias; almost 50% of people undergoing general anesthesia will likewise experience arrhythmias. Arrhythmias are caused by abnormal pacemaker activity or abnormal impulse propagation and manifest themselves as too slow a heart rate (*bradycardia* [<60 bpm]), too fast a heart rate (*tachycardia* [100–220 bpm], *flutter* [220–350 bpm], *fibrillation* [>350 bpm]), or loss of regular rhythm (e.g., *premature ventricular contractions*; PVCs). Atrial fibrillation is not life threatening, so long as the AV node does not transmit the rapid stimulus rate from the atria to the ventricles; but if the rapid rate is successfully transmitted to the ventricles, the resulting ventricular fibrillation is rapidly fatal. Any arrhythmia that functionally decouples the heart's electrical-mechanical pump efficiency, such as ventricular fibrillation, has the capacity to be lethal.

Drug-based treatment of arrhythmias is targeted against the generation and propagation of aberrant electrical activity within the heart's conduction system. Understandably, such drugs target the voltage-gated ion channels, particularly the voltage-gated Na^+ channel. Drug design of therapeutics for the treatment of arrhythmias is centered about four classes of compounds:

1. Class I—"membrane stabilizing drugs" to reduce cardiac electrical excitability: molecules that are sodium channel blockers, usually based on local anesthetic molecular structure
2. Class II—"sympathoplegic drugs" that reduce heart responsiveness to sympathetic autonomic nervous system excitation: molecules that reduce adrenergic stimulation of the heart, usually β-adrenergic blocking agents
3. Class III—drugs that prolong the action potential duration: molecules that either block outward K^+ currents or augment inward Na^+ currents
4. Class IV—drugs that slow cardiac electrical conduction: molecules that block cardiac Ca^{2+} channel currents

Classes I, III, and IV all involve transmembrane ion channels; Classes I and III involve Na^+ channels. Class I compounds are designed to block cardiac Na^+ channels in a voltage-dependent manner, similar to local anesthetics. Not surprisingly, many of these Class I agents are either local anesthetics or are structurally based on local anesthetics. Class I compounds include procainamide (**7.15**), disopyramide (**7.16**), amiodarone (**7.17**), lidocaine (**7.5**), tocainide (**7.18**), mexiletine (**7.19**), and flecainide (**7.20**). The majority of these compounds possess two or three of the fundamental structural building blocks found within local anesthetics. Propranolol (**7.21**) is the prototypic Class II agent. Class III compounds include molecules that block outward K^+ channels, such as sotalol (**7.22**) and dofetilide (**7.23**), and molecules that enhance an inward Na^+ current, such as

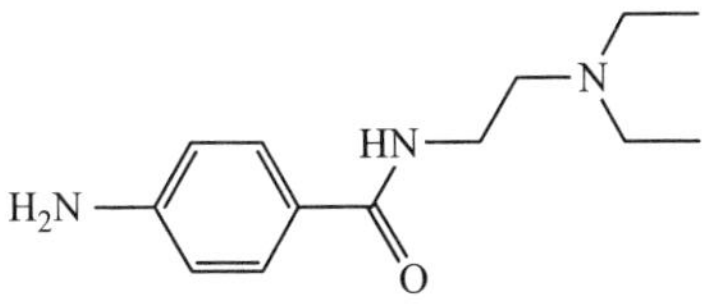

Procainamide (7.15)

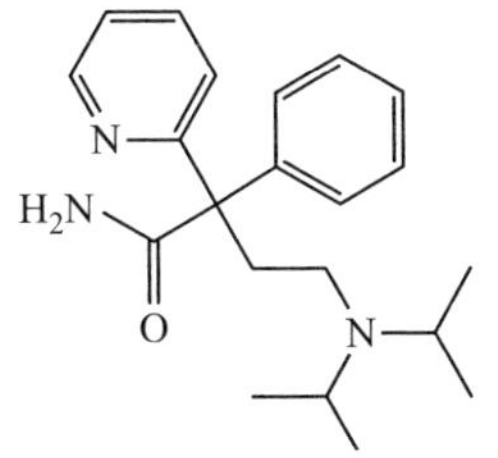

Disopyramide (7.16)

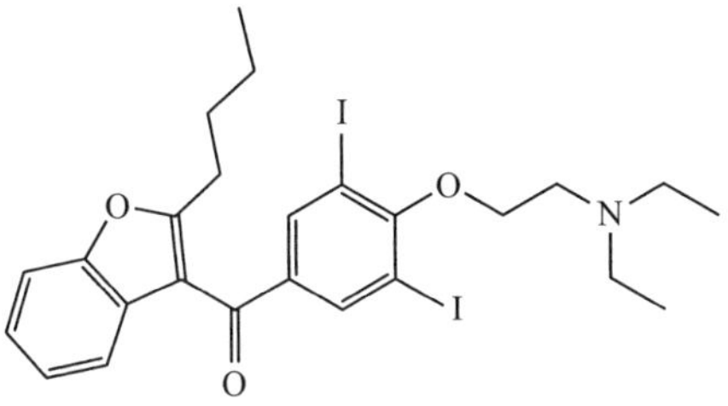

Amiodarone (7.17)

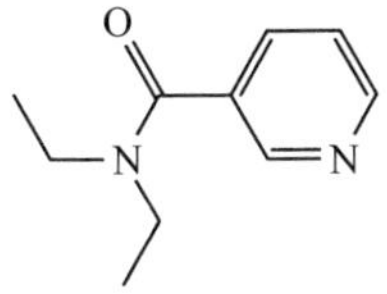

Tocainide (7.18)

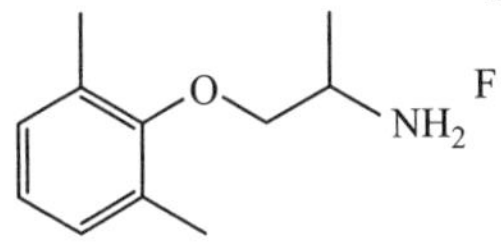

Mexiletine (7.19)

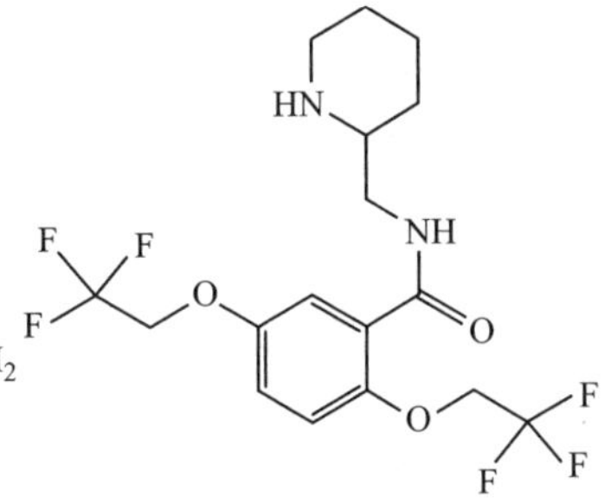

Flecainide (7.20)

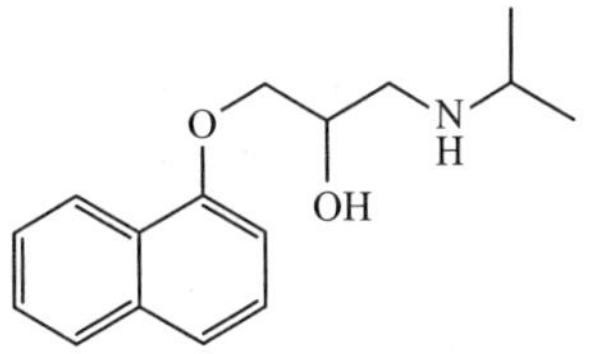

Propranolol (7.21)

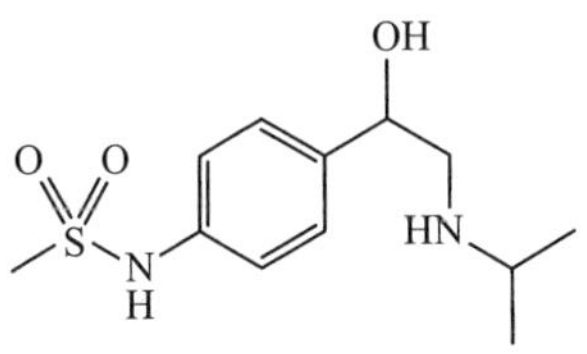

Sotalol (7.22)

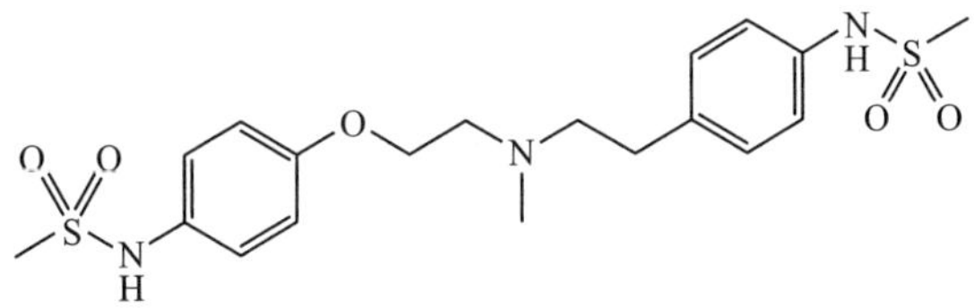

Dofetilide (7.23)

ibutilide (**7.24**). Class IV compounds include traditional Ca^{2+} channel blockers such as verapamil (**7.25**) and diltiazem (**7.26**).

7.4.4 Targeting the Sodium Channel Protein: Anticonvulsants

Seizures are the brain's equivalent to cardiac arrhythmias. Like arrhythmias, seizures are not a disease but rather a symptom of some underlying pathology. The human brain

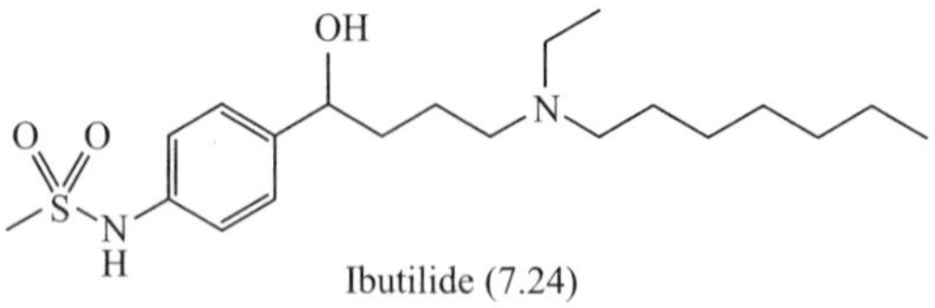

Ibutilide (7.24)

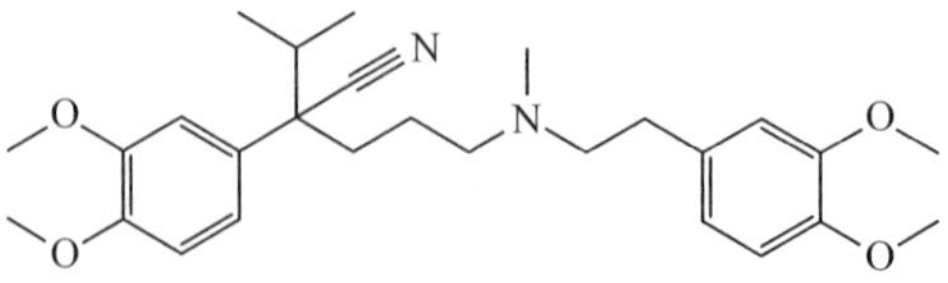

Verapamil (7.25)

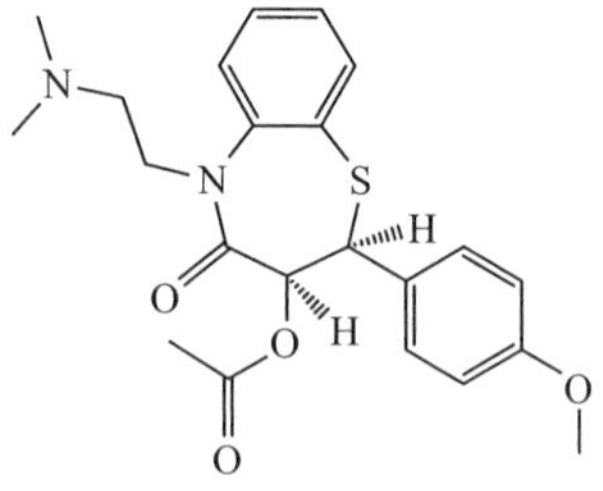

Diltiazem (7.26)

contains 100 billion neurons. When a neuron is injured it may either "hypofunction," producing paralysis, or "hyperfunction," producing seizures. Any insult to a neuron within the brain's outer cortex layer has the capacity to produce seizures. Since anticonvulsants are designed to block the aberrant spread of electrical activity within the brain (much like antiarrhythmics are designed to block the aberrant spread of electrical activity within the heart), most molecules traditionally developed as anticonvulsants function as Na$^+$ channel blockers. The parallel similarities between anticonvulsants and antiarrhythmics—and even local anesthetics—are supported by a variety of other data. Some anticonvulsants (e.g., phenytoin) have been exploited for potential antiarrhythmic activity; some antiarrhythmics (e.g., mexiletine) have been studied as potential anticonvulsants. Certain anticonvulsants (e.g., carbamazepine) may produce cardiac arrhythmias as side effects. Lidocaine, which is both a local anesthetic and an antiarrhythmic, has been used clinically as an anticonvulsant to treat seizures during *status epilepticus* (prolonged [>30 min] uncontrolled seizure activity); during status epilepticus, the structural integrity of the blood–brain barrier is compromised, enabling lidocaine to enter the brain, which it would not normally be able to achieve.

The majority of the time-honored anticonvulsant molecules (phenytoin, carbamazepine, valproic acid) exert their therapeutic effects via blockade of the voltage-gated Na$^+$ channel. Other traditional anticonvulsant drugs, such as phenobarbital (which normally functions by binding to the GABA-A chloride channel), bind and inhibit the voltage-gated Na$^+$ channel when given in large doses (e.g., 20 mg/kg loading dose) for neurological emergencies such as status epilepticus. A significant number of recently introduced anticonvulsant drugs (lamotrigine, topiramate) also inhibit the voltage-gated Na$^+$ channel protein. However, it is important to remember that there are a small number of anticonvulsants that do not block the Na$^+$ channel; these latter agents include vigabatrin, gabapentin, lorazepam, and clobazam.

Structural and theoretical chemistry studies of phenytoin and carbamazepine suggest that they bind to the Na$^+$ channel via a pharmacophore that consists of an aromatic ring and an amide linkage. This pharmacophore consists of two of the three structural features found in local anesthetics. The ionizable group, which is characteristic of local anesthetics, precludes the ability to diffuse across the blood–brain barrier.

7.4.4.1 The Clinical–Molecular Interface: Epilepsy and Voltage-Gated Na⁺ Channels

Case History. "My seizures start without any warning. All of a sudden, I hear Barney Rubble talking to Fred Flintstone. *The Flintstones* was my favourite television cartoon when I was a child, and I really loved Barney Rubble. I can hear them talking. Barney always says 'Hello neighbour … What are you up to today, Fred?' Fred replies 'Hiya Barney, ol' Pal …' As soon as I hear Barney Rubble, I know that I am going to have a seizure. My co-workers say that I stand motionless. I stare straight ahead, sometimes making smacking sounds with my lips. It lasts about one minute. Afterwards, I blink my eyes a few times and shake my head. About 20 minutes later I am back to normal."

Initially, this individual had been given phenobarbital, the GABAergic drug; it had done nothing for his seizures. However, when he received carbamazepine, the Na^+ channel antagonist, his seizures stopped immediately. As long as he was on this medication he had no seizures.

7.4.5 Targeting the Potassium Channel Protein: Agonists

Although K^+ channels share significant structural similarities with voltage-gated Na^+ and Ca^{2+} channels, the latter have received greater attention than the K^+ channel in drug design. This probably arises from the fact that K^+ channel agents are designed as agonists (or openers), whereas the Na^+ and Ca^{2+} channel agents are designed as antagonists (or blockers). (The Na^+ and Ca^{2+} ion channels transport cations into electrically excitable cells; the K^+ channel transports cations out of electrically excitable cells.) Conceptually, it is easier to design antagonists than agonists, since it is simpler to "break machines rather than make them work better."

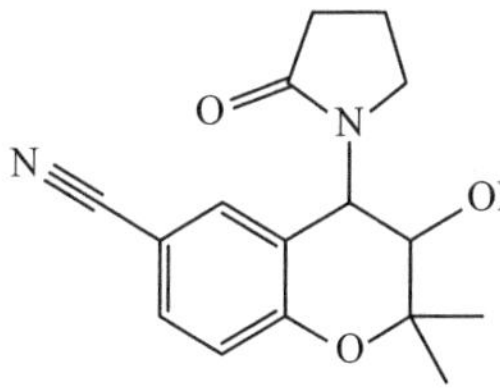

Cromakalim (7.27) 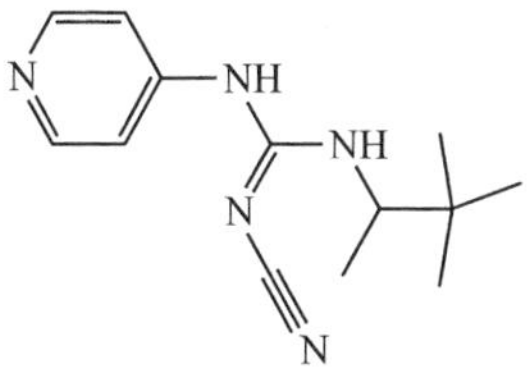   Pinacidil (7.28) 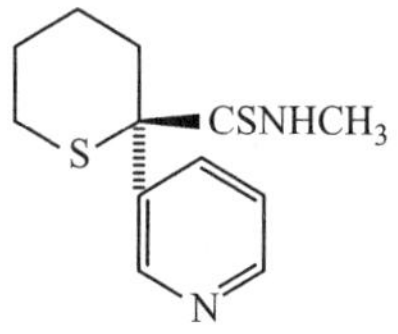   Aprikalim (7.29)

Despite these challenges, the area of K^+ channel openers (PCOs) is emerging as an active area of drug design. Over the past 5–10 years, eight novel structural classes of PCOs have received systematic development: benzopyrans (e.g., cromakalim, **7.27**), cyanoguanidines (e.g., pinacidil, **7.28**), thioformamides (e.g., aprikalim, **7.29**), pyridyl nitrates (e.g., nicorandil, **7.30**), benzothiadiazines (e.g., diazoxide, **7.31**), pyrimidine sulphates (e.g., minoxidil sulphate, **7.32**), tertiary carbinols, and dihydropyridines. These various classes have been subjected to analog preparation with compound optimization via structure–activity studies.

Nicorandil (7.30)   Diazoxide (7.31)   Minoxidil Sulphate (7.32)

From a clinical perspective, some of these PCO classes have attracted initial attention. Diazoxide and minoxidil have been evaluated as antihypertensive agents. These PCOs open K⁺ channels in the plasma membranes of vascular smooth muscle cells, causing vascular vasodilation, thereby lowering blood pressure. Cromakalim has been investigated as a smooth muscle bronchodilator for the treatment of human asthma. Nicorandil was launched in Japan in 1984 for the treatment of angina because of its perceived ability to promote vasodilation of coronary arteries. Developmental work on these and other PCOs is continuing for indications ranging from hypertension, asthma, urinary incontinence, psychosis, epilepsy, pain, and alopecia (hair loss).

7.4.6 Targeting the Calcium Channel Protein: Antagonists

The adult human body contains about 1100 grams of calcium. The plasma concentration of Ca^{2+} is normally about 2.5 mmol/L (5 mEq/L, 10 mg/dL). The Ca^{2+} cation is extremely important to the molecular processes of human physiology. Ca^{2+} plays a central role in the release of neurotransmitter molecules into the synaptic cleft during messenger-based interneuronal communication. Transmembrane Ca^{2+} flux is then itself part of the neurotransmitter message when the transmitter receptor is linked to a ligand-gated cation channel such as the NMDA channel. When the nervous system connects to muscles, Ca^{2+} is once again a crucial intermediate. Ca^{2+} is a key player during the contraction of muscles, whether they are *striated* muscles (e.g., *skeletal muscle* in the arms or legs, or *cardiac muscle* within the heart) or *smooth muscle* (e.g., enabling constriction of arteries). Within the endocrine system, various hormones (e.g., parathyroid hormone, calcitonin) exert control over Ca^{2+} homeostasis. Accordingly, Ca^{2+} concentrations influence cardiac function, affecting both muscle contractility and heart rate. When a person breathes too fast ("overbreathing," causing *hyperventilation*) the resulting reduction in arterial carbon dioxide concentration (*hypocapnia*) yields an elevation of blood pH (*respiratory alkalosis*) which cause a concomitant reduction in plasma Ca^{2+} levels (*hypocalcemia*), producing a clinical phenomenon known as *tetany*; one of the hallmarks of tetany is *Trousseau's sign*, in which spasm of the hand and arm muscles causes flexion of the wrist and thumb with extension of the fingers. Ca^{2+} is also involved in blood clotting, playing a role in the cascade or proteins that biosynthesize a blood clot. Ca^{2+} also plays a structural role in the inorganic chemistry of bone. Finally, during pathological processes (such as ischemia, producing stroke or a heart attack) an influx of Ca^{2+} and an increase in intracellular Ca^{2+} signals the so-called "final common pathway leading to cellular death."

Clearly, Ca^{2+} is an important regulator and mediator of endogenous molecular processes, both functionally and structurally. Thus, from the perspective of a medicinal chemist, Ca^{2+} metabolism would seem to be an ideal target for pharmacological manipulation. However, the fact that Ca^{2+} is involved in so many processes likewise means that drugs that influence calcium also run a substantial risk of producing side effects. Since ion channels are the proteins that transport Ca^{2+} across cell membranes, they are the logical initial targets for drug design. The task of the drug designer is aided by the fact that there are many types of calcium channel. Currently, four principal types of voltage-gated calcium channel are recognized:

1. L-type Ca^{2+} channel (in muscles and neurons)
2. N-type Ca^{2+} channel (in brain)
3. T-type Ca^{2+} channel (in neurons and cardiac cells)
4. P-type Ca^{2+} channel (in cerebellar Purkinje neurons)

Each type of channel is a unique protein which, in principle, could be uniquely targeted for purposes of drug design. Q- and R-type channels have also been described on the basis of polypeptide toxin binding studies, but have not been extensively exploited for purposes of drug design.

The L-type channel is the dominant one in cardiac and smooth muscle. As will be discussed below, many classes of drug work at the level of this channel. Drug design targeting the L-type channel has successfully created many clinically useful chemical entities as therapeutics. The T-type channel also has potential clinical utility, but drug design targeting this channel has been substantially less successful. Within the heart, T-type channels are found, especially in the sinoatrial and atrioventricular nodes, which are responsible for the generation and transmission of the electrical impulse that establishes the heart beat. Mibefradil (**7.33**) is a combined T-type, L-type blocker (with T-type specificity) that was developed for cardiovascular indications but was withdrawn from continued clinical development because of cardiac toxicity and its numerous interactions with other drugs, arising from inhibition of cytochrome P-450 dependent enzymes. Certain anticonvulsant drugs, such as valproic acid (**7.34**) and ethosuximide (**7.35**), seem to work specifically against primary generalized absence seizures by blockade of T-type channels within the thalamic region of the brain. Flunarizine (**7.36**), a weak T-type blocker (initially developed as an H1-antihistamine), is an antimigraine agent with some activity against seizures. N-type channels are a logical target for drug design because of their role in neurotransmitter release. Naturally occurring toxins, such as ω-CTX-GVIA, a conotoxin extracted from several marine snails of the genus *Conus*, are selective antagonists of the N-type channel. Attempts to develop clinically useful peptidomimetic agents based on these peptide toxins are an active area of drug develop. For example, in 2005, zinconotide, an N-type Ca^{2+} channel antagonist began to be used clinically for the treatment of severe, refractory, chronic pain. The P-type channel is blocked by Aga-IVA, a toxin of the funnel web spider, *Agelenopsis aperta*. Clinical utility of P-type antagonists is less apparent.

L-type channel antagonists are the best developed and have been exploited for four primary clinical indications, including:

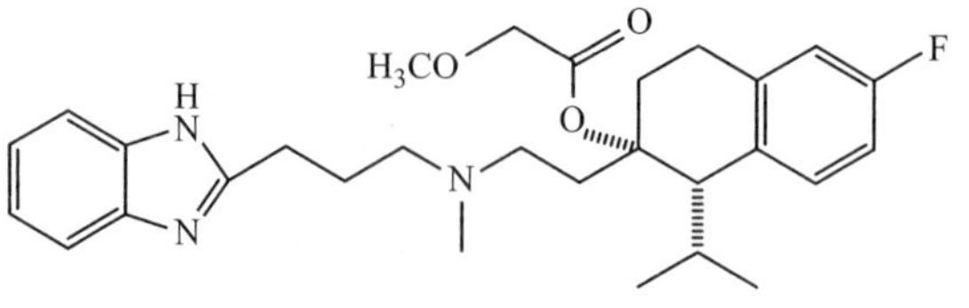

Mibefradil (7.33)

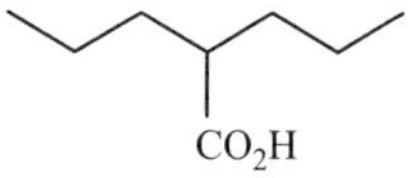

Valproic acid (7.34)

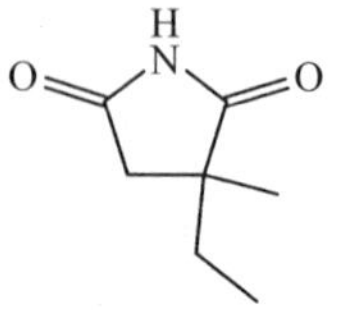

Ethosuximide (7.35)

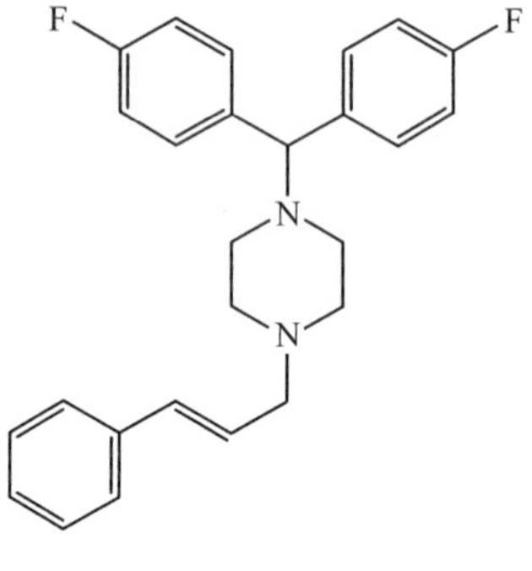

Flunarizine (7.36)

1. *Antiarrythmics,* regulating the timing of heart muscle contraction and thus being useful in the treatment of heart rhythm irregularities which cause palpitations
2. *Hypotensives,* relaxing heart muscle and being used to treat high blood pressure
3. *Anti-anginal agents,* counteracting the chest pain of atherosclerotic coronary artery ischemia
4. *Vasoactive agents,* for disorders such as migraine headache, Raynaud's syndrome (vasospasm of peripheral vessels, producing cold, blue fingers), and Prinzmetal angina (vasospasm of large, surface coronary arteries in the heart, distinct and separate from the atherosclerotic angina described above)

As evidenced by this list, the treatment of cardiovascular diseases by Ca^{2+} channel antagonists is the most important. The contraction of cardiac muscle is based on the interaction of the proteins *actin* and *myosin,* which converts the energy of ATP into mechanical work. ATP hydrolysis is mediated by the enzyme adenosine triphosphatase (ATPase), which requires the binding of Ca^{2+} ions to regulatory proteins, the tropomyosin–troponin complex. When Ca^{2+} is pumped out of the cytosol, contraction ceases and the muscle relaxes. The cardiac troponin complex is a substrate for a cAMP-dependent protein kinase, leading to desensitization that counteracts catecholaminergic stimulation. The uptake of Ca^{2+} from the cytosol into the sarcoplasmic reticulum of cardiac muscle is regulated by an ATP-dependent Ca^{2+} pump, which allows for Ca pooling until Ca^{2+} ions are needed again. During the excitation–contraction coupling of heart muscle there is a *fast influx* of Na^+ ions, causing a rapid action potential, and a *slow influx* of Ca^{2+}, causing a plateau phase of the action potential. The slow Ca^{2+} channel causes depolarization and a rise in Ca^{2+} concentration, and triggers Ca^{2+} release from the endoplasmic reticulum. In addition, there is also a 3 Na^+: 1 Ca^{2+} port–antiport system, which moves three positive charges out of the cell for each Ca^{2+}; the resulting negative inside charge of the resting cell therefore favours Ca efflux. Binding of an L-type channel antagonist produces a reduction in cardiac muscle contractility throughout the heart and a decrease in the sinus node pacemaker rate and in atrioventricular node conduction velocity; it also produces a prolonged relaxation of the smooth muscles within arterial walls.

The 1,4-dihydropyridines enjoy widespread clinical use for a variety of cardiovascular problems, with angina and arterial hypertension being the two most common. One of the 1,4-dihydropyridines, nimodipine (**7.37**), has a greater effect on cerebral arteries than on other arteries. Consequently, nimodipine is sometimes indicated for treating spasm of blood vessels in the brain following rupture of an intracranial aneurysm that produces a *subarachnoid hemorrhage*. The side effects of the L-type Ca^{2+} channel blockers include flushing, nasal congestion, tachycardia, headache, constipation, shortness of breath, and excessively slow heart rate (*bradycardia*). Given the mechanism of action of these agents, such side effects are to be expected. Also, as expected, 1,4-dihydropyridines do not tend to produce skeletal muscle weakness. Skeletal muscle is not depressed by Ca^{2+} channel antagonists because it uses intracellular stores of Ca^{2+} to enable excitation–contraction coupling and is not dependent upon Ca^{2+} being supplied by a transmembrane ion influx.

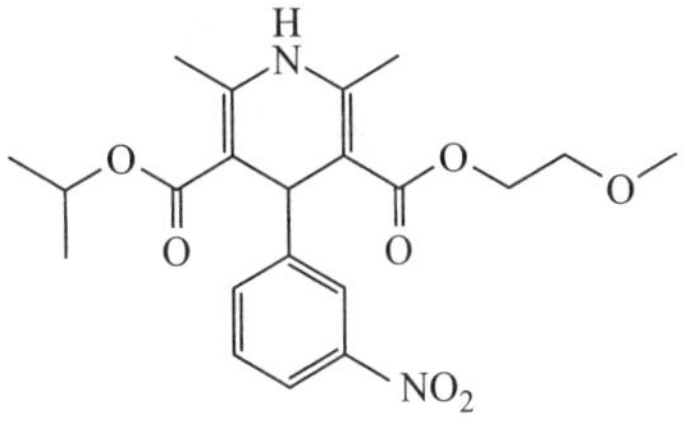

Nimodipine (7.37) 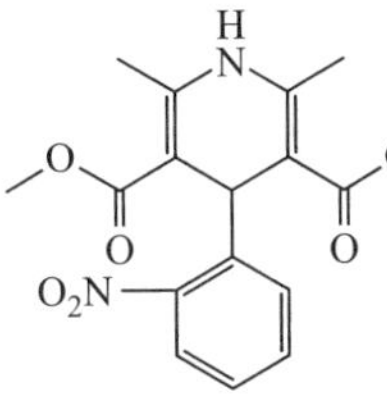   Nifedipine (7.38) 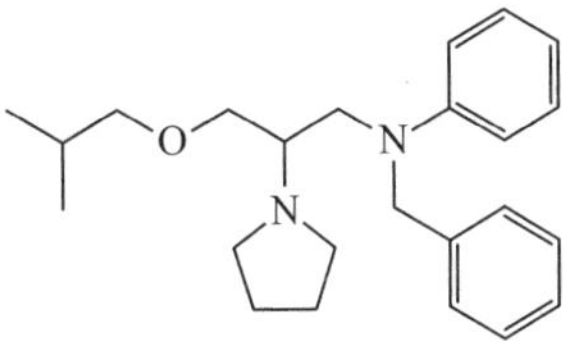   Bepridil (7.39)

From a chemical perspective, there are four classes of L-type Ca^{2+} channel blockers that have been developed for clinical indications:

1. 1,4-Dihydropyridines (e.g., nifedipine, **7.38**)
2. Phenylalkylamines (e.g., verapamil, **7.25**)
3. Benzothiazepines (e.g., diltiazem, **7.26**)
4. Diaminopropanol ethers (e.g., bepridil, **7.39**)

Nifedipine is just one of many 1,4-dihydropyridines; in contrast, the remaining three classes have only one representative agent. Nifedipine is selective for vascular smooth muscle and is therefore an excellent hypotensive. However, it can cause *tachycardia* (i.e., an excessive increase in heart rate), and is therefore prescribed with β-adrenergic blockers. Verapamil and diltiazem have a direct effect on the heart, do not cause tachycardia, and are therefore the ideal antianginal agents. Phenylalkylamines need a 1- to 2-week lag period until their antianginal effect is evident. Bepridil has a relatively non-selective action.

The 1,4-dihydropyridines are the best studied Ca^{2+} channel antagonists. A number of analogs within this series have been developed: amlodipine (**7.40**), felodipine (**7.41**), isradipine (**7.42**), nicardipine (**7.43**), nifedipine (**7.38**), nimodipine (**7.37**), nitrendipine (**7.44**), and nisoldipine (**7.45**). From a pharmacokinetic perspective, these agents vary widely: nisoldipine has only 5% oral bioavailability, whereas amlodipine has 80–90% oral bioavailability; nimodipine has an elimination half-life of 1–2 hours whereas amlodipine's half-life is 40–50 hours. The 1,4-DHPs, in general, undergo extensive first-pass metabolism; in many cases, the dihydropyridine ring is initially oxidized to a pyridine analog. Pharmacodynamically, the 1,4-DHPs bind to the α-1 subunit of the L-type Ca^{2+} channel protein. This protein readily changes shape in response to the trans-membrane voltage gradient and exists in three primary conformations: "resting" (can be stimulated to open by a membrane depolarization), "open" (permitting Ca^{2+} to enter), and "inactive" (temporarily refractory to depolarization-induced opening). 1,4-DHPs preferentially bind to the L-type channel protein when it is in either the "open" or the "inactive" conformation; such conformation-dependent binding is referred to as "use-dependent binding." Due to allosteric interactions between receptors, verapamil inhibits the binding of 1,4-DHPs to the Ca^{2+} channel, whereas diltiazem enhances 1,4-DHP binding.

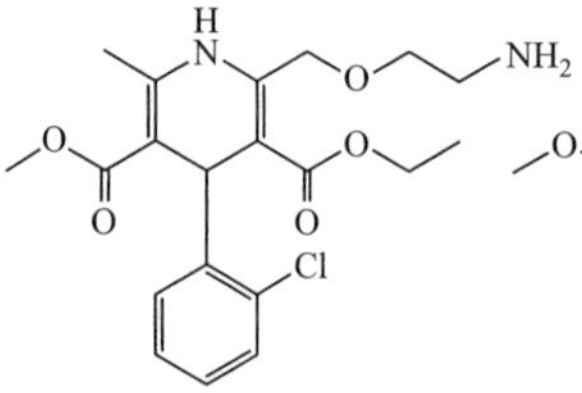

Amlodipine (7.40) 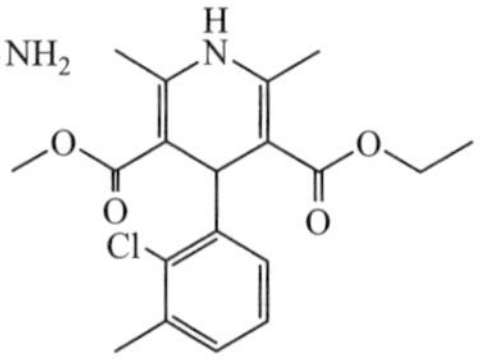   Felodipine (7.41) 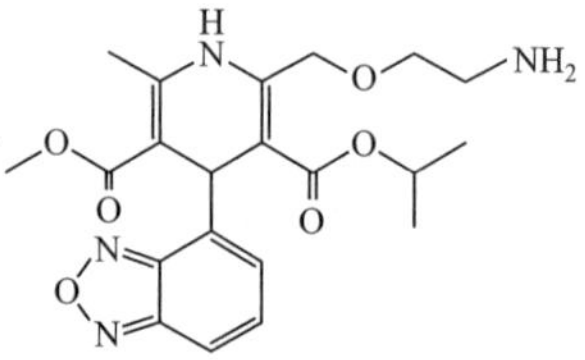   Isradipine (7.42)

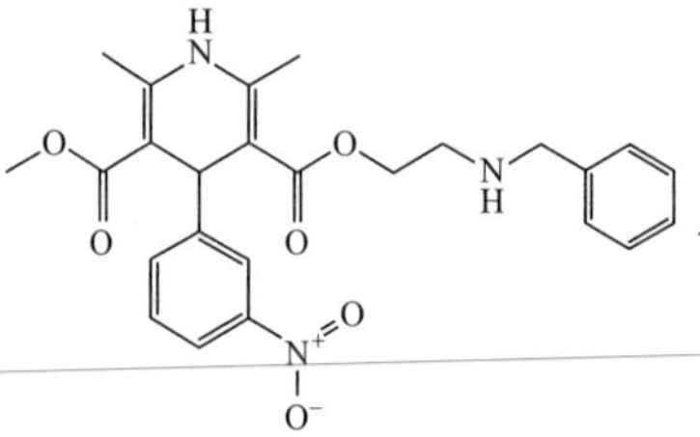

Nicardipine (7.43) 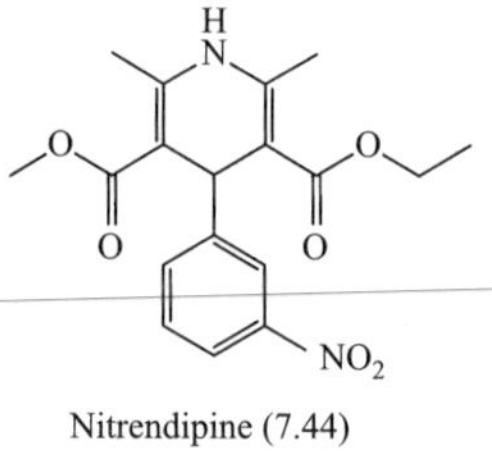   Nitrendipine (7.44) 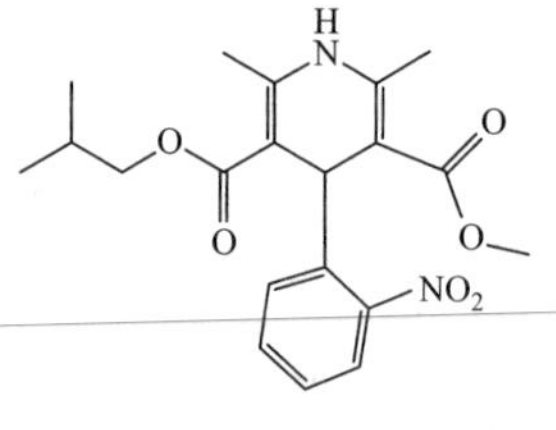   Nisoldipine (7.45)

These various pharmacokinetic and pharmacodynamic properties have enabled the identification of four basic structure–activity rules for the 1,4-DHP antagonists:

1. The 1,4-DHP ring is essential for bioactivity; the use of reduced (pyridine) or oxi-dized (piperidine) ring systems reduces activity; substitution at the N1 position reduces activity
2. Ester groups at the C3 and C5 positions optimize activity; when the C3 and C5 esters are nonidentical, the C4 carbon becomes chiral; such asymmetrical compounds exhibit enhanced selectivity for specific blood vessels

3. A substituted phenyl ring at the C4 position optimizes activity; substitution with heteroaromatic rings (e.g., pyridine) produces similar efficacy but enhances toxicity; substitution with nonplanar alkyl or cycloalkyl groups reduces activity

4. Phenyl ring substitution is crucial for size and location; *ortho-* or *meta-*substituents confer optimal activity by providing sufficient bulk to "lock" the conformation such that the C4 phenyl ring is perpendicular to the DHP ring

Extensive theoretical molecular modelling studies have permitted these SARs to be deduced.

Structural studies have also enabled Ca^{2+} channel agonists to be discovered. For example, Ca^{2+} channel agonists were also found among dihydropyridines. BAY K 8644 (**7.46**; methyl-1,4-dihydro-2,6-dimethyl-3-nitro-4-(-2-trifluoromethylphenyl)-pyridine-5-carboxylate) and PN 202-791 (**7.47**; isopropyl-1,4-dihydro-2,6-dimethyl-3-nitro-4-(2,1,3-benzoxadiazol-4-yl)-pyridine-5-carboxylate) cause vasoconstriction and also positive inotropy (an increase in the force of contraction). The latter occurs because the increase in Ca^{2+} influx prolongs the plateau phase of the cardiac action potential. In theory, such compounds are potential drugs for the treatment of congestive heart failure. Both these compounds are extremely stereoselective: the S enantiomers are agonists, whereas the R enantiomers are antagonists of calcium channels. Such opposite effects are unusual, because normally one enantiomer is the eutomer, the other one the inactive distomer, not an antagonist. Structural chemistry studies have shown that when the C3 ester is replaced with an electron-withdrawing group such as NO_2, the resulting compounds (such as PN 202-791) are channel activators rather than antagonists.

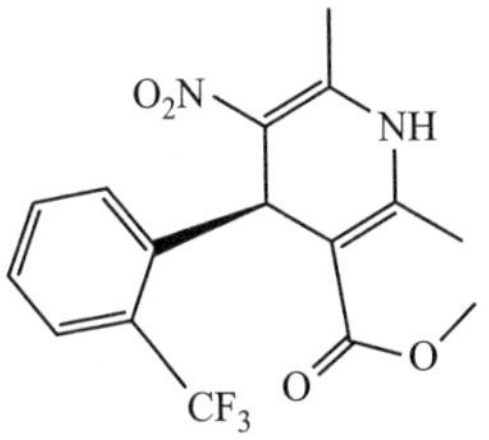

BAY K 8644 (7.46) 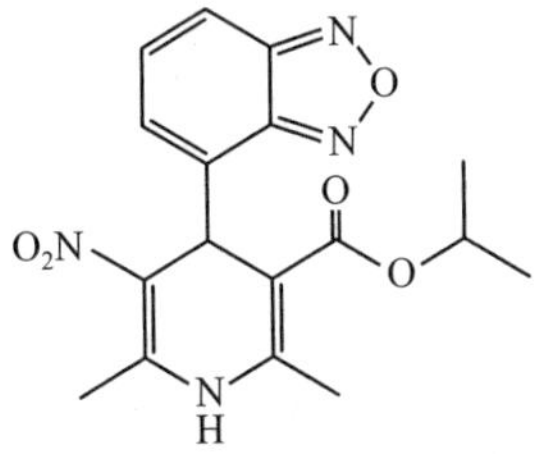   PN 202-791 (7.47)

The 1,4-dihydropyridine "nucleus" has been quite successful as a platform about which to design drugs. The ability of this nucleus to be absorbed and distributed throughout the body is well understood. Because of these data, the 1,4-DHP nucleus is emerging as a preferred platform in drug design. 1,4-DHPs are being used in the design of therapeutics that do not bind to Ca^{2+} channels and that do not exhibit efficacy in cardiovascular disorders. Also, arising from the importance of Ca^{2+} ion to so many other processes in both human and non-human physiology, Ca^{2+} channel blockers are being evaluated for disorders other than those of a cardiovascular nature. For example, Ca^{2+} channel blockers have been studied as a potential therapeutic approach to the treatment of chloroquine-resistant malaria. Given the international spread of malaria arising from the problem of global warming, new, creative approaches to the treatment of malaria will be desperately needed in coming years.

One final anecdote involving 1,4-DHP Ca^{2+} channel blockers concerns their role in the discovery that grapefruit juice can markedly increase oral absorption of drugs, even to potentially toxic levels. Unlike other citrus fruit juices, grapefruit juice can participate in a food–drug interaction that causes a drug to be absorbed in concentrations much greater than anticipated. This observation was first noted for felodipine and then was extended to all of the 1,4-DHP-type Ca^{2+} channel blockers. In addition to the 1,4-DHPs, grapefruit juice can affect a variety of other medications including antihistamines (e.g., loratidine), AIDs drugs (e.g., saquinavir), lipid-lowering agents for atherosclerosis (e.g., simvastatin, lavastatin, atorvastatin), neurological drugs (e.g., carbamazepine, diazepam, midazolam), and immunosuppressants (e.g., cyclosporine, tacrolimus). Grapefruit juice contains furanocoumarins such as 6',7'-dihydroxybergamottin that are natural product inhibitors of intestinal enzymes such as CYP3A4, which oxidizes a broad spectrum of drugs within the small intestine.

7.4.7 Ion Channel Active Agents: Symptomatic but not Curative Drugs

Over the past several decades, drugs developed to be antagonists to voltage-gated ion channels have heralded significant therapeutic advances. Na^+ and Ca^{2+} channel antagonists have demonstrated considerable clinical utility in the treatment of cardiovascular and neurological disorders. Although they have been useful, their value has been more "symptomatic" than "curative" or even "disease stabilizing." Ion channel active agents for cardiac arrhythmias suppress the arrhythmia but do not prevent the development of arrhythmias following cardiac injury. Likewise, ion channel active agents for seizures suppress seizures but do not prevent the development of epilepsy after a brain injury. This observation presents challenges for the selection of druggable targets when one is designing drugs for a particular disorder. Should the drug designer select a target with a higher chance of success, such as electrically excitable tissue, which will enable the development of symptomatic agents, or should the designer target a receptor with a lower likelihood of success, which may afford a curative approach? Recent research in the area of epilepsy has begun to address this drug design dilemma.

The currently available anticonvulsant drugs are "symptomatic" agents that suppress the symptoms of epilepsy (i.e., seizures) while failing to contend with the underlying pathological process that initially caused (or continues to cause) the predisposition to seizures. After an injury such as a depressed skull fracture with associated brain hemorrhage, there is a significant probability of developing epilepsy approximately 1.5–3 years after the injury. The currently available anticonvulsant, if administered at the time of the injury, will do nothing to prevent the ultimate development of epilepsy some two years later. The drug can simply be used to suppress the seizures once they have occurred.

These failings of traditional anticonvulsant drugs to influence the natural history of epilepsy have been demonstrated repeatedly. As shown by various well-controlled post-traumatic epilepsy studies, neither phenytoin nor carbamazepine prophylaxis has any influence on the later development of epilepsy after head trauma. Thus the currently available drugs do not seem to prevent the progressive pathology that underlies a developing seizure disorder. Indeed, limited evidence suggests that these drugs may be achieving quite the contrary. Phenytoin prophylaxis, for instance, has been associated with neurobehavioral deterioration in a variety of neuropsychological measures in post-traumatic epilepsy patients. Some would argue that the search for new agents for epilepsy should probably not use existing ion channel active anticonvulsant drugs as a starting point in the design process. Central to the discovery of more definitive therapeutics, which positively influence the natural history of epilepsy in a curative sense (and not merely mask the symptoms), will be the evolution of concepts concerning the pathogenesis of epilepsy and thus the related molecular targets for drug design.

From a clinical perspective, an important first step in this conceptual evolution of identifying new targets for drug design is to differentiate between the notions of "ictogenesis" and "epileptogenesis." A seizure is a single discrete clinical *event* caused by an excessive electrical discharge from a collection of neurons. Seizures are merely the symptom of epilepsy. Epilepsy, on the other hand, is a dynamic and frequently progressive *process* characterized by an underlying sequence of pathological transformations whereby normal brain is altered, becoming susceptible to spontaneous, recurrent seizures. Ictogenesis (the initiation and propagation of a seizure in time and space) is a rapid electrical/chemical event occurring over seconds or minutes. Epileptogenesis (the gradual process whereby normal brain is transformed into a state susceptible to spontaneous, episodic, time-limited recurrent seizures through the initiation and maturation of an "epileptogenic focus") is a slow biochemical/histological process that occurs insidiously over months to years. Ictogenesis and epileptogenesis have unique biochemical differences; not surprisingly, therapeutics targeting these two processes may have definite differences.

Ictogenesis is a fast, short-term event divided into the rapidly sequential phases of initiation and elaboration; elaboration arises from the extension of the seizure in both time and space. Ictogenesis involves excessive brain electrical discharges propagated by a cascade of chemical events that are initiated by the sequential opening of voltage-gated Na^+ channels with subsequent involvement of K^+ channels and the Ca^{2+} channel-mediated release of neurotransmitters. Logically, diverse mechanisms of action exist for anti-ictogenic (anticonvulsant, anti-seizure) drugs. However, the central role of the transmembrane voltage-gated Na^+ channel in ictogenesis has resulted in the majority of the current anticonvulsant drugs (e.g., phenytoin, carbamazepine, valproate, lamotrigine) being targeted against this receptor site.

Epileptogenesis, unlike ictogenesis, is a gradual two-phase process showing dynamic changes over the course of time: Phase 1, the initiation of the epileptogenic focus; and Phase 2, the maturation of an active epileptogenic focus. Phase 1 epileptogenesis refers to the events that take place prior to the occurrence of the first seizure. There may be a considerable delay of months to years between the occurrence of the brain injury (e.g., stroke, meningitis, trauma) and the onset of spontaneous, recurrent seizures. During this latent period, epileptogenesis is evolving, culminating in active epilepsy in which recurrent seizures occur. Phase 2 epileptogenesis refers to the events that take place after the first seizure(s) has occurred. This also is a long, protracted process in which seizures may become more frequent, more severe, more refractory to treatment, or phenomenologically different in their clinical manifestations.

The cascade of histological/biochemical events that characterize epileptogenesis differs from those of ictogenesis. At the histological level, epileptogenesis involves cellular alterations (brain scarring, referred to as *mesial temporal sclerosis*) in a variety of brain structures. The macroscopic features of these alterations include fibrous gliosis (i.e., "brain scarring") with cellular shrinkage and atrophy. At the biochemical level, various theories of epileptogenesis have been put forth, including the *mossy fibre sprouting hypothesis* and the *dormant basket cell hypothesis*. The mossy fibre sprouting hypothesis postulates an upregulation of excitatory coupling between neurons mediated by N-methy-D-aspartate [NMDA] glutamatergic receptors, which are activated in chronic epileptic brain under circumstances that would not lead to activation in normal brain. In contrast, the dormant basket cell hypothesis suggests a downregulation of inhibitory coupling between neurons such that the connections which normally drive γ-aminobutyric acid [GABA] releasing inhibitory interneurons are disturbed, thereby rendering them functionally dormant.

Although such glutamatergic and GABAergic processes are obvious participants in the molecular mechanism of epileptogenesis, there are other molecular claimants to the throne. For example, in the kindling animal model of epileptogenesis (in which repetitive, subconvulsive, electrical stimulation evokes progressively prolonged electrographic/ behavioral responses that culminate in generalized seizures), neurotrophic peptides, such as nerve growth factor [NGF], play a facilitative role in neuronal synaptic reorganization, thereby contributing to the evolution of the epileptogenic state. Clearly, the future development of antiepileptogenic agents must exploit the full range of targets, extending from amino acid neurotransmitters to peptidic neuromodulators.

Although well exemplified by ion channel active agents, the problem of "symptomatic" versus "curative" druggable targets in drug design is by no means restricted to these classes of compounds. The treatment of Alzheimer's dementia with cholinesterase enzyme inhibitors, rather than with agents such as anti-amyloid compounds, is an analogous example of symptomatic drugs being developed in preference to "disease stabilizing" or curative drugs.

7.5 TARGETING CELL MEMBRANE PROTEINS: LIGAND-GATED ION CHANNELS

Voltage-gated ion channels are transmembrane proteins that change their conformation in response to changes in the transmembrane electrical potential gradient. Ligand-gated ion channels, on the other hand, are transmembrane proteins that change their conformation in response to binding to a particular ligand. Ligand-gated ion channel proteins are also sometimes called *ionotropic receptors*, because once the receptor is occupied by the ligand the resulting change in protein conformation opens a channel that enables ions to flow through. The term ionotropic receptor distinguishes these proteins from *metabotropic receptors*, in which the binding of the ligand to the receptor exerts the biological effect via G proteins.

There are two main families of ligand-gated ion channel proteins that act as ionotropic receptors. One family includes the nicotinic acetylcholine receptor, the GABA-A receptor, the glycine receptor, and a class of serotonin receptor. The other family comprises various types of ionotropic glutamate receptors. Since these various ligand gated ion channels are activated by neurotransmitters, the medicinal chemistry of these proteins is presented in detail in chapter 4.

7.6 TARGETING CELL MEMBRANE PROTEINS: TRANSMEMBRANE TRANSPORTER PROTEINS

The voltage-gated and ligand-gated ion channels enable the transmission of information from cell to cell and along any given cell through the selective permeation of cellular membranes to particular ions. Transport proteins, on the other hand, tend to subserve a support or maintenance function by restoring chemical balance and metabolism within a cell. The transporter proteins are membrane proteins that carry either ions or molecules across membranes, typically in an energy-dependent fashion, usually using ATP as the source of energy. These are "workhorse proteins" that function to restore and maintain cellular metabolism and chemistry.

One of the most important families of transport proteins is the energy-consuming pump family. These proteins literally pump ions across cellular membranes, requiring energy to do so. In electrically excitable tissues, such as neural or cardiac tissue, the electrical signal is transmitted in the form of the "action potential," which involves the sequential opening of voltage-gated ion channels along the course of the cellular membrane. However, once the action potential has passed, it is the function of the Na/K ATPase ("sodium pump") protein to pump the ions back to where they belong so that the cell is once again responsive to another action potential. Pump proteins, such as the Na/K ATPase protein, are targets of drug design, as discussed in detail in section 7.6.1.

The other major class of transporter protein is the carrier protein. A prototypic example of a carrier protein is the *large neutral amino acid transporter.* An important function of the LNAA transporter is to transport molecules across the blood–brain barrier. As discussed previously, most compounds cross the BBB by passive diffusion. However, the brain requires certain compounds that are incapable of freely diffusing across the BBB; phenylalanine and glucose are two major examples of such compounds. The LNAA serves to carry phenylalanine across the BBB and into the central nervous system. Carrier proteins, such as the LNAA transporter, can be exploited in drug design. For example, highly polar molecules will not diffuse across the BBB. However, if the pharmacophore of this polar molecule is covalently bonded to another molecule which is a substrate for the LNAA, then it is possible that the pharmacophore will be delivered across the BBB by "hitching a ride" on the transported molecule.

Another carrier-type protein is the Na^+/glucose cotransporter protein. This carrier must be "loaded" with both Na^+ and glucose in order to fulfil its function, which involves the absorption of both Na^+ and glucose from the bowel. Knowledge of the molecular machinery of this protein has been instrumental in developing a simple but life-saving therapy for the treatment of cholera. There is no specific antibiotic treatment for cholera, since it is due to an exotoxin produced by the bacterium *Vibrio cholerae.* In cholera, the mechanism of death involves severe dehydration (up to 20 L per day via the bowel). The World Health Organization (WHO) promotes the use of an oral rehydration solution consisting of NaCl (3.5 g), $NaHCO_3$ (2.5 g), KCl (1.5 g), and glucose (20 g) per litre of water. By the simple measure of incorporating glucose into the rehydration solution, thereby enabling water and Na^+ to be cotransported across the bowel wall, the potentially lethal dehydration is successfully corrected although the frequent discharge of stool is not prevented.

7.6.1 The Clinical–Molecular Interface: Congestive Heart Failure and the Na⁺/K⁺ ATPase Transporter Protein

Congestive heart failure (CHF) is a common clinical problem in which an abnormality of myocardial function (*myocardium* is the muscle tissue that constitutes the heart) is responsible for the inability of the heart to deliver adequate quantities of blood to the tissues of the body at rest or during normal activity. Thus, CHF occurs when the cardiac output is inadequate to provide the oxygen and nutrients required by the body for its normal metabolism. It occurs because of the heart's failure to act as an efficient mechanical pump, with the primary defect residing at the level of the heart's excitation–contraction coupling machinery. CHF is a serious medical problem with a 5-year mortality of 50%. Its clinical manifestations are varied. The inability of the heart to effectively handle the fluids and blood, which it normally pumps, produces the characteristic clinical symptoms of CHF. Since the heart is overwhelmed by fluid, it "backs up" fluid into the lungs, producing *dyspnea* (shortness of breath) and *orthopnea* (shortness of breath when in the recumbent position). This backed-up fluid in the lungs can be heard with a stethoscope (as wet-sounding *pulmonary rales*) or seen on a chest X-ray. The person with CHF will also experience *edema* as both a symptom and a sign, either in the legs (in ambulatory patients) or over the sacral region at the base of the back (in bed-bound patients). The inability of the heart to meet the perfusional demands of the body will result in symptoms in which the afflicted individual will have fatigue, weakness, anorexia, confusion, and other relatively nonspecific complaints.

At the molecular level, the Na⁺/K⁺ ATPase molecule is a key player in the signs, symptoms, and treatment of CHF. Na⁺/K⁺ ATPase is a membrane-bound transporter protein, often referred to as the "sodium pump." Although transmembrane in structure, much of the Na⁺/K⁺ ATPase protein extends from the extracellular surface. Na⁺/K⁺ ATPase is a dimer constructed from two catalytic α subunits and two inert β subunits, with the β subunits seemingly functioning by holding the α subunits in a bioactive conformation. Several different forms of this protein, each consisting of multiples of α and β subunits, have been identified; the binding sites for Na⁺, K⁺ and ATP all exist on the α subunit. In addition, different isoforms of the subunits have also been identified (three α, two β), thus providing different versions of the molecule with varying affinities in various tissues in the body.

The Na⁺/K⁺ ATPase protein functions as an enzyme crucial to cardiac physiology. During each contraction of the heart, there is an influx of Na⁺ ion and an efflux of K⁺ ion at the cellular level (analogous to the molecular events during generation of an action potential within neurons, described in chapter 4). Before the next contraction of the heart, Na⁺/K⁺ ATPase must restore the ionic concentration gradient by pumping Na⁺ back into the cell against a concentration gradient; this action requires energy which is obtained from the hydrolysis of adenosine triphosphate (ATP) to adenosine diphosphate (ADP), also accomplished by Na⁺/K⁺ ATPase. Thus, Na⁺/K⁺ ATPases operate to "pump out" Na⁺ ions that have entered or leaked into a cell and to "pump in" K⁺ ions that have leaked out of the cell; through this molecular function, the Na⁺/K⁺ ATPase protein maintains the transmembrane gradients for K⁺ and Na⁺ ions and thus the normal electrical excitability of the cell membrane. Since membrane excitability is crucial both to the spread of electrical activity throughout the heart and to the contraction of muscle cells within the heart, it is not surprising that the Na^+/K^+ ATPase protein is a key molecular player in the clinical phenomenology of CHF.

In view of the central importance of the Na^+/K^+ ATPase protein to CHF, it is reasonable that drugs that target Na^+/K^+ ATPase may be clinically useful in the treatment of CHF. The *cardiac glycosides* are such agents. Cardiac glycosides (also called cardiotonic glycosides, cardiosteroids, or digitalis-like compounds) are an important class of naturally occurring drugs. They may be isolated from either plant sources (e.g., *Digitalis purpurea* [foxglove], *Digitalis lanata, Strophanthus gratus, Strophanthus kombe*) or more rarely from animal sources (e.g., skin glands of certain poisonous toads); the glycosides isolated from plants are *cardenolides*, whilst those from the toad are *bufadienolides*. The cardiac glycosides exert their therapeutic effects via inhibition of the function of the Na^+/K^+ ATPase protein, thereby increasing cardiac output and altering the electrical function of the heart. This therapeutic goal is achieved primarily by an augmentation of cardiac contractility (producing a so-called *positive inotropic action*).

Because of their significant clinical importance, the interaction between cardiac glycoside and Na^+/K^+ ATPase has been studied in detail at the molecular–atomic level of structural resolution, using computer modeling and computer-aided drug design. Such studies have revealed that the α dimer of Na^+/K^+ ATPase contains a deep cleft that constitutes the cardiac glycoside binding zone. These *in silico* molecular modeling studies have also shown that cardiac glycosides are not rigid molecules; rather, they are dynamic entities. Therefore, considerations of conformation are crucial to the drug–receptor interaction.

Chemically, cardiac glycosides are composed of two segments: the sugar and the non-sugar (or *aglycone*) moieties. This structural arrangement is shown in figure 7.4. The aglycone segment is a steroid nucleus with a unique combination of fused rings that differentiates these cardiosteroids from other steroids. The A–B and C–D rings are *cis* fused, while the B–C rings are in a *trans* configuration. Frequently, there are two angular methyl groups at C-10 and C-13. The lactone ring at C-17 is the other important structural feature. In cardenolides, this ring at C-17 is a five-membered α,β-unsaturated lactone ring, while in bufadienolides it is a six-membered lactone ring with two conjugated double bonds, forming an α-pyrone structure. The hydroxyl group at the C-3 site of the aglycone is conjugated to either a monosaccharide sugar moiety or to a polysaccharide via β-1,4-glucosidic covalent linkages. The number and type of sugar varies from glycoside to glycoside, with the most commonly occurring sugars being D-glucose, D-digitoxose, L-rhamnose, or D-cymarose. The nature of the sugar does influence biological properties. For example, adding an OH group to the $5'-CH_3$ of rhamnose produces mannose and substantially changes bioactivity. Stereochemically, these sugars exist predominantly in the β-conformation—another variable which influences bioactivity. Structure–activity studies indicate that both the steroid ring system and the lactone ring are optimal, but not essential, for bioactivity as mediated via a pharmacodynamic interaction; pregnane steroids, which lack the C-17 lactone ring, exhibit acceptable binding to Na^+/K^+ ATPase. The sugar moiety at the C-3 position influences pharmacokinetic properties such as absorption and half-life. As discussed

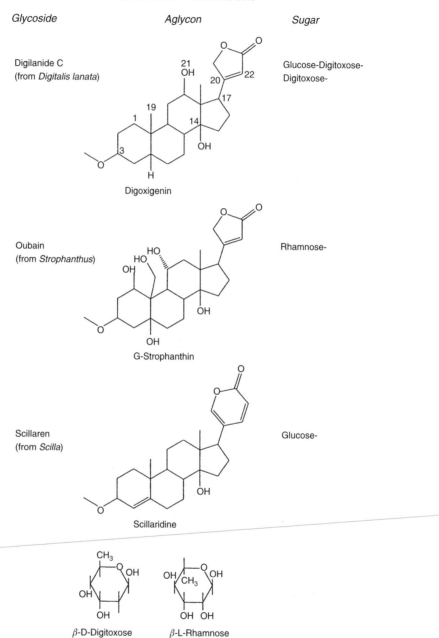

Glycoside — *Aglycon* — *Sugar*

Digilanide C
(from *Digitalis lanata*)

Digoxigenin

Glucose-Digitoxose-
Digitoxose-

Oubain
(from *Strophanthus*)

G-Strophanthin

Rhamnose-

Scillaren
(from *Scilla*)

Scillaridine

Glucose-

β-D-Digitoxose   β-L-Rhamnose

Figure 7.4 Molecular structures of digitalis-like molecules.

above, the cardiac glycosides are conformationally flexible molecules. The two regions of maximal conformational flexibility are the point of connection between C-3 and the sugar moiety and the bond connecting the C-17 side group of the steroid ring D.

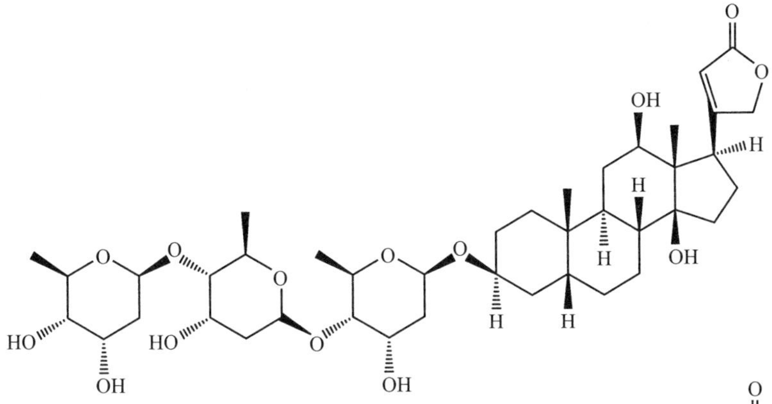

Digoxin (7.48)

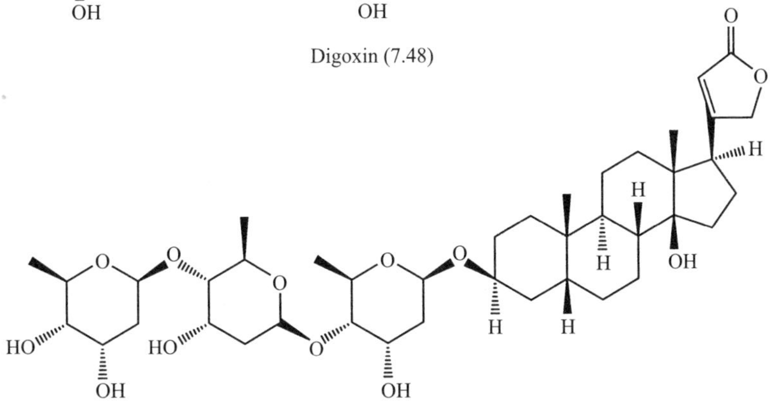

Digitoxin (7.49)

The cardiac glycosides that are employed clinically include digoxin (**7.48**), digitoxin (**7.49**), lanatoside (**7.50**), ouabain, and deslanatoside. Such agents are clinically indicated for the treatment of chronic congestive heart failure and to improve cardiac performance in people suffering from cardiac arrhythmias such as atrial fibrillation or atrial flutter. Clinically, cardiac glycosides are among the most difficult drug molecules to administer. A person with CHF, being treated with these agents, must first be "loaded" or "digitalized" and then treated with a carefully tailored and monitored maintenance dose. The patient's age, kidney function, liver function, and thyroid function will all influence these loading and maintenance doses. These problems are further compounded by numerous interactions between cardiac glycosides and other drug molecules. Finally, the use of cardiac glycosides is associated with significant and ever-prevalent risk of toxicity.

Since the Na^+/K^+ ATPase protein is so ubiquitous throughout the body, it is not surprising that cardiac glycosides have significant toxicities associated with their clinical use. The signs of intoxication include cardiac arrhythmias (including life-threatening ventricular fibrillation), CNS disturbances (e.g., confusion, agitation, hallucinations, distorted color vision [*xanthopsia*]), and gastrointestinal problems (e.g., anorexia, nausea, vomiting, and diarrhea). Since the difference between a therapeutic dose and a toxic dose is so small for cardiac glycosides, these agents are said to have a *narrow therapeutic margin*.

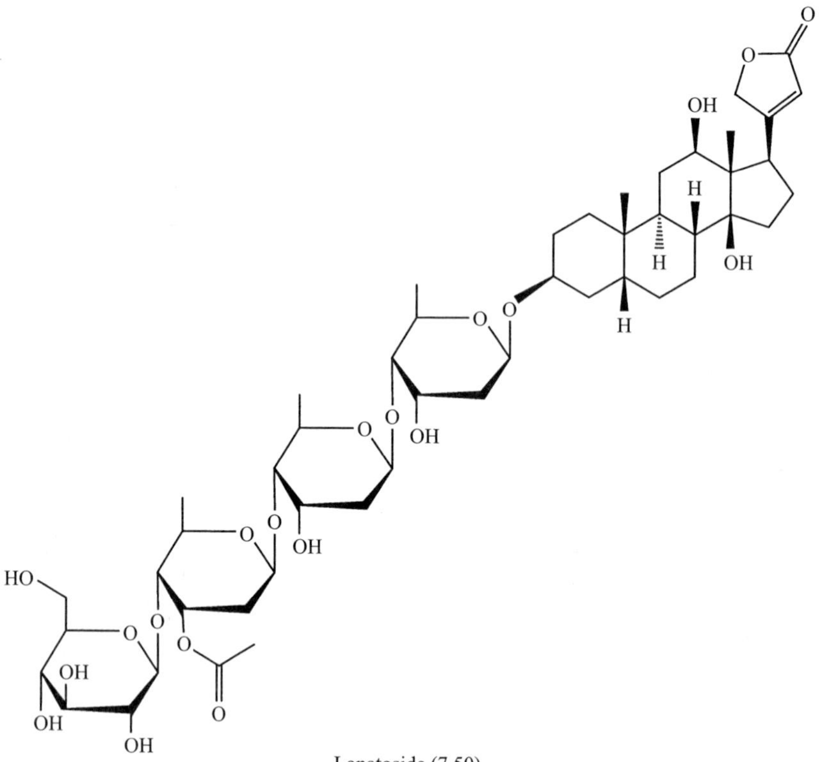

Lanatoside (7.50)

It should be emphasized that this narrow therapeutic margin is directly related to the mechanism of action which involves inhibition of Na$^+$/K$^+$ ATPase. For instance, the CNS side effects of cardiac glycosides are due to binding to neural Na$^+$/K$^+$ ATPases; this, in turn, leads to disturbances of colour vision as well as to stimulation of the area postrema within the brain, leading to nausea and vomiting. Accordingly, the pharmacophore and toxicophore are congruent, and re-engineering the molecule to remove toxicities becomes essentially impossible.

Since the Na$^+$/K$^+$ ATPase enzyme is seemingly so crucial to heart function and thus to CHF, it is tempting to speculate that there must exist chemicals within the human heart (*endogenous ligands*) which bind to and regulate Na$^+$/K$^+$ ATPase. The high affinity of cardiac glycosides for Na$^+$/K$^+$ ATPase and the fact that Na$^+$/K$^+$ ATPases occur naturally in plants and animals suggests that this endogenous ligand may in fact be an endogenous cardiac glycoside. Although the ring stereochemistry is quite different from that of typical cholesterol derivatives, the cardiac glycosides are centered on a steroid nucleus, much like the adrenocortical hormones. Over the past decade, some researchers, using sensitive immunochemical techniques, have suggested that the cardiac glycoside ouabain is synthesized in the human adrenals and perhaps in the human brain. These speculations remain somewhat controversial, but nevertheless underline the importance of the Na$^+$/K$^+$ ATPase protein to the molecular–clinical interface in the pathogenesis and treatment of CHF.

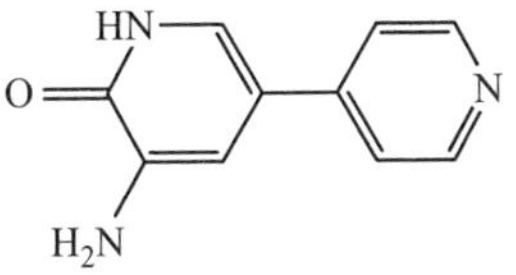

Amrinone (7.51)

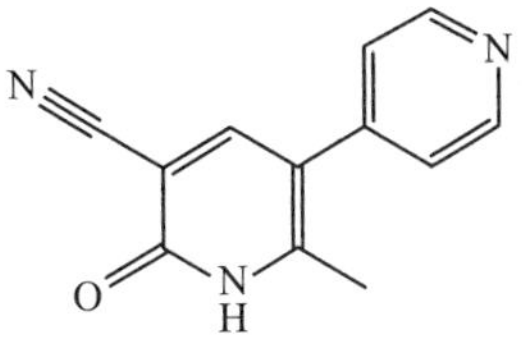

Milrinone (7.52)

Cardiac glycosides are the time-honored treatment for CHF, having been used for centuries. Nonetheless, since the administration of cardiac glycosides is beset with problems of toxicity, new druggable targets and new drugs are currently being evaluated as potential new leads in the pharmacotherapeutic treatment of CHF. Initial work has focused on less toxic compounds with positive ionotropic effects on the heart. Drugs that inhibit phosphodiesterases, the family of enzymes that inactivate cAMP and cGMP, have been evaluated. Chemically, bipyridine derivatives, including amrinone (**7.51**) and milrinone (**7.52**), have been developed. Although efficacious, their use is plagued with problems of toxicity, leading to their relegation to only limited intravenous use for an acute exacerbation of heart failure. Another possible therapeutic avenue is afforded by selective β_1 adrenergic receptor agonists that have the ability to exert a positive inotropic effect via a messenger-mediated mechanism. Dobutamine (**7.53**) is a selective β_1 agonist with a demonstrated capacity to increase cardiac output. Finally, it is possible to treat CHF using non-cardiac strategies. Since CHF involves the inability of the heart to cope with fluid loads, diuretics and ACE inhibitors have been shown to have some therapeutic efficacy.

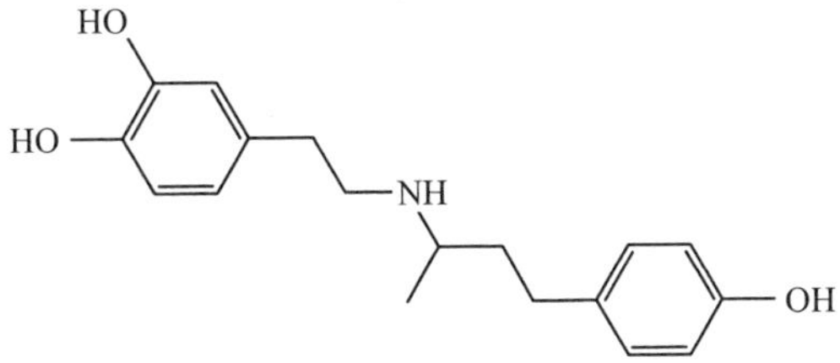

Dobutamine (7.53)

7.7 TARGETING CELLULAR CYTOPLASMIC STRUCTURES

The cytoplasm contains a number of functional organelles. Probably the most important of these are the mitochondria, the energy-producing units within the cell. Drugs that target mitochondria are discussed in section 7.7.1. Other organelles include the rough endoplasmic reticulum, Golgi apparatus, and lysosomes. Traditionally, structures such as these have been relatively ignored in drug design when compared to the attention lavished on the cell membrane and the cell nucleus. Furthermore, the importance of these organelles in human disease processes is only recently being more fully appreciated. For example, recent studies have suggested that anti-Golgi autoantibodies may play a role in connective tissue diseases such as Sjogren's syndrome and systemic lupus erythmatosus.

The cytoplasm also contains three classes of cytoplasmic structural proteins: microtubules, microfilaments, and intermediate filaments. Each one of these classes is composed of a host of different proteins: microtubules (tubulin, microtubule-associated protein [MAP], kinesin, dynein), microfilaments (actin, profilins, moesin, gelsolin, spectrin, tropomyosin, myosin), and intermediate filament proteins (acidic keratin, basic keratin, vimentin, desmin, peripherin, nestin). In principle, and increasingly in practice, each of these proteins could be a target for drug design. Microtubules, for example, play a role in intracellular transport and the mitotic spindle. Thus, proteins associated with microtubules are a reasonable target in the design of anti-cancer agents discussed below.

7.7.1 Targeting Cytoplasmic Structures: Mitochondria

Mitochondria are energy-producing intracellular organelles. They are thought to have arisen by the process of *endosymbiosis* of bacteria; that is to say, since primordial eukaryotic cells lacked the ability to use oxygen, they benefited when aerobic bacteria colonized them. Eventually, these bacteria became an integral part of the cell and ultimately evolved into mitochondria. Mitochondria have multiple functions, including energy production via ATP and the electron-transport chain as well as various biochemical processes (pyruvate oxidation, Krebs cycle, amino acid metabolism). They are unique in that they are the only organelle, other than the cell nucleus, with their own DNA. Moreover, the structure of mitochondrial DNA (mtDNA) differs from that of nuclear DNA. Inherited mitochondrial disorders are transmitted through the maternal line; the mother transmits her mtDNA through the ovum, but the sperm do not.

The clinical symptoms of mitochondrial diseases are highly varied and include seizures, vomiting, deafness, dementia, stroke-like episodes, and short stature. Although there are many types of mitochondrial disorders, four of the most common types are as follows: Kearns–Sayre syndrome, Leber's hereditary optic atrophy, MELAS (mitochondrial encephalopathy, lactic acidosis and stroke-like episodes) and MERRF (myoclonic epilepsy with ragged red fibres).

Since mitochondria are essential to cell health, mitochondrial diseases tend to be severe but, thankfully, relatively uncommon. Accordingly, the medicinal chemistry of mitochondrial disorders is still in its infancy. There are no truly effective drug therapies for mitochondrial disorders, but several agents have been reported to be of some benefit in some individuals. These agents include ubiquinone (coenzyme Q10), carnitine, and riboflavin. These compounds may assist the ailing mitochondria to better complete their metabolic tasks. However, mitochondrial medicinal chemistry is an area of research in need of additional attention.

Since mitochondria are energy factories, they are essential to cellular life. This fact can be usefully exploited in drug design to enable selective killing of unwanted cell types. For example, the mitochondria of certain parasites are fundamentally different from those of the host human cells. Accordingly, it is possible to selectively kill such parasites by targeting the biochemical uniqueness of their mitochondria. Certain 4-hydroxyquinoline derivatives are effective antiparasitic agents that use this mechanism.

7.8 TARGETING CELL NUCLEUS STRUCTURES

The final nonmessenger cellular target for drug design is the cell nucleus—the storage site for the cell's hereditary information. The nucleus is surrounded by a double membrane, the outer layer being from the endoplasmic reticulum and the inner layer from the nucleus itself. Within the nucleus is the important organic acid, deoxyribonucleic acid (DNA). Also in the nucleus is the nucleolus, which is largely composed of ribonucleic acid (RNA).

The nucleic acids DNA and RNA are the chemical carriers of the cell's genetic information. Coded in a cell's DNA is all of the information that determines the molecular nature of that cell, that controls cell growth and division, and that directs the biosynthesis of the enzymes and structural proteins required for all cellular functions. Not surprisingly, nucleic acids represent potentially important targets in drug design. Nucleic acids are biopolymers composed of nucleotides that are joined together to form a long chain. Each nucleotide is constructed from a nucleoside bonded to a phosphate group; each nucleoside is composed of an aldopentose sugar ($2'$-deoxyribose in DNA, ribose in RNA) linked to either a heterocyclic purine or pyrimidine base. In DNA, there are four different heterocyclic bases: two are substituted purines (adenine and guanine), two are substituted pyrimidines (cytosine and thymine). In RNA, thymine is replaced by uracil. In both DNA and RNA the heterocyclic base in bonded to the $C1'$ atom of the pentose sugar, while the phosphoric acid is bonded to the $C5'$ sugar atom via a phosphate ester. Nucleotides join together in DNA and RNA by forming a phosphate ester bond between the $5'$-phosphate group on one nucleotide and the $3'$-hydroxyl group on the sugar of another nucleotide.

In 1953, Watson and Crick made their Nobel Prize winning proposal concerning the secondary structure of DNA. DNA consists of two polynucleotide strands coiled around each other in a double helix. These two complementary strands run in opposite directions and are held together by hydrogen bonds between specific pairs of bases: adenine bonded to thymine, guanine bonded to cytosine. The two strands of the double helix coil in a manner that results in a *major groove* (1.2 nm wide) and a *minor groove* (600 pm wide). Crick's "central dogma" asserted that the function of DNA was to store information and pass it on to RNA, which would in turn use this information to direct the synthesis of proteins. RNA, in the form of messenger RNA (mRNA), carries the information from the DNA to ribosomes for the purpose of protein synthesis. The specific ribonucleotide sequence in mRNA forms a message that determines the order in which different amino acids are to be coupled during creation of the protein; each *codon* (or "word") along the mRNA chain consists of a three-ribonucleotide sequence that is specific for a given amino acid. From this dogma, three fundamental processes emerged:

1. *Replication*: the process by which identical copies of DNA can be made so that information is preserved and handed down from cell to cell
2. *Transcription*: the process by which the genetic information contained in DNA is read and carried out of the nucleus on RNA
3. *Translation*: the process by which the genetic information being carried by RNA is decoded and used to build proteins

These three fundamental processes are central targets in drug design.

The nucleus of every human cell possesses 46 chromosomes (23 pairs), each chromosome consisting of one large DNA molecule. Each chromosome is composed of several thousand DNA segments called *genes*; the sum of all genes in a human cell is the human *genome*. Often a gene will begin in one small section of meaningful DNA called an *exon*, and then be interrupted by a seemingly nonsensical segment called an *intron*.

The nucleic acids DNA and RNA have justifiably stood at the centre of contemporary biology and biochemistry for the past 50 years. Their remarkable structure and the ever-increasing insight into their intricate functions triggered the major scientific revolution labeled "molecular biology." Since medicinal chemistry and molecular pharmacology are at the confluence of physical chemistry and molecular biology, nucleic acids have been investigated and recognized as the targets of several major groups of drugs. Some antibiotics, numerous antiparasitic agents, many antineoplastic (antitumour) drugs, and most of the antiviral compounds exert their varied actions on different phases of nucleic acid function.

From the perspective of a drug designer, a fundamental question concerns the relevance of nucleic acids as a druggable target for rational drug design. Despite the overwhelming importance of DNA and RNA to cell heredity, the number of pathological processes in which nucleic acids play a central role is surprisingly low. Cancer is the major human disease in which nucleic acid biochemistry is crucial. Since cancer arises from uncontrolled cellular reproduction and proliferation, it is understandable that nucleic acids would exert an important influence in the etiology and pathogenesis of cancer.

The other important disease states in which DNA/RNA targets are crucial for drug design are infections. However, there are fundamental differences between targeting nucleic acids for cancer treatment and targeting them to treat a viral or bacterial infection. Cancer is a disease of "self"; the genome of the cancer cell is like the genomes of every other cell in the patient's body—targeting the cancer cell is analogous to targeting the other "healthy" cells in the body. Infections, on the other hand, involve "nonself." The genome of the virus or bacterium is different from the genome of the patient's body (reflecting the fact that the bacterium is a different life form from the human that it infects). If the differences between genomes can be understood and exploited at a molecular level, it is possible to engineer a drug specific for the virus (or bacterium) that will hopefully spare the genome of the human host organism suffering from the infection. In recognition of this fundamental difference, drug design for cancer (i.e., for drugs targeting human genome nucleic acid sequences) is presented in this chapter. Drugs targeting the nonself genome of exogenous pathogens, such as viruses, bacteria, fungi or parasites, are discussed in chapter 9. However, given that the basic nucleotide building blocks are the same for humans as for viruses, many of the drug design concepts are similar when designing "anti-nucleic acid drugs" for either endogenous cancer or exogenous infections.

As a putative receptor, nucleic acids are suitable molecules. Nucleic acids, unlike alkyl chain lipids, are not "bland"; rather, they are reasonably complicated molecules in terms of possessing heteroatoms and hydrogen-bonding donors and acceptors. Such complexity affords a diversity of opportunities for designing molecules capable of unique interactions with DNA or RNA. Given the importance of nucleic acids to heredity and to the control of cellular protein synthesis, drug design that targets nucleic acids should be reserved for major pathologies such as cancer. On the basis of molecular mechanisms, drugs that act upon nucleic acids can be classified in the following way:

1. Drugs interfering with DNA replication
   a. Intercalating cytostatic agents
      Actinomyces
      Anthracyclines
   b. Alkylating cytostatic agents
      Bis(chloroethyl)amines (nitrogen mustards)
      Nitrosoureas
      Aziridines
      Alkylsulphonates
   c. Antimetabolites interfering with DNA synthesis
      Folate antagonists
      Purine antimetabolites
      Pyrimidine antimetabolites
   d. Antibacterial agents interfering with DNA topoisomerase
2. Drugs interfering with transcription and translation
   a. Cytostatic platinum complexes and bleomycin
   b. Antisense oligomers
3. Drugs interfering with mitosis
   a. Vinca alkaloids
   b. Taxane alkaloids

7.8.1 Drugs Interfering with DNA Replication

There are several mechanisms by which a drug molecule can interfere with DNA replication. These range from agents that insert into the DNA structure, altering its geometry, to compounds that inhibit enzymes crucial to DNA metabolism.

7.8.1.1 Intercalating Cytostatic Agents

Intercalating drugs associate strongly with the DNA in the cell nucleus by slipping between two base pairs of the double helix and forming charge-transfer complexes with the nucleotides. This interaction is geometrically controlled, and it has been shown that some compounds (e.g., daunomycin, **7.54**, proflavine, **7.55**) intercalate at the *major groove* of DNA whereas others (e.g., actinomycin D, **7.56**, or ethidium, **7.57**) do so only at the *minor groove* of the helix. There are even indications that some intercalating agents are selective for certain base sequences. By intercalating into the DNA structure, such agents modify DNA conformation, thereby altering its function.

Intercalation has been studied thoroughly on oligonucleotide models using both experimental (X-ray crystallography) and theoretical (molecular modeling calculations) approaches. Such work has studied the complex formed between a model oligonucleotide and ethidium (**7.57**). The ethidium molecule shown forms a charge-transfer stack and also interacts with the phosphate anions, forming salt bonds with the latter

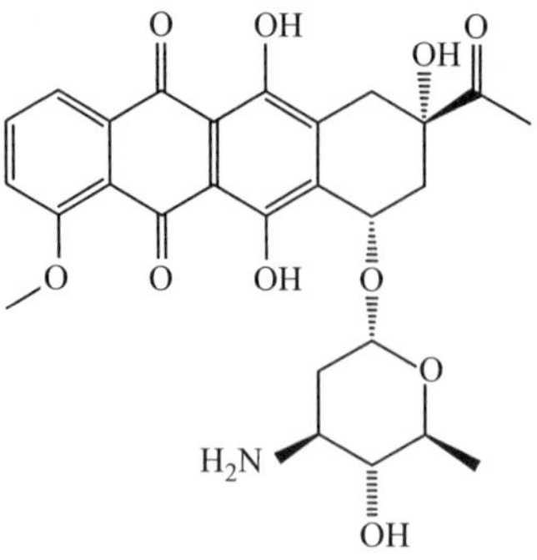

Daunomycin (7.54)

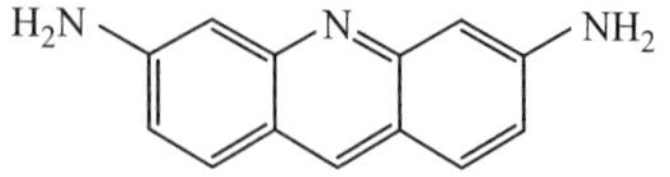

Proflavine (7.55)

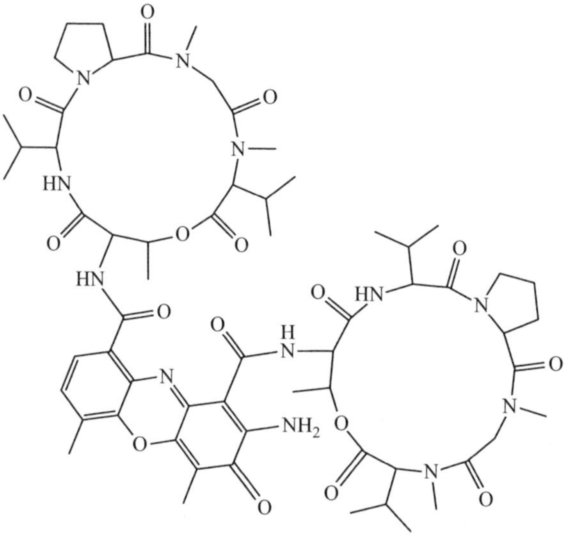

Actinomycin D (7.56)

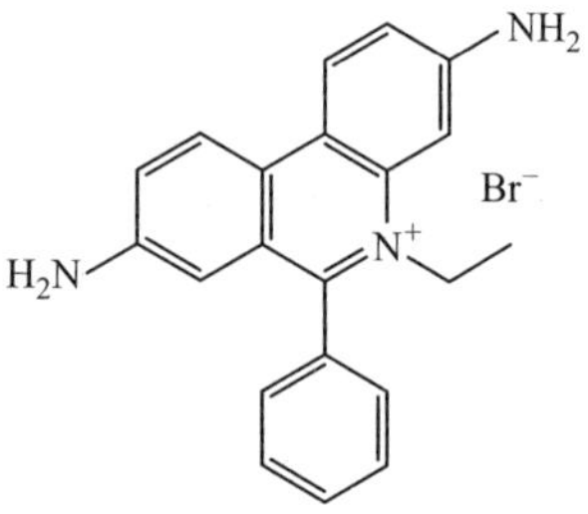

Ethidium (7.57)

through its amino groups. Above (and below) this intercalated ethidium molecule lies another one, which is simply stacked and does not interact with the phosphate in an ionic bond. The phenyl and ethyl groups of the interacting ethidium molecule lie outside (in the minor groove of) the double helix; only the planar tricyclic ring interacts with the nucleotides. As a result, the base pairs of the helix are twisted by 10° and separated by 0.67 nm, and the helix unwinds by −26° at the intercalation site. Since this distorts the double helix, the replication and transcription of genes are compromised.

Ethidium is a model compound for these studies. In human pharmacology, two classes of natural products (actinomyces, anthracyclines) provide prototypic molecules as DNA intercalating agents. (Planar molecules tend to insert well into the stacked nucleic acids, which accounts for the carcinogenic potential of polyaromatic hydrocarbons; see figure 7.5.)

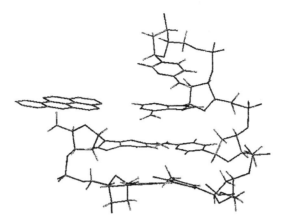

Figure 7.5 Intercalation of carcinogens into the stacked bases of nucleic acids. Both carcinogens and certain anti-neoplastic agents share a common mechanism of action. They have the capacity to insert themselves into the nucleic acid structure, causing geometrical distortions that preclude the ability of the nucleic acid to complete its function in the processes of transcription and translation. Flat aromatic molecules have the ability to intercalate between the stacked bases.

Actinomycin Antitumor Antibiotics. Actinomycin D (**7.56**), produced by the fungus *Actinomyces antibioticus*, is an effective tumor-inhibiting antibiotic. It is a phenoxazone derivative with two cyclic pentapeptide side chains pointing up and down from the plane of the heterocyclic nucleus. The peptides form hydrogen bonds with each other. Actinomycin D forms a strong complex with double-stranded DNA ($K_D = 5 \times 10^{-6}$ M) at G–C pairs, binding to the 2-amino group of the guanine. The flat heterocyclic ring is intercalated. The compound selectively inhibits the ribosomal RNA chain-elongation step during transcription. Although a very effective drug in certain malignancies (testicular tumors, disseminated cancers), it is a very toxic compound. F8-AMD (**7.58**) is a structurally related antitumor antibiotic.

Anthracycline Antitumor Antibiotics. The group of *anthracycline antibiotics*, used in the treatment of several forms of cancer, includes doxorubicin (**7.59**) and daunomycin (**7.54**), which differ by only one hydroxyl group. Both are aminoglycosides of anthraquinones produced by some *Streptomyces* species, and are related to the antibacterial tetracyclines. The four-membered ring system intercalates into DNA, entering from the major groove. The sugar moiety of the drug is ion-bonded through its amino group to the phosphate backbone of DNA. Daunomycin is used primarily in acute leukemia, but adriamycin is effective in solid tumors also. The 4-desmethoxy derivatives of both compounds are much more potent. The principal drawback of these active cancer chemotherapeutic agents is their severe cardiotoxicity, acting through the inhibition of cardiac Na$^+$, K$^+$-ATPase. Idarubicin (**7.60**) and epirubicin (**7.61**) are more recent analogs in this class of compounds.

From a molecular structural perspective, it is the presence of planar aromatic "building blocks" within these molecules that enable them to intercalate into the DNA. However, this planar aromatic structure is also responsible for many molecules being

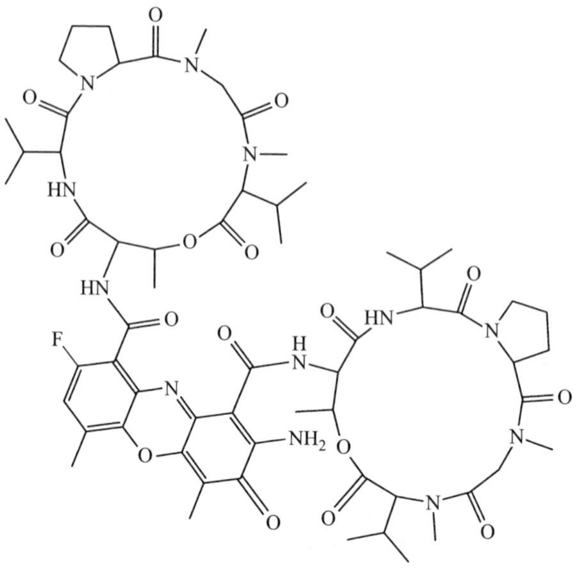

F8-AMD (7.58)

carcinogenic (cancer-causing) substances through precisely the same mechanism. There is a fine line between anticancer compounds and cancer-causing compounds. Indeed, one of the potential long-term side effects of anticancer agents is their ability to induce cancer in a different organ system years after the original cancer has been treated.

7.8.1.2 Alkylating Agents

These antitumor agents are compounds that form carbonium ions or other reactive electrophilic groups. Such compounds bind covalently to DNA, and either crosslink the two strands of the helix or otherwise interfere with replication or transcription. Since these processes are more prevalent in rapidly dividing malignant cells than in normal tissues, alkylating agents can control and in some cases even eliminate tumors. However, their selectivity is limited and they have many and serious side effects.

Bis(chloroethyl)amines (the "Nitrogen Mustards"). Alkylating agents were initially developed from "sulphur mustard," the horrific and infamous "mustard gas" of World War I—a lethal vesicant and cell poison. Its nitrogen analog, the nitrogen mustard mechlorethamine (**7.62**), was first used as an antitumor agent in 1942, with some success; numerous derivatives were subsequently developed. The rationale for this, if any, was to use carrier molecules that are natural products, in the hope that they could direct the active, nitrogen-mustard component of the compound to a selective metabolic site in a tumor. Melphalan (phenylalanine mustard or **7.63**) and chlorambucil (**7.64**) are just two examples of many such compounds. Melphalan is synthesized by treating L-N-phthalimido-*p*-aminophenylalanine ethyl ester with ethylene oxide, followed by phosphorus oxychloride, followed by acid-catalyzed hydrolysis; chlorambucil is

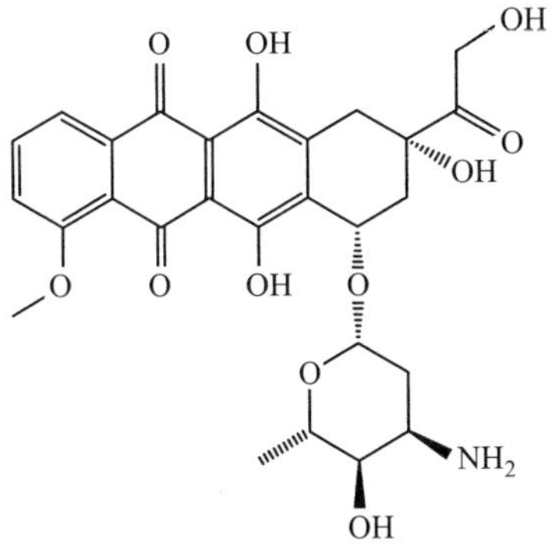

Doxorubicin (7.59)

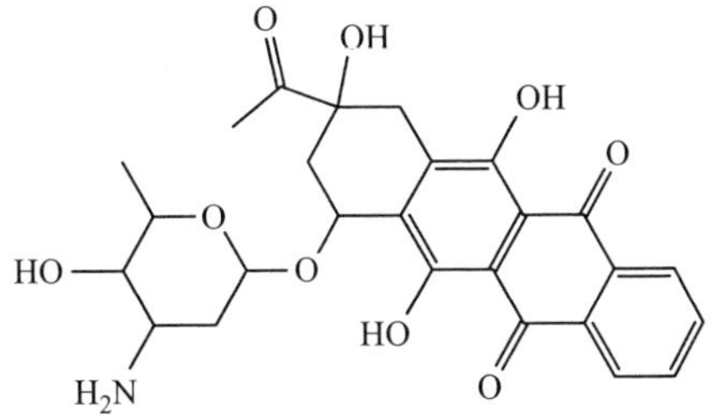

Idarubicin (7.60)

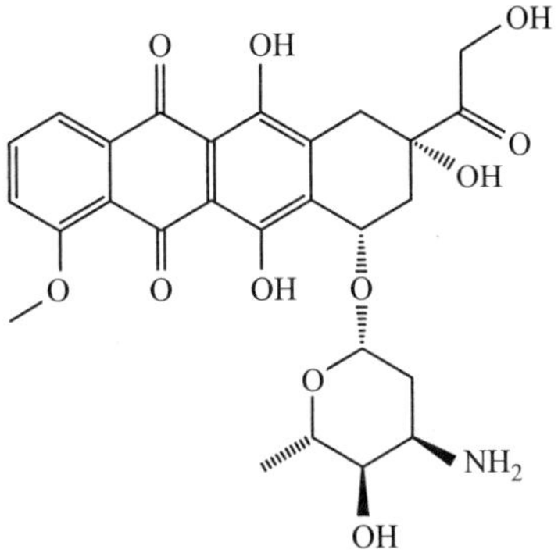

Epirubicin (7.61)

synthesized by treating *p*-aminophenyl butyric acid with ethylene oxide, followed by thionyl chloride. Although the nitrogen mustards did not fulfil expectations regarding selectivity, they are nevertheless useful oral drugs when employed in conjunction with tumor surgery. Ovarian and breast carcinomas, lymphadenoma, and multiple myeloma are the malignancies most successfully treated with these drugs, especially in combination with mitosis inhibitors.

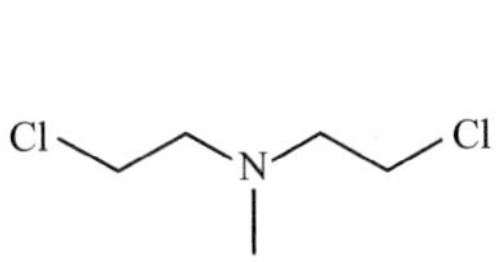

Mechlorethamine (7.62)

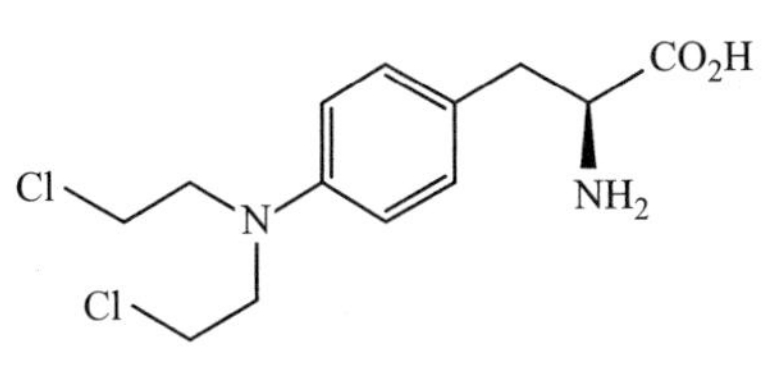

Melphalan (7.63)

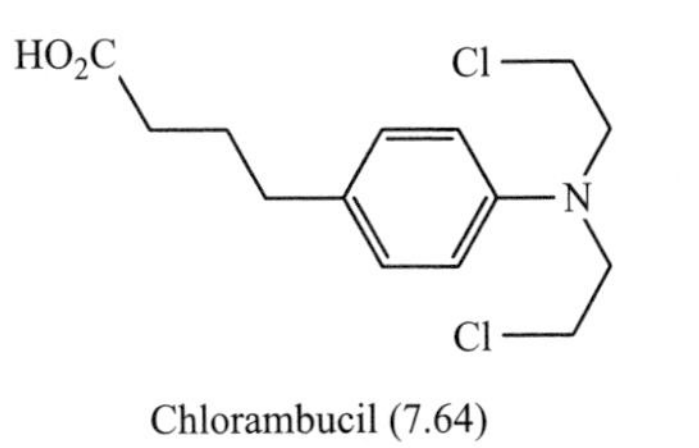

Chlorambucil (7.64)

A nitrogen mustard prodrug, cyclophosphamide (**7.65**), was synthesized in the hope of exploiting the high level of phosphoramidase enzymes in certain tumor cells. Cyclophosphamide is relatively nontoxic but is metabolized in the liver, not the tumor, to form the active drug, the phosphoramide mustard (**7.66**). Whilst not without side effects, cyclophosphamide is a relatively successful drug in a number of carcinomas and lymphomas. Ifosfamide (**7.67**) is an analog of cyclophosphamide; it is structurally related to the nitrogen mustards except that the two chloroethyl arms are not attached to the same nitrogen.

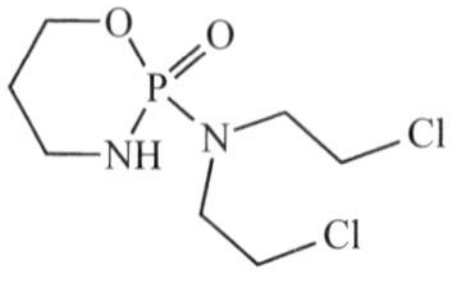

Cyclophosphamide (7.65) 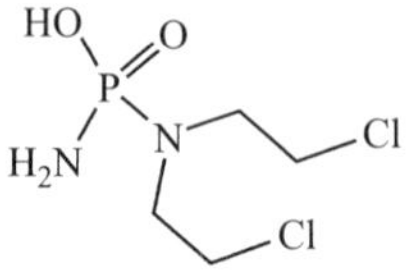   Phosphoramide (7.66) 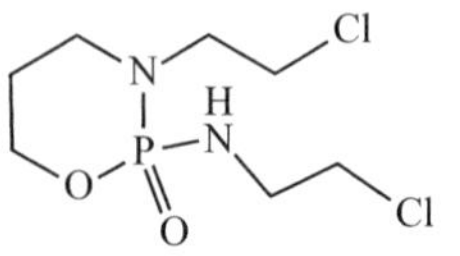   Ifosfamide (7.67)

The mode of action of these compounds is nonspecific, because the active species, the resonance-stabilized carbonium ion, reacts with any nucleophilic centre, including water. Consequently, there is a tremendous waste of drug on the way to the site of action, through hydrolysis alone; this waste is slowed with the aromatic compounds like melphalan. The principal target of the nitrogen mustards seems to be the 7-nitrogen of guanine in DNA, which crosslinks the two strands. This action prevents unwinding, causes deguanylation and base-mispairing, and compromises the template function of DNA. Linking within the same strand and binding to nucleoprotein or the phosphate anion are also possible effects and can lead to functional damage in rapidly proliferating cells, like miscoding and point mutations. The chemistry of the alkylating process is shown in figure 7.6.

Nitrosoureas. The nitrosoureas, represented by carmustine (**7.68**, BCNU), lomustine (**7.69**, CCNU), and semustine (**7.70**, methyl-CCNU), are more recent discoveries. These compounds are relatively easily prepared; for example, carmustine is synthesized by treating 1,3-bis(2-chloroethyl)urea with sodium nitrite and formic acid. These drugs combine the N-NO group with a monofunctional mustard. They function by crosslinking through alkylation of DNA. The compounds are effective against some brain tumors and certain lung carcinomas, both of which tend to respond poorly to chemotherapy. It is their relatively unique lipid-soluble properties that enable these compounds (unlike many chemotherapeutics) to cross the blood–brain barrier. Streptozotocin (**7.71**) is a naturally occurring glucosamine nitrosourea derivative that shows antileukemic activity as well as antibiotic effects. The nitrosoureas can carbamoylate proteins (e.g., on lysine) by forming isocyanates, whereas the chloroethyl carbonium ion formed could potentially crosslink the strands of DNA.

Alkylsulphonate Esters. Methanesulfonate esters such as busulfan (**7.72**) produce clinical remission in chronic myelogenous leukaemia. Busulfan acts through an S_N2 nucleophilic displacement and presumably crosslinks DNA, since the methanesulfonate

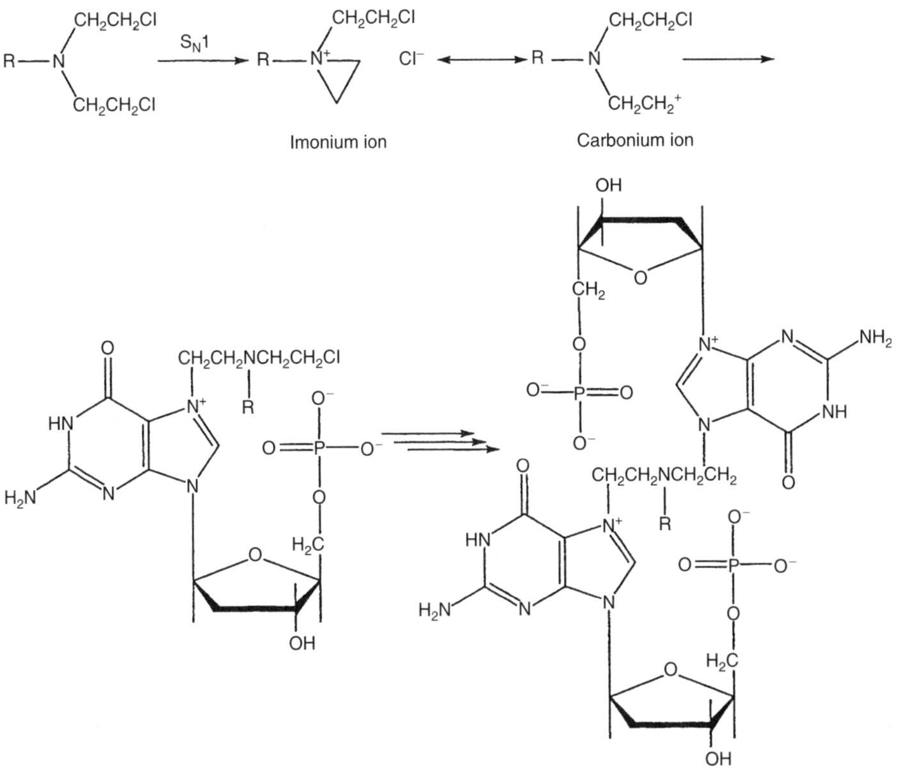

Figure 7.6 Alkylating agents as antineoplastic drugs: the compounds covalently link to both strands of the DNA, literally holding them together. This prevents the DNA from uncoiling, a crucial early step in the process of DNA replication. This stops cellular proliferation in a rapidly growing cell.

ion in an excellent leaving group. Busulfan is synthesized by treating 1,4-butanediol with methanesulfonyl chloride in the presence of pyridine.

Aziridines. This class of compounds includes agents such as thiotepa (**7.73**) and triethylenemelamine (**7.74**). These compounds contain reactive three-membered nitrogen-containing heterocycles that react with nucleophiles to relieve ring strain. The aziridine group is protonated to provide a reactive aziridinium ion that is known to alkylate DNA. Thiotepa is synthesized by treating trichlorophosphine sulphide with aziridine. Altretamine (**7.75**, hexamethylmelamine) is a structural analog of triethylenemelamine that is useful as an alkylating agent in the treatment of ovarian carcinoma.

A variety of other therapeutic compounds have mechanisms that probably involve alkylation. Procarbazine and dacarbazine are the two most successful agents in this group. Dacarbazine is a synthetic compound that functions as an alkylating agent, following metabolic activation by liver microsomal enzymes, to yield diazomethane. It has been used successfully in the treatment of sarcomas.

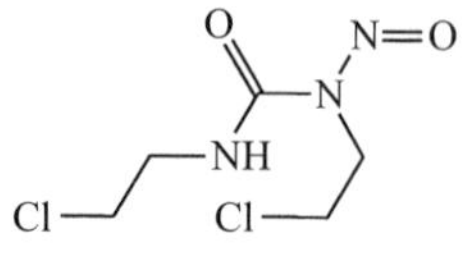

Carmustine (7.68)

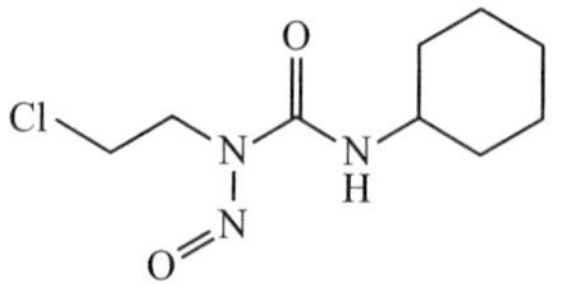

Lomustine (7.69)

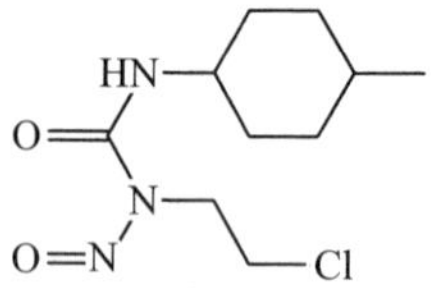

Semustine (7.70)

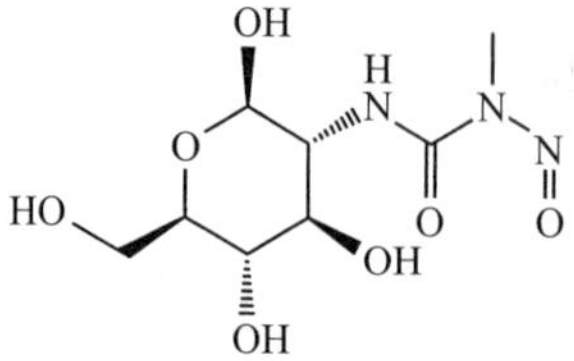

Streptozotocin (7.71)

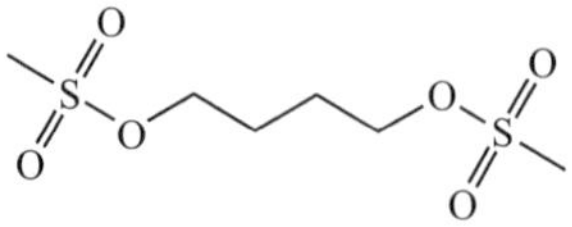

Busulfan (7.72)

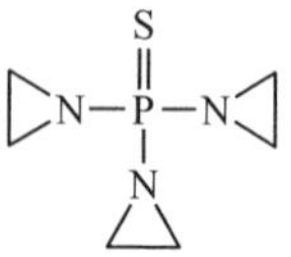

Thiotepa (7.73)

7.8.1.3 Antimetabolites

Antimetabolite inhibitors of DNA synthesis act by the competitive or allosteric inhibition of a number of different enzymes participating in purine or pyrimidine biosynthesis. Actually, some such compounds interfere with as many as 10–12 different enzymes—although admittedly to a different degree.

Folic Acid Antagonists. Folic acid antagonists block the biosynthesis of purine nucleotides. Methotrexate (**7.76**) is the prototypic folic acid antagonist and functions by binding to the active catalytic site of dihydrofolate reductase, thereby interfering with the synthesis of the reduced form that accepts one-carbon units; lack of this cofactor blocks the synthesis of purine nucleotides. As well as being used in the treatment of cancer, methotrexate has been used in the management of rheumatoid arthritis, psoriasis, and even asthma.

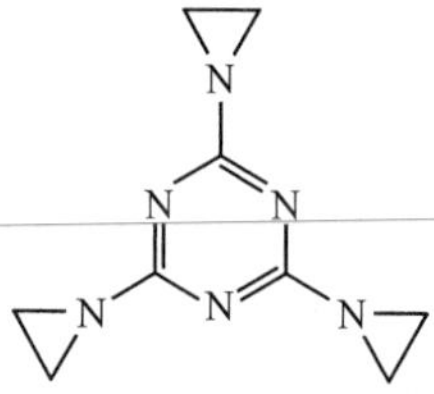

Triethylenemelamine (7.74)

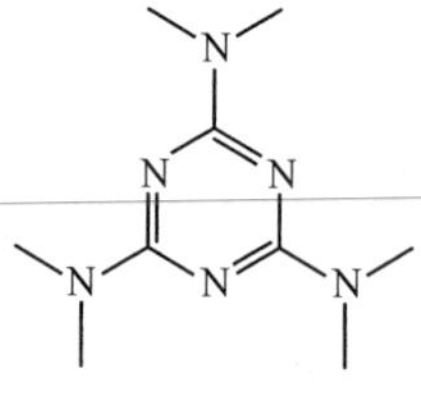

Altretamine (7.75)

Purine Antimetabolites. Purine synthesis can be blocked by 6-mercaptopurine (**7.77**) and 6-thioguanine (**7.78**). Both require conversion to the mononucleotide in a "lethal synthesis"—a mechanism distinguished from the formation of suicide substrates in that the enzyme that transforms the inactive pro-drug to the active inhibitor is different from the enzyme that is being blocked. K_{cat} inhibitors are formed and bound by the same enzyme. Both thiopurines primarily block the amidotransferase in the first step of purine synthesis as "pseudo-feedback" inhibitors. Additionally, the transformations of inosinic

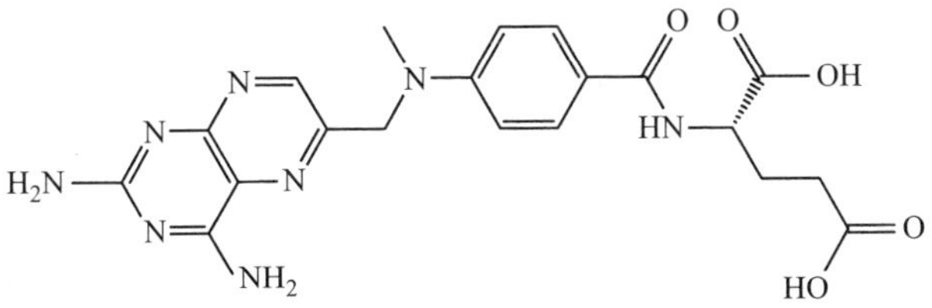

Methotrexate (7.76)

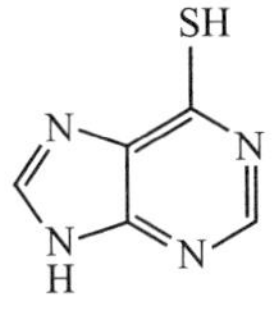

6-Mercaptopine (7.77)

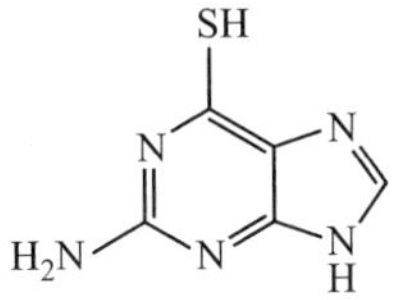

6-Thioguanine (7.78)

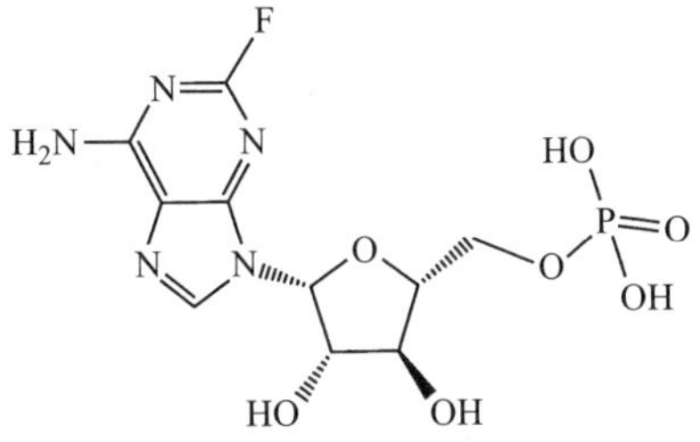

Fludarabine phosphate (7.79)

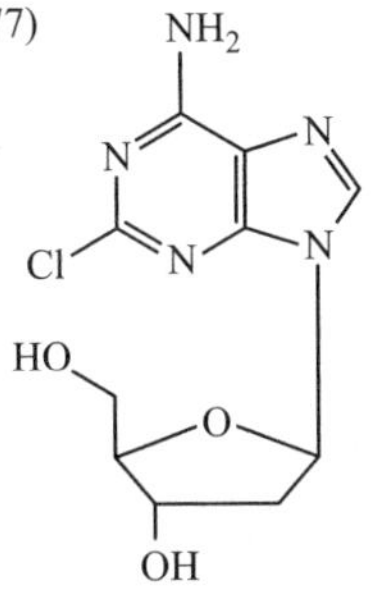

Cladribine (7.80)

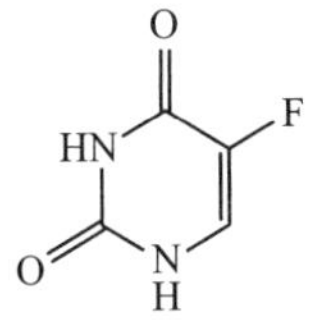

Fluorouracil (7.81)

acid to AMP and GMP are also inhibited. Both 6-mercaptopurine and 6-thioguanine have been used in the treatment of acute leukemia. Both compounds are easily prepared; for example, 6-thioguanine is readily prepared by treating guanine with phosphorus pentasulfide in pyridine. Fludarabine phosphate (**7.79**, 2-fluoro-arabinofuranosyladenine monophosphate) and cladribine (**7.80**, 2-chlorodeoxyadenosine) are other compounds that function as purine antimetabolites.

Pyrimidine Antimetabolites. The commonly used drug fluorouracil (**7.81**, 5-FU) is the prototype pyrimidine antimetabolite. 5-FU undergoes biotransformation to ribosyl nucleotide metabolites, including 5-fluoro-2′-deoxyuridine-5′-phosphate (FdUMP) which in turn forms a covalent bond to the complex formed by the thymidylate synthase enzyme and its $N^{5,10}$-methylenetetrahydrofolate cofactor; this inhibits the synthesis of thymine nucleotides and terminates DNA synthesis by the process of "thymineless death." Traditionally, this compound is prepared by condensing S-ethylisothiouronium bromide with the potassium salt (enolate) of ethyl-2-fluoro-2-formylacetate. Capecitabine (**7.82**) is a fluoropyrimidine carbamate prodrug that is converted to 5-FU. In another pyrimidine antimetabolite, cytarabine (**7.83**, Ara-C, 1-β-D-arabofuranosyl-cytosine), the ribose moiety of cytidine is replaced by the epimeric arabinose. This drug inhibits DNA polymerases after its bioconversion to ara-CTP, competing with CTP. The adenine analog Ara-A acts in a similar fashion. Both drugs are used in leukemia only; they are inactive against solid tumors.

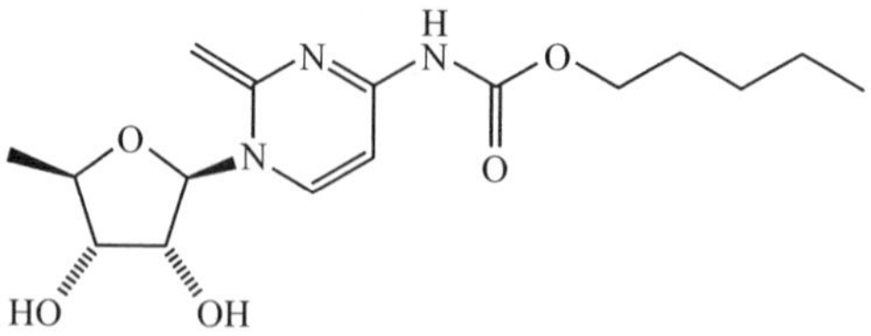

Capecitabine (7.82)

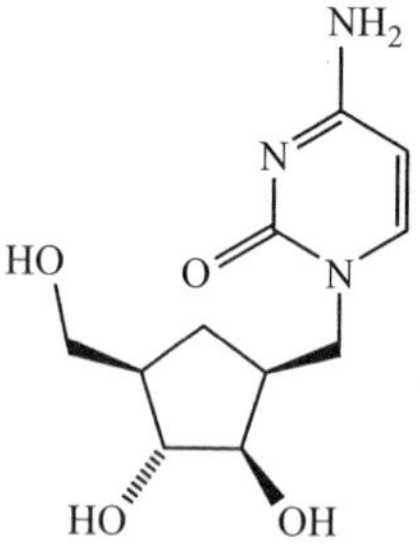

Cytarabine (7.83)

7.8.1.4 DNA Topoisomerase Inhibitors

DNA is a topologically constrained molecule because the ends of the circular double helix are fixed in space, allowing the formation of higher-order structures called "supercoils." During replication, the double helix must unwind, and unwinding introduces additional positive "supertwist." *Topoisomerase I* removes this supertwist by breaking and resealing single strands to produce a "relaxed" DNA. Since the reaction is thermodynamically favorable, no ATP is needed. *Topoisomerase II ("gyrase")* catalyzes the passing of two DNA strands through breaks, and thus produces a "negative supertwist" which further promotes parental strand separation at the replication fork; ATP is needed for this reaction. Since the replicating fork rotates at a speed of about 100 revolutions per second, interference with these enzymes rapidly stops DNA replication. Several natural products have demonstrated efficacy in inhibiting topoisomerase enzymes.

   Camptothecin Topoisomerase Inhibitors. The camptothecins (e.g., topotecan, **7.84**, irinotecan, **7.85**) are natural products that function as topoisomerase I enzyme inhibitors.

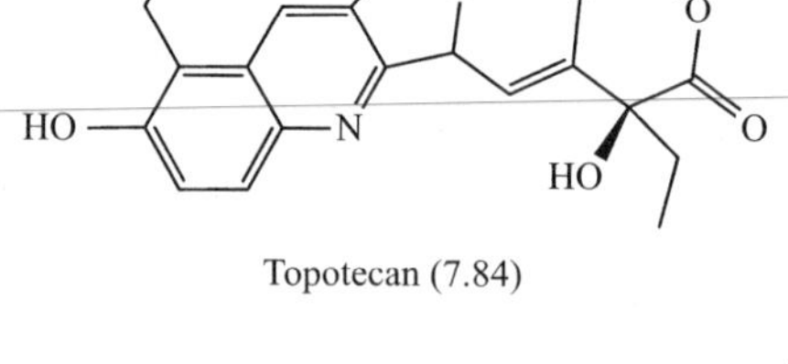

Topotecan (7.84)

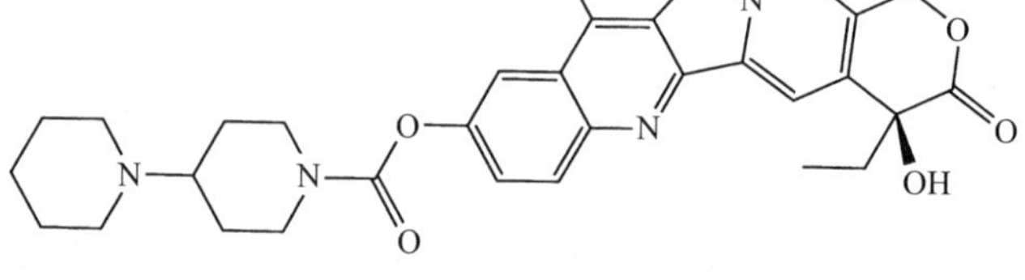

Irinotecan (7.85)

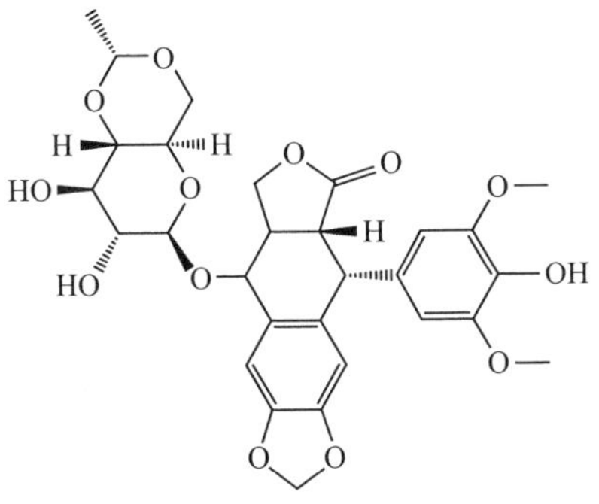

Etoposide (7.86) 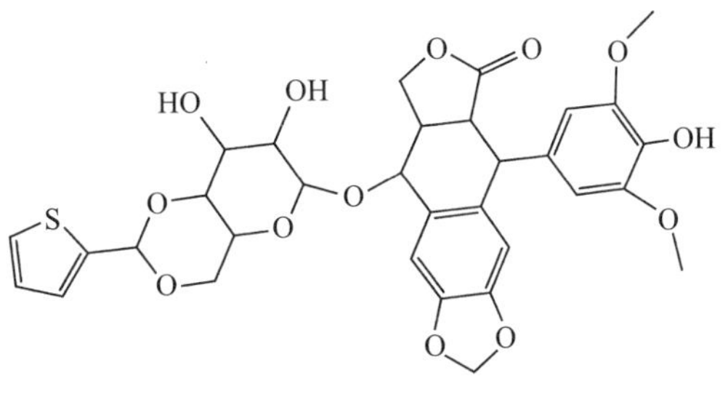   Teniposide (7.87)

They cause substantial DNA damage in tumor cells, preventing tumor growth. Topotecan has been evaluated in metastatic ovarian cancer. Irinotecan is a prodrug that is metabolized to a topoisomerase I inhibitor; it has been used in the treatment of colon and rectal cancer.

Podophyllotoxin Topoisomerase Inhibitors. Two compounds, etoposide (**7.86**) and teniposide (**7.87**) are semisynthetic analogs of podophyllotoxin, a natural product extracted from the root of the mayapple (*Podophyllum peltatum*). These compounds inhibit topoisomerase II, causing DNA strand breakage, and have demonstrated bioactivity in leukemia and lymphoma.

7.8.2 Drugs Interfering with Transcription and Translation

The intercalating drugs and nucleic acid synthesis inhibitors discussed in the preceding sections interfere indirectly with every phase of nucleic acid action because the DNA is rendered nonfunctional. Other drugs, discussed below, affect the regulation of protein synthesis even when the genome, the DNA structure, is intact. Such activity can be the result of interference either with transcription of messenger RNA or with translation of the mRNA to protein.

7.8.2.1 Natural Product Anti-Transcription Agents

Among the cytostatic agents in this group are the *bleomycins*, a very complex group of natural product glycopeptides produced by *Streptomyces verticillus*. Among many biochemical effects, bleomycins seem to cause the scission and breakage of DNA, rendering the process of transcription ultimately impossible. They also seem to inhibit DNA ligase, an important DNA replication and repair enzyme. Bleomycin therefore interferes with the transcription as well as replication of nucleic acids. Since there are no bleomycin-degrading enzymes in the skin, squamous-cell carcinomas of the neck and head respond well to this drug. However, toxic effects also prevail in skin tissue. The unique activity and toxicity of the bleomycins can be utilized in the combination treatment of malignancies, in which drugs with different modes of action and nonoverlapping toxicities are used, often with remarkable success.

7.8.2.2 Synthetic Anti-Transcription Agents

Another group of agents that interfere with DNA transcription are the *platinum complexes*. The *cis*-dichlorodiammine-platinum (II) complex (**7.88**, cisplatin), but not the *trans* isomer, is very active against testicular tumors, epidermoid carcinomas, and ovarian tumours. Cisplatin is prepared by treating potassium chloroplatinite with ammonia. It binds strongly to DNA by intrastrand binding (i.e., not crosslinking) to oligoguanine sequences, unwinds the duplex, and reduces the length of the DNA molecule. The *trans* isomer is selectively removed from the DNA. These platinum agents are therefore similar, but not identical, to alkylating agents. By binding to DNA, this compound ultimately inhibits the transcription process.

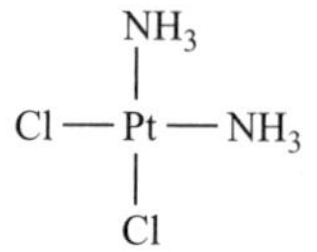

$$Cl - \underset{\underset{Cl}{|}}{\overset{\overset{NH_3}{|}}{Pt}} - NH_3$$

Cisplatin (7.88)

Many platinum analogs, and analogs based on other metal centres, have been prepared and evaluated. Carboplatin (**7.89**, *cis*-diammine(1,1 cyclobutanedicarboxalato)platinum) has markedly reduced gastrointestinal and renal toxicity in comparison with cisplatin. It is prepared by the treatment of *cis*-Pt(NH$_3$)$_2$I$_2$ with silver sulphate, followed by the barium salt of 1,1-cyclobutanedicarboxylic acid.

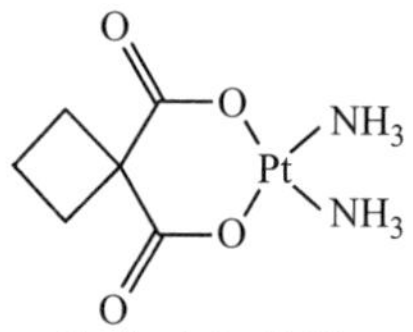

Carboplatin (7.89)

7.8.2.3 Antisense Oligomer Anti-Translation Agents

Antisense oligomers provide a rational approach to the inhibition of translation. Single-strand messenger RNA (mRNA) demonstrates high affinity, sequence-specific binding to a complementary oligonucleotide sequence (via hydrogen bonding of the Watson–Crick pairs). This complementary agent is termed an antisense oligomer. Antisense oligomers can block RNA translation into protein. Since they are analogs of endogenous natural products, antisense oligomers are readily susceptible to enzymatic degradation. This problem can be addressed by the bioisosteric replacement of "building block" segments within the antisense oligomer. For instance, the sugar–phosphate backbone has been replaced with N-(2-aminoethyl)glycine units. Similarly, the sugar moieties have been replaced using 2'-fluoro-2'-deoxysugars. From a clinical perspective, these agents are still in their developmental stages.

7.8.3 Cytostatic Agents Interfering with Mitosis

During the metaphase stage of mitosis, the daughter chromosomes normally begin to migrate toward the poles of the cell; they are pulled toward the poles by *microtubules*, which are assembled at one of their ends and disassembled at the other. The microtubules are hollow tubes built from 13 dimers of two different kinds of protein (α and β) in a helical manner. Their outer diameter is about 24 nm. Some cytostatic agents bind to this αβ dimer, preventing its incorporation into the growing microtubule. Several alkaloid natural products function as antimitotic agents.

7.8.3.1 Vinca Alkaloid Antimitotics

The best known drugs acting as antimitotics are the vinca alkaloids, vincristine (**7.90**) and vinblastine (**7.91**). They are very complex indole derivatives that nevertheless have been synthesized. Both are quite effective in various leukemias and in Hodgkin's lymphoma, but show considerable neurotoxicity. Vinblastine and vincristine bind specifically to the microtubular protein *tubulin* in dimeric form, precipitating depolymerization of the microtubules and functionally acting as a "mitotic poison." Vinorelbine (**7.92**) is a semisynthetic vinca alkaloid functionally identical to vinblastine.

7.8.3.2 Taxane Alkaloid Antimitotics

The alkaloid ester paclitaxel (**7.93**) is a natural product derived from the western yew (*Taxus brevifolia*) and European yew (*Taxus baccata*). This agent functions as a mitotic spindle by causing excessive tubulin polymerization. Paclitaxel has demonstrated considerable clinical utility in ovarian and breast cancer,

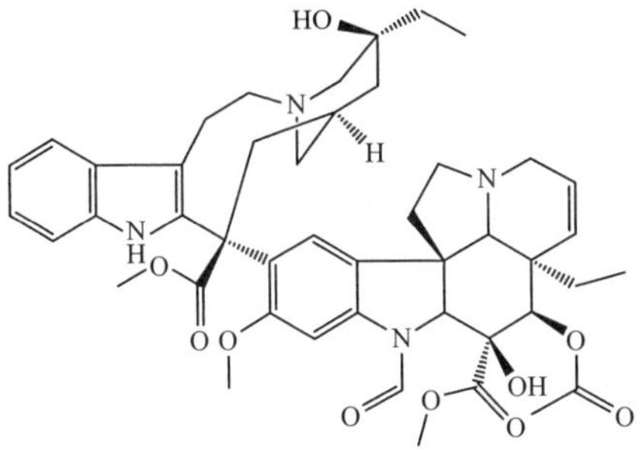

Vincristine (7.90)

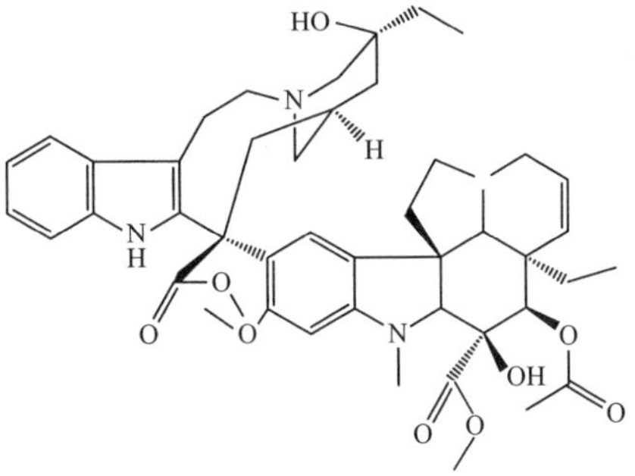

Vinblastine (7.91)

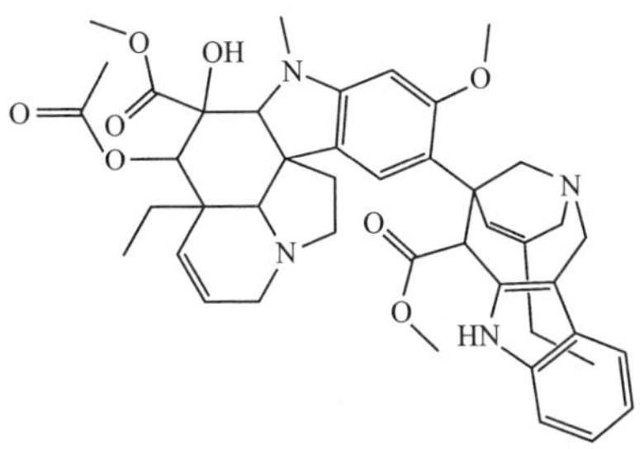

Vinorelbine (7.92)

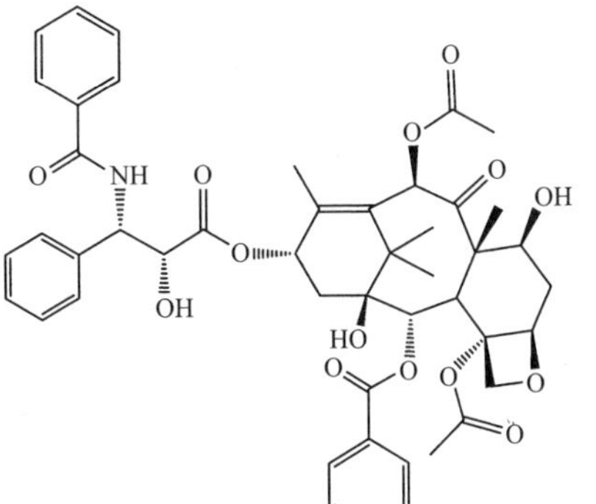

Paclitaxel (7.93)

7.8.4 Targeting Nucleic Acids: Drug Design Approaches

7.8.4.1 Designing Cell-Cycle-Specific Agents

In designing drugs to target the cell nucleus, it is imperative to remember that the cell nucleus is a biochemically dynamic structure. For reproductively active cells, nucleic acid biosynthesis occurs in stages within the cellular nucleus. This gives rise to the concept of a *cell cycle* in which the cell exits in four distinct phases: S (DNA synthesis), M (mitosis, cell division), G_1 (a resting phase in which cellular components for DNA synthesis may be produced), and G_2 (a resting phase in which cellular components for mitosis may be synthesized). The G_1 phase contains a G_0 phase, in which the cell is not active in cell division. The major classes of drugs used in cancer chemotherapy may be either cell-cycle-specific (CCS) or cell-cycle-nonspecific (CCNS), depending upon their mode of interference with the replication–transcription–translation cascade. CCS agents include antimetabolites (e.g., methotrexate, 5-FU, thioguanine), podophyllin alkaloids (e.g., etoposide, teniposide) and plant alkaloids (e.g., vincristine, paclitaxel); CCNS agents include alkylating agents, cisplatin, and antibiotic anticancer drugs (e.g., daunorubicin, doxorubicin).

7.8.4.2 Designing Rational Polypharmacy Agents

In drug design, it is sometimes possible to design new drugs to act synergistically with other existing drugs. This would lead to rational polytherapy—the purposeful therapeutic combination of two or more drugs with complementary mechanisms of action. Typically, monotherapy (the use of one "ideal" drug for the disease in question) is preferable. However, in complicated disease states, such as those presented by cancer, polytherapy (also called polypharmacy) is frequently desirable. The fact that the anticancer agents all work by interfering with DNA/RNA, but do so by differing and noncompetitive molecular mechanisms, enables such agents to be combined in a rational (and hopefully synergistic) manner. A large number of "combination cancer chemotherapeutic regimens" are in common clinical use. A partial listing of such regimens includes:

1. Breast cancer
    a. Doxorubicin + cyclophosphamide + paclitaxel
    b. Doxorubicin + cyclophosphamide + methotrexate + fluorouracil
2. Bladder cancer
    Methotrexate + vinblastine + doxorubicin + cisplatin
3. Testicular cancer
    Bleomycin + etoposide + cisplatin
4. Ovarian cancer
    Paclitaxel + carboplatin
5. Lung cancer
    Cyclophosphamide + doxorubicin + vincristine + etoposide + cisplatin

7.8.5 Emerging Trends in Cancer Drug Design

The diverse compounds discussed above as putative therapies for cancer tend to exhibit significant toxicities that arise from their inability to differentiate between the DNA of the tumor and the DNA of the host patient. Accordingly, several new directions in cancer chemotherapy drug design are exploiting non-nucleic acid strategies. These new directions include:

1. Angiogenesis inhibitors
2. Proteases
3. Signal transduction inhibitors
4. Hormonal manipulation
5. Photodynamic therapy
6. Immunotherapy

These new directions can be used in combination with traditional anti-nucleic acid approaches.

7.8.5.1 Angiogenesis Inhibitors

One characteristic of cancerous tumors is their rapid rate of growth. This accelerated growth rate requires that the tumor establish its own blood supply to fuel its growth needs. *Angiogenesis* is the formation of new blood vessels; uncontrolled angiogenesis is a driving factor in solid tumor growth. Angiogenesis inhibitors should therefore, in principle, halt uncontrolled tumor growth. The search for angiogenesis inhibitors was initiated by seeking "endogenous angiogenesis inhibitors" in human tissues that typically lack blood vessels, such as cartilage. It was subsequently discovered that peptides (such as $CDPGYIGSR–NH_2$) based on the laminin structure inhibit solid tumor growth via an anti-angiogenesis mechanism. Synthetic heparin substitutes, such as sulphated cyclodextrins, achieve the same goal. The search for natural product and synthetic angiogenesis inhibitors is now an active area of ongoing research.

7.8.5.2 Protease Inhibitors

Another characteristic of cancerous tumors is their capacity to infiltrate adjacent tissues and to colonize secondary sites (*metastasis*). To exhibit these biological capabilities, tumor cells degrade the extracellular matrix that surrounds them through a process that is mediated by proteases. An abnormally regulated elaboration of proteases appears to be a characteristic of malignant cell lines. The extracellular matrix is degraded by a complex array of metalloproteinases, including collagenases and gelatinases, which are themselves turned on by tumor-secreted cysteine proteases and serine proteases. Compounds that inhibit cysteine proteases or serine proteases (e.g., nafamostat, **7.94**) are being evaluated as putative anticancer agents.

7.8.5.3 Kinase Inhibitors as Signal Transduction Inhibitors

To achieve rapid, uncontrolled growth with potential for metastasis, the tumor cell has elaborate intracellular biosynthetic capabilities that enable it to fulfil its "vicious" growth characteristics. *Protein tyrosine kinases* are important to signal transduction in initiating aberrant cellular transformation in malignant cells. Microbial products such as staurosporine (**7.95**) have a demonstrated ability to inhibit kinases and may be a good starting point for rational drug design. In addition to kinases, other cellular processes are involved with signal transduction in cancerous cells and are thus targets for drug design.

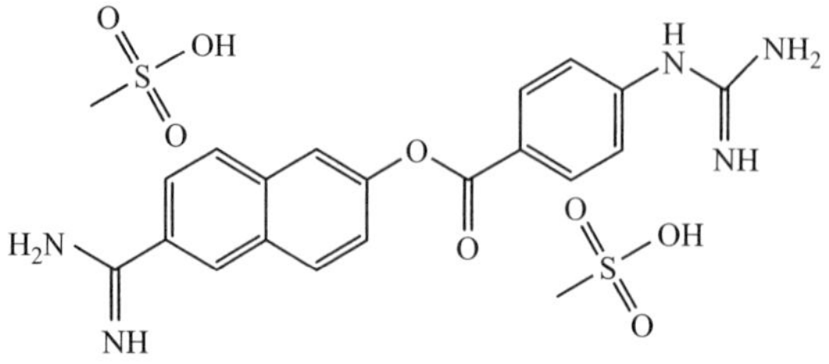

Nafamostat (7.94) 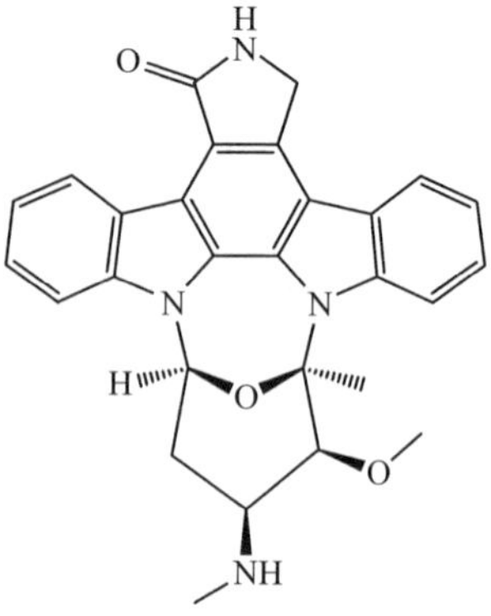   Staurosporine (7.95)

Ras protein, for example, is central amongst the many *oncogene* ("cancer-producing gene") encoded proteins that relay signals from the exterior of the cell to the cell nucleus. In normal cells, ras protein controls the "switch" for normal cellular growth and differentiation by cycling GTP and GDP by a hydrolytic process. Oncogenic ras proteins have lost this hydrolytic capacity, thereby leaving the cellular growth "switched on." Modification of ras protein represents a logical point of molecular attack in drug design.

7.8.5.4 Hormonal Therapy

Tumors that occur in biochemically unique organs offer an opportunity for biochemically unique approaches—the potential to design drugs with reduced toxicity. Hormonal manipulation has been employed successfully for a number of years to treat cancer in gender-specific organ systems. This approach is predicated on the observation that such tumors express hormonal receptors on their cell surfaces and that the tumor's biochemistry is thus susceptible to hormone manipulation. The clinical usefulness of antiestrogens in treating breast cancer and antiandrogens for prostate cancer is well recognized. Gonadrotropin-releasing hormone (GnRH) agonists have also been used in hormone-responsive cancers. A logical extension of hormonal therapy is to block the biosynthesis of the hormones that may be influencing tumor growth. *Aromatase enzyme inhibitors* are an excellent example of this approach. A major step in estrogen biosynthesis involves the aromatase-catalyzed conversion of androstenedione to estrone; blocking the aromatase enzyme lowers the level of circulating estradiol. Two aromatase inhibitors, anastrazole (**7.96**) and letrozole (**7.97**), have been evaluated in the treatment of metastatic breast cancer. Recently, some clinical trials have suggested the use of anastrazole before tamoxifen for post-menopausal women with metastatic breast cancer.

7.8.5.5 Photodynamic Therapy

Since not all tissues are as biochemically unique as hormone-secreting organs, other approaches have been explored in an attempt to achieve tumor-specific killing while sparing other tissues. Photodynamic therapy is such an approach. This type of therapy is based upon the accumulation of a drug molecule within the tumor, followed by a light-mediated

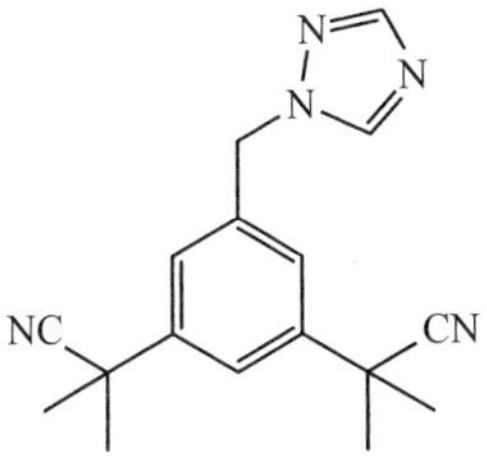

Anastrazole (7.96) 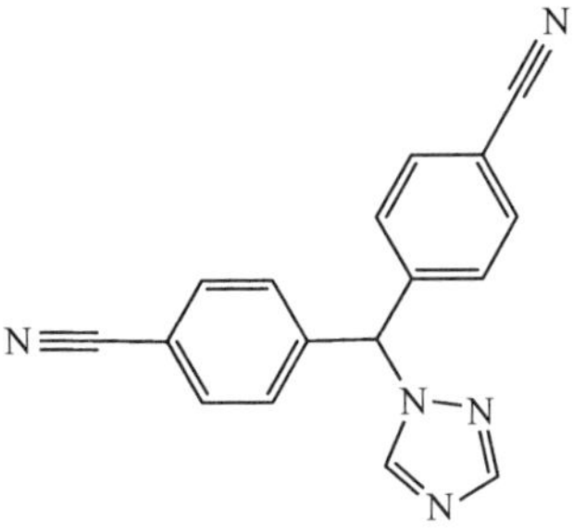   Letrozole (7.97)

conversion of the drug from a nontoxic form to a tumor-killing form. This conversion is achieved uniquely within the tumor cells by shining the activating light only upon the tumor. Elsewhere within the body the drug remains in an inactive, nontoxic form.

Porfimer sodium is an agent used in photodynamic therapy. Porfimer is an oligomeric mixture of porphyrin units linked by ester and ether linkages. The porfimer selectively accumulates in tumor tissues and is cleared from other tissues. Laser illumination of the tumor at 630 nm wavelength then causes the porfimer molecule to enter an excited state in which singlet oxygen is produced by radical reactions. As this reaction propagates (by spin transfer from porfimer to molecular oxygen), hydroxyl radicals are produced. These are locally toxic to the tumor.

7.8.6 The Clinical–Molecular Interface: The Cell Nucleus and Cancer Treatment

Cancer is the common term for any of a group of diseases characterized by abnormal cellular growth that typically produces malignant tumors.

Typically, a tumor is regarded as the pathological hallmark of cancer. The term tumor, however, is a nonspecific term for any swelling or localized enlargement. Tumors that may potentially be cancerous are more correctly categorized as *neoplasms.* A neoplasm is an abnormal mass of tissue whose growth is uncoordinated and exceeds that of normal tissues and whose presence persists in an excessive manner following the termination of the stimulus that provoked the initial formation of the abnormal tissue mass. A neoplasm may be either benign or malignant ("cancerous"). A neoplasm is a benign tumour when it is a slow-growing, localized, expansible mass, enclosed within a capsule; benign tumours have well-formed and differentiated cellular elements (i.e., subcellular structures, such as organelles and the cell nucleus, appear morphologically normal and resemble those of cells from normal tissues). A neoplasm is a malignant tumor when it is a fast-growing, non-encapsulated, infiltrative, erosive growth of cells capable of being transported from its site of origin and being implanted in other sites (with this cellular relocation process being referred to as metastasis); malignant tumors have poorly formed and poorly differentiated cellular elements (i.e., subcellular structures have altered morphology and do not resemble the typical nucleus and organelles of normal cells). Malignant tumor cells may be described as anaplastic when their subcellular structure is more or less undifferentiated and when the cells have lost all resemblance to normal counterpart cells. Anaplasia of cellular morphology is an unmistakable hallmark of malignancy in a neoplasm. Likewise, metastasis also unequivocally marks a tumor as malignant, since benign neoplasms cannot metastasize. Decreased cohesiveness of individual cells within the tumor mass facilitates metastasis by enabling individual cells to readily leave their site of origin and to be transplanted to other sites. Loss of contact inhibition enables malignant cells, either from the site of primary origin or from any secondary metastatic site, to grow in an erosive fashion, progressively impinging upon and ultimately destroying contiguous normal cells and tissues.

Neoplasms can be broadly categorized as either "simple" or "compound"; simple neoplasms are composed of one single neoplastic cell type, whereas compound neoplasms contain more than one neoplastic cell type. Both simple and compound neoplasms may be either benign or malignant. There are two types of malignant simple neoplasms: carcinomas and sarcomas. A carcinoma is a malignant neoplasm that is derived from an epithelial cell line. Epithelium is the cellular layer covering all inner and outer free surfaces (cutaneous, serous, or mucous) within the body, including the skin, gastrointestinal tract, genitourinary tract, respiratory tract, and the associated glands and structures derived therefrom. The type of carcinoma reflects the organ of origin: skin (squamous cell carcinoma), hair follicle (basal cell carcinoma), respiratory tract (bronchogenic carcinoma), urinary bladder (transitional cell carcinoma). When the carcinoma arises from the epithelium that lines a gland, then the resulting neoplasm is called an adenocarcinoma. Thus, a malignant tumor arising from a gland lining the large intestine would be an adenocarcinoma of the colon. Since mammary glands are embryologically derived from modified sweat glands, breast cancer tumors are carcinomas.

A malignant sarcoma is a neoplasm derived from a mesenchymal cell line. Mesenchyme is that portion of the body that produces the muscles, connective tissues (e.g., tendons, cartilage, bone, fat), fluid distribution systems (i.e., blood and lymph vessels), and blood cells (e.g., erythrocytes [red blood cells], leukocytes [white blood cells: granulocytes, monocytes]). The type of sarcoma reflects the organ of origin: fat (liposarcoma), bone (osteogenic sarcoma), lymph vessel (lympangiosarcoma), muscle (rhabdomyosarcoma), blood cell (leukemia). The malignancies of blood cells (leukemias, multiple myeloma) do not produce the localized tumor masses typically associated with neoplasms. From a biological perspective, sarcomas are distinct and different from carcinomas.

Collectively, carcinomas and sarcomas are much more frequent than the rather uncommon compound neoplasms that are populated by more than one neoplastic cell line. Compound tumors are frequently derived from "uncommitted cells" which are *totipotential* or *pluripotential*, i.e., capable of differentiating into a variety of different cell types. Accordingly, compound neoplasms characteristically develop within embryonic cells (primitive cell "rests," sequestered and left over after the person's embryogenetic development) or cells in gonads (reproductive cells waiting to pass their full complement of genetic information to an offspring). A common type of benign compound neoplasm is the *teratoma*, which is principally encountered in the gonads; a single teratoma tumor may contain many types of tissue resembling fat, skin, teeth, hair follicles, gut epithelium, or muscle. One or more of these individual cell lines within a compound neoplasm can become malignant. A skin-type epithelial cell within a teratoma could therefore become malignant, producing a squamous cell carcinoma within a teratoma and thus producing a tumor referred to as a *teratocarcinoma*.

Clinically, the signs and symptoms of cancer are as varied and diverse as the types of neoplasm. If the site of origin of the neoplasm (*primary tumor*) is in an easily observed area (e.g., tongue, skin of face), then the tumor will be detected early. If the site of origin is in a hidden area (e.g., ovaries), then the tumor may grow undetected and be diagnosed only when metastatic spread produces *secondary tumors* in other more easily detected anatomical regions. The primary tumor may produce symptoms by virtue of its location (primary brain tumor causing seizures) or by means of its erosive, infiltrative behavior (pulmonary adenocarcinoma eroding a blood vessel, causing bleeding into the lung and *hemoptysis* [coughing up blood]). Alternatively, symptoms may arise from the secondary metastatic tumors being present in vital organs such as brain or liver. Metastases to bone are common and can cause the bone to break (*pathological fractures*) or can cause severe pain. It is possible for a pea-sized primary tumor to give rise to hundreds of secondary tumors ranging in size from golf-ball dimensions to very small tumors that can be seen only with a microscope ("micromets").

The process of metastasis usually occurs when the cancerous cells are spread either through the bloodstream or via the lymphatic system. Carcinomas are more frequently spread by the lymphatic system, while sarcomas typically are disseminated by embolization through blood vessels; however, either tumor type can be spread by either mode of dissemination. The spread of a breast carcinoma through the lymphatic system to the lymph nodes of the axilla ("armpit") is a well-recognized mode of spread. Organs that receive a voluminous blood supply are frequent sites for blood-borne metastatic spread for either carcinomas or sarcomas. Understandably, secondary tumors are thus common in the liver, lungs, and brain. When a primary tumor has produced many secondary metastases, the overall *tumor burden* is high, and the afflicted individual experiences weight loss and wasting (producing the phenomenon referred to as *cachexia*). Alternatively, tumors can produce widespread effects on the host through the aberrant synthesis of hormones or other chemical factors that influence brain or bone marrow function; such systemic chemical effects are referred to as *paraneoplatic syndromes.* An example of such a syndrome occurs when carcinomas of the lung produce the hormone ACTH.

Understanding the diverse clinical phenomenology of cancer at a molecular level is challenging. Cancer is a fundamental disease of cells and of the macromolecular constituents of cells. Since cancer is characterized by an alteration in the control mechanisms that govern cell proliferation and differentiation, the nucleic acids (DNA, RNA) are the central players in the molecular cascade of cancer. A wealth of data support the key role played by the nucleic acids. Ever since Sir Percival Pott's recognition of environmentally induced scrotal cancer in chimney sweeps (in 1775!), the importance of chemical carcinogenesis has been appreciated. Prototypic chemical carcinogens include polycyclic aromatic hydrocarbons and aromatic amines that intercalate into nucleic acids, thereby altering the conformation and function of DNA and RNA. Carcinogenic nitrosamines are transformed into reagents that donate methyl and ethyl groups to RNA and DNA. Carcinogenic metals such as beryllium and cadmium participate in electrophilic reactions with the basic nitrogen atoms contained within RNA and DNA.

Recognizing that nucleic acids constitute the clinical–molecular interface in cancer represents a major challenge in drug design. Nucleic acids are crucial not only to the growth of the tumor but also to the overall wellbeing of the patient in general. The tumor grows in an independent manner that is not controlled or regulated by the usually operative control mechanisms in the body; in this way, the tumor is behaving like a "nonself" exogenous parasite. However, unlike an exogenous parasite, the tumor is in fact an endogenous structure composed of the exact same molecules and building blocks as the host that it is destroying. Therefore, a tumor is "self acting as nonself." Drugs engineered to attack tumor nucleic acid targets will correspondingly also target (in principle) all other cells within the host's body. This introduces immense complexities when endeavoring to design cancer chemotherapeutic agents with optimal cancer-killing efficacy but minimal toxicity. Since the toxicities are primarily mediated by the therapeutic mechanism of action, separation of pharmacophore from toxicophore becomes a seemingly insurmountable problem. Also, tumors show varying sensitivities to the anticancer therapies that are available (see table 7.1).

The most reasonable approach for addressing this dilemma is to exploit differences in cell growth kinetics between cancer cells and host cells. A hallmark of cancerous cell growth is the rapidity of its cellular proliferation. At any given time, a malignant tumor should have more cells undergoing mitosis and replication than other tissues in the host.

Table 7.1 Varying Responsiveness of Tumors to Anticancer Drugs

Tumor	Treatment
Chemoresistant tumors—not responsive to chemotherapy	
Adrenocortical carcinoma	Combination chemotherapy
Gliomas (brain)	Surgery
Hypernephroma (kidney)	Surgery, chemotherapy, immunotherapy
Malignant melanoma (skin)	Surgery
Pancreatic adenocarcinoma	Surgery
Pancreatic islet cell carcinoma	Surgery, combination chemotherapy
Squamous cell bronchogenic carcinoma (lung)	Surgery and radiotherapy ± chemotherapy
Moderately chemosensitive	
Bladder carcinoma	Combination chemotherapy (local and systemic)
Breast carcinoma	Surgery, radiotherapy, chemotherapy
Bronchogenic carcinoma (lung)	Combination chemotherapy
Endometrial carcinoma (uterus)	Hormonal and cytotoxic chemotherapy, surgery
Head and neck carcinomas	Chemotherapy and surgery/radiotherapy
Multiple myeloma (bone marrow)	Combination chemotherapy
Ovarian carcinoma	Combination chemotherapy, surgery
Prostate carcinoma	Hormonal therapies
Chemosensitive—Tumors responsive to chemotherapy	
Acute leukemias	Combination chemotherapy
Burkitt's lymphoma	Surgery, radiotherapy, chemotherapy
Choriocarcinoma	Methotrexate, dactinomycin
Hodgkin's lymphoma	Combination chemotherapy, radiotherapy
Non-Hodgkin's lymphoma	Combination chemotherapy, radiotherapy
Retinoblastoma (eye)	Radiotherapy, cyclophosphamide
Rhabdomyosarcoma (muscle)	Cyclophosphamide
Testicular carcinoma	Surgery, radiotherapy, chemotherapy
Wilms' tumor (children)	Surgery, radiotherapy, chemotherapy

(Adapted from D. G. Grahame-Smith, J. K. Aronson (2002). *Clinical Pharmacology and Drug Therapy*, 3rd Edn. New York: Oxford University Press. With permission.)

This observation opens a window of opportunity, enabling a partially selective targeting of tumor cells in preference to host cells. By designing agents which attack nucleic acids at particular times during the cell cycle, it is possible to devise molecules with improved specificity for tumor cells.

Selected References

Membrane Ion Channel Targets

M. S. Bretscher (1985). The molecules of the cell membrane. *Sci. Am. 253*: 100–108.

W. A. Catterall (1985). The electroplax sodium channel revealed. *Trends Neurosci. 8*: 39–41.

D. Chapman (Ed.) (1984). *Biomembrane Structure and Function, Topics in Molecular and Structural Biology*, vol. 4. Weinheim: Verlag Chemie.

M. J. Coghlan, W. A. Carroll, M. Gopalakrishnan (2001). Recent developments in the biology and medicinal chemistry of potassium channel modulators: update from a decade of progress. *J. Med. Chem. 44*: 1627–1653.

M. Concepcion, B. G. Covino (1984). Rational use of local anesthetics. *Drugs 27*: 256–270.

N. D. P. Cosford, P. T. Meinke, K. A. Stauderman, S. D. Hess (2002). Recent advances in the modulation of voltage-gated ion channels for the treatment of epilepsy. *Current Drug Targets: CNS & Neurological Disorders 1*: 81–104.

C. R. Craig, R. E. Stitzel (Eds.) (1986). *Modern Pharmacology*, 2nd ed. Boston: Little, Brown.

C. de Dove (1984). *A Guided Tour of the Living Cell*. San Francisco: Scientific American Books.

D. A. Doyle, J. M. Cabral, R. A. Pfuetzner, A. Kuo, J. M. Gulbis, S. L. Cohen, B. T. Chait, R. MacKinnon (1998). The structure of the potassium channel: molecular basis of K^+ conduction and selectivity. *Science 280*: 69–77.

H. Glossman, D. R. Ferry, A. Goll, J. Striessnig, G. Zernig (1985). Calcium channels and calcium channel drugs: recent biochemical and biophysical findings. *Arzneimittelforschung 35*: 1917–1935.

T. Godfraind, R. Miller, M. Wibo (1986). Calcium antagonism and calcium entry blockade. *Pharmacol. Rev. 38*: 321–416.

D. B. Goldstein (1984). The effect of drugs on membrane fluidity. *Annu. Rev. Pharmacol. Toxicol. 24*: 43–64.

V. K. Gribkoff, J. E. Starrett, Jr. (2002). Advances in technologies for the discovery and characterization of ion channel modulators: Focus on potassium channels. *Annu. Rep. Med. Chem. 37*: 237.

A. K. M. Hammarstroem, P. W. Gage (2002). Hypoxia and persistent sodium current. *Eur. Biophys. J. 31*: 323–330.

B. Hille (1984). *Ionic Channels of Excitable Membranes*. Sunderland, MA: Simmer.

B. Hille, C. M. Armstrong, R. MacKinnon (1999). Ion channels: from idea to reality. *Nat. Med. 5*: 1105–1109.

F. Hucho, S. Stengelin, G. Bandini (1979). Effector binding sites and ion channels in excitable membranes. In: F. Gualtieri, M. Gianella, C. Melchiorre (Eds.). *Recent Advances in Receptor Chemistry*. New York: Elsevier/North Holland, pp. 37–58.

Y. Jiang, A. Lee, J. Chen, M. Cadene, B. T. Chait, R. MacKinnon (2002). Crystal structure and mechanism of a calcium-gated potassium channel. *Nature 417*: 515–522.

T. H. Large, M. W. Smith (2001). Screening strategies for ion channel targets. In: *Drugs and the Pharmaceutical Sciences Handbook of Drug Screening*, vol. 114, pp. 313–333.

D. J. Madge (1998). Sodium channels: recent developments and therapeutic potential. *Annu. Rep. Med. Chem. 33*: 51–60.

R. Mannhold (1984). Calmodulin—structure, function and drug action. *Drugs Future 9*: 677–691.

H. Meyer (1983). Structure–activity relationships in calcium antagonists. In: L. R. Opie (Ed.). *Calcium Antagonists and Cardiovascular Disease* New York: Raven Press.

M. Noda, S. Shimizu, T. Tanabe, T, Takai, T. Kayano, T. Ikeda, H. Takahashi, H. Nakayama, Y. Kanaoka, N. Minamino (1984). Primary structure of *Electrophorus electricus* sodium channel deduced from cDNA sequence. *Nature 312*: 121–127.

M. A. Perillo (2002). The drug–membrane interaction: its modulation at the supramolecular level. *Recent Res. Dev. Biophys. Chem. 2*: 105–121.

H. J. Reiser, M. E. Sullivan (1986). Antiarrhythmic drug therapy: new drugs and changing concepts. *Fed. Proc. 45*: 2206–2208.

E. Rucker (1987). Structure, function and assembly of membrane proteins. *Science 235*: 959–961.

G. Schneider, W. Neidhart, G. Adam (2001). Integrating virtual screening methods to the quest for novel membrane protein ligands. *Current Medicinal Chemistry: Central Nervous System Agents 1*: 99–112.

R. R. Schofield, M. G. Darlison, N. Fujita, D. R. Burt, F. H. Stephenson, H. Rodrigues, L. M. Rhee, J. Ramachandran, V. Reale, T. A. Glencorse, P. H. Seeburg, E. A. Barnard (1987). Sequence and functional expression of the GABA receptor shows a ligand-gated receptor super-family. *Nature 328*: 221–227.

M. Spedding (1985). Calcium antagonist subgroups. *Trends Pharmacol. Sci. 6*: 109–114.

M. I. Steinberg, W. B. Lancefield, D. W. Robinson (1986). Class I and III antiarrhythmic drugs. *Annu. Rep. Med. Chem. 21*: 95–108.

C. F. Stevens (1987). Channel families in the brain. *Nature 328*: 198–199.

D. J. Triggle, R. A. Janis (1987). Calcium channel ligands. *Annu. Rev. Pharmacol. Toxicol. 27*: 347–369.

E. Wehinger, R. Gross (1986). Calcium modulators. *Annu. Rep. Med. Chem. 21*: 85–94.

Cell Nucleus Targets

G. L. Chen, L. F. Liu (1986). DNA topoisomerases as therapeutic targets in cancer chemotherapy. *Annu. Rep. Med. Chem. 21*: 257–262.

S. T. Crooke, A. W. Prestayko (Eds.) (1981). *Cancer and Chemotherapy, vol. 3: Antineoplastic Agents.* New York: Academic Press.

T. W. Doyle, T. Kaneko (1985). Antineoplastic agents. *Annu. Rep. Med. Chem. 20*: 163–172.

J. B. Gibbs (2000). Mechanism-based target identification and drug discovery in cancer research. *Science 287*: 1969–1973.

T. Morisaki, M. Katano (2003). Mitochondria-targeting therapeutic strategies for overcoming chemoresistance and progression of cancer. *Curr. Med. Chem. 10*: 2517–2521.

R. Z. Murray, C. Norbury (2000). Proteasome inhibitors as anti-cancer agents. *Anti-Cancer Drugs 11*: 407–417.

D. C. Myles (2002). Emerging microtubule stabilizing agents for cancer chemotherapy. *Annu. Rep. Med. Chem. 37*: 125.

S. Neidle (1979). The molecular basis of action of some DNA-binding drugs. In: G. P. Ellis and G. B. West (Eds.). *Progress in Medicinal Chemistry*, Vol. 16. Amsterdam: Elsevier/North Holland, pp. 151–221.

W. B. Pratt, R. W. Ruddon (1979). *The Anticancer Drugs.* New York: Oxford University Press.

W. E. Ross (1985). DNA topoisomerases as targets for cancer therapy. *Biochem. Pharmacol. 34*: 4191–4195.

A. C. Sartorelli, D. G. Johns (Eds.) (1974, 1975). *Antineoplastic and Immunosuppressive Agents*, Parts 1 and 2. New York: Springer.

K. Shimizu, N. Oku (2004). Cancer anti-angiogenic therapy. *Biol. Pharmaceut. Bull. 27*: 599–605.

F. Takusagawa, R. Carlson, R. F. Weaver (2001). Anti-leukemia selectivity in actinomycin analogues. *Bioorg. Med. Chem. 9*: 719–725.

8

Nonmessenger Targets for Drug Action II

Endogenous macromolecules

8.1 ENDOGENOUS MACROMOLECULES: RELEVANT BIOCHEMISTRY

If targeting messenger molecules such as neurotransmitters, hormones, or immunomodulators fails to address the disease under study, the next approach is to target either nonmessenger endogenous structures, such as the cell membrane or cell nucleus (see chapter 7), or nonmessenger endogenous macromolecules such as proteins (enzymes) or lipids (prostaglandins) (this chapter). These endogenous macromolecules are the catalysts and molecular machinery that enable the cell to perform its normal metabolic functions; accordingly, they afford numerous druggable targets.

At the molecular level, the human body is constructed from a diverse array of molecules and macromolecules that can be broadly categorized into seven general groupings:

1. Water
2. Amino acids, peptides, proteins
3. Lipids
4. Carbohydrates
5. Nucleosides, nucleotides, nucleic acids
6. Heterocycles
7. Minerals and inorganic salts

These seven groups of compounds, and the diversity of compounds within each group, are the molecular building blocks that enable structure and function. A brief overview of the relevant biochemistry for each class is pertinent to an appreciation of their role in medicinal chemistry and drug design.

8.1.1 Water

Water is the single most common molecule in the human body. Indeed, the human body is 73–76% water. Although it is the most ubiquitous molecule, it is the molecule that is most conspicuously ignored during drug design. Water is a powerful *molecular symbiont.* The presence of water profoundly influences the shape and function of all other molecules and macromolecules within the body. Water is the solvent that bathes all molecular events. It is essential in determining the acid–base properties of all biomolecular processes. Water lines all receptor sites, and an incoming drug must either incorporate the water into the receptor binding process or displace it.

8.1.2 Amino Acids, Peptides, and Proteins

Unlike water, amino acids and peptides have been exploited as drug design targets for many years. *Amino acids*, as implied by their name, are difunctional molecules: they possess a basic amino group and an acidic carboxyl group. Arising from this difunctional structure, amino acids are said to be *amphoteric*, since they can react as either acids or bases. Moreover, amino acids can undergo an intramolecular acid–base reaction and exist primarily in the form of a dipolar ion or *zwitterion*. The most important amino acids, both structurally and functionally, are the *α-amino acids*, which have a single carbon atom (the "α-carbon") separating the amino group from the carboxyl group. (β-Amino acids have two carbons separating the amino and carboxyl groups, while γ-amino acids have a three-carbon separation.)

8.1.2.1 Amino Acids, Peptides, and Proteins: Molecular Structure

There are twenty naturally occurring α-amino acids; 19 of them are primary amines, differentiated only by the side-chain substituent appended to the α-carbon (see figure 8.1). Proline is a secondary amine whose nitrogen and α-carbon atoms are part of a five-membered pyrrolidine ring. With the exception of glycine, the α-carbons of the amino acids are stereogenic centers of chirality. α-Amino acids are fundamental building blocks that can join together into long chains by forming amide bonds between the amino terminus of one amino acid and the carboxyl terminus of another. The long repetitive sequence of –N-CH-C(=O)- units that form the chain is called the protein *backbone*. The *side-chains* of each amino acid then bristle away from this backbone structure. The various amino acids that constitute the chain are referred to as *residues*. Chains with fewer than fifty amino acids are *peptides*, while those with more than fifty are *proteins*.

Peptide and protein chains twist and turn upon themselves, producing elaborate and complex structures that enable bioactivity. *Primary structure* refers to the amino acid sequence of the protein; *secondary structure* (consisting of *α-helices, β-sheets,* and *β-turns*) describes how segments of the protein backbone orient into regular patterns via intramolecular hydrogen bonding along the backbone; *tertiary structure* describes how the entire protein molecule coils into a three-dimensional shape by means of multiple intramolecular interactions involving both backbone and side-chain atoms; *quaternary structure* refers to the process whereby many individual protein molecules come together, through various intermolecular interactions, to form a larger aggregate structure.

8.1.2.2 Types of Proteins

Proteins may be structural, functional, or catalytic. Structural proteins are frequently *fibrous proteins* (insoluble polypeptide chains arranged side by side in long filaments)

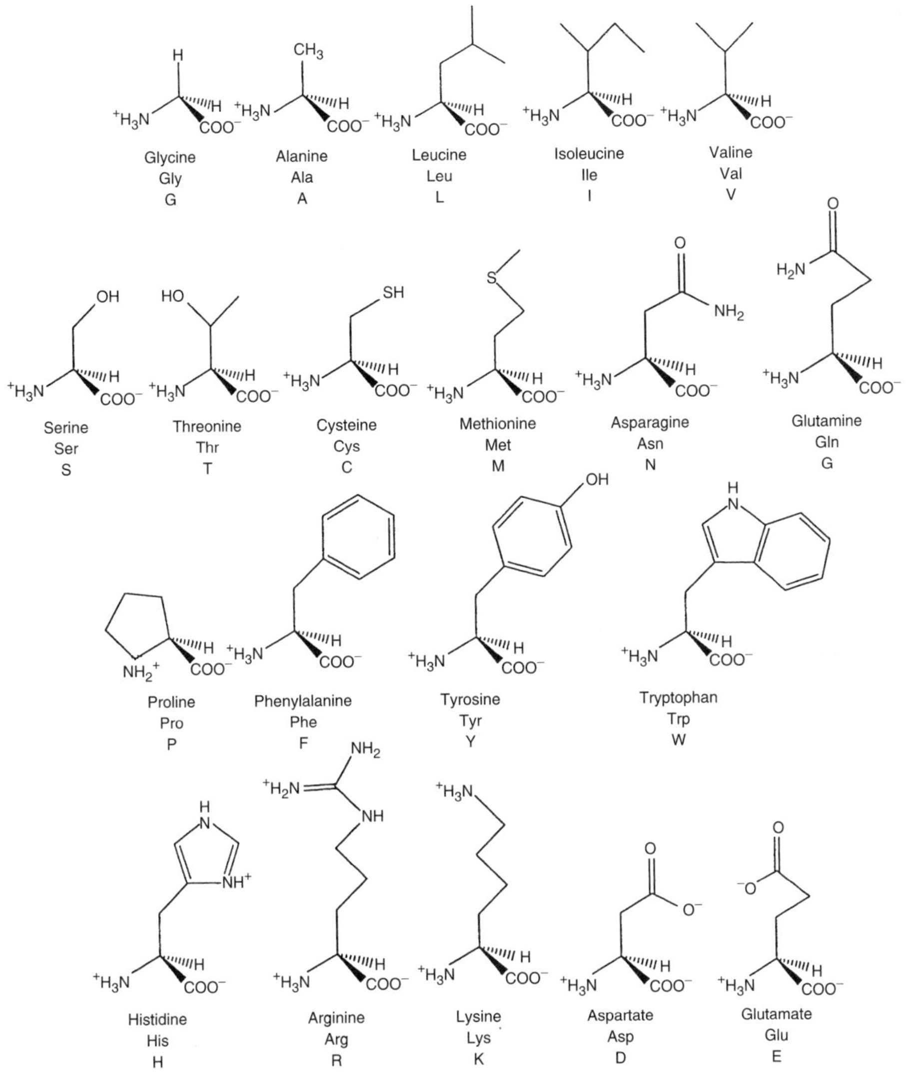

Figure 8.1 Structures of the twenty naturally occurring amino acids.

and include collagens (tendons, connective tissue), elastins (ligaments, blood vessels), keratins (fingernails), and myosins (muscle tissue). Functional proteins are typically *globular proteins* (soluble, compactly coiled, roughly spherical in shape) and include hemoglobin (involved in oxygen transport), immunoglobulins (involved in immune responses), and insulin (involved in glucose metabolism). Catalytic proteins, also called *enzymes*, function by bringing reactant molecules together, positioning them in a favorable orientation for reactivity, and furnishing any required acidic or basic sites.

8.1.2.3 Functional Catalytic Proteins: Enzymes

More than 3000 different enzymes have been identified and characterized. For purposes of drug design, these enzymes may be grouped into six main classes, according to the reaction that they catalyze; in turn, within each group, there are a variety of different enzymes:

1. Hydrolases
     Amidases
     Caspases
     Esterases
     Glycosidases
     Lipases
     Nucleases
         Exonucleases
         Endonucleases
     Peptidases
     Proteases
2. Isomerases
     Cis/trans-isomerases
     Epimerases
     Mutases
     Racemases
     Rotamases
3. Ligases
     Carboxylases
     DNA ligase
     RNA ligase
     Synthetases
4. Lyases
     Aldolases
     Decarboxylases
5. Oxidoreductases
     Catalases
     Dehydrogenases
     Oxidases
     Oxygenases
     Peroxidases
     Reductases
6. Transferases
     Acetyltransferases
     Kinases
     Polymerases
     Transaldolases
     Transaminases
     Transketolases
     Transmethylases

Hydrolases are enzymes that catalyze cleavage reactions (or the reverse fragment condensations); isomerases are enzymes that catalyze intramolecular rearrangements; ligases split C-C, C-N, C-O, C-S, or C-halogen bonds without hydrolysis or oxidation; lyases non-hydrolytically cleave groups from their substrates, concomitantly resulting in the formation of a double bond; oxidoreductases catalyze oxidation and reduction reactions involving the transfer of hydrogen atoms or electrons; and transferases transfer functional groups containing C, N, P, or S atoms from one substrate to another.

Since enzymes are key catalysts in many metabolic processes, they have traditionally been (and will continue to be) among the most important targets for drug design.

8.1.3 Nucleic Acids

Two nucleic acids, deoxyribonucleic acid (**8.1**, DNA) and ribonucleic acid (**8.2**, RNA), constitute, respectively, the molecules that store hereditary information and those that transcribe and translate such hereditary information, thus enabling the directed synthesis of varied but specific proteins throughout the cell and the entire organism. These nucleic acids are biopolymers, composed of monomeric building units called *nucleotides* (figure 8.2). Just as proteins can be hydrolyzed to amino acids and polysaccharides can be hydrolyzed to monosaccharides, nucleic acids can be hydrolyzed to nucleotides.

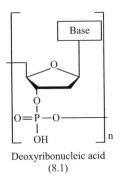
Deoxyribonucleic acid
(8.1)

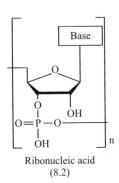
Ribonucleic acid
(8.2)

Mild hydrolysis of nucleic acids yields the monomeric nucleotides. Subsequent complete hydrolysis of a nucleotide furnishes three structural subunits:

1. A 5-carbon monosaccharide (either D-ribose [in RNA] or 2-deoxy-D-ribose [in DNA])
2. A heterocyclic base (either a purine or a pyrimidine)
3. A phosphate ion

A structural variant of a nucleotide is a *nucleoside*, which is a nucleotide with the phosphate group removed.

In nucleotide structure, the central component is the 5-carbon furanoside monosaccharide, either D-ribose (in RNA) or 2-deoxy-D-ribose (in DNA). The four heterocyclic bases found in DNA-based nucleotides are adenine, guanine, cytosine, and thymine:

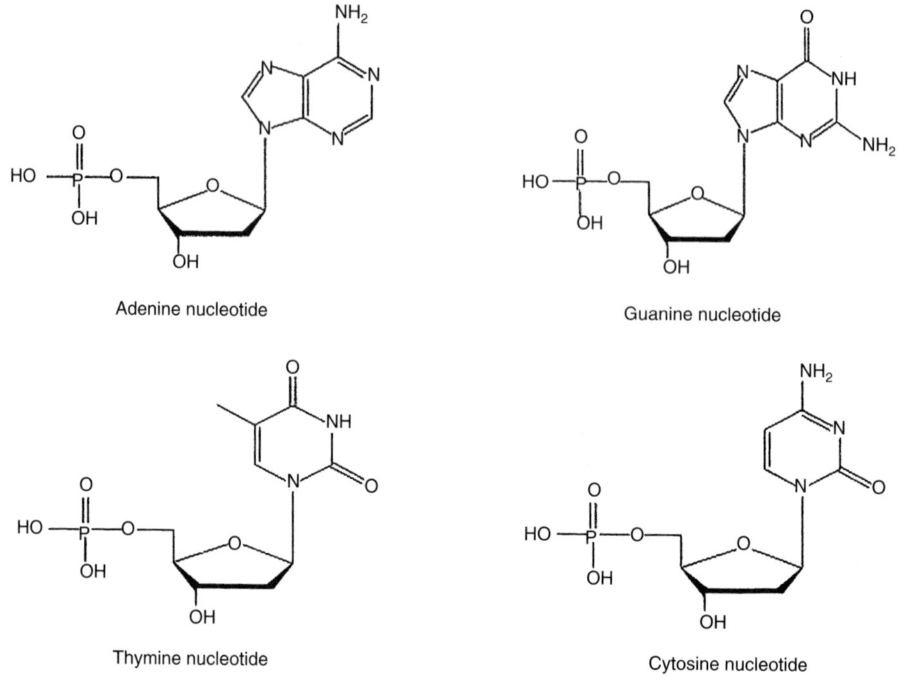

Figure 8.2 Nucleotide: molecular structure.

adenine and guanine are purines; cytosine and thymine are pyrimidines. In RNA-based nucleotides, uracil replaces thymine. These heterocyclic bases are attached through an N-glycosidic linkage to C1′ of the ribose or deoxyribose. The phosphate group is present as a phosphate ester and is attached at either the C5′ or C3′ position. Through these various linkages, the nucleotides are assembled to form the polymeric nucleic acid.

The primary structure of DNA is the base sequence along the DNA chain that contains the encoded genetic information. The secondary structure of DNA is the famous double helix model of Watson and Crick in which two nucleic acid chains are held together by hydrogen bonds between base pairs on opposite strands: adenine pairs with thymine, guanine pairs with cytosine. When the two strands of the double helix coil, two types of groove are produced, a major groove 1.2 nm wide and a minor groove 600 pm wide. In drug design, such structural data is crucial in designing anti-tumor drugs for the treatment of cancer and in understanding the carcinogenic toxicity of flat, polycyclic aromatic molecules.

8.1.3.1 Nucleic Acids: Biological Functions

The structure of DNA/RNA enabled Crick to formulate the "central dogma of molecular genetics." By this dogma, three fundamental processes take place in the transfer of genetic information. *Replication* is the process by which identical copies of DNA are made so that information can be preserved and transferred from generation to generation.

Transcription is the process by which the DNA-based genetic message is read and carried out of the nucleus to an organelle called the ribosome, at which protein synthesis occurs. *Translation* is the process by which the genetic messages are decoded and used to construct proteins. To permit the biochemistry of this dogma, many types of RNA are required: messenger RNA (mRNA) carries genetic messages from DNA to ribosomes; ribosomal RNA (rRNA) is complexed with protein to provide the physical makeup of the ribosomes; and transfer RNA (tRNA) transports amino acids to the ribosomes where they are joined to produce proteins. The specific nucleotide sequence in mRNA forms a message that determines the order in which the amino acid residues are assembled. Each word, or codon, along the mRNA chain consists of a defined sequence of three nucleotides that is specific for a given amino acid.

Nucleotides and nucleosides have emerged as important molecules in medicinal chemistry. In the 1950s, Elion and Hitchings discovered that 6-mercaptopurine had antitumor properties. This pioneering discovery opened the door for many subsequent studies of nucleotide derivatives as therapeutics. Acyclovir (**8.3**), a nucleoside that lacks two carbon atoms of its ribose ring, is effective in the treatment of herpes infections. Allopurinol, a purine derivative, is useful in the treatment of gout.

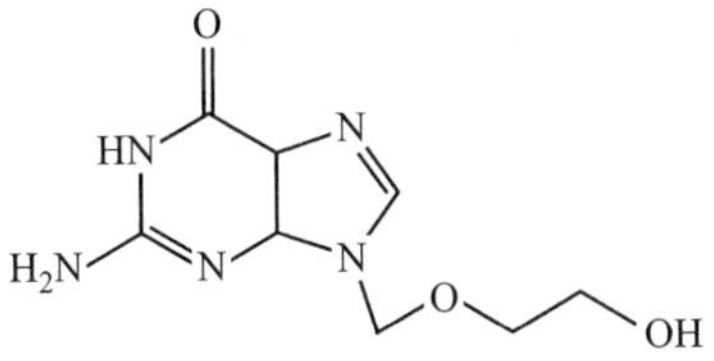

Acyclovir (8.3)

8.1.4 Lipids

Lipids are naturally occurring organic molecules, isolated from animal or plant cells by extraction with nonpolar organic solvents. This definition defines lipids in terms of a physical property (solubility) and differs from structural definitions used for proteins or carbohydrates. Not surprisingly, lipids are highly varied in their structure; from the medicinal chemistry perspective, there are five classes of lipids:

1. Waxes, fats, and oils
2. Phospholipids
3. Prostaglandins
4. Terpenes
5. Steroids

Each of these classes has some relevance to medicinal chemistry and drug design.

8.1.4.1 Lipids: Waxes, Fats, and Oils

Waxes, fats, and oils contain ester linkages that can be hydrolyzed. Waxes are mixtures of esters of long-chain carboxylic acids with long-chain alcohols. Animal fats and vegetable oils, the most widely occurring lipids, are collectively referred to as triacylglycerols (or triglycerides) because they are triesters of glycerol with three long-chain carboxylic acids. Hydrolysis of a fat or oil in an aqueous alkaline solution (saponification) produces glycerol and three fatty acids. There are more than 100 naturally occurring fatty acids. Polyunsaturated fatty acids, such as linoleic or linolenic acid, contain more than one double bond; saturated fatty acids contain no double bonds. The primary biological function of triacylglycerols in animals is to function as an energy storage depot; when triacylglycerols are metabolized they yield more than twice as many calories as does an equal mass of a protein or carbohydrate.

Lipid biochemistry will receive increasing attention in medicinal chemistry because of the epidemic of obesity occurring in developed countries with its associated co-morbidities such as diabetes and accelerated atherosclerosis, leading to strokes and heart attacks. There is compelling evidence that too much saturated fat in the diet can lead to the development of heart disease, and perhaps cancer as well. Moreover, there is also accumulating evidence that "*trans*" fats can likewise lead to an increased risk of cardiovascular disease. Polyunsaturated oils tend to react by auto-oxidation, causing them to become rancid and have a short shelf life; accordingly, partial hydrogenation is performed to convert the oil into an appealing semi-solid with a prolonged shelf life. Regrettably, a problem with partial hydrogenation is that the catalyst isomerizes some of the unreacted double bonds from the natural *cis* configuration to the unnatural *trans* configuration.

8.1.4.2 Lipids: Phospholipids

Phospholipids are diesters of phosphoric acid (H_3PO_4), just as fats and oils are esters of carboxylic acids. There are two main types of phospholipids: *phosphoglycerides* and *sphingolipids*. Phosphoglycerides, organized into a lipid bilayer, comprise the major lipid component of cell membranes. Since this bilayer is the principal barrier to the passage of molecules into and out of a cell, phosphoglyceride structure is of interest in medicinal chemistry and drug design. Phosphoglycerides contain a glycerol backbone linked by ester bonds to two fatty acids and one phosphoric acid. The two most important phosphoglycerides are the lecithins and the cephalins.

Sphingolipids are the second major group of phospholipids. Rather than a glycerol backbone, these lipids have sphingosine (a substituted dihydroxyamine) for their backbone structure. One particular sphingolipid, *sphingomyelin*, is a major lipid constituent of neuronal axons within the brain and nervous tissue. The sphingolipids, together with proteins and polysaccharides, constitute *myelin*, the lipid-based coating of neurons within the brain. Since the need to develop drugs for demyelinating diseases, such as multiple sclerosis, is a continuing drug design target, interest in sphingolipids is of relevance to medicinal chemistry. General structures of some key lipids are given in figure 8.3.

8.1.4.3 Lipids: Prostaglandins and Eicosanoids

Prostaglandins are C20 carboxylic acids, derived from arachidonic acid, that contain a 5-membered ring, at least one double bond and several oxygen-containing functional groups. The 1982 Nobel Prize in medicine was awarded to Bergström, Samuelson, and Vane for their pioneering work in prostaglandin chemistry. Prostaglandins are found in almost all animal tissues, and exert an extraordinary range of biological effects, resulting

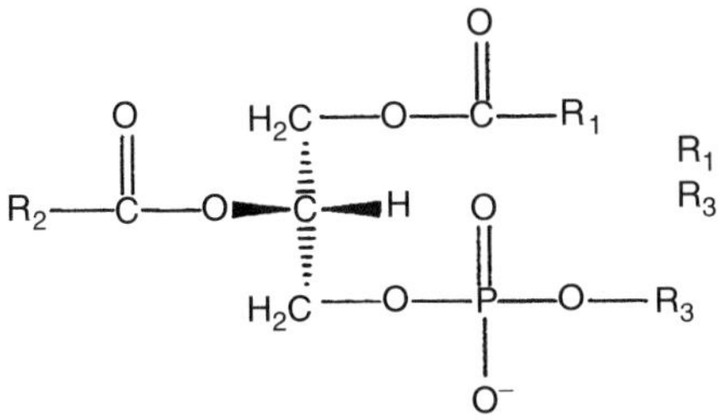

Generalized Phosphoglyceride Structure

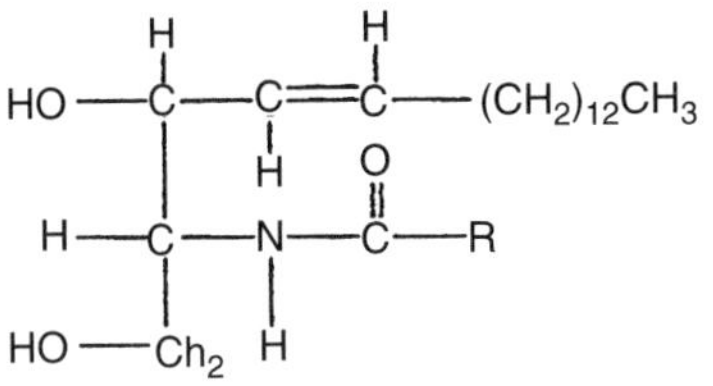

R₁ & R₂ are hydrophobic
R₃ is hydrophilic

R is a hydrocarbon

Generalized Sphingolipid Structure

Figure 8.3 Generalized structures of typical phospholipids.

in their importance in drug design. Prostaglandins affect blood pressure, blood clotting, gastric secretion, kidney function, inflammation, and uterine contractility. One of the prostaglandins (PGE_1) is a potent pyrogen (fever-inducing agent), involved in the symptoms of many pathological processes.

Prostaglandin biosynthesis is also relevant to drug design. In nature, prostaglandins are biosynthesized from arachidonic acid, a C20 unsaturated fatty acid. The synthetic conversion of arachidonic acid to a prostaglandin is catalyzed by the cyclooxygenase (COX) enzyme, which is itself an important target in the design of anti-inflammatory drugs ranging from acetylsalicylic acid (**8.4**) to ibuprofen (**8.5**) to rofecoxib (**8.6**). In addition, prostaglandin-like molecules, including the *leukotrienes*, thromboxanes, and prostacyclin, are also important drug design targets for inflammatory disorders such as asthma. Collectively, these are referred to as the *eicosanoids*.

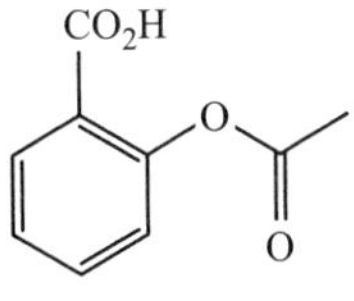

Acetylsalicyclic acid
(8.4)

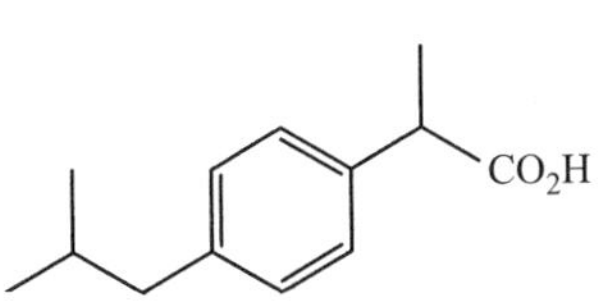

Ibuprofen (8.5)

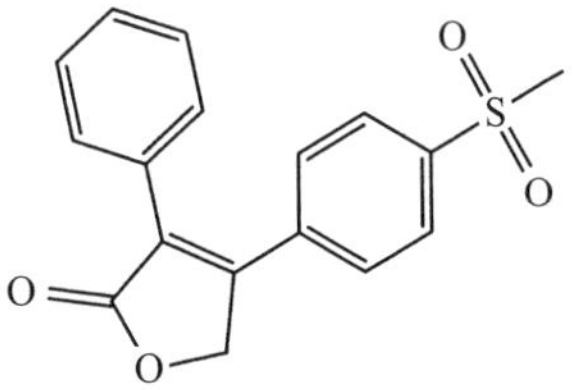

Rofecoxib (8.6)

8.1.4.4 Lipids: Terpenes

Terpenes are a diverse class of small organic molecules. Some are open-chain molecules, some contain rings, some are purely hydrocarbons (classical terpenes), and some contain oxygen-based functional groups (terpenoids). The 1939 Nobel Prize in chemistry was awarded to Leopold Ruzicka for his pioneering work on the chemistry of terpenes, agents that have been used as therapeutics since antiquity. By steam distilling various plant materials, mixtures of odoriferous compounds, known as essential oils, were obtained; terpenes are the most important constituents of these oils. Although they exhibit tremendous diversity, all terpenes are structurally related according to the isoprene rule of Ruzicka. They arise from a head-to-tail joining of 5-carbon isoprene (2-methyl-1,3-butadiene) units; carbon-1 is the head, carbon-4 is the tail. Terpenes are then classified by the number of constituent isoprene units: monoterpenes are 10-carbon compounds biosynthesized from two isoprene units; sesquiterpenes are 15-carbon molecules from three isoprene units. α-Pinene contains two isoprene building blocks assembled into a complex cyclic structure.

   Historically, terpenes have enjoyed a role as symptom-relieving therapeutics. Camphor and carvone (spearmint oil) have been used as soothing agents in the past history of pharmaceutical compounds. More recently, the role of terpenes in biochemistry has been more fully appreciated. β-Carotene, for example, may be cleaved into two units of vitamin A. Other terpenes, such as geraniol, are now recognized as pheromones (a chemical secreted by one individual of a given species in order to elicit a response in another individual of the same species). Pheromones are of interest in medicinal chemistry for a variety of indications; for instance, attractant pheromones can be used for the control of insects known to spread human disease.

8.1.4.5 Lipids: Steroids

Steroids are important lipids whose structures are based on a tetracyclic system. Most steroids function as hormone chemical messengers, and thus these molecules have been discussed in detail in chapter 5. Structurally, steroids are heavily modified triterpenes that are biosynthesized starting from the acyclic hydrocarbon squalene and progressing through cholesterol to the final steroid product; Bloch and Cornforth, who were awarded Nobel Prizes in medicine (1964), contributed greatly to the elucidation of this remarkable biosynthetic transformation.

   Cholesterol is the prototypic steroid lipid. It was first isolated in 1770; in the 1920s, the German chemists Windaus and Wieland deduced the structure for cholesterol, receiving Nobel Prizes for their work in 1927 and 1928. Cholesterol is important in medicinal chemistry, not only for its role in atherosclerosis, but also because it is an important lipid in membrane structure.

   The remaining steroid lipids constitute two main classes of steroid hormones: sex hormones and adrenocortical hormones. The sex hormones include androgens (testosterone, androsterone), estrogens (estrone, estradiol), and progestins (progesterone). The adrenocortical hormones include mineralocorticoids (aldosterone) and glucocorticoids (hydrocortisone).

8.1.5 Carbohydrates

Carbohydrates are naturally occurring compounds that typically have the general formula $C_x(H_2O)_y$, reflecting the fact that they appear to be "hydrates of carbon."

Carbohydrates are the most abundant organic component of plants. Structurally, carbohydrates are usually polyhydroxy aldehydes or polyhydroxy ketones (or compounds that hydrolyze to yield polyhydroxy aldehydes and ketones). Since carbohydrates contain carbonyl groups and hydroxyl groups, they exist primarily as acetals or hemiacetals.

8.1.5.1 Carbohydrates: Molecular Structure

Analogous to peptides, carbohydrates frequently exist as polymers of simpler building blocks. The simplest building block, called a *monosaccharide*, is a simple carbohydrate that cannot be hydrolyzed to simpler carbohydrates. A carbohydrate that can be hydrolyzed to two monosaccharides is called a *disaccharide*. A *trisaccharide* contains three fundamental building blocks. Generalizing, *oligosaccharides* contain 3–10 monosaccharides, while *polysaccharides* (also known as *glycans*) contain more than 10 monosaccharide polymeric units. A polysaccharide that is a polymer of a single monosaccharide is called a *homopolysaccharide*; a polysaccharide composed from different monosaccharides is a *heteropolysaccharide*. A homopolysaccharide composed only of glucose monomeric units is called a *glucan*. In carbohydrate nomenclature, all monosaccharides and most polysaccharides have names ending in the suffix -ose.

Monosaccharides are the simplest, fundamental building block of carbohydrates. Structurally, monosaccharides are classified in accord with two designations: (i) the number of carbon atoms in the molecule, and (ii) the presence of either an aldehyde or a ketone functional group. The number of carbon atoms in a monosaccharide (usually 3–7) is designated as tri-, tetr-, and so on. For example, a three-carbon monosaccharide is a triose, while a six-carbon monosaccharide is a hexose. A monosaccharide containing an aldehyde group is called an aldose; one containing a keto group is termed a ketose. These various designation systems are frequently used in conjunction; thus, a five-carbon monosaccharide with a ketone is called a ketopentose, while a three-carbon monosaccharide with an aldehyde is called an aldotriose.

8.1.5.2 Bioactive Carbohydrates and Stereochemistry

The nomenclature and classification of monosaccharides is further complicated by the presence of chiral carbons within monosaccharides. In the late nineteenth century it was ascertained that the configuration of the last chiral carbon in each of the naturally occurring monosaccharides is the same as that for (+)-glyceraldehyde. This configuration was designated as "D" and it was determined that all naturally occurring monosaccharides were in the D configuration.

Since stereochemistry is crucial to modern drug design, it is imperative that the multiple conventions for representing stereochemistry should not be confused. There are four conventions for representing stereochemical properties: (+)/(−); *d/l*; D/L; R/S. A stereoisomer that rotates plane-polarized light to the right is said to be *dextrorotatory*; its mirror image compound, which rotates plane-polarized light to the left, is *levorotatory*. This experimental direction of light rotation is specified in the name by (+) for dextrorotatory and (−) for levorotatory; alternatively, the symbol *d* can be used for dextrorotatory and *l* for levorotatory. Therefore, the (+)/(−) and *d/l* systems are equivalent. They do not tell us anything about the structure of the molecule; they merely tell us that a solution of the molecule rotates plane-polarized light either to the right or left. An improvement upon this designation is the D/L method for determining *relative configuration*. The D/L system uses the stereoisomers of glyceraldehyde as configurational standards: (+)-glyceraldehyde is designated as having the D configuration, and (−)-glyceraldehyde is designated as having the L configuration. Other molecules are then compared to these two glyceraldehyde standards and are designated as either D or L (relative to glyceraldehydes), depending on which isomer they superimpose upon best. The R/S system is yet another improvement, since it determines *absolute configuration*. In this system, the four groups of atoms attached to a central chiral carbon are prioritized in accord with the *Cahn–Ingold–Prelog sequence rules*; the lowest priority group is directed towards the "rear" of the molecule and a curved arrow unites the other groups from highest to lowest priority. If the arrow is clockwise, the configuration is R; if the arrow is counterclockwise, the configuration is S.

The nomenclature and structure of monosaccharides is additionally complicated by further stereochemical issues arising from the fact that monosaccharides may exist in either "open-chain" or "cyclic" forms. In water solution, a monosaccharide can undergo an intramolecular reaction to produce cyclic hemiacetals: either five-membered ring hemiaceteals (called *furanoses*) or six-membered ring hemiacetals (called *pyranoses*) can exist. An extensive body of experimental evidence reveals that one of the most common monosaccharides, D-glucose, exists as an equilibrium between an open-chain structure and two cyclic forms. The two cyclic forms are hemiacetals produced by an intramolecular reaction of the –OH group at the C5 atom with the aldehyde group. This cyclization creates a new stereocentre at the C1 position; the two cyclic forms are diastereomers that differ in the configuration of C1. Diastereomers of this type are called *anomers*, and the hemiacetal carbon atom is referred to as the *anomeric carbon* atom. Each D-glucose anomer is designated as either an *α-anomer* or a *β-anomer*, depending upon the location of the –OH group on the C1 atom.

8.1.5.3 Types of Carbohydrates

There are many different monosaccharides and oligosaccharides. *Glucose* (also called *dextrose*) is the most important monosaccharide; it is sometimes called *blood sugar* because it is the principal monosaccharide of human blood. *Fructose* (also called *levulose*) is the sweet-tasting sugar found in honey and fruit. *Ribose* and *deoxyribose* are monosaccharides that form part of the polymeric backbone of nucleic acids. *Maltose*, a disaccharide used in baby foods, is composed of two D-glucose building blocks. *Lactose*, a naturally occurring disaccharide found in mammalian milk, is composed of two different monosaccharides, D-glucose and D-galactose. *Sucrose* (common table sugar) is also composed of two different monosaccharides, D-glucose and D-fructose. *Glycogen* is a polysaccharide, found primarily in liver and muscles, that is used as a storehouse for glucose in animal systems. Structures of some common carbohydrates are given in figure 8.4.

Carbohydrates are relevant in medicinal chemistry. The outdated notion that carbohydrates serve only as energy sources in animals and structural materials in plants is no longer tenable. Complex carbohydrates are biochemically important. Carbohydrates, when joined via glycosidic linkages to either proteins or lipids (to produce *glycoproteins* or

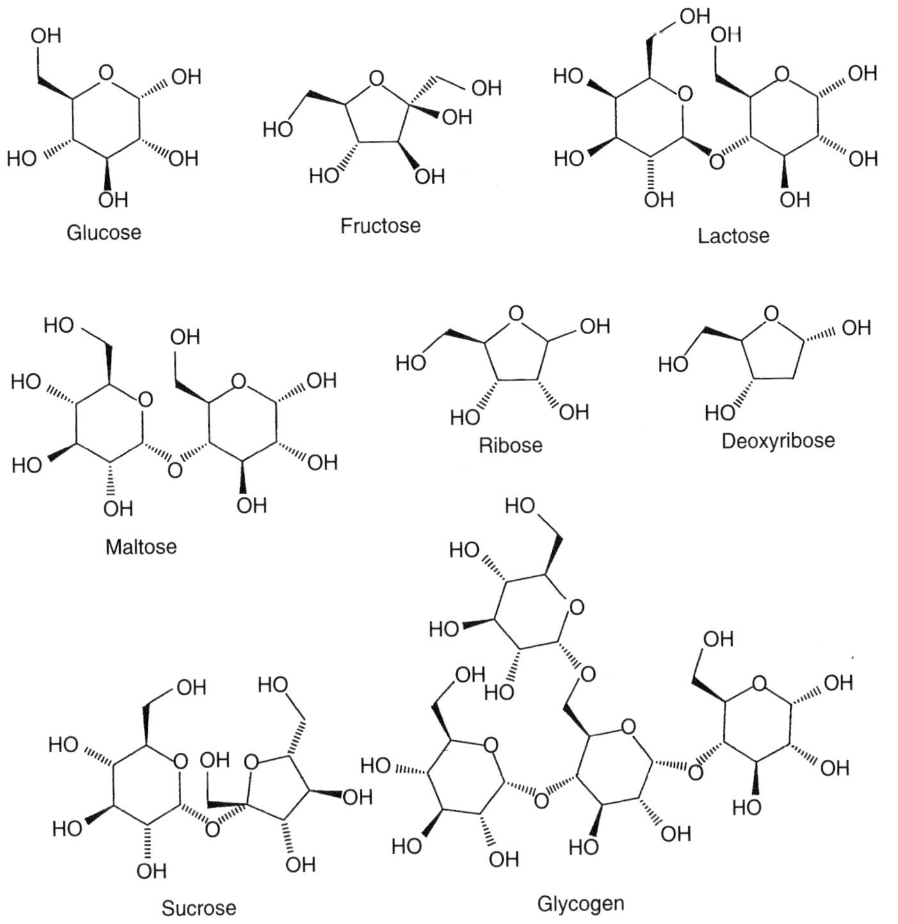

Figure 8.4 Structures of common carbohydrates.

glycolipids) constitute important cell surface macromolecules that are important drug targets and antigens central to immune-mediated disease fighting systems. In addition, carbohydrates may themselves constitute drugs. Heparin, a polysulphated polysacccharide, is used as a therapeutic to prevent blood clotting. Streptomycin (**8.7**) is an important carbohydrate antibiotic.

8.1.6 Heterocycles

Heterocyclic compounds, or *heterocycles*, are molecules with rings that contain more than one type of atom; they are to be distinguished from *carbocycles*, which are molecules with rings that contain only carbon atoms. The heterocyclic compounds of greatest interest to medicinal chemists have carbon rings containing one or two heteroatoms— atoms other than carbon.

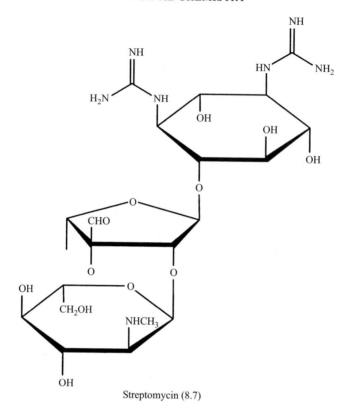

Streptomycin (8.7)

8.1.6.1 Classification of Heterocycles

Structurally and functionally, there is a bewildering array of different heterocycles. These may be classified as follows:

Simple heterocyclic compounds
  Three-membered heterocycles containing one heteroatom
    Oxiranes (epoxides), thiiranes (episulfides), aziridines
  Three-membered heterocycles containing more than one heteroatom
    Oxaziridines, dioxiranes, diazirines
  Four-membered heterocycles containing one heteroatom
    Oxetanes, thietanes, azetidines
  Four-membered heterocycles containing more than one heteroatom
    Dithietanes, dithietes, diazetidines
  Five-membered non-aromatic heterocycles containing one heteroatom
    Pyrrolidine
  Five-membered non-aromatic heterocycles containing more than one heteroatom
    1-Pyrazolines, 1,3-dioxolanes, 2-pyrazolines
  Five-membered aromatic heterocycles containing one heteroatom
    Pyrrole, furan, thiophene

Five-membered aromatic heterocycles containing more than one heteroatom
    Pyrazoles, oxazoles, thiazoles
Six-membered non-aromatic heterocycles containing one heteroatom
    Piperidine
Six-membered non-aromatic heterocycles containing more than one heteroatom
    Morpholines, 1,4-dioxanes, 1,4-oxathianes
Six-membered aromatic heterocycles containing one heteroatom
    Pyridine
Six-membered aromatic heterocycles containing more than one heteroatom
    Pyrimidines, pyrazines
Two-ring aromatic heterocycles containing one heteroatom
    Indole, benzofuran, benzothiophene
    Quinoline, isoquinoline
Two-ring heterocycles containing more than one heteroatom
    Purines
Complex heterocyclic compounds
  Porphyrins and bile pigments
    Hemoglobins, myoglobins, cytochromes
  Alkaloids
    Tropane alkaloids, isoquinoline alkaloids, indole alkaloids

Many of the simple heterocycles occur naturally within human biochemistry. For example, the amino acids proline, histidine, and tryptophan contain, respectively, a pyrrolidine, an imidazole, and an indole ring. The nucleic acids contain purine and pyrimidine rings. Vitamins are heterocyclic compounds: vitamin B_6 (**8.8**) is a substituted pyridine; vitamin B_1 (**8.9**) contains a pyrimidine ring. Simple heterocycles are therefore important to human biochemistry and thus to drug design.

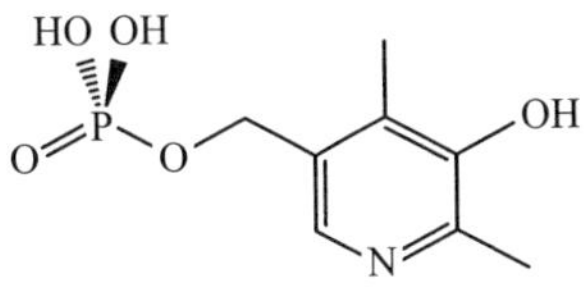

Vitamin B_6 (8.8) 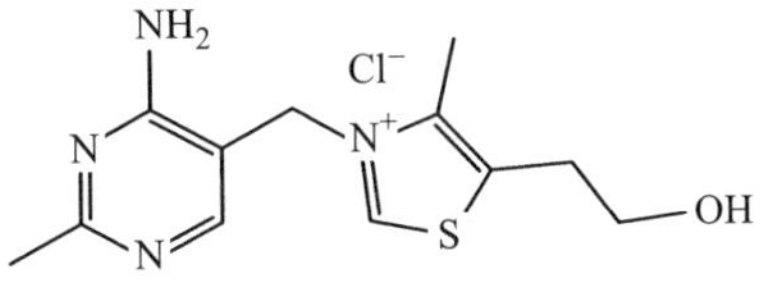   Vitamin B_1 (8.9)

The complex heterocycles are likewise important to drug design. *Porphyrins* are cyclic compounds formed by the linkage of four pyrrole rings through methylene bridges. The general name of porphyrin is used to denote a compound constructed from substituted porphins. *Porphin*, with the four nitrogen atoms of the pyrrole rings pointing toward the center of its large ring system, complexes efficiently with metal ions; in heme, this ion is Fe(II), and in chlorophyll it is magnesium (II). The porphyrin ring system is very stable and has aromatic character. In nature, the metalloporphyrins are conjugated to proteins (globins) to form a number of important biological macromolecules. Hemoglobin serves as a transport mechanism for oxygen within the blood; myoglobin is a respiratory protein found in the muscle cells of vertebrates and invertebrates; cytochromes act as electron transfer agents in oxidation–reduction reactions. Because of the importance of these macromolecules to drug action and because of their involvement in various diseases (e.g., porphyria), porphyrins are of interest in drug design and medicinal chemistry.

The *alkaloids* are also relevant to drug design. Alkaloids are complex heterocyclic compounds that contain nitrogen and thus have base-like (hence the term "alkaloid") properties; they are extremely structurally diverse. Nicotine is one of the simplest alkaloids. Oxidation of nicotine produces nicotinic acid, a vitamin that is incorporated into the important coenzyme nicotinamide adenine dinucleotide, commonly referred to as NAD^+ (oxidized form). The neurotransmitter serotonin is an alkaloid containing the aromatic indole ring system.

8.1.7 Metals

Traditionally, medicinal chemistry and drug design have lain firmly within the domain of organic chemists, not inorganic chemists. Accordingly, the potential role of inorganic salts and organometallic substances has been relatively neglected. Despite concerns about long-term toxicities, the inclusion of metal atoms into the drug design repertoire dramatically increases the diversity of atomic building blocks beyond the time-honoured reliance upon C, O, N, and S. Arguably, the therapeutic potential of organometallic agents as antitumor and antimicrobial drugs has not been fully exploited; their potential role as therapies in a wide range of other disorders has been totally ignored.

8.1.7.1 Classification of Bioactive Metals

Metals that are potentially biologically active, either therapeutically or toxicologically, may be divided into the following groups, based upon their electron configuration and position in the periodic table of the elements:

Main group metals
    Group 1A—the alkali metals: Li, Na, K, Rb, Cs, Fr
    Group 2A—the alkaline earth metals: Be, Mg, Ca, Sr, Ba, Ra
    Group 3A—Al, Ga, In, Tl
    Group 4A—Sn, Pb
    Group 5A—Bi
Transition metals
    Period 4 (4s3d4p)—Sc, Ti, V, Cr, Mn, Fe, Co, Ni, Cu, Zn
    Period 5 (5s4d5p)—Y, Zr, Nb, Mo, Tc, Ru, Rh, Pd, Ag, Cd
    Period 6 (6s5d6p)—La, Hf, Ta, W, Re, Os, Ir, Pt, Au, Hg

The main group metals are the most important, given the role of Na^+, K^+, and Ca^{2+} in bioelectrical excitability. The transition metals also have biological relevance. A formal definition of transition metals is that they have partially filled *d* or *f* orbitals in either their free (uncombined) atoms or one or more of their ions. Transition metals may be divided into *d*-block and *f*-block elements; the *f*-block is further divided into the lanthanide and actinide series. Since *f*-block metals are not of great significance to medicinal chemistry they will not be discussed further. The *d*-block metals have some biological relevance, particularly Fe and Zn.

8.1.7.2 Metals in Biology and Medicine

Of the approximately 25 elements which are currently regarded as important if not essential for life, four can be classified as bulk metal ions: Na, K, Mg, and Ca, and ten as trace metal ions: Fe, Cu, Mn, Zn, Co, Mo, Cr, Sn, V and Ni. The nonmetallic elements are H, B, C, N, O, F, Si, P, S, Cl, Se and I. Over the past years, various evidence has been put forth concerning the roles of Cd, As, Pb, and Al as potential trace elements. These metals play a wide range of functional and structural roles within human biochemistry.

The seven principal mineral macronutrient elements (Ca, Mg, Na, K, P, S, Cl) constitute 70–80% of all the inorganic material in the body. Calcium is present in the body in larger amounts than any other mineral element. A typical 70 kg adult male contains 1.5 kg of calcium; 98–99% of this is in the skeleton, where it is maintained as deposits of calcium phosphates. Not surprisingly, calcium plays an important structural role because of its central involvement in bone structure. Functionally, Ca^{2+} is important to electrical transmission within the brain and heart and to muscle contraction in both skeletal and smooth muscle. Na^+ and K^+ ions are likewise crucial to the functioning of excitable tissues such as neural, cardiac, and muscle tissues. Of the trace elements, the important role of iron in the processes of cellular respiration, via hemoglobin, myoglobin, and cytochromes, is well appreciated. Copper is present in two key enzymes of aerobic metabolism: cytochrome c oxidase, which is responsible for the major part of oxygen consumed, and cytosolic superoxide dismutase, which catalytically scavenges the toxic free radical superoxide ion generated during aerobic metabolism. Zinc is an essential component of a number of enzymes present in animal tissues, including alcohol dehydrogenase, alkaline phosphatase, carbonic anhydrase, procarboxypeptidase, and cytosolic superoxide dismutase. The enzyme pyruvate carboxylase, involved in gluconeogenesis, contains tightly bound manganese.

Not only do metals play central roles in human biochemistry, but also various disease states are recognized as arising from either deficiencies or excesses of various metal ions. Individuals with cancer spreading into their bones may experience life-threatening hypercalcemia from excessive blood concentrations of Ca^{2+}. Elevated or reduced levels of Na^+ (hypernatremia, hyponatremia) or K^+ (hyperkalemia, hypokalemia) can affect electrical excitability in the brain or heart, leading to seizures or cardiac arrhythmias. Iron deficiency leads to anemia; iron excess causes hemosiderosis, leading to liver and pancreatic damage. Pronounced zinc deficiency in humans, resulting in dwarfism and hypogonadism, has been described in geographical areas with zinc-deficient soil. Excess copper produces a neurological disease known as Wilson's disease, resulting in disordered and abnormal body movement.

8.1.8 Endogenous Macromolecules as Drugs and Druggable Targets

These seven groups of endogenous nonmessenger macromolecules offer a plethora of leads as both drugs and druggable targets (see figure 8.5). In a somewhat arbitrary but

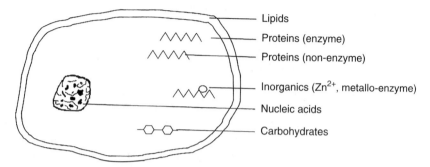

— Lipids
— Proteins (enzyme)
— Proteins (non-enzyme)
— Inorganics (Zn^{2+}, metallo-enzyme)
— Nucleic acids
— Carbohydrates

Figure 8.5 Macromolecular targets for drug design. The mammalian cell contains numerous macromolecules including lipids, enzymatic proteins, non-enzymatic proteins, carbohydrates, nucleic acids, other heterocycles, and water. All of these molecules present important opportunities for drug design.

essentially descending order of importance, these drugs and druggable targets may be classified as follows:

1. Proteins, peptides, amino acids
    Enzyme proteins
        a. As drugs
            Alteplase (for ischemic stroke)
            Retavase, tenecteplase (for acute myocardial infarction)
        b. As druggable targets
            Classical enzyme targets (acetylcholinesterase, monoamine oxidase)
            Emerging targets (kinases, caspases)
    Non-enzyme proteins
        a. As drugs
            Insulin, growth hormone
        b. As druggable targets
            Protein folding disorders (β-amyloid, prions)
2. Nucleic acids, nucleotides, nucleosides
        a. As drugs
            Antimetabolites
            Antisense oligonucleotides
        b. As druggable targets
            Traditional antineoplastic/antiviral agents targeting nuclear DNA
            Agents targeting ribozymes and RNA targets
3. Lipids
        a. As drugs
            Steroids (estrogens, progestins, androgens, corticosteroids)
            Terpenes (pheromones)
        b. As druggable targets
            Anti-prostaglandin agents (cyclooxygenase inhibitors)
            Anti-leukotriene agents (5-lipoxygenase inhibitors)

4. Carbohydrates
    a. As drugs
        Simple carbohydrate therapeutics (mannitol)
        Complex carbohydrate therapeutics (heparin)
    b. As druggable targets
        Membrane glycoproteins
        Membrane glycolipids
5. Heterocycles
    a. As drugs
        Vitamins, indole alkaloids
    b. As druggable targets
        Porphyrins (hemoglobins, cytochromes)
6. Metals
    a. As drugs
        Bismuth, gold, lithium, platinum, zinc
    b. As druggable targets
        Organo-metallic macromolecules
            Metallo-enzyme
            Zinc finger
        Inorganic salts as targets
            Bone (calcium hydroxyapatite)
7. Water
    a. As druggable target
        Antifreeze proteins for organ preservation

Each of these will be discussed in varying detail throughout the remainder of this chapter.

8.2 PROTEINS AS DRUGS AND DRUG DESIGN
TARGETS: ENZYMES

Enzymology occupies a central role in all disciplines that involve biochemical principles. It is therefore understandable that medicinal chemistry has assimilated aspects of enzymology, especially those that explain the mode of action of drugs and help in their rational design. Additionally, the principles and concepts of enzymology have helped to shape our contemporary ideas on drug receptors and the molecular mode of their function. In previous chapters, we have encountered drugs that are associated with some effect on an enzyme and could therefore be discussed in this chapter. For example, the renin–angiotensin system could be discussed here, since most drugs connected with that blood pressure-regulating system act on some enzyme. However, we will restrict our considerations to a limited number of representative examples.

A considerable number of enzymes occupy a central and crucial role in the activity of drugs. Dihydrofolate reductase, an enzyme involved in purine and amino acid biosynthesis, is the target of antibacterial sulfanilamides, which act both as bacteriostatics and antimalarials. These drugs act on the enzyme in different ways, some being so-called antimetabolites (i.e., reversible enzyme inhibitors). Some diuretics act on carbonic anhydrase, which regulates proton equilibria in the kidney. Another ubiquitous enzyme is adenylate cyclase; many drug receptors are coupled to it directly and exert their activity by regulating cAMP production.

Enzymology is a large and complex subdivision of biochemistry, and no attempt will be made to cover it in this book. The reader who requires a review of enzyme kinetics and mechanisms is referred to the many currently available and excellent textbooks of biochemistry.

8.2.1 Mechanisms of Enzyme Inhibition

Enzyme inhibitors inhibit the action of enzymes either reversibly or irreversibly. Since enzymes are such pervasive, powerful biological catalysts, inhibitors can act as potent drugs. Broadly categorized, enzyme inhibitors may be either irreversible or reversible.

Irreversible inhibitors combine or destroy a functional group on the enzyme so that it is no longer active. They often act by covalently modifying the enzyme. Thus a new enzyme needs to be synthesized. Examples of irreversible inhibitors include acetylsalicyclic acid, which irreversibly inhibits cyclooxygenase in prostaglandin synthesis. Organophosphates (e.g., malathion, **8.10**) irreversibly inhibit acetylcholinesterase. *Suicide inhibitors* (mechanism-based inactivators) are a special class of irreversible inhibitors. They are relatively unreactive until they bind to the active site of the enzyme, and then they inactivate the enzyme.

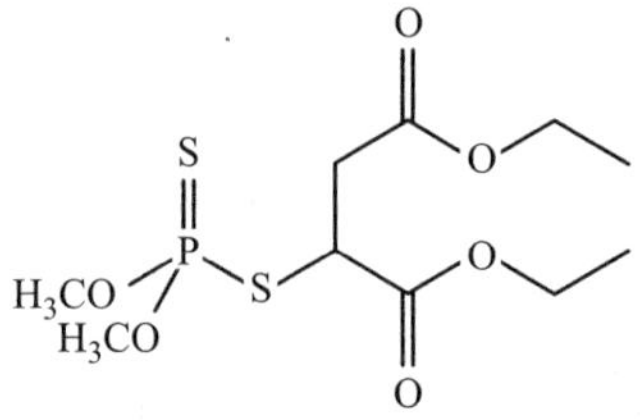

Malathion (8.10)

There are four types of reversible enzyme inhibitors:

1. *Competitive inhibitors.* These compete with the normal substrate for the enzyme's binding site. This does not affect the maximum rate of the reaction but does mean that more substrate has to be supplied. For example, to treat cases of methanol ingestion, ethanol is used to compete with methanol. Most drugs that act as reversible inhibitors are competitive inhibitors.
2. *Non-competitive inhibitors.* These inhibitors bind to the enzyme or the enzyme–substrate complex at a site other than the active site. This results in a decrease in the maximum rate of reaction, but the substrate can still bind to the enzyme. An analogous concept is that of *allosteric inhibition.* The site of binding of an allosteric inhibitor is distinct from the substrate binding site. In this case, the inhibitor is not a steric analog of the substrate and instead binds to the *allosteric site* (the phenomenon was termed thus by Monod and Jacob).

3. *Uncompetitive inhibitors.* These bind only to the enzyme–substrate complex, not to free enzyme. This results in a decrease in the maximum rate of reaction and means that less enzyme is available to bind substrate.
4. *Transition state analog inhibitors.* These are compounds that resemble the substrate portion of the hypothetical transition state of the enzymatic reaction.

Two specific enzyme-inhibitor mechanisms deserve special discussion because they are the basis of action for several important drugs. They are the transition-state analogs and the "suicide" substrates.

8.2.1.1 Transition-State Analogs

Transition-state analogs are inhibitors that mimic the transition-state structure of the substrate of an enzyme, which, by definition, has the highest energy—that is, the least stable conformation. Since the transition state of an enzyme substrate is the form that is most tightly bound, its analog should have a higher affinity and specificity for the enzyme than any substrate in the "ground state." Hence, binding constants of 10^{-15} M for the transition state can be expected for substrates with K_Ds of only 10^{-3}–10^{-5} M.

There are two problems with this potentially very powerful concept. First, the mechanism of the enzymatic reaction must be known in order to mimic the transition state of the substrate, since the structural specificity of the reaction is quite high. Second, a stable analog of the labile transition state is, by implication, very difficult to prepare. Often one must be content with a metastable intermediate analog of the substrate. Nevertheless, there are several successful applications of this interesting principle. Penicillin (**8.11**) is a transition-state analog of a distorted Gly–D–Ala–D–Ala peptide involved in the crosslinking of glycol-peptides constituting the cell walls of bacteria.

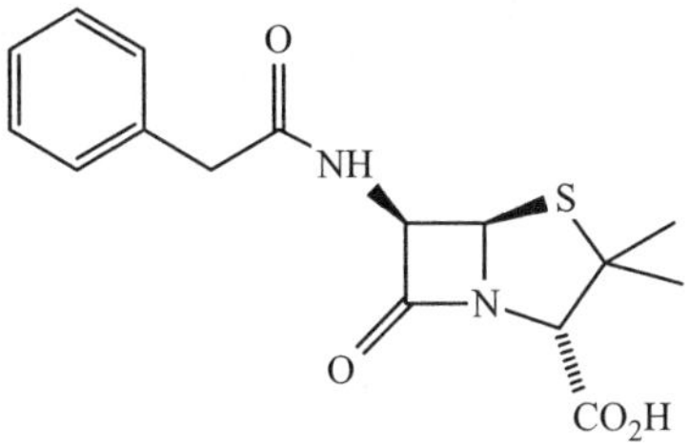

Penicillin (8.11)

Suicide Enzyme Inhibitors. "Suicide" substrates are irreversible enzyme inhibitors that bind covalently. The reactive anchoring group is catalytically activated by the enzyme itself through the enzyme–inhibitor complex. The enzyme thus produces its own inhibitor from an originally inactive compound, and is perceived to "commit suicide."

To design a K_{cat} substrate, the catalytic mechanism of the enzyme as well as the nature of the functional groups at the enzyme active site must be known. Conversely, successful K_{cat} inhibition provides valuable information about the structure and mechanism of an enzyme. Compounds that form carbanions are especially useful in this regard. Pyridoxal phosphate-dependent enzymes form such carbanions readily because the coenzyme can stabilize the anion by resonance delocalization on the heterocyclic ring (see figure 8.6). Additionally, flavine-dependent oxidases such as monoamine oxidase can be attacked by acetylenes as suicide substrates.

8.2.2 Enzyme Targets: Acetylcholinesterase

Acetylcholinesterase (AChE) is the enzyme that hydrolyzes and thereby deactivates acetycholine (ACh) after it binds to the receptor. The enzyme is present in peripheral and central synaptic sites, in erythrocytes, and in the placenta.

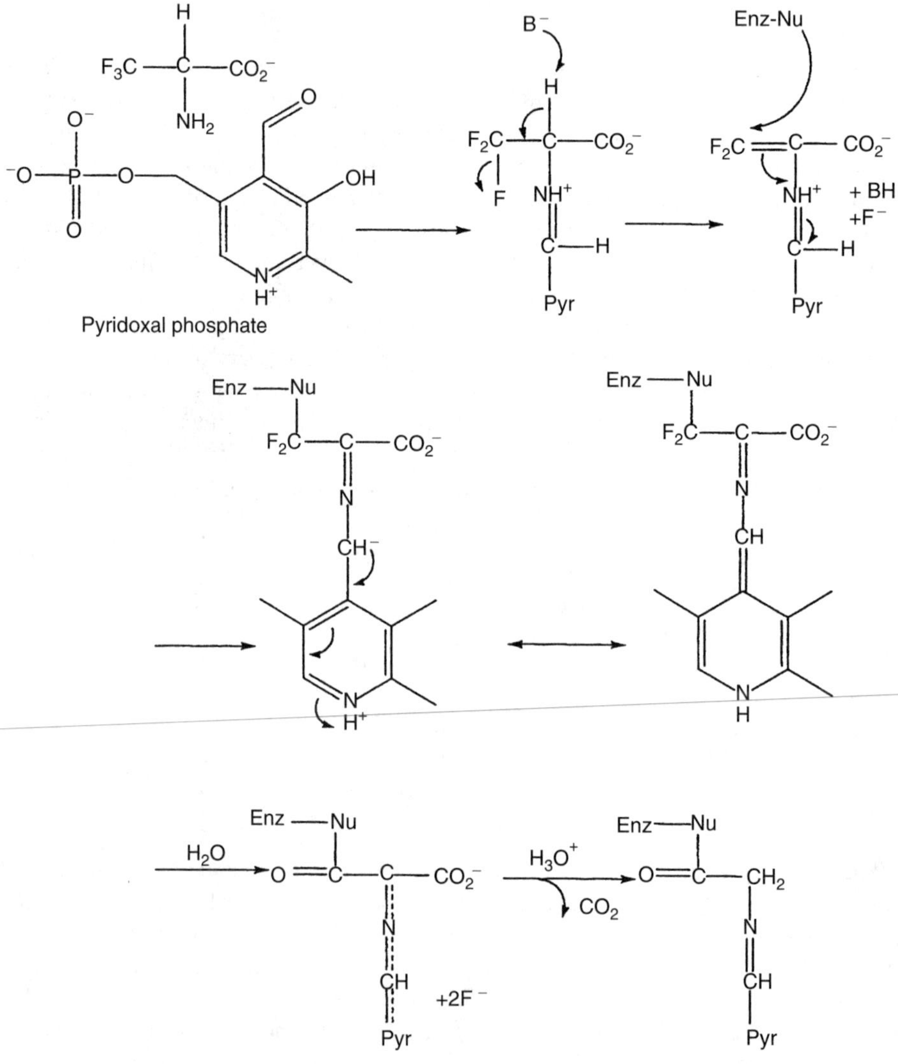

Figure 8.6 Trifluoroalanine is transformed by a pyridoxal coenzyme into a covalently bound "suicide substrate."

8.2.2.1 Physicochemical Properties of Acetylcholinesterase

The purification of AChE and the elucidation of its physicochemical properties were achieved by using the enzyme isolated from the electric eel (*Electrophorus electroplax*) the richest source of AChE, as well as from brain and erythrocytes. With high ionic strength solutions (1 M NaCl or 2 M $MgCl_2$), the extraction is selective and can be facilitated by treatment of the electroplax with collagenase. The basic unit of the enzyme is a tetramer with a molecular weight of 320,000; each of the protomers contains an active site. Normally, three such tetrameric units are linked through disulfide bonds to a 50×2 nm stem. This stem is a collagen triple helix that seems to have a structural role only. In neural and electroplax tissue, the enzyme "trees" (stems with tetramers) are anchored in the basement membrane or neurolemma—a porous, collagen-rich structure. However, because many tissues (erythrocytes, peripheral ganglia) do not have a basement membrane, attachment of the AChE must occur in another, unknown fashion. Analysis of the amino acid composition of the enzyme shows that it bears a close similarity to the AChR in its high proportion of acidic amino acids.

In contrast to the AChR, AChE does not bind bungarotoxin or sulfhydryl reagents. It is inhibited by excess substrate (3×10^{-6} M), and the K_D of electric eel AChE is about 10^{-4} M. The specific activity of the enzyme is one of the highest known: 750 nmol/mg-hr, with a turnover time of 30–60 msec and a turnover number of $2–3 \times 106$. It is therefore one of the most efficient and fastest enzymes known.

The active site of AChE has the composition of most serine esterases. It includes a charge-relay system of histidine and serine, and an acidic center, probably glutamate, which binds the choline cation. The serine hydroxyl is rendered more nucleophilic through the proton-acceptor role of histidine and is capable of executing a nucleophilic attack on the carbonyl carbon of ACh. A tetrahedral transition state is reached, resulting in serine acetylation and the desorption of free choline. The acetyl group is taken over by histidine as an N-acetate, which is then easily hydrolyzed, regenerating the enzyme active site. The choline is taken up into the nerve ending by an active transport system and reused for ACh synthesis. Finally, the acetate goes into the ubiquitous acetate pool of intermediary metabolism.

On the basis of studies of the hydrolysis rate of succinyl-methylcholine isomers, Stenlake proposed the hypothesis that the AChR and AChE bind two different faces of ACh, which implies that they have opposite sterochemistries at their respective anionic subsites. Simultaneous attachment to the receptor and the enzyme is therefore impossible, although a fast sequential attack of the enzyme on the receptor-bound neurotransmitter is still feasible since no ACh reorientation is necessary for this. However, there is no evidence that such a sequence of events occurs *in vivo*.

Another cholinesterase, called serum cholinesterase or butyrylcholinesterase, is found in serum and the liver. It plays an important role in drug metabolism.

8.2.2.2 Anticholinesterase Inhibitors—General Properties

Anticholinesterase drugs are compounds that block AChE and inhibit the destruction of released ACh. The resultant higher neurotransmitter levels then increase the biological response. Anticholinesterases can therefore be considered as indirect cholinergic agents. Acetycholinesterase inhibitors can act by either of two mechanisms:

1. As classical competitive enzyme inhibitors, they have a high affinity for the active site but are not substrates. The enzyme is occupied by the inhibitor for relatively long periods, and therefore cannot handle ACh efficiently, as a result of the saturation phenomenon.
2. The inhibitor acylates the serine hydroxyl of AChE, forming an ester more stable than acetate, such as a carbamate or phosphate. The hydrolysis of these esters takes a long time even if they are not irreversible, as was formerly thought. Acetycholine cannot then be hydrolyzed, since the active site is covalently occupied.

A representative of the first group of inhibitors is edrophonium (**8.12**), a short-acting drug that binds to the anionic site of the enzyme and also forms a hydrogen bond with the imidazole nitrogen of the active site. Ambenonium (**8.13**) also does not react covalently with the enzyme but has a much longer duration of action than edrophonium.

The second group of AChE inhibitors is represented by the alkaloid physostigmine (**8.14**, eserine) isolated from the seeds of *Physostigma venenosum*, the Calabar bean, which has been used in West Africa in witch ordeals. Synthesic analogs such as neostigmine (**8.15**) and its congeners are also active. They have a very high affinity for the enzyme and will carbamoylate the serine hydroxyl of the active site. Because the half-life of the dimethyl-carbamoyl ester is 20–30 minutes, as compared to only microseconds for the acetylated enzyme, AChE is inhibited for several hours after a single dose of such drugs.

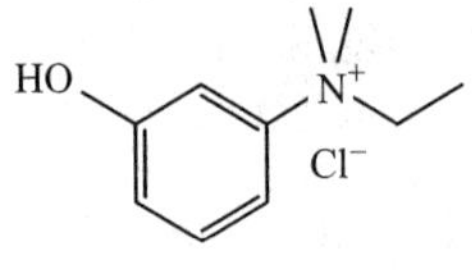

Edrophonium (8.12) 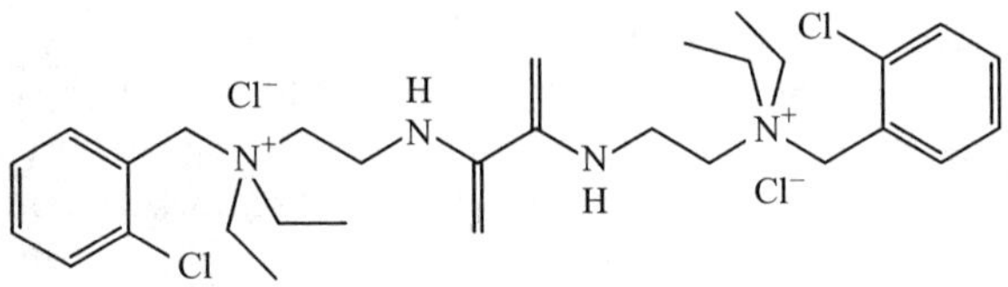   Ambenonium (8.13)

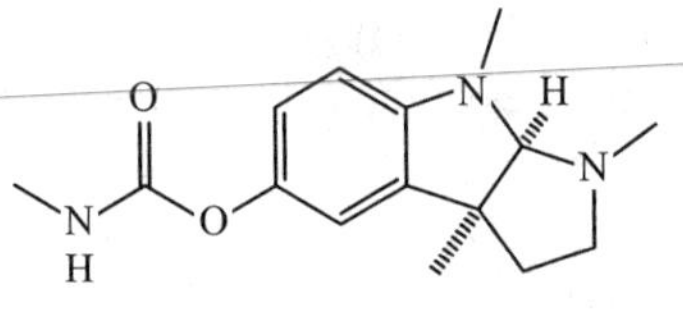

Physostigmine (8.14) 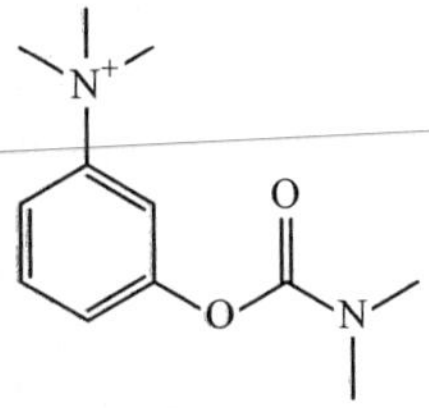   Neostigmine (8.15)

8.2.2.3 Acetylcholinesterase Inhibitors: Use in Myasthenia Gravis

Both groups of AChE inhibitors are used therapeutically. One use of anticholinesterase drugs is in *myasthenia gravis*. This is an autoimmune disease caused by the development of antibodies against the patient's own ACh receptors, accompanied by disturbed neuromuscular transmission. The disturbance is caused by a reduction in the number of nerve terminals and an increase in the width of the synaptic cleft. Normally, nicotinic

ACh receptors are destroyed by endocytosis via coated pits and proteolysis in lysosomes. In myasthenia gravis, the receptors are crosslinked by antireceptor antibodies, which facilitate the rate-limiting endocytosis step; receptor destruction occurs in less than half the normal time, resulting in net receptor loss. The chronic disease is characterized clinically by such muscular weakness and abnormal fatigue that patients cannot even keep their eyes open. Acetylcholinesterase inhibitors increase the ACh concentration and excitation of the neuromuscular junction, resulting in increased strength and endurance. As expected, AChE inhibitors are also potent curare antidotes because the increased ACh levels displace the blocker more readily.

8.2.2.4 Acetylcholinesterase Inhibitors: Clinical Use in Alzheimer's Disease

Most recently, acetylcholinesterase inhibitors have been used to afford symptomatic relief in Alzheimer's disease (AD). In AD, neurons are progressively killed by some neurotoxic factor, presumably mediated by β-amyloid. As the neurons die, they decrease their production of acetylcholine; as acetylcholine production decreases, symptoms such as decreased memory and confusion become apparent. Administering an acetylcholinesterase inhibitor prolongs the effective half-life of the acetylcholine that is being biosynthesized, thereby producing a symptomatic improvement. Since the acetylcholinesterase inhibitors are not dealing with the underlying neuropathological process (i.e., β-amyloid mediated toxicity) they are not curative. Once the disease has progressed to the point where neurons are no longer producing effective concentrations of acetylcholine, these agents are no longer of value. The acetylcholinesterase inhibitors currently in clinical use for AD include donepezil (**8.16**), rivastigmine (**8.17**), and galantamine (**8.18**).

It is important to emphasize that dementia and Alzheimer's disease are not synonymous terms. All Alzheimer's disease is dementia, but not all dementia is Alzheimer's

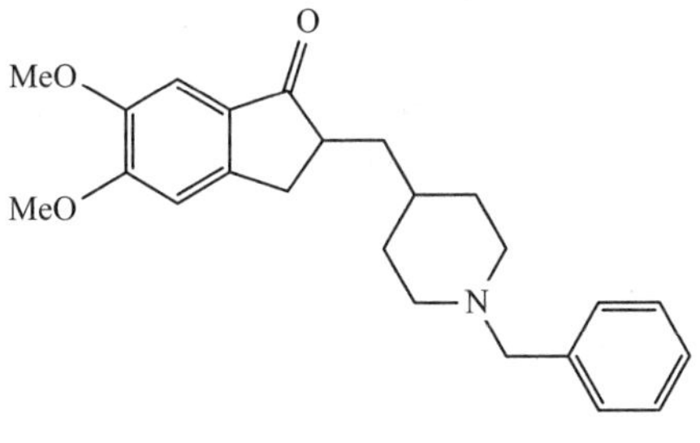

Donepezil (8.16) 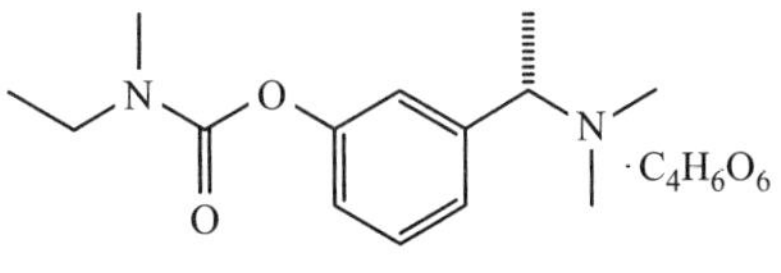   Rivastigmine (8.17)

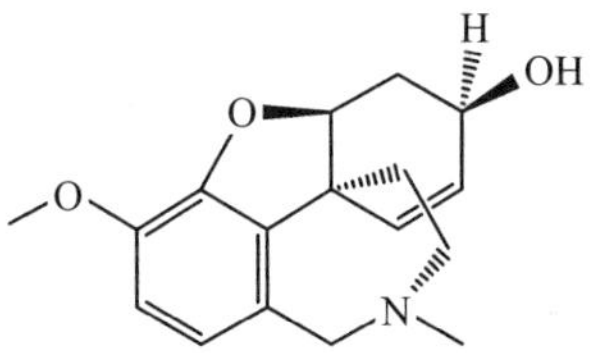

Galantamine (8.18)

disease. Dementia is a nonspecific term applied to any chronic, progressive, non-reversible neurodegenerative process that results in diminished intellectual function; AD is merely one type of dementia. Dementia can also occur secondary to multiple strokes affecting the brain, producing so-called *multi-infarct dementia* or *vascular dementia*. It is also important to differentiate dementia from *delirium*. Delirium (also sometimes called *encephalopathy*) is a reversible reduction in intellectual functioning, typically manifesting as confusion or agitation. In the elderly, delirium not infrequently arises from overmedication or excessive medication. Acetylcholinesterase inhibitors are of no value in the treatment of delirium, and in fact may make it worse. Interestingly, for reasons that remain unclear, acetylcholinesterase inhibitors, such as donepezil, do demonstrate weak activity in vascular dementia.

8.2.2.5 Acetylcholinesterase Inhibitors: Clinical Use in Glaucoma

Another use for acetylcholinesterase inhibitors is in *glaucoma*, in which high intraocular pressure can lead to permanent damage to the optic disk, resulting in blindness. The local instillation of physostigmine (**8.14**) or echothiophate (**8.19**) solution in the eye results in a long-lasting decrease in the intraocular pressure as well as myosis (contraction of the pupil).

8.2.2.6 Acetylcholinesterase Inhibitors: Industrial Use as Insecticides

Uncharged carbamates, such as carbaryl (**8.20**, sevin), can penetrate the CNS of insects (which do not use AChE in their neuromuscular junction) and they act quite selectively as insecticides with a low toxicity to mammals (median lethal dose [LD_{50}] in the rat = 540 mg/kg, p.o.). Many useful insecticides can thus be found in this group. Malathion (**8.21**) is a pro-drug, since the thiophosphate must be bioactivated to the phosphate form—a transformation carried out by insects but not mammals. Additionally, the ester groups of malathion are rapidly hydrolyzed in higher organisms to water-soluble and

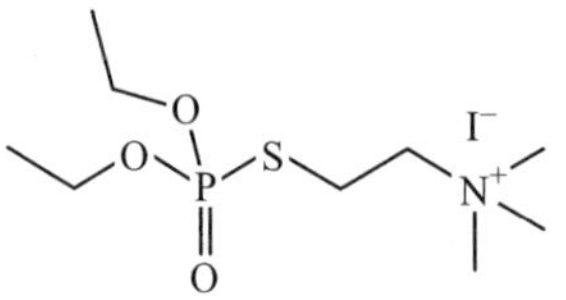

Echothiophate (8.19)

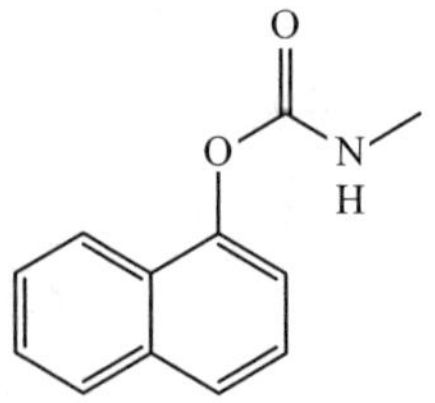

Carbaryl (8.20)

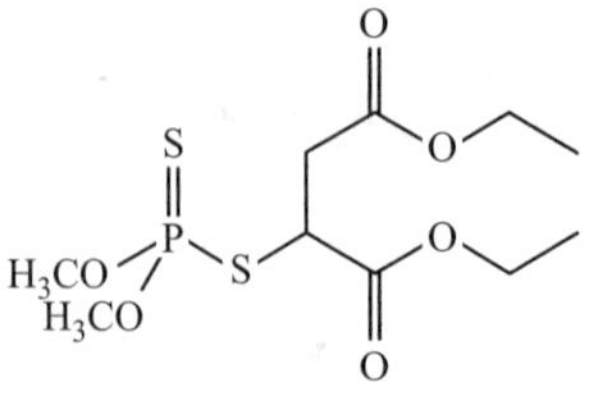

Malathion (8.21)

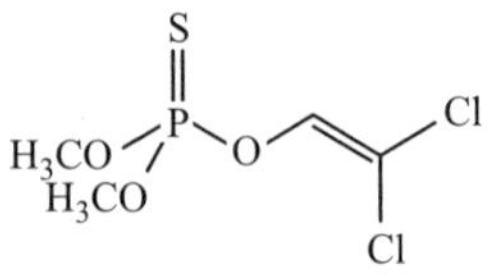

Dichlorvos (8.22)

inactive compounds; however, insects do not have the hydrolases that would deactivate the insecticide. Because of this double safety feature, the selectivity of malathion is high and its mammalian toxicity low: the LD_{50} for rats is 1500 mg/kg, p.o. The volatile dichlorvos (**8.22**) is used in fly-killer strips. Whereas the vapors of this compound are rapidly hydrolyzed by mammals, they are cumulative and highly toxic for insects.

Because of accidental exposure to insecticides by children or by agricultural workers, an appreciation of the toxicities of insecticide anticholinesterase inhibitors is medically relevant.

8.2.2.7 Acetylcholinesterase Inhibitors: Terrorism and Use as Nerve Gas Agents

Even longer-acting covalent AChE inhibitors are the organophosphate esters, such as diisopropyl-phosphorofluoridate (**8.23**), developed from "nerve gases" discovered but never used in World War II. These compounds are known for their extreme toxicity and rapid dermal absorption. The covalently binding AChE inhibitors, and especially the organophosphates, can be highly toxic, and poisoning from their improper use could occur. An antidote, pralidoxime (**8.24**; 2-N-methylpyridinium-2-aldoxim iodide), was

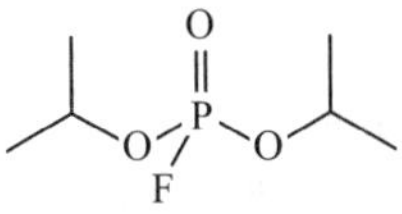

Diisopropyl-phosphorofluoridate (8.23)

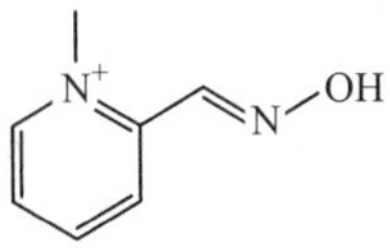

Pralidoxime (8.24)

designed on the basis of knowledge of their mode of action. The quaternary pyridinium ion of this compound binds to the anionic site of the enzyme, removes the phosphate of the inhibitor from the serine active site in the form of an oxime-phosphonate, and regenerates AChE. Atropine (**8.25**) is also administered to victims of organophosphate poisoning to relieve peripheral and central muscarinic symptoms due to excessive ACh.

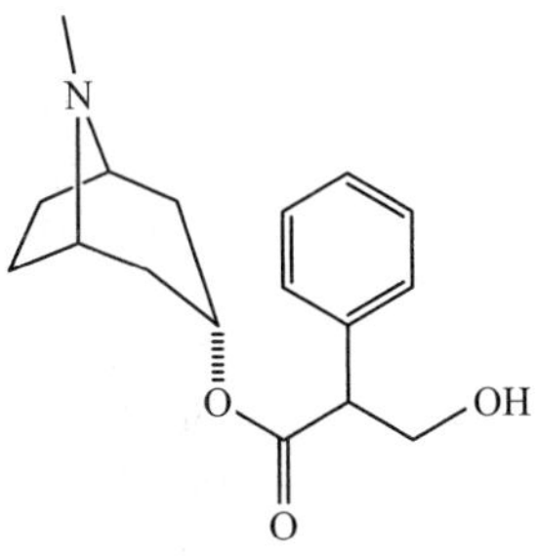

Atropine (8.25)

8.2.3 Enzyme Targets: Adenosine Triphosphatase

Adenosine triphosphatase (ATPase, E.C. 3.6.1.3) comprises a group of extremely wide-spread membrane-bound enzymes. All cell membranes contain the Na^+–K^+-dependent enzyme, which, among its other roles, is extremely important in the establishment of the mitochondrial proton gradient and in bacterial permease systems that transport amino acids. As a result of the proton gradient in mitochondria, or a very steep Na^+–K^+ gradient, ATP synthesis rather than ATP hydrolysis can take place, according to the Mitchell chemiosmotic hypothesis. This model suggests that Na^+–K^+-activated ATPase located in neural membranes is responsible for maintaining transmembrane ionic asymmetry, or ion gradients, across these membranes. Adenosine triphosphatase is phosphorylated by ATP and then dephosphorylated by K^+. The free energy of ATP hydrolysis is used to transport sodium and potassium against their respective gradients in a 3 Na^+: 2 K^+: 1 ATP ratio, but Na^+ must be inside and K^+ outside the cell in order to activate the enzyme. A rich source for ATPase isolation is the electroplax of *Torpedo* or *Electrophorus*.

8.2.3.1 Physicochemical Properties of ATPase

The Na^+–K^+–ATPase has been purified and sequenced. It is a dimeric protein with an α subunit of 100 kD and a β subunit of 55 kD. The α subunit contains the ATP hydrolysis subsite in which an aspartate accepts the γ phosphate of ATP. The function of the glycoprotein β subunit is not clear. ATPases from different sources (electroplax, sheep kidney, rat, bacterial K^+ pump) have been shown, through cDNA cloning, to have a high degree of homology, indicating their ancient common evolutionary history. Determination of the primary sequence, proteolytic digestion, and labeling experiments have enabled investigators to elucidate the folding of the protein within the cell membrane. There are eight transmembrane domains and a large cytosolic portion. ATP binds to Lys-501, and Asp-369 is phosphorylated. The cardiac glycoside binding site, which is partially located on the outside of the α subunit, inhibits the ATP-driven ion transport and the Na^+-dependent conformational change. Investigators also discovered that the mammalian brain and heart contain distinct isozymes with different α subunits showing different steroid binding capabilities. The sarcoplasmic reticulum of muscle cells contains a Ca^{2+}-dependent ATPase, responsible for Ca^{2+} concentration changes in the muscle that are intimately connected with the contraction process.

8.2.3.2 Cardiac Steroid Glycosides and Na^+-K^+/ATPase

Myocardial cell membrane ATPase, the enzyme present in heart muscle, is the site of action of the cardiac steroid glycosides, which have a specific action on the heart muscle. These drugs increase the force of contraction of the muscle (positive inotropic effect) as well as its conductivity and automaticity. They are also valuable in treating congestive heart failure, in which the circulatory needs of organs are no longer satisfied, and heart arrhythmias, in which the rhythm of the cardiac contractions is upset. The effect of the drug is that the force of contraction increases and the heart rate is slowed (chronotropic effect). Consequently, the cardiac output is elevated while the size of the heart decreases.

The exterior of the myocardial enzyme, situated in the plasmalemma, is considered to be the specific binding site of cardiac steroid glycosides. It is believed that the positive inotropic effect is due to the inhibition of enzyme dephosphorylation and

Na^+–K^+ exchange, which occurs before each heart contraction. The inhibition of Na^+ extrusion increases the Na^+ concentration, which in turn triggers greater Ca^{2+} mobilization and controls the contraction of the heart. If, however, the Na^+ pump is further inhibited, toxic effects appear; indeed, cardiac steroids show considerable toxicity that, unfortunately, is due to the same mechanism as their beneficial effect. Nevertheless, it has been shown that the inotropic effect of these drugs is possible without pump inhibition. The mechanisms involved are complex, and there are several possible interpretations. The major difficulties in firmly correlating steroid glycoside action with Na^+–K^+–ATPase activity are the lack of a specific antagonist and the fact that many other compounds inhibit the enzyme without showing cardiac activity. In order to form a complex with a drug, the ATPase must be in the proper conformation. This conformation exists only immediately after the enzyme has transported Na^+ and is ready to bind K^+.

Cardiac steroids, or *cardenolides*, are steroid glycosides. Their effect has been known since the time of the ancient Egyptians. In more recent times, the foxglove (*Digitalis purpurea*) and its effect were described in 1785 by William Withering, who knew of its use in folk medicine.

Types of Cardiac Glycoside. Three groups of plants produce cardenolides: the *Digitalis* species, growing in temperate climates; the *Strophanthus* species, of tropical provenance; and *Scilla* (sea onion or squill), a Mediterranean plant.

The *Digitalis* glycosides are the most widely used cardenolids. The aglycones (the steroid parts) of the molecule differ from the usual steroid structure in several points. The anellation of the A–B and C–D rings is *cis* (Z), the 3-OH is axial (β), and all of these steroids carry a 14β-OH group. The C17 side chain is an unsaturated lactone ring. The sugar part, binding to the 3-OH, is a tri- or tetrasaccharide consisting mainly of digitoxose (2,6-dideoxy-β-D-allose) and glucose. The *strophanthin* aglycones have a 5β-OH group in addition to other hydroxyls, up to a maximum of six in ouabain. The 19-methyl is replaced by an aldehyde or primary alcohol and the sugars are the unusual rhamnose or cymarose. The *squill* aglycones carry a six-membered lactone ring with two double bonds, and are closely related to some toad venoms (bufotalin). None is used therapeutically since all are highly toxic.

Structure–Activity Correlations. The structure–activity relationships of cardenolides have been thoroughly investigated, and have undergone considerable revision on the basis of crystallographic work and potential energy calculations. The correlations are summarized as follows:

1. The A–B *cis* fused rings, the axial methyl group, and ring C form the rigid essential backbone of the structure.
2. Ring D has conformational flexibility, influenced by the nature of the 17β side group. The C–D cis junction and 14β conformation are essential.
3. The 14-OH group, previously considered essential, can be omitted in some structures or can be replaced by a 14β-NH_2 group.
4. The Δ^{20-22} double bond serves to properly orient the carbonyl oxygen. The ring itself is not as essential as previously thought, and some derivatives with side chains instead of a ring have even higher activity. However, the side chains must be coplanar, and are much less flexible than one would expect.

5. The activity of a compound depends to a very great extent on the position of the 23-carbonyl oxygen, which is held quite rigidly by ring D and the double bond. The standard is the position of the carbonyl relative to the rigid backbone of digitoxigenin. In synthetic analogs, every 0.22 nm deviation from this position causes a loss of activity by one order of magnitude from a maximum IC_{50} of 49 nM. This correlation is highly significant, as the regression coefficient of $r^2 = 0.993$ indicates.

6. Removal of the sugar portion allows epimerization of the 3β-OH group, with a decrease in activity and an increase in toxicity due to changes in polarity.

More recently, a polypeptide of 49 amino acids was isolated from a sea anemone. This compound, anthopleurine A (**8.26**), is 30 times more potent than the *Digitalis* glycosides and is less toxic. The discovery stimulated speculation over the possibility of a native inotropic peptide receptor, in a situation similar to the endorphin–opiate relationship, in which a plant alkaloid fortuitously fits a peptide receptor. Also an endogenous digitalis-like activity was discovered in rat, guinea pig, and bovine heart homogenates, which inhibited Na^+–K^+–ATPase and had an affinity for the steroid receptor one to two orders of magnitude higher than digoxin. This *cardiodigin* may be the long-sought endogenous cardioactive factor.

GVSCLCDSDGPSVRGNTLSGTLWLYPSGCPSGWHNCKAHGPTIGWCCKQ

Anthopleurine A (8.26)

8.2.3.3 Miscellaneous ATPase Drugs

The H^+–K^+–ATPase acts as the proton pump in the parietal cells of the stomach mucosa. It transports protons and Cl^- ions into the stomach via a K^+ antiport. Substituted benzimidazoles, such as omeprazole (**8.27**) inhibit this enzyme and are 2–12 times as active as cimetidine (**8.28**) in inhibiting gastric stimulation.

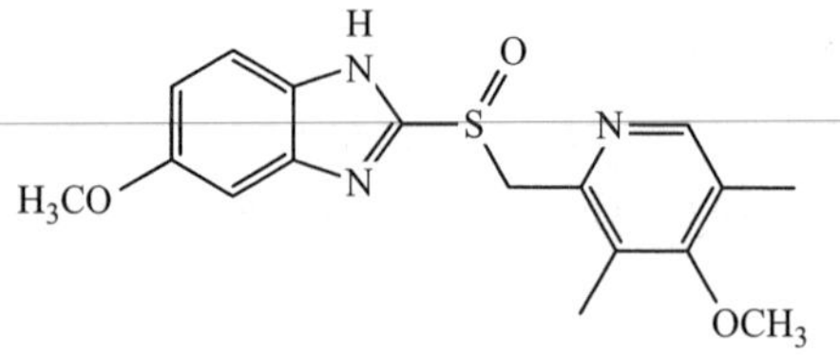

Omeprazole (8.27) 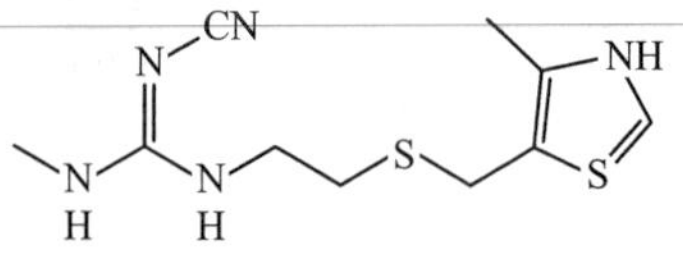　　　　　　　　　　Cimetidine (8.28)

8.2.4 Enzyme Targets: Carbonic Anhydrase

Carbonic anhydrase (E.C. 4.2.1.1) is an enzyme located in the renal tubular epithelium and in red blood cells. It catalyzes the seemingly simple reaction

$$H_2O + CO_2 \rightarrow H_2CO_3 \rightarrow HCO_3^- + H_3O^+$$

which would be shifted far to the left without an enzyme.

In the red blood cell this reaction plays an important role in CO_2 transport from the tissues to the lungs. In the kidney, the protons of the H_3O^+ are exchanged for Na^+ ions, which are reabsorbed, while HCO_3 is decomposed through a shift of the equilibrium to the left. Carbonic anhydrase therefore plays a crucial role in maintaining the ion and water balance between the tissues and urine.

When carbonic anhydrase inhibitors block the enzyme in the kidney, H_2CO_3 formation— and consequently the availability of H_3O^+ (i.e., protons)—decreases. Since the Na^+ ions in the filtrate cannot be exchanged, sodium is excreted, together with large amounts of water, as a result of ion hydration and osmotic effects. The result is *diuresis,* accompanied by a dramatic increase in urine volume. There is also failure to remove HCO_3^- ions because there is no H_3O^+ to form H_2CO_3, which would decompose to $CO_2 + H_2O$. Therefore, the normally slightly acidic urine becomes alkaline. The strong carbonic anhydrase inhibitors also increase K^+ excretion, an undesirable effect.

Carbonic anhydrase has also been discovered in the eye and the CNS. Consequently, its inhibitors have found use in the treatment of glaucoma, in reducing the high intraocular pressure which can lead to blindness in this disease. They are also effective, in combination with anticonvulsant drugs, in controlling certain forms of *epilepsy.*

8.2.4.1 Carbonic Anhydrase: Enzyme Structure

The structure of carbonic anhydrase has been completely elucidated. Both the amino acid sequence and the three-dimensional structure of the crystalline enzyme are known. Actually, there are two isozymes, a low- and a high-activity form having been isolated from human erythrocytes, with the latter designated the C form (HCA-C).

The large molecule consists of a single peptide chain; 35% β-sheet and 20% helical structure are found in the folded structure. The active site is a 1.2 nm deep conical cavity in the central pleated sheet, with a Zn^{2+} ion located at its bottom. Three histidine residues hold the Zn^{2+}, which also binds an H_2O molecule. The active-site cavity is divided into hydrophilic and hydrophobic halves. The inhibitors of the enzyme replace the water on the Zn^{2+} ion and also block the fifth coordination site where CO_2 should bind.

8.2.4.2 Sulfonamide Carbonic Anhydrase Inhibitors

The development of sulfonamide carbonic anhydrase inhibitors was based on the observation that antibacterial sulfanilamides produce alkaline urine. This discovery led to the development of acetazolamide (**8.29**), a thiadiazole derivative. It is not an ideal drug because it promotes K^+ excretion and causes a very high urine pH. Since chloride ions are not excreted simultaneously, systemic acidosis also results. Much more useful are the chlorothiazide (**8.30**) derivatives, which are widely used as oral diuretic drugs. These compounds differ from one another mainly in the nature of the substituent on C3;

Acetazolamide (8.29)   Chlorothiazide (8.30) they are much weaker carbonic anhydrase inhibitors than acetazolamide, and may have another mode of action in addition to carbonic anhydrase inhibition. They are widely used in edema, hypertension, and cardiac insufficiency, in which a decrease in the amount of tissue-bound or circulating water is imperative. Some newer derivatives, like polythiazide (**8.31**), are three orders of magnitude more active and have a duration of action of up to 24 hours. Furosemide (**8.32**) is formally a sulfanilamide but not a carbonic anhydrase inhibitor. It not only inhibits Na^+ reabsorption in the loop of Henle—a part of the nephron—but probably also functions as an inhibitor of Na^+–K^+–ATPase, which has a role in renal sodium transport as it does in other organs.

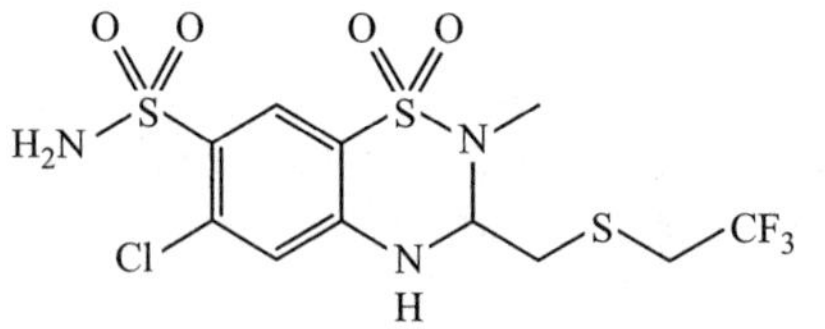

Polythiazide (8.31)

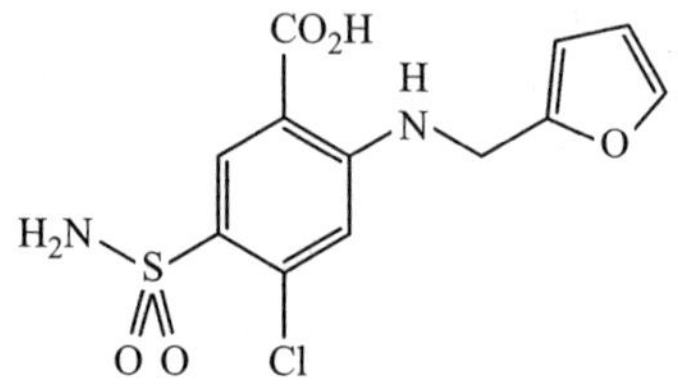

Furosemide (8.32)

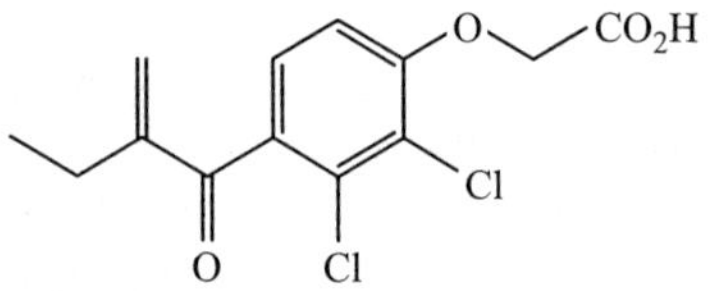

Ethacrynic acid (8.33)

8.2.4.3 Miscellaneous Diuretic Drugs

Other diuretics do not act through carbonic anhydrase inhibition and are not sulfanylamides. Ethacrynic acid (**8.33**) is based on some older Hg-containing diuretics, which block enzymatic Na transport by binding to enzyme —SH groups. The unsaturated ketone of ethacrynic acid can react in a similar way, binding —SH, whereas the carboxyl group ensures the concentration of the compound in the kidneys. The halogens increase the electrophilic nature of the unsaturated ketone. Chloride ion elimination is increased together with Na^+ excretion, but K^+ and HCO_3 elimination are low and the urine pH stays at 6. Furosemide and ethracrynic acid are powerful drugs that are used in patients resistant to other diuretics.

8.2.5 Enzyme Targets: Thymidylate Synthase

Thymidylate synthase (E.C. 2.1.1.45) is the enzyme that methylates UMP to thymidine, using methylene tetrahydrofolate as the carbon carrier. The enzyme can be inhibited directly by analogues of uracil such as 5-fluorouracil (**8.34**, 5-FU). The antimetabolite must be in the 5-fluorodeoxyuridine monophosphate (FdUMP) form to become active, and the capability of cells to achieve this transformation is a major determinant of their sensitivity to such drugs.

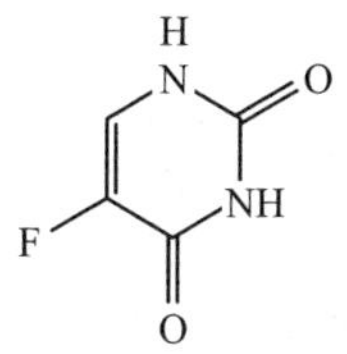

5-Fluorouracil (8.34)

It was previously thought that 5-FU inhibits the enzyme by classical competitive inhibition. However, it was found that 5-FU is a transition-state substrate, and it forms a covalent complex with tetrahydrofolate and the enzyme in the same way that the natural substrate does. The reaction, however, will not go to completion, since the fluorouridine derived from the antimetabolite remains attached to the enzyme, and the latter becomes irreversibly deactivated. Recovery can occur only through the synthesis of new enzyme. Fluorouracil is used in the treatment of breast cancer and has found limited use in some intestinal carcinomas. Unfortunately, this drug has the side effects usually associated with antimetabolites. Its prodrug, fluorocytosine (**8.35**, which is also an antifungal agent) is better tolerated.

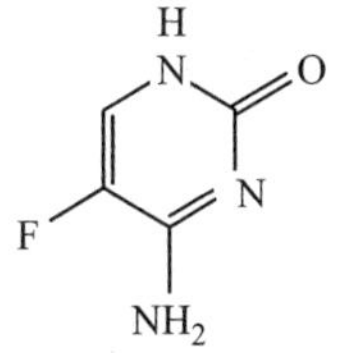

Fluorocytosine (8.35)

8.2.6 Enzyme Targets: Monoamine Oxidase

Monoamine oxidase (MAO) (E.C. 1.4.3.4) is an enzyme found in all tissues and almost all cells, bound to the outer mitochondrial membrane. Its active site contains flavine adenine dinucleotide (FAD), which is bound to the cysteine of a –Ser–Gly–Gly–Cys–Tyr sequence. Ser and Tyr in this sequence suggest a nucleophilic environment, and histidine is necessary for the activity of the enzyme. Thiol reagents inhibit MAO. There are at least two classes of MAO binding sites, either on the same molecule or on different isozymes. They are designated as MAO-A, which is specific for 5-HT (serotonin) as a substrate, and MAO-B, which prefers phenylethylamine. Similarly, MAO inhibitors show a preference for one or the other active site, as discussed below.

Monoamine oxidase catalyzes the deamination of primary amines and some secondary amines, with some notable exceptions. Aromatic amines with unsubstituted α-carbon atoms are preferred, but aromatic substituents influence the binding of these substrates. For example, *m*-iodobenzylamine is a good substrate, whereas the *o*-iodo analog is an inhibitor. The mechanism of deamination is as follows: hydrolysis of the Schiff base that results from loss of a hydride ion on an α-proton yields an aldehyde, which is then normally oxidized to the carboxylic acid. Aromatic substrates are probably preferred because they can form a charge-transfer complex with the FAD at the active site, properly orienting the amino group of the substrate and decreasing the energy of the transition state.

Monoamine oxidase has some important physiological roles:

1. It inactivates many of the neurotransmitters in the synaptic gap or in the synapse if the latter are not protected by synaptic vesicles. The metabolism of NE, DA, 5-HT, tyramine, and histamine is thus taken care of by MAO as well as by some other enzymes.
2. It detoxifies exogenous amines and may even help to maintain the blood–brain barrier, since it is also localized in the walls of the blood vessels.

8.2.6.1 Monoamine Oxidase Inhibitors

Monoamine oxidase inhibitors (MAOIs) are useful as thymoleptic (antidepressant) drugs, especially since the action of some of these agents is very rapid, as compared to the lag period of days or even weeks shown by tricyclic antidepressants. All MAOIs act by increasing the available concentration of the neurotransmitters NE and 5-HT which, because they are not metabolized, accumulate in the synaptic gap and exert an increased postsynaptic effect. The drugs show hypotensive activity as a side effect, and some MAOIs are used as hypotensive drugs.

There are four structural types of MAOI. These are hydrazines, cyclopropylamines, propargylamines, and carbolines.

Hydrazines. The hydrazines have only historic significance. The entire group of MAOIs was discovered through the euphoric side effect of isoniazid (**8.36**, isonicotinyl-hydrazide), a successful antituberculotic drug introduced in 1952. Iproniazid (**8.37**) is the corresponding isopropyl derivative. All of the hydrazides are highly hepatotoxic, and are no longer available.

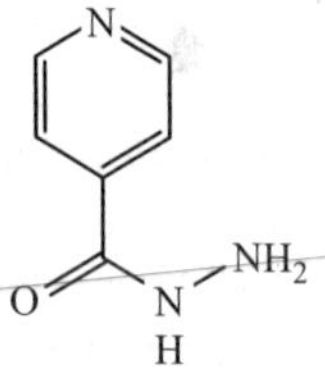

Isoniazid (8.36) 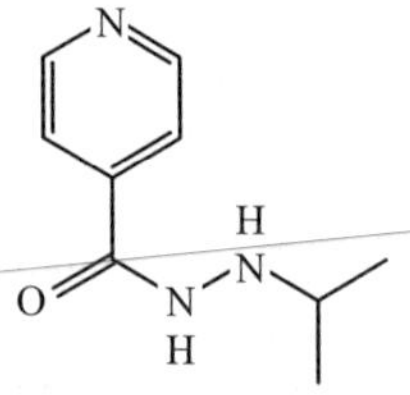   Iproniazid (8.37) 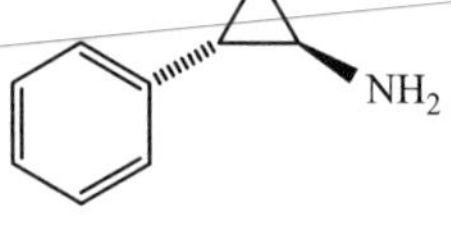   Tranylcypromine (8.38)

Cyclopropylamines. Tranylcypromine (**8.38**) (*trans*-phenyl-cyclopropylamine) can be regarded as a ring-closed derivative of amphetamine, and therefore provides rapid stimulation as well as protracted effect. Like all MAOI drugs, it can produce severe or even fatal hypertensive crises if taken together with foods containing tyramine, such as cheese. Preventing the destruction of such a pressor amine produces a sudden increase in blood pressure, which is especially dangerous in individuals exhibiting high blood pressure. Therefore, depressive patients taking tranylcypromine must practice diet control and avoid tyramine-containing foods. Reversible MAOIs do not show this "cheese effect."

Propargylamines. The propargylamines are K_{cat} inhibitors or suicide substrates of MAO, forming covalent derivatives with the flavine group of the enzyme. Pargyline (**8.39**), while an MAOI, is used as a hypotensive agent, even though it is incompatible with certain foods and sympathomimetic drugs (i.e., adrenergic agonists). It may act by negative feedback on norepinephrine synthesis, and has a long duration of action.

Carbolines. Harmine (**8.40**) and related carboline alkaloids are reversible MAO-A inhibitors, but are not used therapeutically. Deprenyl (**8.41**) is a selective MAO-B inhibitor and produces an increase in DA levels, but does not influence NE or 5-HT concentrations. It has been proposed as an antidepressant in aging males.

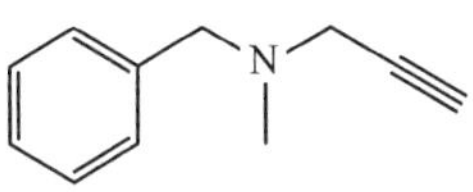

Pargyline (8.39) 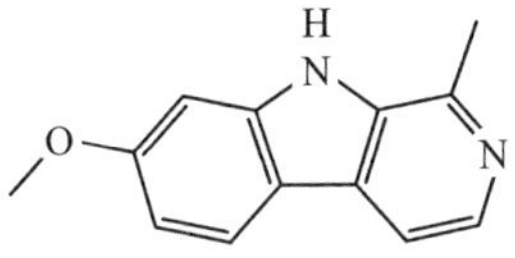   Harmine (8.40) 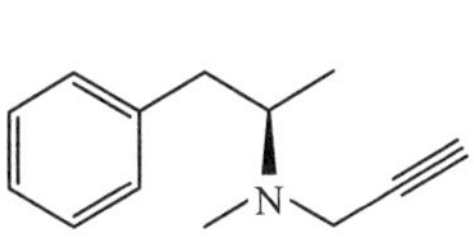   Deprenyl (8.41)

8.2.7 Emerging Enzyme Targets: Caspases and Kinases

As evidenced by the ongoing discussion in this chapter, enzymes have historically represented very important targets in drug design. This trend will undoubtedly continue. Although many enzymes continue to be targeted for purposes of drug design, two families of enzymes are emerging as important sources for future drug design: kinases and caspases. Both of these enzyme families trigger cascades of molecular events that exert profound effects on cellular functioning; both regulate crucial biological processes. Both families are therefore strong candidates for drug design. However, there are some major hurdles that must be confronted. Since there are many different kinases and many different caspases, the drug design process must strive for selectivity, developing a compound that inhibits one particular kinase (or caspase) without necessarily inhibiting the entire family. Furthermore, since both of these enzymes are integral to fundamental biochemical processes, care must be taken to ensure that the enzyme inhibitors are not excessively toxic.

8.2.7.1 Kinases

For regulating crucial biological processes, living organisms depend upon a family of enzymes called protein kinases. Although they were first recognized in the 1950s, it is only recently that their potential in drug discovery has been more fully exploited. Current estimates suggest that there are several thousand kinase enzymes. They work by attaching phosphate groups to other proteins, thereby activating cellular processes, including transcription of new proteins. Protein phosphorylation is recognized as the most important factor in the regulation of protein function by switching cellular activities from one state to another, thereby regulating gene expression, cell proliferation, and cell differentiation. All kinases have "on" switches and "off" switches that keep them under tight control. Protein kinase A, for example, uses a regulatory subunit—a second protein molecule—to control the activity of its catalytic subunit. Not surprisingly, kinases are widely studied as druggable targets.

Protein tyrosine kinases (PTKs) are enzymes that phosphorylate specific tyrosine residues within the sequence of a wide variety of proteins that transmit signals central to cellular processes. Certain PTKs seem to be involved in *tumorigenesis*, transforming normal cells to cells with a neoplastic phenotype. Accordingly, small-molecule inhibitors of PTKs are being developed as therapeutics. Initial work focused on anilinoquinazoline congeners. Additional work confirmed that most PTK inhibitors have common chemical features, including a substituted fused bicyclic ring system combined with another aromatic ring located off the central bicyclic moiety.

Another important family of kinases for drug discovery is the mitogen-activated protein kinases (MAPKs). These are proline-directed serine/threonine kinases that activate their substrates by dual-phosphorylation. MAPK enzymes are activated by a variety of signals including growth factors and cytokines, discussed in chapter 6. The MAPK family plays a critical role in cell cycle progression. Small molecule inhibitors of MAPK may have utility in the treatment of cancer.

Yet another interesting family of kinases is the cyclin-dependent kinases (CDKs). The cell cycle incorporates the coordinated activities of cellular "checkpoint molecules" to ensure the accurate replication of the genome during cellular division. Since the CDKs are an integral part of these checkpoint controls, they are compelling targets for drug design. In early work, flavopiridol (**8.42**) has been one of the most widely studied compounds as a CDK inhibitor. Such inhibitors may have a therapeutic role in the management of cancer.

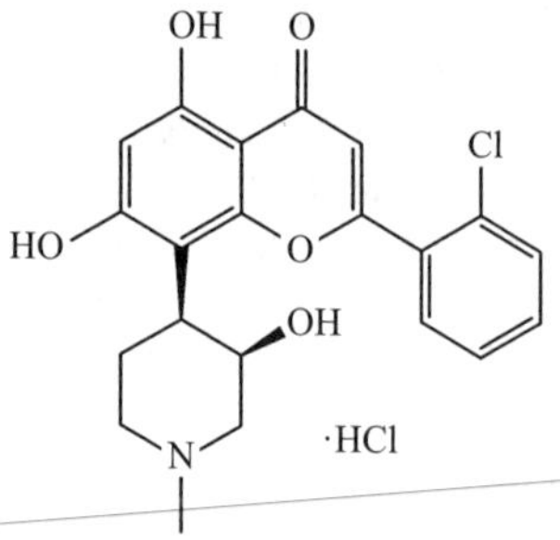

Flavopiridol (8.42)

8.2.7.2 Caspases

Apoptosis is a potentially extremely important target in drug design. Apoptosis and necrosis are the two fundamental biochemical pathologies that cause cells to die. Necrosis is death from injury (e.g., burn injury to skin); apoptosis is predestined death from "old age" and degeneration. Apoptosis and necrosis may be compared as shown in table 8.1.

Apoptosis is pre-programmed cell death. It is why cells (and eventually humans) eventually die with age. Apoptosis is the blueprint for death contained within the genetic code of our cells. Being able to modulate apoptosis would be immensely powerful.

Table 8.1 Comparison of Apoptosis and Necrosis

Characteristic	Apoptosis	Necrosis
1. Source of death-inducing trigger	Internal	External
2. Reversibility	No	Sometimes
3. Appearance of dying cells	Shrink by 50%	Swell during death process
4. Integrity of cell membrane	Usually intact	Frequently ruptured
5. Morphology of Nucleus	Fragmented	No or little change
6. Structure of chromatin	Condensed	Minor changes only

There are some conditions, such as cancer, in which it would be desirable to "turn on" apoptosis, hastening the death of the malignant cell. There are other degenerative conditions, such as Alzheimer's disease, in which it would be desirable to "turn off" apoptosis, prolonging the life of cells. Caspases are a family of proteins that are one of the main effectors of apoptosis.

Because they are central to apoptosis, the caspases have emerged as important prospects for drug discovery targets. The term caspase comes from *cysteine* prote*ase* with *asp*artic acid specificity. The family is defined by sequence homology and conservation of catalytic and substrate recognition residues. Caspases are present in all cells as latent enzymes, called *zymogens* because they are inactive pro-forms of proteins. In order to become biologically active, caspases must be activated. Typically, activation occurs when two caspase proteins associate with each other (*homodimerization*) and then use their own catalytic enzymatic portions to cleave themselves into subunits via cleavage at internal aspartates (*autoproteolysis*). Alternatively, a caspase enzyme can be cleaved into subunits by another caspase. Since all caspases are thiol-proteases, they cleave at the carboxyl terminal of aspartate residues. Functionally, there are two groups of caspases:

1. Caspases involved in apoptosis
2. Caspases involved in immunoregulation via cytokines

This section will focus on the pro-apoptotic caspases. Cytokines and their involvement in immune regulation are discussed in chapter 6.

The caspases that are involved in apoptosis may be further classified as either *initiators* or *effectors*. Induction of apoptosis via *death receptors* results in the activation of an initiator caspase. Caspases-8, -9, and -10 are initiators because they initiate the cascade of biochemical events that culminates in apoptosis. Caspases-3, -6, and -7 propagate this cascade (the so-called *doomsday signal*), thus functioning as effectors. Although there is some overlap, caspases-1, -4, -5, -11, -12, and -13 are involved in processing cytokines, thus influencing immunoregulation. The various caspases are listed in table 8.2.

Crystal structures for a number of these caspases have been determined. Activated caspase-3 has twofold symmetry. Caspase-3 has a protein core composed of a 12-stranded β-sheet, surrounded by ten α-helices. There are two active sites, each residing on an opposite protein side. Each active site is composed of a large subunit,

Table 8.2 Caspases Relevant to Drug Design

Caspace	Alternate nonmenclature	Function	Substrates
Caspase-1	ICE	Processing of interleukins (inflammation)	Interleukin-18
			Lamins
			pre-Interleukin-1β
Caspase-2	Ich-1	Apoptosis	Golgin-160
Caspase-3	CPP32	Apoptosis	β-Catenin
			Calpastatin
			Caspase-6
			Caspase-7
			Caspase-9
			DNA–PK
			FAK
			Fodrin
			Gas2
			Gelsolin
			HnRNP proteins
			ICAD
			Lamins
			MDM2
			PARP
			Presenilin2
			Topoisomerase I
Caspase-4	Ich-2, ICE II		Caspase-1
Caspase-5	ICE III, TY	Inflammation/Apoptosis	
Caspase-6	Mch2	Apoptosis	Caspase-3
			FAK
			Keratin-18
			NuMA
			PARP
Caspase-7	Mch3, ICE-LAP3, CMH-1	Apoptosis	Calpastatin
			EMAP II
			FAK
			PARP
			SREB1
Caspase-8	FLICE, MACH, Mch5	Apoptosis	Caspase-3
			Caspase-4
			Caspase-6
			Caspase-7
			Caspase-9
			Caspase-10
			Caspase-13
			PARP

(Continued)

Table 8.2 Continued

Caspace	Alternate nonmenclature	Function	Substrates
Caspase-9	Apaf-3, ICE-LAP6, Mch6	Apoptosis	Caspase-3
			Caspase-7
			PARP
			pro-Caspase-9
Caspase-10	FLICE-2, Mch4	Apoptosis	Caspase-3
			Caspase-4
			Caspase-6
			Caspase-7
			Caspase-8
			Caspase-9
Caspase-11	Ich_3, ICE-B	Inflammation and apoptosis	
Caspase-12	ICE_C	Apoptosis	
Caspase-13	ERICE	Inflammation	

from one of the procaspase units of the homodimer, and a small unit from the other procaspase.

The activated caspases kill cells by a process of selectively directed proteolysis. The caspases catalyze the cleavage and breakdown of a select group of proteins within the cell. This highly focused cellular destruction produces a disassembly of the cell into membrane-limited fragments called *apoptotic bodies.* These bodies are engulfed and digested within phagocytes. The entire process is focused and tightly controlled, preventing the release of protein breakdown products into the surrounding environment; in turn, this precludes the activation of an inflammatory response.

Small-molecule inhibitors of caspases would have obvious use as therapeutics. The current medicinal chemistry research literature is rich in studies attempting to achieve this important design goal. In early work, both reversible and irreversible peptide-based inhibitors of various caspases have been developed. Peptidomimetic ketones were also devised; for example, acyloxymethyl ketones were designed and developed as potent, time-dependent irreversible caspase inhibitors.

8.2.8 Enzyme Cofactors as Targets in Drug Design: Vitamins

Vitamins are low-molecular-weight organic compounds that, in almost all cases, function as *coenzymes*—that is, molecules that "assist" an enzyme in the completion of its catalytic task. As such, they are indispensable in a multitude of biochemical reactions; however, higher organisms cannot biosynthesize many of them. Those that must be supplied in the human diet are classified in the vitamin category because their absence leads to deficiency diseases. Vitamins therefore have therapeutic importance; hence they are of interest to the medicinal chemist, even though they belong to a mature and slow-moving field.

The knowledge of some vitamins reaches back into folk medicine: North American aboriginal peoples treated scurvy (later recognized as vitamin C deficiency) with cedar leaf tea (*Thuja*), and, from the seventeenth century on, the British Navy issued lime juice to its sailors to prevent scurvy. (This is the origin of the slang word "limey" for the British.) The use of fish-liver oils for the treatment of rickets (later recognized as vitamin D deficiency), a bone-growth disorder, has been known since the nineteenth century. These uses all predate the recognition of the essential function of these substances and the term "vitamine," coined by Funk in 1912.

On the basis of their solubility and polarity, vitamins are divided into two categories: the water-soluble and the fat-soluble vitamins. The former cannot be stored in any biocompartment and must be ingested regularly. In that context, it is important to caution about the many and periodically recurring fads and myths embraced by the lay public and press. Since vitamins are widely known but misunderstood, there is a need to educate the public about their misuse, as well as in other areas of pseudoscientific misunderstanding.

8.2.8.1 Vitamin B_1 (Thiamine)

Vitamin B_1 (**8.43**, thiamine) is a pyrimidinyl methyl thiazolium derivative that occurs in bran (mostly rice hulk), beans, nuts, egg yolk, yeast, and vegetables. In its pyrophosphate form it is a coenzyme of pyruvate dehydrogenase, which oxidatively decarboxylates pyruvate to form acetyl-coenzyme A. Consequently, it is a very important coenzyme, catalyzing the connecting step between glycolysis and the Krebs cycle. The carbon between the sulfur and nitrogen of the thiazole ring of the vitamin is very acidic and adds to the carbonyl of pyruvate, leading to decarboxylation. Thiamine is also a coenzyme of transketolase, which transfers two-carbon fragments between carbohydrates in the pentose phosphate pathway (pentose shunt), a multipurpose metabolic reaction sequence.

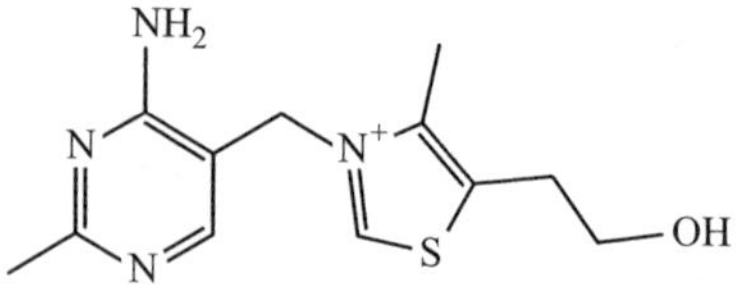

Vitamin B_1 (8.43)

Vitamin B_1 deficiency, known as *beriberi*, was historically seen primarily in people of southeast Asia, for dietary reasons. The symptoms of beriberi are neurological disorders (weakness, paralysis, painful neuritis), diarrhea, loss of appetite, dermatitis, and anemia. These symptoms are due mainly to the accumulation of pyruvate and lactate.

Some analogs of thiamine are biologically active, but have not attained wide use because thiamine itself is cheaply available.

8.2.8.2 Vitamin B_2 (Riboflavin)

Vitamin B_2 (**8.44**, riboflavin) is a benzopteridine derivative carrying a ribityl (reduced ribose) side chain. It occurs in almost all foods, the largest amounts being found in eggs, meat, spinach, liver, yeast, and milk. Riboflavin is one of the major electron carriers as a component of flavine-adenine dinucleotide (FAD), which is involved in carbohydrate and fatty acid metabolism. A hydride ion and a proton are added to the pyrazine ring of

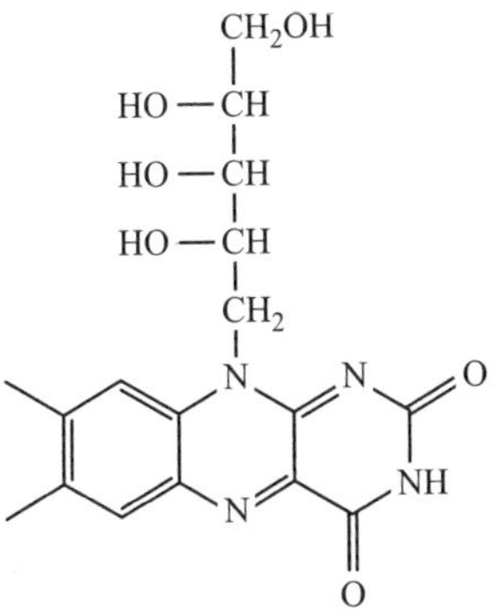

$$CH_2OH$$

HO — CH

HO — CH

HO — CH

$$CH_2$$

Vitamin B$_2$ (8.44)

the vitamin, with the result that FAD carries a pair of electrons and two hydrogens. The monoamine oxidases (see section 8.2.6) are also flavoenzymes.

Vitamin B$_2$ deficiency is seldom seen, and its symptoms of leg ulceration (cheilosis), skin symptoms, a purplish tongue, eye disturbances, and photophobia are vague.

8.2.8.3 Nicotinamide and Nicotinic Acid (Niacin)

Nicotinamide (**8.45**) and nicotinic acid (**8.46**, niacin)—which have also been referred to as vitamin B$_3$ or B$_5$—are simple pyridine-3-carboxylic acid derivatives occurring in liver, yeast, and meat. In the form of nicotinamide-adenine dinucleotide (NAD$^+$) or its phosphorylated form (NADP$^+$), nicotinamide is the most important electron carrier in intermediary metabolism. Unlike FAD, it adds a hydride ion (i.e., one pair of electrons and one hydrogen) only.

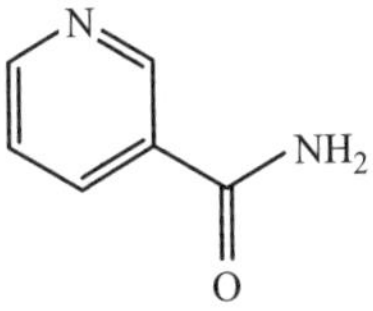

Nicotinamide (8.45)

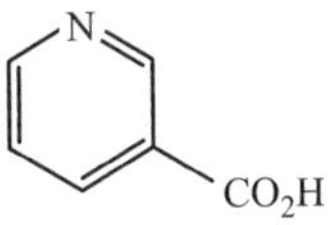

Nicotinic acid (8.46)

Nicotinamide deficiency leads to *pellagra* [*pelle agra* (Italian) = rough skin], which manifests itself in dermatitis (skin rash), pigmentation, a red and inflamed tongue, diarrhea, and weakness. People who consume large amounts of corn in an unbalanced diet are prone to develop the disease.

8.2.8.4 Vitamin B$_6$ (Pyridoxine)

Vitamin B$_6$ (**8.47**, pyridoxine) is a pyridine-alcohol, but its biologically active forms are pyridoxal 5-phosphate and the corresponding pyridoxamine. Like all the members of the vitamin B complex, it occurs in yeast, bran, wheat germ, and liver. It is a coenzyme of

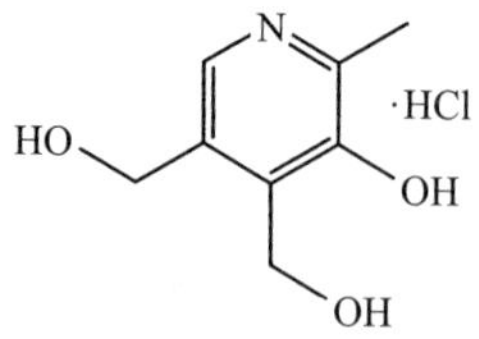

Vitamin B$_6$ (8.47)

a remarkable number of enzymes, most of which are involved in amino acid metabolism. By forming a Schiff base, pyridoxal phosphate acts as an electron sink and weakens all three bonds about the α-carbon of amino acids. It therefore functions in transamination, decarboxylation, deamination, racemization, and aldol cleavage, and can also influence the β and γ carbons of an amino acid. Glutamic acid decarboxylase (chapter 4) is also a pyridoxal enzyme and is involved in GABA synthesis. If it is inhibited by hydrazine-type compounds through Schiff-base formation, the resulting lack of GABA may lead to a seizure.

The typical B$_6$ avitaminosis is very similar to riboflavine and niacin deficiency and is manifested by eye, mouth, and nose lesions as well as neurological symptoms.

8.2.8.5 Pantothenic Acid

Pantothenic acid (**8.48**), a hydroxyamide, occurs mainly in liver, yeast, vegetables, and milk, but also in just about every other food source, as its name implies [*pantos* (Greek) = everywhere]. It is part of coenzyme A, the acyl-transporting enzyme of the Krebs cycle and lipid syntheses, as well as a constituent of the acyl carrier protein in the fatty-acid synthase enzyme complex.

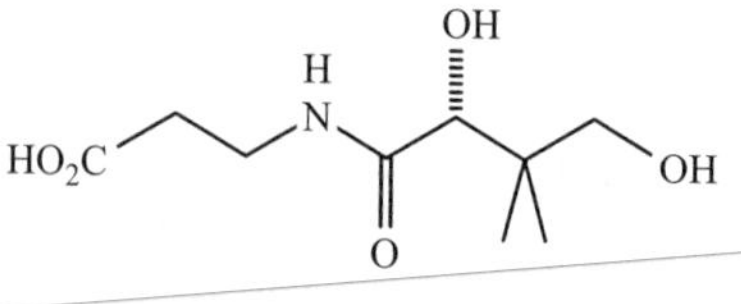

Pantothenic acid (8.48)

There is no record of pantothenic acid deficiency in humans, since all food contains sufficient quantities of this vitamin. Experimentally, however, neurological, gastrointestinal, and cardiovascular symptoms result from a diet lacking in pantothenic acid.

8.2.8.6 Biotin

Biotin (**8.49**), a thiophene-lactam, occurs in yeast, liver, kidney, eggs, vegetables, and nuts. It functions as a cocarboxylase in a number of biochemical reactions. It binds CO_2 in the form of an unstable carbamic acid on one of the lactam nitrogens. The carbamate carboxyl is then donated easily.

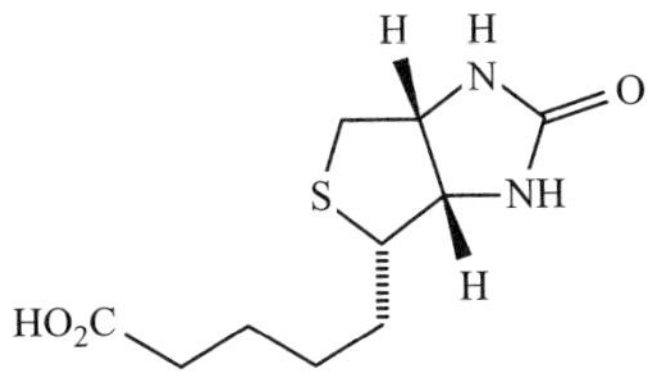

Biotin (8.49)

Biotin deficiency can be triggered only experimentally, using diets rich in raw egg white. The latter contains avidin, a 70 kD protein that binds biotin in an inactive form. The deficiency leads to dermatitis and hair loss in rats.

8.2.8.7 Vitamin B₁₂ (Cobalamin)

Vitamin B_{12} (**8.50**, cobalamin) is an extremely complex molecule consisting of a corrin ring system similar to heme. The central metal atom is cobalt, coordinated with a ribofuranosyl-dimethylbenzimidazole. Vitamin B_{12} occurs in liver, but is also produced by many bacteria and is therefore obtained commercially by fermentation. The vitamin is a catalyst for the rearrangement of methylmalonyl-CoA to the succinyl derivative in the degradation of some amino acids and the oxidation of fatty acids with an odd number of carbon atoms. It is also necessary for the methylation of homocysteine to methionine.

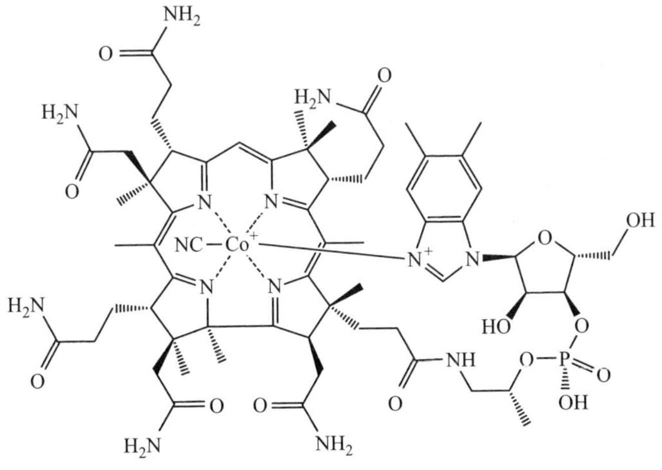

Vitamin B_{12} (8.50)

Lack of B_{12} leads to *pernicious anemia*, an uncommon but potentially fatal disease if untreated. In 1926, Minot and Murphy discovered that raw liver could keep patients with the disease alive. The vitamin itself was isolated only in 1948. As little as $3–6 \times 10^{-6}$ g is curative, and the large amounts of the vitamin taken by some fadists are therefore totally unnecessary and without benefit. Some patients with inflammatory bowel diseases, such as Crohn's disease, may experience pernicious anemia if the inflammatory disease affects their ileum (a portion of the small intestine). In these patients, the uptake of the vitamin is impaired because they lack the so-called "intrinsic factor"—a gastric glycoprotein necessary for the absorption of the vitamin. The administration of very large amounts (milligrams) of vitamin B_{12} forces the uptake of sufficient cobalamin to cover the need, but this is only a treatment and not a cure for pernicious anemia.

8.2.8.8 Vitamin C (Ascorbic Acid)

Vitamin C (**8.51**, ascorbic acid) is a carbohydrate derivative. It occurs in citrus fruits, green pepper, and fresh vegetables, including onions, but is rapidly hydrolyzed by cooking. Only primates and the guinea pig are unable to synthesize ascorbic acid. Vitamin C is necessary for the hydroxylation of proline to hydroxyproline, reducing the Fe atom in collagen hydroxylase. Hydroxyproline is a major amino acid in the collagen that is present in all fibrous tissues, the intracellular matrix, and capillary walls. Vitamin C is also used in steroid hydroxylation in the adrenal gland, and finds a role in tyrosine metabolism.

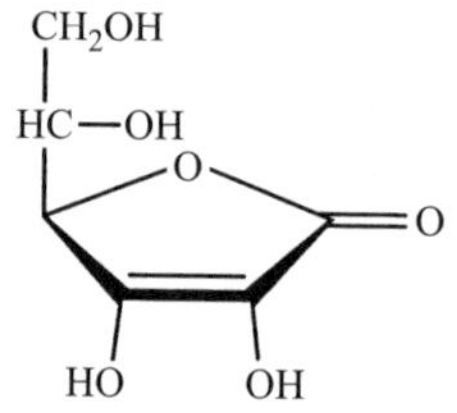

Vitamin C (8.51)

    In vitamin C avitaminosis, or *scurvy,* the joints become painful and the gums bleed and deteriorate, resulting in tooth loss. Gangrene and infections may also occur, and wounds do not heal properly. These symptoms all result from impaired collagen synthesis and require 3–4 months to develop.

    Increased doses of vitamin C are sometimes given after surgery or burns to promote healing by increasing collagen synthesis. There is controversy about its purported activity of alleviating the common cold and other viral diseases when given in extremely large doses. Experiments with radioisotope-labeled ascorbic acid have shown that the normal body pool is about 20 mg/kg, which can be maintained by an intake of approximately 100 mg/day (140 mg for smokers). The consumption of several grams per day—as advocated by believers in "megavitamin" hypotheses—results in the excretion of excess vitamin C. As opponents of megavitamin administration sometimes cynically observe, the only obvious effect of megavitamin therapy is the production of brightly coloured urine!

8.2.8.9 Vitamin A (Retinol)

Vitamin A (**8.52**, retinol) occurs in plants as the provitamin carotene; it is a highly unsaturated terpenoid, with all of its double bonds *trans* (E) to one another. Fish liver oils,

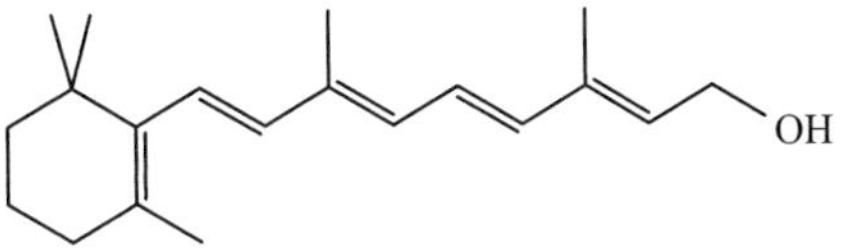

Vitamin A (8.52)

milk, and eggs contain vitamin A itself, a cleavage product of β-carotene. It is necessary for the growth of young animals since it regulates bone-cell formation and the shape of bones. It also plays a vital role in vision, where *cis*-retinal (**8.53**), an isomer of retinol, is formed upon irradiation with light and becomes attached to the protein rhodopsin as a Schiff base. Light energy is thus transformed into atomic motion, resulting in hyperpolarization of the plasma membrane of the rods in the retina. A single photon can block the Na$^+$ permeability of rod membranes, amplifying the response a million fold. In general metabolism, vitamin A also influences glycogen synthesis.

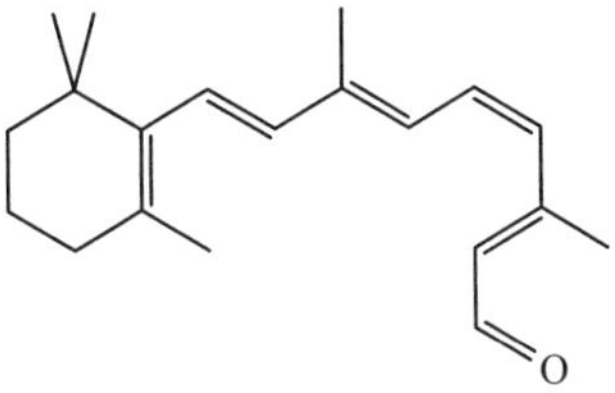

Cis-Retinal (8.53)

Avitaminosis A results in the loss of night vision (*nyctalopia*). Furthermore, the removal of vitamin A from the diet causes the cornea of the eye to "dry out" (*xerophthalmia*). However, excessive intake of vitamin A can result in severe and even fatal toxicity.

The past twenty years have witnessed considerable progress in the synthesis and use of other retinoid-like molecules related to vitamin A. The aromatic retinoid etretin (**8.54**) and its ester etretinate (**8.55**) had some effectiveness in the treatment of psoriasis, a disorder of skin. 13-*cis*-Retinoic acid (isotretinoin) produces sebaceous gland atrophy and could prove useful in the treatment of severe *acne vulgaris*. Although these compounds have toxic side effects and are not in regular use, they have opened up new therapeutic possibilities. Retinoic acid (tretinoin, **8.56**) has been employed in the treatment of acne.

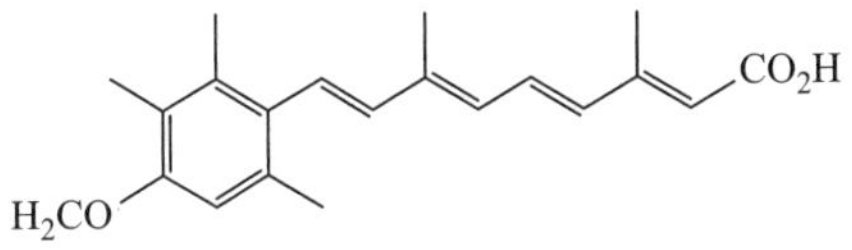

Etretin (8.54)

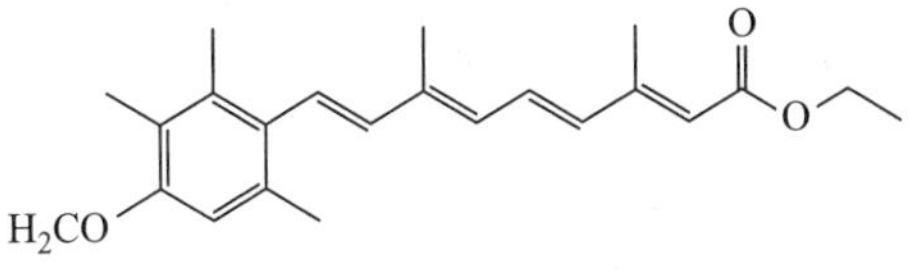

Etretinate (8.55)

8.2.8.10 Vitamin D (Calciferol)

Vitamin D (**8.57**, calciferol) is found in fish-liver oils and milk. It is also produced in human skin from a steroid previtamin by sunlight. Ergosterol is another previtamin D compound and undergoes photo-rearrangement to vitamin D_2, whereas cholesterol and cholesterol derivatives produce vitamin D_3. The two vitamins have the same activity in humans. These photo-rearrangements, resulting in the cleavage of the steroid B ring, are used industrially to manufacture these vitamins.

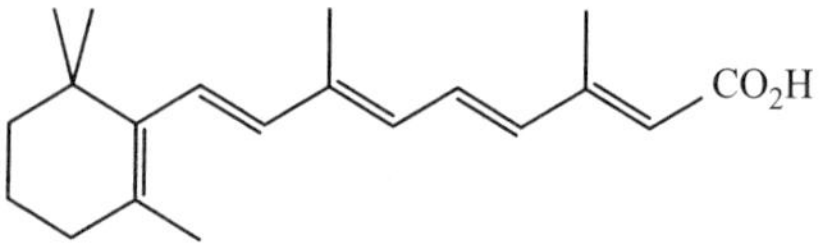

Retinoic acid (8.56) 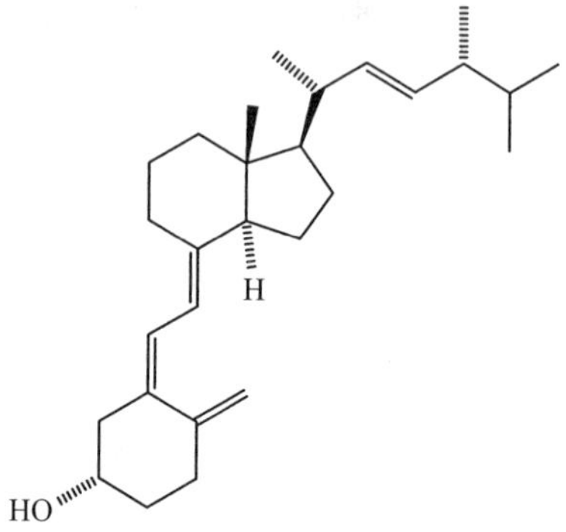   Vitamin D (8.57)

The active forms of the D vitamins are $1\alpha,25$-dihydroxy-vitamin D_3 and 25-hydroxy-vitamin D_3. They are formed by enzymatic hydroxylation in the liver microsomes and then in the kidney mitochondria by a ferredoxin flavoprotein and cytochrome P-450. The 1,25-dihydroxy vitamin is then transported to the bone, intestine, and other target organs (kidneys, parathyroid gland). Consequently, it can be considered a hormone since it is produced in one organ but used elsewhere. It mobilizes calcium and phosphate and also influences the absorption of these ions in the intestine, thus promoting bone mineralization. The hormone is also active in relieving hypoparathyroidism and postmenopausal osteoporosis, which, for example, results in the brittle bones of elderly women.

Overall calcium metabolism is regulated by the parathyroid gland (parathormone) and the thyroid hormone calcitonin. Parathormone regulates the synthesis of $1,25$-$(OH)_2D_3$, which manages bone and kidney calcium metabolism as well as promoting Ca^{2+} intake in the gut. The role of calcitonin is less clear, although it works against parathormone. Serum calcium represses parathormone synthesis, completing a feedback loop.

Vitamin D deficiency, known as *rickets,* was described in children as early as 1645 by Webster, in "gloomy, sunless England." Women wearing extensive black veils also can show osteoporosis, since they are not exposed to sunshine at all. In many Western countries, milk is fortified with about 400 units of vitamin D_3 (10 µg/liter), the minimum daily requirement for children or adults. Therefore, with a proper diet and sunshine, rickets should not occur in children. Pregnant women have a higher requirement, but overdosage of the vitamin leads to toxic symptoms. Although rickets is not a current clinical problem, osteoporosis ("thinning of the bones") has emerged as a very relevant clinical problem, especially in older women.

Because of the importance of osteoporosis, the area of vitamin D research has become exceptionally active in the generally quiescent vitamin field. Vitamin D analogs such as calcitriol (**8.58**, 1α,25,25-trihydroxycholecalciferol) and other compounds have been suggested as new drugs for these conditions.

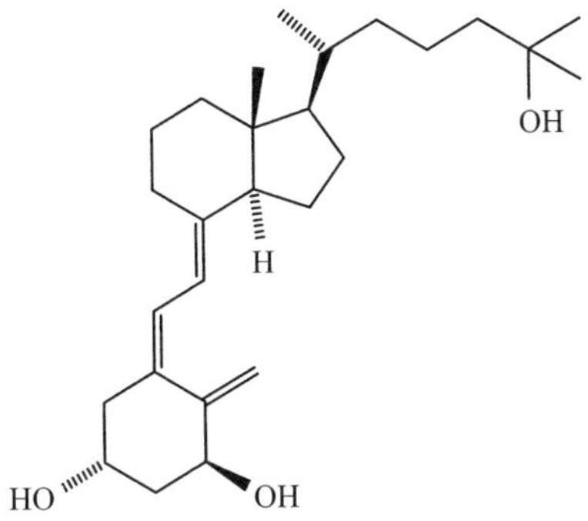

Calcitrol (8.58)

8.2.8.11 Vitamin E (Tocopherol)

Vitamin E (**8.59**, tocopherol), a chromane, occurs in just about all vegetables as well as in oils, grain, milk, meat, and yeast. Its principal known action is the maintenance of normal pregnancy in rats kept on a special diet. Vitamin E-deficient rats reabsorb their fetuses even though they are developing normally in every other respect. There is no evidence for this in humans. The vitamin has antioxidant properties, probably stabilizing vitamin A and unsaturated fatty acids and preventing free-radical reactions. Free radicals, according to some hypotheses, may be involved in aging. Accordingly, some researchers have advocated the ingestion of large doses of vitamin E (2000 IU/day) for the treatment of Alzheimer's disease. However, large doses such as this (especially when co-administered with an anti-inflammatory agent such as acetylsalicylic acid) may predispose to bleeding problems. Vitamin E deficiency is unknown in humans, but there are periodic fads promoting the intake of vitamin E.

8.2.8.12 Vitamin K₁

Vitamin K₁ (**8.60**) is a phytyl-naphthoquinone occurring in the green leaves of most plants. The several related active compounds differ in the length of the phytyl side chain.

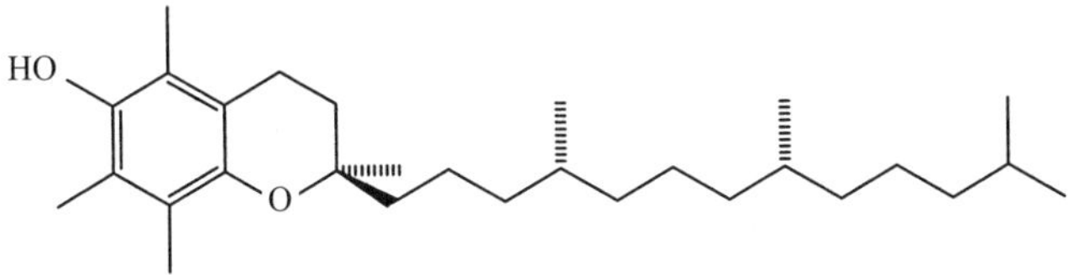

Vitamin E (8.59)

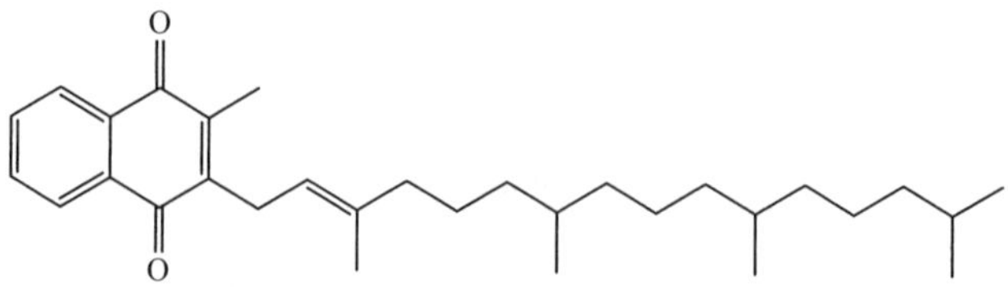

Vitamin K₁ (8.60)

In the absence of the vitamin, the blood-clotting time increases, since the post-translational carboxylation of several glutamate residues in prothrombin and other factors involved in blood clotting is impaired. In humans, this disorder is unknown except in cases of faulty vitamin K absorption, since the normal diet covers the need generously.

Vitamin K antagonists, such as dicoumarol (**8.61**, a natural product) and warfarin (**8.62**), are used as anticoagulants in human therapy (thrombosis, atherosclerosis) and as rat poisons that lead to internal bleeding and death in rodents. Heparin, a polysaccharide consisting of 2-0-sulfonated glucuronic acid and 2-N,6-0-disulfonated glucosamine, is also a widely used anticoagulant, but its effect is connected not with Vitamin K but with enzyme inhibition.

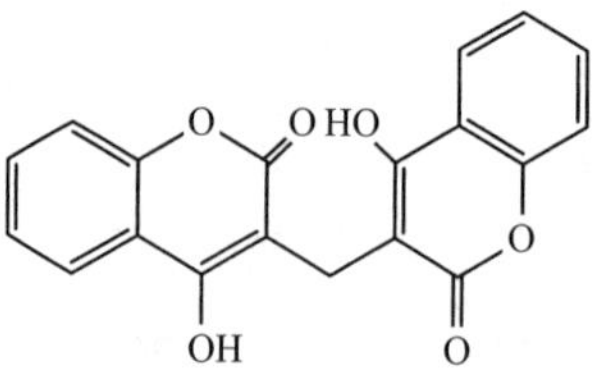

Dicoumarol (8.61) 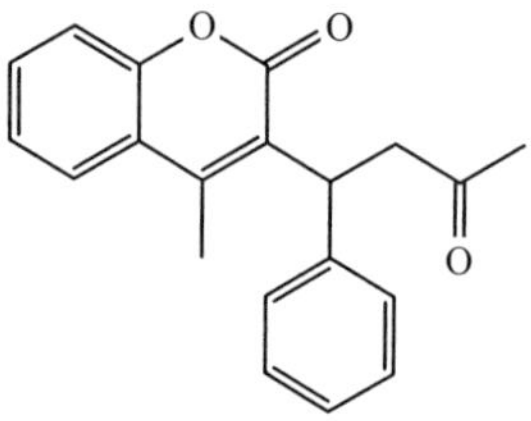   Warfarin (8.62)

8.2.9 Molecular–Clinical Interface: Human Disease and Enzyme Cofactors

8.2.9.1 Case Studies

In a single 12 hour period, one of the authors saw the following five sequential patients in the emergency department of an urban teaching hospital.

Patient 1. This 32-year-old markedly obese female had undergone a gastric stapling operation to decrease the size of her stomach and thus promote weight loss. Although she had recovered well from the surgical procedure, in the ensuing 3–4 weeks she had only been able to take fluids, vomiting all solid foods. Nevertheless, since she was rapidly losing weight, she did not complain. Suddenly, over the course of 24 hours, she rapidly developed confusion (not recognizing her children or spouse), ataxia (staggering gait when walking and inability to stand), and diplopia (double vision). A diagnosis of *Wernicke's syndrome* secondary to acute thiamine deficiency was made. Following the administration of 50 mg of thiamine intravenously, her confusion and diplopia resolved within hours.

Patient 2. This 68-year-old malnourished chronic alcoholic was found wandering in a state of confusion and disorientation. In the emergency department, he demonstrated a severe disturbance of memory in which new information could not be stored. The normal temporal sequence of his established long-term memories was disrupted. When asked to recall events of the past week, he was easily suggestible and gave a semi-fictionalized account of the circumstances (i.e., he demonstrated *confabulation*). A diagnosis of *Korsakoff syndrome* secondary to chronic thiamine deficiency was made. His mental state responded partially, but incompletely, to prolonged thiamine administration.

Patient 3. This 24-year-old male presented with a recent increase in seizures (arising from an exacerbation of his underlying epilepsy) as well as "pins and needles" and "uncomfortable numbness" in his feet, lower legs, and hands. Based on "advice" that he had obtained from an internet chatroom on the World-Wide Web, he had decided to take "megadoses" of pyridoxine for several months and then to stop his regular anti-convulsant drugs. A diagnosis of peripheral neuropathy secondary to pyridoxine toxicity was made. He ultimately returned to normal following discontinuation of the pyridoxine and reinstitution of his carbamazepine anticonvulsant drug.

Patient 4. This 79-year-old male presented with marked agitation and bloody urine. In the previous week, he had experienced a spontaneous nosebleed. A diagnosis of Alzheimer's disease had been made 18 months previously and the patient was being "managed" by his four children, a family physician, and a geriatrician. He was receiving donepezil (cholinesterase enzyme inhibitor), Vitamin E (2000 IU/day), acetylsalicylic acid (325 mg/day) and Gingko biloba ("as a complementary alternative therapy"). A diagnosis of a urinary tract hemorrhage, possibly precipitated by the administration of large doses of vitamin E, was made. The family refused to reduce the dose of vitamin E, since "everyone knows it is a harmless vitamin." The patient died of a brain hemorrhage three months later.

Patient 5. This 74-year-old female presented with three identical episodes of sudden-onset left-sided body paralysis lasting 5–10 minutes, occurring over the past 24 hours. Over the past four months she had experienced intermittent episodes of an "irregularly irregular" heart rate. Multiple electrocardiograms over the past 12 weeks had demonstrated an intermittent atrial fibrillation heart rhythm disorder. A diagnosis of transient ischemic attacks (TIAs, "mini-strokes") secondary to blood clots arising from atrial fibrillation was made. The patient had her blood "thinned" with warfarin, an anticoagulant that works as a vitamin K antagonist. The TIAs subsequently resolved completely.

8.3 PROTEINS AS DRUGS AND DRUG DESIGN TARGETS: NON-ENZYMES

8.3.1 Protein Folding Disorders: Neurodegenerative Diseases

A number of neurodegenerative disorders are characterized by neuronal damage that is induced by neurotoxic, aggregation-prone proteins. Various neurodegenerative diseases, including Alzheimer's disease ("caused" by β-amyloid deposits), prion disease

(PrP^{SC} protein), Parkinson's disease (α-synuclein), Pick's disease (tau), Huntington's disease (polyglutamine-containing protein), and familial amyotrophic lateral sclerosis (SOD1), share a single mechanistic feature—the aggregation and deposition of an abnormal protein product. Following the identification of various genes implicated in the molecular pathogenesis of these diseases, it has become clear that the major effect of mutations in these genes is the abnormal processing and accumulation of misfolded protein products, leading to neuronal inclusions and plaques. These insights into toxic protein accumulation and abnormal protein deposition may permit the rational design of effective therapies. To demonstrate these design principles, two diseases will be discussed: prion-induced dementia and Alzheimer's dementia.

8.3.1.1 Protein Folding Diseases: Prion-Induced Dementia

Prion diseases have attracted immense attention over the past decade. This attention was prompted, in part, by the outbreak of "mad cow disease" in the United Kingdom, an outbreak that ultimately involved humans. More recently, the identification of cows with mad cow disease in North America has emphasized the widespread nature of this problem.

The most common prion disease is sporadic Creutzfeldt–Jakob Disease (CJD). Less common prion-based diseases include Gerstmann–Straussler–Scheinker disease, fatal familial insomnia, and familial CJD. Unlike other neurodegenerative diseases, prion-based disorders are uniquely characterized by the property of transmissibility. Clinically, CJD is characterized by a rapidly progressive dementia accompanied variably by early-onset seizures, insomnia, disordered movements, and psychiatric disturbances; the disease is uniformly fatal. Pathologically, the brains of people who have died from prion diseases show neuronal loss, gliosis (i.e., scarring) and spongiform vacuolation (i.e., the brain contains many microscopic holes, giving it a "sponge-like" appearance; hence the alternative name of *spongiform encephalopathy* for prion diseases). Histochemically, another pathological feature of prion disease is the abnormal accumulation of an amyloid-like material composed of prion protein (PrP), which is encoded by a single gene on the short arm of chromosome 20.

The abnormal deposits found in the brains of CJD victims consist of an abnormal isoform of PrP. Prion protein is normally found in cells. Detailed structural studies show that normal cellular PrP (PrP^{C}) is a soluble protein whose conformation is rich in α-helices with very little β-sheet. The PrP protein extracted from the brains of CJD victims (i.e., PrP^{SC}) is identical in primary amino acid sequence to the normal PrP (PrP^{C}). However, PrP^{SC} has a much greater content of β-sheet conformation with little α-helical structure. Thus PrP^{SC} is neurotoxic because of its three-dimensional structure. When the prion protein is predominantly in an α-helical conformation it is nontoxic; when the prion protein is predominantly in a β-sheet conformation, it kills neurons. The prion protein is thus made neurotoxic not by its amino acid composition but by its conformation. This concept is both fascinating and terrifying. Prion diseases are transmissible; thus prions are infectious agents. However, prions are not like bacteria or viruses, or other infectious microbes—they are simply protein molecules. Prions are *not* microbes with cell membranes and nucleic acids; they are not "living things." Indeed, prions are not even infectious molecules, they are infectious molecular shapes.

Like any other protein, the molecular structure of the prion is subject to conformational flexibility and to various thermal-induced fluctuations between varying conformational states. However, if these fluctuations permit the PrP^{SC} conformation to be attained, then this abnormal conformer promotes the widespread conversion of PrP^C to PrP^{SC}, leading to the precipitous deposition of the abnormal protein throughout the brain (mirrored by the rapid and relentlessly downhill clinical course). This pathological self-propagating shape conversion of α-helical PrP^C to β-sheet PrP^{SC} may in principle be initiated by a "seed" PrP^{SC} molecule in the neurotoxic conformation. This explains the transmissibility of prion diseases and accounts for how susceptible humans exposed to beef from an animal with mad cow disease develop variant Creutzfeldt–Jakob disease.

The concept of abnormal proteins in CJD may provide insights useful for drug design. The pioneering (and Nobel Prize winning) work of Prusiner has enabled the preliminary identification of prototype agents as therapies for CJD. Preliminary work identified two classes of compounds with therapeutic potential: polysulphated molecules and tricyclic molecules (e.g., phenothiazines, aminoacridines). These compounds bind to PrP and endeavor to inhibit the PrP^C to PrP^{SC} cascade of conformational change.

8.3.1.2 Protein Folding Diseases: Alzheimer's Dementia

Alzheimer's disease (AD) is the most common neurodegenerative disease, affecting 2.2 million North Americans. Clinically, AD is characterized by a slowly progressive dementia in which loss of short-term memory is a frequent early manifestation. Pathologically, AD is characterized by the presence of two lesions: the *plaque*, which is an extracellular deposit composed of β-amyloid protein, and the *tangle*, which is an intracellular deposit of tau protein. An important gene that has been implicated in the cause of AD codes for the amyloid precursor protein (APP).

Strong evidence suggests that the main constituent of the plaque, β-amyloid peptide (Aβ), exerts a prominent role in the cause, initiation and progression of AD. Thus AD appears to arise from the abnormal deposition of a protein. Aβ is derived from proteolytic cleavage of amyloid precursor protein (APP), an integral membrane protein. APP is cleaved by the sequential actions of three unique proteases, called α-, β-, and γ-secretases. Each secretase cleaves at a unique site (see figure 8.7).

8.3.2 Targeting Proteins Endogenous to Other Species

It is also possible to design drugs around proteins that are exogenous to humans but endogenous to other species. As discussed in chapter 3, exogenous molecules may be pursued as potential lead compounds in drug discovery. By designing drugs using proteins that are endogenous to other species, it is sometimes possible to exploit targets in humans that do not have messenger molecules. Although there are many examples of such proteins, two illustrative examples are presented:

1. Fish antifreeze proteins
2. Snail conotoxins

Both of these "proteins from the sea" offer opportunities for drug design.

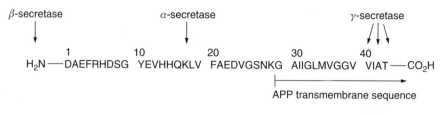

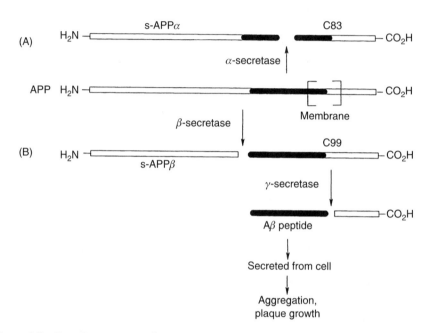

Figure 8.7 Top: Sequence of Aβ1–43 and sites of secretase cleavage. γ-Secretase has low specificity, cleaving the amyloid precursor protein (APP) anywhere between residues 39 to 43 of Aβ. The transmembrane portion of APP is indicated. Bottom: processing of amyloid precursor protein (APP): (A) "Normal" cleavage within Aβ region by α-secretase; (B) pathogenic cleavage of APP by β- and γ-secretase, liberating Aβ, which can become incorporated into growing plaques. (Note: Aβ = β-Amyloid Peptide.)

8.3.2.1 Fish Antifreeze Proteins and Drug Design

Antifreeze proteins are a group of unique macromolecules that prevent marine bony fish from freezing in their icy ocean habitat. This class of proteins was discovered by DeVries of Stanford University in the early 1960s. Many types of antifreeze proteins have been isolated and characterized from different fish. The first type is a glycoprotein antifreeze made from a repeating tripeptide unit (Ala–Ala–Thr)$_n$ with a disaccharide moiety attached to the threonyl residue. Next, there are three antifreeze protein (AFP) macromolecules: type I is an alanine-rich, amphiphilic α-helix; Type II is a protein with five disulfide bridges and a secondary structure characterized by numerous reverse β-turns; Type III is composed of a variety of different proteins of varying length and amino acid composition. These proteins hydrogen-bond to the prism face of ice crystals, inhibiting crystalline growth. These peptides are 300 times more effective in preventing freezing than conventional organic solvent antifreezes. From the perspective of drug design, antifreeze proteins could be exploited for purposes of cryopreservation, prolonging the effective "shelf-life" of organs removed for donation and transplantation. Also, understanding the mechanism of antifreeze proteins may help drug design for human diseases involving *biocrystallization.* These disorders include gout, kidney stones, and gallstones. If one could exploit the prism face of a crystal as a potential receptor, and then design drugs to bind to the crystal face, it might be possible to retard crystal growth, thus preventing stone formation.

8.3.2.2 Snail Conotoxins and Drug Design

Conotoxins are small (10–30 amino acids), disulphide-rich, conformationally constrained peptides produced by marine mollusks such as cone snails. The fish-hunting snails, in particular *Conus geographus*, have been extensively studied. Depending upon the arrangement of disulphide bonds and the number of residues between cysteines, five or more classes of conotoxin can be structurally identified:

1. Alpha class (e.g., alpha conotoxin G1)
2. Mu class (e.g., mu conotoxin GIIIA)
3. Omega class (e.g., omega conotoxin GVIA)
4. Delta class (e.g., delta conotoxin TxVIA)
5. Kappa class (e.g., kappa conotoxin PVIIA)

These various conotoxins enable the snail to paralyze fish that it is hunting. The conotoxins bind to various voltage-gated ion channels, unlike endogenous ligands found in humans. Thus the conotoxins present a unique opportunity to design drugs capable of binding to voltage-gated ion channels. Mu conotoxins can block the voltage-gated Na^+ channel, while omega conotoxins can inhibit the voltage-gated Ca^{2+} channel. The conotoxins can also differentiate between types of voltage-gated Ca^{2+} channels, including the L-, N-, and P-type channels. Because they are conformationally constrained bioactive peptides, they are a good point for the initiation of drug design.

8.4 NUCLEIC ACIDS AS DRUGS AND DRUG DESIGN TARGETS

Nucleic acids are obvious drug design targets, offering important receptors around which to design drugs. Drugs targeting various aspects of nucleic acid replication, transcription, and translation have obvious applications as agents with which to treat cancer or infectious diseases; these are presented in chapters 7 and 9, respectively, and will not be repeated here.

Nucleic-acid-related molecules (nucleotides, nucleosides, purines, pyrimidines) may also be used as drugs themselves (and not only as drug receptors). Once again, as discussed in chapters 7 and 9, this is most relevant in the areas of cancer and infectious disease, with purine/pyrimidine analogs being exploited as antimetabolites. 5-Fluorouracil is a well-described antineoplastic agent. Analogously, 5-fluorocytosine is used as an antifungal agent. Analogs of purines and pyrimidines have been used for other applications as well. For example, analogs of uracil have been studied as sedatives and as *anxiolytics* (drugs to treat anxiety). The strengths and weaknesses of such a design approach are apparent. Analogs of purines or pyrimidines will have predictable pharmacokinetic properties; however, they may also have undesirable pharmacodynamic properties, being carcinogenic, mutagenic, or teratogenic by interfering with endogenous nucleic acids.

8.4.1 Ribozymes as Drug Design Targets

There are other ways in which nucleic-acid-related compounds could be exploited as therapeutics. A new, emerging area concerns the application of RNA as a drug. The discovery of catalytic RNA (*ribozymes*) by Cech and Altman was a fundamental advance in nucleic acid chemistry. According to traditional "double helix dogma," RNA was a passive information-transmitting molecule. The identification of ribozymes enabled the conceptual advance that RNA can also act as a catalyst for the following biochemical processes:

1. RNA splicing
2. RNA cleavage
3. DNA cleavage
4. Peptide bond cleavage
5. Transfer of phosphate groups

Ribozymes therefore show promise as therapeutic agents with which to downregulate RNA activity. For instance, a ribozyme-based drug could be used to attack the mRNA coding for a protein associated with a particular disease; this attack would prevent the protein's expression by rendering the mRNA untranslatable.

Engineering drugs on the basis of RNA is beset with the same problems as engineering drugs on the basis of peptides. Unmodified RNA is metabolically vulnerable and unstable within a biological milieu. Therefore, just as there is peptidomimetic chemistry to make peptide-like drugs to mimic peptides, so too is "nucleotidomimetic" chemistry beginning to emerge. There are seven obvious positions at which to modify a nucleotide in order to make it more "drug-like" and less "nucleotide-like":

1. 2′ Sugar position
2. 3′ Sugar position
3. 4′ Sugar position
4. 5′ Sugar position
5. Replacing the base with base bioisosteres
6. Replacing the phosphodiester backbone with bioisosteres
7. Using non-nucleotide linkers between nucleotides

There are many examples of each of these possibilities (see figure 8.8). In the case of the 2′ sugar position, most of the 2′-substititutions have been achieved with 2′-O-alkyl, 2′-amino, and 2′-fluoro replacements. Purine base substitutions have been performed with 2-aminopurine, xanthosine, and isoguanosine. Phosphodiester replacements have been effected using phosphorothioate substitutions. As for the non-nucleotide linkers, propanediol linkers have been employed. Using these various bioisosteric substitutions,

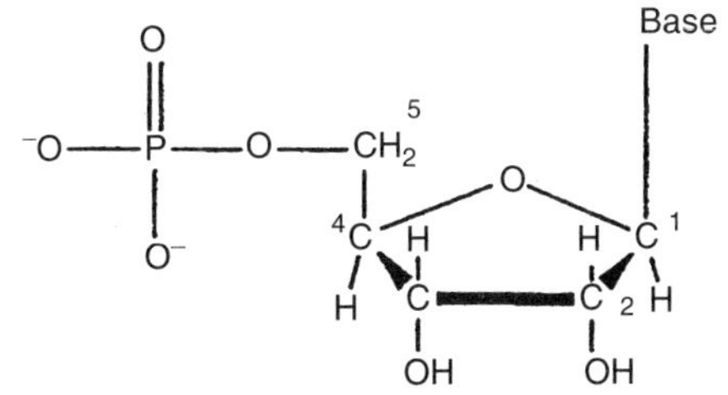

Figure 8.8 Locations of strategic modification positions for drug design in ribonucleotides.

a variety of analogs have been prepared and evaluated. Ribozyme-based drugs may in theory be used against a diversity of disorders, including cancer and viral infections.

8.4.2 Antisense Oligonucleotides as Potential Therapeutics

Synthetic oligonucleotide ligands, designed to bind to RNA receptors in a sequence-specific fashion by Watson–Crick base pair recognition, are another nucleotide-based emerging drug discovery paradigm. This approach to the discovery of drug design leads is called the *antisense technology*. More than 20 oligonucleotides with anti-inflammatory, antineoplastic, and antiviral properties have been preliminarily evaluated in human clinical trials. Many of these are full-length compounds ranging from 15 to 20 bases. However, bioisosteric equivalents are frequently needed to ensure survivability within the biological milieu. The first generation of antisense oligonucleotides were phospho-rothioate oligodeoxynucleotides (PSs). Second-generation antisense oligonucleotides have incorporated 2′-O-methyl and 2′-O-methoxyethyl modifications.

8.5 LIPIDS AS DRUGS AND DRUG DESIGN TARGETS

In 1934, von Euler in Sweden discovered a group of polyunsaturated fatty acids that had a powerful effect on smooth muscle and blood pressure. They were isolated from seminal fluid and the prostate, and were named *prostaglandins*. Their structure was elucidated by the Samuelsson group in Stockholm, who, in 1975, also discovered even more potent fatty acid metabolites, the *thromboxanes* and *prostacyclin*. The effect of these compounds on blood platelet aggregation and the contraction of blood vessels connected them to the etiology of stroke and cardiovascular disease. Additionally, it was discovered that both steroidal and nonsteroidal anti-inflammatory agents act through the prostaglandin system, adding further impetus to the research in this area. Consequently, the field of bioactive lipids became a center of intense interest.

As the research area expanded, the *leukotrienes* were discovered next. The leukotrienes are potent lipid mediators associated with asthma and allergic reactions. In contrast to prostaglandins, leukotrienes are made predominantly in inflammatory cells, like leukocytes, macrophages, and mast cells.

Prostaglandins, thromboxane, and the leukotrienes are lipids that are collectively called *eicosanoids,* since they are all derived from the C20 fatty acid, arachidonic acid [*eicosa* (Gr.) = twenty]. Over the past twenty years, the eicosanoids have emerged as important molecules around which to target drug design and development.

8.5.1 Prostaglandins and Thromboxanes

8.5.1.1 Prostaglandins: Structure and Biosynthesis

The biosynthesis of prostaglandins and thromboxanes starts from arachidonic acid. Arachidonic acid, obtained from its phospholipid form by the action of phospholipase A, is cyclized to prostaglandin endoperoxide (**8.63**) in the form of PGG (a side-chain peroxide), from which PGH_2 (a side-chain hydroxyl) is obtained. Interleukin-1, a cytokine (see chapter 6) produced by leukocytes and possessing multiple immunological roles, mediates inflammation by increasing phospholipase activity and thus prostaglandin synthesis. The first reaction in prostaglandin biosynthesis is catalyzed by PG cyclooxygenase in the presence of O_2 and heme. The second reaction requires tryptophan, probably as a source of electrons.

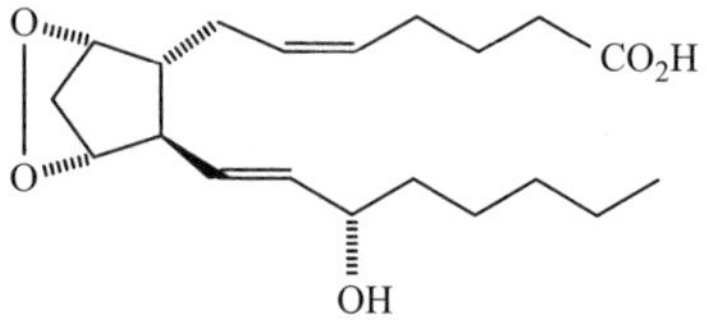

Prostaglandin Endoperoxide (8.63)

The biosynthetic pathway is given in figure 8.9. The endoperoxide PGH_2 then undergoes a variety of rapid changes. It can be isomerized to the various ketol derivatives, the so-called "primary" prostaglandins designated PGD_2, PGE_2, and PGF_2. The endoperoxide is also transformed into the extremely unstable and potent thromboxane A_2 (**8.64**, TXA_2) in blood platelets (thrombocytes). This compound has a half-life of only about 30 seconds, and its isolation and characterization were therefore an experimental *tour de force* of the Samuelsson group. Thromboxane A_2 is rapidly inactivated to the stable but inactive TXB_2.

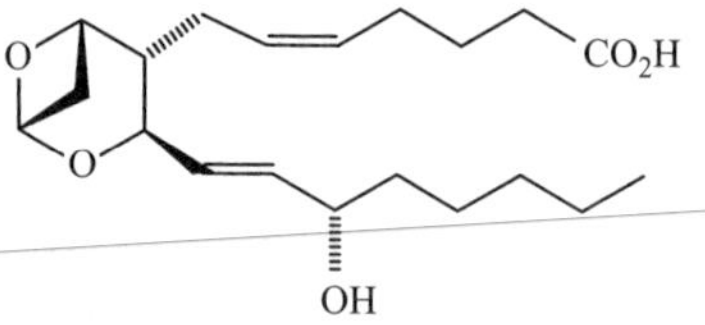

Thromboxane A_2 (8.64)

Another and equally important substance produced from the endoperoxide is prostacyclin (**8.65**, PGI2), which is synthesized in the walls of blood vessels. It has an additional tetrahydrofuran ring which is easily opened and deactivated. Prostacyclin has a half-life of less than 10 minutes.

Prostaglandins, prostacyclin, and thromboxane are considered *autocoids*, synthesized in many different organs and acting locally. They are not stored like neurotransmitters or conventional hormones but are continuously synthesized and released immediately into the circulation, where they are usually deactivated after only one passage through the lungs. The synthesis depends on the availability of the starting material, arachidonic acid, and is modulated by cAMP. Prostaglandin A, PGB, and PGC are inactive degradation products.

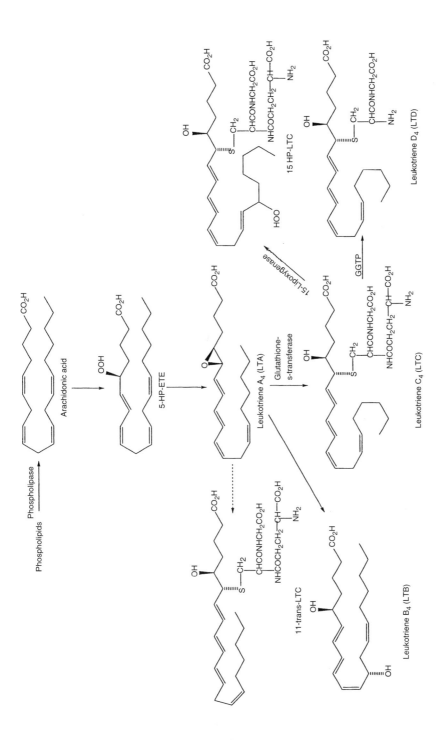

Figure 8.9 Prostaglandins and leukotrienes are potent eicosanoid lipid mediators, derived from phospholipase-released arachidonic acids, that are involved in numerous homeostatic biological functions and inflammation. They are generated by cyclooxygenase isozymes and 5-lipoxygenase, respectively, and their biosynthesis and pharmacological actions are inhibited by clinically relevant nonsteroidal anti-inflammatory drugs.

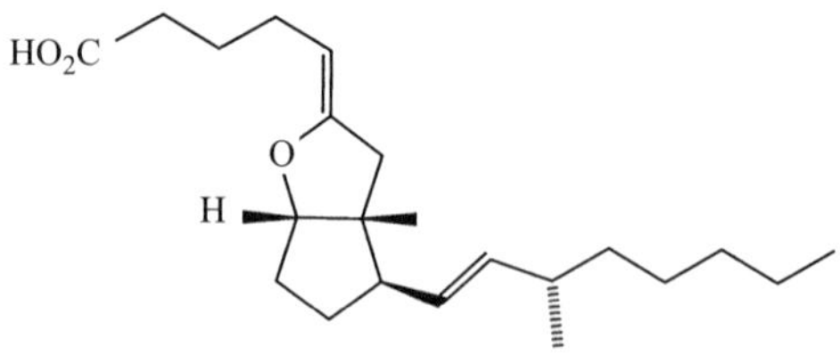

Prostacyclin (8.65)

8.5.1.2 Prostaglandins: Pharmacological Effects

The pharmacological effects of the prostaglandins and TXA_2 comprise many different activities—in fact too many. The lack of specificity of their activities implies a number of side effects which preclude the clinical application of several highly active natural prostaglandins, necessitating the development of selective synthetic compounds. The following effects of prostaglandins/eicosanoids are known and are summarized in table 8.3.

Table 8.3 Biologic Activities Associated with the Eicosanoids

Substance	Observed biologic Activity
PGD_2	Weak inhibitor of platelet aggregation
PGE_1	Bronchodilatation
	Inhibitor of fat breakdown
	Inhibitor of platelet aggregation
	Stimulates contraction of gastrointestinal smooth muscle
	Vasodilation
PGE_2	Elevates body temperature set-point in anterior hypothalamus
	Protects stomach lining acid degradation
	Reduces secretion of stomach acid
	Renal vasodilatation in kidneys
	Stimulates uterine smooth-muscle contraction
PGF_2	Stimulates uterine smooth-muscle contraction
PGI_2	Potent inhibitor of platelet aggregation
	Potent vasodilator
PGJ_2	Inhibits cell proliferation
	Stimulates osteogenesis (bone formation)
TXA_2	Potent inducer of platelet aggregation
	Potent vasoconstrictor
	Stimulates release of serotonin from platelets
LTB_4	Increases leukocyte chemotaxis and aggregation
LTC/D_4	Bronchoconstrictive
5- or 12-HPETE	Inhibits platelet aggregation
	Aggregates leukocytes
	Promotes leukocyte chemotaxis

Vasodilation and Constriction. PGE$_2$ and especially PGI$_2$ (prostacyclin) are powerful, short-acting vasodilators, probably involved in blood pressure regulation. Prostaglandin F$_2$ and TXA$_2$, on the other hand, are potent vasoconstrictors.

Blood Platelet Aggregation. Blood platelet aggregation is an all-important mechanism in normal blood-clot (thrombus) formation, and therefore also highly significant in the pathophysiology of cardiovascular disease, stroke, coronary occlusion, and other circulatory catastrophes. Thromboxane A$_2$, formed in platelets, promotes their aggregation, whereas PGI$_2$ has the opposite effect, just as in their vaso-activity. It seems that a very efficient homeostatic control system exists: the endoperoxide PGH$_2$, a precursor of both compounds, is converted in the platelets to TXA$_2$ but is used to produce PGI$_2$ in the blood vessel wall, which does not have thromboxane synthetase. Prostacyclin dilates the vessel and increases the cAMP concentration, which in turn reverses the platelet aggregation caused by TXA$_2$ (which inhibits adenylate cyclase).

Oxytocic Activity. The oxytocic activity of prostaglandins is used clinically. Prostaglandin E$_2$ can induce labor at term in pregnant women, while PGF$_2$ and its methyl ester are used for terminating pregnancies in the second trimester when administered by the intrauterine (intraamniotic) route. The activity of PGF$_2$ and its ester is probably due to a direct effect on uterine muscle, since in late pregnancy progesterone is already being produced by the placenta rather than by the *corpus luteum*. In earlier pregnancy, the PGF$_2$ causes abortion by luteolysis and a decreased production of progesterone. Abortions initiated in this way are safe, but gastrointestinal side effects (vomiting, diarrhea) are not uncommon.

Other Effects. Other effects of the prostaglandins include *bronchodilation* by the PGE series and constriction by PGF2, as well as antiulcer and antisecretory effects of some synthetic analogs in the stomach. It has been suggested that alcohol facilitates arachidonic acid release and subsequent PGE$_2$ synthesis, and is indeed responsible for hangover headaches. The analgesic tolfenamic acid (**8.66**) is alleged to have a preventive effect ... hope springs eternal.

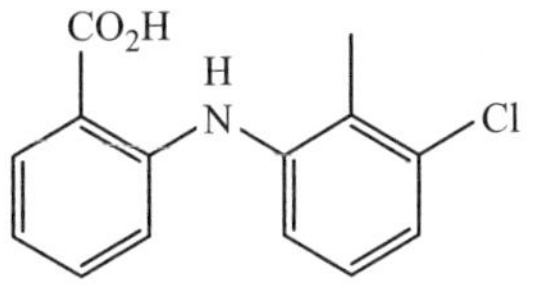

Tolfenamic acid (8.66)

8.5.1.3 Prostaglandin Receptors

Prostaglandin receptors have been demonstrated, even if the extent of binding does not always correlate with physiological activity. Prostaglandin binding sites have been reported in adipocytes, the corpus luteum, blood platelets, and the uterus, skin, stomach, and liver. The K_D values for these vary from 10^{-8} to 10^{-11} M, indicating a high affinity;

however, correlation of these values with the activation of adenylate cyclase is not very convincing. Prostaglandin receptors have also been found in plasma membranes of the corpus luteum, with a K_D of 5–8 × 10^{-8} M. There are at least nine known prostaglandin receptor forms, as well as several additional splice variants. The prostaglandin receptors belong to three clusters (on the basis of homology) within a particular subfamily of the G-protein receptor superfamily of seven transmembrane spanning proteins.

8.5.1.4 Prostaglandins: Structure–Activity Relationships

Studies of the structure–activity correlations and synthetic modifications of prostaglandins and thromboxane are extremely active fields. Although a large number of analogs have been synthesized, pharmacological information on these analogs is still unfolding.

Expansion or contraction of the cyclopentane ring or its replacement by heteroaromatic rings provides less active compounds in the E$_2$ and F$_2$ series. The replacement of the ring oxygen in prostacyclin gives more stable analogs, which show about half of the platelet aggregation-inhibiting effect of PGI$_2$ but act as a vasoconstrictor instead of a vasodilator. Analogs containing a nitrogen atom at the same position are very stable and good mimics of PGI$_2$.

The carboxylic acid side chain has also been modified extensively. The phenoxy derivative is 10 times more active than PGE as an inhibitor of platelet aggregation, but is a weak smooth-muscle spasmogen. The sulfonamide is 10–30 times more active than PGE$_2$ as an antifertility agent, and has few side effects. This compound has also undergone co-chain (carbinol chain) modification. Successful derivatives were mainly found among the 15- and 16-methyl PGE$_2$ analogs: carboprost (8.67) possesses very high uterotonic activity and is used to induce abortions. They probably inhibit the dehydrase that inactivates the prostaglandins by removing the 15-OH group. Misoprostol (8.68) is a modified prostaglandin analog that shows potent gastric antisecretory and gastroprotective effects. It has been used to treat and prevent gastric ulcers.

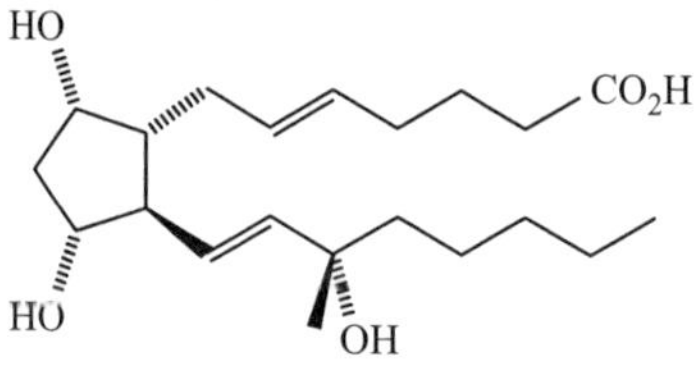

Carboprost (8.67) 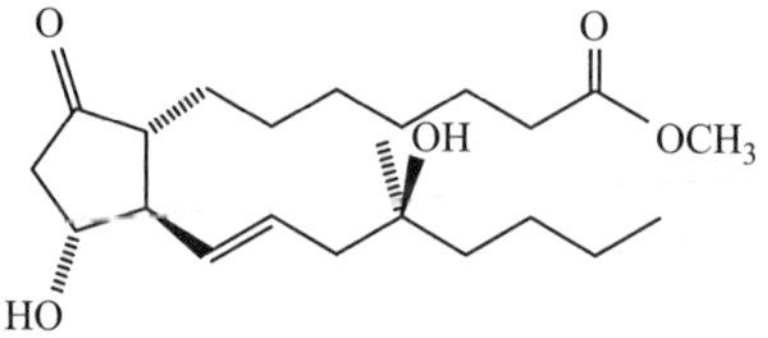   Misoprostol (8.68)

8.5.2 Anti-Prostaglandins as Anti-Inflammatory Agents and Minor Analgesics

Anti-inflammatory agents are believed to act by disrupting the arachidonic acid cascade. These drugs are widely used for the treatment of minor pain and arthritis. Some of them are antipyretics (drugs that reduce fever) in addition to having analgesic and anti-inflammatory actions. Some of these agents are widely available in over-the-counter (OTC) preparations. Collectively, these are referred to as nonsteroidal anti-inflammatory drugs (NSAIDs). They may be classified as follows:

1. NSAIDs: Non-selective cyclo-oxygenase inhibitors
   a. Arylanthranilic acids (mefenamic acid, meclofenamate)
   b. Arylbutyric acids (nabumetone)
   c. Arylpropionic acids (ibuprofen, ketoprofen, fenoprofen, naproxen, ketorolac)
   d. Indene derivatives (sulindac)
   e. Indole derivatives (indomethacin)
   f. Naphthylacetic acid derivatives (nabumetone)
   g. Oxicams (piroxicam, meloxicam, tenoxicam)
   h. Phenylacetic acid derivatives (diclofenac)
   i. Phenylalkanoic acid derivatives (flurbiprofen)
   j. Pyrazolone derivatives (phenylbutazone, azapropazone)
   k. Pyrrolealkanoic acid derivatives (tolmetin)
   l. Salicylate derivatives (aspirin, diflunisal)

2. NSAID: Selective COX-2 inhibitors
   a. Coxibs (celecoxib, rofecoxib)

This list provides examples for each category. There are other analogs as well, reflecting the widespread use of these agents.

The nonsteroidal anti-inflammatory agents block the cyclooxygenase enzyme that catalyzes the conversion of arachidonic acid to the prostaglandins PGG_2 and PGH_2. Since these two cyclic endoperoxides are the precursors of all other prostaglandins, the implications of cyclooxygenase inhibition are significant. Prostaglandin E_1 is known to be a potent *pyrogen* (fever-causing agent), and PGE_2 causes pain, edema, erythema (reddening of the skin), and fever. The prostaglandin endoperoxides (PGG_2 and PGH_2) can also produce pain, and inhibition of their synthesis can thus account for the action of the nonsteroidal anti-inflammatory agents.

Most of the nonsteroidal anti-inflammatory drugs (NSAIDs) are carboxylic acids. Aspirin (**8.69**) (acetylsalicylic acid, ASA) has been used since the turn of the last century to reduce pain and fever, but the parent compound, salicylic acid, has been known and used since antiquity, owing to its common occurrence as a glycoside in willow bark. Acetylation merely decreases its irritating effect. Among the numerous other salicylates known and used, flufenisal (**8.70**) has a longer duration of activity and fewer side effects than aspirin. Mefenamic acid (**8.71**) and flufenamic acid (**8.72**) are derivatives of anthranilic acid, while ibuprofen (**8.73**) and naproxen (**8.74**) are derivatives of phenylacetic and naphthylacetic acids, respectively.

Among indole derivatives, indomethacin (**8.75**) is very widely used despite side effects. Its indene analog sulindac (**8.76**) is a pro-drug, the active form being its —SH derivative. Piroxicam (**8.77**) is a long-lasting anti-rheumatoid agent but can have serious gastrointestinal side effects. The once widely used phenylbutazone (**8.78**) derivatives have too many side effects and have fallen into disrepute.

Among the nonselective cyclooxygenase (COX) inhibitors, there is no clear-cut statistical evidence for the superiority of one or another of these useful drugs. Individual patients may do better with some than with others, and there are differences in side effects, primarily gastric bleeding and renal toxicity, which can be especially serious with the prolonged administration of high doses—necessary in chronic diseases such as

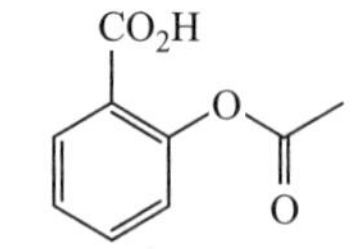

Aspirin (8.69)

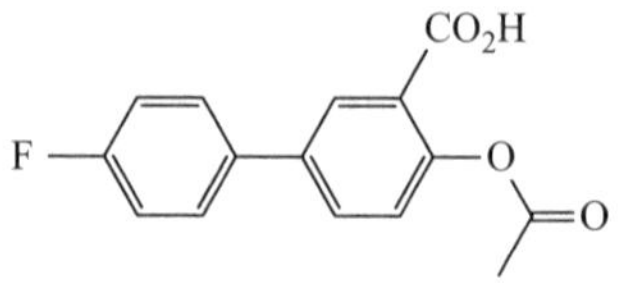

Flufenisal (8.70)

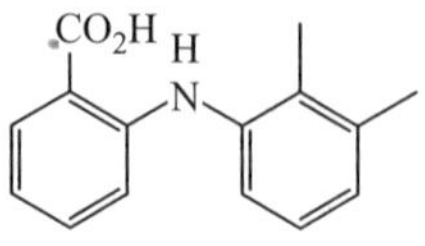

Mefenamic acid (8.71)

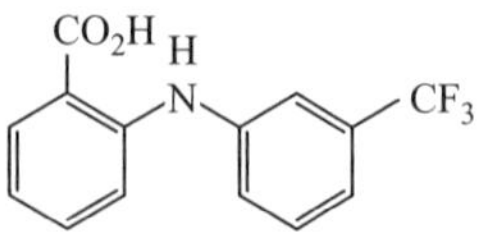

Flufenamic acid (8.72)

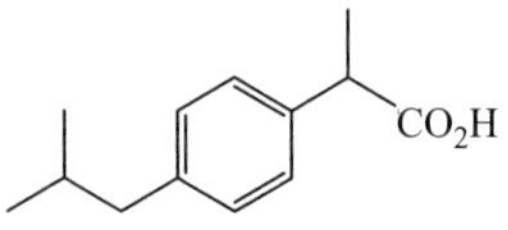

Ibuprofen (8.73)

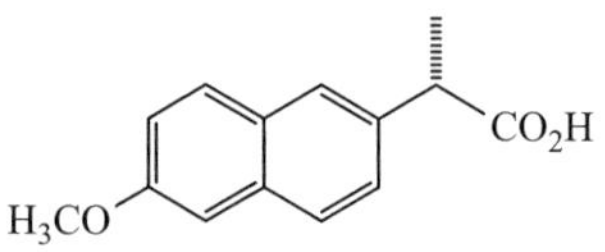

Naproxen (8.74)

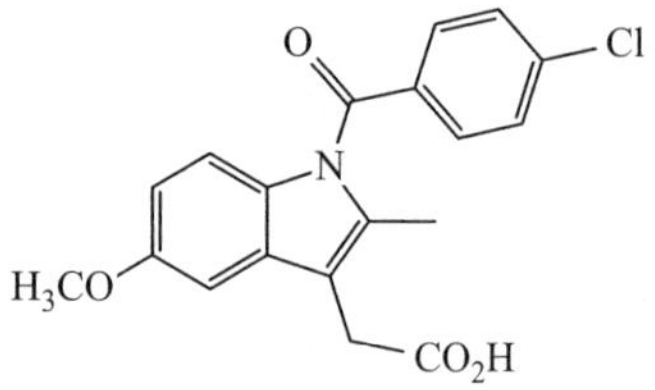

Indomethacin (8.75)

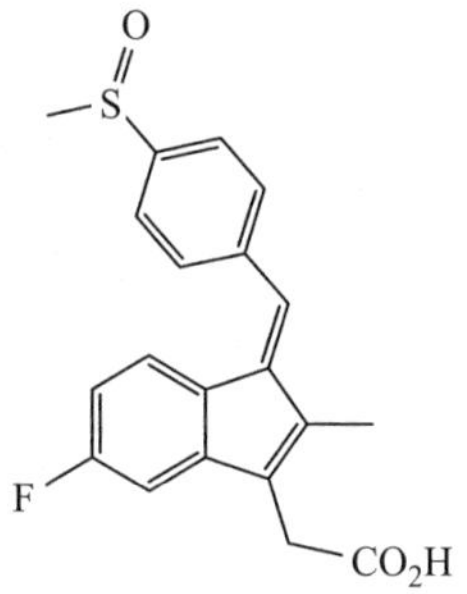

Sulindac (8.76)

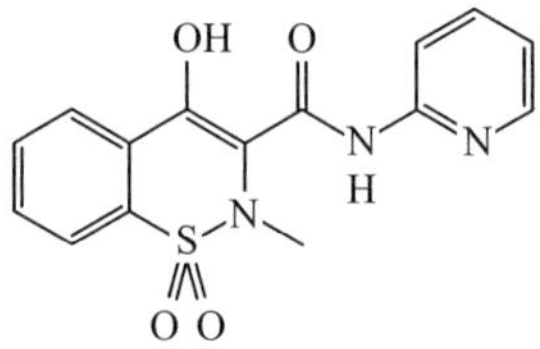

Piroxicam (8.77)

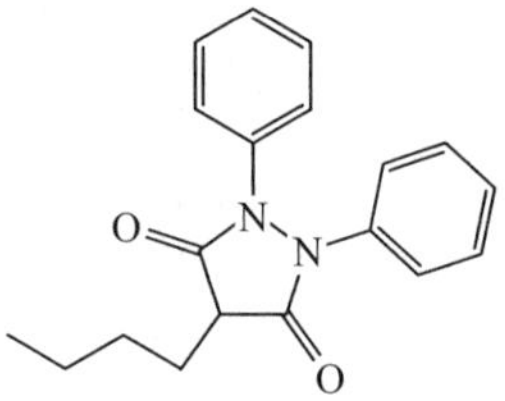

Phenylbutazone (8.78)

rheumatoid arthritis. Some of these compounds are powerful enough to be effective against the major pain caused by malignancies.

In theory the selective COX-2 inhibitors may offer some advantages. There are at least two forms of the COX enzyme: COX-1 and COX-2. COX-1 is located in most tissues, including stomach and kidneys. COX-2, on the other hand, is undetectable in most tissues under physiological conditions. COX-2 is induced at sites of inflammation, secondary to the effects of cytokines. Accordingly, specific COX-2 inhibitors, such as celecoxib (**8.79**) and rofecoxib (**8.80**), are less likely to cause stomach ulceration and gastrointestinal bleeding. Nevertheless, they still have the capacity to cause fluid retention and renal damage. More significantly, in 2004 it was recognized that COX-2 inhibitors may pose an increased cardiovascular risk. For example, rofecoxib was suddenly withdrawn from the market when it was noted that there was an increased relative risk for confirmed cardiovascular events, including heart attack and stroke, beginning after 18 months of treatment, compared to placebo.

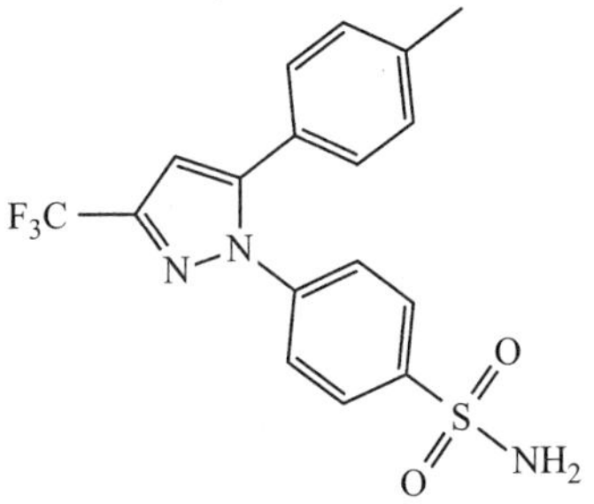

Celecoxib (8.79)

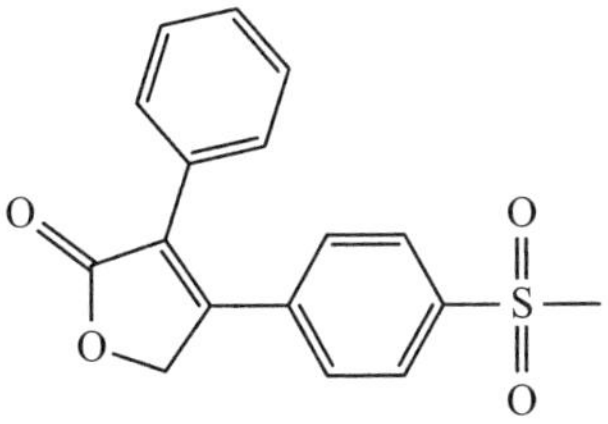

Rofecoxib (8.80)

8.5.3 Anti-Leukotrienes as Anti-Inflammatory Agents for Asthma

Like the other eicosanoids, the leukotrienes are also involved in inflammatory diseases. Accordingly, leukotriene pathway inhibitors are emerging as important agents. Two fundamental approaches to the design of leukotriene pathway inhibitors have been pursued:

1. Inhibitors of 5-lipoxygenase (to prevent leukotriene biosynthesis)
2. Leukotriene receptor antagonists (to prevent leukotriene binding)

Zileuton (**8.81**) is an inhibitor of 5-lipoxygenase; montelukast (**8.82**) and zafirlukast (**8.83**) are inhibitors of the leukotriene LTD_4 receptor. All of these agents have demonstrated efficacy in the treatment of asthma, a common chronic inflammatory disease of the airways.

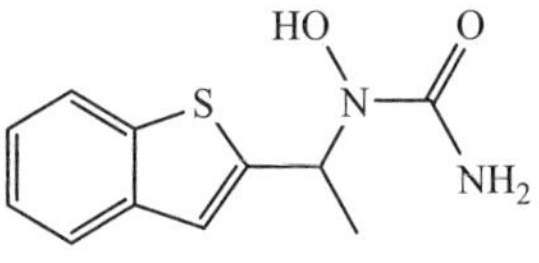

Zileuton (8.81)

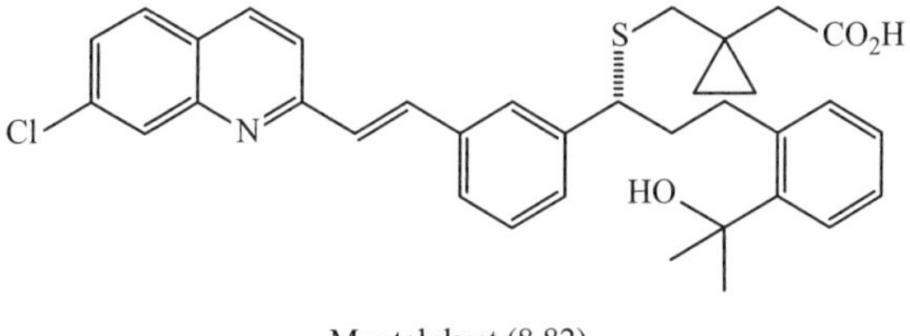

Montelukast (8.82)

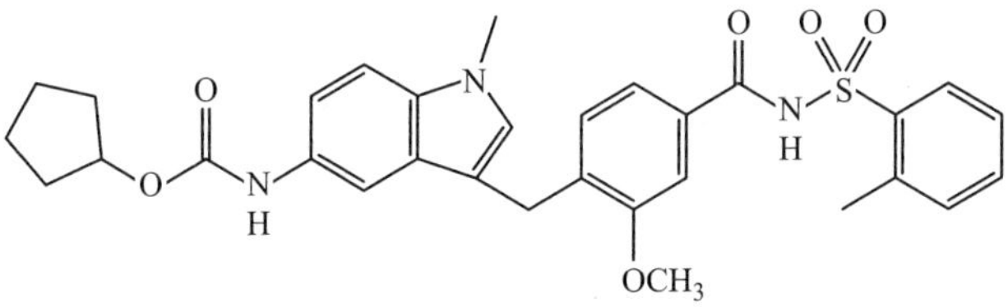

Zafirlukast (8.83)

8.6 CARBOHYDRATES AS DRUGS AND DRUG DESIGN TARGETS

Carbohydrates are underexploited as receptors against which to target drugs and under-utilized as leads in drug discovery. The role of carbohydrates has regrettably been generalized to functioning primarily as energy stores. The functional and structural role of carbohydrates has been somewhat neglected. Moreover, the capacity of carbohydrates to provide large numbers of molecules with diversity and complexity has not been fully explored. For example, when two identical amino acids or nucleotides are connected, they can produce just one dipeptide or dinucleotide; but when two identical saccharides are joined, 11 different disaccharides are possible. In modern medicinal chemistry, the ability to generate families with molecular diversity is an asset; carbohydrates possess such ability.

Over the years, few carbohydrates have emerged as drug candidates or drug additives. Of the sugar alcohols, sorbitol (**8.84**) produces sweet and viscous solutions that are used in the formulation of pharmaceutical preparations such as cough syrups. Mannitol (**8.85**), on the other hand, is used as an actual drug. Since mannitol creates an osmotic gradient within the proximal tubules, distal tubules, and collecting ducts of the kidney, it functions as a potent diuretic. This strong diuretic property has resulted in mannitol being used to treat *increased intracranial pressure.* The volume of the human skull is fixed; the bones of the skull are not expandable. When a brain hemorrhage or some other brain pathology uses volume within the skull, the contents of the skull become more crowded, producing increased pressure within the cranium. A diuretic, such as mannitol, dehydrates the brain temporarily, making more room within the skull and decreasing intracranial pressure. Starch, a high-molecular-weight carbohydrate composed of amylose (20%) and amylopectin (80%), is used as an excipient additive in drug tablet formulation. Ferrous gluconate (**8.86**) is a carbohydrate salt of iron that is used (for its iron content, rather than its carbohydrate content) to treat iron deficiency anemia.

The single most important drug molecule based on a carbohydrate structure is heparin. Heparin is a heterogeneous mixture of sulfated mucopolysaccharides; it is composed of

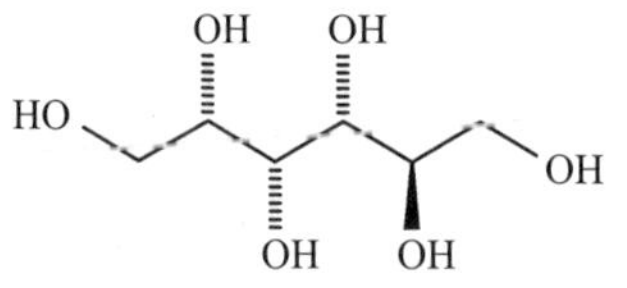

Sorbitol (8.84) 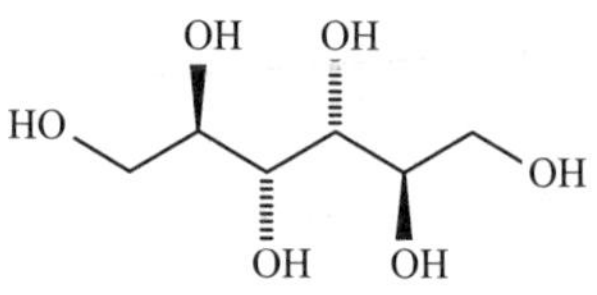   Mannitol (8.85)

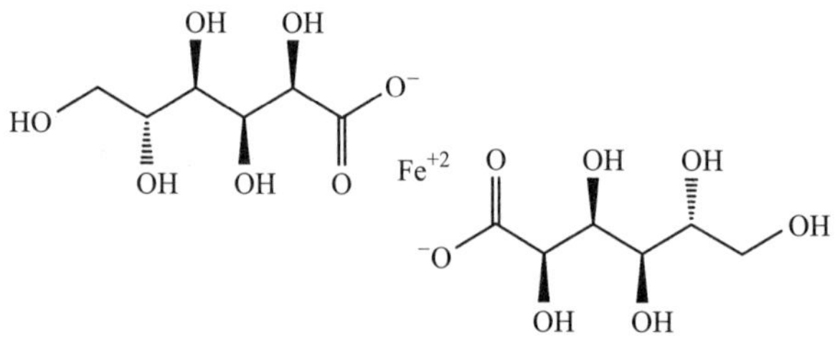

Ferrous Gluconate (8.86)

α-D-glucuronic acid, α-L-iduronic acid, α-D-glucosamine, and *N*-acetyl-α-D-glucosamine. These monosaccharide units are partially sulfated and linked into a polymer via 1–4 linkages. The pharmacological activity of heparin is dependent upon a protein, antithrombin III (AT III), which occurs naturally within the blood plasma. Antithrombin III inhibits enzymes that are called clotting factor proteases. In doing so, antithrombin III "thins the blood," acting as an anticoagulant. Heparin binds to antithrombin III, producing a conformational change that better exposes the antithrombin active site; in doing so, heparin catalyzes the antithrombin–protease reaction without itself being consumed. Through this mechanism, heparin acts as a potent, intravenously administered anticoagulant (warfarin is the principal orally administered anticoagulant). Low-molecular-weight (LMW) heparins are fragments or fractions of commercial-grade heparin. LMW heparins have higher anticoagulant potency in whole blood than standard heparin. Enoxaparin and dalteparin are two LMW heparins that are clinically employed.

Although carbohydrates have not traditionally enjoyed a favored status as drug discovery leads, there is room for optimism concerning the future. Research on carbohydrates is undergoing considerable growth. Computational chemistry (molecular mechanics and *ab initio* molecular orbital calculations), as applied to carbohydrate chemistry, is improving. Likewise, progress is being made in the synthesis of peptides. Many carbohydrates of current medical interest are expressed on cell surfaces as glycoconjugates, including glycolipids, glycosaminoglycans, and glycoproteins. These various cell-surface glycoconjugates help to promote cell–cell recognition and cell–cell adhesion.

In cell–cell adhesion, *lectins* are important; these are carbohydrate binding proteins. *Selectins* are a family of glycoprotein lectins that are implicated in the adhesion of white blood cells and platelets to the lining of blood vessels. As such, they play a role in blood clotting and inflammation. Drugs that target selectin receptors are being studied as putative immunological and anti-inflammatory agents. Since adhesion is also important to the spread of cancer cells through the process of metastasis, drugs that target specific cell-surface carbohydrates may also be of therapeutic value in the future.

8.7 HETEROCYCLES AS DRUGS AND DRUG DESIGN TARGETS

Drug molecules should be small, conformationally constrained structures that are rich in functional groups (e.g., ammonium, carboxylate) and/or hetero-atoms (e.g., N, O, S) that are capable of establishing energetically favorable intermolecular interactions between the drug and its receptor. Not surprisingly, heterocyclic compounds are ideal drug candidates. As discussed at the beginning of this chapter, heterocycles are cyclic organic molecules that contain heteroatoms other than simply carbon and hydrogen. The organic chemistry classification of heterocycles was given in section 8.1.6. From a medicinal chemistry perspective, heterocycles may be functionally divided as follows:

1. Endogenous heterocycles
2. Natural product exogenous (exogenous to humans, but endogenous to other life forms) heterocycles
3. Semisynthetic heterocycles
4. Synthetic heterocycles

Each of these four categories is rich in drug discovery lead candidates.

This chapter focuses primarily on endogenous molecules as nonmessenger targets for drug design. There are many such molecules. As discussed earlier, amino acid derivatives, lipids (eicosanoids), nucleoside/nucleotide derivatives, and carbohydrates all afford heterocyclic leads for drug design. The discussion will not be repeated here.

In addition to endogenous heterocycles, there are also medically important exogenous heterocycles. Nature is a great source of molecular diversity, especially for bioactive molecules. Nature provides a rich source of peptidic (penicillin), lipid (terpenes), and other (alkaloid) heterocyclic natural products. These compounds are produced in plants or nonhuman animals, but may exert profound biological effects when administered to humans.

Heterocycles which are not biosynthesized in humans, but which are natural products produced by other life forms, are very important in the history of drug design. This is particularly true of alkaloids containing a piperidine ring. These include coniine (**8.87**, extracted from poison hemlock, *Conium maculatum*, a member of the Umbelliferae carrot family), atropine (from *Atropa belladonna* and other genera of the Solanaceae plant family; the plant was called *belladonna* ["beautiful woman"] since it was used by

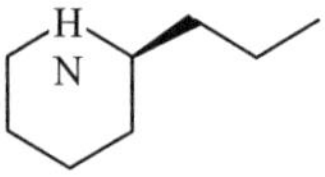

Coniine (8.87)

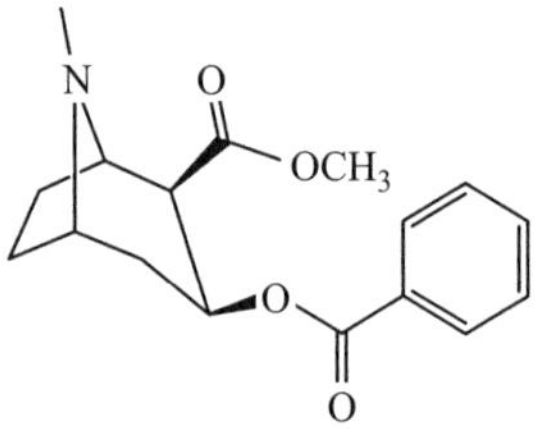

Cocaine (8.88)

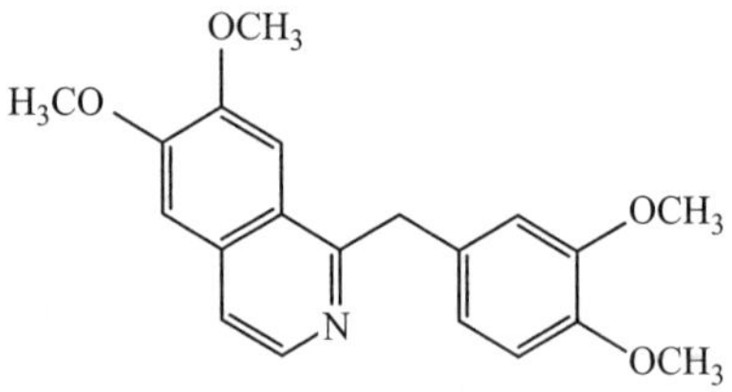

Papaverine (8.89)

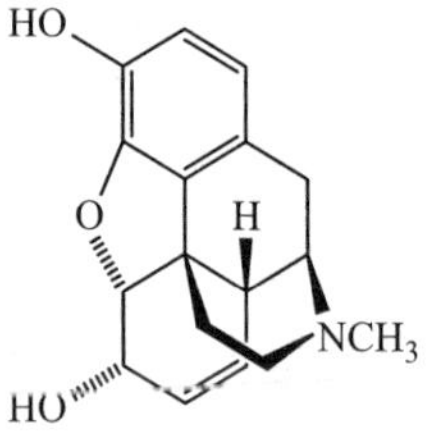

Morphine (8.90)

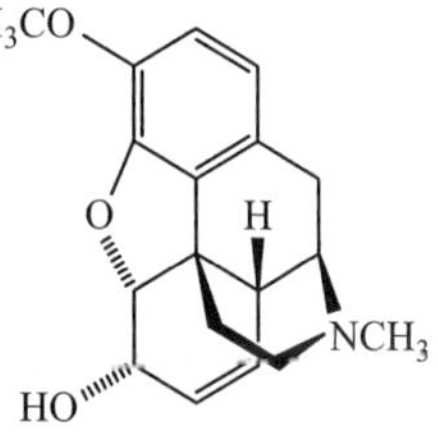

Codeine (8.91)

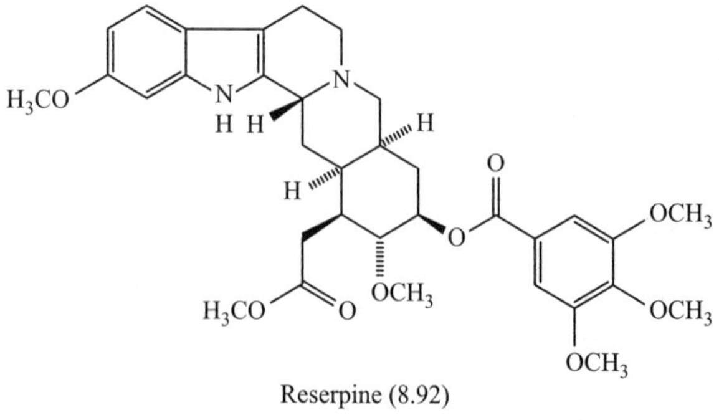

Reserpine (8.92)

women to make their ocular pupils look darker and thus supposedly more attractive), and cocaine (**8.88**, from *Erythroxylon coca*). In medicinal chemistry, cocaine was the starting point in the design of many local anesthetic agents, including lidocaine and procaine. Alkaloids such as atropine and cocaine contain a pyrrolidine ring that is bridged by three carbon atoms between the second and fifth carbons; hence, they are sometimes referred to as tropane alkaloids. Alkaloids containing an isoquinoline or reduced isoquinoline ring are likewise medically important. Papaverine (**8.89**), morphine (**8.90**), and codeine (**8.91**) are all alkaloids obtained from the opium poppy, *Papaver somniferum*. Papaverine has an isoquinoline ring; morphine and codeine contain a partially reduced isoquinoline ring. These compounds have played a central role in the design of analgesic agents for the treatment of pain. Alkaloids containing indole rings have also been used medically. Reserpine (**8.92**), obtained from the Indian snakeroot plant (*Rauwolfia serpentina*), was used in aboriginal medicine for centuries as a "tranquilizer" and in modern medicine as an agent to treat systemic arterial hypertension (high blood pressure).

Semisynthetic heterocycles are also important drug molecules. These compounds attempt to capture the best of both worlds, being synthetic derivatives of natural products. The use of a natural product in the preliminary stages of the synthesis enables the elimination of numerous costly synthetic steps. The subsequent synthetic modifications enable further fine tuning of the natural product pharmacophore. There are a number of semisynthetic penicillin derivatives available. Similarly, there are also semisynthetic hormone analogs, especially of estrogens and gestagens.

Finally, there are numerous purely synthetic heterocycles. These are discussed in many places throughout this book. The hydantoin ring of phenytoin and the barbiturate ring of phenobarbital are good examples of these. There are a variety of drugs containing pyrrolidine, furan, pyrazole, pyridine, and indole rings.

8.7.1 Porphyria and Diseases Influencing Heterocycle-Based Drug Design

Porphyrins are endogenous heterocycles formed by the linkage of four pyrrole rings through methylene bridges. As discussed previously, porphyrins play an important role in crucial biomolecules, such as hemoglobin or myoglobin. *Porphyria* is a disorder of porphyrin metabolism that may be either inherited or acquired; thus, porphyria embraces a group of diseases, each with unusual and characteristic manifestations, which have in common the excessive excretion of one or more of the porphyrins and/or porphyrin biosynthetic precursors.

Of the various types of porphyria, one type (*acute intermittent porphyria*, AIP) is relevant to medicinal chemistry and drug design. Clinically, AIP is characterized by periodic attacks of intense abdominal pain, severe nausea and vomiting, psychotic behavior, disturbances of neuromuscular transmission (which may even lead to quadriplegia), and occasionally death. In AIP, the basic defect lies in an enzyme called porphobilinogen deaminase. Acute and potentially life-threatening attacks of AIP can be brought on by certain drugs. The drugs that precipitate porphyria are inducers of hepatic cytochrome P-450, an important heme-containing enzyme found in the liver. Drugs that have been implicated in porphyria exacerbations are structurally diverse and include barbiturates, carbamazepine, chloramphenicol, diazepam, diclofenac, diphenhydramine, enalapril, erythromycin, flucloxacillin, furosemide, imipramine, nifedipine, phenytoin, and sulfonamides.

Porphyria is not the only disorder that may be inadvertently precipitated by the administration of a drug. *Malignant hyperthermia* is a serious, life-threatening complication of general anesthesia with halothane, methoxyflurane, and succinylcholine. It occurs in 1 in 20,000 people. Clinically, it is characterized by high body temperature (41°C), muscle rigidity, and cardiovascular collapse.

8.8 INORGANIC SUBSTANCES AS DRUGS AND DRUG DESIGN TARGETS

Traditionally, drug design has been practiced predominantly by organic chemists; hence, most drug molecules are created primarily from four elements—carbon, oxygen, hydrogen, and nitrogen—with the occasional use of sulphur. Regrettably, inorganic substances (as either inorganic salts or organo-metallic complexes) have been largely ignored as either drug targets or drugs themselves. The neglect of inorganic substances as therapeutics arises in part from concerns regarding toxicity and in part from the fact that most medicinal chemists are organic chemists rather than inorganic chemists. Arguably, this neglect of inorganic substances is regrettable since inorganics may offer novel avenues to new therapeutics, especially for indications in which the drug is administered for a short acute period (e.g., antibiotics) rather than for prolonged chronic use.

When they are used medically, inorganic substances are given more commonly as salts and only rarely, if ever, as the elemental metal. An inorganic salt consists of the metallic cation and the counter anion. Within the human organism, inorganic salts play a variety of normal roles, both functional and structural. As discussed in section 8.1.7, from a functional perspective the metallic cations Na^+, K^+, and Ca^{2+} are essential to the transmembrane transmission of information via voltage-gated ion channels. Metallic ions (e.g., Zn^{2+}, Cu^{2+}) are also found in some enzymes, termed metallo-enzymes; such enzymes employ trace amounts of metal ion to aid their catalytic function. The important role of these trace metallic ions was appreciated in the early days of TPN (total parenteral nutrition) when individuals received complete nutritional support via intravenous solutions; the essential role of metallic ions was demonstrated when they were eventually included in the TPN solutions. Metallic ions also play crucial structural roles. This is most evident in the role of Ca^{2+} salts in the structure of bone. Clearly, there are many opportunities to exploit inorganic substances as either drugs or drug receptors.

8.8.1 Inorganic Substances as Drugs

Inorganic salts of metals such as gold, lithium, and bismuth have demonstrated utility as therapeutics for a variety of disorders.

8.8.1.1 Antimony

Trivalent antimonials (**8.93**) and pentavalent antimonials (**8.94**) are sometimes used as antiparasitic agents against leishmaniasis and schistosomiasis. Although the mechanism

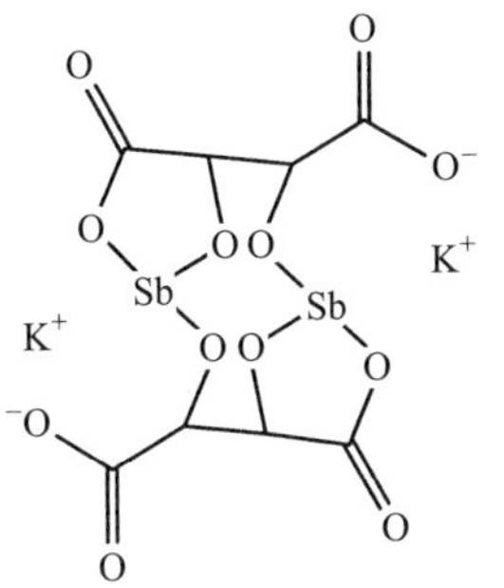

Trivalent Antimonials (8.93)

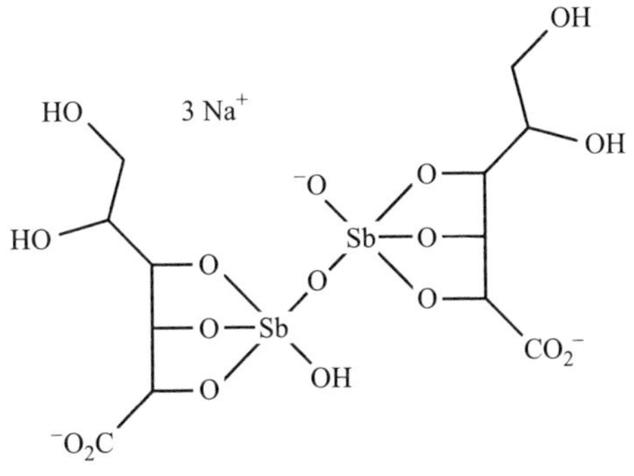

Pentavalent Antimonials (8.94)

of action is unknown, they may function by inhibition of the enzyme phosphofructokinase within the parasite.

8.8.1.2 Bismuth

Bismuth compounds have been used in the treatment of peptic ulcer disease. They function by selective binding to the ulcer, coating it and shielding it from the effects of gastric acid. Bismuth may also have activity against bacteria such as *Helicobacter pylori*, shown to be a causative factor in peptic ulcer disease of the stomach. Bismuth has been administered as bismuth subsalicylate (**8.95**) or tripotassium dicitrato bismuthate.

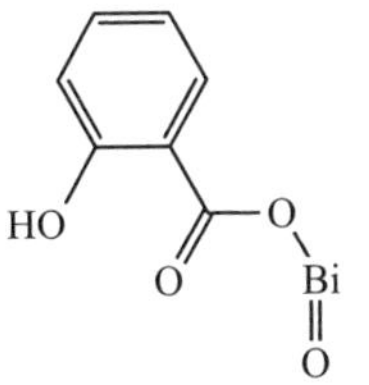

Bismuth Subsalicylate
(8.95)

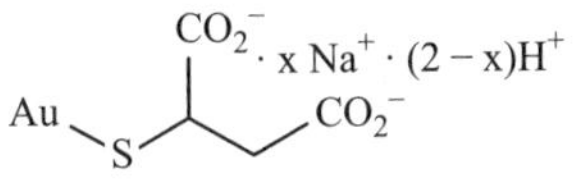

Sodium Aurothiomalate
(8.96)

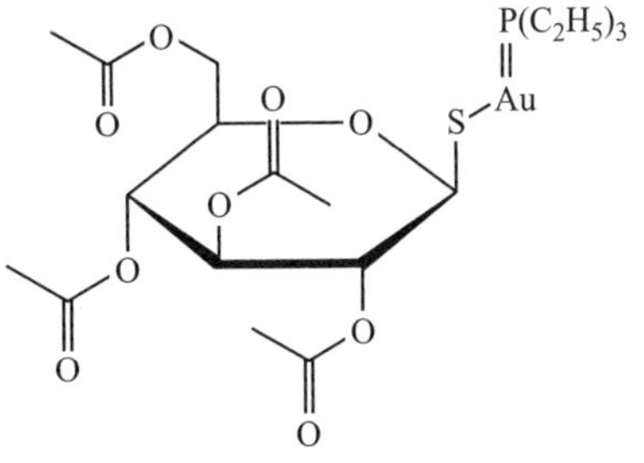

Auranofin (8.97)

8.8.1.3 Gold

Gold salts are used occasionally in the treatment of severe active rheumatoid arthritis. Its use is not widespread because of toxicity concerns. Although the precise mechanism of action is unknown, gold is taken up by macrophages, inhibiting phagocytosis and lysosomal enzyme activity; in doing so, gold suppresses inflammation and immune reactions. Clinically, two gold salts are used: sodium aurothiomalate (**8.96**) and auranofin (**8.97**). The former is administered intramuscularly, whereas the latter is given orally. More than 40% of people receiving gold therapy will have side effects, with 10% of patients experiencing serious side effects which include diarrhea, oral ulcerations, kidney damage with protein in the urine (proteinuria), and bone marrow suppression (aplastic anemia).

8.8.1.4 Lithium

Lithium salts are used in the treatment of bipolar affective disorder (i.e., manic depression) and occasionally in mania (but its slow onset of action is somewhat of a disadvantage in this case). Its mechanism of action is still open to debate, but lithium has effects on brain monoamines, on neuronal transmembrane sodium flux, and on cellular phosphatidylinositides related to second messenger systems. Lithium is administered in two salt forms, lithium carbonate (**8.98**) and lithium citrate (**8.99**). Side effects are common and include diarrhea, kidney failure, and drowsiness with tremor.

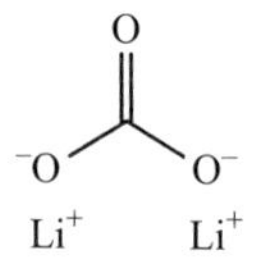

Lithium Carbonate (8.98) 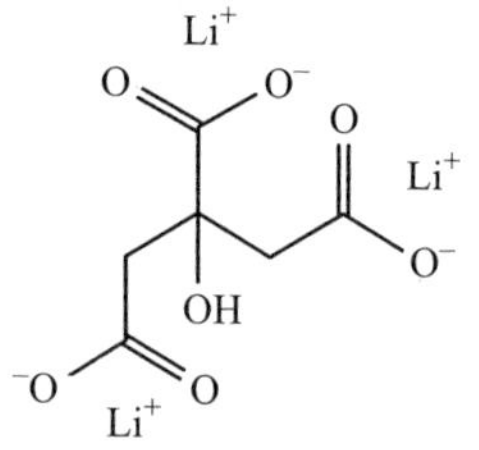   Lithium Citrate (8.99) 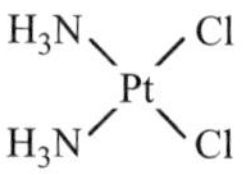   Cis-Platinum (8.100)

8.8.1.5 Platinum

Platinum complexes (e.g., *cis*-platinum, **8.100**) are used as antineoplastic agents in the treatment of cancer. Their use is described in chapter 7.

8.8.1.6 Selenium

Selenium sulfide (**8.101**) is used dermatologically as an antiseborrheal agent. Selenium acts as a cytostatic agent to decrease epidermal turnover rates in the skin.

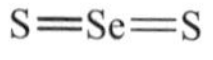

Selenium Sulfide   Zinc Acetate   Zinc Sulfate   Zinc Pyrithione
(8.101)   (8.102)   (8.103)   (8.104)

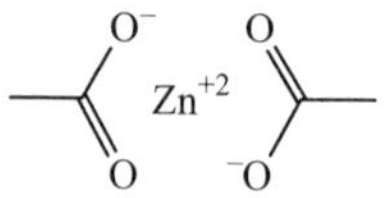

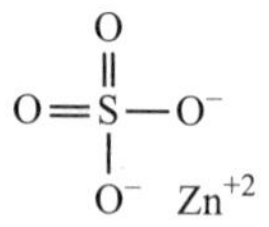

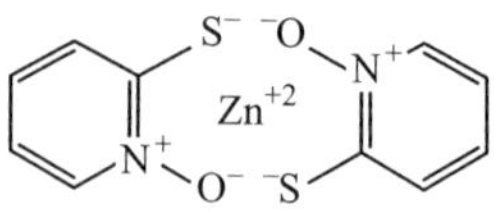

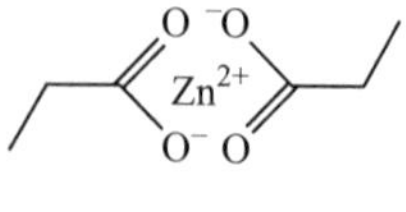

Zinc Propionate (8.105) 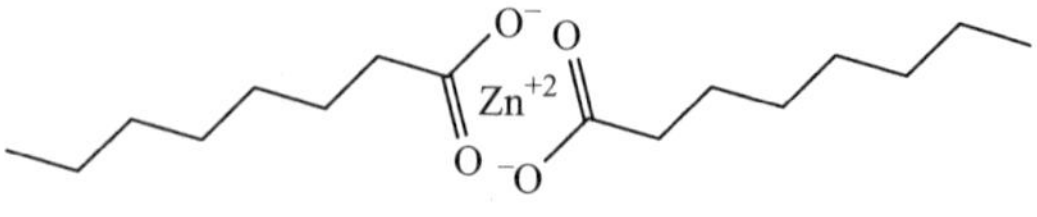   Zinc Caprylate (8.106)

8.8.1.7 Zinc

Zinc is used for a variety of indications. Zinc acetate (**8.102**) or, rarely, zinc sulfate (**8.103**) have been used orally to treat Wilson's disease, a recessively inherited disorder of copper metabolism, characterized by brain and liver dysfunction arising from excessive deposits of copper. Zinc pyrithione (**8.104**) is used in shampoos to treat seborrhea. Zinc propionate (**8.105**) and zinc caprylate (**8.106**) have been used as topical antifungal agents.

8.8.1.8 Other Metals

Of historical interest, arsenical antibiotics (**8.107**) were once used to treat syphilis and mercurial diuretics (**8.108**) were once used to treat edema and to promote water excretion. Due to their high toxicity, neither is currently used.

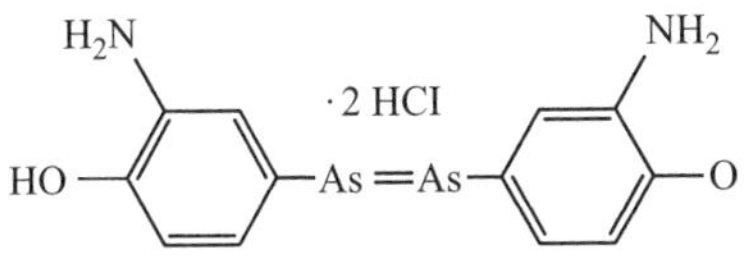

Arsphenamine (8.107) 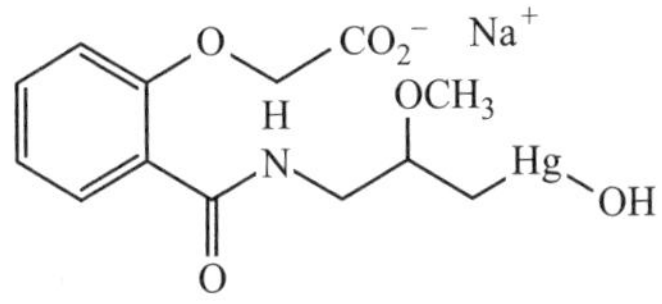   Mersalyl (8.108)

8.8.2 Inorganic Substances as Drug Receptors

There are several structures in which inorganic substances may act as receptors for drugs.

8.8.2.1 Endogenous Organo-Metallic Macromolecules as Drug Receptors

Metallo-Enzymes. These are enzymes that have a metal atom at their active site. This atom may play a role as a point of contact with the drug pharmacophore. A good example of this is provided by the discovery of angiotensin-converting enzyme (ACE) inhibitors for the treatment of high blood pressure. The ACE enzyme contains a zinc atom. Drug design of enzyme inhibitors, such as captopril, has targeted the zinc through the binding process.

Zinc Fingers. A zinc finger is part of a protein that binds to DNA, forming a complex stabilized in part by the presence of a zinc cation. A zinc finger typically has two β-sheets, each with a cysteine amino acid residue, and an α-helix with two histidine residues. These four amino acids form the chelating moiety that holds the zinc ion. Since zinc fingers are proteins that bind to DNA, their manipulation has the capacity to influence genome expression. Zinc fingers have been considered as targets for breast cancer therapy development.

8.8.2.2 Inorganic Salts as Drug Receptors

About 98% of the 1.5 kg of calcium and 85% of the 1 kg of phosphorus in the human adult are found in bone. Bone is composed of two distinct tissue structures: cortical (compact) bone and trabecular (cancellous) bone. More than 80% of the skeleton is composed of cortical bone, a dense tissue that is 90% calcified. Inorganic and organic components are both present in bone; the highly dense, crystalline inorganic component is hydroxyapatite. The principal calcium salt contained in the hydroxyapatite crystalline lattice is $Ca_{10}(PO_4)_6(OH)_2$. Although primarily an inorganic material, bone is dynamic in its structure, undergoing a continuous process of resorption and formation. Approximately 3% of cortical bone is remodeled yearly. Two different cells assist this remodeling process: *osteoblasts* permit new bone to be laid down; *osteoclasts* enable old bone to be reabsorbed.

A variety of diseases can affect bone and its structure. *Paget's disease*, for example, is a disorder arising from abnormal osteoclasts, characterized by excessive bone resorption followed by replacement of the normal mineralized bone with structurally weak, poorly mineralized tissue. However, the most important bone disease is *osteoporosis*. This is a skeletal bone disease characterized by microarchitectural deterioration of bony tissue and loss of bone mass, yielding increased susceptibility to bone fracture and bone fragility. In the United States, osteoporosis results in 1.5 million bone fractures annually, with 250,000 of these being hip fractures that sometimes ultimately culminate in patient death. There is a variety of therapies for the prevention and treatment of osteoporosis.

Bisphosphonates. Bisphosphonates are synthetic compounds designed to function as mimics of pyrophosphate, in which the oxygen atom in P-O-P is replaced with a carbon atom, creating a non-hydrolyzable backbone structure. The bisphosphonates selectively bind to the hydroxyapatite portion of the bone, decreasing the number of sites along the bone surface at which osteoclast-mediated bone resorption can occur. This permits the osteoblasts to lay down well-mineralized new bone without competition from osteoclasts. Clinically employed bisphosphonates include etidronate (**8.109**), tiludronate (**8.110**), risedronate (**8.111**), alendronate (**8.112**), and pamidronate (**8.113**).

Inorganic Calcium Salts. Adequate intake of Ca^{2+} salts in adolescence and early adulthood should increase bone mineral density, subsequently, in theory, reducing risk

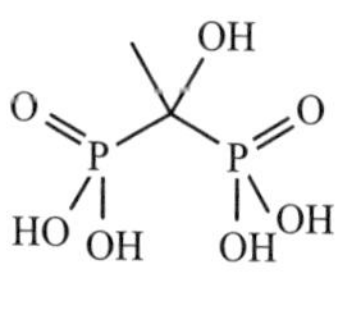

Etidronate (8.109) 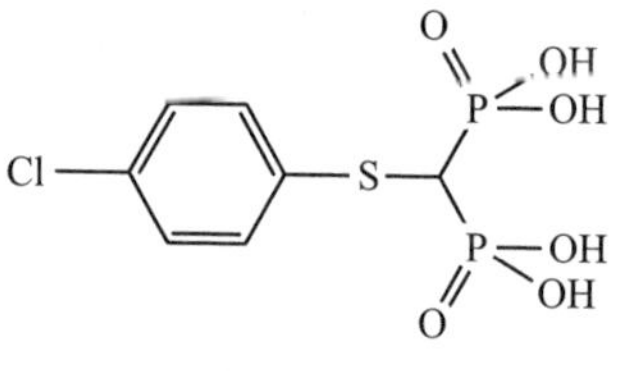   Tiludronate (8.110) 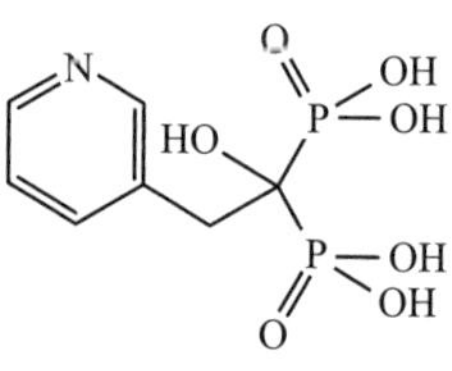   Risedronate (8.111)

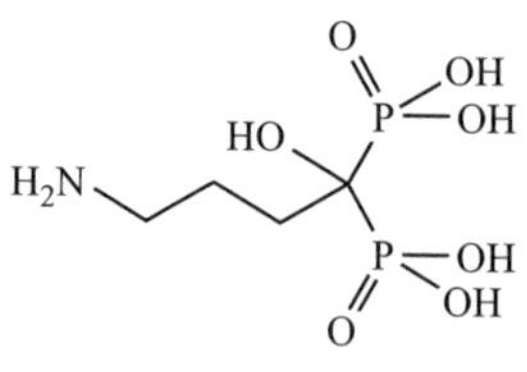

Alendronate (8.112) 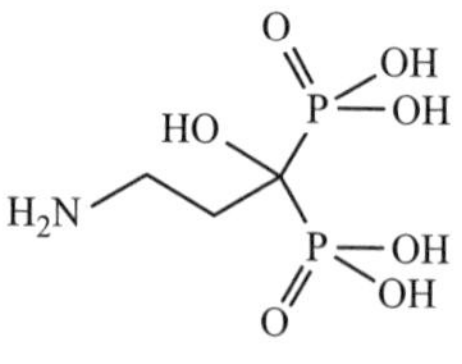   Pamidronate (8.113) 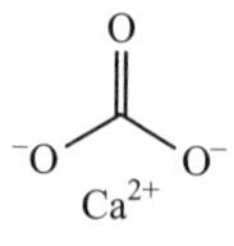   Calcium Carbonate (8.114)

of future osteoporosis. Approximately 1200–1500 mg of elemental calcium per day is suggested for teenagers. Calcium supplementation alone is not sufficient to reverse the bone loss of established osteroporosis. Some of the inorganic calcium salts that have been used clinically include calcium carbonate (**8.114**), tricalcium phosphate (**8.115**), calcium citrate (**8.116**), and calcium gluconate (**8.117**).

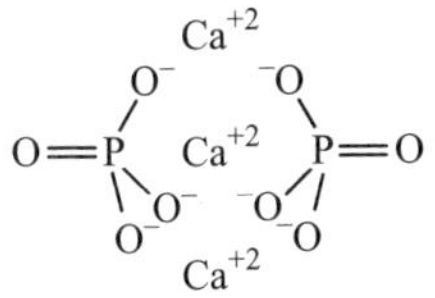

Tricalcium Phosphate (8.115) 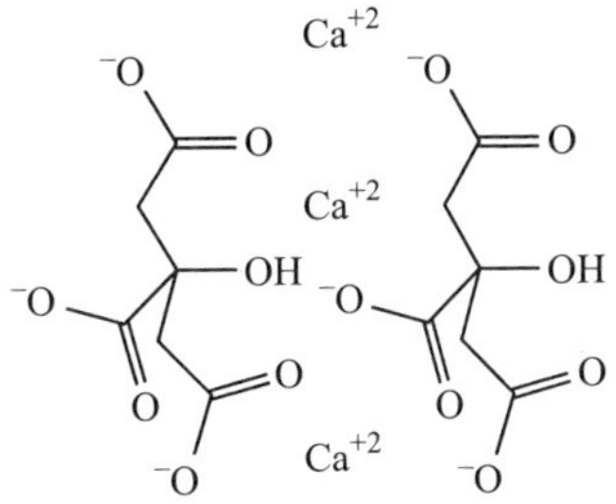   Calcium Citrate (8.116)

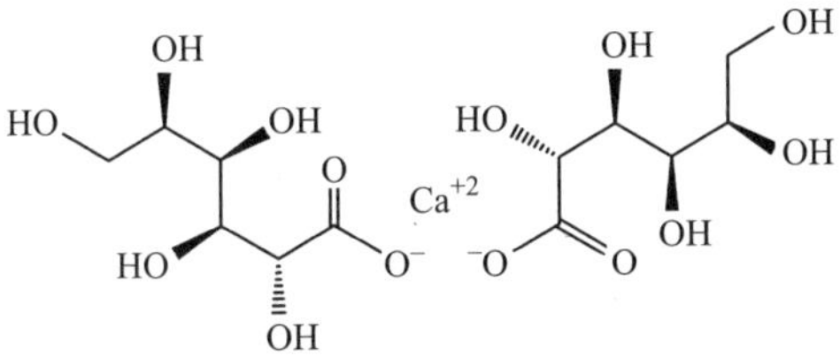

Calcium Gluconate (8.117)

Selective Estrogen Receptor Modulators (SERMs). Antiestrogen compounds such as tamoxifen (**8.118**) and raloxifene (**8.119**) exhibit agonist activity in some tissues (e.g., bone) and antagonist activity in other tissues (e.g., breast, uterus). Since osteoporosis is more prominent in post-menopausal women, possibly secondary to a failure of estrogen-mediated osteoclast inhibition, SERMs may play a role in the treatment of osteoporosis. Raloxifene has shown some positive results in clinical trials. Alternatively, *estrogen replacement therapy* (ERT) with estrogen analogs may be pursued. Such therapies have

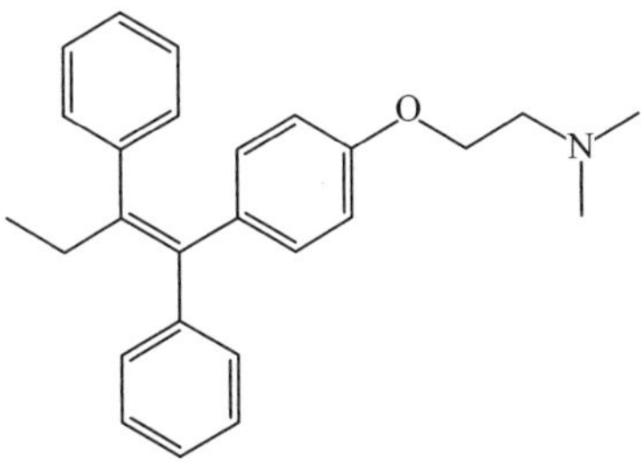

Tamoxifen (8.118) 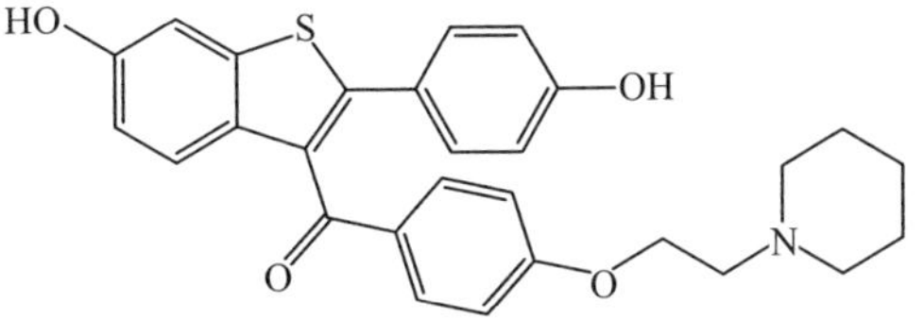   Raloxifene (8.119)

to be balanced against the risk of breast cancer and other shortcomings associated with chronic estrogen replacement.

Selected References

Proteins as Drug Design Targets: Enzymes

D. C. Evans, D. P. Hartley, R. Evers (2001). Enzyme induction—Mechanisms, assays, and relevance to drug discovery and development. *Ann. Rep. Med. Chem. 38*: 315.

A. Fersht (1985). *Enzyme Structure and Mechanism,* 2nd ed. San Francisco: Freeman.

M. Garcia-Viloca, J. Gao, M. Karplus, D. Truhlar (2004). How enzymes work: analysis by modern rate theory and computer simulations. *Science 303*: 186–195.

M. J. Jung (1978). Selective enzyme inhibitors in medicinal chemistry. *Annu. Rep. Med. Chem. 13*: 249–260.

R. N. Lindquist (1975). The design of enzyme inhibitors: Transition state analogs. In: E. J. Ariëns (Ed.). *Drug Design*, vol. 5. New York: Academic Press, pp. 23–80.

T. M. Penning (1983). Design of suicide substrates. *Trends Pharmacol. Sci. 4*: 212–217.

R. R. Rando (1974). Mechanism based irreversible enzyme inhibitors. *Annu. Rep. Med. Chem. 9*: 234–243.

M. Sandler (Ed.) (1980). *Enzyme Inhibitors as Drugs*. London: Macmillan.

C. T. Walsh (1984). Suicide substrates, mechanism-based enzyme inactivators: recent developments. *Annu. Rev. Biochem. 53*: 493–535.

R. Wolfenden (1976). Transition state analog inhibitors and enzyme catalysis. *Annu. Rev. Biophys. Bioeng. 5*: 271–306.

Enzyme Targets: Acetylcholinesterase

A. Khalid, Zaheer-uls-Haq, S. Anjum, M. Riaz Khan, Atta-ur-Rahman, M. Iqbal Choudhary (2004). Kinetics and structure–activity relationship studies on pregnane-type steroidal alkaloids that inhibit cholinesterases. *Bioorg. Med. Chem. 12*: 1995–2003.

J. Lindstrom (1986). Acetylcholine receptors: structure, function, synthesis, destruction and antigenicity. In: A. G. Engel, B. Q. Banker (Eds.). *Myology*. New York: McGraw-Hill, pp. 769–790.

P. Taylor, M. Schumacher, K. MacPhee-Quigley, T. Friedman, S. Taylor (1987). The structure of acetylcholinesterase: relationships to its function and cellular disposition. *Trends Neurosci. 10*: 92–96.

Enzyme Targets: Adenosine Triphosphatase

J. A. Bristol, D. B. Evans (1981). Cardiotonic agents for the treatment of heart failure. *Annu. Rep. Med. Chem. 16*: 91–102.

S. F. Campbell, J. C. Danilewicz (1978). Agents for the treatment of heart failure. *Annu. Rep. Med. Chem. 13*: 92–102.

L. Cantley (1986). Ion transport systems sequenced. *Trends Neurosci. 9*: 1–3.

P. Chene (2002). ATPases as drug targets: learning from their structure. *Nat. Rev. Drug Discov. 1*: 665–673.

P. Chene (2003). The ATPases: a new family for a family-based drug design approach. *Expert Opin. Ther. Targ. 7*: 453–461.

E. Erdman, K. Werdan, L. Brown (1985). Multiplicity of cardiac glycoside receptors in the heart. *Trends Pharmacol. Sci. 6*: 293–295.

T. Godfraind, A. de Pover, G. Castaneda Hernandez, M. Fagoo (1982). Cardiodigin: endogenous digitalis-like material from mammalian heart. *Arch. Int. Pharmacodyn. 258*: 165–167.

P. Lindberg, P. Norberg, T. Alminger, A. Brändström, B. Wallmark (1986). The mechanism of action of the gastric acid secretion inhibitor omeprazole. *J. Med. Chem. 29*: 1327–1329.

H. Lullmann, T. Peters, A. Ziegler (1979). Kinetic events determining the effect of cardiac glycosides. *Trends Pharmacol. Sci. 1*: 102–106.

K. Repke (1985). New developments in cardiac glycoside structure–activity relationships. *Trends Pharmacol. Sci. 6*: 275–278.

D. C. Rohrer, D. S. Fullerton, K. Yoshioka, A. H. L. Frome, K. Ahmed (1979). Functional receptor mapping for modified cardenolides: use of the Prophet system. In: E. C. Olson, R. E. Christoffersen (Eds.). *Computer Assisted Drug Design*. Washington, DC: American Chemical Society, pp. 259–279.

R. Thomas, J. Boutagy, A. Gelbart (1974). Synthesis and biological activity of semisynthetic digitalis analogs. *J. Pharmacol. Sci. 63*: 1649–1683.

B. Wetzel, N. Hanel (1984). Cardiotonic agents. *Annu. Rep. Med. Chem. 19*: 71–80.

Enzyme Targets: Carbonic Anhydrase

E. J. Cragoe, Jr (Ed.) (1983). *Diuretics: Chemistry, Pharmacology and Medicine*. New York: Wiley.

S. P. Gupta (2003). Quantitative structure–activity relationships of carbonic anhydrase inhibitors. *Prog. Drug Res. 60*: 171–204.

H. R. Jacobsen, J. K. Kokko (1976). Diuretics: sites and mechanism of action. *Annu. Rev. Pharmacol. Toxicol. 16*: 201–226.

K. K. Kannan, I. Vaara, B. Notstrand, S. Lovgren Borell, K. Fridborg, M. Petef (1977). Structure and function of carbonic anhydrase: comparative studies of sulfonamide binding to human erythrocyte carbonic anhydrases B and C. In: G. C. K. Roberts (Ed.). *Drug Action at the Molecular Level*. Baltimore: University Park Press, pp. 73–91.

R. L. Smith, O. W. Woltersdorf, Jr, E. J. Cragoe, Jr (1976, 1978). Diuretics. *Annu. Rep. Med. Chem. 11*: 71–79; *13*: 61–70.

C. T. Supuran, D. Vullo, G. Manole, A. Casini, A. Scozzafava (2004). Designing of novel carbonic anhydrase inhibitors and activators. *Curr. Med. Chem. Cardiovasc. Hematol. Agents 2*: 49–68.

Enzyme Targets: Monoamine Oxidase

F. P. Bymaster, R. K. McNamara, P. V. Tran (2003). New approaches to developing antidepressants by enhancing monoaminergic neurotransmission. *Expert Opin. Invest. Drugs 12*: 531–543.

J. van Dijk, J. Hartog, F. C. Hillen (1978). Non-tricyclic antidepressants. In: G. P. Ellis, G. B. West (Eds.). *Progress in Medicinal Chemistry*, vol. 15. Amsterdam: Elsevier/North Holland, pp. 261–320.

J. Knoll (1982). Selective inhibition of B-type monoamine oxidase in brain: a drug strategy to improve quality of life in senescence. In: J. A. Keverling Buisman (Ed.). *Strategy in Drug Research*. Amsterdam: Elsevier.

R. A. Maxwell, H. L. White (1978). Tricyclic and monoamine oxidase inhitor antidepressants: structure–activity relations. In: L. L. Iversen, S. D. Iversen, S. H. Snyder (Eds.). *Handbook of Psychopharmacology*, vol. 14. New York: Plenum Press, pp. 83–155.

M. Naoi, W. Maruyama, G. M. Nagy (2004). Dopamine-derived salsolinol derivatives as endogenous monoamine oxidase inhibitors: occurrence, metabolism and function in human brains. *Neurotoxicology 25*: 193–204.

R. R. Ramsay, M. B. Gravestock (2003). Monoamine oxidases: to inhibit or not to inhibit. *Mini-Rev. Med. Chem. 3*: 129–136.

M. Strolin Benedetti, P. Dostert (1985). Stereochemical aspects of MAO interactions: reversible and selective inhibitors of monoamine oxidase. *Trends Pharmacol. Sci. 6*: 246–251.

M. Tuncel, V. C. S. Ram (2003). Hypertensive emergencies: etiology and management. *Am. J. Cardiovasc. Drugs 3*: 21–31.

Enzyme Targets: Proteases and Kinases

S. Chakravarty, S. Dugar (2002). Inhibitors of p38α MAL kinase. *Annu. Rep. Med. Chem. 37*: 177.

Y. Dai, S. Grant (2003). Cyclin-dependent kinase inhibitors. *Curr. Opin. Pharmacol. 3*: 362–370.

S. D. Kimball, K. R. Webster (2001). Cell cycle kinases and checkpoint regulation in cancer. *Annu. Rep. Med. Chem. 36*: 139.

M. E. Noble, J. Endicott, L. Johnson (2004). Protein kinase inhibitors. *Science 303*: 1800–1803.

J. T. Randolph, D. A. DeGoey (2004). Peptidomimetic inhibitors of HIV protease. *Curr. Top. Med. Chem. 4*: 1079–1095.

P. A. Renhowe (2001). Growth factor receptor kinases in cancer. *Annu. Rep. Med. Chem. 36*: 109.

C. T. Supuran, A. Casini, A. Scozzafava (2003). Protease inhibitors of the sulfonamide type: anticancer, antiinflammatory, and antiviral agents. *Med. Res. Rev. 23*: 535–558.

Enzyme Cofactors: Vitamins

H. F. DeLuca, H. K. Schnoes (1984). Vitamin D: metabolism and mechanism of action. *Annu. Rep. Med. Chem. 19*: 179–190.

D. Hornig (1982). Requirement of vitamin C in man. *Trends Pharmacol. Sci. 3*: 294–296.

L. J. Machlin (Ed.) (1984). *Handbook of Vitamins: Nutritional, Biochemical and Clinical Aspects*. New York: Marcel Dekker.

D. P. McDonnell, D. L. Mangelsdorf, J. W. Pike, M. R. Haussler, B. W. O'Malley (1987). Molecular cloning of complementary DNA encoding the avian receptor for vitamin D. *Science 235*: 1214–1217.

L. Ovesen (1984). Vitamin therapy in the absence of obvious deficiency. What is the evidence? *Drugs 27*: 148–170.

S. Vertuani, A. Angusti, S. Manfredini (2004). The antioxidants and pro-antioxidants network: an overview. *Curr. Pharm. Des. 10*: 1677–1694.

Proteins as Drug Design Targets: Non-Enzymes

F. E. Cohen, J. W. Kelly (2003). Therapeutic approaches to protein-misfolding diseases. *Nature 426*: 905–908.

C. M. Dobson (2003). Protein folding and misfolding. *Nature 426*: 884–890.

G. M. Hirschfield, P. N. Hawkins (2003). Amyloidosis: new strategies for treatment. *Int. J. Biochem. Cell Biol. 35*: 1608–1613.

A. Y. Kornilova, M. S. Wolfe (2001). Secretase inhibitors for Alzheimer's disease. *Annu. Rep. Med. Chem. 38*: 41.

J. P. Taylor, J. Hardy, K. Fischbeck (2002). Toxic proteins in neurodegenerative disease. *Science 296*: 1991–1995.

P. A. Temussi, L. Masino, A. Pastore (2003). From Alzheimer to Huntington: Why is a structural understanding so difficult? *EMBO J. 22*: 355–361.

D. Selkoe (1999). Translating cell biology into therapeutic advances in Alzheimer's disease. *Nature 399* (Suppl.): A23–A31.

D. Selkoe (2003). Folding proteins in fatal ways. *Nature 426*: 900–905.

Lipids as Drug Design Targets

P. G. Adaikan, S. R. Kottegoda (1985). Prostacyclin analogs. *Drugs Future 10*: 765–774.

M. R. Bell, F. H. Barzold, R. C. Winneker (1986). Chemical control of fertility. *Annu. Rep. Med. Chem. 21*: 169–177.

K. Brune, K. D. Rainsford (1980). New trends in the understanding and development of anti-inflammatory drugs. *Trends Pharmacol. Sci. 1*: 95–97.

D. A. Clark, A. Marfat (1982). Structure elucidation and the total synthesis of leukotrienes. *Annu. Rep. Med. Chem. 17*: 291–300.

R. A. Coleman, P. P. A. Humphrey, I. Kennedy, P. Lumley (1984). Prostanoid receptors—the development of a working classification. *Trends Pharmacol. Sci. 5*: 303–306.

C. A. Dinarello (1984). Interleukin-1: an important mediator of inflammation. *Trends Pharmacol. Sci. 5*: 420–422.

C. Funk (2001). Prostaglandins and leukotrienes: advances in eicosanoid biology. *Science 294*: 1871–1875.

J. G. Gleason, C. D. Perchonock, T. J. Torphy (1986). Pulmonary and antiallergy agents. *Annu. Rep. Med. Chem. 21*: 73–83.

R. R. Gorman (1978). Prostaglandins, thromboxanes and prostacyclin. *Int. Rev. Biochem. 20*: 91–107.

P. Gund, J. D. Andose, J. B. Rhodes, G. M. Smith (1980). Three-dimensional molecular modeling and drug design. *Science 208*: 1625–1731.

F. D. Hart, E. C. Huskinson (1984), Non-steroidal antiinflammatory drugs: current status and rational therapeutic use. *Drugs 27*: 232–255.

S. Kaivola, J. Paratainen, T. Osterman, H. Tiomonen (1983). *Cephalagia 3*: 31–36; as quoted in *Trends Pharmacol. Sci. 6*: 435 (1995).

W. Kreutner, M. I. Siegel (1984). Biology of leukotrienes. *Annu. Rep. Med. Chem. 19*: 241–251.

M. Kucher, V. Rejholec (1986). Antithrombotic agents. *Drugs Future 11*: 687–701.

W. E. M. Lands (1985). Mechanism of action of antiinflammatory drugs. *Adv. Drug Res. 14*: 147–164.

A. Lupulescu (1996). Prostaglandins, their inhibitors and cancer. *Prostaglandins, Leukotrienes and Essential Fatty Acids 54*: 83–94.

R. Nickander, F. G. McMahon, A. S. Ridollo (1979). Non-steroidal anti-inflammatory agents: *Annu. Rev. Pharmacol. Toxicol. 19*: 649–690.

K. C. Nicolaou, J. B. Smith (1979). Prostacyclin, thromboxane and the arachidonic acid cascade: *Annu. Rep. Med. Chem. 14*: 178–187.

J. E. Pike, D. R. Morton, Jr (Eds.) (1985). *Advances in Prostaglandin, Thromboxane, and Leukotriene Research*, vol. 14. New York: Raven Press.

K. D. Rainsford (1985). *Anti-inflammatory and Anti-rheumatic Drugs*, 3 vols. Boca Raton: CRC Press.

J. A. Salmon (1997). Inhibition of prostaglandin, thromboxane and leukotriene biosynthesis. *Adv. Drug Res. 15*: 111–167.

B. Samuelsson (1980). The leukotrienes: a new group of biologically active compounds including slow reacting substance. *Trends Pharmacol. Sci. 1*: 227–230.

B. Samuelsson, P. W. Ramwell, R. Paoletti (1976). *Advances in Prostaglandin and Thromboxane Research*, 18 vols. New York: Raven Press.

B. Samuelsson, M. Goldyne, E. Granström, M. Ramberg, S. Hammarström, C. Malmensten (1978). Prostaglandins and thromboxanes. *Annu. Rev. Biochem. 47*: 997–1029.

V. Ullrich, H. Graf (1984). Prostacyclin and thromboxane synthase as P-450 enzymes. *Trends Pharmacol. Sci. 5*: 352–355.

M. C. Venuti (1985). Platelet-activating factor: multifaceted biochemical and physiological mediator. *Annu. Rep. Med. Chem. 20*: 193–202.

9

Nonmessenger Targets for Drug Action III

Exogenous pathogens and toxins

9.1 EXOGENOUS PATHOGENS AS TARGETS
FOR DRUG DESIGN

The fundamental mechanistic causes of human disease and pathology are multiple, but they can be comprehensively categorized into the following ten groups:

1. Traumatic (pathology from injury)
2. Toxic (pathology from poisons)
3. Hemodynamic/vascular (pathology from disorders of blood vessels)
4. Hypoxic (pathology from inadequate supply or excessive demand for oxygen by a tissue)
5. Inflammatory (pathology from abnormal inflammatory response in the body)
6. Infectious (pathology from microbes or infectious agents)
7. Neoplastic (pathology from tumors, cancer)
8. Nutritional (pathology from too much/too little food intake)
9. Developmental (pathology in the chemistry of heredity)
10. Degenerative (pathology from age-related tissue breakdown)

When designing therapeutics for a specific disease, the medicinal chemist must consider the pathological mechanism of the underlying disorder and then design the drug accordingly. For example, *anti-neoplastic* agents are designed to treat malignancies like colon cancer, whilst *anti-inflammatory agents* would be designed to treat diseases such as rheumatoid arthritis. Of these ten fundamental pathological mechanisms, most represent errors or imbalances in endogenous mechanisms. Developmental diseases, for example, arise from inborn errors in metabolism or from genetic disorders. Many common human diseases may be understood in terms of these endogenous pathological mechanisms: a myocardial infarction (heart attack), for instance, usually arises from an occluded coronary artery—occlusion of this artery may be the product of *traumatic* damage to the artery (from years of untreated high blood pressure), leading to *degenerative* atherosclerotic changes in the arterial wall, resulting in reduced *hemodynamic* blood supply to the tissues of the heart, causing oxygen deprivation and *hypoxic* damage to the heart muscles. Rational drug design for diseases arising from these various endogenous pathological mechanisms has already been addressed, using the diverse messenger and nonmessenger design strategies discussed in chapters 4–8.

The remaining two mechanisms of pathology (infectious, toxic) arise primarily from exogenous *pathogens* (where a pathogen is any microorganism or other substance causing disease). These pathogens may be biological (bacteria), chemical (environmental toxins), or physical (radiation); pathogens may be naturally occurring (viruses) or human-made (nerve gas). Pathogens may be *opportunistic* (causing disease only when the host's state of health is diminished; e.g., a fungal pneumonia occurring in a person whose resistance has been lowered by having acquired immune deficiency syndrome [AIDS]) or *indiscriminate* (producing disease in any individual, whether healthy or ill, exposed to the agent; e.g., high levels of radiation).

9.1.1 Types of Exogenous Pathogens

The exogenous pathogens that lead to human disease may be sub-divided into two broad categories:

1. Microbes
    Prions
    Viruses
    Bacteria
    Fungi
    Parasites
2. Toxins
    Biological
    Chemical
    Physical

Of these two categories of pathogens, microbe-induced infectious diseases are probably the most important.

The World Health Organization has identified infectious disease as the greatest challenge to global health. The identification of such an important problem has been made apparent by the current health crises confronting the modern world. The spectre of AIDS, SARS (severe acute respiratory syndrome), and West Nile virus emphasizes the looming importance of infectious disease to the health of our planet. In future years, the summer months may also be accompanied by increasing problems with malaria in North America and Europe as global warming heralds fundamental changes to our planet's ecosystems. In Africa, bacterial meningitis continues to reach epidemic proportions, with concomitant increases in polio and tuberculosis.

It is also of great concern that the traditional antibiotic drugs (the miracle agents of the 20th century) are becoming increasingly ineffective against an ever-enlarging number of bacteria. In the United States, a recent National Academy of Science report identified resistance to conventional antibiotics and the need for new antibiotics as crucial issues. The "To Market, To Market" summaries of the Annual Reports in Medicinal

Chemistry of the American Chemical Society (which document new drugs introduced on a yearly basis) verifies this and reveals that the majority of so-called new antibiotics developed over the past two decades continue to be penicillin and cephalosporin derivatives. These grim realities attest to the worries of microbiologists and epidemiologists who contend that "the bugs are eventually going to get us" and "our current drugs probably won't be up to this challenge." Clearly, the design of drugs that target exogenous pathogens will emerge as an important drug design field in coming years.

9.1.2 Designing Drugs for Exogenous Pathogen-Based Diseases

Designing drugs for exogenous pathogens has some fundamental differences from the process of designing drugs to achieve the manipulation of endogenous processes. Exogenous pathogens represent targets for drug design that are *nonself*. When a drug binds to a receptor in the human heart, the target is *self*; however, when a drug binds to bacteria within lung tissue, then the target is nonself. The toxicity that the drug can inflict upon the surrounding tissues of the receptor microenvironment is quite different between self and nonself. It is desirable for a drug binding to a bacterium to kill that bacterium; it is undesirable for a drug binding to heart tissue to kill cardiac cells.

The types of intermolecular interactions exploited during drug design against nonself exogenous targets may also be different. For example, it is acceptable for a drug to bind to a nonself target by a covalent bond. Whereas drug–receptor interactions via covalent bonding are typically avoided for endogenous targets for reasons related to toxicity, covalent bonds are acceptable when targeting a nonself receptor on an exogenous pathogen. This observation is well exemplified by the example of antibacterial agents, such as penicillin, that covalently link to the bacterial cell wall.

In developing drugs for the treatment of diseases caused by microbes, drug design strategies may differ widely from microbe to microbe. In terms of structural complexity, microbes exist on a structural spectrum (prions, viruses, bacteria, fungi, parasites), with prions being the least complex and parasites the most complex. Drug design for prion and viral diseases is the most challenging, since these microbes are structurally simple. A prion is merely a protein; viruses are composed principally of nucleic acids. Because of the structural overlap between prion proteins and viral nucleic acids and the corresponding macromolecules found in humans, it is difficult to design a drug specific for the microbe. At the other end of the spectrum, parasites have a structural complexity (organs, rudimentary nervous system) that begins to approach the sophistication of human cell lines. Because of this similarity, it may be difficult to identify a target that will enable selective killing of the parasite without causing concomitant harm to the host organism. The structurally intermediate bacteria have sufficient complexity to enable drug design without the complexity overlapping with that of the host biochemistry. Accordingly, antibacterial drug design has traditionally been more successful than drug design targeted against the other microbes.

In designing drugs for exogenous pathogens, it is sometimes possible to minimize or even neglect pharmacokinetic design considerations. For instance, if a drug were being designed to treat intestinal parasites, it would be beneficial to ensure that the drug is restricted to the gastrointestinal tract and is never absorbed. The strategic use of charged or polar substituent groups to influence drug solubility and absorption can be manipulated under such circumstances to reduce the likelihood of drug absorption.

9.1.3 Microbial Genomics and Drug Design

As discussed in chapter 3, the elucidation of both the human genome and numerous pathogen genomes may exert a profound impact on the understanding and treatment of human disease. When designing drugs to contend with exogenous pathogens such as infectious microbes, it is immensely useful to understand the biochemistry of the microbe and how that biochemistry differs from human biochemistry. Such data are useful when designing a drug to kill a microbe without harming the human host. Understanding the genome of a microorganism enables insights concerning its biochemical operation and the identification of a potential druggable target; comparing the microbe's genome to that of a human enables an appreciation of whether that target is shared between the microbe and humans.

Over the past 5–10 years, a number of pathogenic microbial genomes have been determined. Amongst these, one of the most noteworthy has been the full genome of *Plasmodium falciparum*, the most important of the parasites causing malaria. This parasite's 30-million-base-pair genome is spread over 14 chromosomes. The elucidation of this genome was challenging because adenine and thymine, two of the four building blocks of DNA, together make up more than 80% of the organism's genome. Sifting through this enormous collection of genomic data on *Plasmodium falciparum* and comparing it to the human genome for purposes of rational drug design is a Herculean task, made somewhat easier by recent advances in bioinformatics (discussed in chapter 1).

The use of a variety of techniques, ranging from classical drug design to genome-based drug design, will be required for a full and effective attack on the microbes of human disease, extending from simple prions to complex parasites.

9.2 DRUG DESIGN TARGETING PRIONS

Prion diseases have attracted immense attention over the past decade, prompted, in part, by the outbreak of "mad cow disease" in the United Kingdom. The most common prion disease is sporadic Creutzfeldt–Jakob disease (CJD). Clinically, CJD is characterized by a rapidly progressive dementia accompanied variably by early-onset seizures, insomnia, disordered movements, and psychiatric disturbances; the disease is uniformly fatal. Histochemically, the principal pathological feature of prion disease is the abnormal accumulation of an amyloid-like material composed of prion protein (PrP), which is encoded by a single gene on the short arm of chromosome 20.

The abnormal deposits found in the brains of CJD victims consist of an abnormal isoform of PrP. Prion protein is normally found in cells. Detailed structural studies show that normal cellular PrP (PrP^C) is a soluble protein whose conformation is rich in α-helices with very little β-sheet. The PrP protein extracted from the brains of CJD victims (i.e., PrP^{SC}) is identical in primary amino acid sequence to the normal PrP (PrP^C). However, PrP^{SC} has a much greater content of β-sheet conformation with little α-helical structure. Thus PrP^{SC} is neurotoxic because of its three-dimensional structure. When the prion protein is predominantly in an α-helical conformation, it is nontoxic; when it is predominantly in a β-sheet conformation, it kills neurons. The prion protein is thus made neurotoxic not by its amino acid composition but by its conformation. Prion diseases are transmissible; thus prions are infectious agents. However, prions are not like viruses, or other infectious microbes—prions are simply protein molecules; they are not microbes with cell membranes and nucleic acids. Indeed, prions are not even infectious molecules—they are infectious molecular shapes.

Like any other protein, the molecular structure of the prion is subject to conformational flexibility and to various thermal-induced fluctuations between varying conformational states. However, if these fluctuations permit the PrP^{SC} conformation to be attained, then this abnormal conformer promotes the widespread conversion of PrP^C to PrP^{SC}, leading to the precipitous deposition of the abnormal protein throughout the brain (mirrored by the rapid and relentlessly downhill clinical course). This pathological self-propagating shape conversion of α-helical PrP^C to β-sheet PrP^{SC} may in principle be initiated by a "seed" PrP^{SC} molecule in the neurotoxic conformation. This explains the transmissibility of prion diseases and accounts for how susceptible humans exposed to beef from an animal with mad cow disease develop variant Creutzfeldt–Jakob disease.

The concept of abnormal proteins in CJD may provide insights useful for drug design. The pioneering work of Prusiner has enabled the preliminary identification of prototype agents as therapies for CJD. Preliminary work identified two classes of compounds with therapeutic potential: polysulphated molecules and tricyclic molecules (e.g., phenothiazines, aminoacridines). These compounds bind to PrP and endeavor to inhibit the PrP^C to PrP^{SC} cascade of conformational change.

Since prions affect cattle, elk, and other ruminant animals, it may be necessary to design therapies for veterinary applications. However, designing drugs with a veterinary application in cattle is different from designing drugs for domestic pets such as cats or dogs. Cattle are an integral part of the human food chain. Drugs administered to cattle have the potential to be consumed by humans. Under such circumstances, a drug administered to cattle and consumed by humans could become an exogenous chemical pathogen, conceivably causing health care concerns for the human. Arguably, other therapies such as vaccines may have to be considered for the mass treatment of animals susceptible to prion diseases.

9.3 DRUG DESIGN TARGETING VIRUSES

Viruses are on the borderline between inanimate and living matter—one step above prions, but one step below bacteria. Viruses can be crystallized, and the virus particle shows a high degree of structural symmetry; alternatively, viruses can be cultured and grown. Viruses are small (20–300 nm in diameter), containing a molecule of nucleic acid as their genome. The nucleic acid is encased in a protein shell, termed a *capsid*, which is itself composed of protein building blocks, or *capsomeres*. Some viruses have a further membrane-like envelope around the capsomere. A number of viruses have this envelope (pox, herpes, rhabdo, paramyxo, orthomyxo, toga); others do not (papova, adeno, parvo, picorna). The entire infectious viral unit is called a *virion* and is shown schematically in figure 9.1.

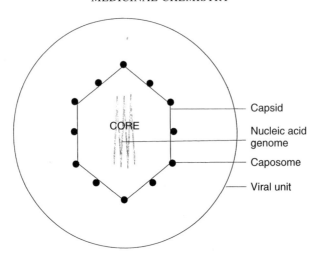

Figure 9.1 The viral infectious unit, or virion, consists of a nucleic acid encased in a protein shell. The various structural components of the virion exhibit differing functional properties and thus afford a variety of targets for antiviral drug design.

Viruses can exist and reproduce only as extreme cellular parasites because they do not possess a complete synthetic machinery for independent survival. Accordingly, viruses are obligate parasites, able to replicate only in living cells. To survive, a virus must first penetrate a host cell. Viruses consist of infectious DNA or RNA (but not both) wrapped in a protective protein coat, which shows helical or spherical symmetry and which may show molecular appendages. The replication and assembly of such complex structures is not a spontaneous self-assembly but is instead directed by enzymes and by "assisting proteins" that collectively form a biosynthetic framework. The replication of viruses is very different from that of higher organisms. In some, such as the DNA viruses, transcription and translation occur, whereas in others, like the single stranded RNA viruses, the RNA is its own messenger and activates an RNA-directed DNA *polymerase*. The latter viruses are known as *retroviruses* because of the reversal of the normal DNA–RNA–protein sequence; the DNA produced in this reverse way then becomes the template for viral RNA and protein. The virus then uses the enzymes, nucleotides, and amino acids of the infected host cell to build the virion. Because of this diversion of starting material and synthetic capacity, the infected cell may die, with virions being released to infect other cells (*lytic* viruses); alternatively, the viral DNA may join the infected cell by recombination, which then continues to produce virions (*lysogenic* viruses). In some instances, the host cell is transformed into a malignant, cancerous cell by the virus, which is obviously a process of great interest to the mechanistic understanding of human disease. The discovery of cancer genes (*oncogenes*) will ultimately shed more light on the process of viral-induced carcinogenesis.

A high proportion of human and animal diseases are caused by viruses, from the common cold to poliomyelitis, rabies, hepatitis, and many others. Table 9.1 provides an overview of viruses causing diseases in humans. In the last two decades of the twentieth century, the emergence of the AIDS epidemic focused attention on the need to develop

Table 9.1 Viruses Causing Disease in Humans

Family agent	Disease
DNA viruses	
Adenovirus	
Human adenovirus	Upper respiratory tract infections
Hepadnavirus	
Hepatitis B virus	Hepatitis
Herpesvirus	
Cytomegalovirus	Infections in neonates
Epstein–Barr virus	Infectious mononucleosis
Herpes simplex 1	Stomatitis, eye infections encephalitis
Herpes simplex 2	Genital herpes, skin eruptions
Varicella zoster	Chicken pox (children), shingles (adults)
Papovavirus	
Papillomavirus	Warts
Polyomavirus (JC virus)	Progressive leucoencephalopathy
Poxvirus	
Variola	Smallpox
RNA viruses	
Arenavirus	
Lassa virus	Hemorrhagic fever
Lymphocytic choriomeningitis virus	Meningitis
Bunyavirus	
Arboviruses	Encephalitis, hemorrhagic fever
Hantavirus	Fever, renal failure
Calicivirus	
Norwalk virus	Gastroenteritis
Coronavirus	Respiratory infection, SARS
Filovirus	
Ebola virus	Hemorrhagic fever
Marburg virus	Marburg disease
Flavivirus	
Flaviviruses	Dengue, encephalitis, hemorrhagic fever, yellow fever
Hepatitis C virus	Hepatitis
Orthomyxovirus	
Influenza virus	Influenza
Paramyxovirus	
Morbillivirus	Measles (rubeola)
Mumps virus	Mumps
Parainfluenza virus	Respiratory infection
Respiratory syncytial virus	Respiratory infection
Picornavirus	
Coxsackie viruses	Variety of symptoms
Enterovirus	Polio

(Continued)

Table 9.1 Continued

Family agent	Disease
Hepatitis A virus	Hepatitis (ususally mild, rarely chronic)
Rhinovirus (over 100 serotypes)	Common cold, pneumonia
Reovirus	
Human rotavirus	Gastroenteritis in infants
Orbivirus	Colorado tick fever
Retrovirus	
Human immunodeficiency viruses (HIV-1, HIV-2)	T-cell leukemia
Rhabdovirus	
Rabies virus	Rabies
Togavirus	
Alphaviruses	Encephalitis, hemorrhagic fevers
Rubivirus	Rubella (German measles)

effective drugs for the treatment of human immunodeficiency virus. The dawning of the 21st century has brought new concerns about seemingly new viruses, including SARS and avian flu. The new century has also brought new concerns about old viruses, such as the *weaponization* of viruses (e.g., smallpox) as tools of bioterrorism and weapons of mass destruction. These worries are further compounded by the dire predictions of epidemiologists who foresee the "long overdue" re-emergence of worldwide killer flu epidemics such as the Spanish flu epidemic of 1918. Although vaccines represent an effective tool against viruses, vaccines can take a long time to produce and their development may be too late to confront a rapidly spreading epidemic. Moreover, not all viruses are amenable to immunotherapy via vaccines. Clearly, the design of antivirals is an important, indeed crucial, medicinal chemistry priority. Regrettably, the development of antiviral drugs has been (and continues to be) a very slow and frustrating field. It is not nearly as well advanced as bacterial chemotherapy.

When designing new chemical entities as putative antivirals, there are seven logical steps in viral biochemistry and replication at which druggable targets suitable for drug design can be identified:

1. Penetration of the virus into the susceptible host cell
   (blocked, non-specifically, by γ-globulins)
2. Uncoating of the viral nucleic acid through shedding of the protein coat
   (blocked by amantadine (**9.1**))
3. Synthesis of early regulatory proteins (nucleic acid polymerases) used to aid nucleic acid synthesis
   (blocked by fomivirsen (**9.2**))
4. Synthesis of either RNA or DNA
   (blocked by purine or pyrimidine analogs)
5. Synthesis of late structural proteins so that the nucleic acids can be "re-coated"
   (blocked by methimazole (**9.3**) or protease inhibitors)

6. Assembly of new virus particles from nucleic acids and proteins (from steps 4 and 5)
   (blocked by rifampin (**9.4**))
7. Release and extrusion of the virus from the host cell
   (blocked by neuraminidase inhibitors)

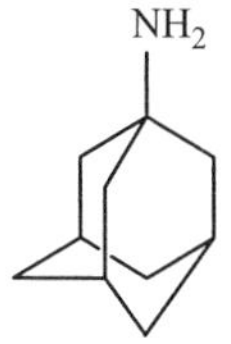

Amantadine (9.1)

5′-GCG TTT GCT CTT CTT CTT GCG-3′
note: racemic phosphorothioate nucleotide junctions

Fomivirsen (9.2)

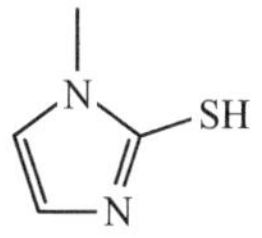

Methimazole (9.3)

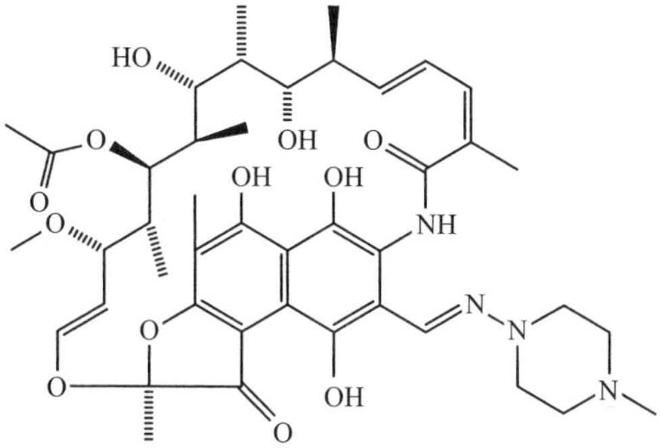

Rifampin (9.4)

Although any of these seven steps could be a druggable target, most of the antiviral agents clinically employed for non-AIDS infections act on the synthesis or assembly of either purines or pyrimidines (steps 3 and 4). For AIDS, reverse transcriptase inhibitors block transcription of the HIV RNA genome into DNA, thereby preventing synthesis of viral mRNA and protein; protease inhibitors act on the synthesis of late proteins (steps 5 and 6).

9.3.1 Antivirals Targeting Purine or Pyrimidine Biosynthesis

In the host cell, viruses induce the formation of enzymes that they themselves cannot produce. The most important group of such enzymes is that of the DNA polymerases, but thymidine kinase is also essential. Interference with these enzymes by either enzyme inhibitors or through fraudulent *antimetabolites* is the basis of the activity of many antiviral drugs. In this respect, antiviral compounds and cytostatics used in the treatment of malignant tumors have much in common and indeed overlap each other in their activity.

Viral DNA polymerase is an important catalyst for the synthesis of viral nucleic acids. DNA polymerase inhibitors have already been encountered as antitumor agents. Ara-A (**9.5**, vidarabine) is a DNA polymerase inhibitor that has demonstrated activity against herpes simplex virus type I (HSV-1) infections, responsible for "cold sores" on

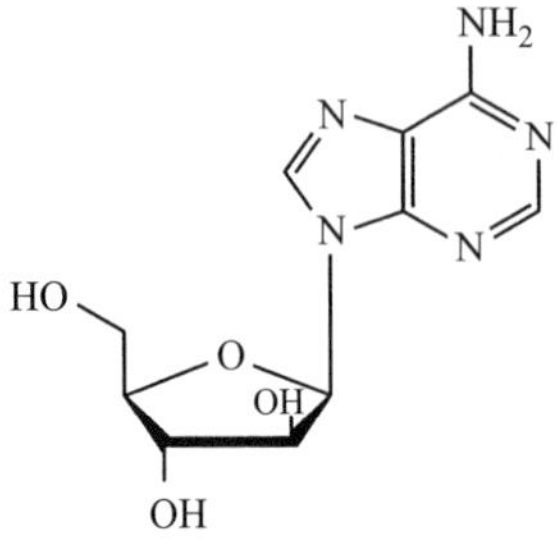

Vidarabine (9.5)

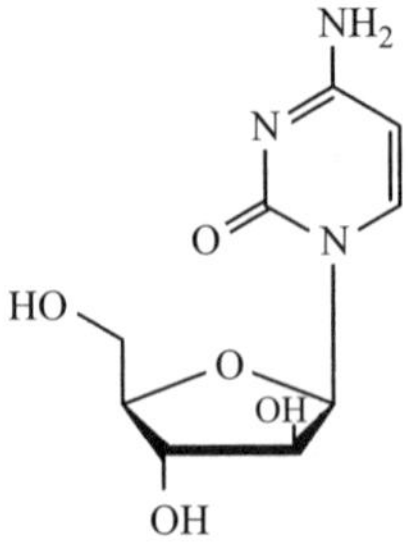

Cytarabine (9.6)

the lip, keratitis of the cornea, and encephalitis. Its analog, Ara-C (**9.6**, cytarabine), is primarily an antineoplastic drug, but its 2'-fluoro-5-iodo derivative has also shown good activity against HSV-1. Another enzyme, thymidylate synthase, is involved in pyrimidine biosynthesis, and this can be exploited by antimetabolites that are mistaken for true nucleotide metabolites. The enzyme-mediated incorporation of such antimetabolites into the viral DNA and RNA will destroy the virus, eliminating its infectious properties. This is in contrast to the mode of action of the antitumor agent 5-fluorouracil, which is a suicide substrate of thymidylate kinase and inactivates the enzyme, thereby interrupting the dTTP supply of the tumor cell.

Idoxuridine (**9.7**) and trifluridine (**9.8**) are antiviral agents that are phosphorylated to their active form in virus-infected cells, and thus show specificity for two reasons: their higher affinity for the viral enzyme, and the higher phosphorylase levels in viral-infected than in normal cells. Both compounds have been used locally on lesions of HSV-1 and HSV-2 (the latter of which causes genital herpes, now reaching epidemic proportions) with fair success. They are rather toxic if administered parenterally, as are all moderately selective antimetabolities.

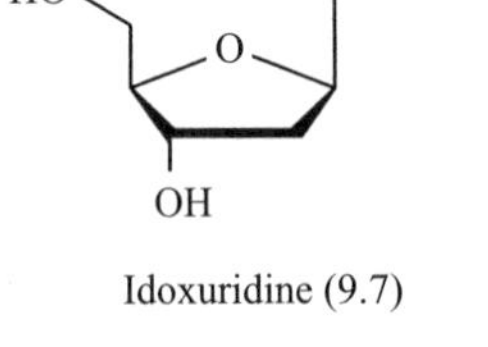

Idoxuridine (9.7)

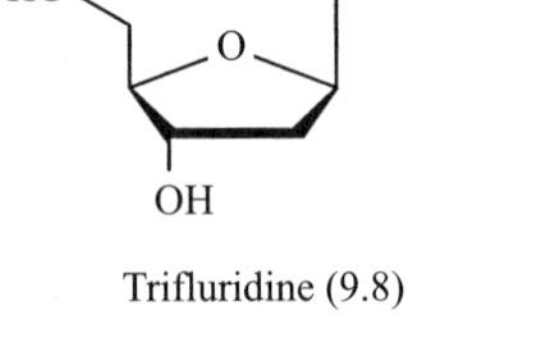

Trifluridine (9.8)

Acyclovir (**9.9**) shows a unique specificity and lack of toxicity in HSV-1, HSV-2, and varicella (chickenpox, shingles) viral infections. A guanine derivative, acyclovir lacks the pentose of similar compounds and is phosphorylated at the alcoholic OH by the viral thymidylate kinase only. Consequently, it is not activated in uninfected cells; additionally,

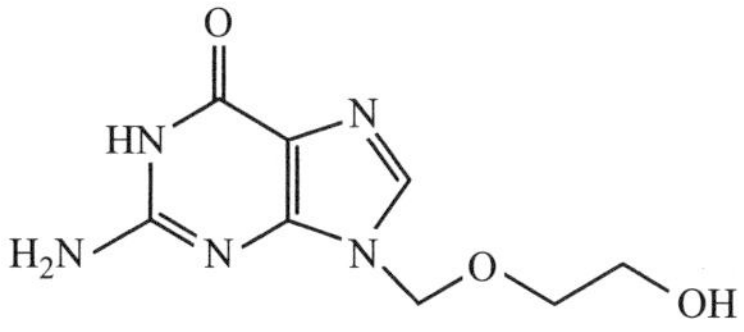

Acyclovir (9.9)

it is a viral DNA polymerase inhibitor but does not readily block the polymerase of the host cell. Therefore, acyclovir is a spectacularly nontoxic drug [LD_{50} (mouse) = 1000 mg/kg, i.p.] and is not degraded metabolically. Valacyclovir (**9.10**) is the L-valyl ester of acyclovir. After oral administration valacyclovir is rapidly converted to acyclovir, attaining serum concentrations 3–5 times higher than those achieved with oral acyclovir.

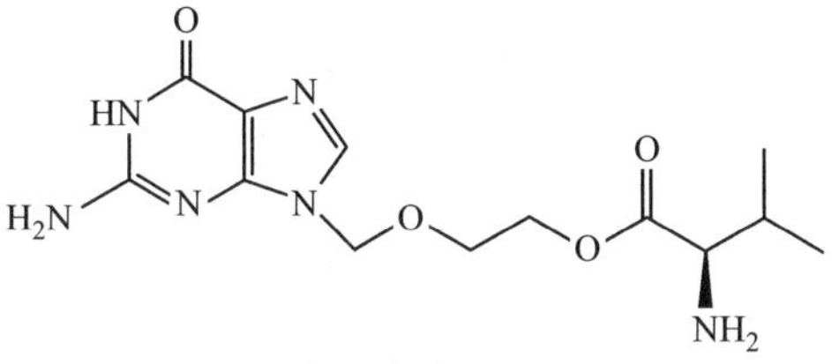

Valacyclovir (9.10)

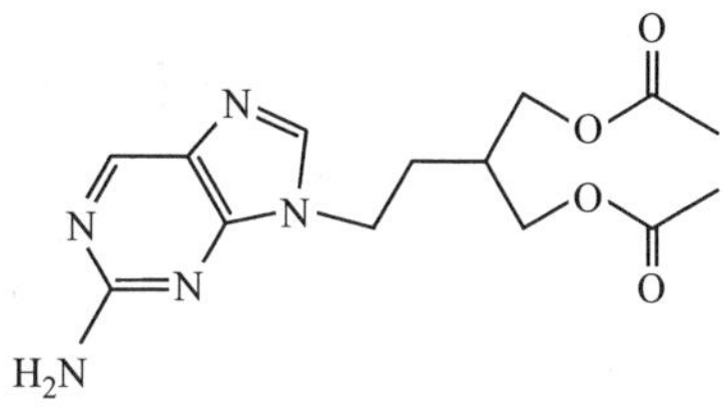

Famciclovir (9.11)

Valacyclovir has been successfully used in doses of 2 grams four times daily. Famciclovir (**9.11**) is the diacetyl ester prodrug of 6-deoxy penciclovir (**9.12**), an acyclic guanosine analog; in doses of 500 mg every 12 hours, famciclovir is an effective viral DNA polymerase inhibitor. Ganciclovir (**9.13**) is an acyclic guanosine analog that requires triphosphorylation prior to being activated as a viral DNA polymerase inhibitor. Cidofovir (**9.14**) is a cytosine nucleotide analog with activity against HSV-1, HSV-2, adenovirus, and poxvirus.

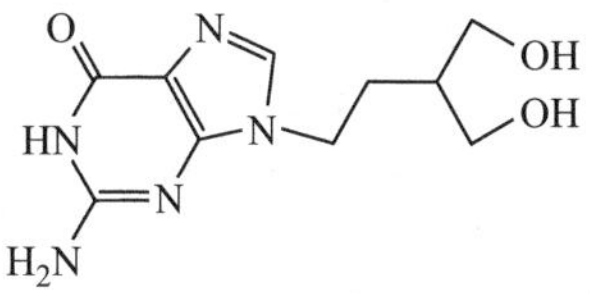

Penciclovir (9.12)

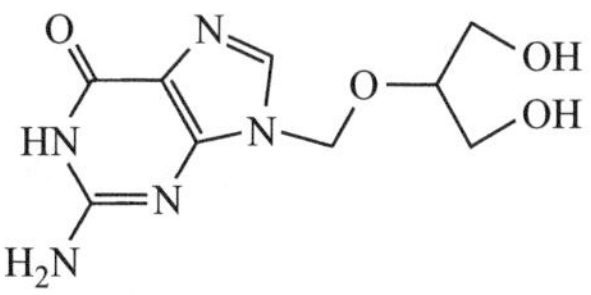

Ganciclovir (9.13)

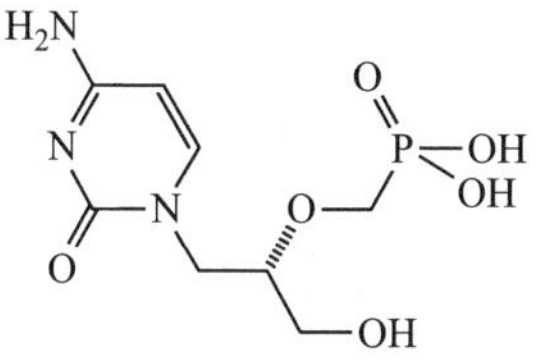

Cidofivir (9.14)

Another anti-herpes compound is the remarkably simple foscarnet (**9.15**, phosphonoformate). This is an inorganic pyrophosphate compound that inhibits viral DNA polymerase and RNA polymerase without requiring activation by phosphorylation. Fomivirsen is an oligonucleotide that inhibits human cytomegalovirus (CMV) via an

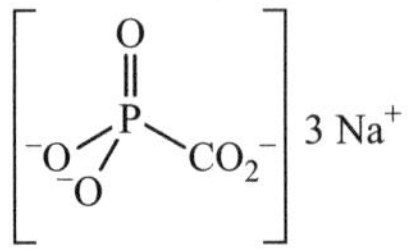

Foscarnet (9.15) 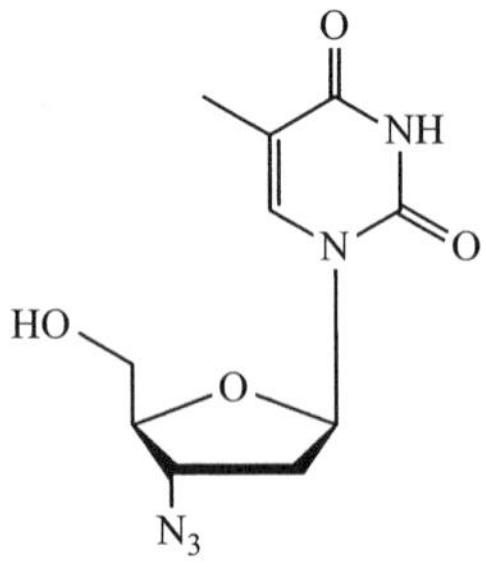   AZT (9.16)

antisense mechanism. It binds to target mRNA, resulting in inhibition of early protein synthesis within the virus-infected cell. Fomivirsen may even be injected directly into the eyeball for the treatment of CMV retinitis.

9.3.2 Antivirals Targeting Antiretroviral Mechanisms

The *acquired immune deficiency syndrome* (AIDS) has become a serious public health problem around the world, and commands much concern in medical and lay circles alike. Its causative agent is a human T-cell lymphotropic virus (HIV) that destroys helper/inducer T cells (discussed in chapter 6) of the immune system and causes mortality by allowing opportunistic infections and malignancies. A compound designed along traditional lines, 3′-azido-3′-deoxythymidine (**9.16**, AZT), emerged as one of the first useful therapeutics. However, since that time many other drugs have been devised.

Drugs designed for the treatment of AIDS function through antiretroviral mechanisms. Antiretroviral agents may be classified into three categories:

1. Competitive reverse transcriptase inhibitors
2. Non-competitive reverse transcriptase inhibitors
3. Protease inhibitors

The first agents to be developed were nucleoside analogs designed to function through the inhibition of the viral reverse transcriptase enzyme. These agents are also referred to as *nucleoside reverse transcriptase inhibitors* (NRTIs). Azidothymidine (AZT) is a deoxythymidine analog and was the first successful drug in this class. It is an effective drug, decreasing the rate of clinical disease and prolonging survival. Didanosine (**9.17**, ddI) is a synthetic analog of deoxyadenosine; lamivudine (**9.18**, 3TC) and zalcitabine (**9.19**, ddC) are cytosine analogs; stavudine (**9.20**, d4T) is a thymidine analog; abacavir (**9.21**) is a guanosine analog. All of these compounds function as nucleoside reverse transcriptase inhibitors.

The second class of agents comprises non-competitive inhibitors of reverse transcriptase. These agents are also referred to as *non-nucleoside reverse transcriptase inhibitors* (NNRTIs). Unlike NRTIs, NNRTIs do not require phosphorylation to be activated and do not compete with nucleoside triphosphates. The NNRTIs bind to a site on the viral reverse transcriptase that is close to but separate from the NRTI receptor site. This binding ultimately results in blockade of RNA- and DNA-dependent DNA

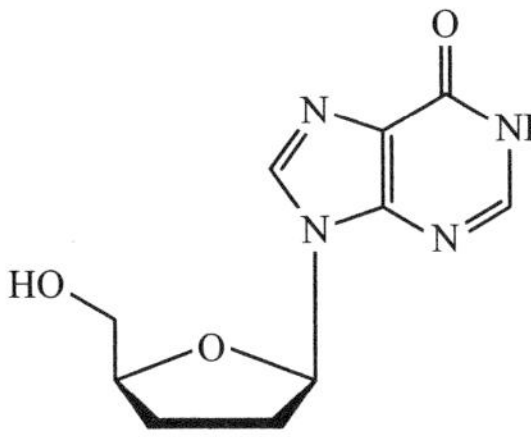

Didanosine (9.17)

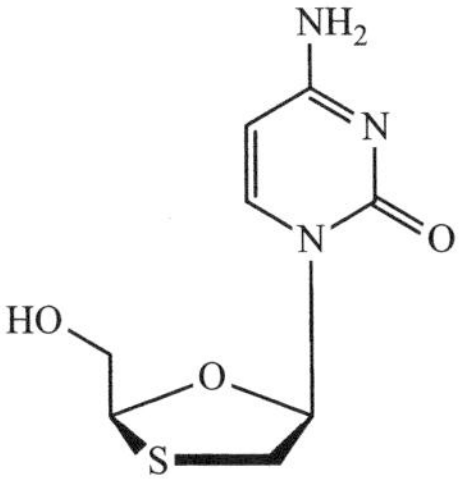

Lamivudine (9.18)

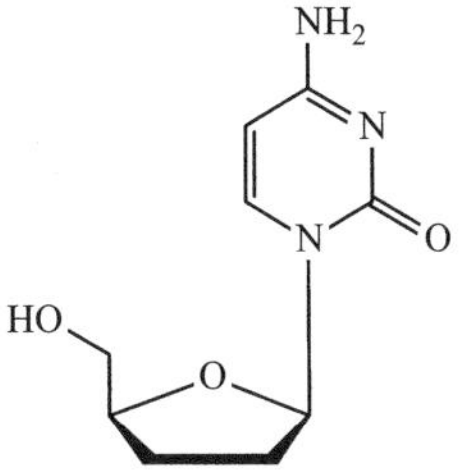

Zalcitabine (9.19)

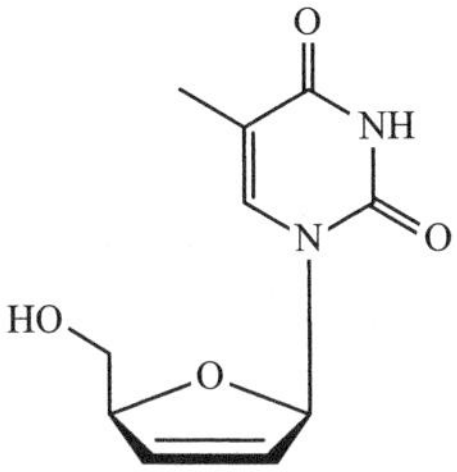

Stavudine (9.20)

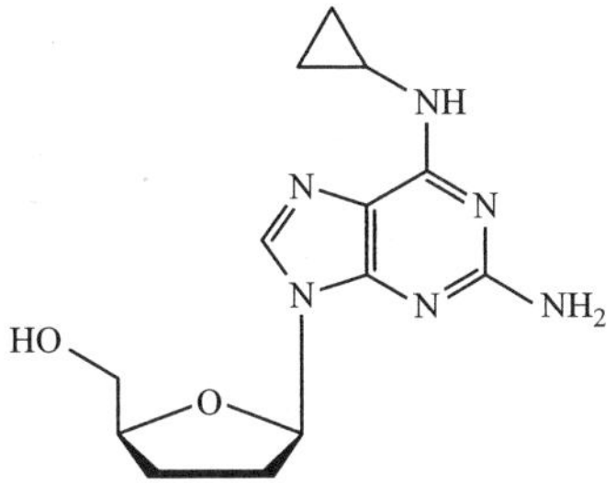

Abacavir (9.21)

polymerase activities. The NNRTIs include nevirapine (**9.22**), delavirdine (**9.23**), and efavirenz (**9.24**).

The third and most recent class of antiretroviral agent is the *protease inhibitor*. During the late stage of the HIV growth cycle, precursor macromolecules are cleaved into structural proteins that constitute an integral part of the mature HIV virion particle. This cleavage process is catalyzed by a protease enzyme; inhibition of this enzyme renders the viral particle non-infectious, preventing the ongoing wave of HIV infection.

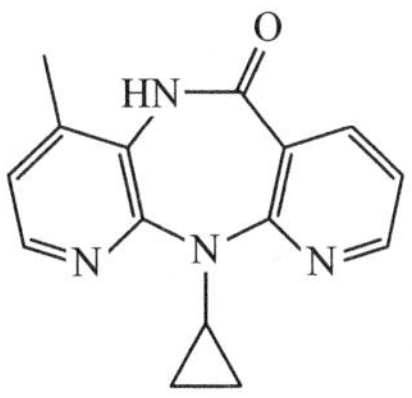

Nevirapine (9.22)

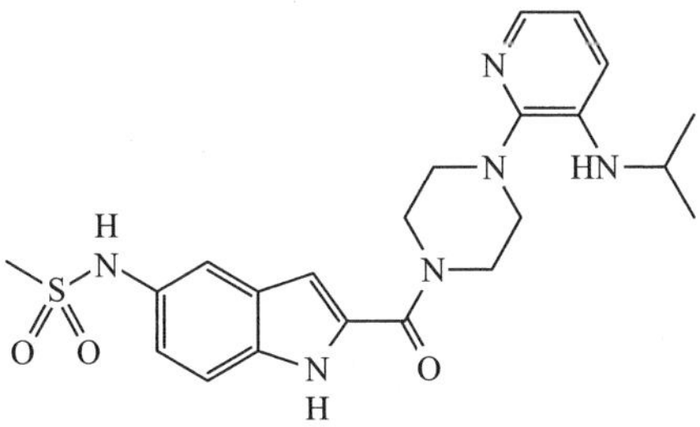

Delavirdine (9.23)

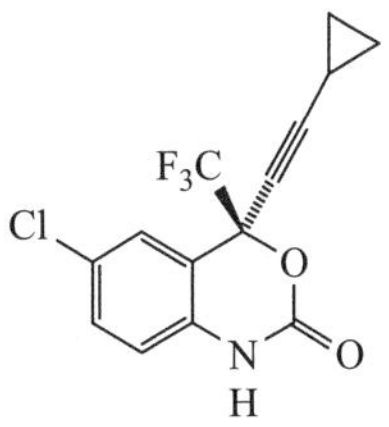

Efavirenz (9.24)

HIV protease is a dimer in which each monomer has an active site containing one or two aspartate residues. Using crystallographic three-dimensional data and molecular modeling studies, drugs have been designed to be transition-state mimetics, aligning at the enzyme active site. Some of the successful protease inhibitors include saquinavir (**9.25**), ritonavir (**9.26**), indinavir (**9.27**), nelfinavir (**9.28**), and amprenavir (**9.29**).

In the clinical setting, the combination of at least two antiretroviral agents is now recommended to ensure enhanced potency and to delay the emergence of resistance.

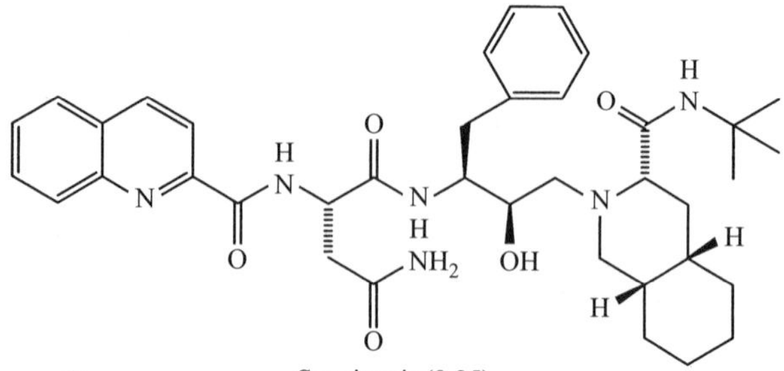

Saquinavir (9.25)

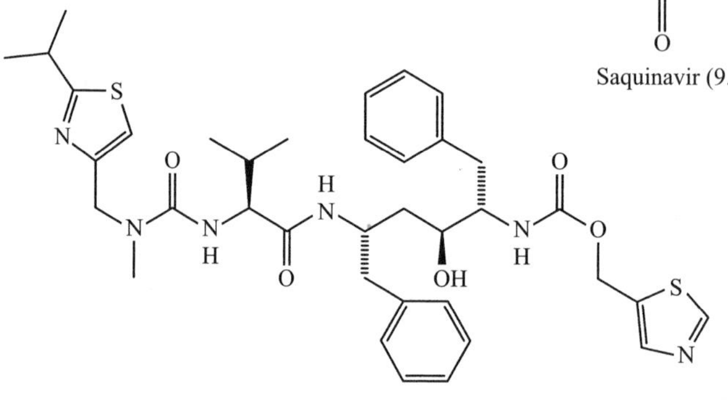

Ritonavir (9.26)

9.3.3 Antivirals Targeting Other Mechanisms

Amantadine (**9.1**, 1-aminoadamantane) and related compounds (e.g., rimantadine (**9.30**)) are available commercially for the prophylactic treatment of influenza A. Amantadine is not active against many strains of the influenza virus, which is a disadvantage when one considers the great variability of the virus. Amantadine is a cyclic amine that inhibits the uncoating of the viral RNA of influenza A. Fortuitously, amantadine is also an antiparkinsonism drug, acting as a cholinergic blocking agent—a totally unrelated effect. Amantadine has even been used to treat fatigue in multiple sclerosis.

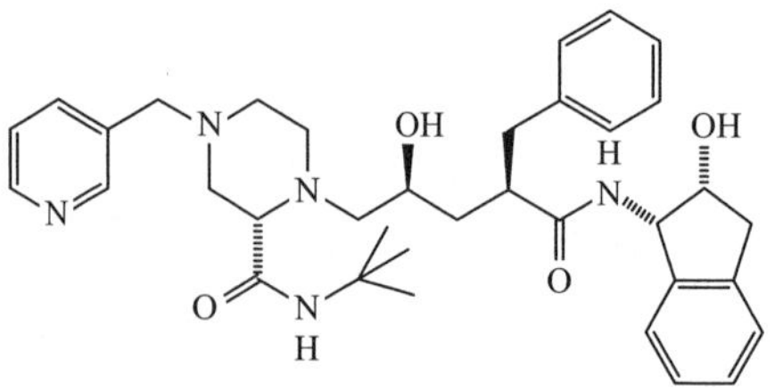

Indinavir (9.27)

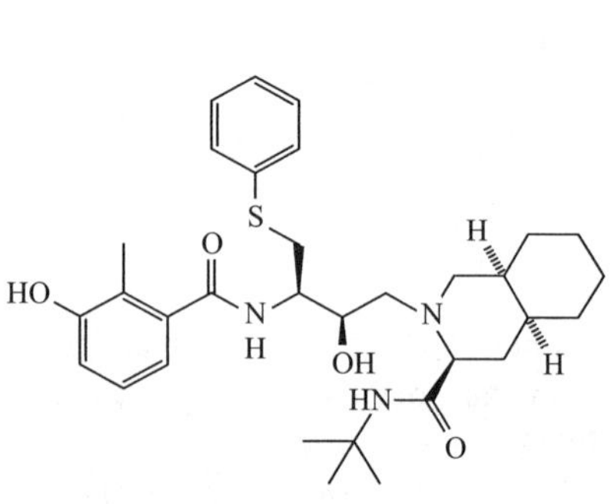

Nelfinavir (9.28)

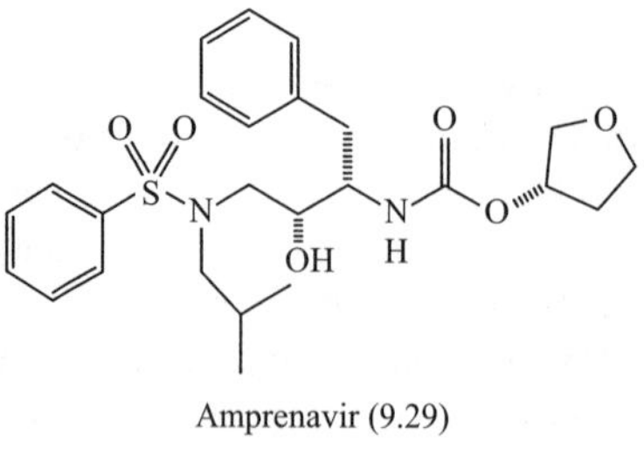

Amprenavir (9.29)

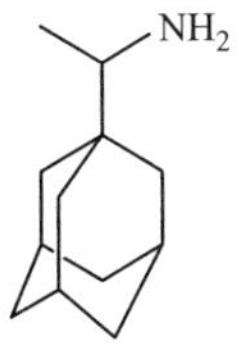

Rimantadine (9.30) 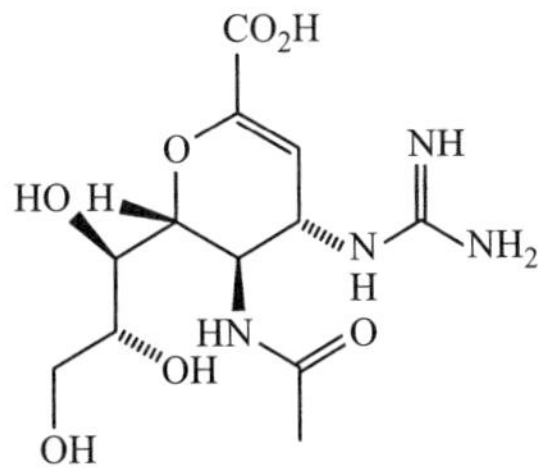   Zanamivir (9.31) 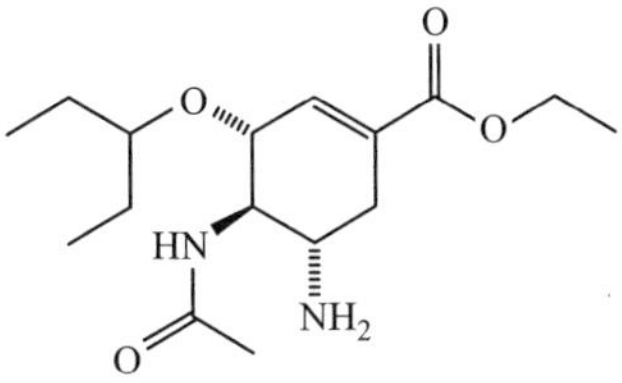   Oseltamivir (9.32)

Zanamivir (**9.31**) and oseltamivir (**9.32**) are neuraminidase inhibitors. Neuraminidase is an essential viral glycoprotein, playing a central role in viral replication and release. These agents work against both influenza A and B. Zanamivir is given intranasally, 10 mg twice daily for 5 days. Peramivir (**9.33**, figure 9.2) is another selective influenza neuraminidase inhibitor that has been under development. The synthesis of peramivir is shown in figure 9.2.

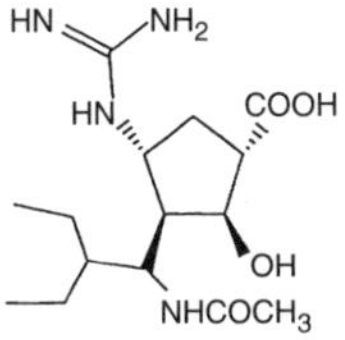

(Peramivir (9.33)
(BCX-1812)

Among drugs with uncertain modes of action, ribavirin (**9.34**) is not exactly a nucleotide since the purine ring is replaced by a triazole. It is active against HSV-1 and -2, hepatitis, and perhaps influenza viruses. It seems to have multiple effects on viral replication, blocking RNA synthesis and mRNA capping. Rifampin, an antibacterial, is effective against DNA viruses like HSV and the smallpox virus. It also prevents the virus-induced transformation of cells to malignant forms. Bleomycin (**9.35**), primarily an antitumor agent, also shows some activity as an antiviral agent.

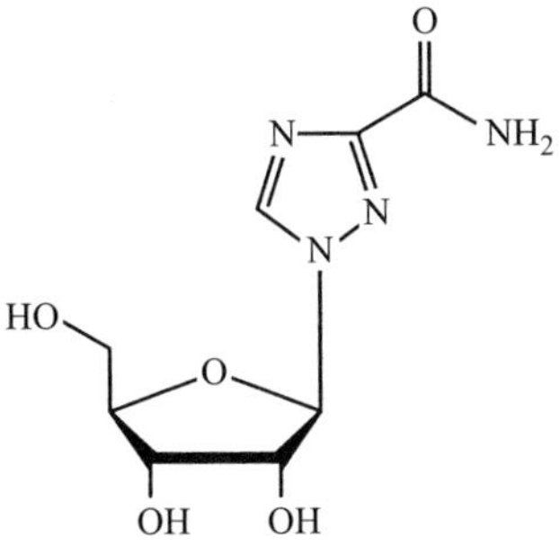

Ribavirin (9.34)

An antiviral (and anticancer) compound very much in the news media is interferon, a peptide consisting of about 150 amino acids (discussed in chapter 6). It is produced by most cells upon viral infection or a challenge by interferon-inducing agents, and protects

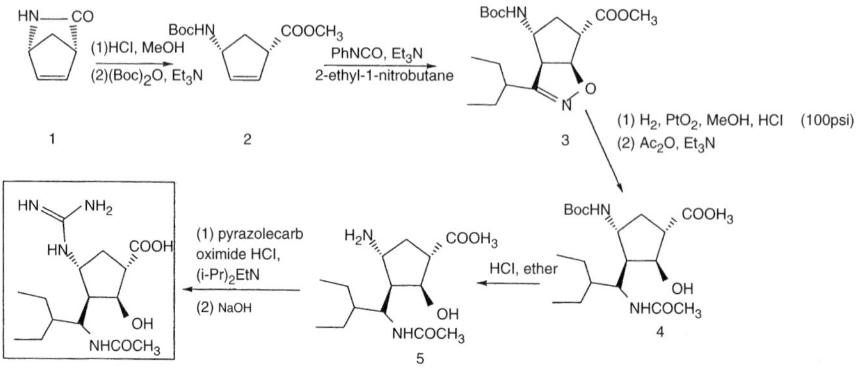

Figure 9.2 Synthesis of Peramivir. The starting material, (−)-lactam, **1**, underwent hydrolysis with methanol in acidic conditions. The amino group of the resulting amino ester was protected with a *t*-butyloxycarbonyl group (Boc) to give **2**. A [3+2] cycloaddition between **2** and the nitrile oxide produced from 2-ethyl-1-nitrobutane, phenyl isocyanate, and triethylamine gave **3** and a mixture of isomers. Compound **3** was isolated, then hydrogenated in a 1:1 methanol:HCl (aq.) mixture in the presence of PtO_2 at 100 psi to give an amine hydrochloride. This was then reacted with acetic anhydride to give the corresponding *N*-acetyl derivative **4**. Subsequent to this, the amine was deprotected in an ether–acid solution to give **5**. Compound **5** was guanylated with pyrazole carboxamidine hydrochloride in DMF in the presence of diisopropylethylamine. Finally, treatment with NaOH gave the desired compound, Peramivir. (Reference: Y. S. Babu, P. Chand, S. Bantia, P. Kotian, A. Dehghani, Y. El-Kattan, T. -H. Lin, T. L. Hutchinson, A. J. Elliot, C. D. Parker, S. L. Ananth, L. L. Horn, G. W. Laver, J. A. Montgomery, (2000). *J. Med. Chem. 43*: 3482.)

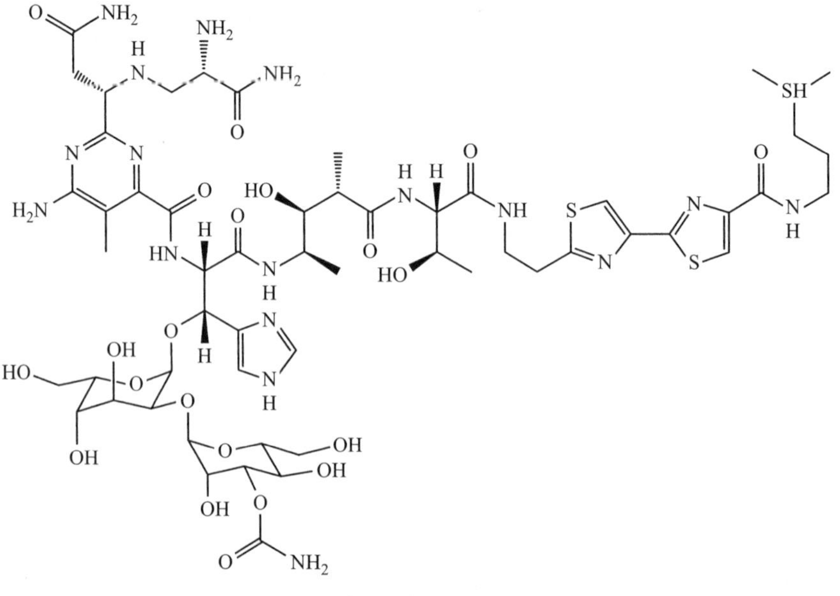

Bleomycin (9.35)

cells against viral infection by altering the plasma membrane (i.e., in a nonspecific manner). It also activates an endonuclease that destroys viral mRNA and a protein kinase that inactivates a protein synthesis initiation factor. Interferon can be isolated very laboriously from leukocytes and other cells, but the gene that encodes its synthesis has been transferred into bacteria by recombinant DNA techniques. Recently, α-interferon has been suggested as a possible therapy for SARS.

9.3.4 Antiviral Design: Evolving Targets

Viruses are important pathogens in human disease. The possibility of a worldwide "viral flu" pandemic that kills 50 million people is a haunting spectre. The need for new and effective antivirals is undeniable. Accordingly, new avenues of antiviral drug design are being pursued. Classes of agents known as *fusion inhibitors* are being devised. A fusion protein is a viral membrane protein that interacts with lipids to induce fusion during viral entry into the host cell. Various classes of fusion proteins are recognized; some have had their three-dimensional structures evaluated by crystallographic studies.

In addition to fusion inhibitors, novel viral enzymes (e.g., integrase) are also being studied; integrase inhibitors may constitute a new direction in antiviral therapy. Viral serine proteases have likewise been identified as attractive antiviral targets. Based upon crystallographic structural data, novel peptidomimetic inhibitors are being developed.

9.4 DRUG DESIGN TARGETING BACTERIA

Traditionally, biologists endeavored to categorize all organisms into two kingdoms: plant or animal. Realizing that this broad generalization could not apply to all species, Haeckel proposed in 1866 that microorganisms be placed in a separate kingdom, the *Protista*. The Protista were further subdivided into *prokaryotes* and *eukaryotes*. Eukaryotes (protozoa, fungi, molds) are more advanced, having their genetic material in a nucleus that is separated from the rest of the cell by a nuclear membrane; prokaryotes (bacteria, blue-green algae) are more primitive and are characterized by having their genetic material in the form of simple filaments of DNA that are not separated from the cytoplasm by a membrane. Bacteria are differentiated from other prokaryotes by the fact that they do not have chlorophyll for purposes of photosynthesis.

Bacteria have been killing humans for millennia. Many diseases (pneumonia, meningitis, gangrene) are caused by bacterial infections. One of the greatest triumphs of medicinal chemistry in the 20th century was the discovery of antibacterial drugs. There are many types of bacteria that can produce disease in humans. Although they can be classified using rigid biological nomenclature systems, this is not always clinically relevant. Accordingly, a somewhat more commonly employed classification system is as follows:

1. Gram-negative enteric bacteria (infectious diarrhea)
   Pseudomonas
   Salmonellae
   Shigellae
   Vibrios

2. Gram-negative small rods (respiratory infections: pneumonia, whooping cough)
   Hemophilus influenzae
   Bordetella pertussis
3. Gram-positive bacilli (food poisoning)
   Clostridium botulinum
   Clostridium tetani
4. Pyogenic cocci (various localized infections)
   Staphylococci (wound infections)
   Streptococci (pharyngitis—"sore throat")
   Pneumococci (pneumonia)
   Neisseria meningitides (meningitis)
   Neisseria gonorrheae (gonorrhea)
5. Gram-positive rods
   Corynebacterium diphtheriae (diphtheria)
6. Mycobacteria (slow chronic infections)
   Mycobacterium tuberculosis (tuberculosis)
   Mycobacterium leprae (leprosy)
7. Spirochetes (chronic diseases, venereal disease)
   Treponema pallidum (syphilis)
8. Rickettsia (fever and rash transmitted by arthropods such as lice)
   Rickettsia prowazekii (typhus)
   Rickettsia rickettsii (Rocky Mountain spotted fever)
9. Chlamydia
   Chlamydia trachomatis (venereal infection)

This classification is based on various criteria, including shape of the bacterium (rod, cocci) and the ability of the bacteria to be stained with a crystal violet-iodine complex in the presence of alcohol (Gram staining). This classification system also contains rickettsiae and chlamydiae; both of these tend to be at the interface with viruses in that they are intracellular bacteria.

Currently there is a wide variety of agents available for the treatment of bacterial infections. A *broad spectrum* agent works against many types of bacteria. A *bacteriostatic* agent does not kill bacteria but does inhibit their reproductive growth; a *bactericidal* agent actually kills bacteria. The term *antibiotic* is frequently used interchangeably for antibacterial. The word antibiotic, proposed by Waksman in 1942, refers to a substance that is able to inhibit the growth or even destroy microorganisms; the term is derived from Vuillemin's concept of *antibiosis* (which literally means "against life"). The designation of antibiotic can thus be applied not only to antibacterials but also to other antimicrobials such as antifungal agents. To avoid confusion, this book will use the more precise term of antibacterial agent.

The rational design of antibacterial agents depends upon the exploitation of a molecular structural feature found in bacteria but not found in humans. There are a number of such targets within bacteria, including the bacterial cell wall, bacterial cell membrane, bacterial protein synthesis, and bacterial nucleic acid synthesis. Various types of antibacterial agents can be categorized into four druggable target groupings according to the general structure of bacteria as shown in figure 9.3:

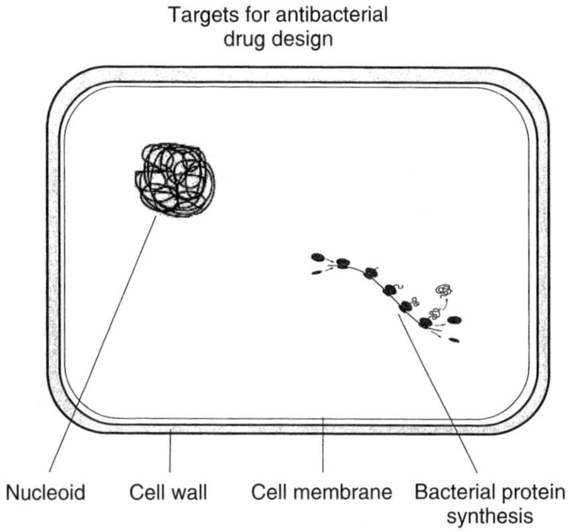

Targets for antibacterial
drug design

Nucleoid   Cell wall   Cell membrane   Bacterial protein
synthesis

Figure 9.3 Targets for antibacterial drugs. The various classes of antibacterial drugs exert their effects at one of the four fundamental structural components of bacteria. Each of these components is vulnerable to drug attack. Penicillin, for example, attacks at the level of the cell wall; chloramphenicol, however, works at the level of bacterial protein synthesis.

1. Bacterial cell wall targets
   Penicillins (bactericidal; inhibit cell wall crosslinking)
      e.g., benzylpenicillin, phenoyxmethylpenicillin, ampicillin, amoxicillin, flu-
         cloxacillin, methicillin, piperacillin
   Cephalosporins (bactericidal; inhibit cell wall crosslinking)
      e.g., cefaclor, cefalexin, cefradine, cefuroxime, cefazolin, cefotaxime, ceftriaxone,
         cefoxitin, cefsulodin, ceftazidime, ceftizoxime
   Monobactams (bactericidal, β-lactam-like activity)
      e.g., aztreonam
   Carbapenems (bactericidal, β-lactam-like activity)
      e.g., imipenem
   Bacitracin (bactericidal; interrupts mucopeptide synthesis)
   Vancomycin (bactericidal; interrupts mucopeptide synthesis)
   Cycloserine (bactericidal; interrupts synthesis of cell wall peptides)
2. Bacterial cell membrane targets
   Polymyxins (bactericidal; disrupt bacterial membrane structural integrity)
3. Bacterial protein synthesis
   Chloramphenicol (bacteriostatic; interrupts protein synthesis at the ribosome)
   Macrolides (bacteriostatic; interrupt protein synthesis at the 50S ribosome subunit)
      e.g., erythromycin, azithromycin, clarithromycin
   Lincomycins (bacteriostatic; interrupt protein synthesis at the 50S subunit)
   Aminoglycosides (bactericidal; interrupt protein synthesis at the 30S subunit)
      e.g., gentamicin, amikacin, kanamycin, neomycin, tobramycin
   Tetracyclines (bacteriostatic; interrupt protein synthesis at the 30S subunit)
      e.g., tetracycline, doxycycline, minocycline

4. Bacterial nucleic acid synthesis
   Sulfonamides (bacteriostatic; inhibit bacterial folic acid synthesis)
   Trimethoprim (bacteriostatic; inhibits bacterial folic acid synthesis)
   Quinolones (bacteriostatic; inhibit DNA gyrase)
      e.g., ciprofloxacin, cinoxacin, enoxacin, norfloxacin
   Rifampin (bactericidal; blocks mRNA synthesis in bacteria, inhibits RNA
      polymerase)

Table 9.2 lists various bacteria and their susceptibility to different classes of antibacterial agents.

9.4.1 Antibacterials Targeting Cell Wall Synthesis

The successful chemotherapeutic management of any host–parasite interaction—whether viral, bacterial, or protozoan—depends upon the exploitation of biochemical differences between the host and the parasite. The greater these differences are, the better the likelihood of finding or designing drugs that exploit them and inhibit some crucial function of the parasite in order to kill it without harming the host cell. This almost utopian goal (Paul Ehrlich's "magic bullet") has been approximated very closely in the case of cell wall synthesis inhibitors, such as antibacterial agents, for the simple reason that a very fundamental difference exists between bacteria and mammalian cells: the former have cell walls and the latter do not. The rigid cell wall of bacteria encloses and strengthens the vulnerable cell membrane, which is subjected to considerable internal osmotic pressure. If the integrity of the cell wall is impaired, the bacterial cell will undergo breakdown (*lysis*) and the bacterium will perish. The antibiotics that inhibit cell wall synthesis cannot find an analogous target in animal cells and are in most cases extremely nontoxic.

Cell walls are complex and variable structures but have a number of common characteristics, discussed in most biochemistry textbooks and numerous monographs. The basic structural unit of the wall is the *muropeptide* [*murus* (Latin) = wall], a repeating disaccharide linked through a lactyl ether to a tetrapeptide. The peptides are, in turn, crosslinked (in *Staphylococcus aureus*) by a pentaglycine chain. The resulting polymer, called *murein,* forms a closed sack around the bacterium and can be dissolved by the enzyme lysozyme. Other glycopeptides, such as teichoic acid, and polypeptides contribute to the antigenic properties of bacteria.

The classical division of bacteria into Gram-positive and Gram-negative groups on the basis of specific staining procedures also depends on cell wall components. The Gram-positive organisms have a rigid cell wall that is covered with an outer layer containing teichoic acids, whereas the wall of Gram-negative bacteria is covered with a smooth, soft lipopolysaccharide. Most penicillins are much more effective against Gram-positive bacteria.

During the biosynthesis of the cell wall, the muropeptide is formed from acetylmuramyl-pentapeptide, which terminates in a D-alanyl-D-alanine. The synthesis of this precursor is inhibited by the antibiotic cycloserine (**9.36**), a compound produced by many *Streptomyces* fungi but which is not used clinically. During the crosslinking of the pentapeptide precursor, the terminal fifth alanine must be split off by a transpeptidase enzyme. This last reaction in cell wall synthesis is inhibited by the β-lactam antibiotics,

Table 9.2 Sensitivities of Some Organisms to Commonly Used Antibiotics

Drug	1	2	3	4	5	6	7	8	9	10	11	12	13	14	15	16	17	18	19	20	21
Ampicillin/Amoxicillin	*	*	*	**	*	*	**	**	*	**							**			**	
Cefoxitin																	*				
Ceftazidime											*		**								
Cefuroxime/cefotaxime cefamandole		*		*			*	*	**	*	**	*									
Chloramphenicol			*			*		*							*				*		
Co-trimoxazole	*	*				*		*	*	**	**	**				*				*	
Erythromycin	*	*	*		*	*			*						**			**	*		
Gentamicin	*	*		*						*	*	*	**								
Flucloxacillin		**											**								
Metronidazole																	**				
Penicillin G/Penicillin V	**	**	**	**	**	**	**									*					
Piperacillin												**	**								
Quinolones													**			**					
Tetracyclines	*		*					*	*	*	*			**	*			*	**		**

** Drug of first choice.

* Drug of second choice.

[1]Staphylococcus aures (penicillin-sensitive); [2]Staphylococcus aureus (penicillin-resistant); [3]Streptococcus pyogenes; [4]Streptococcus pneumoniae; [5]Enterococcus faecalis; [6]Viridans' streptococci; [7]Neisseria meningitidis; [8]Neisseria gonorrhoeae; [9]Haemophilus influenzae; [10]Escherichia coli; [11]Klebsiella; [12]Proteus mirabilis; [13]Pseudomonas aeruginosa; [14]Brucella; [15]Legionella pneumophila; [16]Salmonella typhi; [17]Bacteroides spp. and other anaerobes; [18]Mycoplasma pneumoniae; [19]Chlamydiae; [20]Listeria monocytogenes; [21]Rickettsiae.

(Adapted from D. G. Grahame-Smith, J. K. Aronson (2002). *Clinical Pharmacology and Drug Therapy*, 3rd Edn. New York: Oxford University Press. With permission.)

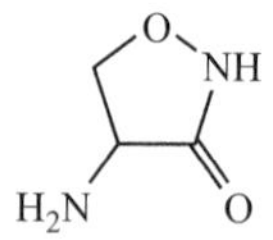

Cycloserine (9.36)

such as the penicillins and cephalosporins, after the bacterium has expended considerable biosynthetic energy. In contrast to this inhibition of the last step of a reaction sequence, the feedback inhibition of enzymatic reactions normally occurs at the first step in a sequence, avoiding any wastage of precursor substances; if it occurs late, biochemical efficiency is seriously jeopardized.

Various investigations have elucidated many details of this process. First, the antibacterial agent has to penetrate the outer membrane of the Gram-negative bacteria, which are less susceptible to antibiotics. This membrane consists of lipopolysaccharides, phospholipids, lipoproteins, and proteins. The β-lactam antibiotics (penicillins, cephalosporins) cross this diffusion-resistant membrane through *porin channels*, trimeric proteins that traverse the membrane. There are about 10^5 channels per bacterial cell, and their diameter is 1.2 nm. Some bacterial genera (e.g., *Pseudomonas*) are insensitive to most β-lactam antibiotics because the majority of their porin channels are not functional. The next hurdle the antibiotic has to surmount involves the *β-lactamase* enzymes in the periplasmic space, between the outer and inner membranes; these can deactivate the antibiotic (Gram-positive bacteria excrete the lactamase into the medium). Beyond that is the peptidoglycan cell wall with the associated penicillin-binding proteins, which are the essential transpeptidases, transglycosylases, and D-alanine carboxykinases involved in cell wall synthesis. Penicillin (**9.37**) inactivates these by acylation of the active sites, as a "suicide substrate." Different transpeptidases have different roles in the cell, and their selective inactivation can lead to cell lysis or production of deformed (spherical or threadlike) cells as well as cell wall synthesis inhibition.

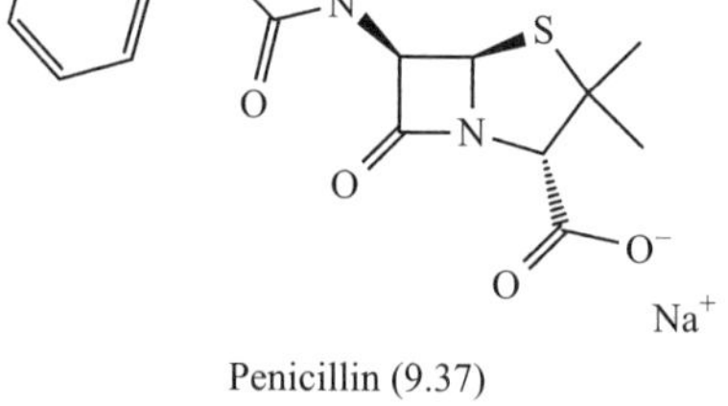

Penicillin (9.37)

9.4.1.1 Penicillins

The penicillins (or penams) were discovered in 1929 by Sir Alexander Fleming, and developed by Florey, Chain, and Abraham at Oxford University. The history of penicillin became a story of legendary proportions, illustrating the case of a serendipitous discovery combined with brilliant development; it also marks the beginning of the modern chemotherapy of infectious diseases.

The penicillins are produced by the molds *Penicillium notatum* and *P. chrysogenum*. Through the use of different culture media or biosynthetic precursors (e.g., phenylacetic acid), a number of biosynthetic ("natural") penicillins have been isolated, frequently distinguished by Roman numerals in the United Kingdom and by letters in the United States. The most important and still used among these is benzylpenicillin or penicillin G, a singularly nontoxic compound highly active against Gram-positive infections such as staphylococcal sepsis, meningitis, and gonorrhea.

Structurally, the β-lactam ring fused with the thiazolidine ring is most unusual, since β-lactam rings were unknown before the discovery of penicillin. Consequently, the elucidation of the structure of penicillin during World War II, a top-secret joint Anglo-American project, was a difficult undertaking, ultimately settled by X-ray crystallography. The penam ring can be considered as a dipeptide composed of a cysteine and a valine residue.

Owing to the strain of the four-membered β-lactam, the ring is easily cleaved by acid hydrolysis and alcoholysis, and by heavy metals such as Zn^{2+}, Cu^{2+}, and Pb^{2+}. The resulting penicilloic acid is inactive and undergoes a complex series of rearrangements. The acid sensitivity of penicillins varies with their structure. For example, phenoxymethyl penicillin is more resistant to acid cleavage than benzylpenicillin, and is therefore more suitable for oral use. Even so, considerably higher peroral doses are required than parenteral ones. Only among the semisynthetic penicillins does one find good acid resistance. The high reactivity of the β-lactam ring is the key to the biological activity of the β-lactam antibiotics. It acts as an irreversible inhibitor of the bacterial transpepticlase because it acylates the enzyme protein near the active site through opening of the lactam ring (see figure 9.4).

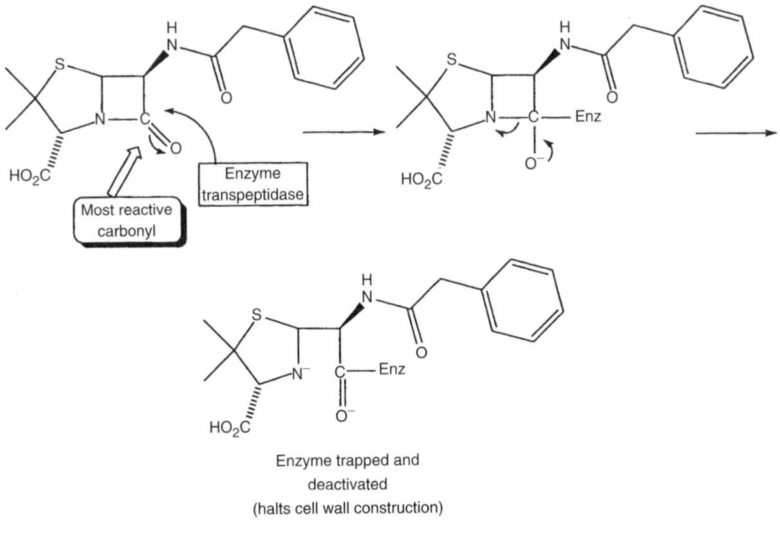

Figure 9.4 Mechanism of penicillin. By means of its highly reactive lactam ring, penicillin is able to deactivate the transpeptidase enzyme. This in turn leads to a halting of cell wall construction within the bacterium, ultimately leading to bacterial death.

Penicillins and Mechanisms of Bacterial Resistance. The most serious threat to antibiotic therapy, and to the use of β-lactam antibiotics in particular, is the emergence of resistant bacterial strains. The primary reason for this resistance is the production of an enzyme, β-lactamase (penicillinase), which in Gram-positive bacteria is excreted into the growth medium but in Gram-negative bacteria remains contained in the cell. Thus, Gram-positive organisms quickly destroy the antibiotic in the surrounding solution by hydrolysis, converting it to the inactive penicilloic acid. Since production of penicillinase enzymes is under plasmid control, resistant bacteria can transfer their resistance. Hence, bacterial species that were in the past easily controlled with penicillin have increasingly become a serious medical problem.

Inhibitors of β-lactamase are known. The synthetic sulfone tazobactam (**9.38**), and clavulanic acid (**9.39**) both have weak antibacterial activity besides β-lactamase inhibitory activity, and they can be used in combination with vulnerable antibiotics.

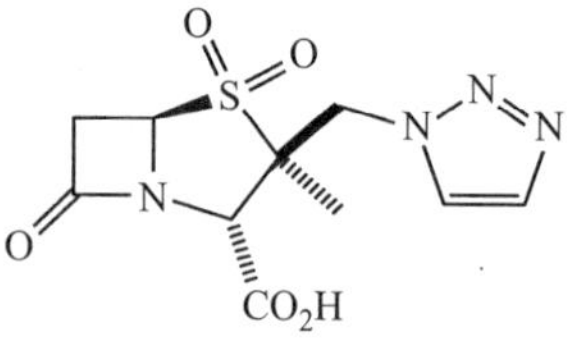

Tazobactam (9.38) 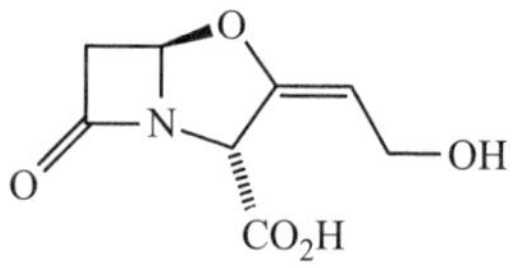   Clavulanic acid (9.39)

Semisynthetic Penicillins. However useful they may be, natural penicillins have several drawbacks. They have a relatively narrow activity spectrum, primarily inhibiting Gram-positive bacteria only. They are acid- and lactamase-sensitive, and in a small percentage of patients they cause allergic side effects. All of these limitations could potentially be overcome by molecular modifications during the biosynthesis of these drugs. Unfortunately, however, the fermentation process used in penicillin production is not very flexible and does not permit the incorporation of very many amide side chains into the molecule.

Although the total synthesis of penicillins was accomplished by Sheehan and his co-workers in 1953, and although some other approaches have also been successful, the syntheses are of limited practical value; nevertheless, they do allow modification of the ring system.

On the other hand, the discovery of the parent amine 6-aminopenicillanic acid (**9.40**, 6-APA) in fermentation products constituted a major breakthrough in penicillin synthesis.

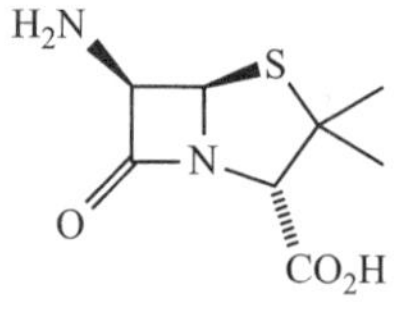

6-APA (9.40)

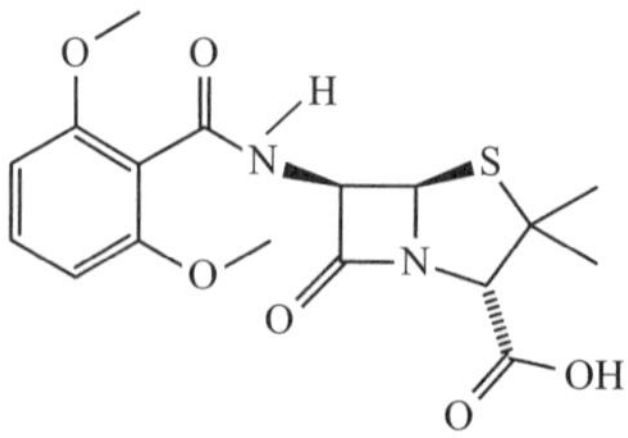

Methicillin (9.41) 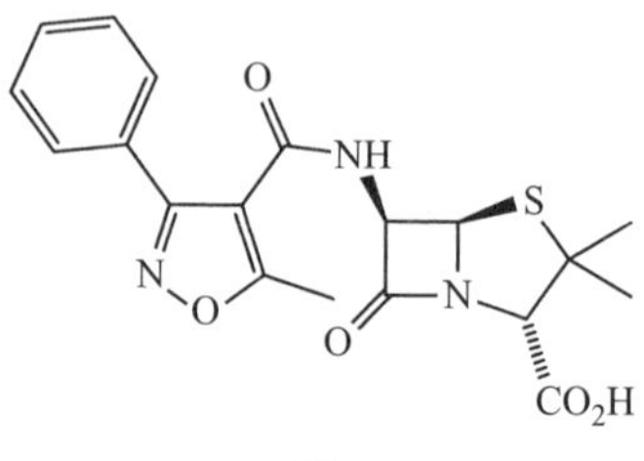   Oxacillin (9.42)

It is formed by acylases that cleave off the side chain of the penicillins, and can also be obtained by the selective chemical cleavage of the amide, leaving the lactam intact. After this, 6-APA can be easily acylated by any carboxylic acid, and this has yielded literally thousands of semisynthetic penicillins in the past 30 years, many showing improved stability and activity. Some of them are lactamase resistant (methicillin (**9.41**), oxacillin (**9.42**) and its halogenated derivatives), whereas others are broad-spectrum antibiotics, like the orally active ampicillin (**9.43**), which also inhibits Gram-negative bacteria but is sensitive to lactamase. Carbenicillin (**9.44**) is particularly active against *Pseudomonas* and *Proteus* infections, which are unaffected by "natural" penicillins. Piperacillin (**9.45**), a broad-spectrum compound, is spectacularly active against *Pseudomonas*.

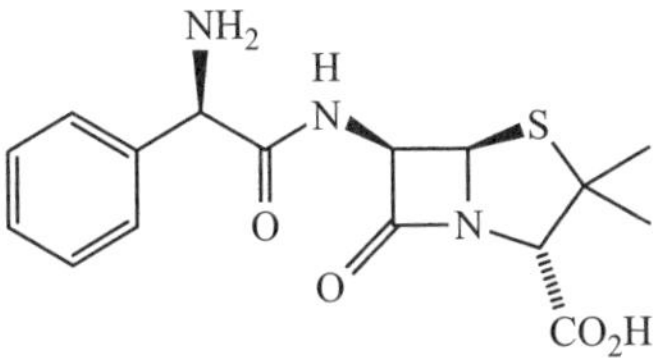

Ampicillin (9.43) 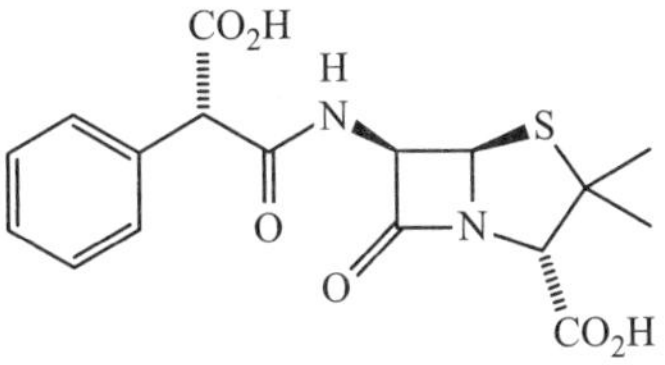   Carbenicillin (9.44)

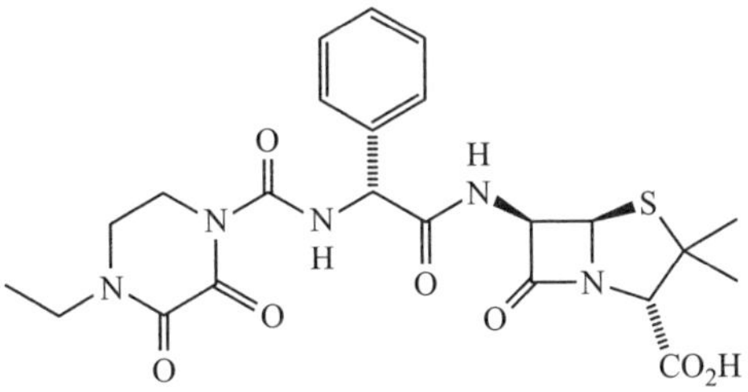

Piperacillin (9.45)

9.4.1.2 Cephalosporins

The cephalosporins, discovered in the 1950s, are produced by various species of the mold *Cephalosporium*. Cephalosporin C (**9.46**) is the prototype of these antibiotics, and its structure shows a close similarity to the penam structure. The 5-thia-1-azabicyclo[4.2.0] octane ring system is therefore called the cepham ring. The parent compound carries the aminoadipate side chain, which can be cleaved to supply the 7-amino-cephalosporanic acid. This amine can easily be acylated and thus forms the basis of many useful derivatives. The 3-acetoxymethyl substituent is also amenable to modifications.

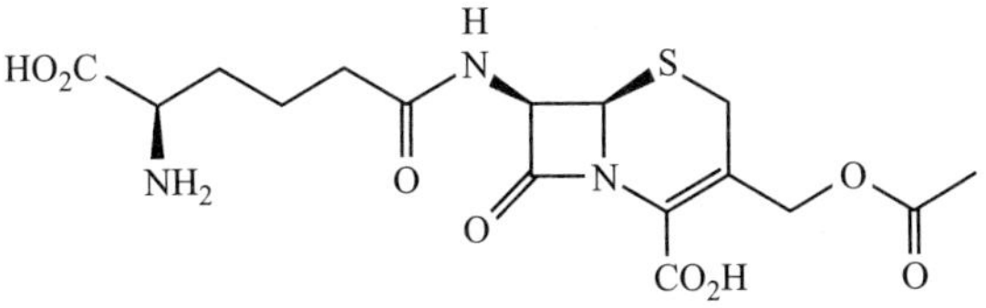

Cephalosporin C (9.46)

Since cephalosporin C is only one-thousandth as active as benzylpenicillin, its use is very limited. However, it is remarkably resistant to enzymatic hydrolysis and becomes highly concentrated in the urine, which makes it useful in urinary tract infections caused by Gram-negative organisms.

Among the semisynthetic derivatives, cephalothin (**9.47**) is the most widely used since it is a broad-spectrum antibiotic resistant to lactamase. Its main drawback is that it must be injected. Cefazolin (**9.48**) and cephaloridine (**9.49**) are metabolized to a lesser extent; cephalexin (**9.50**, analogs to ampicillin) is orally active and has a much higher acid stability than the penicillins. Cefotaxime (**9.51**) and moxalactam (**9.52**) are highly active against meningitis.

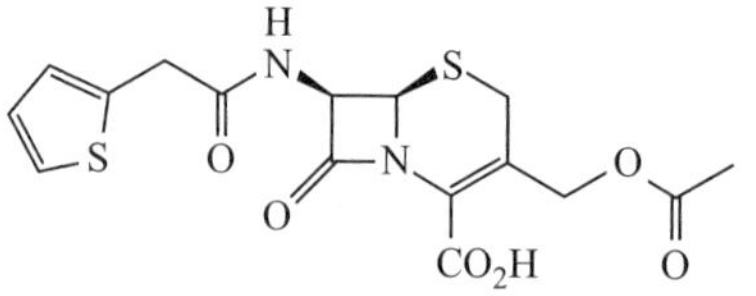

Cephalothin (9.47)

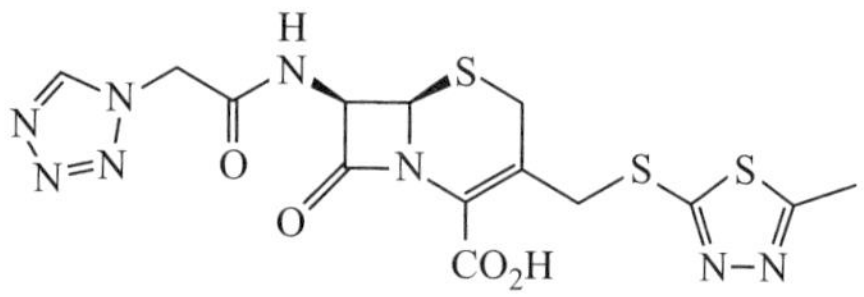

Cefazolin (9.48)

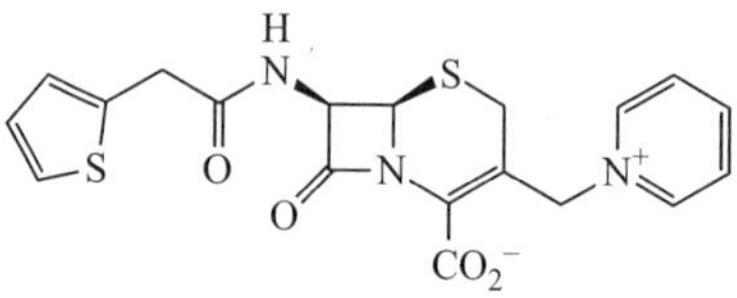

Cephaloridine (9.49)

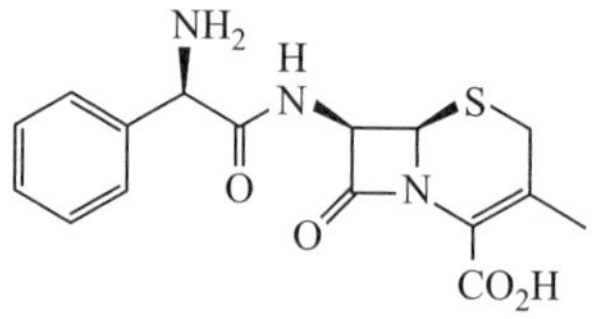

Cephalexin (9.50)

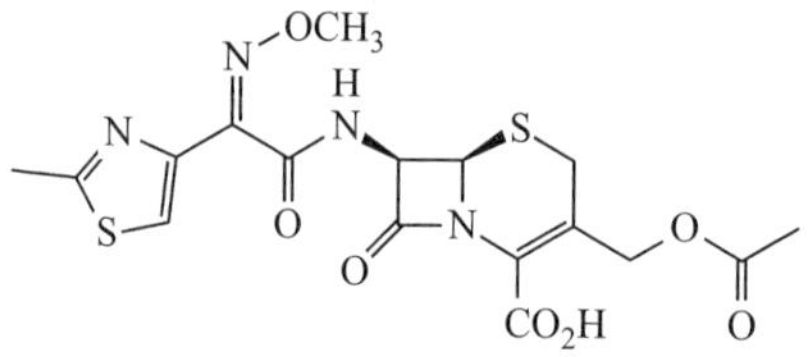

Cefotaxime (9.51)

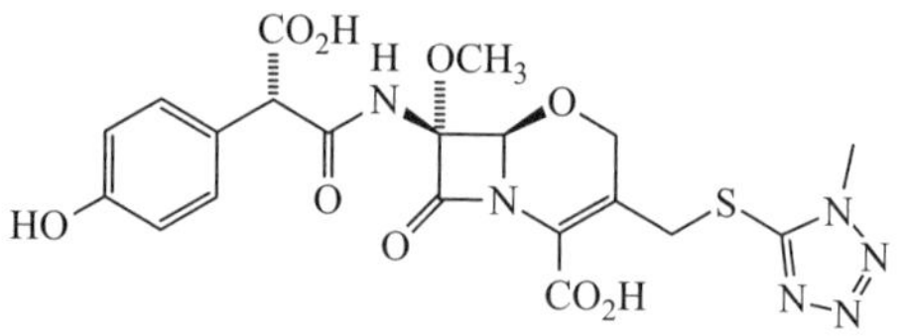

Moxalactam (9.52)

9.4.1.3 Other β-Lactam Antibiotics

Other β-lactam antibiotics have revolutionized our understanding of the structure–activity relationships in this large group of antibiotics. Thienamycin (**9.53**), discovered in 1976, is a broad-spectrum antibiotic of high activity. It is lactamase resistant because of its hydroxyethyl side chain but is not absorbed orally as it is highly polar. Unfortunately,

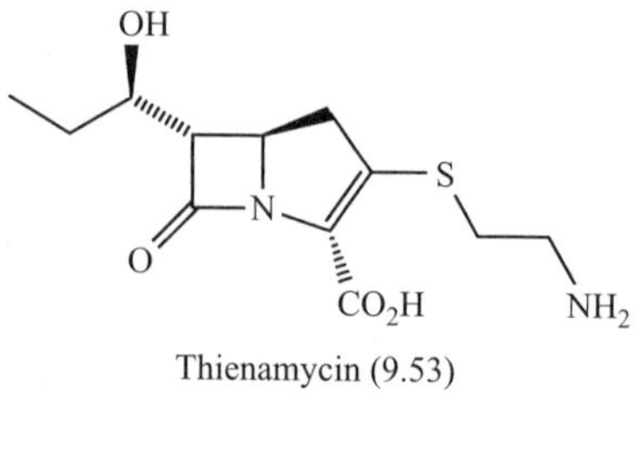

Thienamycin (9.53)

it is very unstable and therefore unlikely to be of use in its native form. The N-formimidyl derivative overcomes this problem.

Clavulanic acid (**9.39**), which is produced by a *Streptomyces* species, has only weak antibiotic activity but is a potent β-lactamase inhibitor. It can therefore protect lactamase-sensitive but otherwise potent antibiotics (e.g., ampicillin) from deactivation.

The monocyclic nocardicins represent the ultimate "simplification" of the β-lactam structure, containing the azetidinone ring by itself, with a side chain resembling that of cephalosporin C. Nocardicin A (**9.54**), the (Z)-oxime, has limited activity against some Gram-negative bacteria. The similar aztreonam (**9.55**) is active against Gram-negative bacteria and *Pseudomonas,* and is lactamase resistant.

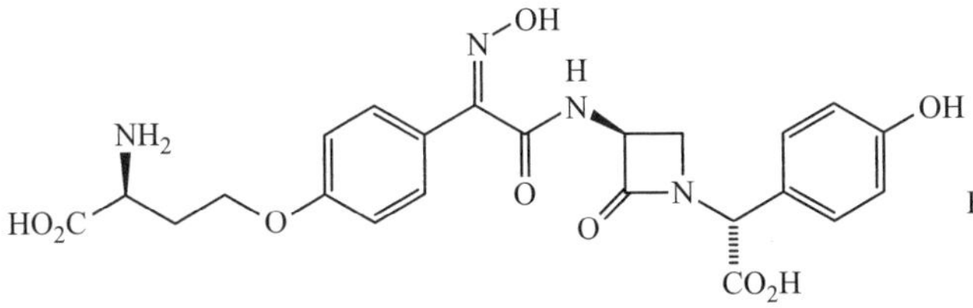

Nocardicin A (9.54)

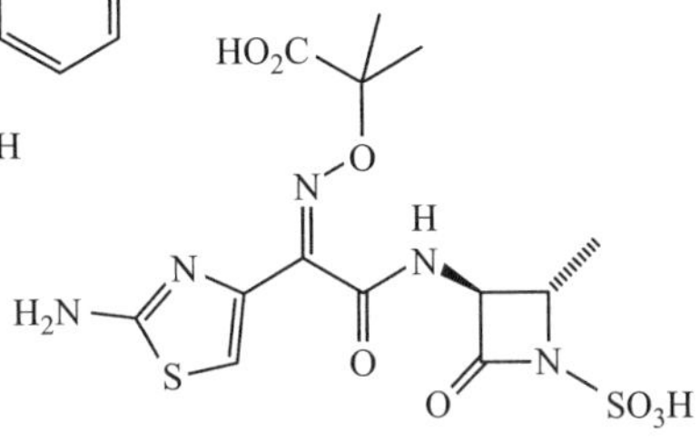

Aztreonam (9.55)

9.4.1.4 β-Lactam Structure–Activity Correlations

Structure–activity correlations in the β-lactam antibiotic field have required drastic re-evaluation in view of the novel structures described above. Apparently, only the intact β-lactam ring is an absolute requirement for activity. The sulfur atom can be replaced (moxalactam) or omitted (thienamycin), and the entire ring itself is, in fact, unnecessary (nocardicin). The carboxyl group, previously deemed essential, can be replaced by a tetrazolyl ring (as a bioisostere), which results in increased activity and lactamase resistance. The amide side chain, so widely varied in the past, is also unnecessary, as shown in the example of thienamycin. There is a considerable literature analyzing the classical structure–activity relationships of the penicillin and cephalosporin groups.

9.4.2 Antibacterials Targeting Bacterial Cell Membranes

Some microorganisms produce compounds that can become incorporated into lipid membranes and will facilitate the transmembrane transport of ions, notably K^+. These natural products are antibacterial, killing bacteria by lethally altering the transmembrane ion flux. Such antibacterial molecules are called *ionophores*, or ion carriers, in contrast to other antibacterials, such as polyene antibiotics, which simply produce leakage through the cell membrane.

Ionophoric antibiotics can function either as "cage" carriers of an ion or as channel formers. The cage carrier encloses an ion and transports it from one side of the membrane to the other, releasing the ion on the other side. A channel former simply provides a polar tunnel that allows the migration of a polar ion across an otherwise impenetrable lipid layer.

An example of an ionophore that is a cage carrier is valinomycin (**9.56**). This cyclic peptide lactone consists of three molecules each of L-valine, D-α-hydroxyisovaleric acid, and L-lactate. The six highly polarized lactone carbonyl oxygens line the inside of the ring, whereas the nonpolar alkyl groups point to the outside of the molecule. Thus

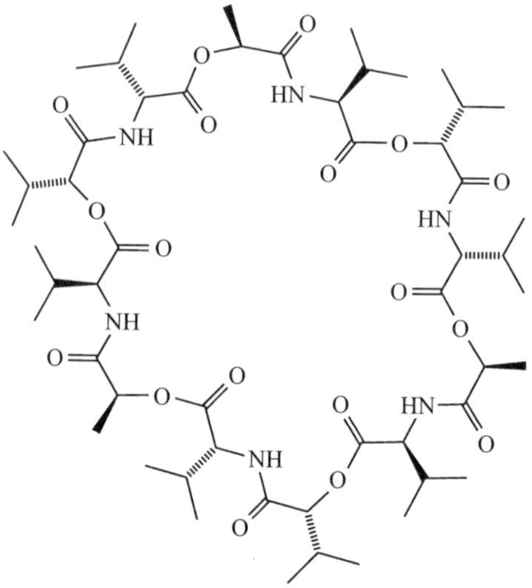

Valinomycin A (9.56)

the polar interior can accommodate a nonhydrated potassium ion and surround it with an apolar bracelet. This complex can then be transported through a membrane in an energy-dependent K^+–H^+ exchange. The selectivity of valinomycin for K^+ over Na^+ is very high, the ratio being about 10^4:1. In this way, valinomycin will increase the K^+ conductivity of lipid membranes at concentrations as low as 10^{-9} M. The high K^+ selectivity is due to the relative ease of dehydration of this ion: with its larger diameter, the potassium ion holds hydrate water less firmly than does sodium; consequently, whereas the hydrated sodium ion does not fit the valinomycin "doughnut," the dehydrated K^+ does bind easily, with the bonding energy providing a further energy advantage for the selective reaction. A hydrated sodium ion is larger than a potassium ion with or without a hydrate envelope.

An example of a channel- or pore-forming antibiotic is gramicidin A (**9.57**), a peptide consisting of 15 amino acids. It induces the transmembrane transport of protons, alkalimetal ions, and thallium ions at concentrations as low as 10^{-10} M, even though it is unable to complex these ions in solution. Gramicidin also forms several dimers with itself.

Val-Gly-Ala-Leu-Ala-Val-Val-Val-Trp-Leu-Trp-Leu-Trp-Leu-Trp-NH-CH$_2$-CH$_2$-OH

Amino Acid Sequence of Gramicidin A (9.57)

Many hypotheses explaining the channel formation induced by gramicidin A have been put forth over the years. For example, two molecules of gramicidin form a "head-to-head" helix, spanning the total width of the cell membrane. The helix creates a pore lined with hydrophilic groups, permitting ion transport across the otherwise impermeable lipid barrier of the membrane. Such a pore can accommodate even large ions, as long as they are dehydrated. The movement of ions in the pore has been simulated *in silico*, using elaborate molecular modeling calculations.

Colicins, bacteriostatic peptides encoded by bacterial plasmids, have been crystallized and investigated by X-ray crystallographic methods. They have a mass of 79,000 daltons and an axial ratio of 1:10, giving these peptides a length of about 20 nm. They are capable of forming a transmembrane channel with a diameter so large that glucose molecules can pass through. Of course, this fatally disrupts the membrane potential of the bacterial cell, with consequent bacteriostatic activity.

Ionophoric antibiotics do not distinguish microbial from mammalian membranes and are therefore therapeutically limited. However, they are excellent tools for studying membrane transport phenomena.

9.4.2.1 Surface-Active Antibacterial Agents

Biological membranes are indispensable to the proper functioning of all cells, including bacteria and fungi. Hence, any agent that disrupts the membrane or otherwise interferes with its integrity or function is a potential threat to the life of the cell.

Aliphatic alcohols are bactericidal because they damage the bacterial membrane, resulting in a rapid loss of the cytoplasmic constituents of the bacterium. At high concentrations, such alcohols cause lysis (dissolution) of the bacterial cell. Because they damage bacterial membranes, phenol (**9.58**) and cresol (**9.59**) are also effective disinfectants and are used in commercial preparations (e.g., Lysol). They not only denature proteins but also act as detergents, owing to the polarity of the phenolic hydroxyl group. The activity of phenols can be increased considerably by attaching an alkyl side chain to the benzene ring, as in *n*-hexylresorcinol (**9.60**), which makes the resulting compound more surface active. Hexachlorophene (**9.61**) and fentichlor (**9.62**) are also very active and are used in disinfectant

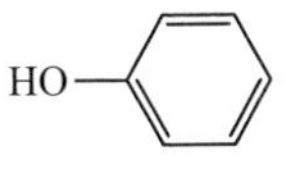

Phenol (9.58) 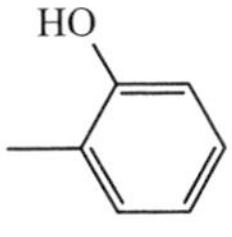   Cresol (9.59)

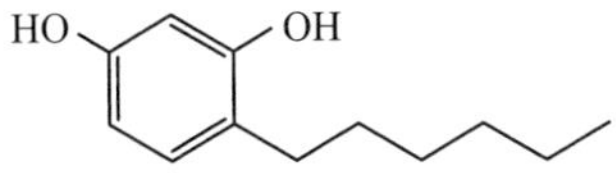

n-Hexylresorcinol (9.60) 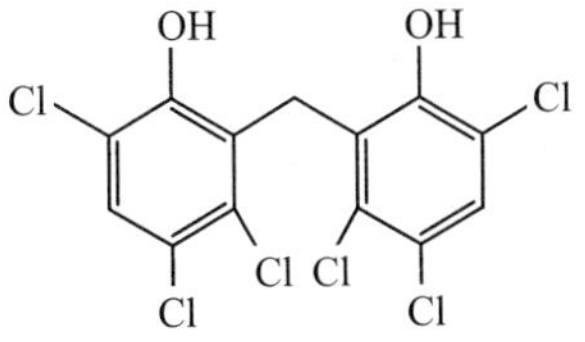   Hexachlorophene (9.61) 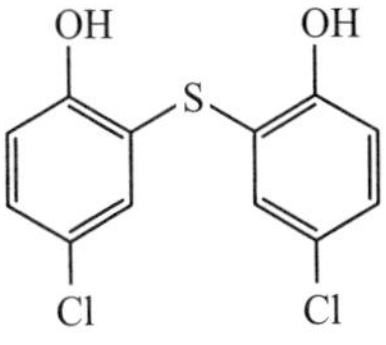   Fentichlor (9.62) soaps. Moreover, since fentichlor has a low oral toxicity, it can be used internally for skin infections. Inhibition of the metabolic electron-transport chain and of amino acid uptake also constitutes the bacteriostatic action of fentichlor.

Cationic detergents such as cetyl-trimethylammonium chloride (**9.63**) are more effective than anionic soaps such as sodium dodecylsolfonate (**9.64**, SDS). Nonionic detergents such as Triton X-100 (**9.65**, octoxynol or (polyethylene glycol)$_{10-}$

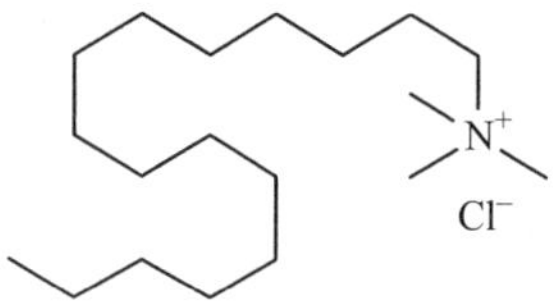

Cetyl-trimethylammonium Chloride (9.63) 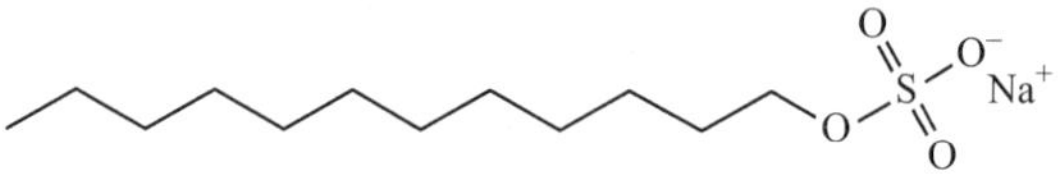   Sodium Dodecylsulfonate (9.64)

p-isooctylphenyl ether) are very mild, and are used to disperse membranes rather than to kill bacteria. Chlorhexidine (**9.66**), a chlorophenyl-biguanidine derivative, is a very effective compound. It has very low mammalian toxicity and is widely used as a wound and burn antiseptic and surgical disinfectant. Because the imino group of the biguanidine moiety becomes protonated by salt formation, the compound is a cationic detergent. Low concentrations (10–100 µg/ml) cause a rapid release of cytoplasmic material from the bacterial cell. At concentrations as low as 1 µg/ml little leakage is seen, but chlorhexidine is still active because it inhibits the membrane-bound ATPase of bacteria.

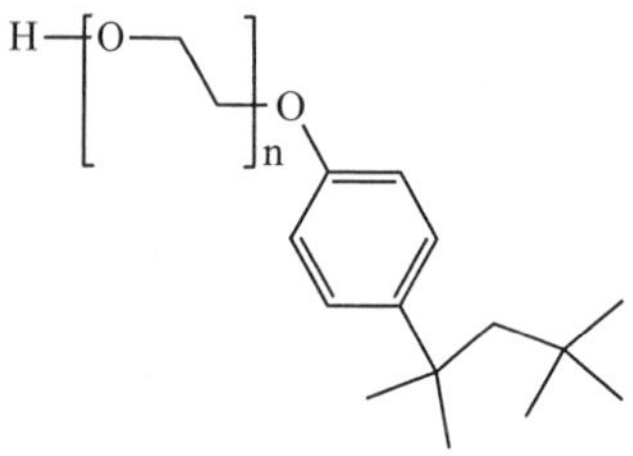

Triton X-100 (9.65)

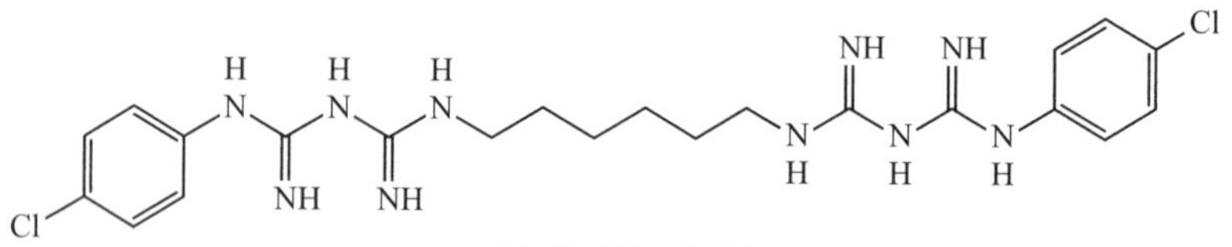

Chlorhexidine (9.66)

9.4.3 Antibacterials Targeting Bacterial Protein Synthesis

Like other cells, bacteria must biosynthesize proteins in order to survive. Proteins are biosynthesized following the arrival of an mRNA message to an organelle in the cytoplasm, called a ribosome. In making proteins, the ribosome employs two of its structural units, the 30S and 50S subunits. Molecules capable of blocking this process within bacteria are putative antibacterials.

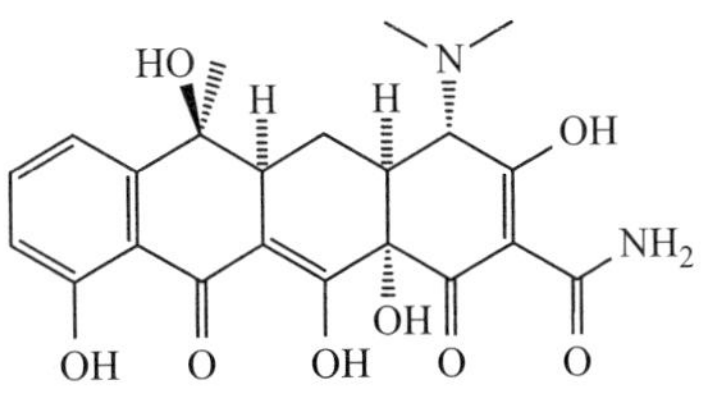

Tetracycline (9.67)

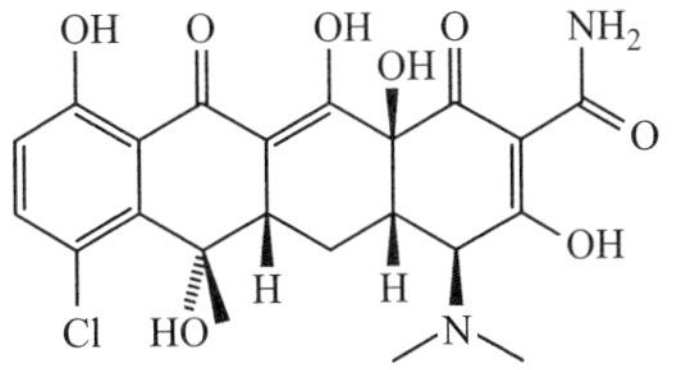

Chlortetracycline (9.68)

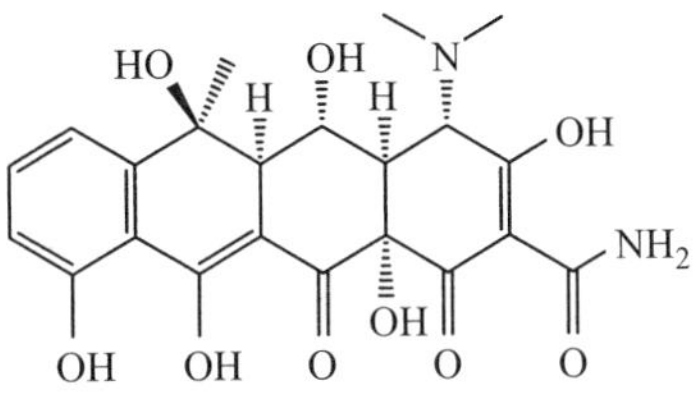

Oxytetracycline (9.69)

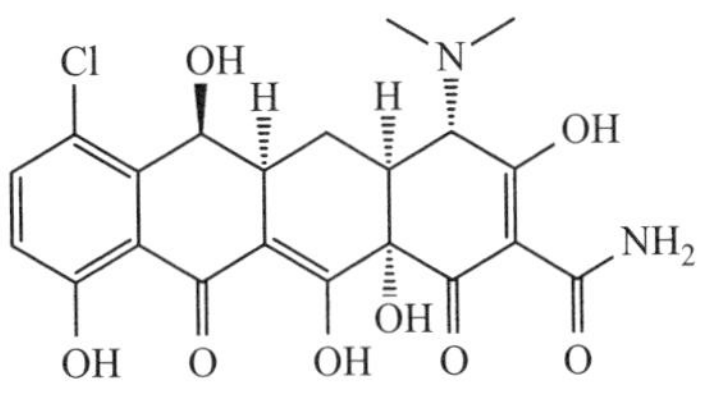

Demeclocycline (9.70)

9.4.3.1 Antibacterials Targeting the Ribosomal 30S Subunit: Tetracyclines

The tetracyclines are closely related to the anthracycline glycoside antitumor agents. Available since the early 1950s, they are broad-spectrum antibiotics active against a wide variety of microorganisms, including some that either are not sensitive or are resistant to β-lactam antibiotics. Tetracyclines interfere with protein synthesis by inhibiting the binding of aminoacyl-tRNA to the 30S subunit of the ribosome in the microorganisms. The release of completed peptides from the ribosome is also blocked. The analogs include tetracycline (**9.67**), chlortetracycline (**9.68**), oxytetracycline (**9.69**), demeclocycline (**9.70**), methacycline (**9.71**), doxycycline (**9.72**), and minocycline (**9.73**).

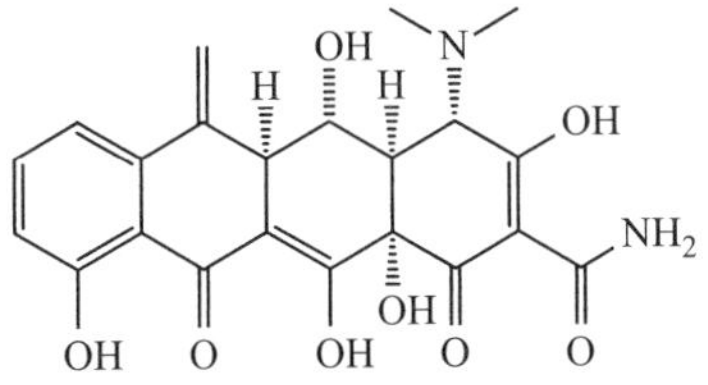

Methacycline (9.71)

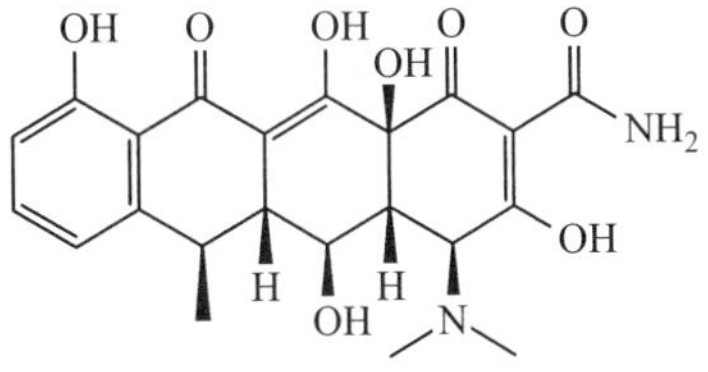

Doxycycline (9.72)

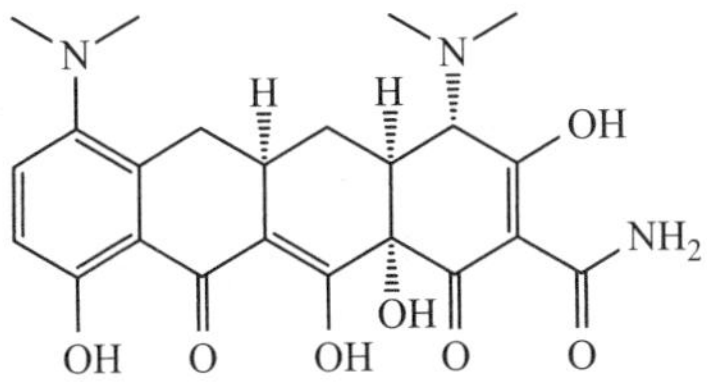

Minocycline (9.73)

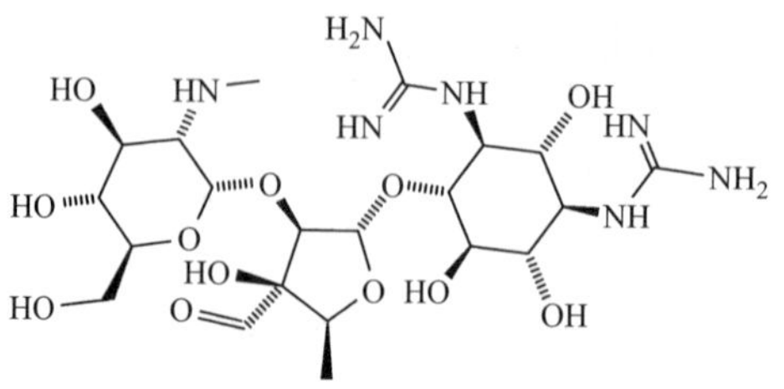

Streptomycin (9.74)

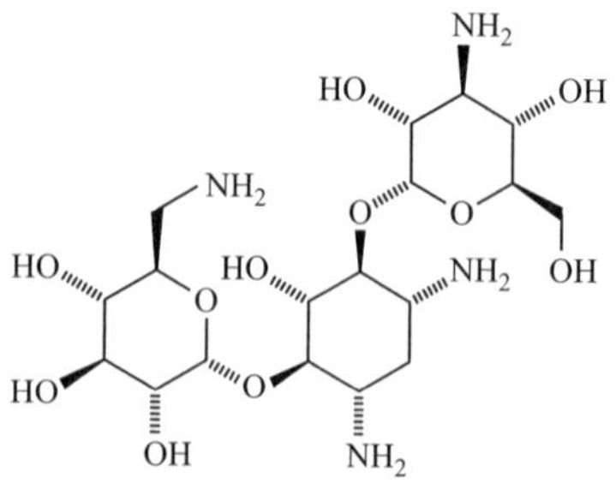

Kanamycin (9.75)

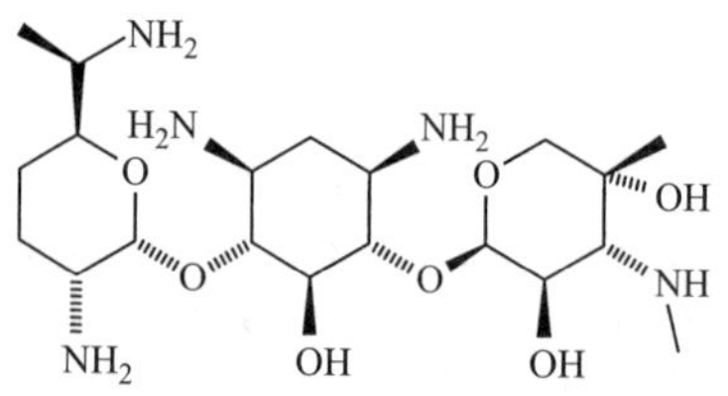

Gentamycin (9.78)

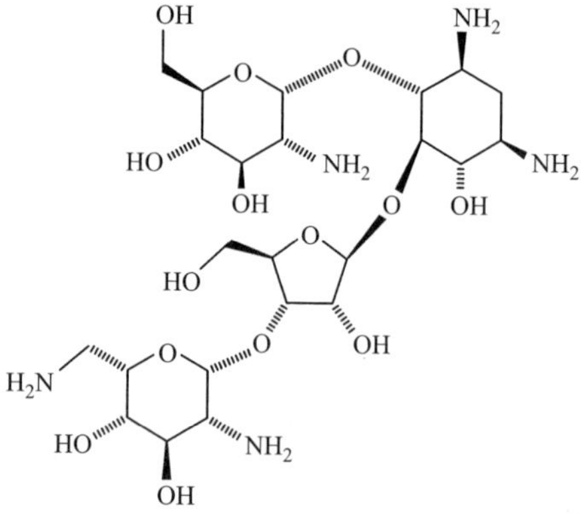

Paromomycin (9.77)

9.4.3.2 Antibacterials Targeting the Ribosomal 30S Subunit: Aminoglycosides

Clinically useful aminoglycosides contain a highly substituted 1,3-diaminocyclohexane central ring and amino sugars (such as an aminohexose) linked glycosidically. Originally they were isolated from the actinomycetes, particularly from the genus *Streptomyces*. Seven important aminoglycoside congeners were isolated from that genus: streptomycin (**9.74**), kanamycin (**9.75**), neomycin (**9.76**), paromomycin (**9.77**), gentamycin (**9.78**),

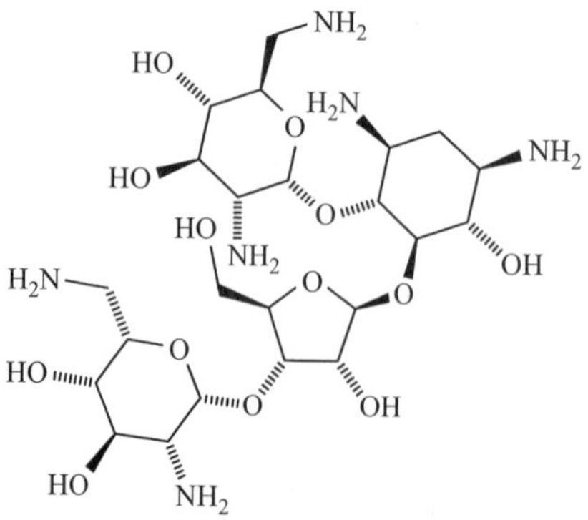

Neomycin (9.76)

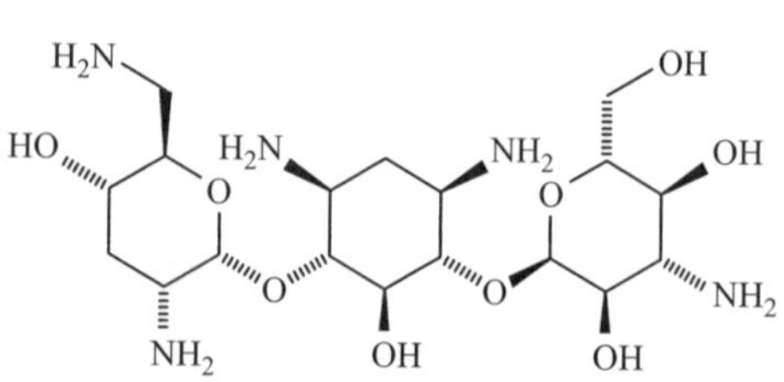

Tobramycin (9.79)

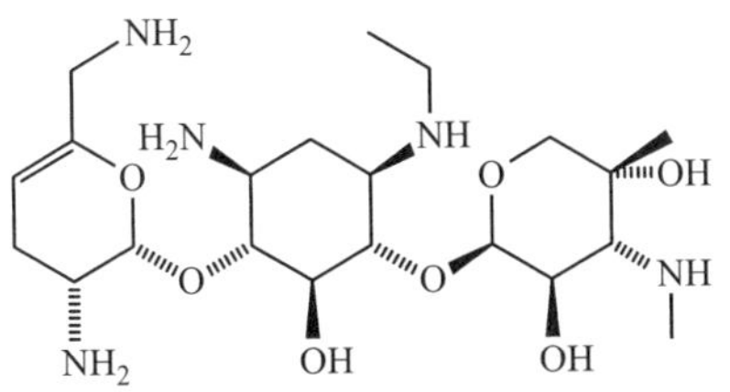

Netilmycin (9.80)

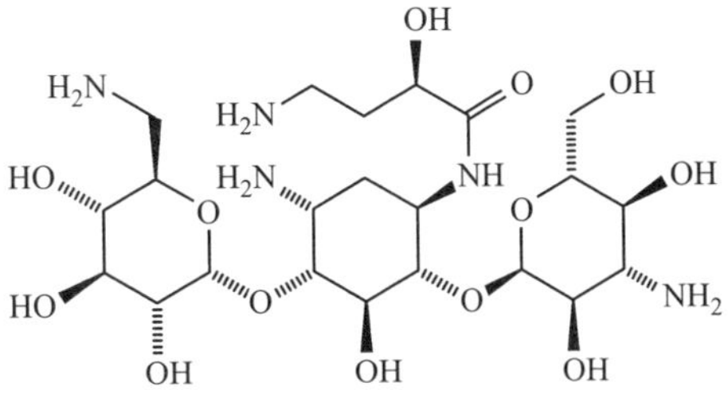

Amikacin (9.81)

tobramycin (**9.79**), and netilmycin (**9.80**). Amikacin (**9.81**), a semi-synthetic analog of kanamycin, has also been used clinically. The great usefulness of aminoglycosides arises from their ability to treat infections caused by Gram-negative bacilli. In this regard, they complement the β-lactam antibacterials and may be used in combination in the empirical treatment of life-threatening infections. Bacterial strains develop resistance to aminoglycosides.

The aminoglycosides decrease the fidelity of translation by binding to the 30S sub-unit of the ribosome. This permits the formation of the peptide initiation complex but prohibits any subsequent addition of amino acids to the peptide. This effect is due to the inhibition of polymerization as well as to the failure of tRNA and mRNA codon recognition. Aminoglycosides are ototoxic (i.e., may produce partial deafness), damaging the auditory nerve. Kanamycin is less toxic. Since aminoglycosides are concentrated in the kidney, they may occasionally cause kidney damage.

Among the aminoglycoside antibiotics, streptomycin and kanamycin are very important. Streptomycin contains a diguanidine derivative of 1,3-diaminoinositol (streptidine), a specific hexofuranose (streptose) carrying a 3-aldehyde group, and N-methyl-2-glucosamine. Kanamycin is somewhat simpler, being derived from a 1,3-diamino-2-deoxyinositol, a 3-glucosamine, and a 6-glucosamine. Both antibiotics have a fairly wide antibiotic spectrum, but practically they enjoy unique use as agents for the treatment of tuberculosis. Tuberculosis was a widely fatal form of chronic pneumonia, sometimes called the *white plague*, that exacted a devastating toll on humanity more than a century ago. Despite misconceptions to the contrary, tuberculosis ("TB") is still not uncommon as a clinical problem.

9.4.3.3 Antibacterials Targeting the Ribosomal 50S Subunit: Chloramphenicol

Chloramphenicol (**9.82**) is a product of a *Streptomyces* species. Its structure is remarkably simple, and it is obtained synthetically rather than by fermentation. It is a broad-spectrum

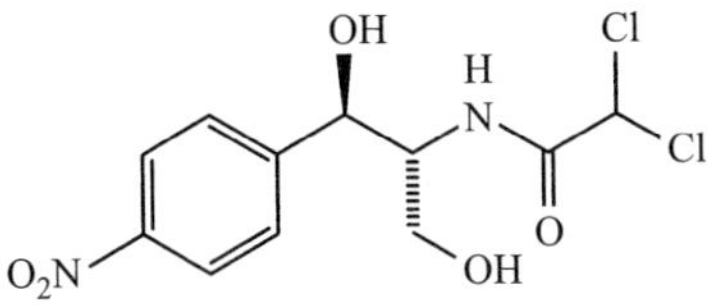

Chloramphenicol (9.82)

antibiotic, but because it can cause fatal blood *dyscrasias* (i.e., bone marrow suppression and failure) its use is largely restricted to microorganisms that cannot be well controlled by other antibiotics. Thus, it has been the drug of choice against typhoid. Chloramphenicol binds to the 50S subunit of the ribosome and inhibits the enzyme peptidyl transferase. This blocks peptide-bond formation between the amino acid-tRNA on the aminoacyl site and the growing peptide chain on the peptidyl site of the ribosome, interrupting translation.

9.4.3.4 Antibacterials Targeting the Ribosomal 50S Subunit: Macrolides

The macrolides are a group of chemically related compounds isolated from the actinomycetes microorganism. Macrolides bind selectively to a specific site on the 50S ribosomal subunit, preventing the translocation step of bacterial protein synthesis. Currently, more than 40 macrolide compounds with antibacterial properties have been identified. Chemically, all macrolide antibacterials share three common features: a large lactone ring, a glycosidically linked amino sugar, and a ketone group; the lactone ring usually has 12, 14, or 16 atoms and is frequently unsaturated. Erythromycin (**9.83**) is

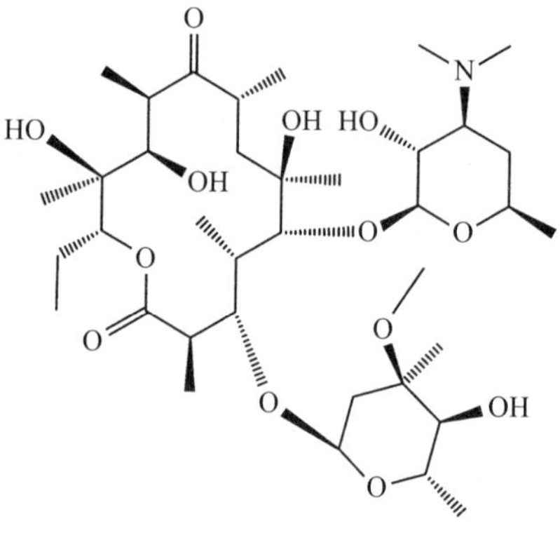

Erythromycin (9.83) 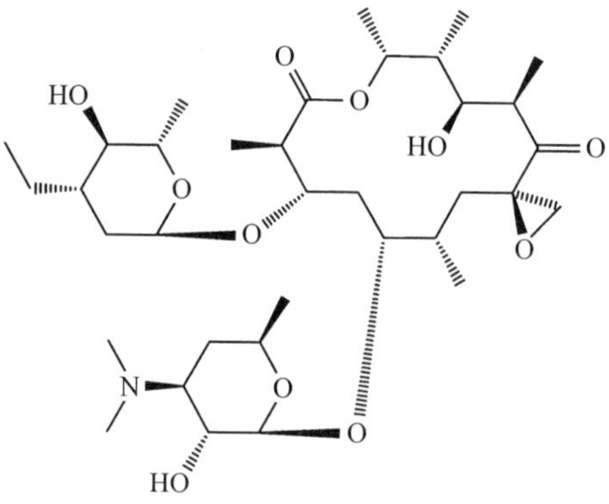   Oleandomycin (9.84)

the most widely used macrolide; oleandomycin (**9.84**) has also been employed clinically over the years. More recently, semisynthetic erythromycin analogs (clarithromycin (**9.85**), azithromycin (**9.86**)) have demonstrated enhanced acid stability, improved bioavailability, and superior pharmacokinetics. Erythromycin acts in a similar way to chloramphenicol and can actually compete with chloramphenicol for the same binding site. It is a nontoxic macrocyclic lactone, widely used against penicillin-resistant *Staphylococcus* strains, and is the drug of choice to cure *Legionnaires' disease*, caused by *Legionella pneumophila*. Gastrointestinal discomfort is a common side effect of the macrolides.

9.4.3.5 Antibacterials Targeting the 50S Ribosomal Subunit: Lincomycins

Lincomycins are similar to macrolides in biochemical mechanism of action and antibacterial spectrum of activity. They are sulphur-containing antibacterials, first

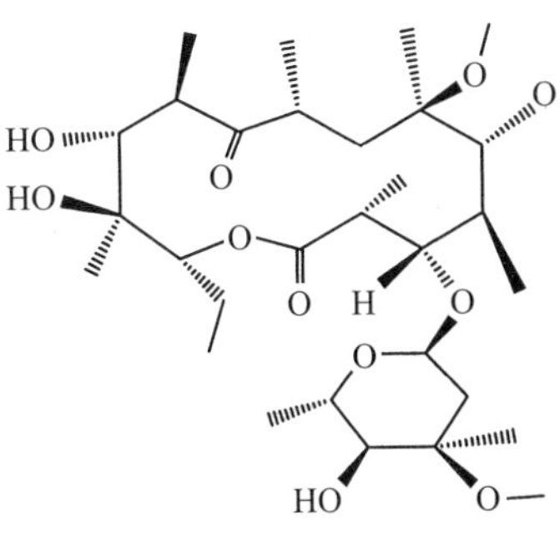

Clarithromycin (9.85)

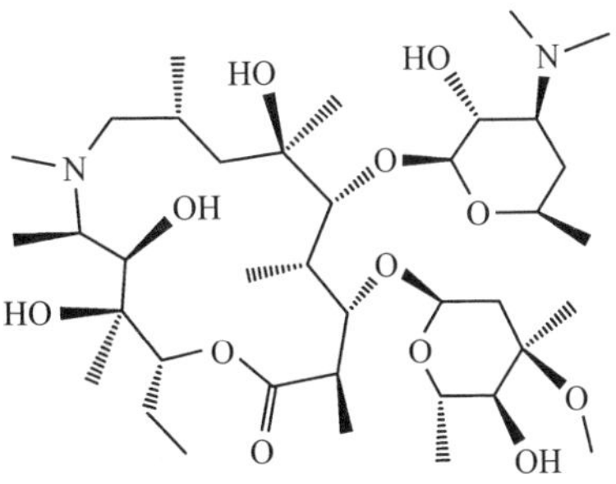

Azithromycin (9.86)

isolated from *Streptomyces lincolnensis*. Lincomycin (**9.87**) is the prototype compound in this class. It consists of a pyrrolidine moiety (*trans*-L-*n*-propylhygric acid) and a sugar moiety (α-thiolincosamide) connected by an amide bond. Replacement of the 7(R)-hydroxy group of lincomycin with a chlorine and subsequent inversion of configuration leads to 7(S)-chloro-7-deoxylincomycin (**9.88**, clindamycin), which has also been used clinically. Lincomycins are primarily active against Gram-positive bacteria, particularly the cocci. The use of clindamycin has been associated with gastrointestinal toxicity, including a very serious form of diarrhea called *pseuodomembranous colitis*.

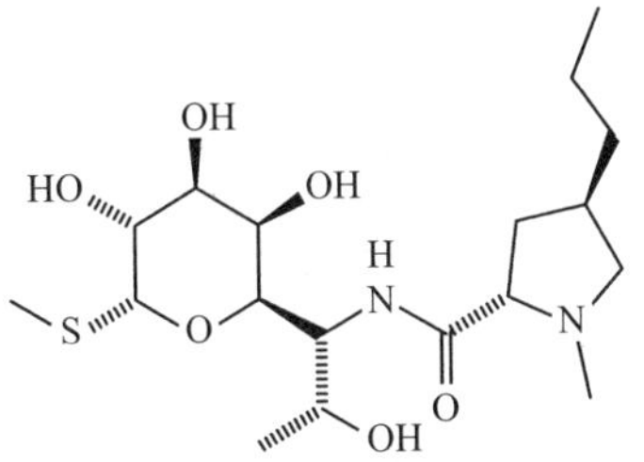

Lincomycin (9.87)

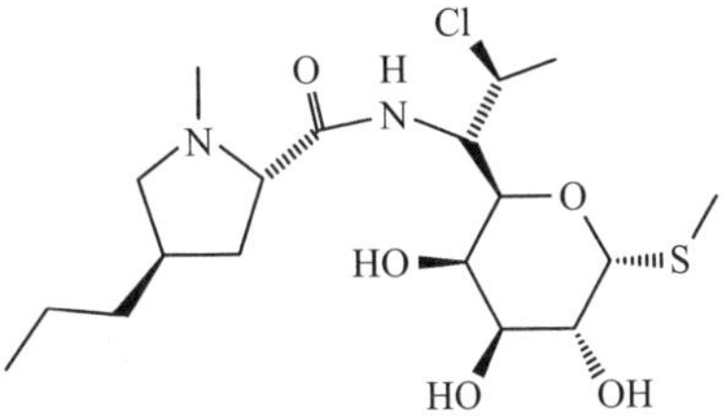

Clindamycin (9.88)

9.4.4 Antibacterials Targeting Bacterial Nucleic Acids

Another logical means of confronting the diseases of bacterial infection is to prevent the bacteria from reproducing. This may be readily achieved by targeting the bacterial nucleic acids.

9.4.4.1 Antibacterials Targeting Dihydrofolate Reductase: Sulfas

Dihydrofolate reductase (E.C. 1.5.1.3) is a relatively small monomeric protein (MW = 18,000–36,000) containing no disulfide bonds. It is present in all cells in different isozyme forms that depend on the organism. The best studied form contains 159 amino acids, with 30% of them in an eight-stranded pleated sheet. The binding site is a 1.5 nm deep cleft across the enzyme, and about a dozen amino acids are involved in binding substrate or inhibitor.

The biochemical role of dihydrofolate reductase (DHFR) is enormously important in a number of biosynthetic pathways. It reduces dihydrofolate to tetrahydrofolate, a cofactor that accepts a one-carbon fragment in several forms and transfers it to substrates during the synthesis of some amino acids, purines, and pyrimidines. Tetrahydrofolate—or its precursor, folate—is a vitamin for humans, and must be acquired from food. It is present abundantly in the green leaves of vegetables [folium (Latin) = leaf]. Bacteria, on the other hand, cannot use external folic acid, relying instead on an obligatory folate synthetic mechanism. In addition, variations in DHFR isozymes in different organisms offer a number of pharmacological targets in accord with the principle of antimetabolite therapy—that is, the use of competitive enzyme inhibitors. Sulfanilamide DHFR inhibitors have found use as antibacterial agents which act on the basis of folate synthesis inhibition. As mentioned above, bacteria must synthesize their own folate. Interference with this process will inhibit nucleic acid biosynthesis and result in bacteriostatic action because cell division ceases.

Discovered in 1935 by Domagk in Germany, these sulfa agents opened the modern era of bacterial chemotherapy. A tremendously rapid development of new and improved derivatives increased the antimicrobial spectrum and therapeutic ratio of these drugs and eliminated many side effects. Although the introduction of penicillin and other antibiotics led to some loss of interest in sulfanilamides, a new emphasis began when their utility in coexistence with other antibiotics was recognized. Research in this area also triggered the formulation of many contemporary concepts of drug action, metabolism, and molecular mechanisms, specifically on competitive enzyme inhibitors.

The mode of action of sulfanilamides became known around 1947, when the structure and biosynthesis of folic acid were elucidated. This compound is built by bacteria from the heterocyclic pteroyl moiety, p-aminobenzoate, and glutamate. p-Aminobenzene-sulfonamide (9.89, sulfanilamide) is a competitive inhibitor of the synthase enzyme, acting as an "antimetabolite" of p-aminobenzoate. Occasionally, the sulfanilamide can even be incorporated into the modified folate, resulting in an inactive compound and thus an inactive enzyme. This theory, proposed by Woods and Fildes in 1940, became the first molecular explanation of drug action.

The structure of sulfonamides is shown for representative purposes. Among the several thousand compounds in existence, about 25–30 have found widespread use. Sulfanilamide itself is, by present-day standards, very inactive. It was the development of heterocyclic derivatives that produced the highly potent sulfathiazole (9.90). When a succinyl or phthalyl group is attached to the aniline nitrogen, the inactive acylanilide derivatives will not be absorbed from the intestinal tract. Slow deacylation by intestinal

Sulfanilamide (9.89)

Sulfathiazole (9.90)

enzymes releases the active free sulfanilamide. Succinyl-sulfathiazole was therefore widely used for intestinal infections. The pyrimidine derivatives, such as sulfamerazine, have a much longer duration of action and a broad antibacterial spectrum, including both Gram-positive and Gram-negative organisms. Sulfisoxazole (**9.91**), although less active, is better tolerated. Among the newer compounds, sulfamethoxine (**9.92**) has a spectacular half-life (about 150 hours) and requires administration only once a week. A sulfone, acedapsone (**9.93**), has found successful use in the treatment and even the clinical cure of the once dreaded leprosy (Hansen's disease), caused by *Mycobacterium leprae*.

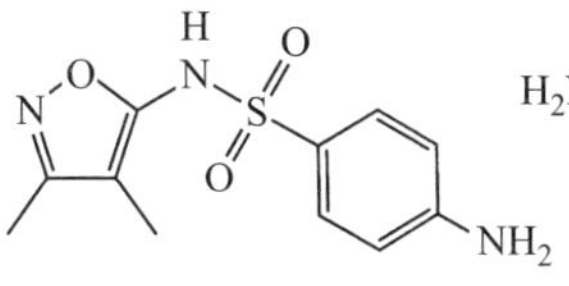

Sulfisoxazole (9.91) 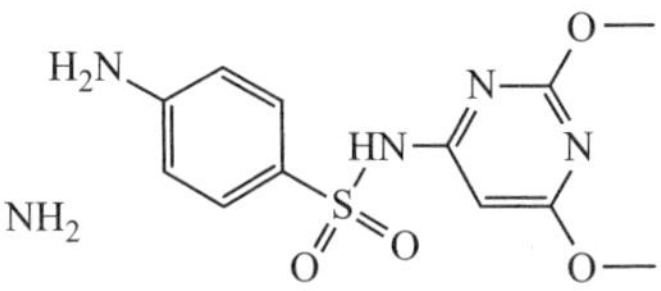   Sulfamethoxine (9.92) 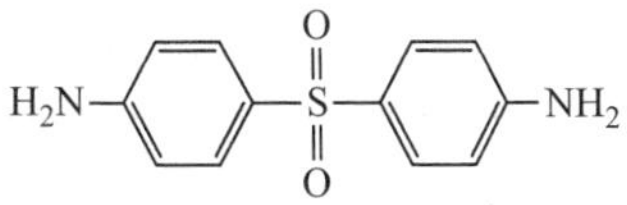   Acedapsone (9.93)

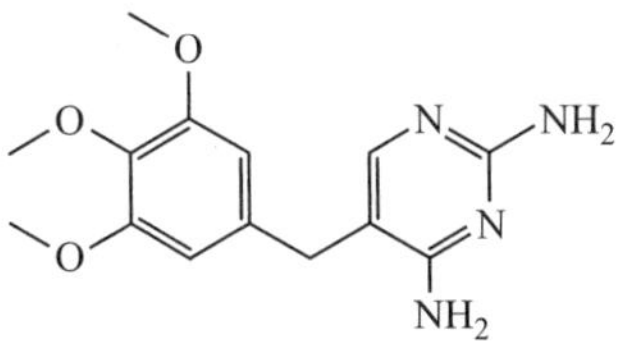

Trimethoprim (9.94) 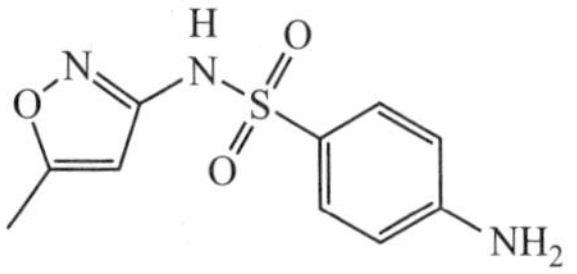   Co-trimoxazole (9.95)

Some DHFR inhibitors are active against the bacterial enzyme even though originally used against the malaria parasite. For instance, trimethoprim (**9.94**) is used in combination with sulfa drugs (co-trimoxazole (**9.95**)), attacking both the synthesis and the proper functioning of DHFR. The therapeutic index, tolerance, delay in the development of resistance, and antimicrobial spectrum of this synergistic combination of two drugs are all spectacularly greater than the corresponding effects of the individual drugs, to the extent that microorganisms unaffected by either drug alone are successfully eliminated by the combination.

Structure–activity correlations for the sulfa-DHFR inhibitors, derived on the basis of some 5000 compounds, have led to the deduction of the following regularities:

1. The amino groups must be in the *para* position to the sulfonamide group, and this amine must be unsubstituted or become unsubstituted *in vivo*.
2. The benzene ring can be substituted in positions 1 and 4 only.
3. Sulfones, carboxamides, or ketones replacing the sulfonamide may retain activity, but often at a lower level.
4. The sulfonamide nitrogen can be monosubstituted only, and heteroaromatic substituents increase the drug activity.

It is noteworthy that sulfanilamide structural modifications have led to other valuable classes of drugs already discussed, including the hypoglycemic sulfonylureas and the diuretic carbonic anhydrase inhibitors.

Correlations between physicochemical parameters and the antibacterial activity of sulfa drugs have been explored for a long time. The value of the acid dissociation constant of the sulfonamide group (pK_a) is essential, the optimum being between 6.0 and 7.4. This pK_a depends on the electronegativity of the $-SO_2-$ group, but the degree of dissociation is also greatly influenced by the intracellular pH. For the optimal balance between intrinsic activity and penetration of the bacterial cell membrane, a 50% ionization is desirable. The hydrophobicity of the drug molecule also plays a role in its activity, as Hansch correlations have shown.

9.4.4.2 Antibacterials Targeting RNA Polymerase: Rifamycins

The rifamycins, a class of antibacterials isolated from *Streptomyces mediterranei*, contain a macrocyclic ring bridging across two nonadjacent positions on an aromatic system. Rifampicin (**9.96**), a semisynthetic derivative of rifamycin, is a drug of choice in the treatment of tuberculosis as well as leprosy, either alone or in combination with other drugs. Rifampicin is much safer than other antituberculotics since it inhibits DNA-directed RNA polymerase in bacteria but not in mammals. Another rifamycin, rifabutin (**9.97**), is a spiroimidazopiperidyl derivative of the rifamycin.

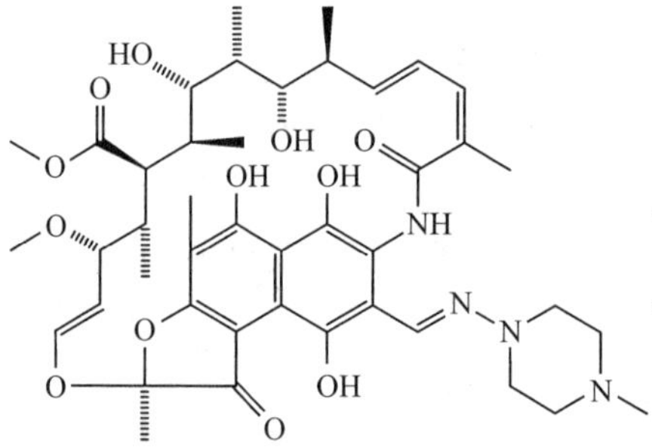

Rifampicin (9.96) 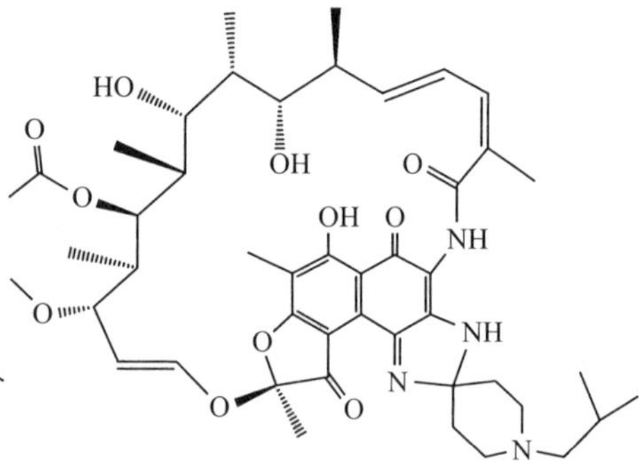   Rifabutin (9.97)

9.4.4.3 Antibacterials Targeting DNA Gyrase: Quinolones

Quinolones are synthetic compounds designed around nalidixic acid, a naphthyridine derivative used to treat urinary tract infections in the 1960s. By replacing various functional groups within the nalidixic acid pharmacophore with bioistosteric substitutions, three structural classes of quinolones were devised:

1. Quinolines (e.g., norfloxacin (**9.98**), ciprofloxacin (**9.99**), lomefloxacin (**9.100**))
2. Naphthyridines (e.g., enoxacin (**9.101**))
3. Cinnolines (e.g., cinoxacin (**9.102**))

These various quinolones have a 1,4-dihydro-4-oxo-3-pyridinecarboxylic acid moiety that is crucial to their antibacterial activity. Potency can be dramatically increased by adding a 6-fluoro substituent, producing analogs collectively referred to as fluoroquinolones.

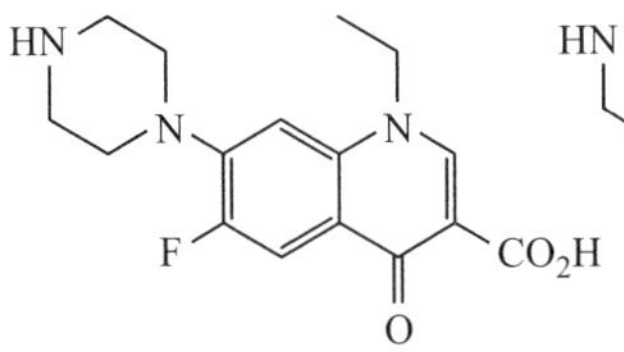

Norfloxacin (9.98)

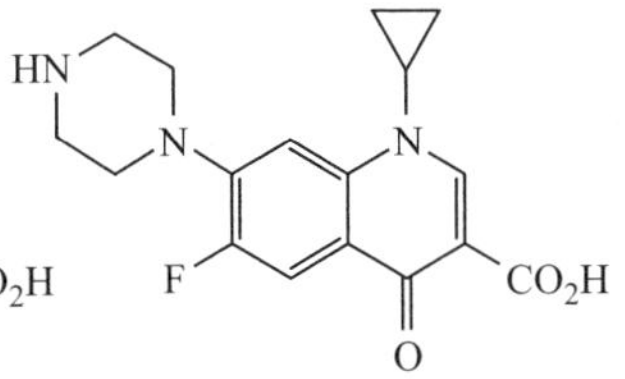

Ciprofloxacin (9.99)

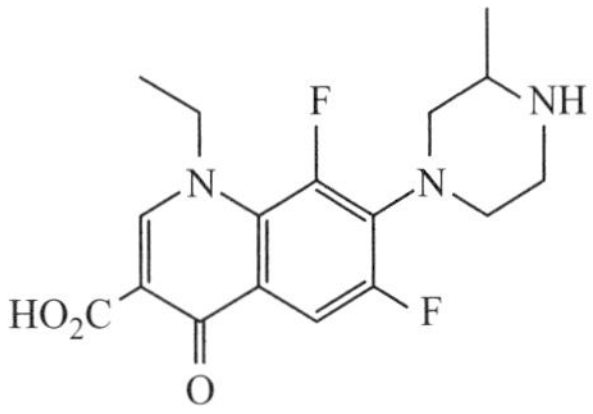

Lomefloxacin (9.100)

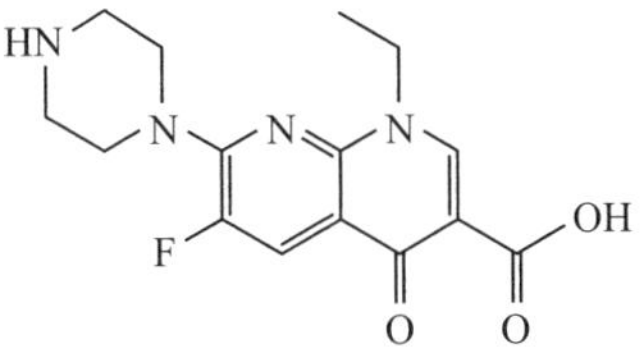

Enoxacin (9.101)

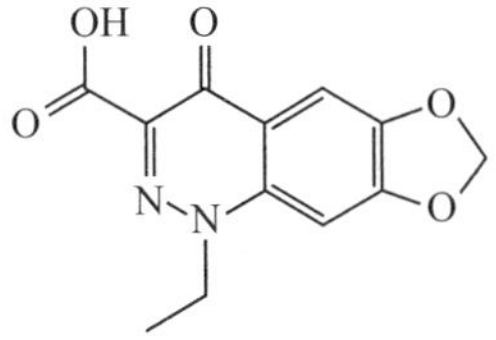

Cinoxacin (9.102)

The fluoroquinolones represent an important therapeutic advance and are extremely useful agents. They are important for the treatment of Gram-negative urinary tract infections, especially those from sulfa-resistant strains of *E. coli*, and for the treatment of serious bacterial infections such as those caused by *Pseudomonas aeruginosa*. On the basis of microbiological considerations, the fluoroquinolones may be further subdivided into three groups:

1. Good Gram-negative activity; poor Gram-positive activity
   norfloxacin
2. Excellent Gram-negative activity; moderate Gram-positive activity
   ciprofloxacin, enoxacin, lomefloxacin, levofloxacin, ofloxacin, perfloxacin
3. Excellent Gram-negative activity; very good Gram-positive activity
   clinafloxacin, gatifloxacin, sparfloxacin

These compounds exert their antibacterial effect by inhibiting the bacterial DNA gyrase (topoisomerase II) enzyme. By inhibiting this enzyme, these drugs prevent the super-coiled bacterial DNA from undergoing normal transcription and translation. Some quinolones also inhibit the topoisomerase IV enzyme, thereby interfering with separation of replicated chromosomal DNA into the respective daughter cells during cell division.

9.5 DRUG DESIGN TARGETING FUNGI

Fungi are non-photosynthetic microorganisms growing as a mass of branching, inter-lacing filaments ("*hyphae*") known as *mycelium*. The entire organism is a *coenocyte* (a multinucleate mass of continuous cytoplasm) confined within a series of branching tubes. The fungal kingdom includes molds, yeasts, rusts, and mushrooms. Many fungi cause plant diseases, but only about 100 of the thousands of known yeasts and molds cause disease in humans. Fungal infections in humans are known as *mycotic* infections. Human mycotic infections may be grouped as superficial (skin [tinea corporis], hair [tinea capitis], nails [tinea unguium; *Microsporum canis, Trichophyton mentagrophytes*]), subcutaneous (abscess in muscle or even bone; *Sporothrix schenckii*) or deep (diffuse, systemic infection, possibly fatal; *Histoplasma capsulatum, Blastomyces brasiliensis*).

In recent years, the incidence and severity of human fungal infections have increased dramatically. The use of powerful immunosuppressive agents for cancer chemotherapy and for organ transplant, combined with the AIDS epidemic, has led to this significant increase. However, in healthy individuals, a normally functioning immune system tends to ward off fungal infections.

Drug design for fungal infections is more challenging than for bacterial infections. Fungi are "more sophisticated" and their cellular structure starts to more closely approximate mammalian cellular structure, leading to design challenges in developing agents that are toxic to fungi but nontoxic to humans. One vulnerable feature of fungal biochemistry is the structure of the membrane. There is a fundamental difference in lipid composition between fungal and mammalian cells. Ergosterol is the predominant membrane sterol lipid in fungi; cholesterol is the principal sterol in human cells. This is a biochemical difference that can be exploited for drug design. Another significant difference is that fungi (like bacteria) have cell walls; mammalian cells do not. Accordingly, possible druggable targets for antifungal drug design include:

1. Fungal membrane disruptors, via mechanical insertion
2. Ergosterol biosynthesis inhibitors, via 14α-demethylase enzyme inhibition
3. Ergosterol biosynthesis inhibitors, via squalene epoxidase enzyme inhibition
4. Ergosterol biosynthesis inhibitors, via Δ^{14}-reductase enzyme inhibition
5. Fungal cell wall disruptors

Of these five targets, those that involve membrane biochemistry (targets 1–4) have proven to be the most successful.

Streptomyces nodosus produces natural products such as amphotericin A and B. These are macrocyclic (large-ring) compounds containing numerous (three to seven) double bonds and multiple hydroxyl groups which are usually located on one side of the molecule. Amphotericin B (**9.103**) and the very similar nystatin (**9.104**) are antifungal

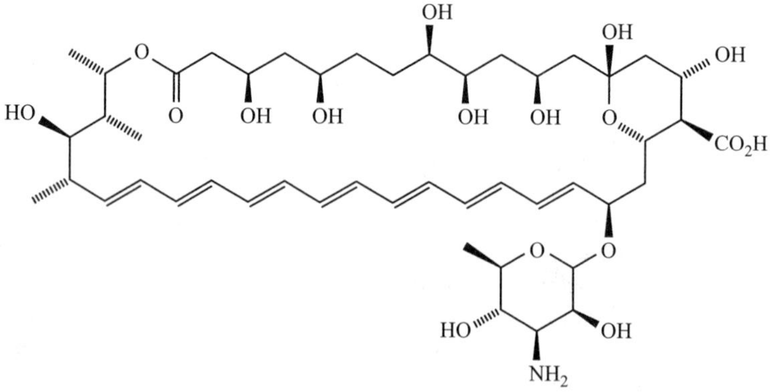

Amphotericin B (9.103)

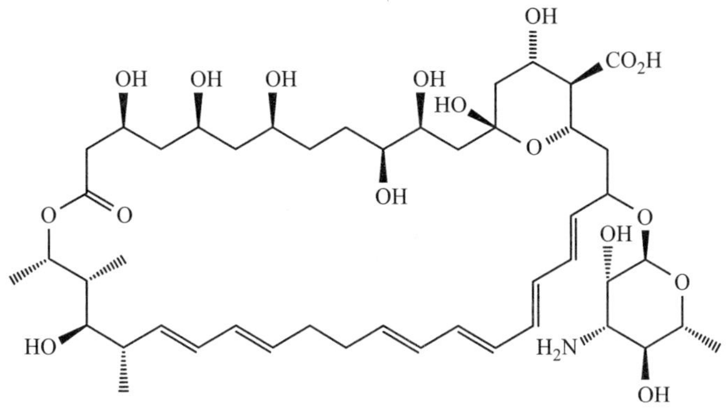

Nystatin (9.104)

agents that interact with sterols in the microbial plasma membrane. From five to ten molecules of these antifungals form a conducting pore or channel through which K$^+$ ions, sugars, and proteins are lost from the microorganism. The inside of the pore is lined with the hydroxyl groups of the macrocyclic molecule; the polyene section interacts with the hydrophobic sterol component of the cell membrane. Because these antifungals interact preferentially with ergosterol, they are fungicidal. With regard to animal membranes they have a strange selectivity, being lethal to flatworms and snails, not affecting bacteria, and exerting limited toxicity in mammals. They are used for such fungal infections as athlete's foot and vaginal candidiasis, and for systemic (internal) fungal infections, which are nearly always fatal if untreated. Thus the therapeutic value of amphotericin B outweighs the drawback of its toxic side effects.

The more recently discovered group of *azole antifungal agents* also act through membrane destabilization by inhibiting ergosterol biosynthesis, via 14α-demethylase enzyme inhibition; the basic nitrogen of the azole forms a bond to the heme iron in the enzyme, preventing the enzyme from oxidizing its normal substrates. Ketoconazole (**9.105**) can be taken orally, whereas clotrimazole (**9.106**), miconazole (**9.107**), and related compounds are applied topically. Other azoles include itraconazole (**9.108**) and fluconazole (**9.109**). The azoles bind very avidly to skin and accumulate in fungi very rapidly. Although contact times may be only 15–20 minutes, the azole remains in the fungus for 120 hours and thus, even in sublethal doses, leads to a decrease in virulence. Another group of sterol synthesis inhibitors (via squalene epoxidase enzyme inhibition) includes the allylamines, represented by naftifine (**9.110**) and terbinafine (**9.111**).

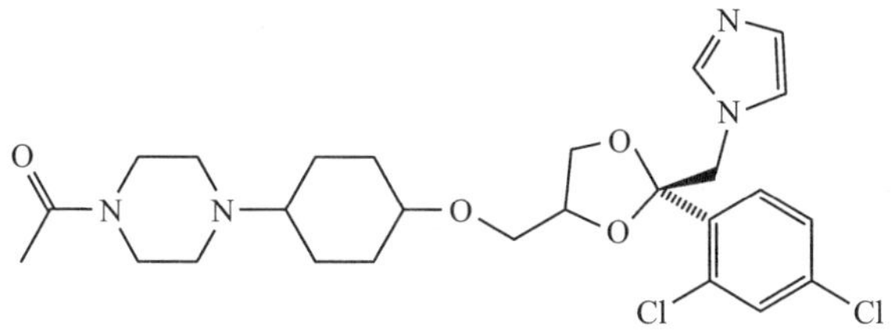

Ketoconazole (9.105)

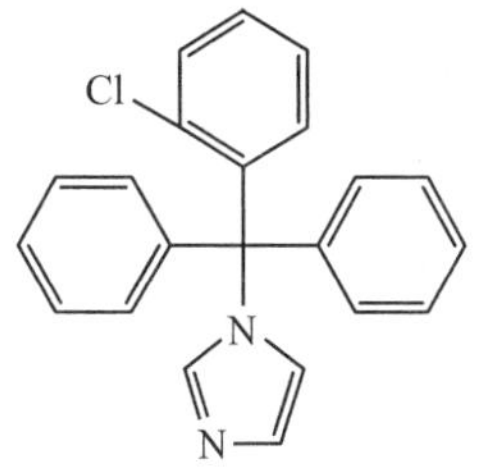

Clotrimazole (9.106)

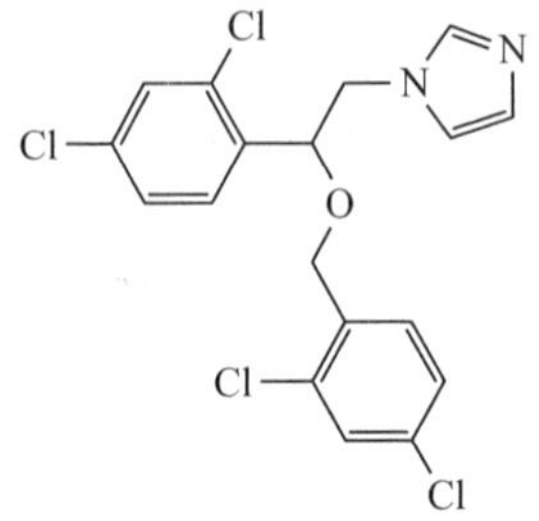

Miconazole (9.107)

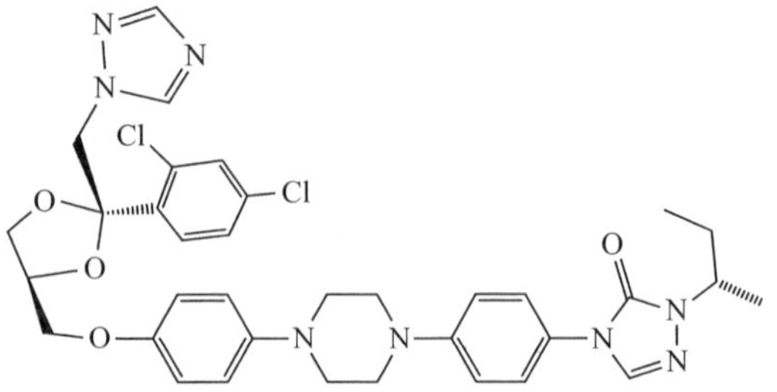

Itraconazole (9.108)

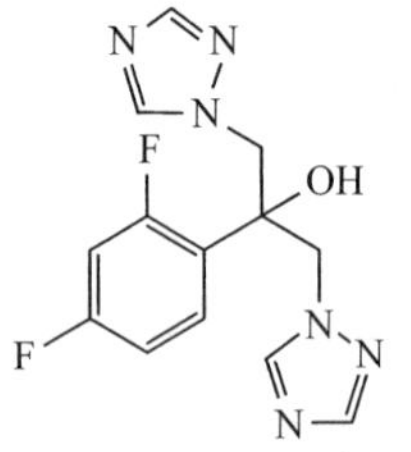

Fluconazole (9.109)

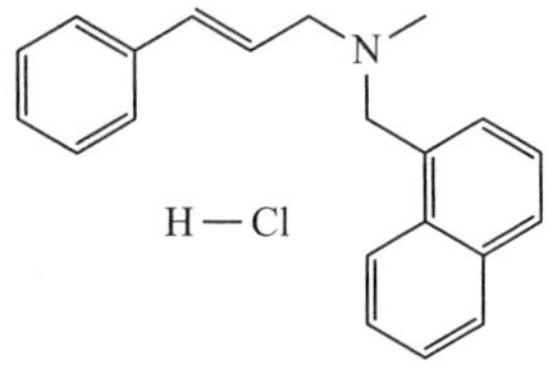

Naftifine (9.110)

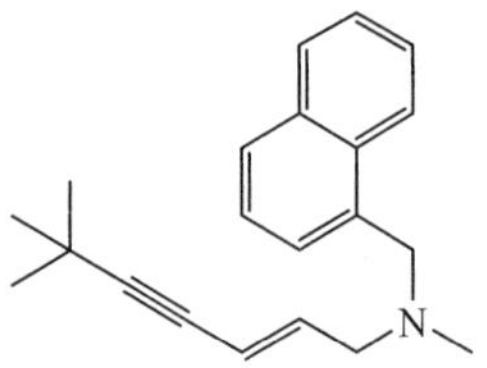

Terbinafine (9.111)

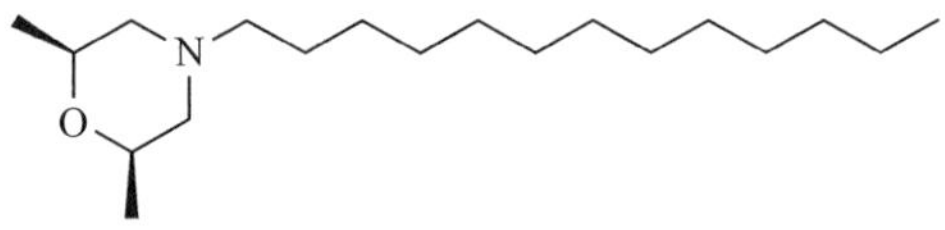

Tridemorph (9.112)

Morpholines such as tridemorph (**9.112**) attack sterol biosynthesis via Δ^{14}-reductase enzyme inhibition and are valuable agricultural fungicides.

9.6 DRUG DESIGN TARGETING PARASITES

Although any microbe that infects a human (virus, bacteria, fungus, etc.) could theoretically be considered to be a parasite, the term parasite tends to be reserved for three types of infectious agents: protozoa, helminths, and arthropods. These may be subdivided as follows:

A. Protozoa (unicellular forms)
   a. Mastigophora (flagellates)
       Giardia (causing giardiasis—cramps, diarrhea, nausea)
       Trichomonas (causing trichomoniasis—painful urination, vaginal discharge)
       Leishmania (causing leishmaniasis—"black fever," gastrointestinal bleeding)
       Trypanosoma (causing Chaga's disease—malaise, fever, heart failure)
   b. Sarcodina (ameboids)
       Entamoeba (causing amebiasis—potentially life-threatening dysentery)

c. Sporozoa
   Plasmodium (causing malaria—high fever, sweating, headache, nausea)
d. Ciliophora
   Balantidium (causing balantidiasis—diarrhea, dysentery, ulcers)
B. Helminths (worms)
   a. Platyhelminths (flatworms)
      Cestode (tapeworms)
         Tenia (beef/pork tapeworm—5 m long worm in bowel)
      Trematode (flukes)
         Schistosoma (blood or bladder flukes—headache, gut symptoms)
   b. Nematode
      Ancylostoma (hookworm—intestinal ulcerations, lung lesions)
      Enterobius (pinworm—perianal itching)
      Ascaris (roundworm—30 cm long worm in small intestine)
      Trichuris (whipworm—5 cm long worm in colon)
C. Arthropods (ectoparasitic infestations; insects)
      Sarcoptes scariei (scabies—severe itching in skin fold areas)
      Pediculus corporis (lice—itching of scalp and skin)
(This is only a partial list of the more common parasitic infestations.)

Parasitic infections affect (and sometimes kill) a huge number of people worldwide. Conservative estimates suggest that more than one billion people have parasitic infections; other epidemiologists suggest a figure of three billion—that is, almost one-half of this planet's human population. In some developing countries more than 80% of the population has a parasitic infection. Parasitic infections constitute one of the most widespread human health problems in the modern world. It is a sad comment on the "drug design culture" that so little is done for so great a problem. Regrettably, the intellectual and technical might of modern drug design is seldom focussed on the issue of parasitic infection.

Designing drugs for parasitic infections presents some undeniable challenges. Worms are surprisingly sophisticated and differentiating helminths from humans is not always easy. As with the other microbes, the key to successful drug design is to identify druggable targets that permit selective parasite killing. This may be achieved using three classes of druggable targets:

1. Enzymes unique to the parasite and not found in humans (e.g., dihydropteroate synthesis enzymes, trypanothione reductase)
2. Enzymes found in both parasites and humans, but essential to parasite life while being nonessential or less essential to human life (e.g., purine nucleoside kinase, ornithine decarboxylase)
3. Non-enzymatic proteins found in both parasites and humans, but exhibiting different pharmacological profiles between the two species (e.g., thiamine transporter)

Using these strategies enables rational drug design of antiparasitic agents.

Because of the importance of malaria, antiparasitic drugs are sometimes divided into two classes: antiparasitic agents for protozoans, especially malaria, and antiparasitic agents for helminthic infestations. Each of these classes will be examined separately.

9.6.1 Antiparasitic Drugs Targeting Protozoans

Although there is a variety of protozoan infestations, malaria is probably the most important on a world-wide scale. Four species of the plasmodium parasite cause malaria in humans: *Plasmodium falciparum, Plasmodium vivax, Plasmodium malariae, and Plasmodium ovale; P. falciparum* is responsible for most of the deaths. Malaria parasites have a very complex life cycle and can exist in many different forms (sporozoite, schizont, merozoite, trophozoite, gametocyte). When a human is bitten by an *Anopheline* mosquito, the parasite is introduced in its sporozoite form. This rapidly invades the liver where it matures first into a schizont and then into a merozoite. This merozoite form of the parasite leaves the liver and then invades red blood cells (erythrocytes). Within erythrocytes the merozoite is sequentially transformed into a trophozoite, then into a schizont, and finally back to a merozoite, which ruptures out of the erythrocyte to inoculate other erythrocytes. Within the erythrocyte, some merozoites also develop into gametocytes, which are taken up by mosquitoes, where they are matured back into infective sporozoites—thereby initiating the infective process all over again. This circuitous life cycle, shown in figure 9.5, offers a number of targets for antimalarial drugs.

There are four major types of compounds that exhibit activity against malaria:

1. Quinoline and acridine antimalarials
2. Folate synthesis antagonists
3. Atovaquone compounds that inhibit mitochondrial electron transport in the parasite
4. Artemisia derivatives that kill parasites via carbon-centered free radicals

Of these four groups of compounds, the quinoline/acridine class is the most widely used.

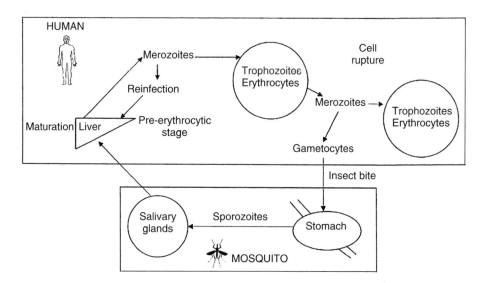

Figure 9.5 Malaria parasites have a complex life cycle, existing in many different forms between humans and mosquitoes. This offers a number of targets for drug design.

Many antimalarial agents have a common structural feature: a quinoline ring or an acridine ring (which is a quinoline ring with an extra benzene ring appended to it).

The other major class of antimalarials are the folate synthesis antagonists. There is a considerable difference in the drug sensitivity and affinity of dihydrofolate reductase enzyme (DHFR) between humans and the *Plasmodium* parasite. The parasite can therefore be eliminated successfully without excessive toxic effects to the human host. DHFR inhibitors block the reaction that transforms deoxyuridine monophosphate (dUMP) to deoxythymidine monophosphate (dTMP) at the end of the pyrimidine-synthetic pathway. This reaction, a methylation, requires N^5, N^{10}-methylene-tetrahydrofolate as a carbon carrier, which is oxidized to dihydrofolate. If the dihydrofolate cannot then be reduced back to tetrahydrofolate (THF), this essential step in DNA synthesis will come to a standstill.

The DHFR-inhibitor antimalarial drugs are competitive inhibitors and therefore structural analogs of folate. They are either diaminobenzyl- or diaminophenyl-pyrimidines (trimethoprim (**9.113**); pyrimethamine (**9.114**)) or triazines such as cycloguanil (**9.115**), which is the active form of chlorguanide (**9.116**), a pro-drug that undergoes an oxidative ring closure *in vivo*. The replacement of the folate 6-OH by an amino group in these analogs increases their affinity for the enzyme by a factor of 1×10^4 to 5×10^4 as compared to folate. Some very insoluble derivatives of cycloguanil show a protective effect for several months after a single intramuscular injection, and may play an important role in the eradication of malaria. Because of the emergence of many resistant strains of *Plasmodium*, combination chemotherapy is imperative. Eradication of the insect vector, the *Anopheles* mosquito, is also an important but elusive goal if endemic malarial areas in Africa and Southeast Asia are to be cleared.

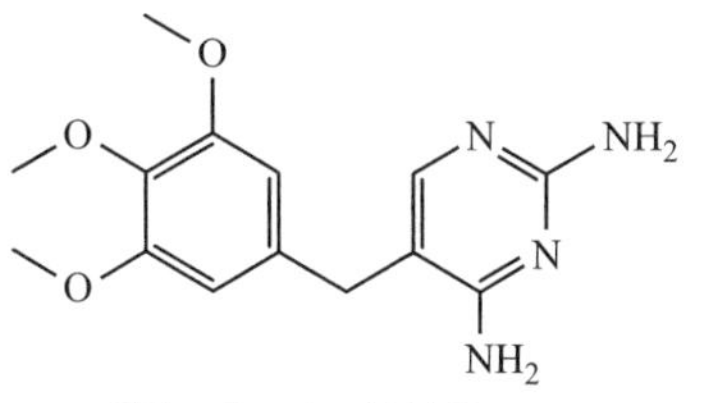

Trimethoprim (9.113)

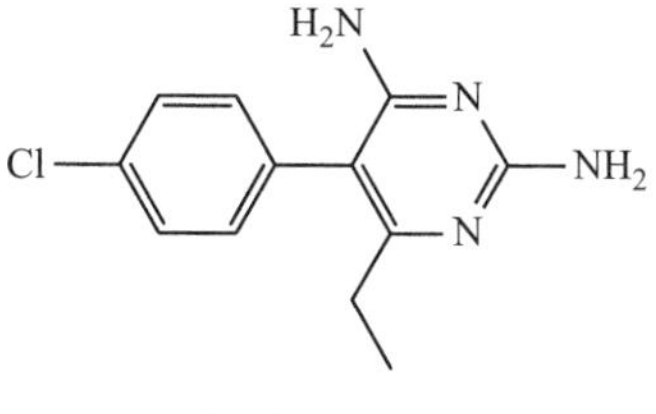

Pyrimethamine (9.114)

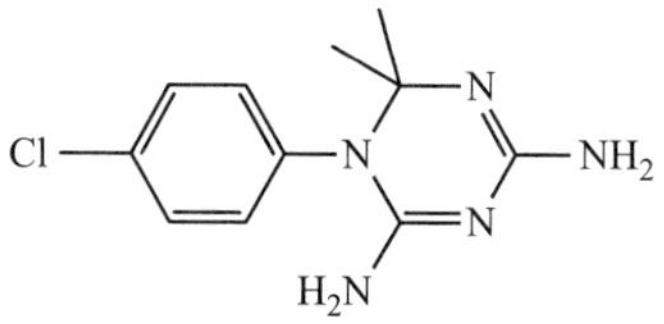

Cycloguanil (9.115)

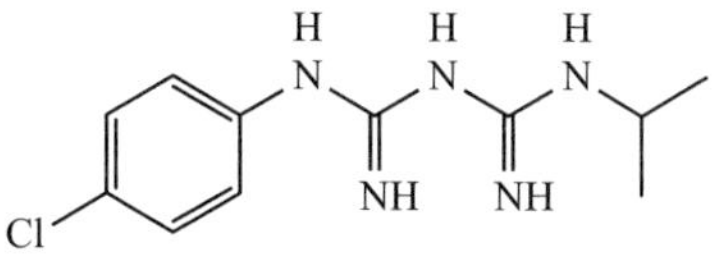

Chlorguanide (9.116)

Malaria is still one of the major health problems in tropical countries, where tens of millions of people suffer from this debilitating disease. Drug design for malaria is a research area with the capacity to blossom in coming years. Due to environmental changes such as global warming, the distribution of this disease may change, leading to malaria occurrences in the southern portions of North America and Europe. This will undoubtedly be a motivating factor to drive new research directions.

9.6.2 Antiparasitic Drugs Targeting Helminths

A variety of structurally diverse compounds is used to eradicate or at least reduce the numbers of helminthic parasites in the intestinal tract or tissues of the body. The search for antihelminthics has been somewhat less systematic than for other microbial pathogens. Most of these agents were discovered by traditional screening programs; their mechanisms of action at a molecular level are frequently unknown.

Albendazole (**9.117**), a benzimidazole carbamate, is a broad-spectrum oral antihelminthic used for cysticercosis, ascariasis, pinworm infestation, and hookworm infestation. It is thought to work by blocking microtubule synthesis in the nematode, subsequently impairing glucose uptake.

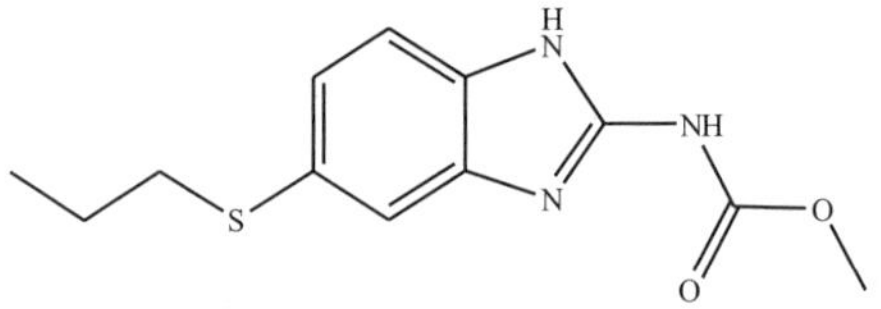

Albendazole (9.117)

Diethylcarbamazine citrate (**9.118**), a synthetic piperazine derivative, is used for the treatment of filariasis and loiasis. Although its mechanism of action is basically unknown, it appears to somehow alter the surface structure of the helminth, rendering it more susceptible to destruction by the host defenses.

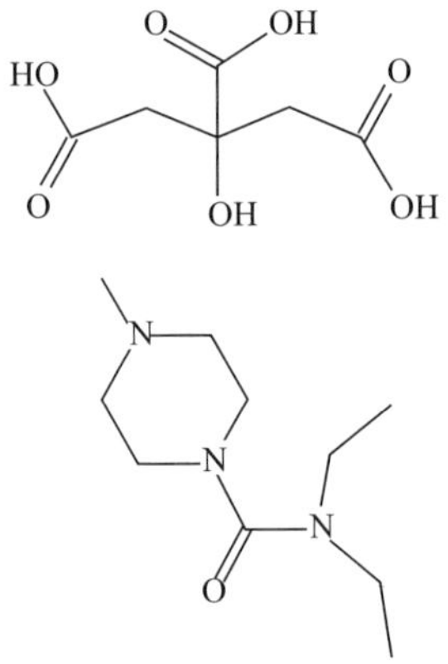

Diethylcarbamazine citrate (9.118)

Ivermectin (**9.119**), a semisynthetic macrocyclic lactone, is the drug of choice for strongyloidiasis and onchocerciasis types of helminth infestation. Ivermectin appears to paralyze the nematode, which may lead to its death but which certainly facilitates its expulsion from the body.

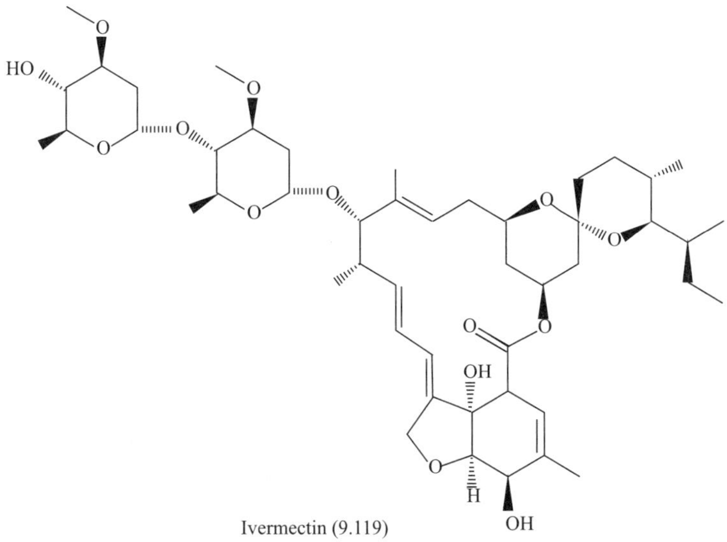

Ivermectin (9.119)

Mebendazole (**9.120**), a synthetic benzimidazole, is a broad-spectrum oral antihelminthic used for ascariasis, pinworm infestation, and hookworm infestation. It is thought to work by blocking microtubule synthesis in the nematode, subsequently impairing glucose uptake.

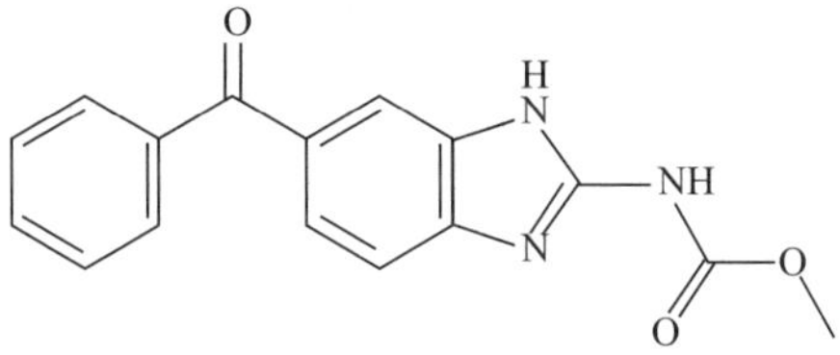

Mebendazole (9.120)

Metrifonate (**9.121**), an organophosphate compound, is a low-cost antihelminthic used to treat *Schistosoma hematobium* infections. The drug appears to alter cholinesterase activity in the nematode, temporarily paralysing the adult worm and causing the worm to die. Because of its activity against the cholinesterase enzyme, this drug has also been evaluated as a treatment for Alzheimer's disease.

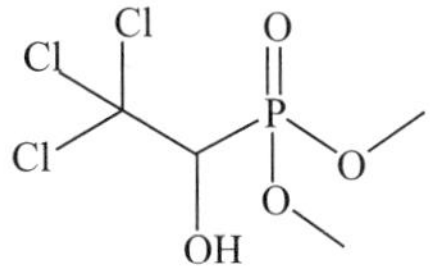

Metrifonate (9.121)

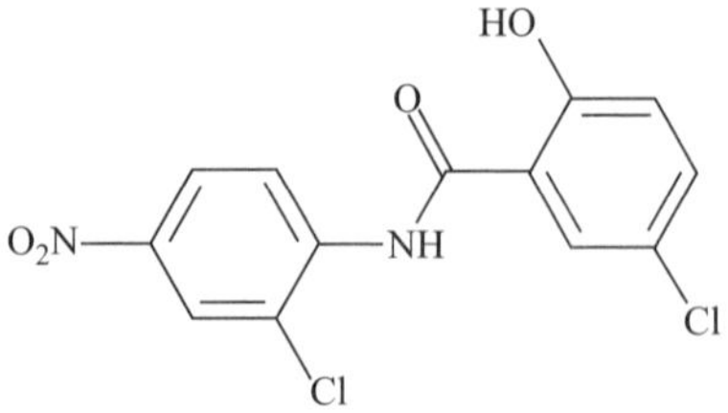

Niclosamide (9.122)

Niclosamide (**9.122**), a salicylamide derivative, is the drug of choice for most tape-worm infestations. Niclosamide inhibits oxidative phosphorylation within the parasite, causing the cestode to be rapidly killed.

Oxamniquine (**9.123**), a semisynthetic tetrahydroquinoline, is the treatment of choice for *Schistosoma mansoni* infections. Its mechanism of action is unknown but appears to entail binding of the drug to the nematode's DNA, somehow causing its death. Alternatively, it may cause worm paralysis and expulsion.

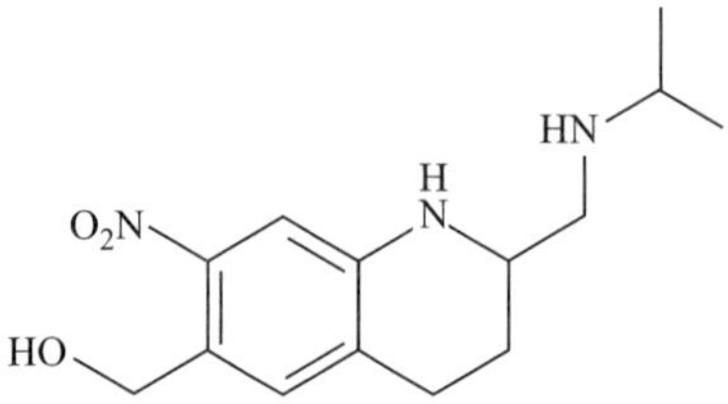

Oxamniquine (9.123)

Praziquantel (**9.124**), a synthetic isoquinoline–pyrazine derivative, is effective for the treatment of schistosome infections; it is also effective against trematode and cestode infestations. This agent increases membrane permeability to Ca^{2+} in the nematode, causing paralysis, dislodgement, and death of the worm.

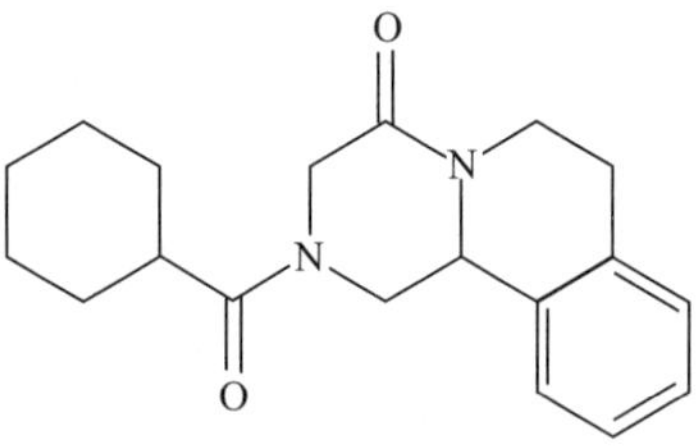

Praziquantel (9.124)

Pyrantel pamoate (**9.125**), a tetrahydropyrimidine derivative, is a broad-spectrum antihelminthic active against pinworm and hookworm. This drug causes release of acetylcholine in the worm, producing stimulation and paralysis, leading to expulsion from the host's intestinal tract.

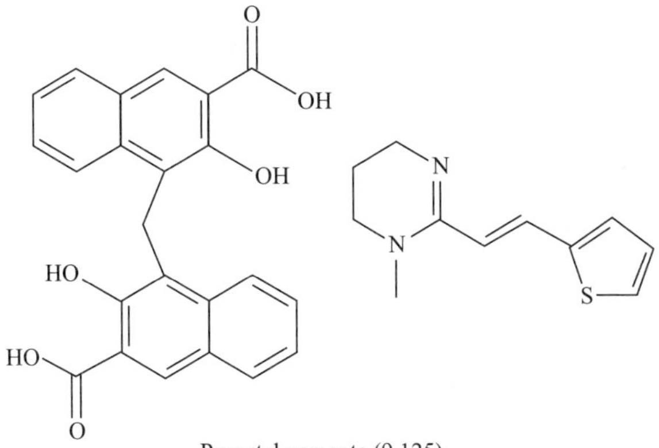

Pyrantel pamoate (9.125)

Thiabendazole (**9.126**), a benzimidazole, is used to treat strongyloidiasis and trichinosis. Within the parasite, thiabendazole inhibits an enzyme called fumarate reductase, resulting in death of the worm. This drug may be carcinogenic or mutagenic to the host.

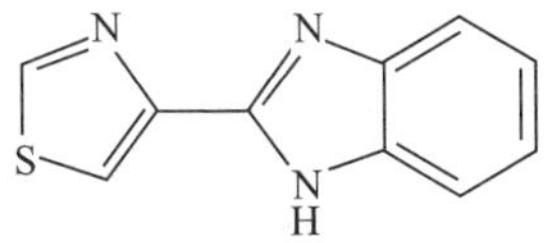

Thiabendazole (9.126)

9.7 THE CLINICAL–MOLECULAR INTERFACE: PNEUMONIA

Infections can occur in virtually every organ and tissue in the human body: encephalitis (brain), otitis media (middle ear), sinusitis (sinus), pharyngitis (throat), bronchitis (bronchi), pneumonitis (lung), myocarditis (heart), hepatitis (liver), gastroenteritis (intestine), cholecystitis (gall bladder), nephritis (kidney), cystitis (urinary bladder), osteitis (bone), and septicemia (blood). The infection may be diffuse (encephalitis of the brain) or localized (brain abscess). It may be caused by any of the infectious agents: meningitis may be caused by viral, bacterial, or fungal pathogens. The pathological consequences of infection may be immediate (meningococcal meningitis, causing rapid death) or delayed (syphilis, causing brain damage twenty years after the first infection). Similarly, the pathological ramifications of infections may be direct or indirect; for example, some researchers have speculated that a chlamydia infection of arteries may damage the arterial wall, causing atherosclerosis (leading to strokes and heart attacks) in later years.

Among the various types of infection, pneumonia is one of the most common. Pneumonia is an infection of lung tissue. Clinically, pneumonia is characterized by fever, chills, shortness of breath, cough, and sputum production. The amount, viscosity, and color of the sputum are directly related to the type of organism causing the pneumonia.

In some people, the associated failure of breathing can lead to death. The infectious agents that cause pneumonia are many and varied, including many types of viruses, bacteria, and fungi. Certain microbes are more prevalent at particular ages; in neonates *E. coli* is a common pathogen, whereas in adults *Pneumococcus* is more common. There is also a distinction between community-acquired and hospital-acquired (*nosocomial*) pneumonias; they may be caused by different microbes. In people who are immunocompromised (due to AIDS or cancer chemotherapy) unusual organisms may cause pneumonia, including viruses (cytomegalovirus), fungi (cryptococcus), or even protozoa (*Pneumocystis carinii*). The treatment of pneumonia is determined by the type of microbial agent.

The treatment of infectious diseases such as pneumonia is dictated more by the nature of the microbe than by the location of the infection. For example, a staphylococcal pneumonia is treated with the same antibacterial drugs as a staphylococcal infection elsewhere in the body. This tends to be different from other forms of pathology. For example, an adenocarcinoma tumor of the lung is treated differently from adenocarcinomas elsewhere in the body. When given for a localized infection, antibiotics are not "smart bombs"; they are not tissue-specific and are widely distributed throughout the entirety of the body. This accounts for some of the side effects seen with antibacterial agents. An antibiotic given for a pneumonia will just as easily kill bacteria in the patient's gastrointestinal tract, permitting undesirable microorganisms to overgrow and produce side effects such as diarrhea.

9.8 THE CLINICAL–MOLECULAR INTERFACE: MENINGITIS AND ENCEPHALITIS

Meningitis is an infection of the meninges, the membranes that cover the surface of the brain. Infectious causes of meningitis may be bacteria, viruses, fungi, or parasites. If untreated, bacterial meningitis is uniformly fatal; with treatment, the mortality rate is 20–25%. Encephalitis is an inflammation, usually secondary to infection, of the substance of the brain. Whereas numerous viral, bacterial, fungal or parasitic agents are capable of producing the encephalitis syndrome, viral causes are the most frequent. The mortality and long-term morbidity rate of encephalitis is dependent upon the causative agent. The following two patients recently presented to an emergency room within minutes of each other.

Patient 1. This 24-year-old male university student was brought to the emergency department at 1600 h by his roommate. He was delirious and had a depressed level of consciousness. Although he had been well the previous day, that morning he had complained of a fever, severe headache, severe neck and back stiffness, nausea, and vomiting. He had become progressively unwell over 7–8 hours. On physical examination he was acutely ill with a temperature of 40°C. He was delirious and had neck rigidity with severe resistance to any attempt to passively flex his neck. A CT scan of his brain was normal. A spinal tap was performed and cerebrospinal fluid (CSF) was removed; it was cloudy.

A clinical diagnosis of bacterial meningitis was made. Therapy was started immediately because of the life-threatening nature of the illness. He was empirically treated with ceftriaxone and vancomycin, intravenously. A short time later, a Gram stain of his CSF revealed that the bacteria were Gram-positive cocci. Accordingly, the empiric antibiotic therapy was left as ceftriaxone and vancomycin. (If the Gram stain had revealed Gram-positive bacilli, then the empiric antibiotics might have been changed to ampicillin plus gentamicin because of the possibility of infection with *L. monocytogenes*.) Later studies confirmed that the meningitis was due to *S. pneumoniae*, and he was treated with ceftriaxone for an additional 12 days. He made a complete and uneventful recovery.

Patient 2. This 36-year-old female was brought to the emergency department at 1610 h by her husband. She had been unwell for several days. Over the preceding two days her family had noted a personality change with bizarre behavior; she had even complained of smelling odors that no one else could smell (i.e., she was having *olfactory hallucinations*). On the preceding day, she had developed episodes in which she would stare blankly for several minutes, sometimes licking her lips, while appearing to be in a dream state with altered consciousness (i.e., she was having *complex-partial seizures*). On the day of her presentation to hospital, she had a severe headache with fever. She collapsed to the ground and had a generalized seizure with body twitching. Physical examination revealed that she was confused. Her speech was abnormal in that she was having trouble forming the words to express herself. A CT scan of the brain showed asymmetrical areas of swelling in both temporal lobes of her brain.

A clinical diagnosis of viral encephalitis secondary to *Herpes simplex* was made. Therapy was started immediately. She was given acyclovir intravenously as an antiviral agent. To address the brain swelling she was given dexamethasone, an anti-inflammatory corticosteroid hormone. Since she was having seizures, she was also intravenously "loaded" with phenytoin, an anticonvulsant drug. She survived this acute illness and was discharged home 15 days later. She was left with mild, but definite, memory difficulties. Fourteen months later, she developed epilepsy, requiring long-term treatment with carbamazepine.

9.9 DRUG DESIGN OF THERAPIES AND ANTIDOTES FOR TOXINS

The industrial revolution brought many advances in technology; it also brought increased problems with pollution and environmental toxins. Humans now live in an environment that is rich in man-made chemicals. Currently, there are more than 65,000 chemicals in common use with approximately 400 new chemicals entering the environment on a yearly basis. Not surprisingly, toxicology may assume a role of increasing importance within the realm of human health.

Toxicology is the science dedicated to the study of the harmful effects of chemical and physical agents on living systems, especially humans. *Occupational toxicology* deals with exposure to chemicals within the workplace environment. *Environmental toxicology*, on the other hand, deals with the deleterious health effects of chemicals and pollutants within the environment as a whole. The magnitude of the toxicity experienced by an exposed human is dependent upon a number of factors, including route of exposure (topical vs. inhaled), duration of exposure (acute vs. chronic), quantity of exposure (low-level vs. high-level), and presence of mixtures (synergistic effects of two toxins, potentiation of one toxin by another). There are also a number of environmental considerations when assessing toxicology. If exposure to a contaminant is sufficiently large that it exceeds the ability of an organism to excrete the substance, then the chemical accumulates within the tissues of that organism, leading to the phenomenon of *bioaccumulation*. If the organism that is bioaccumulating contaminants is an integral part of the food chain, then the concentration of the contaminant may be magnified thousands of times as it passes up the food chain—this is the process of *biomagnification*. Recently, concern has been expressed concerning the possibility of biomagnification of toxins within farmed fish such as salmon.

A wide variety of specific chemicals have been implicated as toxins in occupational and environmental toxicology. Gaseous toxins include sulphur dioxide, nitrogen oxides, carbon monoxide, and ozone. These arise from a diverse range of industrial, recreational, and automotive sources. Halogenated aliphatic hydrocarbons (carbon tetrachloride, chloroform, trichloroethylene, tetrachloroethylene) are extensively used as industrial solvents, cleaning agents, and degreasing agents. The environment also contains many insecticides, including chlorinated hydrocarbons (chlorophenothanes, benzene hexachlorides, cyclodienes, toxaphenes), carbamates (aminocarb, carbofuran, isolan, pyramat), organophosphorus compounds (diazinon, malathion, parathion), and botanical insecticides (rotenone, pyrethrum). In addition to insecticides, various herbicides with "agrotech" applications also pollute our environment, including chlorophenoxy herbicides (2,4-dichlorophenoxyacetic acid) and bipyridyl herbicides (paraquat). The treatment of acute exposure to these various toxins is typically nonspecific and involves removing the victim from the contaminated environment, coupled with general supportive measures. The toxic effects of low-level, long-term exposure to these agents have yet to be fully determined.

Toxicity from inorganic pollutants and heavy-metal intoxication is also a significant public health concern. Lead poisoning is probably the oldest environmental disease in the world; it is credited with the downfall of the Roman Empire. Although lead has been removed from gasoline and paint, it still lurks as an environmental contaminant. Lead can cause damage to the brain, peripheral nervous system, and kidneys. Lead serves absolutely no useful purpose in the human body; no safe limit for exposure to lead exists. Arsenic is also a toxic element. In recent years, arsenic has received widespread commercial application in the manufacture of electronic semiconductors, cotton desiccants, and wood preservatives. In addition, arsenic leached from natural mineral deposits can contaminate groundwater; arsenic in the drinking water in the Ganges delta region of India has emerged as one of the world's largest environmental health problems. Like lead, mercury can also cause diffuse brain injury. In the 1950s, an epidemic of neurologic disease occurred in the Japanese fishing village of Minamata. The resulting Minamata disease was due to mercury poisoning. A number of other heavy metals can also contribute to human disease.

Specific treatments do exist for heavy metal toxicity. *Chelating agents* serve as antidotes for inorganic metal poisonings. *Chelates* (from Greek: chele = claw [of crayfish]) are complexes between the chelating agent and the metal ion. The chelating agent possesses several binding sites (ligands) that act to complex and "inactivate" the heavy metal ion.

The rational design of chelating agents as antidotes requires a careful consideration of acid–base chemistry. Metal ions are Lewis acids, while the chelating agents or ligands are Lewis bases. The concepts of hardness and softness may be used to describe systematically the interaction between them. A hard metal cation is one that retains its valence electrons very strongly; hard cations are not readily polarized and are of small size and high charge. Conversely, a soft cation is relatively large, does not retain its valence electrons firmly, and is easily polarized. Examples of hard cations include Li^+, Na^+, K^+, Ca^{2+}, Fe^{3+}, and Mn^{2+}; examples of soft cations include Cu^+, Ag^+, Au^+, Pd^{2+}, and Cd^{2+}. Chelating ligands containing highly electronegative donor atoms (O, N, F) are difficult to polarize and can be classified as hard bases. Easily polarized ligands, containing phosphorus or sulphur, act as soft bases. Examples of hard ligands include OH-, OR-, ROH, NH_3, Cl-, F- and NO_3-; examples of soft ligands include RSH, RS-, R_2S, R_3P, SCN- and I-. As a general rule, the formation of stable chelates, or complexes, results from interactions between hard acids and hard bases, or between soft acids and soft bases. Hard–soft interactions are weak and less desirable.

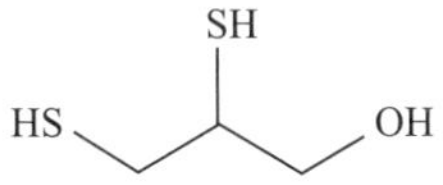

Dimercaprol (9.127)

A number of chelating agents are in clinical use. The chelates formed between the chelating agent and the toxic metal are nontoxic and are excreted via the kidney. The organometallic bond tends to be stable within the concentrated, acidic milieu of the renal tubular urine, thus facilitating the urinary excretion of the metal ion. Ethylenediaminetetraacetate (EDTA) is the prototypic chelating agent. EDTA possesses four carboxylate groups whose geometry enables the chelation of lead. Dimercaprol (**9.127**) (British Anti-Lewisite, BAL) was developed during World War II as an antidote against arsenical chemical warfare agents. Although it chelates a variety of metal ions, dimercaprol tends

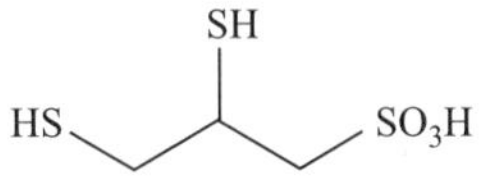

Dimercaptopropane
sulfonate (9.128)

to be chemically unstable and is administered intramuscularly in an oily vehicle. Dimercaptopropane sulfonate (**9.128**) is a chemically stable analog of dimercaprol that may be given orally. D-penicillamine (**9.129**) is a chelating agent that promotes the removal of copper from the body.

Drug design for the treatment of toxins is a discipline that is probably still in its infancy. We continue to introduce new chemicals into our environment; someday we will be confronted with the health issues that will arise. Currently, antidotes are essentially

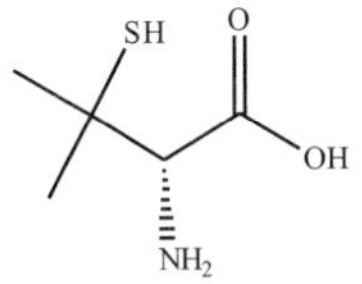

D-Penicillamine (9.129)

restricted to chelating agents for heavy-metal poisoning. However, health effects extend to the numerous organic chemicals in our environment. This has been known for many years. Environmental agents causing cancer were first brought to light by Sir Percival Potts in 1775, when he astutely related the high incidence of scrotal cancer among chimney sweeps to their chronic exposure to soot. Poisons in our environment continue to be a problem. In recent years there has been a growing concern regarding the presence of *endocrine disruptors* in the environment. These are synthetic chemicals that mimic or inhibit hormonal actions, producing estrogen-like or antiandrogenic effects. The resulting modified biochemical and endocrine responses may ultimately be a cause for concern among the human population. Potential problems such as this, and others that have yet to be recognized, may ultimately herald drug design for environmental toxins as an important pursuit in the future of medicinal chemistry.

Selected References

Antiviral Agents

K. R. Beutner (1995). Valacyclovir: A review of its antiviral activity, pharmacokinetics and clinical efficacy. *Antiviral Res. 28*: 281.

D. M. Coen, P. A. Schaffer (2003). Antiherpesvirus drugs: a promising spectrum of new drugs and drug targets. *Nat. Rev. Drug Discov. 2*: 278–288.

E. de Clerq (2001). Antiviral drugs: current state of the art. *J. Clin. Virol. 22*: 73–89.

E. de Clerq, R. T. Walker (1986). Chemotherapeutic agents for herpes virus infections. In: G. P. Ellis, G. B. West (Eds.). *Progress in Medicinal Chemistry*, vol. 23. Amsterdam: Elsevier, pp. 230–255.

R. Dolin (1985). Antiviral chemotherapy and chemoprophylaxis. *Science 227*: 1296–1303.

J. C. Drach (1980). Antiviral agents. *Annu. Rep. Med. Chem. 15*: 149–161.

M. Enserink (2004). Infectious diseases. One year after outbreak, SARS virus yields some secrets. *Science 304*: 1097.

C. Flexner (1998). HIV–protease inhibitors. *New Engl. J. Med. 338*: 1281.

S. Gupta (1986). Therapy of AIDS and AIDS-related syndromes. *Trends Pharmacol. Sci. 7*: 393–397.

S. Kinloch-de Loes (2004). Role of therapeutic vaccines in the control of HIV-1. *J. Antimicrob. Chemother. 53*: 562–566.

H. Mitsuga, S. Boder (1987). Strategies for antiviral therapy in AIDS. *Nature 325*: 773–778.

K. G. Nicholson, J. M. Wood, M. Zambon (2003). Influenza. *Lancet 362*: 1733–1745.

C. M. Perry, A. Wagstaff (1995). Famciclovir: A review of its pharmacological properties and therapeutic efficacy. *Drugs 50*: 396.

L. M. Schang (2002). Cyclin-dependent kinases as cellular targets for antiviral drugs. *J. Antimicrob. Chemother. 50*: 779–792.

T. J. Smith, M. J. Kremer, M. Luo, G. Vriend, E. Arnold, G. Kramer, M. G. Rossman, M. A. McKinley, G. D. Diana, M. J. Otto (1986). The site of attachment in human rhinovirus-14 for antiviral agents that inhibit uncoating. *Science 233*: 1286–1293.

S. D. Young (2003). Recent advances in the chemotherapy of HIV. *Annu. Rep. Med. Chem. 38*: 173.

Antibacterial Agents

S. D. Bentley, K. Chater, A. Tarraga, *et al.* (2002). Complete genome sequence of the actinomycete *Streptomyces coelicolor*. *Nature 417*: 141–147.

J. M. Blondeau (1999). Expanded activity of new fluoroquinolones. *Clin Ther. 21*: 3.

A. G. Brown (1985). Clavulanic acid and related compounds: inhibitors of β-lactamase enzymes. In: S. M. Roberts, B. J. Price (Eds.). *Medicinal Chemistry: The Role of Organic Chemistry in Drug Research*. New York: Academic Press.

H. Busse, C. Wostmann, E. Bakker (1992). The bactericidal action of streptomycin. *J. Gen. Microbiol.* 138: 551.

L. D. Cama, B. G. Christensen (1978). Structure–activity relationships of "non-classical" β-lactam antibiotics. *Annu. Rep. Med. Chem. 13*: 149–158.

I. Chopra (1998). Protein synthesis as a target for antibacterial drugs: current status and future opportunities. *Expert Opinion on Investigational Drugs 7*: 1237–1244.

I. Chopra, P. Hawkey, M. Hinton (1992). Tetracyclines: molecular and clinical aspects. *J. Antimicrob. Chemother. 29*: 245.

B. G. Christensen, R. W. Radcliffe (1976). Total synthesis of β-lactam antibiotics. *Annu. Rep. Med. Chem. 11*: 271–280.

M. Debono, R. S. Gordee (1982) Antibacterial agents. *Annu. Rep. Med. Chem. 17*: 107–117.

M. E. Falages, S. Gorbach (1995). Clindamycin and metronidazole. *Med. Clin. North Am. 79*: 845.

E. F. Gale, E. Cundliffe, P. E. Reynolds, M. H. Richmond, and M. J. Waring (1981). *The Molecular Basis of Antibiotic Action*, 2nd ed. New York: Wiley.

D. M. Gilbert (2001). Making sense of eukaryotic DNA replication origins. *Science 294*: 96–99.

T. D. Gootz (1985). Determinants of bacterial resistance to beta-lactam antibiotics. *Annu. Rep. Med. Chem. 20*: 137–144.

D. W. Green (2002). The bacterial cell wall as a source of antibacterial targets. *Expert Opinion on Therapeutic Targets 6*: 1–19.

E. S. Hamanaka, M. S. Kelly (1983). Antibacterial agents. *Annu. Rep. Med. Chem. 18*: 109–118.

R. Hare (1970). *The Birth of Penicillin*. London: Allen and Unwin.

D. Havlir, P. Barnes (1999). Tuberculosis in patients with HIV infection. *New Engl. J. Med. 340*: 367.

S. Houston, A. Fanning (1994). Current and potential treatment of tuberculosis. *Drugs 48*: 689.

B. K. Hubbard, C. T. Walsh (2003). Vancomycin assembly: nature's way. *Angew. Chem. Int. Ed. 42*: 730–765.

J. A. Kelly, P. C. Moews, J. R. Knox, J. M. Frére, J. M. Glunysen (1982). Penicillin target enzyme and the antibiotic binding site. *Science 218*: 479–481.

A. Maxwell, D. M. Lawson (2003). The ATP-binding site of type II topoisomerases as a target for antibacterial drugs. *Curr. Top. Med. Chem. 3*: 283–303.

R. B. Morin, M. Gorman (Eds.) (1982). *Chemistry and Biology of β-Lactam Antibiotics*. New York: Academic Press.

J. Rosamond, A. Allsop (2000). Harnessing the power of the genome in the search for new antibiotics. *Science 287*: 1973–1977.

A. D. Russel (1983). Design of antimicrobial chemotherapeutic agents. In: J. Smith, H. Williams (Eds.). *Introduction to Principles of Drug Design*. Bristol: Wright.

M. B. Schmid (1998). Novel approaches to the discovery of antimicrobial agents. *Curr. Opin. Chem. Biol. 2*: 529–534.

B. G. Spratt (1994). Resistance to antibiotics mediated by target alterations. *Science 264*: 389.

B. Suh, B. Lorber (1995). Quinolones. *Med. Clin. North Am. 79*: 869.

A. S. Wagman, M. L. MacKichan (2003). Antibacterial treatment of community-acquired respiratory tract infections. *Annu. Rep. Med. Chem. 38*: 183.

C. T. Walsh (1993). Vancomycin resistance: decoding the molecular logic. *Science 261*: 308.

C. T. Walsh (2000). Molecular mechanisms that confer antibacterial drug resistance. *Nature 406*: 775–781.

C. T. Walsh (2004). Polyketide and nonribosomal peptide antibiotics: modularity and versatility. *Science 303*: 1805–1810.

D. J. Waxman, J. L. Strominger (1983). Penicillin-binding proteins and the mechanism of action of β-lactam antibiotics. *Annu. Rev. Biochem. 52*: 825–869.

M. P. Wentland, J. B. Comett (1985). Quinolone antibacterial agents. *Annu. Rep. Med. Chem. 20*: 145–154.

Antifungal Agents

J. Afeltra, P. E. Verweij (2003). Antifungal activity of nonantifungal drugs. *Eur. J. Clin. Microbiol. & Inf. Dis. 22*: 397–407.

A. Albert (1985). *Selective Toxicity*, 7th ed. London: Chapman and Hall.

M. B. Anderson, T. Roemer, R. Fabrey (2003). Progress in anti-fungal drug discovery. *Annu. Rep. Med. Chem. 38*: 163.

S. Arikan (2002). Lipid-based antifungal agents: a concise overview. *Cell. & Mol. Biol. Lett. 7*: 919–922.

D. Berg, K. H. Büchel, M. Plempel, E. Regel (1986). Antimycotic sterol biosynthesis inhibitors. *Trends Pharmacol. Sci. 7*: 233–238.

J. Como, W. Dismukes (1994). Oral azole drugs as antifungal therapy. *New Engl. J. Med. 330*: 263.

D. W. Denning (2003). Echinocandin antifungal drugs. *Lancet 362*: 1142–1151.

B. DiDomenico (1999). Novel antifungal drugs. *Curr. Op. Microbiol. 2*: 509–515.

M. B. Gravestock, J. F. Ryley (1984). Antifungal chemotherapy. *Annu. Rep. Med. Chem. 19*: 127–136.

R. Hay (1990). Fluconazole. *J. Infect. 21*: 1.

C. Kauffman (1996). Role of azoles in antifungal therapy. *Clin. Infect. Dis. 22*: S148.

R. C. Matthews, J. P. Burnie (2004). Recombinant antibodies: a natural partner in combinatorial antifungal therapy. *Vaccine 22*: 865–871.

C. P. Selitrennikoff, M. Nakata (2003). New cell wall targets for antifungal drugs. *Curr. Opin. Invest. Drugs 4*: 200–205.

M. Viviani (1995). Flucytosine—What is the future? *J. Antimicrob. Chemother. 35*: 241.

A. Wong-Beringer, J. Kriengkauykiat (2003). Systemic antifungal therapy: new options, new challenges. *Pharmacotherapy 23*: 1441–1462.

Antiparasitic Agents

K. L. Blair, J. L. Bennett, R. A. Pax (1992). Praziquantel: Physiological evidence for its site of action in *Schistosoma mansoni*. *Parisitology 104*: 59.

I. Bruchhaus, E. Tannich (1993). Primary structure of the pyruvate phosphate dikinase in *Entoemeba histolytica*. *Mol. Biochem. Parisitol. 62*: 153.

K. Chibale (2002). Towards broad spectrum antiprotozoal agents. *ARKIVOC 9*: 93–98.

R. N. Davidson (1998). Practical guide to the treatment of Leishmaniasis. *Drugs 56*: 1009.

C. Doerig (2004). Protein kinases as targets for anti-parasitic chemotherapy. *Biochim. Biophys. Acta 1697*: 155–168.

J. E. Ellis (1994). Coenzyme Q homologs in parasitic protozoa as targets for chemotherapeutic attack. *Parisitol. Today 10*: 296.

M. Foley, L. Tilley (1998). Quinoline antimalarials: Mechanisms of action. *Pharmacol. Ther. 79*: 55.

O. Kayser, A. F. Kiderlen, S. L. Croft (2002). Natural products as potential antiparasitic drugs. *Stud. Nat. Prod. Chem. 26*: 779–848.

H. O. Lobel, P. E. Kozarsky (1997). Update on the prevention of malaria. *J. Amer. Med. Assoc. 278*: 1767.

W. A. Petri, U. Singh (1999). Diagnosis and management of amebiasis. *Clin. Infect. Dis. 29*: 1117.

J. E. Rosenblatt (1999). Antiparisitic agents. *Mayo Clin. Proc. 74*: 1161.

C. C. Wang (1995). Molecular mechanisms and therapeutic approaches to the treatment of African trypansomiasis. *Annu. Rev. Pharmacol. Toxicol. 35*: 93.

L. M. Werbel, D. F. Worth (1980). Antiparasitic agents. *Annu. Rep. Med. Chem. 15*: 120–129.

K. A. Werbovetz (2002). Tubulin as an antiprotozoal drug target. *Mini-Rev. Med. Chem. 2*: 519–529.

N. J. White (1996). The treatment of malaria. *New Engl. J. Med. 335*: 69.

Appendix: Drugs Arranged by Pharmacological Activity

CARDIOLOGY

Angina

BETA-ADRENERGIC BLOCKING AGENTS, SELECTIVE, INTRINSIC SYMPATHOMIMETIC ACTIVITY (ISA)
acebutolol hydrochloride

BETA-ADRENERGIC BLOCKING AGENTS, SELECTIVE, NON-ISA
atenolol
metoprolol tartrate

BETA-ADRENERGIC BLOCKING AGENTS, NONSELECTIVE, ISA
pindolol

BETA-ADRENERGIC BLOCKING AGENTS, NONSELECTIVE, NON-ISA
nadolol
propranolol hydrochloride
timolol maleate

CALCIUM CHANNEL BLOCKERS
amlodipine besylate
diltiazem hydrochloride
nifedipine
verapamil hydrochloride

CORONARY VASODILATORS, NITRATES
isosorbide dinitrate
isosorbide-5-mononitrate
nitroglycerin

Arrhythmias

CARDIAC GLYCOSIDES
digoxin

CLASS I, TYPE 1A ANTIARRHYTHMICS
disopyramide
disopyramide phosphate
procainamide hydrochloride
quinidine bisulfate
quinidine gluconate
quinidine sulfate

CLASS I, TYPE 1B ANTIARRHYTHMICS
lidocaine hydrochloride
mexiletine hydrochloride

CLASS I, TYPE 1C ANTIARRHYTHMICS
flecainide acetate
propafenone hydrochloride

CLASS II, BETA-ADRENERGIC BLOCKING AGENTS
esmolol hydrochloride
propranolol hydrochloride
sotalol hydrochloride

CLASS III
amiodarone hydrochloride
bretylium tosylate
ibutilide fumarate

CLASS IV, CALCIUM CHANNEL
    BLOCKERS
diltiazem hydrochloride
verapamil hydrochloride

VARIOUS ANTIARRHYTHMICS
adenosine

CARDIAC SYMPATHOMIMETICS
dobutamine hydrochloride
dopamine hydrochloride
eninephrine hydrochloride
isoproterenol
methoxamine hydrochloride
norepinephrine bitartrate
phenylephrine hydrochloride

Congestive heart failure

ANGIOTENSIN CONVERTING
    ENZYME (ACE) INHIBITORS
captopril
cilazapril
enalapril maleate
fosinopril sodium
lisinopril
perindopril erbumine
quinapril hydrochloride
ramipril

BETA-ADRENERGIC BLOCKING
    AGENTS, NONSELECTIVE,
    NON-ISA
carvedilol

CARDIAC GLYCOSIDES
digoxin

DIURETICS
amiloride hydrochloride
bumetanide
chlorthalidone
ethacrynate sodium
ethacrynic acid
furosemide
hydrochlorothiazide
metolazone
sprionolactone
triamterene

INOTROPES
dobutamine hydrochloride
dopamine hydrochloride
milrinone lactate

Dyslipidemias

Cholesterol and triglyceride reducers

BILE ACID SEQUESTRANTS
cholestyramine resin
colestipol hydrochloride

FIBRATES
bezafibrate
clofibrate
fenofibrate (micronized)
gemfibrozil

3-HYDROXY-3-METHYLGLUTERYL
    (HMG)-CoA REDUCTASE INHIBITORS
atorvastatin calcium
fluvastatin sodium
lovastatin
pravastatin sodium
simvastatin

NIACIN DERIVATIVES
niacin

Hypertension

Antiadrenergic agents

ALPHA$_1$-ADRENERGIC BLOCKING AGENTS
doxazosin mesylate
prazosin hydrochloride
terazosin hydrochloride dihydrate

ALPHA$_1$- AND ALPHA$_2$-ADRENERGIC
    BLOCKING AGENTS
phentolamine mesylate

ALPHA$_1$- AND BETA-ADRENERGIC BLOCKING
    AGENTS
labetalol hydrochloride

BETA-ADRENERGIC BLOCKING AGENTS,
    SELECTIVE, INTRINSIC
    SYMPATHOMIMETIC ACTIVITY (ISA)
acebutolol hydrochloride

BETA-ADRENERGIC BLOCKING AGENTS, SELECTIVE, NON-ISA
atenolol
bisoprolol fumarate
esmolol hydrochloride
metoprolol tartrate

BETA-ADRENERGIC BLOCKING AGENTS, NONSELECTIVE, ISA
oxprenolol hydrochloride
pindolol

BETA-ADRENERGIC BLOCKING AGENTS, NONSELECTIVE, NON-ISA
nadolol
propranolol hydrochloride
timolol maleate

CENTRALLY ACTING ANTIADRENERGIC AGENTS
clonidine hydrochloride
methyldopa

Arteriolar smooth muscle agents

CALCIUM CHANNEL BLOCKERS
amlodipine besylate
diltiazem hydrochloride
felodipine
nifedipine
verapamil hydrochloride

VASODILATORS
diazoxide
epoprostenol sodium
hydralazine hydrochloride
minoxidil
nitroglycerin
sodium nitroprusside

Diuretics

LOOP DIURETICS
ethacrynate sodium
ethacrynic acid
furosemide

OSMOTIC DIURETICS
mannitol

POTASSIUM-SPARING AGENTS
amiloride hydrochloride
spironolactone
triamterene/hydrochlorothiazide

THIAZIDES AND RELATED AGENTS
chlorthalidone
hydrochlorothiazide
indapamide
indapamide hemihydrate
metolazone

Renin–angiotensin system agents

ANGIOTENSIN CONVERTING ENZYME (ACE) INHIBITORS
benazepril
captopril
cilazapril
enalaprilat
enalapril maleate
fosinopril sodium
lisinopril
perindopril erbumine
quinapril hydrochloride
ramipril
trandolapril

ANGIOTENSIN II RECEPTOR ANTAGONISTS
candesartan cilexetil
eprosartan mesylate
irbsartan
losartan potassium
telmisartan
valsartan

Peripheral vascular disease

PERIPHERAL VASODILATORS
nylidrin hydrochloride
papaverine hydrochloride
pentoxifylline

DERMATOLOGY

Acne

Systemic

ANTIANDROGEN/ESTROGEN COMBINATIONS
cevonorgesterl/ethinyl estradiol
cyproterone acetate/ethinyl estradiol

ANTIBIOTICS
erythromycin
erythromycin estolate
minocycline hydrochloride
tetracycline hydrochloride

RETINOIDS
isotretinoin

Topical

ANTIBIOTICS
clindamycin phosphate
crythromycin

PEROXIDES
benzoyl peroxide

RETINOIDS
isotretinoin
tazarotene
tretinoin

RETINOID ANALOGS
adapalene

VARIOUS ACNE PREPARATIONS FOR
   TOPICAL USE
povidone-iodine
salicylic acid
triclosan

Dermatologic therapy

ANTIHYPERHIDROTICS (ANTIPERSPIRANTS)
methenamine

DEPIGMENTING AGENTS
hydroquinone
mequinol/tretinon

Dermatitis, atopic
(see also Corticosteroid therapy, topical)

IMMUNOMODULATOR, TOPICAL
tacrolimus

DERMATITIS HERPETIFORMIS THERAPY
dapsone
sulfapyridine

EXFOLIANTS
glycolic acid
strontium chloride

Photodamaged skin therapy

tazarotene
tretinion

Rosacea therapy

metronidazole

Scar management

polysiloxane/silicone dioxide

Sunscreens

UVA ABSORBERS (LARGE SPECTRUM)
butyl methoxydibenzoylmethane
oxbenzoneterephthalylidene dicamphor
   sulfonic acid

UVB ABSORBERS (NARROW SPECTRUM)
homosalate
methylbenzylidene camphor
octocrylene
octyl dimethyl PABA (Padimate O)
octyl methoxycinnamate
octyl salicylate
phenylbenzymidazole sulfonic acid

PHYSICAL AGENTS
zinc oxide

Vitiligo therapy

methoxslen

Warts/genital warts and corn preparations
(see also Anogenital Warts)

cantharidin
imiquimod
podofilox
podophyllum resin
salicylic acid

Psoriasis

Psoriasis therapy, systemic

CYTOTOXICS
methotrexate sodium

IMMUNOSUPRESSANTS
cyclosporine

methoxsalen
psoralens

RETINOIDS
acitretin

Psoriasis therapy, topical

ANTRACEN DERIVATIVES
anthralin

PSORALENS
methoxsalen

RETINOIDS
taxaroten

TARS
coal tar
fractar

VITAMIN D DERIVATIVES
calcipotriol

VARIOUS PSORIASIS THERAPY AGENTS, TOPICAL
phenol
salicylic acid

ENDOCRINOLOGY

Diabetes mellitus

Insulins, analogs

VERY RAPID ACTING
insulin aspart
insulin lispro

MIXED
insulin lispro/lispro protamine (25/75)

Insulins, human

RAPID ACTING
insulin regular, biosynthetic

INTERMEDIATE ACTING
insulin lente, biosynthetic
insulin NPH, biosynthetic

LONG ACTING
insulin ultralente, biosynthetic

MIXED (REGULAR/NPH)
insulin (10/90) biosynthetic
insulin (20/80) biosynthetic
insulin (30/70) biosynthetic

insulin (40/60) biosynthetic
insulin (50/50) biosynthetic

Insulins, pork

RAPID ACTING
insulin regular

INTERMEDIATE ACTING
insulin lente
insulin NPH

Oral agents

ALPHA-GLUCOSIDASE INHIBITORS
acarbose

BIGUANIDES
metformin hydrochloride

MEGLITINIDES
nateglinide
repaglinide

SULFONYLUREAS
chlorpropamide
gliclazide
limepiride
glyburide
tolbutamide

THIZAOLIDINEDIONES
pioglitzaone
rosiglitazone

ADJUNCTIVE THERAPY
orlistat

Hormonal therapy

Androgen replacement therapy

ANDROGENS, ORAL
testosterone undecanoate

ANDROGENS, PARENTERAL
nandrolone decanoate
testosterone cypionate
testosterone enanthate

ANDROGENS, TOPICAL
testosterone

ANDROGENS, TRANSDERMAL
testosterone

Estrogen/progestin replacement therapy

ESTROGENS, ORAL
conjugated estrogens
estradiol-17β
estropipate
ethinyl estradiol

ESTROGENS, PARENTERAL
estradiol valerate

ESTROGENS, TRANSDERMAL
estradiol-17β

ESTROGENS, VAGINAL PREPARATIONS
conjugated estrogens
estradiol
estradiol-17β hemihydrate
estrone

PROGESTOGENS
medrogestone
medroxypogesterone acetate
megestrol acetate
norethindrone

ESTROGEN/PROGESTOGEN
    COMBINATIONS, ORAL
conjugated estrogens/
    medroxyprogesterone
ethinyl estradiol/norethindrone acetate
mestranol/norethindrone

ESTROGEN/PROGESTOGEN COMBINATIONS,
    TRANSDERMAL
estradiol-17β/norethindrone

Hypercalcemia and Paget's disease

Bone metabolism regulators

ANTIPARATHYROID HORMONES
calcitonin salmon

BISPHOSPHONATES
alendronate sodium
clodronate disodium
etidronate disodium
pamidronate disodium
risedronate sodium
zoledronic acid

PHOSPHATE PREPARATIONS
sodium acid phosphate

Hypothalamic and pituitary hormones

*(see also Cancer, hormones;
    Endometriosis; Hormonal therapy;
    Thyroid disorders)*

Anterior pituitary hormones

ADRENOCORTICOTROPHIC HORMONES
    (ACTH)
cosyntropin
cosyntropin/zinc hydroxide

GROWTH HORMONES
somatrem
somatropin

Hypothalamic hormones

GONADOTROPIN-RELEASING HORMONE
    (GNRH) ANALOGS
buserelin acetate
goserelin acetate
leuprolide acetate
nafarelin acetate

SOMATOSTATIN AND ANALOGS
octreotide acetate
somatostatin

Posterior pituitary hormones

ANTIDIURETIC HORMONE ANALOGS
desmopressin acetate
vasopressin

OXYTOCICS
oxytocin

Infertility

GONADOTROPINS
follitropin alpha (rDNA origin)
follitropin beta (rec FSH)
gonadotropin (human) chorionic
menotropins

GONADOTROPIN-RELEASING HORMONE
    ANTAGONISTS
ganirelix acetate

OVULATION STIMULANTS, SYNTHETIC
clomiphene citrate
gonadorelin acetate

Osteoporosis

Bone metabolism regulators

BISPHOSPHONATES
alendronate
etidronate
risedronate

ANTIPARATHYROID HORMONES
calcitorin

CALCIUM SUPPLEMENTS
calcium carbonate
calcium gluconate

*Selective estrogen receptor
  modulators (SERMs)*

BENZOTHIOPHENE
raloxifene hydrochloride

Thyroid disorders

ANTITHYROID AGENTS
methimazole
propylthiouracil

THYROID HORMONES
levothyroxine sodium
liothyronine sodium
thyroid

THYROID STIMULATING HORMONE, HUMAN
thyrotropin alfa

GASTROENTEROLOGY

Cholestatic liver disease

GALLSTONE SOLUBILIZING AGENTS
ursodiol

Constipation

BULK FORMING AGENTS
psyllium hydrophilic mucilloid
sterculia gum

HYPEROSMOTIC LAXATIVES
glycerin
sorbitol

LUBRICANT LAXATIVES
mineral oil

OSMOTIC LAXATIVES
lactulose
magnesium citrate
magnesium hydroxide
polyethylene glycol/electroltyes
sodium phosphates

STIMULANT LAXATIVES
bisacodyl
cascara
sennosides

STOOL SOFTENERS
docusate calcium
docusate sodium

Diarrhea therapy and intestinal anti-infective agents

ANTIPERISTALTICS
diphenoxylate hydrochloride/atropine
  sulfate
loperamide hydrochloride

FLORA MODIFIERS
lactobacillus

INTESTINAL ADSORBENTS
attapulgite, activated
bismuth subsalicylate

Intestinal anti-infectives

ANTIBACTERIALS
ciprofloxacin hydrochloride
co-trimoxazole (sulfamethoxazole/
  trimethoprim)
doxycycline hyclate
mentronidazole
vancomycin hydrochloride

ANTIFUNGALS
nystatin

Eating disorders

ANTIANOREXIC/ANTICACHEXIC
megestrol acetate
somatropin

ANTIBULIMICS
fluoxetine hydrochloride

Appetite suppressants

NORADRENERGIC AGENTS
diethylpropion hydrochloride
mazindol
phentermine

NORADRENERGIC AND SEROTONERGIC
    AGENTS
sibutramine

Antiobesity agents

GASTROINTESTINAL LIPASE INHIBITORS
orlistat

Flatulence

ALPHA-D-GALACTOSIDASE ENZYMES
alpha-d-galactosidase

COALESCING AGENTS
simethicone

Gastroesophageal reflux

Antacids

ALUMINUM-CONTAINING PREPARATIONS
aluminum hydroxide

ALUMINUM/MAGNESIUM-CONTAINING
    PREPARATIONS
aluminum hydroxide/magnesium
    hydroxide
magaldrate

CALCIUM-CONTAINING PREPARATIONS
calcium carbonate

MAGNESIUM-CONTAINING PREPARATIONS
magnesium carbonate
magnesium hydroxide

FOAMING AGENTS
alginic acid
sodium alginate

HISTAMINE H$_2$-RECEPTOR ANTAGONISTS
cimetidine
famotidne

nizatidne
ranitidine hydrochloride

PROTON PUMP INHIBITORS
lansoprazole
omeprazole magnesium
pantoprazole sodium
rabeprazole sodium

Gastrointestinal motility disorders

UPPER GASTROINTESTINAL TRACT AGENTS
doperidone maleate
metoclopramide hydrochloride

Gastrointestinal spasticity
(see also Urinary Tract Therapy)

Anticholinergic agents

NATURAL ALKALOIDS, TERTIARY AMINES
atropine sulfate
hyoscine butylbromide
hyoscine hydrobromide
hyoscyamine sulfate

SYNTHETIC AMINES, QUATERNARY
    AMMONIUM PREPARATIONS
pinaverium bromide
propantheline bromide

SYNTHETIC AMINES, TERTIARY AMINE
    PREPARATIONS
dicyclomine hydrochloride

LOWER GASTROINTESTINAL TRACT
    MOTILITY REGULATORS
trimebutine maleate

Inflammatory bowel disease

5-AMINOSALICYLIC ACID DERIVATIVES
5-aminosalicylic acid (mesalamine)
olsalazine sodium
sulfasalazine

BIOLOGICAL RESPONSE MODIFIERS
infliximab

CORTICOSTEROIDS FOR RECTAL USE
betamethasone sodium phosphate
budesonide

hydrocortisone
hydrocortisone acetate

Irritable bowel syndrome

ANTISPASMODICS
dicyclomine
hyoscyamine sulfate
pinaverium bromide
trimebutine

5-HT$_4$ PARTIAL AGONISTS
tegaserod

Nausea and vomiting

ANTIHISTAMINES
dimenhydrinate
hydroxyzine hydrochloride
promethazine hydrochloride

ANTICHOLINERGICS
scopolamine

CANNABINOIDS
dronabinol
nabilone

DOPAMINE ANTAGONISTS
chlorpromazine hydrochloride
metoclopramide hydrochloride
perphenazine
prochlorperazine
prochlorperazine mesylate
trifluoperazine hydrochloride

SEROTONIN (5-HT$_3$) ANTAGONISTS
dolasetron mesylate
granisetron hydrochloride
ondansetron
ondansetron hydrochloride dihydrate

Peptic ulcer

ANTACIDS
aluminum hydroxide
aluminum hydroxide/magnesium
    hydroxide
magaldrate

CYTOPROTECTIVES
sucralfate

HISTAMINE H$_2$-RECEPTOR ANTAGONISTS
cimetidine
famotidine
nizatidine
ranitidie hydrochloride

MUCOSAL PROTECTIVE AGENT
misoprostol

PROTON PUMP INHIBITORS
lansoprazole
omeprazole magnesium
pantoprazole
pantoprazole sodium
rabeprazole

H. PYLORI ERADICATION THERAPY
lansoprazole/clarithromycin/amoxicillin
omeprazole/clarithromycin/amoxicillin
omeprazole/clarithromycin/
    metonidazole

GYNECOLOGY

Contraception
(see also Hormonal therapy)

Oral contraceptives

ESTROGENS AND PROGESTOGENS,
    MONOPHASIC
desogestrel/ethinyl estradiol
ethinyl estradiol/ethynodiol diacetate
ethinyl estradiol/levonorgestrel
ethinyl estradiol/norethindrone
ethinyl estradiol/norethindrone acetate
ethinyl estradiol/norgestimate
ethinyl estradiol/norgestrel
mestranol/norethindrone

ESTROGENS AND PROGESTOGENS,
    BIPHASIC
ethinyl estradiol/norethindrone

ESTROGENS AND PROGESTOGENS,
    TRIPHASIC
ethinyl estradiol/levonorgeestrel
ethinyl estradiol/norethindrone
ethinyl estradiol/norgestimate

PROGESTOGENS
levonorgestrel
norethindrone

Injectable contraceptives

PROGESTOGENS
medroxyprogesterone acetate

Contraceptive implants

PROGESTOGENS
levonorgestrel

Intrauterine contraceptives

PROGESTOGENS
levonorgestrel

Transdermal contraceptives
norelgestromin/ethinyl estradiol

Vaginal contraceptives
nonoxynol-9

Endometriosis

ESTROGEN/PROGESTOGEN-CONTAINING
    PREPARATIONS
mestranol/norethindrone

GONADOTROPIN INHIBITORS
danazol

GONADOTROPIN-RELEASING HORMONE
    (GNRH) ANALOGS
buserelin
goserelin acetate
nafarelin acetate

PROGESTOGENS
medroxyprogesterone acetate
norethindrone acetate

HEMATOLOGY

Anemia therapy and hematopoietics
*(see also Mineral replacement; Vitamin
    therapy)*

ERYTHROPOIESIS STIMULANTS
darbepoetin alfa
epoetin alfa

HEMATOPOIETIC AGENTS
ancestim
filgrastim

Iron preparations

IRON BIVALENT, ORAL PREPARATIONS
ferrous fumarate
ferrous gluconate
ferrous sulfate

IRON COMPLEX, PARENTERAL PREPARATIONS
iron–dextran

IRON TRIVALENT, PARENTERAL PREPARATIONS
iron–sorbitol–citric acid complex

Vitamin B_{12} and folic acid

FOLIC ACID
folic acid

VITAMIN B_{12} DERIVATIVES
cyanocobalamin
hydroxocobalamin

Various anemia agents

ANDROGENIC-ANABOLIC STEROIDS
nandrolone decanoate

Anticoagulant therapy

HEPARINS, STANDARD
heparin sodium

HEPARINS, LOW MOLECULAR WEIGHT
    (LMWH)
dalteparin sodium
enoxaparin sodium
nadroparin calcium
tinzaparin sodium

HEPARINOIDS
danaparoid sodium

VITAMIN K ANTAGONISTS
nicoumalone
warfarin sodium

VARIOUS ANTICOAGULANTS
antithrombin III (human)
lepirudin

Bleeding therapy

Antifibrinolytics

AMINO ACIDS
aminocaproic acid
tranexamic acid

PROTEINASE INHIBITORS
aprotinin

VITAMIN K ANALOGUES
phytonadione

Hemostatics

BLOOD COAGULATION FACTORS
antihemophilic factor (recombinant)
eptacog alfa (activated)
factor IX contentrate (human)
factor IX (recombinant)
moroctocog alfa

LOCAL HEMOSTATICS
thrombin (bovine)
gelatin, absorbable

VARIOUS HEMOSTATICS
desmopressin acetate

Mineral replacement

Iron replacement therapy

IRON BIVALENT, ORAL PREPARATIONS
ferrous fumarate
ferrous gluconate
ferrous sulfate

IRON TRIVALENT, PARENTERAL PREPARATIONS
iron sorbitol citric acid complex

Platelet antiaggregation therapy

ADENOSINE DIPHOSPHATE INHIBITORS
clopidogrel bisulfate
dipyridamole
sulfinpyrazone

GLYCOPROTEIN (GP LLB/LLLA)
   RECEPTOR INHIBITORS
abciximab
eptifibatide
tirofiban hydrochloride

FIBRINOGEN-PLATELET BINDING INHIBITORS
ticlipidine hydrochloride

THROMBOXANE-A$_2$ INHIBITORS
acetylsalicylic acid (ASA)

Thrombolytic therapy

Plasminogen activators

NATURAL ENZYMES
urokinase

PROTEINS, DERIVED FROM BACTERIA
reteplase
streptokinase

PROTEINS, RECOMBINANT
   DNA ORIGIN
alteplase
anistreplase
tenecteplase

VARIOUS ANTITHROMBOTIC AGENTS
anagrelide hydrochloride
fondaparins sodium

IMMUNOLOGY

Corticosteroid therapy, systemic

GLUCOCORTICOIDS
betamethasone acetate
betamethasone sodium phosphate
cortisone acetate
dexamethasone
dexamethasone sodium
   phosphate
hydrocortisone
hydrocortisone sodium succinate
methylprednisolone acetate
methylprednisolone sodium
   succinate
prednisolone sodium phosphate
prednisone
triamcinolone
triamcinolone acetonide
triamcinolone diacetate

MINERALOCORTICOIDS
fludrocortisone acetate

Corticosteroid therapy, topical

Anti-inflammatory agents, topical

CORTICOSTEROIDS, WEAK (GROUP I)
hydrocortisone
hydrocortisone acetate
methylprednisolone

CORTICOSTEROIDS, MODERATELY POTENT
    (GROUP II)
clobetasone 17-butyrate
desonide
flumethasone pivalate
hydrocortisone acetate
hydrocortisone 17-valerate
prednicarbate
triamcinolone acetonide

CORTICOSTEROIDS, POTENT (GROUP III)
amcinonide
betamethasone dipropionate
betamethasone valerate
desoximetasone
diflucortolone valerate
fluocinolone acetonide
fluocinonide
mometasone furoate

CORTICOSTEROIDS, VERY POTENT
    (GROUP IV)
clobetasol 17-propionate
halcinonide
halobetasol propionate

Immunosuppressive therapy

Corticosteroids, systemic

GLUCOCORTICOIDS
betamethasone sodium phosphate
cortisone acetate
dexamethasone
hydrocortisone
hydrocortisone sodium succinate
methylprednisolone
methylprednisolone sodium succinate
prednisolone
prednisone
triamcinolone diacetate

CYCLIC PEPTIDES
cyclosporine

CYTOTOXIC AGENTS
azathioprine

IMMUNE GLOBULINS
anti-thymocyte globulin (equine)

MONOCLONAL ANTIBODIES
basiliximab
daclizumab
muromonab-CD3

SELECTIVE IMMUNOSUPPRESSIVE AGENTS
mycophenolate mofetil
sirolimus
tacrolimus

INFECTIOUS DISEASE

Fungal infections

Antifungals, systemic

ALLYLAMINES
terbinafine hydrochloride

ANTIFUNGAL ANTIBIOTICS
amphotericin B
amphotericin B (lipid-based)
griseofulvin

ECHINOCANDINS
caspofungin acetate

IMIDAZOLES
ketoconazole

PYRIMIDINES
flucytosine

TRIAZOLES
fluconazole
itraconazole

Antifungals, topical
(see also Vaginal therapy)

ALLYLAMINES
naftifine hydrochloride
terbinafine hydrochloride

ANTIFUNGAL ANTIBIOTICS
nystatin

IMIDAZOLES
clotrimazole
econazole nitrate
ketoconazole
miconazole nitrate
oxiconazole nitrate
tioconazole

VARIOUS ANTIFUNGALS,
     TOPICAL
chlorphenesin
ciclopirox olamine
clioquinol
selenium sulfide
tolnaftate
undecylenic acid

HIV (human immunodeficiency virus) infections and related disorders

Prophylactic therapy

CYTOMEGALOVIRUS
ganciclovir sodium

MYCOBACTERIUM AVIUM COMPLEX
azithromycin
clarithromycin
rifabutin

MYCOBACTERIUM TUBERCULOSIS
isoniazid

PNEUMOCYSTIS CARINII
     PNEUMONIA
atovaquone
co-trimoxazole
     (sulfamethoxazole/trimethoprim)
dapsone
pentamidine isethionate

TOXOPLASMA GONDII ENCEPHALITIS
co-trimoxazole
     (sulfamethoxazole/trimethoprim)
dapsone
pyrimethamine

AIDS-associated bacterial infections

MYCOBACTERIUM AVIUM COMPLEX
azithromycin
clarithromycin
ethambutol hydrochloride
rifabutin
rifampin

AIDS-associated fungal infections

CANDIDA SPECIES
amphotericin B
fluconazole
ketoconazole
itraconazole

CRYPTOCOCCUS NEOFORMANS
amphotericin B
fluconazole

AIDS-associated parasitic infections

PNEUMOCYSTIS CARINII PNEUMONIA
atovaquone
co-trimoxazole
     (sulfamethoxazole/trimethoprim)
dapsone
pentamidine isethionate

TOXOPLASMA GONDII
atovaquone
azithromycin
clarithromycin
co-trimoxazole
     (sulfamethoxazole/trimethoprim)
dapsone
pyrimethamine
slfadiazine

Viral infections

CYTOMEGALOVIRUS
ganciclovir sodium
valganciclovir

NON-NUCLEOSIDE REVERSE TRANSCRIPTASE
     INHIBITORS (NNRTIS)
delavirdine mesylate
efavirenz
nevirapine

NUCLEOSIDE ANALOG REVERSE
  TRANSCRIPTASE INHIBITORS (NRTIs)
abacavir suflate
didanosine (ddl)
lamivudine (3TC)
stavudine (d4T)
zalcitabine (ddC)
zidovudine (AZT)
abacavir
  sulfate/lamivudine/zidovudine

PROTEASE INHIBITORS
amprenavir
indinavir sulfate
nelfinavir
ritonavir
ritonavir/lopinavir
saquinavir
saquinavir mesylate

Related disorders therapy

ANEMIA (AZT-INDUCED)
epoetin alfa

KAPOSI'S SARCOMA
alitetinoin
daunorubicin (liposomal)
doxorubicin hydrochloride
  (pegylated liposomes)
interferon alfa-2a
interferon alfa-2b
paclitaxel

WEIGHT LOSS, ANOREXIA, CACHEXIA
megestrol acetate
somatropin

Bacterial infections

Antibiotics

AMINOGLYCOSIDES
amikacin sulfate
gentamicin sulfate
netilmicin sulfate
paromomycin sulfate
streptomycin sulfate
tobramycin sulfate

CARBAPENEMS
imipenem/cilastatin sodium
meropenem

CEPHALOSPORINS, 1ST GENERATION
cefadroxil
cefazolin sodium
cephalexin

CEPHALOSPORINS, 2ND GENERATION
cefaclor
cefotetan disodium
cefoxitin sodium
cefprozil
cefuroxime axetil
cefuroxime sodium

CEPHALOSPORINS, 3RD GENERATION
cefixime
cefotaxime sodium
ceftazidime
ceftazidime pentahydrate
ceftizoxime sodium
ceftriaxone sodium

CEPHALOSPORINS, 4TH GENERATION
cefepime hydrochloride

FLUOROQUINOLONES
ciprofloxacin
ciprofloxacin hydrochloride
gatifloxacin
levofloxacin
moxifloxacin hydrochloride
norfloxacin
ofloxacin

GLYCOPOPTIDES
vancomycin hydrochloride

LINCOSAMIDES
clindamycin hydochloride
clindamycin palmitate hydrochloride
clindamycin phosphate
lincomycin hydrochloride monohydrate

MACROLIDES
azithromycin dihydrate
clarithromycin

erythromycin
erythromycin stearate
spiramycin

OXAZOLIDINONES
linezolid

PENICILLINS, AMINOPENICILLINS
amoxicillin trihydrate
ampicillin
ampicillin sodium
bacampicillin hydrochloride
pivampicillin
pivmecillnam hydrochloride

PENICILLINS, ANTIPSEUDOMONAL
piperacillin sodium
ticarcillin disodium

PENICILLINS, PENICILLINASE SENSITIVE
penicillin G benzathine
penicillin G sodium
phenoxymethyl penicillin (penicillin V)
phenoxymethyl penicillin
    (penicillin V) benzathine
phenoxymethyl penicillin
    (penicillin V) potassium

PENICILLINS, PENICILLINASE-RESISTANT
cloxacillin sodium

PENICILLINS, B -LACTAMASE INHIBITOR
    COMBINATIONS
amoxilillin trihydrate/clavulanate
    potassium
poperacillin sodium/tazobactam
    sodium
ticarcillin disodium/clavulanate
    potassium

STREPTOGRAMINS
quinupristin/dalfopristin

SULFONAMIDES
sulfamethoxazole

SULFONAMIDE COMBINATIONS
co-trimazine
    (sulfadiazine/trimethoprim)

co-trimoxazole
    (sulfamethoxazole/trimethoprim)
erythromycin
    ethylsuccinate/sulfsoxazole

TRIMETHOPRIM AND DERIVATIVES
trimethoprim

TETRACYCLINES
doxycycline hyclate
minocycline hydrochloride
tetracycline hydrochloride

VARIOUS ANTIBIOTICS
bacitracin
chloramphenicol
colistimethate sodium
fosfomycin tromethamine
fusidic acid
metronidazole
polymyxin B sulfate
rifabutin
sodium fusidate

ANTIBACTERIALS, TOPICAL
bacitracin
bacitracin zinc
chloramphenicol
chlorhexidine acetate
chlorhexidine gluconate
chlortetracycline hydrochloride
clioquinol
framycetin sulfate
fusidic acid
gentamicin sulfate
mupirocin
mupirocin calcium
neomycin sulfate
polymyxin B sulfate
silver sulfadizaine
sodium fusidate
tetracycline hydrochloride

Parasitic infestations

ANTHELMINTICS
mebendazole
praziquantel

pyrantel pamoate
pyrvinium pamoate

Protozoal infections

Amebicides

AMINOGLYCOSIDES
paromomycin sulfate

8-HYDROXYQUINOLINE DERIVATIVES
iodoquinol

NITROIMIDAZOLE DERIVATIVES
metronidazole

Malaria therapy

BIGUANIDES
proguanil

CINCHONA ALKALOIDS
quinidine gluconate injection
quinine sulfate

FOLIC ACID ANTAGONISTS
pyrimethamine

QUINOLINE DERIVATIVES
chloroquine phosphate
hydroxychloroquine sulfate
mefloquine hydrochloride
primaquine phosphate

COMBINATION PRODUCTS
atovaquone/proguanil

Pneumocystis carinii therapy
atovaquone
co-trimoxazole
    (sulfamethoxazole/trimethoprim)
pentamidine isethionate

Tuberculosis

AMINOSALICYLIC ACID DERIVATIVES
para-aminosalicylate sodium
    (PAS sodium)

ANTIBIOTICS
cycloserine
rifampin
streptomycin sulfate

HYDRAZIDES
isoniazid

COMBINATION ANTITUBERCULOSIS AGENTS
isoniazid/pyrazinamide/rifampin

VARIOUS ANTITUBERCULOSIS AGENTS
ethambutol hydrochloride
pyrazinamide

Vaginal infections

Antibacterials, vaginal

LINCOSAMIDES
clindamycin phosphate

Antifungals, vaginal

ANTIBIOTICS
nystatin

IMIDAZOLES
clotrimazole
econazole nitrate
metronidazole
miconazole nitrate

TRIAZOLES
terconazole

Trichomonas therapy, vaginal

NITROIMIDAZOLES
metronidazole

Viral infections

CYCLIC AMINES
amantadine hydrochloride

NEURAMINIDASE INHIBITORS
oseltamivir
zanamivir

NON-NUCLEOSIDE REVERSE TRANSCRIPTASE
    INHIBITORS (NNRTIs)
delavirdine mesylate
efavirenz
nevirapine

NUCLEOSIDES
acyclovir
acyclovir sodium

famciclovir
ganciclovir sodium
rabavirin
valacyclovir hydrochloride
valganciclovir

NUCLEOSIDE ANALOG REVERSE
  TRANSCRIPTIASE INHIBITORS
  (NRTIs)
abacavir sulfate
didanosine (ddl)
lamivudine (3TC)
stavudine (d4T)
zalcitabine (ddC)
zidovudine (AZT)

PROTEASE INHIBITORS
amprenavir
indinavir sulfate
nelfinavir
ritonavir
ritonavir/lopinavir
saquinavir
saquinavir mesylate

Viral infections, topical

NUCLEOSIDES
acyclovir
idoxuridine
trifluridine

NEPHROLOGY

Edema

LOOP DIURETICS
bumetanide
ethacrynate sodium
ethacrynic acid
furosemide

OSMOTIC DIURETICS
mannitol

POTASSIUM SPARING AGENTS
amiloride hydrochloride
spironolactone
triamterene/hydrochlorothiazide

THIAZIDES AND RELATED AGENTS
chlorthalidone
hydrochlorothiazide
indapamine
indapamide hemihydrate
metolazone

VARIOUS DIURETICS
pamabrom

Hyperkalemia

Potassium exchange resins

CATION EXCHANGE RESINS
sodium polystyrene sulfonate

ION EXCHANGE RESINS
calcium polystyrene sulfonate

Hyperphosphatemia

Phosphate binders

ALUMINUM-CONTAINING PREPARATIONS
aluminum hydroxide gel

POLYMERIC PHOSPHATE BINDER
sevelamer hydrochloride

NEUROLOGY

Dementia

Alzheimer's disease

CHOLINESTERASE INHIBITORS
donepezil hydrochloride
galantamine hydrobromide
rivastigmine tartrate

Idiopathic dementia

ADJUNCTIVE THERAPY
ergoloid mesylates

Epilepsy

BARBITURATES AND DERIVATIVES
phenobarbital
primidone

BENZODIAZEPINES
clobazam
clonazepam

diazepam
lorazepam
nitrazepam

CARBOXYLIC ACID DERIVATIVES
divalproex sodium
valproic acid

GAMMA AMINOBUTYRIC ACID (GABA)
   DERIVATIVES
gabapentin
vigabatrin

HYDANTOIN DERIVATIVES
fosphenytoin sodium
phenytoin

IMINOSTILBENE DERIVATIVES
carbamazepine
oxcarbazepine

SUCCINIMIDE DERIVATIVES
ethosuximide
methsuximide

VARIOUS ANTICONVULSANTS
lamotrigine
magnesium sulfate
paraldehyde
topiramate

Migraine

BARBITURATES
butalbital

BETA-ADRENERGIC BLOCKING AGENTS
propranolol hydrochloride

CALCIUM CHANNEL BLOCKING AGENTS,
   SELECTIVE
flunarizine hydrochloride

ERGOT ALKALOIDS
dihydroergotamine mesylate
methysergide maleate

SEROTONIN (5-HT) RECEPTOR
   AGONISTS
naratriptan hydrochloride
rizatriptan benzoate

sumatriptan
zolmitriptan

VARIOUS ANTIMIGRAINE PREPARATIONS
feverfew
pizotifen

Multiple sclerosis

BIOLOGICAL RESPONSE MODIFIERS
glatiramer acetate
interferon beta-1a
interferon beta-1b

Muscle spasticity

Centrally acting agents

ALPHA₂-ADRENERGIC AGONISTS
tizanidine

ANTIHISTAMINE DERIVATIVES
orphenandrine citrate

BENZODIAZEPINES
diazepam

CARBAMIC ACID EATERS
carisoprodol
methocarbamol

GAMMA AMINOBUTYRIC ACID (GABA)
   DERIVATIVES
baclofen

TRICYCLIC DERIVATIVES
cyclobenzaprine hydrochloride

Direct acting agents
dantrolene sodium

Neuromuscular blocking agents

DEPOLARIZING AGENTS, CHOLINE
   DERIVATIVES
succinylcholine chloride

NONDEPOLARIZING AGENTS, QUATERNARY
   AMMONIUM COMPOUNDS
atracurium besylate
cisatracurium besylate
doxacurium chloride
mivacurium chloride

pancuronium bromide
rocuronium bromide
vecuronium bromide

Neuromuscular paralytic agents
botulinum toxin type A

Myasthenia gravis

CHOLINESTRASE INHIBITORS
neostigmine bromide
pyridostigmine bromide

Narcolepsy

CNS STIMULANTS
dexamphetamine
methylphenidate
modafinil

Neuralgia

TRIGEMINAL NEURALGIA
carbamazepine

TOPICAL THERAPY
capsaicin

Parkinson's disease

ANTICHOLINERGIC AGENTS
benztropine mesylate
biperiden hydrochloride
ethopropazine hydrochloride
procyclidine hydrochloride
trihexyphenidyl hydrochloride

Dopaminergic agents

DOPAMINE AGONISTS
bromocriptine mesylate
pergolide mesylate
pramipexole dihydrochloride
ropinirole hydrochloride

DOPAMINE PRECURSORS
levodopa

DOPAMINE PRECURSORS AND DECARBOXY-
LASE INHIBITORS
levodopa/benserazide hydrochloride
levodopa/carbidopa

MONOAMINE OXIDASE (MAO)
INHIBITORS, SELECTIVE
(TYPE B)
selegiline hydrochloride

VARIOUS DOPAMINERGIC AGENTS
amantadine hydrochloride

COMT INHIBITORS
entacapone

ONCOLOGY

Cancer

Alkylating agents

ALKYL SULPHONATES
busulfan

ETHYLENEIMINES
thiotepa

IMIDAZOTETRAZINES
temozolomide

NITROGEN MUSTARD ANALOGS
chlorambucil
cyclophosphamide
estramustine sodium phosphate
ifosfamide
mechlorethamine hydrochloride
melphalan

NITROSOUREAS
carmustine
lomustine
streptozocin

PLATINUM-CONTAINING
COMPOUNDS
carboplatin
cisplatin

Antimetabolites

CYTIDINE ANALOGS
gemcitabine hydrochloride

FOLIC ACID ANALOGS
methotrexate sodium
raltitrexed disodium

PURINE ANALOGS
cladribine
fludarabine phosphate
mercaptopurine
pentostatin
thioguanine

PYRIMIDINE ANALOGS
capecitabine
cytarabine
cytarabine liposomal
fluorouracil

UREA DERIVATIVES
hydroxyurea

Cytotoxic antibiotics

ANTHRACYCLINES
daunorubicin
doxorubicin hydrochloride
doxorubicin hydrochloride pegylated
   liposomal
epirubicin hydrochloride
idarubicin hydrochloride
valrubicin

ACTINOMYCINS
dactinomycin

VARIOUS CYTOTOXIC ANTIBIOTICS
bleomycin sulfate
mitomycin
mitotane
mitoxantrone hydrochloride

Plant alkaloids and other natural products

CAMPTOTHECIN DERIVATIVES
irinotecan hydrochloride

EPIPODOPHYLLOTOXINS
etoposide
teniposide

TAXANES
docetaxel
paclitaxel

VINCA ALKALOIDS AND ANALOGS
vinblastine sulfate
vincristine sulfate
vinorelbine tartrate

Hormones

ESTROGENS
diethylstilbestrol sodium diphosphate

GONADOTROPIN-RELEASING HORMONE
   ANALOGS
buserelin acetate
goserelin acetate
leuprolide acetate

PROGESTOGENS
medroxyprogesterone acetate
megestrol acetate

Hormone antagonists

ANTIANDROGENS
bicalutamide
cyproterone acetate
flutamide
nilutamide

ANTIESTROGENS
tamoxifen citrate

NONSTEROIDAL AROMATASE INHIBITORS
anastrozole
exemestane
letrozole

PROTEIN KINASE INHIBITOR
imatinib mesylate

VARIOUS ANTINEOPLASTICS
altretamine
amsacrine
l-asparaginase
dacarbazine
melanoma theraccine
porfimer sodium
procarbazine hydrochloride
topotecan hydrochloride
tretinoin (all-*trans* retinoic acid),
   systemic

PSYCHIATRY

Anxiety disorders

AZASPIRODECANEDIONE DERIVATIVES
buspirone hydrochloride

BENZODIAZEPINES
alprazolam
bromazepam
chlordiazepoxide hydrochloride
ciazepam
clorazepate dipotassium
lorazepam
oxazepam

VARIOUS ANXIOLYTICS
hydroxyzine hydrochloride
paroxetine
trifluoperazine hydrochloride
venlafaxine

Attention deficit hyperactivity disorder

STIMULANTS
dexamphetamine sulfate
methylphenidate hydrochloride

Depression
(see also Obsessive-compulsive disorder;
Panic disorder; Eating disorders)

MONOAMINE OXIDASE (MAO) INHIBITORS,
NONSELECTIVE (TYPES A,B)
phenelzine sulfate
tranylcypromine sulfate

MONOAMINE OXIDASE (MAO) INHIBITORS,
SELECTIVE (TYPE A)
moclobemide

NONSELECTIVE MONOAMINE REUPTAKE
INHIBITORS
amitriptyline hydrochloride
clomipramine hydrochloride
desipramine hydrochloride
doxepin hydrochloride
imipramine hydrochloride
maprotiline hydrochloride
nortriptyline hydrochloride
trimipramine maleate

SELECTIVE SEROTONIN REUPTAKE
INHIBITORS (SSRIs)
citalopram hydrobromide
fluoxetine hydrochloride
fluvoxamine maleate

paroxetine hydrochloride
sertaline hydrochloride

SEROTONIN–NOREPINEPHRINE REUPTAKE
INHIBITORS
venlafaxine hydrochloride

VARIOUS ANTIDEPRESSANTS
bupropion hydrochloride
nefazodone hydrochloride
trazodone hydrochloride
L-tryptophan

Mania

IMINOSTILBENE DERIVATIVES
carbamazepine

LITHIUM SALTS
lithium carbonate
lithium citrate

VARIOUS ADJUNCTIVE AGENTS
L-tryptophan

Obsessive-compulsive disorder
(see also depression)

SELECTIVE SEROTONIN REUPTAKE INHIBITORS
fluoxetine hydrochloride
fluvoxamine maleate
paroxetine hydrochloride
sertraline hydrochloride

TRICYCLIC DERIVATIVES
clomipramine hydrochloride

Panic disorder

BENZODIAZEPINES
alprazolam

SELECTIVE SEROTONIN REUPTAKE INHIBITORS
paroxetine hydrochloride
sertraline

Psychoses

BENZISOXAZOLE DERIVATIVES
risperidone

BUTYROPHENONE DERIVATIVES
haloperidol
haloperidol decanoate

DIBENZODIAZEPINE DERIVATIVES
clozapine

DIBENZOTHIAZEPINE DERIVATIVES
quetiapine fumarate

DIBENZOXAZEPINE DERIVATIVES
loxapine

DIPHENYLBUTYLPIPERIDINE DERIVATIVES
pimozide

PHENOTHIAZINES, ALIPHATIC
chlorpromazine hydrochloride
methotrimeprazine maleate
promazine hydrochloride

PHENOTHIAZINES, PIPERAZINE
fluphenazine decanoate
fluphenazine enanthate
fluphenazine hydrochloride
perphenazine
prochlorperazine
thioproperazine mesylate
trifuoperazine hydrochloride

PHENOTHIAZINES, PIPERIDINE
mesoridazine besylate
pericyazine
pipotiazine palmitate
thioridazine hydrochloride

THIENOBENZODIAZEPINE DERIVATIVES
olanzapine

THIOXANTHENE DERIVATIVES
flupenthixol decanoate
flupenthixol dihydrochloride
thiothixene
zuclopenthixol acetate
zuclopenthixol decanoate
zuclopenthixol dihydrochloride

RESPIROLOGY

Allergies

Allergenic extracts

IMMUNOTHERAPY
grass tyrosine adsorbate, modified
ragweed tyrosine adsorbate
tree tyrosine adsorbate, modified

ANAPHYLAXIS THERAPY
epinephrine

Antihistamines

ALKYLAMINES
brompheniramine maleate
chlorpheniramine maleate

ETHANOLAMINES
clemastine hydrogen fumarate
diphenhydramine hydrochloride

PHENOTHIAZINE DERIVATIVES
promethazine hydrochloride
trimeprazine tartrate

PIPERAZINE DERIVATIVES
cetirizine hydrochloride
cyclizine lactate
hydroxyzine hydrochloride
meclizine hydrochloride

PIPERIDINE DERIVATIVES
azatadine maleate
cyproheptadine hydrochloride
desloratadine
fexofenadine hydrochloride
loratadine

Asthma
*(see also Corticosteroid therapy,
   systemic)*

Adrenergics, inhalants

ALPHA- AND BETA-ADRENERGIC AGONISTS
epinephrine bitartrate
epinephrine hydrochloride, racemic

BETA-ADRENERGIC AGONISTS,
   NONSELECTIVE
orciprenaline sulfate

BETA-2-ADRENERGIC AGONISTS, SELECTIVE
fenoterol hydrobromide
formoterol fumarate
formoterol fumarate dihydrate
salbutamol
salbutamol sulfate
salmeterol xinafoate
terbutaline sulfate

Adrenergics, systemic

ALPHA- AND BETA-ADRENERGIC AGONISTS
epinephrine

BETA-ADRENERGIC AGONISTS, NONSELECTIVE
isoproterenol hydrochloride
orciprenaline sulfate

BETA-2-ADRENERGIC AGONISTS, SELECTIVE
fenoterol hydrobromide
salbutamol
salbutamol sulfate
terbutaline sulfate

*Bronchial anti-inflammatory agents,
   inhalants*

CORTICOSTEROIDS
beclomethasone dipropionate
budesonide
fluticasone propionate

NONSTEROIDAL AGENTS
nedocromil sodium

COMBINATION INHALANTS
fenoterol hydrobromide/ipratropium
   bromide
formoterol fumaratedihydrate/ budesonide
salbutamol sulfate/ipratropium bromide
salmeterol xinafoate/fluticasone
   propionate

*Bronchial anti-inflammatory
   agents, systemic*

CORTICOSTEROIDS
hydrocortisone sodium succinate
methylprednisolone sodium succinate
prednisone
triamcinolone

LEUKOTRIENE RECEPTOR ANTAGONISTS,
   SYSTEMIC
montelukast sodium
zafirlukast

XANTHINES, SYSTEMIC
aminophylline
oxtriphylline
theophylline

Pulmonary hypertension

ENDOTHELIN RECEPTOR ANTAGONIST
bosentan

PROSTAGLANDINS
epoprostenol sodium

Respiratory distress syndrome

LUNG SURFACTANTS
beractant
bovine lipid extract surfactant

MUCOLYTICS
acetylcysteine
dornase alfa, recombinant

RHEUMATOLOGY

Gout and hyperuricemia

Gout therapy

ANTIMITOTICS
colchicine

CORTICOSTEROIDS
dexamethasone
dexamethasone sodium phosphate
hydrocortisone sodium succinate
methylprednisolone acetate
prednisone
trimcinolone

NONSTEROIDAL ANTI-INFLAMMATORY
   DRUGS (NSAIDs)
indomethacin
phenylbutazone
sulindac

Hyperuricemia therapy

URICOSURICS
probenecid
sulfinpyrazone

XANTHINE OXIDASE INHIBITORS
allopurinol

Rheumatoid arthritis

BIOLOGICAL RESPONSE MODIFIERS
anakinra
etanercept
infliximab

CORTICOSTEROIDS
betamethasone sodium phosphate
cortisone acetate
dexamethasone
dexamethasone sodium phosphate
methylprednisolone acetate
prednisolone
prednisone
triamcinolone
triamcinolone diacetate

Disease-modifying drugs (DMARDs)

CYTOTOXICS
azathioprine
methotrexate sodium

GOLD PREPARATIONS
aurothioglucose
sodium arothiomalate

OTHERS DMARDS
cyclosporine
hydroxychloroquine sulfate
leflunomide
penicillamine
sulfasalazine

*Various agents for rheumatic
    disease therapy*

JOINT AND MUSCULAR PAIN THERAPY, TOPICAL
capsaicin
menthol
triethanolamine salicylate

UROLOGY

Erectile dysfunction

PHOSPHODIESTERASE TYPE 5 (PDE5)
    INHIBITORS
sildenafil citrate

PROSTAGLANDINS
alprostadil

Prostatic hyperplasia

ALPHA₁-ADRENERGIC BLOCKING AGENT
alfuzosin
doxazosin

tamsulosin hydrochloride
terazosin hydrochloride

5 ALPHA-REDUCTASE INHIBITORS
finasteride

Urinary tract therapy

Urinary analgesics

AZO DYES
phenazopyridine hydrochloride

*Urinary antiseptics and
    anti-infectives
(see also Infectious diseases)*

METHENAMINE SALTS
methenamine (hexamine)
methenamine mandelate

NITROFURAN DERIVATIVES
nitrofurantion
nitrofurantion monohydrate

QUINOLONES
nalidixic acid

Urinary antispasmodics

ANTICHOLINERGICS
hyoscine butylbromide
hyoscyamine sulfate

SMOOTH MUSCLE RELAXANTS
flavoxate hydrochloride
oxybutynin chloride
tolterodine L-tartrate

Urinary retention

PARASYMPATHOMIMETIC AGENTS
bethanechol chloride

MISCELLANEOUS

Anesthesia, general

BARBITURATES
methohexital sodium
thiopental sodium

HALOGENATED HYDROCARBONS
desflurane

enflurane
isoflurane
sevoflurane

OPOID ANESTHESTICS
alfentanil hydrochloride
fentanyl citrate
remifentanil hydrochloride
sufentanil citrate

VARIOUS GENERAL ANESTHETICS
ketamine hydrochloride
midazolam hydrochloride
propofol

Anesthesia, local

AMIDES
articaine hydrochloride
bupivacaine hydrochloride
lidocaine
lidocaine/prilocaine
mepivacaine hydrochloride
prilocaine hydrochloride
ropivacaine hydrochloride

ESTERS OF AMINO BENZOIC ACID
benzocaine
chloroprocaine hydrochloride
cocaine hydrochloride
procaine hydrochloride
tetracaine
tetracaine hydrochloride

Fever
*(see also Malignant
  hyperthermia)*

ANTIPYRETICS
acetaminophen
acetylsalicylic acid (ASA)
ibuprofen

Hair regrowth

Systemic

5 ALPHA-REDUCTASE INHIBITORS
finasteride

TOPICAL
minoxidil

Hepatitis
*(see also Viral infection therapy,
  Immunization therapy)*

HEPATITIS B THERAPY
interferon alfa-2a
interferon alfa-2b

HEPATITIS C THERAPY
interferon alfa-2b/ribavirin
peginterferon-2b
peginterferon alfa-2b/ribavirin

Insomnia and sedation

Hypnotics and sedatives

ALDEHYDES AND DERIVATIVES
chloral hydrate
paraldehyde

ANTIHISTAMINES
diphenhydramine
doxylamine succinate

BARBITURATES
amobarbital sodium
pentobarbital sodium
phenobarbital
secobarbital sodium

BENZODIAZEPINES
alprazolam
bromazepam
clorazepate dipotassium
diazepam
flurazepam hydrochloride
lorazepam
midazolam hydrochrloide
nitrazepam
oxazepam
temazepam
trizolam

CYCLOPYRROLONES
zopiclone

PYRAZOLOPYRIMIDINES
zaleplon

VARIOUS HYPNOTICS AND SEDATIVES
propofol
valerian root

Pain

Nonsteroidal anti-inflammatory drugs (NSAIDs)

ACETIC ACID DERIVATIVES (INCLUDING INDOLE DERIVATIVES)
diclofenac potassium
diclofenac sodium
etodolac
indomethacin
ketorolac tromethamine
sulindac
tolmetin sodium

CYCLOOXYGENASE-2 (COX-2) INHIBITORS
celecoxib
meloxicam
rofecoxib

FENAMATES
floctafenine
mefenamic acid

OXICAMS
meloxicam
piroxicam
tenoxicam

PROPIONIC ACID DERIVATIVES
fenoprofen calcium
flurbiprofen
ibuprofcn
ketoproen
naparoxen
naproxen sodium
oxaprozin
tiaprofenic acid

SALICYLIC ACID DERIVATIVES
acetylsalicylic acid (ASA)
diflunisal
choline magnesium trisalicylate
choline salicylate
triethanolamine salicylate

VARIOUS NONSTEROIDAL ANTI-INFLAMMATORY AGENTS
nabumetone

Opioids

BENZOMORPHAN DERIVATIVES
pentazocine hydrochloride
pentazocine lactate

DIPHENYLPROPYLAMINE DERIVATIVES
propoxyphene hydrochloride
propoxyphene napsylate

MORPHINAN DERIVATIVES
butorphanol tartrate
nalbuphine hydrochloride

NATURAL OPIUM ALKALOIDS
codeine phosphate
hydromorphone hydrochloride
morphine hydrochloride
morphine sulfate
oxycodone hydrochloride
oxymorphone hydrochloride

PHENYLHEPTYLAMINE DERIVATIVES
methadone

PHENYLPIPERIDINE DERIVATIVES
alfentanil hydrochloride
fentanyl citrate
meperidine hydrochloride (pethidine)
sufentanil citrate

Para-aminophenol derivatives

ANILIDES
acetaminophen

Poisoning management

ACETAMINOPHEN ANTIDOTES
acetylcysteine

BENZODIAZEPINE ANTAGONISTS
flumazenil

CHELATING AGENTS
deferoxamine mesylate

CYANIDE ANTIDOTES
sodium thiosulfate

DIGOXIN ANTIBODIES
digoxin immune Fab (ovine)

Ethylene glycol antidote

ALCHOHOL DEHYDROGENASE INHIBITOR
fomepizol

HEPARIN ANTAGONISTS
protamine sulfate

Nonspecific therapy for overdoses

ADSORBENTS
activated charcoal

EMETICS
ipecac

OPOID ANTAGONISTS
naloxone hydrochloride
naltexone hydrochloride

Organic phosphorus insecticides antidote

CHOLINESTERASE REACTIVATORS
pralidoxime chloride

Smoking cessation

Nicotine replacement therapy

NICOTINE GUM
nicotine polacrilex

NICOTINE TRANSDERMAL
nicotine

VARIOUS SMOKING CESSATION AIDS
bupropion hydrochloride

Vitamin therapy
*(see also Anemia therapy and
    Hematopoietics: Mineral replacement)*

Vitamin A (fat soluble)

VITAMIN A
retinol

Vitamin B complex (water soluble)

FOLIC ACID
folic acid

VITAMIN B$_1$
thiamine hydrochloride

VITAMIN B$_2$
riboflavin

VITAMIN B$_3$
niacin
niacinamide

VITAMIN B$_5$
pantothenic acid (calcium
    pantothenate)

VITAMIN B$_6$
pyridoxine hydrochloride

VITAMIN B$_{12}$
cyanocobalamin
hydroxocobalamin

Vitamin C (water soluble)

ASCORBIC ACID
ascorbic acid

Vitamin D analogs (fat soluble)

VITAMIN D$_2$ ANALOGS
ergocalciferol (calciferol)

VITAMIN D$_3$ ANALOGS
alfacalcidol
calcitriol
cholecalciferol

VARIOUS VITAMIN D ANALOGS
dihydrotachysterol
doxercalciferol

Vitamin E (fat soluble)

TOCOPHEROLS
alpha tocopherol

Vitamin K (fat soluble) analogs

VITAMIN K$_1$
phytonadione

Index